Specific energy

$1 \text{ kJ/kg} = 0.42992 \text{ Btu/lbm}$
$1 \text{ Btu/lbm} = 2.326 \text{ kJ/kg}$
$1 \text{ lbf-ft/lbm} = 2.98907 \times 10^{-3} \text{ kJ/kg}$
$= 1.28507 \times 10^{-3} \text{ Btu/lbm}$

Power

$1 \text{ W} = 1 \text{ J/s} = 1 \text{ Nm/s}$
$1 \text{ kW} = 1 \text{ kJ/s} = 3412.14 \text{ Btu/h}$
$1 \text{ Btu/s} = 1.055\ 056 \text{ kW}$
$1 \text{ hp (metric)} = 0.735\ 499 \text{ kW}$
$1 \text{ hp (UK)} = 0.7457 \text{ kW}$
$= 550 \text{ lbf-ft/s}$
$= 2544.43 \text{ Btu/h}$
$1 \text{ lbf-ft/s} = 1.355\ 818 \text{ W}$
$= 4.62624 \text{ Btu/h}$
$1 \text{ ton of refrigeration} = 12\ 000 \text{ Btu/h}$
$= 3.516\ 85 \text{ kW}$

Heat capacity

$1 \text{ Btu/lbm-R} = 4.1868 \text{ kJ/kg-K}$
$1 \text{ kJ/kg-K} = 0.238\ 846 \text{ Btu/lbm-R}$

Pressure

$1 \text{ Pa} = 1 \text{ N/m}^2 = 1 \text{ kg/m-s}$
$1 \text{ bar} = 1.0 \times 10^5 \text{ Pa} = 100 \text{ kPa}$
$1 \text{ atm} = 101.325 \text{ kPa}$
$= 1.01325 \text{ bar}$
$= 14.696 \text{ lbf/in.}^2$
$= 760 \text{ mm Hg } [0°C]$
$= 10.332\ 56 \text{ m H}_2\text{O } [4°C]$
$1 \text{ lbf/in.}^2 = 6.894\ 757 \text{ kPa}$
$1 \text{ mm Hg } [0°C] = 0.133\ 322 \text{ kPa}$
$1 \text{ m H}_2\text{O } [4°C] = 9.806\ 38 \text{ kPa}$
$1 \text{ in. Hg } [0°C] = 3.386\ 38 \text{ kPa} = 0.49115 \text{ lbf/in.}^2$
$1 \text{ in. H}_2\text{O } [4°C] = 0.249\ 082 \text{ kPa} = 0.036126 \text{ lbf/in.}^2$
$1 \text{ torr} = 1 \text{ mm Hg } [0°C] = 0.133\ 322 \text{ kPa}$

Specific volume

$1 \text{ cm}^3/\text{g} = 0.001 \text{ m}^3/\text{kg}$
$1 \text{ m}^3/\text{kg} = 16.018\ 45 \text{ ft}^3/\text{lbm}$
$1 \text{ ft}^3/\text{lbm} = 0.062\ 428 \text{ m}^3/\text{kg}$

Temperature

$TC = TK - 273.15 = (TF - 32)/1.8$
$TR = 1.8\ TK$
$TF = TR - 459.67 = 1.8\ TC + 32$

Force

$1 \text{ lbf} = 4.448\ 222 \text{ N}$
$1 \text{ N} = 0.224809 \text{ lbf}$

Universal Gas Constant

$\overline{R} = N_0 k = 8.31451 \text{ kJ/kmol-K}$
$= 1.98589 \text{ Btu/lbmol-R}$
$= 1.98589 \text{ kcal/kmol-K}$
$= 1545.36 \text{ lbf-ft/lbmol-R}$
$= 10.7317 \text{ (lbf/in.}^2)\text{-ft}^3/\text{lbmol-R}$
$= 0.73024 \text{ atm-ft}^3/\text{lbmol-R}$
$= 82.0578 \text{ atm-L/kmol-K}$

Specific kinetic energy (V^2)

$1 \text{ m}^2/\text{s}^2 = 0.001 \text{ kJ/kg}$
$1 \text{ kJ/kg} = 1000 \text{ m}^2/\text{s}^2$
$1 \text{ ft}^2/\text{s}^2 = 1/(32.17405 \times 778.16934) \text{ Btu/lbm}$
$= 3.9941 \times 10^{-5} \text{ Btu/lbm}$
$1 \text{ Btu/lbm} = 25037 \text{ ft}^2/\text{s}^2$

Specific potential energy (Zg)

$1 \text{ m-g}_{std} = 9.80665 \times 10^{-3} \text{ kJ/kg}$
$1 \text{ ft-g}_{std} = 1.0 \text{ lbf-ft/lbm}$
$= 0.001285 \text{ Btu/lbm}$
$= 0.002989 \text{ kJ/kg}$

Gravitation

$g = 9.80665 \text{ m/s}^2$
$= 32.17405 \text{ ft/s}^2$

FUNDAMENTALS OF CLASSICAL THERMODYNAMICS

FOURTH EDITION

GORDON J. VAN WYLEN
Hope College

RICHARD E. SONNTAG
CLAUS BORGNAKKE
University of Michigan

John Wiley & Sons, Inc.
New York   Chichester   Brisbane   Toronto   Singapore

Acquisitions Editor Cliff Robichaud
Marketing Manager Susan Elbe
Senior Production Editor Savoula Amanatidis
Text Designer Dawn L. Stanley
Cover Designer Carolyn Joseph
Illustration Coordinator Anna Melhorn
Manufacturing Manager Andrea Price
Cover Photo Norman Chigier

ABOUT THE COVER   The photograph on the cover of this textbook was taken by Professor Norman Chigier, William J. Brown Professor of Mechanical Engineering at Carnegie Mellon University. The high speed photograph is of a natural gas emerging from a nozzle and mixing with the surrounding air. The color photograph shows temperature variations and flow patterns. The highest temperature regions of the flame, around 2000K, are located where the air-fuel ratio is stoichiometric. The gas jet is initially laminar and becomes turbulent farther downstream. Entrainment of air from the surroundings can be seen to be associated with the engulfment of large vortex structures. The most luminous regions of the flame have solid carbon particles that radiate at high temperatures to the surroundings at lower temperatures. These carbon particles can be burnt in the flame in the presence of oxygen or emitted as smoke. Several hundred intermediate chemical reactions occur when methane and oxygen react in high temperature flames and cause ionization of the gas. High speed visualization of flame structures is leading to more detailed understanding of the chemical and physical processes of combustion.

This book was typset in Times New Roman by General Graphic Services and printed and bound by R.R. Donnelley in Willard. The cover was printed by Phoenix Color Corporation.

Library of Congress Cataloging in Publication Data:
Van Wylen, Gordon J.
      Fundamentals of classical thermodynamics/Gordon J. Van Wylen,
         Richard E. Sonntag, Claus Borgnakke.—4th ed.
               p.   cm.
            Includes index.
            ISBN 0-471-59395-8 (cloth)
            1. Thermodynamics.   I. Sonntag, Richard Edwin.   II. Borgnakke, C.
      (Claus) III. Title
      TJ265.V23   1993
      536'.7—dc20   93-35997
                           CIP

Printed in the United States of America

10 9 8 7 6 5 4 3 2

PREFACE

In this fourth edition we have retained the basic objective of the first three editions: to present a comprehensive and rigorous treatment of classical thermodynamics while retaining an engineering perspective and, in doing so, to lay the groundwork for subsequent studies in such fields as fluid mechanics, heat transfer, and statistical thermodynamics, and also to prepare the student to effectively use thermodynamics in the practice of engineering.

We have deliberately directed our presentation to students. New concepts and definitions are presented in the context where they are first relevant. The first thermodynamic properties to be defined (Chapter 2) are those that can be readily measured: pressure, specific volume, and temperature. In Chapter 3, tables of thermodynamic properties are introduced, but only in regard to these measurable properties. Internal energy and enthalpy are introduced in connection with the first law, entropy with the second law, and the Helmholtz and Gibbs functions in the chapter on availability. Many examples have been included in the book to assist the student in gaining an understanding of thermodynamics, and the problems at the end of each chapter have been carefully sequenced to correlate with the subject matter and provide some progression in difficulty.

We have attempted to cover fairly comprehensively the basic subject matter of classical thermodynamics, and believe that the book provides adequate preparation for study of the application of thermodynamics to the various professional fields as well as for study of more advanced topics in thermodynamics, such as those related to materials, surface phenomena, plasmas, and cryogenics. We also recognize that a number of colleges offer a single introductory course in thermodynamics for all departments, and we have tried to cover those topics that the various departments might wish to have included in such a course. However, since specific courses vary considerably in prerequisites, specific objectives, duration, and background of the students, we have arranged the material, particularly in the later chapters, so that there is considerable flexibility in the amount of material that may be covered.

The principal changes of our philosophy and direction in this edition are twofold: an increased awareness and emphasis on environmental issues and concerns throughout the text and in the text problems; and a recognition of the steadily growing use of computers in the study of thermodynamics and also in the solution of thermodynamic problems. The use of computers is reflected in our inclusion of the entire appendix material on thermodynamic properties, Tables A.1 through A.18, on a computer disk and also in the inclusion there of FORTRAN subroutines for the properties of substances listed in these tables, Tables A.1 through A.7, A.12, A.13, and A.15. This allows a considerable flexibility for studying many problems in thermodynamics in terms of the variable quantities involved, and for creating many new design and open-ended problems, which is one principal change in detail in this edition. Others include reorganization and expansion of Chapter 9 on power and refrigeration systems, new development of the material on thermodynamic availability in Chapter 8, optional inclusion of three- instead of two-parameter generalized correlations in Chapter 10, and a chemical equilibrium program solver to enable analysis and solution of complex problems in Chapter 13. There is also an increased emphasis on rate processes, such as the introduction to the potential for power output from reversible heat engines in Chapter 6 and also in the Design Problems throughout the text. In addition to these changes, there has been a general updating of text material, definitions, and thermodynamic data, some of which have been carried over from the latest edition of our book, *Introduction to Thermodynamics, Classical and Statistical*. For example, the new temperature scale, ITS-90, is included in Chapter 2, a conceptual approach to the introduction of thermodynamic tables has been presented in Chapter 3, and new ideal-gas tables consistent with the most recent JANAF tables are introduced in Chapter 5, as well as new air tables that are consistent with these tables. In addition, new analysis and the correlation between thermodynamic and ideal-gas temperature scales are included in Chapter 6. Thermochemical properties and equilibrium constants are made internally consistent with the earlier-mentioned gas tables, and the appendix tables and data have been thoroughly revised and extended.

Throughout the book we have attempted to maintain an engineering perspective, particularly through the choice of examples and problems. One of the major changes from the previous edition is in the problems. Although a number of the earlier problems have been retained, most are new or have been significantly revised. A number of new problems requiring a computer-based solution have been added throughout the book, beginning in Chapter 2, so that instructors who wish to incorporate these problems into the course may do so at any time throughout the term. Several different classes of applications of thermodynamics are introduced in the first chapter, and these then serve as a basis for many of the problems in later chapters. We have retained a chapter on systems (Chapter 9) because we find that many students enjoy this subject, and it can also serve to effectively strengthen the student's understanding of the first and second laws of thermodynamics, and introduce him on her to engineering design and practice. Many of the computer-related problems are in this chapter, as are problems concerning a number of modern and relevant applications: heat pumps, cogeneration, topping and bottoming cycles, two-stage systems, combined cycles, and many others. We have also included comments and problems that relate thermodynamics to environmental concerns.

Chapter 14, which includes an introduction to compressible flow and flow through blade passages, has been retained for those students who would not otherwise undertake extensive studies in fluid flow. We recognize that in many cases the material in this chapter will not be included in the thermodynamics courses, but may instead be covered elsewhere in the curriculum.

In regard to symbols used in this text, we have tried to maintain consistency throughout the book. In a limited number of cases, we have used a given symbol for more than one purpose. We believe, however, that the context will clarify the meaning of the symbol in these cases.

Our philosophy regarding units in this edition has been to organize the book so that the course or sequence can be taught entirely in SI units (Le Système International d'Unités). Thus, all the text examples are in SI units, as are the complete problem sets and the thermodynamic tables. In recognition, however, of the continuing need for engineering graduates to be familiar with English Engineering units, we have included an introduction to this system in Chapter 2. We have also repeated a sufficient number of examples, problems, and tables in these units, which should allow for suitable practice for those who wish to use these units. For dealing with English units, there is the problem of distinguishing between force and mass units. The symbols lbf and lbm have been used for pound force and pound mass to emphasize this distinction in the English system.

Such symbols as lbf/in.^{2} have been used for pressure (rather than psi) and ft^3/lbm for specific volume (rather than cu ft/lb) in order to stress the fundamental units involved in the various parameters. The force–mass conversion constant g_c is treated as being implicit, to enable equations including kinetic or potential energy terms to be handled in a consistent manner with regard to units. In dealing with SI units, we have commonly used the basic units for pressure (pascal) and volume (cubic meter), although we have used the liter extensively as a convenient volume unit. Others may wish to use the bar more extensively as the pressure unit, and we trust that such flexibility in these units will present no particular difficulties to the student. Concerning the extensive properties, a lowercase letter (v, u, h, s) designates the property per unit mass (either pound mass or kilogram); an uppercase letter (V, U, H, S) the property for the entire system; a lowercase letter with a bar ($\bar{v}$, $\bar{u}$, $\bar{h}$, $\bar{s}$) the property per unit mole (either the pound mole or lb mol, or the kilogram mole or kmol, in this text); and an uppercase letter with a bar ($\bar{V}$, $\bar{U}$, $\bar{H}$, $\bar{S}$) the partial molal property in a mixture. Following this pattern, we have found it convenient to designate the total heat transfer as Q, the heat transfer per unit mass of the system as q, the total work as W, and the work per unit mass of the system as w.

Furthermore, we represent the rate of flow across a system boundary or control surface by a dot over the given quantity. Thus, $\dot{Q}$ represents a rate of heat transfer across the system boundary; $\dot{W}$ the rate at which work crosses the system boundary (that is, the power); and $\dot{m}$ the mass rate of flow across a control surface ($\dot{n}$ is used when the mass rate of flow is expressed in moles per unit time). We realize that we have departed from the usual mathematical use of a dotted symbol, in which the dotted symbol typically refers to a derivative with respect to time. However, we have used the dotted symbol to indicate only a flow of heat and work across a system boundary and a flow of heat, work, and mass across a control surface, and not in any other context. We believe that these designations have contributed to a simple and consistent use of symbols for this book.

We acknowledge with appreciation the suggestions, counsel, and encouragement of many colleagues, both at the University of Michigan and elsewhere. This assistance has been very helpful to us during the writing of this edition, as it was with the earlier editions of the book. Several secretaries have aided us immeasurably throughout, and we acknowledge this assistance with many thanks. Both undergraduate and graduate students have been of particular assistance, for their perceptive questions have often caused us to rewrite or rethink a given portion of the text, or to try to develop a better way of presenting the material in order to anticipate such questions or difficulties. We must single out two graduate students at the University of Michigan who have been of particular help with this project, especially in the development of the computer programs, former doctoral students Dr. Young Moo Park and Dr. Kyoung Kuhn Park. We appreciate their many valuable contributions to this material. Finally, for each of us, the encouragement and patience of our wives and families have been indispensable, and have made this time of writing pleasant and enjoyable, in spite of the pressures of the project.

Our hope is that this book will contribute to the effective teaching of thermodynamics to students who face very significant challenges and opportunities during their professional careers. Your comments, criticism, and suggestions will also be appreciated.

GORDON J. VAN WYLEN
RICHARD E. SONNTAG
CLAUS BORGNAKKE
Ann Arbor, Michigan
July 1993

CONTENTS

SYMBOLS

a	acceleration
a	activity
a, A	specific Helmholtz function and total Helmholtz function
AF	air-fuel ratio
c	velocity of sound
c	mass fraction
C_D	coefficient of discharge
C_p	constant-pressure specific heat
C_v	constant-volume specific heat
C_{po}	zero-pressure constant-pressure specific heat
C_{vo}	zero-pressure constant-volume specific heat
e, E	specific energy and total energy
f	fugacity
$\bar{f}_i$	fugacity of component i in a mixture
F	force
FA	fuel-air ratio
g	acceleration due to gravity
g, G	specific Gibbs function and total Gibbs function
g_c	a constant that relates force, mass, length, and time
h, H	specific enthalpy and total enthalpy
i	electrical current
I	irreversibility
J	proportionality factor to relate units of work to units of heat
k	specific heat ratio: C_p/C_v
K	equilibrium constant
KE	kinetic energy
L	length
m	mass
$\dot{m}$	mass rate of flow
M	molecular weight
M	Mach number
n	number of moles
n	polytropic exponent

P	pressure
P_i	partial pressure of component i in a mixture
PE	potential energy
P_r	relative pressure as used in gas tables
q, Q	heat transfer per unit mass and total heat transfer
$\dot{Q}$	rate of heat transfer
Q_H, Q_L	heat transfer with high-temperature body and heat transfer with low-temperature body; sign determined from context
R	gas constant
$\bar{R}$	universal gas constant
s, S	specific entropy and total entropy
S_{gen}	entropy generation
$\dot{S}_{\text{gen}}$	rate of entropy generation
t	time
T	temperature
u, U	specific internal energy and total internal energy
v, V	specific volume and total volume
v_r	relative specific volume as used in gas tables
vf	volume fraction
V	velocity
V$_r$	relative velocity
w, W	work per unit mass and total work
$\dot{W}$	rate of work, or power
w_{rev}	reversible work between two states assuming heat transfer with surroundings
x	quality
x	liquid-phase or solid-phase mole fraction
y	vapor-phase mole fraction
Z	elevation
Z	compressiblity factor
Z	electrical charge

SCRIPT LETTERS

$\mathscr{A}$	area
$\mathscr{C}$	number of components
$\mathscr{E}$	electrical potential
$\mathscr{H}$	magnetic field intensity
$\mathscr{M}$	magnetization
$\mathscr{P}$	number of phases
$\mathscr{S}$	surface tension
$\mathscr{T}$	tension
$\mathscr{V}$	variance

GREEK LETTERS

α	residual volume
α_P	volume expansivity
β	coefficient of performance for a refrigerator
β'	coefficient of performance for a heat pump
β_S	adiabatic compressibility
β_T	isothermal compressibility
η	efficiency
μ	chemical potential
μ_J	Joule-Thomson coefficient
ν	stoichiometric coefficient
ρ	density
Φ	equivalence ratio
ϕ	relative humidity
ϕ	availability for a control mass
ψ	availability associated with a steady-state, steady-flow process
ω	acentric factor
ω	humidity ratio or specific humidity

SUBSCRIPTS

c	property at the critical point
c.v.	control volume
e	state of a substance leaving a control volume
f	formation
f	property of saturated liquid
fg	difference in property for saturated vapor and saturated liquid
g	property of saturated vapor
i	state of a substance entering a control volume
i	property of saturated solid
ig	difference in property for saturated vapor and saturated solid
r	reduced property
s	isentropic process
0	property of the surroundings
0	stagnation property

SUPERSCRIPTS

$^{-}$	bar over symbol denotes property on a molal basis (over V, H, S, U, A, G, the bar denotes partial molal property)
$^{\circ}$	property at standard-state condition
$*$	ideal gas
$*$	property at the throat of a nozzle
L	liquid-phase
rev	reversible
S	solid-phase
V	vapor-phase

SOME INTRODUCTORY COMMENTS 1

In the course of our study of thermodynamics, a number of the examples and problems presented refer to processes that occur in equipment such as a steam power plant, a fuel cell, a vapor-compression refrigerator, a thermoelectric cooler, a rocket engine, and an air separation plant. In this introductory chapter a brief description of this equipment is given. There are at least two reasons for including such a chapter. First, many students have had limited contact with such equipment, and the solution of problems will be more meaningful when they have some familiarity with the actual processes and the equipment. Second, this chapter will provide an introduction to thermodynamics, including the use of certain terms (which will be more formally defined in later chapters), some of the problems to which thermodynamics can be applied, and some of the things that have been accomplished, at least in part, from the application of thermodynamics.

Thermodynamics is relevant to many other processes than those cited in this chapter. It is basic to the study of materials, chemical reactions, and plasmas. The student should bear in mind that this chapter is only a brief and necessarily very incomplete introduction to the subject of thermodynamics.

1.1 THE SIMPLE STEAM POWER PLANT

A schematic diagram of a simple steam power plant is shown in Fig. 1.1. High-pressure superheated steam leaves the boiler, which is also referred to as a steam generator, and enters the turbine. The steam expands in the turbine and, in doing so, does work, which enables the turbine to drive the electric generator. The low-pressure steam leaves the turbine and enters the condenser, where heat is transferred from the steam (causing it to condense) to the cooling water. Because large quantities of cooling water are required, power plants are frequently located near rivers or lakes. This transfer of heat to the water in lakes and rivers leads to the thermal pollution problem, which has been studied extensively in recent years. During our study of thermodynamics we will gain an understanding of why this heat transfer is necessary and how it can be minimized. When the supply of cooling water is limited, a cooling tower may be used. In the cooling tower some of the cooling water evaporates in such a way that it lowers the temperature of the water that remains as a liquid.

The pressure of the condensate leaving the condenser is increased in the pump and thus enables the condensate to flow into the steam generator. In many steam generators an economizer is used. An economizer is simply a heat exchanger in which heat is trans-

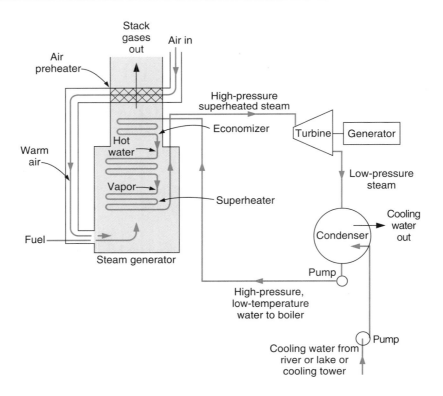

FIGURE 1.1 Schematic diagram of a steam power plant.

ferred from the products of combustion (just before they leave the steam generator) to the condensate. The temperature of the condensate is increased, but no evaporation takes place. In other sections of the steam generator, heat is transferred from the products of combustion to the water, causing it to evaporate. The temperature at which evaporation occurs is called the saturation temperature. The steam then flows through another heat exchanger, known as a superheater, where the temperature of the steam is increased well above the saturation temperature.

In many power plants the air that is used for combustion is preheated in the air preheater by transferring heat from the stack gases as they are leaving the furnace. This air is then mixed with fuel—which might be coal, fuel oil, natural gas, or other combustible material—and combustion takes place in the furnace. As the products of combustion pass through the furnace, heat is transferred to the water in the superheater, the boiler, the economizer, and to the air in the air preheater. The products of combustion from power plants are discharged to the atmosphere, which is one of the facets of the air pollution problem we now face.

A large power plant has many other pieces of equipment, some of which will be considered in later chapters.

Figure 1.2 shows a steam turbine and the generator that it drives. Steam turbines vary in capacity from less than 10 kilowatts to more than 1 000 000 kilowatts.

Figure 1.3 shows a large steam generator. The flow of air and products of combustion are indicated. The condensate, also called the boiler feedwater, enters at the economizer inlet, and the superheated steam leaves at the superheater outlet.

The number of operating nuclear power plants has increased substantially in recent

FIGURE 1.2 A large steam turbine. (Courtesy General Electric Co.)

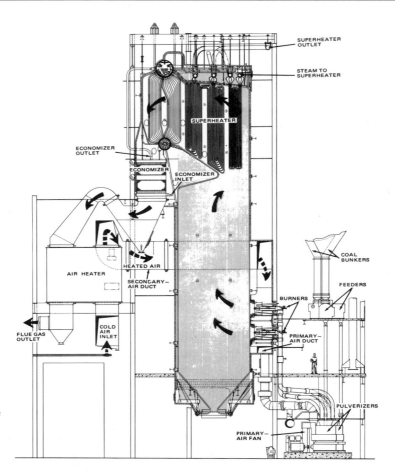

FIGURE 1.3 A large steam generator (Courtesy Babcock and Wilcox Co.).

years. In these power plants the reactor replaces the steam generator of the conventional power plant, and the radioactive fuel elements replace the coal, oil, or natural gas.

There are several different reactor designs in current use. One of these is the boiling-water reactor, such as the system shown in Fig. 1.4. In other nuclear power plants a secondary fluid circulates from the reactor to the steam generator, where heat is transferred from the secondary fluid to the water, which in turn goes through a conventional steam cycle. Safety considerations and the necessity to keep the turbine, condenser, and related equipment from becoming radioactive are always major considerations in the design and operation of a nuclear power plant.

1.2 FUEL CELLS

When a conventional power plant is viewed as a whole, as shown in Fig. 1.5, fuel and air enter the power plant and products of combustion leave the unit. There is also a transfer of heat to the cooling water, and work is done in the form of the electrical energy leaving the power plant. The overall objective of a power plant is to convert the availability (to do work) of the fuel into work (in the form of electrical energy) in the most efficient manner, taking into consideration cost, space, safety, and environmental concerns.

We might well ask whether all the equipment in the power plant, such as the steam generator, the turbine, the condenser, and the pump, is necessary. Is it possible to produce electrical energy from the fuel in a more direct manner?

The fuel cell accomplishes this objective. Figure 1.6 shows a schematic arrangement of a fuel cell of the ion-exchange membrane type. In this fuel cell hydrogen and oxygen react to form water. Let us consider the general features of the operation of this type of fuel cell.

The flow of electrons in the external circuit is from anode to cathode. Hydrogen enters at the anode side, and oxygen enters at the cathode side. At the surface of the ion-exchange membrane the hydrogen is ionized according to the reaction

$$2H_2 \rightarrow 4H^+ + 4e^-$$

The electrons flow through the external circuit and the hydrogen ions flow through the membrane to the cathode, where the following reaction takes place.

$$4H^+ + 4e^- + O_2 \rightarrow 2H_2O$$

There is a potential difference between the anode and cathode, and thus there is a flow of electricity through a potential difference; this, in thermodynamic terms, is called work. There may also be a transfer of heat between the fuel cell and the surroundings.

At the present time the fuel used in fuel cells is usually either hydrogen or a mixture of gaseous hydrocarbons and hydrogen. The oxidizer is usually oxygen. However, current development is directed toward the production of fuel cells that use hydrocarbon fuels and air. Although the conventional (or nuclear) steam power plant is still used in large-scale power-generating systems, and conventional piston engines and gas turbines are still used in most transportation power systems, the fuel cell may eventually become a serious competitor. The fuel cell is already being used to produce power for space and other special applications.

Thermodynamics plays a vital role in the analysis, development, and design of all power-producing systems, including reciprocating internal-combustion engines and gas turbines. Considerations such as the increase of efficiency, improved design, optimum operating conditions, environmental pollution, and alternate methods of power generation involve, among other factors, the careful application of the fundamentals of thermodynamics.

1.3 THE VAPOR-COMPRESSION REFRIGERATION CYCLE

A simple vapor-compression refrigeration cycle is shown schematically in Fig. 1.7. The refrigerant enters the compressor as a slightly superheated vapor at a low pressure. It then leaves the compressor and enters the condenser as a vapor at some elevated pressure, where the refrigerant is condensed as heat is transferred to cooling water or to the surroundings. The refrigerant then leaves the condenser as a high-pressure liquid. The pressure of the liquid is decreased as it flows through the expansion valve, and as a result, some of the liquid flashes into cold vapor. The remaining liquid, now at a low pressure and temperature, is vaporized in the evaporator as heat is transferred from the refrigerated space. This vapor then reenters the compressor.

In a typical home refrigerator the compressor is located in the rear near the bottom of the unit. The compressors are usually hermetically sealed, that is, the motor and com-

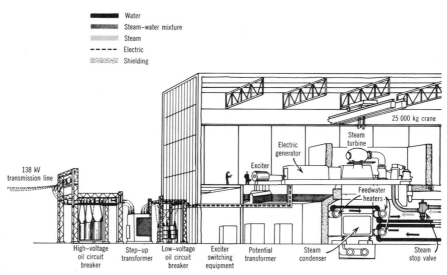

FIGURE 1.4 Schematic diagram of the Big Rock Point nuclear plant of Consumers Power Company at Charlevoix, Michigan (Courtesy Consumers Power Company).

pressor are mounted in a sealed housing, and the electric leads for the motor pass through this housing. This seal prevents leakage of the refrigerant. The condenser is also located at the back of the refrigerator and is arranged so that the air in the room flows past the condenser by natural convection. The expansion valve takes the form of a long capillary tube, and the evaporator is located around the outside of the freezing compartment inside the refrigerator.

Figure 1.8 shows a large centrifugal unit that is used to provide refrigeration for an air-conditioning unit. In this unit, water is cooled and then circulated to provide cooling where needed.

1.4 THE THERMOELECTRIC REFRIGERATOR

We may well ask the same question about the vapor-compression refrigerator that we asked about the steam power plant—is it possible to accomplish our objective in a more direct manner? Is it possible, in the case of a refrigerator, to use the electrical energy (which goes to the electric motor that drives the compressor) to produce cooling in a more direct manner, and to avoid the cost of the compressor, condenser, evaporator, and all the related piping?

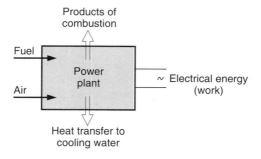

FIGURE 1.5 Schematic diagram of a power plant.

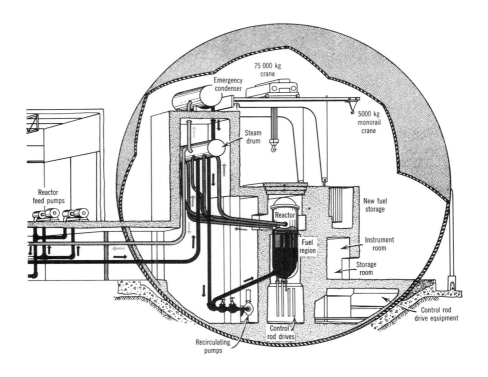

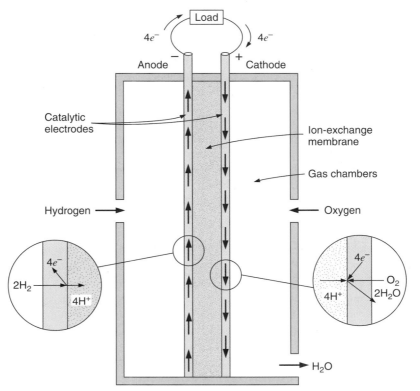

FIGURE 1.6 Schematic arrangement of an ion-exchange membrane type of fuel cell.

7

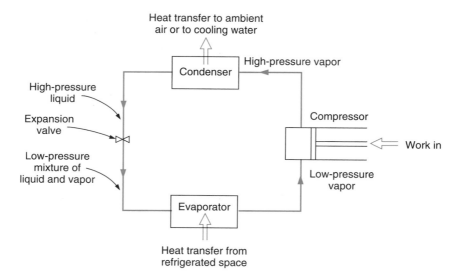

FIGURE 1.7 Schematic diagram of a simple refrigeration cycle.

The thermoelectric refrigerator is such a device. This is shown schematically in Fig. 1.9. The thermoelectric device, like the conventional thermocouple, uses two dissimilar materials. There are two junctions between these two materials in a thermoelectric refrigerator. One is located in the refrigerated space and the other in ambient surroundings. When a potential difference is applied, as indicated, the temperature of the junction located in the refrigerated space will decrease and the temperature of the other junction will increase. Under steady-state operating conditions, heat will be transferred from the refrigerated space to the cold junction. The other junction will be at a temperature above the ambient, and heat will be transferred from the junction to the surroundings.

A thermoelectric device can also be used to generate power by replacing the refrigerated space with a body that is at a temperature above the ambient. Such a system is shown in Fig. 1.10.

The thermoelectric refrigerator cannot yet compete economically with the conventional vapor-compression units. However, in certain special applications, the thermoelectric refrigerator is already in use and, in view of research and development efforts under way in this field, it is quite possible that thermoelectric refrigerators will be much more extensively used in the future.

1.5 THE AIR SEPARATION PLANT

One process of great industrial significance is the air separation plant, in which air is separated into its various components. The oxygen, nitrogen, argon, and rare gases so produced are used extensively in various industrial, research, space, and consumer goods applications. The air separation plant can be considered an example from two major fields, the chemical process industry and cryogenics. Cryogenics is a term applied to technology, processes, and research at very low temperatures (in general, below 150 K). In both chemical processing and cryogenics, thermodynamics is basic to an understanding of many phenomena that occur and to the design and development of processes and equipment.

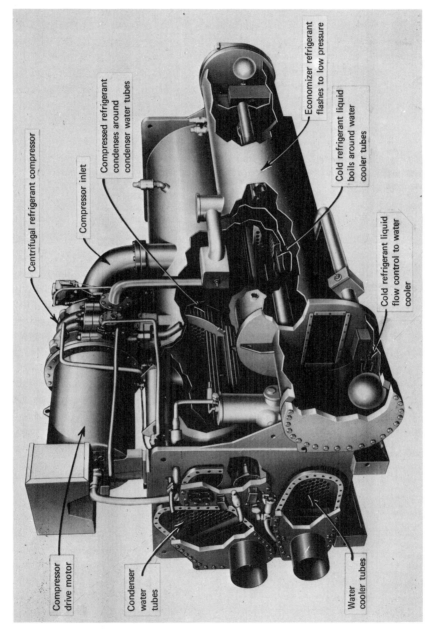

FIGURE 1.8 A refrigeration unit for an air-conditioning system (Courtesy Carrier Air Conditioning Co.)

Centrifugal refrigerant compressor

Compressor inlet

Compressed refrigerant condenses around condenser water tubes

Economizer refrigerant flashes to low pressure

Cold refrigerant liquid boils around water cooler tubes

Cold refrigerant liquid flow control to water cooler

Compressor drive motor

Condenser water tubes

Water cooler tubes

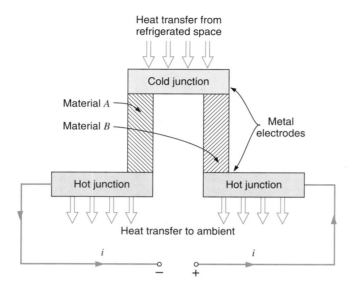

FIGURE 1.9 A thermo-electric refrigerator.

A number of different designs of air separation plants have been developed. Consider Fig. 1.11, which shows a somewhat simplified sketch of a type of plant that is frequently used. Air from the atmosphere is compressed to a pressure of 2 to 3 MPa. It is then purified, particularly to remove carbon dioxide (which would plug the flow passages as it solidifies when the air is cooled to its liquefaction temperature). The air is then compressed to a pressure of 15 to 20 MPa, cooled to the ambient temperature in the aftercooler, and dried to remove the water vapor (which would also plug the flow passages as it freezes).

The basic refrigeration in the liquefaction process is provided by two different processes. In one process the air in the expansion engine expands. During this process the air does work and as a result the temperature of the air is reduced. In the other refrig-

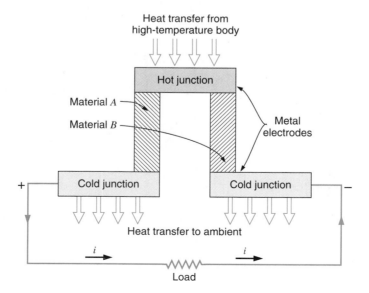

FIGURE 1.10 A thermoelectric power generation device.

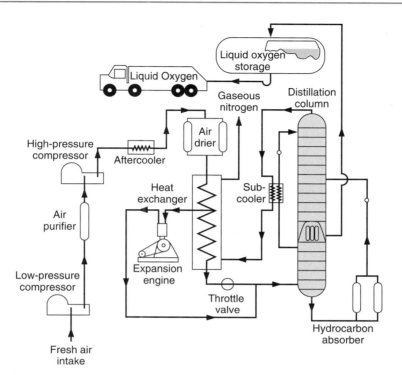

FIGURE 1.11 A simplified diagram of a liquid oxygen plant.

eration process air passes through a throttle valve that is so designed and so located that there is a substantial drop in the pressure of the air and, associated with this, a substantial drop in the temperature of the air.

As shown in Fig. 1.11, the dry, high-pressure air enters a heat exchanger. The air temperature drops as it flows through the heat exchanger. At some intermediate point in the heat exchanger, part of the air is bled off and flows through the expansion engine. The remaining air flows through the rest of the heat exchanger and through the throttle valve. The two streams join (both are at the pressure of 0.5 to 1 MPa) and enter the bottom of the distillation column, which is referred to as the high-pressure column. The function of the distillation column is to separate the air into its various components, principally oxygen and nitrogen. Two streams of different composition flow from the high-pressure column through throttle valves to the upper column (also called the low-pressure column). One of these streams is an oxygen-rich liquid that flows from the bottom of the lower column, and the other is a nitrogen-rich stream that flows through the subcooler. The separation is completed in the upper column. Liquid oxygen leaves from the bottom of the upper column, and gaseous nitrogen leaves from the top of the column. The nitrogen gas flows through the subcooler and the main heat exchanger. It is the transfer of heat to this cold nitrogen gas that causes the high-pressure air entering the heat exchanger to become cooler.

Not only is a thermodynamic analysis essential to the design of the system as a whole, but essentially every component of such a system, including the compressors, the expansion engine, the purifiers and driers, and the distillation column, operates according to the principles of thermodynamics. In this separation process we are also concerned with the thermodynamic properties of mixtures and the principles and procedures by which these mixtures can be separated. This is the type of problem encountered in the refining of petroleum and many other chemical processes. It should also be noted that cryogenics is particularly relevant to many aspects of the space program, and a thorough knowledge of thermodynamics is essential for creative and effective work in cryogenics.

1.6 THE GAS TURBINE

The basic operation of a gas turbine is similar to the steam power plant, except that air is used instead of water. Fresh atmospheric air flows through a compressor that brings it to a high pressure. Energy is then added by spraying fuel into the air and igniting it so the combustion generates a high temperature flow. This high-temperature, high-pressure gas enters a turbine, where it expands down to the exhaust pressure, giving a shaft work out in the process. The turbine shaft work is used to drive the compressor and other devices, such as an electric generator that may be coupled to the shaft. The energy that is not used for shaft work comes out in the exhaust gases, so these have either a high temperature or a high velocity. The purpose of the gas turbine determines the design so that the most desirable energy form is maximized. An example of a large gas turbine for stationary power generation is shown in Fig. 1.12; the unit has 16 stages of compression, 4 stages in the turbine and is rated at 150 MW. Notice that since the combustion of fuel uses the oxygen in the air, the exhaust gases cannot be recirculated as the water is in the steam power plant.

A gas turbine is often the preferred power generating device where a large amount of power is needed, but only a small physical size is possible. Examples are jet engines, turbofan jet engines, offshore oilrig power plants, ship engines, helicopter engines, smaller local power plants, or peak load power generators in larger power plants. Since the gas turbine has relatively high exhaust temperatures, it can also be arranged so the exhaust gases are used to heat water that runs in a steam power plant before the exhaust to the atmosphere.

In the examples mentioned previously, the jet engine and turboprop application utilizes part of the power to discharge the gases at high velocity. This is what generates the thrust of the engine that moves the airplane forward. The gas turbines in these applications are therefore designed differently than for the stationary power plant, where the en-

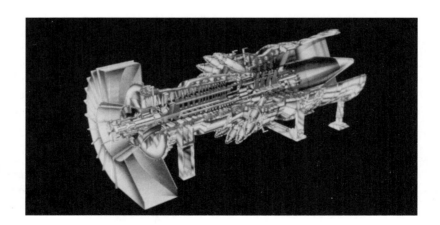

FIGURE 1.12 A 150-MW gas turbine (Courtesy Westinghouse Electric Corporation).

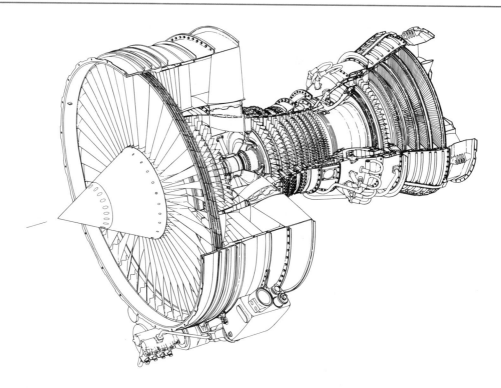

FIGURE 1.13 A turbofan jet engine (Courtesy General Electric Aircraft Engines).

ergy is taken out as shaft work to an electric generator. An example of a turbofan jet engine used in a commercial airplane is shown in Fig. 1.13. The large front end fan also blows air past the engine providing cooling and gives additional thrust.

1.7 THE CHEMICAL ROCKET ENGINE

The advent of missiles and satellites brought to prominence the use of the rocket engine as a propulsion power plant. Chemical rocket engines may be classified as either liquid propellant or solid propellant, according to the fuel used.

Figure 1.14 shows a simplified schematic diagram of a liquid propellant rocket. The oxidizer and fuel are pumped through the injector plate into the combustion chamber where combustion takes place at high pressure. The high-pressure, high-temperature products of combustion expand as they flow through the nozzle, and as a result they leave the nozzle with a high velocity. The momentum change associated with this increase in velocity gives rise to the forward thrust on the vehicle.

The oxidizer and fuel must be pumped into the combustion chamber, and some auxiliary power plant is necessary to drive the pumps. In a large rocket this auxiliary power plant must be very reliable and have a relatively high power output, yet it must be light in weight. The oxidizer and fuel tanks occupy the largest part of the volume of an actual rocket, and the range and payload of a rocket are determined largely by the amount of oxidizer and fuel that can be carried. Many different fuels and oxidizers have been considered and tested, and much effort has gone into the development of fuels and oxidizers that will give a higher thrust per unit mass rate of flow of reactants. Liquid

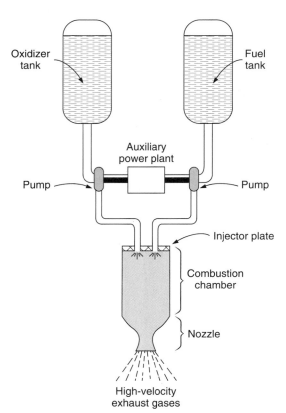

FIGURE 1.14 Simplified schematic diagram of a liquid-propellant rocket engine.

oxygen is frequently used as the oxidizer in liquid-propellant rockets, and liquid hydrogen is frequently used as the fuel.

Much work has also been done on solid-propellant rockets. They have been very successfully used for jet-assisted takeoffs of airplanes, military missiles, and space vehicles. They are much simpler in both the basic equipment required for operation and the logistic problems involved in their use, but they are more difficult to control.

1.8 ENVIRONMENTAL ISSUES

In the first seven sections of this chapter, we have introduced and discussed a number of devices, each of which produces certain effects for the use and convenience of humankind. For example, the construction and operation of the steam power plant creates electricity, which is so deeply entrenched in our society that we take its ready availability for granted. In recent years, however, it has become increasingly apparent that we need to consider seriously the effects of such an operation on our environment. Combustion of hydrocarbon fuels releases carbon dioxide into the atmosphere, where its concentration is increasing. Carbon dioxide, as well as other gases, absorbs infrared radiation from the surface of the earth, holding it close to the planet and creating the "greenhouse effect," which in turn is believed to cause global warming and critical climatic changes around the earth. Power plant combustion, particularly of coal, releases sulfur dioxide, which is absorbed in clouds and later falls as acid rain in many areas. Combustion processes in power plants and gasoline and diesel engines generate pollutants other than the carbon dioxide and sulfur oxides mentioned. Species such as carbon monoxide, nitric oxides, and partly burned fuels together with particulates all contribute to atmospheric pollution and are regulated by law for many applications. Refrigeration and air-conditioning systems, as well as other industrial processes, use certain chlorofluorocarbons that eventually find their way to the upper atmosphere and destroy the protective ozone layer.

These are only some of the many environmental problems caused by our efforts to produce goods and effects intended to improve our way of life. During our study of thermodynamics, which is the science of the conversion of energy from one form to another, we must continue to reflect on these issues. We must consider how we can eliminate or at least minimize damaging effects, as well as use our natural resources efficiently and responsibly.

2 SOME CONCEPTS AND DEFINITIONS

One excellent definition of thermodynamics is that it is the science of energy and entropy. Since we have not yet defined these terms, an alternate definition in already familiar terms is: thermodynamics is the science that deals with heat and work and these properties of substances that bear a relation to heat and work. Like all sciences, the basis of thermodynamics is experimental observation. In thermodynamics these findings have been formalized into certain basic laws, which are known as the first, second, and third laws of thermodynamics. In addition to these laws, the zeroth law of thermodynamics, which in the logical development of thermodynamics precedes the first law, has been set forth.

In the chapters that follow, we will present these laws and the thermodynamic properties related to these laws, and apply them to a number of representative examples. The objective of the student should be to gain both a thorough understanding of the fundamentals and an ability to apply these fundamentals to thermodynamic problems. The examples and problems further this twofold objective. It is not necessary for the student to memorize numerous equations, for problems are best solved by the application of the definitions and laws of thermodynamics. In this chapter some concepts and definitions basic to thermodynamics are presented.

2.1 A THERMODYNAMIC SYSTEM AND THE CONTROL VOLUME

A thermodynamic system is a device or combination of devices containing a quantity of matter that is being studied. To define this more precisely a control volume is chosen, so that it contains the matter and devices inside a control surface. Everything external to the control volume is the surroundings, with the separation given by the control surface. The surface may be open or closed to mass flows and it may have flows of energy in terms of heat transfer and work across it. The boundaries may be movable or stationary. In the case of a control surface that is closed to mass flow, so that no mass can escape or enter the control volume, it is called a **control mass** containing the same amount of matter at all times.

Selecting the gas in the cylinder of Fig. 2.1 as a control volume by placing a control surface around it, we recognize this as a control mass. If a Bunsen burner is placed under the cylinder, the temperature of the gas will increase and the piston will rise. As the piston rises, the boundary of the control mass moves. As we will see later, heat and

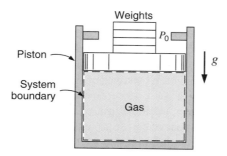

FIGURE 2.1 Example of a control mass.

work cross the boundary of the control mass during this process, but the matter that comprises the control mass can always be identified and remains the same.

An isolated system is one that is not influenced in any way by the surroundings. This means that no mass, heat, or work cross the boundary of the system. In many cases a thermodynamic analysis must be made of a device, such as an air compressor, which has a flow of mass into it, out of it, or both, as shown schematically in Fig. 2.2. The procedure followed in such an analysis is to specify a control volume that surrounds the device under consideration. The surface of this control volume is the control surface, which may have mass, momentum, as well as heat and work, cross it.

Thus the more general control surface defines a control volume, where mass may flow in or out, with a control mass as the special case of no mass flow in or out, so the control mass contains a fixed mass at all times, hence its name. The difference in the formulation of the analysis is considered in detail in Chapter 5. The terms *closed system* (fixed mass) and *open system* (involving a flow of mass) are sometimes used to make this distinction. Here, we use the term *system* as a more general and loose description for a mass, device, or combination of devices that then is more precisely defined, when a control volume is selected. The procedure that will be followed in the presentation of the first and the second laws of thermodynamics is first to present these laws for a control mass and then to extend the analysis to the more general control volume.

2.2 MACROSCOPIC VERSUS MICROSCOPIC POINT OF VIEW

An investigation into the behavior of a system may be undertaken from either a microscopic or a macroscopic point of view. Let us briefly describe a system from a microscopic point of view. Consider a system consisting of a cube 25 mm on a side and con-

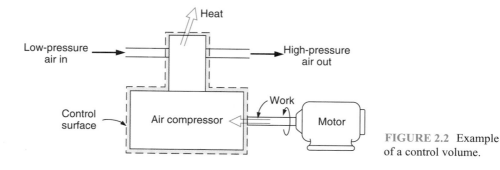

FIGURE 2.2 Example of a control volume.

taining a monatomic gas at atmospheric pressure and temperature. This volume contains approximately 10^{20} atoms. To describe the position of each atom, we need to specify three coordinates; to describe the velocity of each atom, we specify three velocity components.

Thus, to describe completely the behavior of this system from a microscopic point of view we must deal with at least 6×10^{20} equations. Even with a large digital computer, this is a quite hopeless computational task. However, there are two approaches to this problem that reduce the number of equations and variables to a few that can be computed relatively easily. One approach is the statistical approach, in which, on the basis of statistical considerations and probability theory, we deal with "average" values for all particles under consideration. This is usually done in connection with a model of the atom under consideration. This is the approach used in the disciplines known as kinetic theory and statistical mechanics.

The other approach that reduces the number of variables to a few that can be handled is the macroscopic point of view of classical thermodynamics. As the word macroscopic implies, we are concerned with the gross or average effects of many molecules. These effects can be perceived by our senses and measured by instruments. However, what we really perceive and measure is the time-averaged influence of many molecules. For example, consider the pressure a gas exerts on the walls of its container. This pressure results from the change in momentum of the molecules as they collide with the wall. From a macroscopic point of view, however, we are not concerned with the action of the individual molecules but with the time-averaged force on a given area, which can be measured by a pressure gage. In fact, these macroscopic observations are completely independent of our assumptions regarding the nature of matter.

Although the theory and development in this book is presented from a macroscopic point of view, a few supplementary remarks regarding the significance of the microscopic perspective are included as an aid to the understanding of the physical processes involved. Another book in this series, *Introduction to Thermodynamics: Classical and Statistical,* by R. E. Sonntag and G. J. Van Wylen, includes thermodynamics from the microscopic and statistical point of view.

A few remarks should be made regarding the continuum. From the macroscopic view, we are always concerned with volumes that are very large compared to molecular dimensions and, therefore, with systems that contain many molecules. Because we are not concerned with the behavior of individual molecules, we can treat the substance as being continuous, disregarding the action of individual molecules; this is called a continuum. The concept of a continuum, of course, is only a convenient assumption that loses validity when the mean free path of the molecules approaches the order of magnitude of the dimensions of the vessel, as, for example, in high-vacuum technology. In much engineering work the assumption of a continuum is valid and convenient, and goes hand in hand with the macroscopic view.

2.3 PROPERTIES AND STATE OF A SUBSTANCE

If we consider a given mass of water, we recognize that this water can exist in various forms. If it is a liquid initially, it may become a vapor when it is heated or a solid when it is cooled. Thus, we speak of the different phases of a substance. A phase is defined as a quantity of matter that is homogeneous throughout. When more than one phase is present, the phases are separated from each other by the phase boundaries. In each phase the substance may exist at various pressures and temperatures or, to use the thermodynamic term, in various states. The state may be identified or described by certain observable, macroscopic properties; some familiar ones are temperature, pressure, and density. In later chapters other properties will be introduced. Each of the properties of a substance in a given state has only one definite value, and these properties always have the same value for a given state, regardless of how the substance arrived at the state. In fact, a property can be defined as any quantity that depends on the state of the system and is independent of the path (that is, the prior history) by which the system arrived at the given state. Conversely, the state is specified or described by the properties, and later we will consider the number of independent properties a substance can have, that is, the minimum number of properties that must be specified to fix the state of the substance.

Thermodynamic properties can be divided into two general classes, intensive and extensive properties. An intensive property is independent of the mass; the value of an extensive property varies directly with the mass. Thus, if a quantity of matter in a given state is divided into two equal parts, each part will have the same value of intensive properties as the original, and half the value of the extensive properties. Pressure, temperature, and density are examples of intensive properties. Mass and total volume are examples of extensive properties. Extensive properties per unit mass, such as specific volume, are intensive properties.

Frequently we will refer not only to the properties of a substance but to the properties of a system. When we do so we necessarily imply that the value of the property has significance for the entire system, and this implies equilibrium. For example, if the gas that comprises the system (control mass) in Fig. 2.1 is in thermal equilibrium, the temperature will be the same throughout the entire system, and we may speak of the temperature as a property of the system. We may also consider mechanical equilibrium, which is related to pressure. If a system is in mechanical equilibrium, there is no tendency for the pressure at any point to change with time as long as the system is isolated from the surroundings. There will be a variation in pressure with elevation because of the influence of gravitational forces, although under equilibrium conditions there will be no tendency for the pressure at any location to change. However, in many thermodynamic problems, this variation in pressure with elevation is so small that it can be neglected. Chemical equilibrium is also important and will be considered in Chapter 13. When a system is in equilibrium regarding all possible changes of state, we say that the system is in thermodynamic equilibrium.

2.4 PROCESSES AND CYCLES

Whenever one or more of the properties of a system change, we say that a change in state has occurred. For example, when one of the weights on the piston in Fig. 2.3 is removed, the piston rises and a change in state occurs, for the pressure decreases and the specific volume increases. The path of the succession of states through which the system passes is called the process.

Let us consider the equilibrium of a system as it undergoes a change in state. The moment the weight is removed from the piston in Fig. 2.3, mechanical equilibrium does

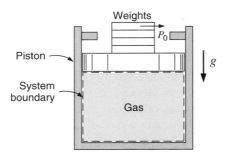

FIGURE 2.3 Example of a system that may undergo a quasi-equilibrium process.

not exist, and as a result the piston is moved upward until mechanical equilibrium is again restored. The question is this: Since the properties describe the state of a system only when it is in equilibrium, how can we describe the states of a system during a process if the actual process occurs only when equilibrium does not exist? One step in the answer to this question concerns the definition of an ideal process, which we call a quasi-equilibrium process. A quasi-equilibrium process is one in which the deviation from thermodynamic equilibrium is infinitesimal, and all the states the system passes through during a quasi-equilibrium process may be considered equilibrium states. Many actual processes closely approach a quasi-equilibrium process, and may be so treated with essentially no error. If the weights on the piston in Fig. 2.3 are small and are taken off one by one, the process could be considered quasi-equilibrium. On the other hand, if all the weights were removed at once, the piston would rise rapidly until it hit the stops. This would be a nonequilibrium process, and the system would not be in equilibrium at any time during this change of state.

For nonequilibrium processes, we are limited to a description of the system before the process occurs and after the process is completed and equilibrium is restored. We are unable to specify each state through which the system passes or the rate at which the process occurs. However, as we will see later, we are able to describe certain overall effects that occur during the process.

Several processes are described by the fact that one property remains constant. The prefix iso- is used to describe this process. An isothermal process is a constant-temperature process, an isobaric (sometimes called isopiestic) process is a constant-pressure process, and an isochoric process is a constant-volume process.

When a system in a given initial state goes through a number of different changes of state or processes and finally returns to its initial state, the system has undergone a cycle. Therefore, at the conclusion of a cycle, all the properties have the same value they had at the beginning. Steam (water) that circulates through a steam power plant undergoes a cycle.

A distinction should be made between a thermodynamic cycle, which has just been described, and a mechanical cycle. A four-stroke cycle internal-combustion engine goes through a mechanical cycle once every two revolutions. However, the working fluid does not go through a thermodynamic cycle in the engine, since air and fuel are burned and changed to products of combustion that are exhausted to the atmosphere. In this text the term cycle will refer to a thermodynamic "cycle" unless otherwise designated.

2.5 UNITS FOR MASS, LENGTH, TIME, AND FORCE

Since we are considering thermodynamic properties from a macroscopic perspective, we are dealing with quantities that can, either directly or indirectly, be measured and counted. Therefore, the matter of units becomes an important consideration. In the remaining sections of this chapter we will define certain thermodynamic properties and the basic units. The relation between force and mass is often a difficult matter for students, and is considered in this section in some detail.

Force, mass, length, and time are related by Newton's second law of motion, which states that the force acting on a body is proportional to the product of the mass and the acceleration in the direction of the force.

$$F \propto ma$$

The concept of time is well established. The basic unit of time is the second (s), which in the past was defined in terms of the solar day, the time interval for one complete revolution of the earth relative to the sun. Since this period varies with the season of the year, an average value over a one-year period is called the mean solar day, and the mean solar second is 1/86 400 of the mean solar day. (The measurement of the earth's rotation is sometimes made relative to a fixed star, in which case the period is called a sidereal day.) In 1967, the General Conference of Weights and Measures (CGPM) adopted a definition of the second as the time required for a beam of cesium-133 atoms to resonate 9 192 631 770 cycles in a cesium resonator.

For periods of time less than a second, the prefixes milli, micro, nano, or pico, as listed in Table 2.1, are commonly used. For longer periods of time, the units minute (min), hour (h), or day (day) are frequently used. It should be pointed out that the prefixes in Table 2.1 are used with many other units as well.

The concept of length is also well established. The basic unit of length is the meter (m). For many years the accepted standard was the International Prototype Meter, the distance between two marks on a platinum–iridium bar under certain prescribed conditions. This bar is maintained at the International Bureau of Weights and Measures, Sevres, France. In 1960, the CGPM adopted a definition of the meter as a length equal to 1 650 763.73 wavelengths in a vacuum of the orange-red line of krypton-86. Then in 1983, the CGPM adopted a more precise definition of the meter in terms of the speed of light (which is now a fixed constant): the meter is the length of the path traveled by light in a vacuum during a time interval of 1/299 792 458 of a second.

The fundamental unit of mass is the kilogram (kg). As adopted by the first CGPM in 1889 and restated in 1901, it is the mass of a certain platinum–iridium cylinder maintained under prescribed conditions at the International Bureau of Weights and Measures.

TABLE 2.1
Unit Prefixes

Factor	Prefix	Symbol	Factor	Prefix	Symbol
10^{12}	tera	T	10^{-3}	milli	m
10^{9}	giga	G	10^{-6}	micro	μ
10^{6}	mega	M	10^{-9}	nano	n
10^{3}	kilo	k	10^{-12}	pico	p

A related unit that is used frequently in thermodynamics is the mole (mol), defined as an amount of substance containing as many elementary entities as there are atoms in 0.012 kg of carbon-12. These elementary entities must be specified, and may be atoms, molecules, electrons, ions, or other particles or specific groups. For example, one mole of diatomic oxygen, having a molecular weight of 32 (compared to 12 for carbon), is a mass of 0.032 kg. The mole is often termed a gram mole, since it is an amount of substance in grams numerically equal to the molecular weight. In this text, when using the metric SI system we will find it preferable to use the kilomole (kmol), the amount of substance in kilograms numerically equal to the molecular weight, rather than the mole.

The system of units in use presently throughout most of the world is the metric International System, commonly referred to as SI units (from Le Système International d'Unités). In this system, the second, meter, and kilogram are the basic units for time, length, and mass, respectively, as just defined, and the unit of force is defined directly from Newton's second law.

Therefore, a proportionality constant is unnecessary, and we may write that law as an equality:

$$F = ma$$

The unit of force is the newton (N), which by definition is the force required to accelerate a mass of one kilogram at the rate of one meter per second per second.

$$1\,N = 1\,kg\,m/s^2$$

It is worth noting that SI units derived from proper nouns use capital letters for symbols; others use the lowercase letters. The liter, with the symbol L, is an exception.

The traditional system of units used in the United States is the English Engineering System. In this system the unit of time is the second, which has been discussed earlier. The basic unit of length is the foot (ft), which at present is defined in terms of the meter as

$$1\,ft = 0.3048\,m$$

The inch (in.) is defined in terms of the foot

$$12\,in. = 1\,ft$$

The unit of mass in this system is the pound mass (lbm). It was originally the mass of a certain platinum cylinder kept in the Tower of London, but now it is defined in terms of the kilogram as

$$1\,lbm = 0.453\,592\,37\,kg$$

A related unit is the pound mole (lb mol), which is an amount of substance in pounds mass numerically equal to the molecular weight of that substance. It is important to distinguish between a pound mole and a mole (gram mole).

In the English Engineering System of Units, the concept of force is not defined from Newton's second law, but is instead established as an independent quantity. Thus, for this system, unlike other systems of units, a conversion constant g_c is necessary in Newton's second law. The unit for force is defined in terms of an experimental procedure as follows. Let the standard pound mass be suspended in the earth's gravitational field at a location where the acceleration due to gravity is 32.1740 ft/s². The force with which the standard pound mass is attracted to the earth (the buoyant effects of the atmosphere on the standard pound mass must also be standardized) is defined as the unit for force and is termed a pound force (lbf). We must be careful to distinguish between a lbm and a lbf, and we do not use the term pound alone. Since we have independently defined the units for force, mass, length, and time, we have the relation, from Newton's second law,

$$F = \frac{ma}{g_c}$$

or

$$1 \text{ lbf} = \frac{1 \text{ lbm} \times 32.174 \text{ ft} / \text{s}^2}{g_c}$$

or

$$g_c = 32.174 \frac{\text{lbm ft}}{\text{lbf s}^2}$$

Note that the conversion constant g_c has both a numerical value and dimensions. We also note that for gravitational accelerations that are not too different from the standard value (32.174 ft/s^2 or 9.806 65 m/s^2), the lbf is approximately numerically equal to the lbm. This is the reason that people often drop the designation, which can lead to confusion and inconsistencies. To illustrate, let us calculate the force due to gravity on a one-pound mass at a location where the acceleration due to gravity is 32.14 ft/s^2 (about 10 000 ft above sea level).

$$F = \frac{ma}{g_c} = \frac{1 \text{ lbm} \times 32.14 \text{ ft} / \text{s}^2}{32.174 \text{ lbm ft} / \text{lbf s}^2}$$
$$= 0.999 \text{ lbf}$$

We further note that the pound force could very well have been defined as

$$1 \text{ lbf} = 32.174 \text{ lbm ft} / \text{s}^2$$

in a manner analogous to that for the newton in the SI system. It follows that

$$1 = 32.174 \frac{\text{lbm ft}}{\text{lbf s}^2}$$

Comparing this with the conversion constant g_c, we have

$$g_c = 32.174 \frac{\text{lbm ft}}{\text{lbf s}^2} = 1$$

Since a pure number can be substituted into an equation at any point, we can therefore substitute this unit conversion constant g_c as necessary. Throughout the remainder of this text, we will follow this philosophy, and will not include the constant g_c explicitly in our equations, even when our attention is focused on the English Engineering system of units.

The term "weight" is often used with respect to a body, and is sometimes confused with mass. Weight is really correctly used only as a force. When we say a body weighs so much, we mean that this is the force with which it is attracted to the earth (or some other body), that is, the product of its mass and the local gravitational acceleration. The mass of a substance remains constant with elevation, but its weight varies with elevation.

2.6 ENERGY

One of the very important concepts in a study of thermodynamics is the concept of energy. Energy is a fundamental concept, such as mass or force and, as is often the case with such concepts, is very difficult to define. Energy has been defined as the capability to produce an effect. Fortunately the word "energy" and the basic concept that this word represents are familiar to us in everyday usage, and a precise definition is not essential at this point.

It is important to note that energy can be stored within a system and can be transferred (as heat, for example) from one system to another. In a study of statistical thermodynamics we would examine, from a molecular view, the ways in which energy can be stored. Because it is helpful in a study of classical thermodynamics to have some notion of how this energy is stored, a brief introduction is presented here.

Consider as a system a certain gas at a given pressure and temperature contained within a tank or pressure vessel. When considered from the molecular view, we identify three general forms of energy.

1. Intermolecular potential energy, which is associated with the forces between molecules.
2. Molecular kinetic energy, which is associated with the translational velocity of individual molecules.
3. Intramolecular energy (that within the individual molecules), which is associated with the molecular and atomic structure and related forces.

The first of these forms of energy, the intermolecular potential energy, depends on the magnitude of the intermolecular forces and the position the molecules have relative to each other at any instant of time. It is impossible to determine accurately the magnitude of this energy because we do not know either the exact configuration and orientation of the molecules at any time, or the exact intermolecular potential function. However, there are two situations for which we can make good approximations. The first situation is at low or moderate densities. In this case the molecules are relatively widely spaced, so that only two-molecule or two- and three-molecule interactions contribute to the potential energy. At these low and moderate densities, techniques are available for determining, with reasonable accuracy, the potential energy of a system composed of reasonably simple molecules. The second situation is at very low densities; under these conditions the average intermolecular distance between molecules is so large that the potential energy may be assumed to be zero. Consequently, we have in this case a system of independent particles (an ideal gas) and, therefore, from a statistical point of view, we are able to concentrate our efforts on evaluating the molecular translational and internal energies.

The translational energy, which depends only on the mass and velocities of the molecules, is determined by using the equations of mechanics—either quantum or classical.

The intramolecular internal energy is more difficult to evaluate because, in general, it may result from a number of contributions. Consider a simple monatomic gas such as helium. Each molecule consists of a helium atom. Such an atom possesses electronic energy as a result of both orbital angular momentum of the electrons about the nucleus and angular momentum of the electrons spinning on their axes. The electronic energy is commonly very small compared with the translational energies. (Atoms also possess nuclear

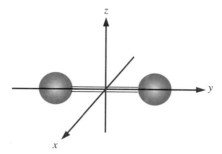

FIGURE 2.4 The coordinate system for a diatomic molecule.

energy, which, except in the case of nuclear reactions, is constant. We are not concerned with nuclear energy at this time.) When we consider more complex molecules, such as those comprised of two or three atoms, additional factors must be considered. In addition to having electronic energy, a molecule can rotate about its center of gravity and thus have rotational energy. Furthermore, the atoms may vibrate with respect to each other and have vibrational energy. In some situations there may be an interaction between the rotational and vibrational modes of energy.

In the evaluation of the energy of a molecule, reference is often made to the degree of freedom, f, for these energy modes. For a monatomic molecule, such as helium, $f = 3$, and this represents the three directions, x, y, and z in which the molecule can move. For a diatomic molecule, such as oxygen, $f = 6$. Three of these are the translation of the molecule as a whole in the x, y, and z directions, and two are for rotation. The reason why there are only two modes of rotational energy is evident from Fig. 2.4, where we take the origin of the coordinate system at the center of gravity of the molecule, and the y-axis along the molecule's internuclear axis. The molecule will then have an appreciable moment of inertia about the x-axis and the z-axis, but not about the y-axis. The sixth degree of freedom of the molecule is vibration, which relates to stretching of the bond joining the atoms.

For a more complex molecule such as H_2O, there are additional vibrational degrees of freedom. Figure 2.5 shows a model of the H_2O molecule. From this diagram it is evident that there are three vibrational degrees of freedom. It is also possible to have rotational energy about all three axes. Thus, for the H_2O molecule, there are nine degrees of freedom ($f = 9$), three translational, three rotational, and three vibrational.

This general discussion can be summarized by referring to Fig. 2.6. Let heat be transferred to the water. During this process the temperature of the liquid and vapor (steam) will increase, and eventually all the liquid will become vapor. From the macroscopic view we are concerned only with the energy that is transferred as heat, the change in properties, such as temperature and pressure, and the total amount of energy (relative to some base) that the H_2O contains at any instant. Thus, questions about how energy is

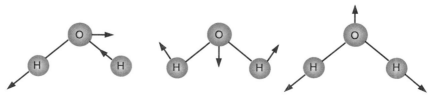

FIGURE 2.5 The three principal vibrational modes for the water molecule.

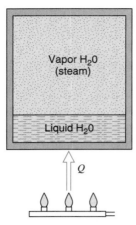

FIGURE 2.6 Heat transfer to water.

stored in the H_2O do not concern us. From a microscopic viewpoint we are concerned about the way in which energy is stored in the molecules. We might be interested in developing a model of the molecule so that we could predict the amount of energy required to change the temperature a given amount. Although the focus in this book is on the macroscopic or classical viewpoint, it is helpful to keep in mind the microscopic or statistical perspective as well, as the relationship between the two helps us in understanding basic concepts such as energy.

2.7 SPECIFIC VOLUME

The specific volume of a substance is defined as the volume per unit mass, and is given the symbol v. The density of a substance is defined as the mass per unit volume, and is therefore the reciprocal of the specific volume. Density is designated by the symbol ρ. Specific volume and density are intensive properties.

The specific volume of a system in a gravitational field may vary from point to point. For example, if the atmosphere is considered a system, the specific volume increases as the elevation increases. Therefore, the definition of specific volume involves the specific volume of a substance at a point in a system.

Consider a small volume δV of a system, and let the mass be designated δm. The specific volume is defined by the relation

$$v = \lim_{\delta V \to \delta V'} \frac{\delta V}{\delta m}$$

where $\delta V'$ is the smallest volume for which the mass can be considered a continuum. Volumes smaller than this will lead to the recognition that mass is not evenly distributed in space, but concentrated in particles as molecules, atoms, electrons, etc. This is tentatively indicated in Fig. 2.7, where in the limit of a zero volume the specific volume may be infinite (the volume does not contain any mass) or very small (the volume is part of a nucleus).

Thus, in a given system, we should speak of the specific volume or density at a point in the system, and recognize that this may vary with elevation. However, most of

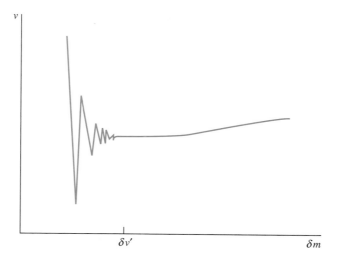

FIGURE 2.7 The continuum limit for the specific volume.

the systems that we consider are relatively small, and the change in specific volume with elevation is not significant. Therefore, we can speak of one value of specific volume or density for the entire system.

In this text, the specific volume and density will be given either on a mass or on a mole basis. A bar over the symbol (lowercase) will be used to designate the property on a mole basis. Thus, $\bar{v}$ will designate molal specific volume and $\bar{\rho}$ will designate the molal density. In SI units, those for specific volume are m^3/kg and m^3/mol (or $m^3/kmol$); for density the corresponding units are kg/m^3 and mol/m^3 (or $kmol/m^3$). In English units, those for specific volume are ft^3/lbm and ft^3/lb mol; the corresponding units for density are lbm/ft^3 and lb mol/ft^3.

Although the SI unit for volume is the cubic meter, a commonly used volume unit is the liter (L), which is a special name given to a volume of 0.001 cubic meters, that is, 1 L = 10^{-3} m^3.

2.8 PRESSURE

When dealing with liquids and gases, we ordinarily speak of pressure; for solids we speak of stresses. The pressure in a fluid at rest at a given point is the same in all directions, and we define pressure as the normal component of force per unit area. More specifically, if $\delta\mathcal{A}$ is a small area, $\delta\mathcal{A}'$ the smallest area over which we can consider the fluid a continuum, and δF_n the component of force normal to $\delta\mathcal{A}$, we define pressure, P, as

$$P = \lim_{\delta\mathcal{A} \to \delta\mathcal{A}'} \frac{\delta F_n}{\delta\mathcal{A}}$$

where the lower limit corresponds to sizes as mentioned for the specific volume, shown in Fig. 2.7. The pressure P at a point in a fluid in equilibrium is the same in all directions. In a viscous fluid in motion, the variation in the state of stress with orientation becomes an important consideration. These considerations are beyond the scope of this book, and we will consider pressure only in term of a fluid in equilibrium.

The unit for pressure in the International System is the force of one newton acting on a square meter area, which is called the pascal (Pa). That is,

$$1 \text{ Pa} = 1 \text{ N} / \text{m}^2$$

In general, the unit for pressure that is consistent with the English units is pounds force per square foot (lbf/ft^2). On the other hand, in common parlance and general experimental work, pressures are often measured in pounds force per square inch (lbf/in.^{2}). Therefore, the student should be careful in numerical calculations to introduce the conversion 144 in.^{2} = 1 ft^2 as necessary.

Two other units, not part of the International System, continue to be widely used. These are the bar, where

$$1 \text{ bar} = 10^5 \text{ Pa} = 0.1 \text{ MPa}$$

and the standard atmosphere, where

$$1 \text{ atm} = 101\,325 \text{ Pa} = 14.696 \text{ lbf} / \text{in.}^2$$

which is slightly larger than the bar. In this text, we will normally use the SI unit, the pascal, and especially the multiples of kilopascal and megapascal. The bar will be utilized often in the examples and problems, but the atmosphere will not be used, except in specifying certain reference points.

In most thermodynamic investigations we are concerned with absolute pressure. Most pressure and vacuum gauges, however, read the difference between the absolute pressure and the atmospheric pressure existing at the gauge. This referred to as gauge pressure. This is shown graphically in Fig. 2.8, and the following examples illustrate the principles. Pressures below atmospheric and slightly above atmospheric, and pressure differences (for example, across an orifice in a pipe) are frequently measured with a manometer, which contains water, mercury, alcohol, oil, or other fluids. From the princi-

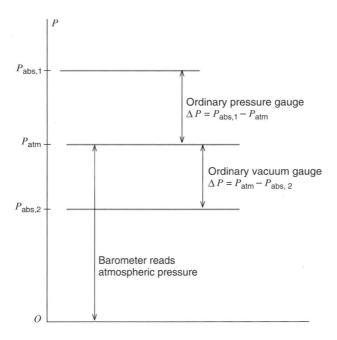

FIGURE 2.8
Illustration of terms used in pressure measurement.

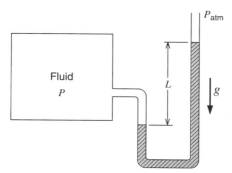

FIGURE 2.9 Example of pressure measurement using a column of fluid.

ples of hydrostatics one concludes that for a difference in level of L meters, the pressure difference in pascals is calculated by the relation

$$\Delta P = \rho L g$$

where ρ is the density of the fluid, and g is the local acceleration due to gravity. The accepted standard value for gravitational acceleration is

$$g = 9.806\ 65\ \text{m}/\text{s}^2 = 32.174\ \text{ft}/\text{s}^2$$

but the value varies with location and elevation. The use of a manometer is illustrated in Fig. 2.9.

For distinguishing between absolute and gauge pressure in this text, the term pascal or lbf/in.^{2} will always refer to absolute pressure. Any gauge pressure will be indicated as such.

2.9 EQUALITY OF TEMPERATURE

Although temperature is a familiar property, an exact definition of it is difficult. We are aware of "temperature" first of all as a sense of hotness or coldness when we touch an object. We also learn early that when a hot body and a cold body are brought into contact, the hot body becomes cooler and the cold body becomes warmer. If these bodies remain in contact for some time, they usually appear to have the same hotness or coldness. However, we also realize that our sense of hotness or coldness is very unreliable. Sometimes very cold bodies may seem hot, and bodies of different materials that are at the same temperature appear to be at different temperatures.

Because of these difficulties in defining temperature, we define equality of temperature. Consider two blocks of copper, one hot and the other cold, each of which is in contact with a mercury-in-glass thermometer. If these two blocks of copper are brought into thermal communication, we observe that the electrical resistance of the hot block decreases with time and that of the cold block increases with time. After a period of time has elapsed, however, no further changes in resistance are observed. Similarly, when the blocks are first brought in thermal communication, the length of a side of the hot block decreases with time, but the length of a side of the cold block increases with time. After a period of time, no further change in length of either of the blocks is perceived. In addition, the mercury column of the thermometer in the hot block drops at first and that in the cold block rises, but after a period of time no further changes in height are observed. We may say, therefore, that two bodies have equality of temperature if, when they are in thermal communication, no change in any observable property occurs.

2.10 THE ZEROTH LAW OF THERMODYNAMICS

Now consider the same two blocks of copper and another thermometer. Let one block of copper be brought into contact with the thermometer until equality of temperature is established, and then remove it. Then let the second block of copper be brought into contact with the thermometer. Suppose that no change in the mercury level of the thermometer occurs during this operation with the second block. We then can say that both blocks are in thermal equilibrium with the given thermometer.

The zeroth law of thermodynamics states that when two bodies have equality of temperature with a third body, they in turn have equality of temperature with each other. This seems very obvious to us because we are so familiar with this experiment. Because the principle is not derivable from other laws, and because it precedes the first and second laws of thermodynamics in the logical presentation of thermodynamics, it is called the zeroth law of thermodynamics. This law is really the basis of temperature measurement. Every time a body has equality of temperature with the thermometer, we can say that the body has the temperature we read on the thermometer. The problem remains how to relate temperatures that we might read on different mercury thermometers or obtain from different temperature-measuring devices, such as thermocouples and resistance thermometers. This observation suggests the need for a standard scale for temperature measurements.

2.11 TEMPERATURE SCALES

Two scales are commonly used for measuring temperature, namely the Fahrenheit (after Gabriel Fahrenheit, 1686–1736) and the Celsius. The Celsius scale was formerly called the centigrade scale but is now designated the Celsius scale after Anders Celsius (1701–1744), the Swedish astronomer who devised this scale.

The Fahrenheit temperature scale is used with the English Engineering system of units, and the Celsius scale with the SI unit system. Until 1954 both of these scales were based on two fixed, easily duplicated points—the ice point and the steam point. The temperature of the ice point is defined as the temperature of a mixture of ice and water that is in equilibrium with saturated air at a pressure of 1 atm. The temperature of the steam point is the temperature of water and steam, which are in equilibrium at a pressure of 1 atm. On the Fahrenheit scale these two points are assigned the numbers 32 and 212, respectively, and on the Celsius scale the points are 0 and 100, respectively. Why Fahrenheit chose these numbers is an interesting story. In searching for an easily reproducible point, Fahrenheit selected the temperature of the human body and assigned it the number 96. He assigned the number 0 to the temperature of a certain mixture of salt, ice, and salt solution. On this scale the ice point was approximately 32. When this scale was slightly revised and fixed in terms of the ice point and steam point, the normal temperature of the human body was found to be 98.6 F.

In this text the symbols F and °C will denote the Fahrenheit and Celsius scales, respectively. The symbol T will refer to temperature on all temperature scales.

At the tenth CGPM in 1954, the Celsius scale was redefined in terms of a single fixed point and the ideal-gas temperature scale. The single fixed point is the triple point of water (the state in which the solid, liquid, and vapor phases of water exist together in equilibrium). The magnitude of the degree is defined in terms of the ideal-gas temperature scale, which is discussed in Chapter 6. The essential features of this new scale are a single fixed point and a definition of the magnitude of the degree. The triple point of water is assigned the value of 0.01°C. On this scale the steam point is experimentally found to be 100.00°C. Thus, there is essential agreement between the old and new temperature scales.

We have not yet considered an absolute scale of temperature. The possibility of such a scale comes from the second law of thermodynamics and is discussed in Chapter 6. On the basis of the second law of thermodynamics, a temperature scale that is independent of any thermometric substance can be defined. This absolute scale is usually referred to as the thermodynamic scale of temperature. However, it is very complicated to use this scale directly, and therefore, a more practical scale, the International Temperature Scale, which closely represents the thermodynamic scale, has been adopted.

The absolute scale related to the Celsius scale is the Kelvin scale (after William Thomson, 1824–1907, who is also known as Lord Kelvin), and is designated K (without the degree symbol). The relation between these scales is

$$K = °C + 273.15$$

In 1967, the CGPM defined the kelvin as 1/273.16 of the temperature at the triple point of water. The Celsius scale is now defined by this equation instead of by its earlier definition.

The absolute scale related to the Fahrenheit scale is the Rankine scale and is designated R. The relation between these scales is

$$R = F + 459.67$$

2.12 THE INTERNATIONAL TEMPERATURE SCALE OF 1990

In 1989 the International Committee on Weights and Measures adopted a revised International Temperature Scale of 1990, the ITS-90, which is described as follows. This scale, similar to earlier ones of 1927, 1948, and 1968, has been refined, extended in range, and made to conform more closely to the thermodynamic temperature scale. It is based on a number of fixed and easily reproducible points that are assigned definite numerical values of temperature, and on specified formulas relating temperature to the readings on certain temperature-measuring instruments for the purpose of interpolation between the defining fixed points. These defining points and a summary of the interpolation techniques are listed here for the sake of completeness, although the student will have limited need for them at this time.

The defining fixed-point temperatures in International Kelvin Temperatures are as given in Table 2.2.

The means available for measurement and interpolation lead to a division of the temperature scale into four general ranges.

1. The range from 0.65 to 5.0 K is based on vapor pressure measurements for ^{3}He and ^{4}He, the latter divided into two parts. Each has a specified 12-term equation for the vapor-pressure as a function of temperature.

TABLE 2.2

^{3}He and ^{4}He	VP	3 to 5
e-H$_2$	TP	13.8033
e-H$_2$ (or He)	VP (or CVGT)	≈17
e-H$_2$ (or He)	VP (or CVGT)	≈20.3
Ne	TP	24.5561
O$_2$	TP	54.3584
Ar	TP	83.8058
Hg	TP	234.3156
H$_2$O	TP	273.16
Ga	MP	302.9146
In	FP	429.7485
Sn	FP	505.078
Zn	FP	692.677
Al	FP	933.473
Ag	FP	1234.93
Au	FP	1337.33
Cu	FP	1357.77

VP indicates vapor pressure point; CVGT indicates constant volume gas thermometer point; TP indicates triple point (equilibrium temperature at which the solid, liquid and vapor phases coexist); FP indicates freezing point and MP indicates melting point (the FP and the MP are equilibrium temperatures at which the solid and liquid phases coexist under a pressure of 101 325 Pa, one standard atmosphere). The isotopic composition is that naturally occurring.

2. The range from 3.0 to 24.5561 K is based on measurements using a helium constant-volume gas thermometer, which has been calibrated at three points: one point in the range from 3 to 5 K, based on helium vapor pressure measurements; one at the triple point of equilibrium-hydrogen; and one at the triple point of neon.

3. The range from 13.8033 to 1234.93 K is based on fixed points in the table and on measurements of resistance ratios of certain platinum resistance thermometers calibrated at specified sets of the fixed points, and using equations and corrections in each of four subranges of temperature. There are also alternatives for further subdividing these ranges.

4. The range above 1234.93 K is based on measurements of the intensity of visible-spectrum radiation compared with that of the same wavelength at the freezing point of either silver, gold, or copper, and on Planck's equation for blackbody radiation.

PROBLEMS

2.1 A 20-kg iron ball moving at 100 km/h should be decelerated at a constant rate of 10 m/s^2. What is the force required?

2.2 A 1200-kg car accelerates from rest with g/4 for a period of 10 s. What force is needed and what is the final velocity?

2.3 Two kilomoles of diatomic oxygen gas are enclosed in a 10-kg steel container. A force of 2 kN now accelerates this system. What is the acceleration?

2.4 The "standard" acceleration (at sea level and 45° latitude) due to gravity is 9.80665 m/s². If a force of 700 N is required to hold a mass at rest in this gravitational field, find the mass.

2.5 A crane lifts a mass of 200 kg at a location where the local gravitational acceleration is 9.5 m/s². Find the required force.

2.6 On the moon the gravitational acceleration is approximately one-sixth that on the surface of the earth. A 5-kg mass is "weighed" with a beam balance on the surface on the moon. What is the expected reading? If this mass is weighed with a spring scale that reads correctly for standard gravity on earth (see Problem 2.4), what is the reading?

2.7 A washing machine has 2 kg of clothes spinning at a rate that generates an acceleration of 12 m/s². What is the force needed to hold the clothes?

2.8 One kilogram of diatomic nitrogen (N_2 molecular weight 28) is contained in a 500-L tank. Find the specific volume on both a mass and mole basis (v and $\bar{v}$).

2.9 A 15-kg steel gastank holds 300 L of liquid gasoline, having a density of 800 kg/m³. What force is needed to accelerate this combined system at a rate of 4 m/s²?

2.10 A vertical hydraulic cylinder has a 150-mm diameter piston with hydraulic fluid inside the cylinder and an ambient pressure of 1 bar. Assuming standard gravity, find the piston mass that will create a pressure inside of 1250 kPa.

2.11 A barometer to measure absolute pressure shows a mercury column height of 725 mm. The temperature is such that the density of the mercury is 13 550 kg/m³. Find the ambient pressure.

2.12 A differential pressure gauge mounted on a vessel shows 1.25 MPa and a local barometer gives atmospheric pressure as 0.96 bar. Find the absolute pressure inside the vessel.

2.13 The absolute pressure in a tank is 85 kPa and the local ambient absolute pressure is 97 kPa. If a U-tube with mercury, density 13 550 kg/m³, is attached to the tank to measure the vacuum, what column height difference would it show?

2.14 A 5-kg piston in a cylinder with diameter of 100 mm is loaded with a linear spring and the outside atmospheric pressure of 100 kPa. The spring exerts no force on the piston when it is at the bottom of the cylinder and for the state shown the pressure is 400 kPa with volume 0.4 L. The valve is opened to let some air in, causing the piston to rise 2 cm. Find the new pressure.

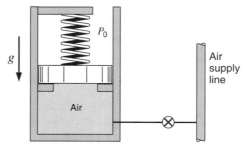

FIGURE P2.14

2.15 A U-tube manometer filled with water, density 1000 kg/m³, shows a height difference of 25 cm. What is the gauge pressure? If the right branch is tilted to make an angle of 30° with the horizontal, as shown in Fig. P2.15, what should the length of the column in the tilted tube be relative to the U-tube?

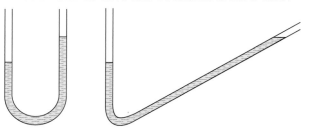

FIGURE P2.15

2.16 A piston/cylinder with cross-sectional area of 0.01 m² has a piston mass of 101 kg resting on the stops, as shown in Fig. P2.16. With an outside atmospheric pressure of 100 kPa, what should the water pressure be to lift the piston?

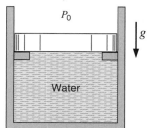

FIGURE P2.16

2.17 The difference in height between the columns of a manometer is 200 mm with a fluid of density 900 kg/m³. What is the pressure difference? What is the height difference if the same pressure difference is measured using mercury, density 13 600 kg/m³, as manometer fluid?

2.18 Two reservoirs, A and B, open to the atmosphere, are connected with a mercury manometer. Reservoir A is moved up/down so the two top surfaces are level at h_3 as shown in Fig. P2.18. Assuming that you know ρ_A, ρ_{Hg} and measure the heights h_1, h_2, and h_3, find the density ρ_B.

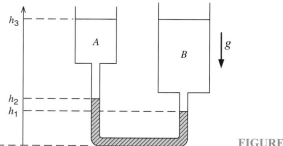

FIGURE P2.18

2.19 The density of mercury changes approximately linearly with temperature as

$$\rho_{Hg} = 13595 - 2.5T \quad kg/m^3 \quad T \text{ in Celsius}$$

so the same pressure difference will result in a manometer reading that is influenced by temperature. If a pressure difference of 100 kPa is measured in the summer at 35°C and in the winter at −15°C, what is the difference in column height between the two measurements?

2.20 Liquid water with density ρ is filled on top of a piston in a cylinder with cross-sec tional area A and height H. Air is let in under the piston so it pushes up, spilling the water over the edge. Deduce the formula for the air pressure as a function of piston height from the bottom, h.

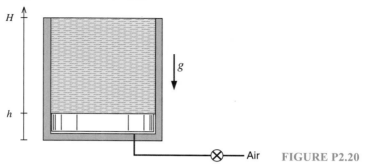

FIGURE P2.20

2.21 A piston, $m_p = 5$ kg, is fitted in a cylinder, $A = 15$ cm², that contains a gas. The setup is in a centrifuge that creates an acceleration of 25 m/s². Assuming standard atmospheric pressure outside the cylinder, find the gas pressure.

2.22 A piece of experimental apparatus is located where $g = 9.5$ m/s² and the temperature is $-2°C$. An air flow inside the apparatus is determined by measuring the pressure drop across an orifice with a mercury manometer (see Problem 2.19 for density) showing a height difference of 200 mm. What is the pressure drop in kPa?

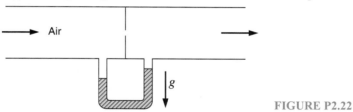

FIGURE P2.22

2.23 Repeat the previous problem if the flow inside the apparatus is liquid water, $\rho \approx 1000$ kg/m³, instead of air. Find the pressure difference between the two holes in the bottom of the channel.

2.24 Two piston/cylinder arrangements, A and B, have their gas chambers connected by a pipe. Cross-sectional areas are $A_A = 75$ cm² and $A_B = 25$ cm² with the piston mass in A being $m_A = 25$ kg. Outside pressure is 100 kPa and standard gravitation. Find the mass m_B so that none of the pistons have to rest on the bottom.

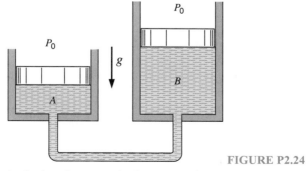

FIGURE P2.24

2.25 At the beach, atmospheric pressure is 1025 mbar. You dive 10 m down in the ocean and you later climb a hill up to 100 m elevation. Assume the density of water is about 1000 kg/m³ and the density of air is 1.18 kg/m³. What pressure do you feel at each place?

2.26 In the city water tower, water is pumped up to a level 35 m above ground in a pressurized tank with air at 125 kPa over the water surface. This is illustrated in Fig. P2.26. Assume the water density is 1000 kg/m³, standard gravity and find the pressure required to pump more water in at ground level.

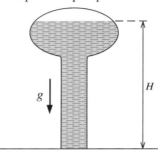

FIGURE P2.26

2.27 Two cylinders are connected by a piston as shown in Fig. P2.27. Cylinder A is used as a hydraulic lift and pumped up to 500 kPa. The piston mass is 15 kg and there is standard gravity. What is the gas pressure in cylinder B?

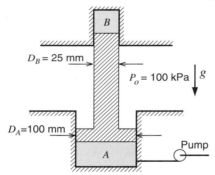

FIGURE P2.27

2.28 Two cylinders are filled with liquid water, $\rho \approx 1000$ kg/m³, and connected by a line with a closed valve. A has 100 kg and B has 500 kg of water, their cross-sectional areas are $A_A = 0.1$ m² and $A_B = 0.25$ m² and the height h is 1 m. Find the pressure on each side of the valve. The valve is opened and water flows to an equilibrium. Find the final pressure at the valve location.

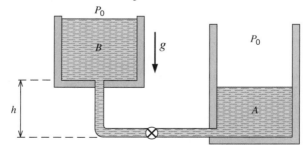

FIGURE P2.28

2.29 Using the freezing and boiling point temperatures for water in both Celsius and Fahrenheit scales, develop a conversion formula between the scales. Find the conversion formula between Kelvin and Rankine temperature scales.

2.30 A particular platinum versus rhodium-platinum thermocouple is to be calibrated and used in the temperature range described as range 3 in the discussion of the ITS-90 in Section 2.12. During calibration, the thermocouple EMF readings in microvolts are 5858, 9147, and 10 333 at the freezing points of aluminum, silver and gold, respectively. Assume a polynomial relation of the form

$$EMF = C_0 + C_1 T + C_2 T^2$$

and determine the temperature if the thermocouple gives a reading of 7500 µV.

ENGLISH UNIT PROBLEMS

2.31E A 2500-lbm car accelerates from rest with g/4 for a period of 10 s. What force is needed and what is the final velocity?

2.32E Two pound moles of diatomic oxygen gas are enclosed in a 20-lbm steel container. A force of 2000 lbf now accelerates this system. What is the acceleration?

2.33E A crane lifts a mass of 400 lbm at a location where the local gravitational acceleration is 31 ft/s². Find the required force.

2.34E One pound-mass of diatomic nitrogen (N_2 molecular weight 28) is contained in a 100-gal tank. Find the specific volume on both a mass and mole basis (v and $\bar{v}$).

2.35E A 30-lbm steel gas tank holds 10 ft³ of liquid gasoline, having a density of 50 lbm/ft³. What force is needed to accelerate this combined system at a rate of 15 ft/s²?

2.36E A differential pressure gauge mounted on a vessel shows 185 lbf/in.² and a local barometer gives atmospheric pressure as 0.96 atm. Find the absolute pressure inside the vessel.

2.37E A U-tube manometer filled with water, density 62.3 lbm/ft³, shows a height difference of 10 in. What is the gauge pressure? If the right branch is tilted to make an angle of 30° with the horizontal, as shown in Fig. P2.15, what should the length of the column in the tilted tube be relative to the U-tube?

2.38E A piston/cylinder with cross-sectional area of 0.1 ft² has a piston mass of 200 lbm resting on the stops, as shown in Fig. P2.16. With an outside atmospheric pressure of 1 atm, what should the water pressure be to lift the piston?

2.39E The density of mercury changes approximately linearly with temperature as

$$\rho_{Hg} = 851.5 - 0.086T \quad lbm/ft^3 \quad T \text{ in degrees Fahrenheit}$$

so the same pressure difference will result in a manometer reading that is influenced by temperature. If a pressure difference of 14.7 lbf/in.² is measured in the summer at 95 F and in the winter at 5 F, what is the difference in column height between the two measurements?

2.40E A piston, m_p = 10 lbm, is fitted in a cylinder, A = 2.5 in.², that contains a gas. The setup is in a centrifuge that creates an acceleration of 75 ft/s². Assuming standard atmospheric pressure outside the cylinder, find the gas pressure.

2.41E At the beach, atmospheric pressure is 1025 mbar. You dive 30 ft down in the ocean and you later climb a hill up to 300 ft elevation. Assume the density of water is about 62.3 lbm/ft³ and the density of air is 0.0735 lbm/ft³. What pressure do you feel at each place?

2.42E In the city water tower, water is pumped up to a level 100 ft above ground in a pressurized tank with air at 19 lbf/in.^{2} over the water surface. This is illustrated in Fig.P2.26. Assume the water density is 62.4 lbm/ft^3, standard gravity and find the pressure required to pump more water in at ground level.

COMPUTER, DESIGN, AND OPEN ENDED PROBLEMS

2.43 Write a program to list corresponding temperatures in °C, K, F, and R from −50°C to 100°C in increments of 10 degrees.

2.44 Write a program that will input pressure in kPa or atm or lbf/in.^{2} and write the pressure out in: kPa, atm, bar, and lbf/in.^{2}.

2.45 Write a program to do the temperature correction on a mercury barometer reading (see Problem 2.19). Input reading and temperature and output corrected reading at 20°C and pressure in kPa.

2.46 Write a program to solve the thermocouple calibration problem discussed in Problem 2.30. Any three sets of calibration readings can be given, and the result is a polynomial equation (parabola) with specified coefficients that can be used at any temperature in the calibrated range.

2.47 Make a list of different weights and scales that are used to measure mass directly or indirectly. Investigate the ranges of mass and the accuracy that can be obtained.

2.48 Thermometers are based on several principles. Expansion of a liquid with a rise in temperature is used in many applications. Electrical resistance, thermistors, and thermocouples are common in instrumentation and remote probes. Investigate a variety of thermometers and make a list of their range, accuracy, advantages and disadvantages.

2.49 Find the information for a resistance, thermistor, and thermocouple based thermometer suitable for the range of temperatures from 0°C to 200°C. For each of the three types list the accuracy and response of the transducer (output per degree change). Is any calibration or corrections necessary when it is used in an instrument?

2.50 A thermistor is used as a temperature transducer. Its resistance changes with temperature approximated as

$$R = R_o \exp\left[\alpha(1/T - 1/T_o)\right]$$

where it has the resistance, R_o, at the temperature, T_o. Select the constants as R_o = 3000Ω, T_o = 298 K and compute α so it has the resistance of 200 Ω at 100°C. Write a program to convert a measured resistance, R, into information about the temperature. Find information for actual thermistors and plot the calibration curves with the above formula and the recommended correction given by the manufacturer.

2.51 Investigate possible transducers for the measurement of temperature in a flame with temperatures near 1000 K. Are any available for a temperature of 2000 K?

2.52 Devices to measure pressure are available as differential or absolute pressure transducers. Make a list of 5 different differential pressure transducers to measure pressure differences in the order of 100 kPa. Note their accuracy, response (linear or ?), and price.

2.53 A micromanometer uses a fluid with density 1000 kg/m^3 and it is able to measure the height difference with an accuracy of ±0.5 mm. Its range is a maximum height difference of 0.5 m. Investigate if any transducers are available to replace the micromanometer.

2.54 An experiment involves the measurements of temperature and pressure of a gas flowing in a pipe at 300°C and 250 kPa. Write a report with a suggested set of transducers (at least two alternatives for each) and give the expected accuracy and cost.

3 PROPERTIES OF A PURE SUBSTANCE

In the previous chapter we considered three familiar properties of a substance—specific volume, pressure, and temperature. We now turn our attention to pure substances and consider some of the phases in which a pure substance may exist, the number of independent properties a pure substance may have, and methods of presenting thermodynamic properties.

3.1 THE PURE SUBSTANCE

A pure substance is one that has a homogeneous and invariable chemical composition. It may exist in more than one phase, but the chemical composition is the same in all phases. Thus, liquid water, a mixture of liquid water and water vapor (steam), and a mixture of ice and liquid water are all pure substances; every phase has the same chemical composition. On the other hand, a mixture of liquid air and gaseous air is not a pure substance, because the composition of the liquid phase is different from that of the vapor phase.

Sometimes a mixture of gases, such as air, is considered a pure substance as long as there is no change of phase. Strictly speaking, this is not true. As we will see later, we should say that a mixture of gases such as air exhibits some of the characteristics of a pure substance as long as there is no change of phase.

In this text the emphasis will be on simple compressible substances. This term designates substances whose surface effects, magnetic effects, and electrical effects are insignificant when dealing with the substances. On the other hand, changes in volume, such as those associated with the expansion of a gas in a cylinder, are very important. Reference will be made, however, to other substances for which surface, magnetic, and electrical effects are important. We will refer to a system consisting of a simple compressible substance as a simple compressible system.

3.2 VAPOR–LIQUID–SOLID-PHASE EQUILIBRIUM IN A PURE SUBSTANCE

Consider as a system 1 kg of water contained in the piston-cylinder arrangement shown in Fig. 3.1a. Suppose that the piston and weight maintain a pressure of 0.1 MPa in the cylinder, and that the initial temperature is 20°C. As heat is transferred to the water, the

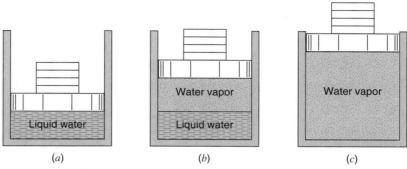

FIGURE 3.1 Constant-pressure change from liquid to vapor phase for a pure substance.

temperature increases appreciably, the specific volume increases slightly, and the pressure remains constant. When the temperature reaches 99.6°C, additional heat transfer results in a change of phase, as indicated in Fig. 3.1b. That is, some of the liquid becomes vapor, and during this process both the temperature and pressure remain constant, but the specific volume increases considerably. When the last drop of liquid has vaporized, further transfer of heat results in an increase in both temperature and specific volume of the vapor, as shown in Fig. 3.1c.

The term saturation temperature designates the temperature at which vaporization takes place at a given pressure. This pressure is called the saturation pressure for the given temperature. Thus, for water at 99.6°C the saturation pressure is 0.1 MPa, and for water at 0.1 MPa the saturation temperature is 99.6°C. For a pure substance there is a definite relation between saturation pressure and saturation temperature. A typical curve, called the vapor-pressure curve, is shown in Fig. 3.2.

If a substance exists as liquid at the saturation temperature and pressure, it is called saturated liquid. If the temperature of the liquid is lower than the saturation temperature for the existing pressure, it is called either a subcooled liquid (implying that the temperature is lower than the saturation temperature for the given pressure) or a compressed liquid (implying that the pressure is greater than the saturation pressure for the given temperature). Either term may be used, but the latter term will be used in this text.

When a substance exists as part liquid and part vapor at the saturation temperature, its quality is defined as the ratio of the mass of vapor to the total mass. Thus, in Fig. 3.1b, if the mass of the vapor is 0.2 kg and the mass of the liquid is 0.8 kg, the quality is 0.2 or 20%. The quality may be considered an intensive property and has the symbol x. Quality has meaning only when the substance is in a saturated state, that is, at saturation pressure and temperature.

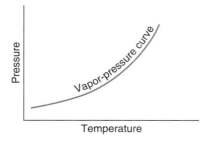

FIGURE 3.2 Vapor-pressure curve of a pure substance.

If a substance exists as vapor at the saturation temperature, it is called saturated vapor. (Sometimes the term "dry saturated vapor" is used to emphasize that the quality is 100%.) When the vapor is at a temperature greater than the saturation temperature, it is said to exist as superheated vapor. The pressure and temperature of superheated vapor are independent properties, since the temperature may increase while the pressure remains constant. Actually, the substances we call gases are highly superheated vapors.

Consider Fig. 3.1 again. Let us plot on the temperature–volume diagram of Fig. 3.3 the constant-pressure line that represents the states through which the water passes as it is heated from the initial state of 0.1 MPa and 20°C. Let state *A* represent the initial state, *B* the saturated-liquid state (99.6°C), and line *AB* the process in which the liquid is heated from the initial temperature to the saturation temperature. Point *C* is the saturated-vapor state, and line *BC* is the constant-temperature process in which the change of phase from liquid to vapor occurs. Line *CD* represents the process in which the steam is super-heated at constant pressure. Temperature and volume both increase during this process.

Now let the process take place at a constant pressure of 1 MPa, starting from an initial temperature of 20°C. Point *E* represents the initial state, in which the specific volume is slightly less than that at 0.1 MPa and 20°C. Vaporization begins at point *F*, where the temperature is 179.9°C. Point *G* is the saturated-vapor state, and line *GH* the constant-pressure process in which the steam is superheated.

In a similar manner, a constant pressure of 10 MPa is represented by line *IJKL,* for which the saturation temperature is 311.1°C.

At a pressure of 22.09 MPa, represented by line *MNO*, we find, however, that there is no constant-temperature vaporization process. Instead, point *N* is a point of inflection with a zero slope. This point is called the critical point. At the critical point the saturated-liquid and saturated-vapor states are identical. The temperature, pressure, and specific volume at the critical point are called the critical temperature, critical pressure, and critical volume. The critical-point data for some substances are given in Table 3.1. More extensive data are given in Table A.8 in the Appendix.

A constant-pressure process at a pressure greater than the critical pressure is represented by line *PQ*. If water at 40 MPa, 20°C, is heated in a constant-pressure process in a cylinder as shown in Fig. 3.1, there will never be two phases present and the state shown

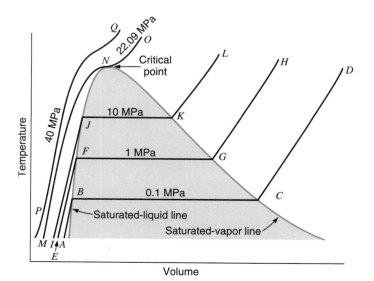

FIGURE 3.3
Temperature–volume diagram for water showing liquid and vapor phases (not to scale).

TABLE 3.1
Some Critical-Point Data

	Critical Temperature, °C	Critical Pressure, MPa	Critical Volume, m³/kg
Water	374.14	22.09	0.003 155
Carbon dioxide	31.05	7.39	0.002 143
Oxygen	−118.35	5.08	0.002 438
Hydrogen	−239.85	1.30	0.032 192

in Fig. 3.1*b* will never exist. Instead, there will be a continuous change in density and at all times there will be only one phase present. The question then is when do we have a liquid and when do we have a vapor? The answer is that this is not a valid question at supercritical pressures. We simply term the substance a fluid. However, rather arbitrarily, at temperatures below the critical temperature we usually refer to it as a compressed liquid and at temperatures above the critical temperature as a superheated vapor. It should be emphasized, however, that at pressures above the critical pressure we never have a liquid and vapor phase of a pure substance existing in equilibrium.

In Fig. 3.3, line *NJFB* represents the saturated-liquid line and line *NKGC* representsw the saturated-vapor line.

Let us consider another experiment with the piston–cylinder arrangement. Suppose that the cylinder contains 1 kg of ice at −20°C, 100 kPa. When heat is transferred to the ice, the pressure remains constant, the specific volume increases slightly, and the temperature increases until it reaches 0°C, at which point the ice melts and the temperature remains constant. In this state the ice is called a saturated solid. For most substances the specific volume increases during this melting process, but for water the specific volume of the liquid is less than the specific volume of the solid. When all the ice has melted, a further heat transfer causes an increase in temperature of the liquid.

If the initial pressure of the ice at −20°C is 0.260 kPa, heat transfer to the ice results in an increase in temperature to −10°C. At this point, however, the ice passes directly from the solid phase to the vapor phase in the process known as sublimation. Further heat transfer results in superheating of the vapor.

Finally, consider an initial pressure of the ice of 0.6113 kPa and a temperature of −20°C. Through heat transfer let the temperature increase until it reaches 0.01°C. At this point, however, further heat transfer may cause some of the ice to become vapor and some to become liquid, for at this point it is possible to have the three phases in equilibrium. This point is called the triple point, which is defined as the state in which all three phases may be present in equilibrium. The pressure and temperature at the triple point for a number of substances are given in Table 3.2.

This whole matter is best summarized by the diagram of Fig. 3.4, which shows how the solid, liquid, and vapor phases may exist together in equilibrium. Along the sublimation line the solid and vapor phases are in equilibrium, along the fusion line the solid and liquid phases are in equilibrium, and along the vaporization line the liquid and vapor phases are in equilibrium. The only point at which all three phases may exist in equilibrium is the triple point. The vaporization line ends at the critical point because there is no distinct change from the liquid phase to the vapor phase above the critical point.

Consider a solid in state *A*, as shown in Fig. 3.4. When the temperature increases

TABLE 3.2
Some Solid-Liquid-Vapor Triple-Point Data

	Temperature, °C	Pressure, kPa
Hydrogen (normal)	−259	7.194
Oxygen	−219	0.15
Nitrogen	−210	12.53
Carbon dioxide	−56.4	520.8
Mercury	−39	0.000 000 13
Water	0.01	0.6113
Zinc	419	5.066
Silver	961	0.01
Copper	1083	0.000 079

but the pressure (which is less than the triple-point pressure) is constant, the substance passes directly from the solid to the vapor phase. Along the constant-pressure line *EF*, the substance passes from the solid to the liquid phase at one temperature, and then from the liquid to the vapor phase at a higher temperature. Constant-pressure line *CD* passes through the triple point, and it is only at the triple point that the three phases may exist together in equilibrium. At a pressure above the critical pressure, such as *GH*, there is no sharp distinction between the liquid and vapor phases.

Although we have made these comments with rather specific reference to water (only because of our familiarity with water), all pure substances exhibit the same general behavior. However, the triple-point temperature and critical temperature vary greatly from one substance to another. For example, the critical temperature of helium, as given

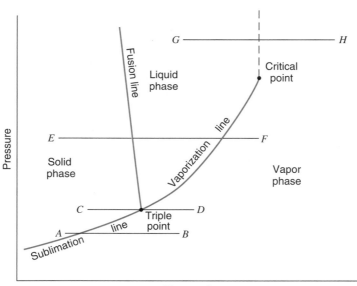

FIGURE 3.4
Pressure–temperature diagram for a substance such as water.

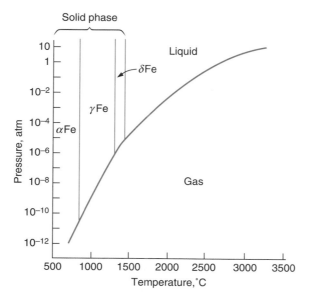

FIGURE 3.5 Estimated pressure–temperature diagram for iron (From *Phase Diagrams in Metallurgy* by F. N. Rhines, copyright 1956, McGraw-Hill Book Company; used by permission).

in Table A.8, is 5.3 K. Therefore, the absolute temperature of helium at ambient conditions is over 50 times greater than the critical temperature. On the other hand, water has a critical temperature of 374.14°C (647.29 K), and at ambient conditions the temperature of water is less than one-half the critical temperature. Most metals have a much higher critical temperature than water. When we consider the behavior of a substance in a given state, it is often helpful to think of this state in relation to the critical state or triple point. For example, if the pressure is greater than the critical pressure, it is impossible to have a liquid phase and a vapor phase in equilibrium. Or, to consider another example, the states at which vacuum-melting a given metal is possible can be ascertained by a consideration of the properties at the triple point. Iron at a pressure just above 5 Pa (the triple-point pressure) would melt at a temperature of about 1535°C (the triple-point temperature).

It should also be pointed out that a pure substance can exist in a number of different solid phases. A transition from one solid phase to another is called an allotropic transformation. Figure 3.5 is a pressure–temperature diagram for iron that shows three solid phases, the liquid phase, and the vapor phase. Figure 3.6 shows a number of solid phases for water. A pure substance can have a number of triple points, but only one triple point has a solid, liquid, and vapor equilibrium. Other triple points for a pure substance can have two solid phases and a liquid phase, two solid phases and a vapor phase, or three solid phases.

3.3 INDEPENDENT PROPERTIES OF A PURE SUBSTANCE

One important reason for introducing the concept of a pure substance is that the state of a simple compressible pure substance (that is, a pure substance in the absence of motion, gravity, and surface, magnetic, or electrical effects) is defined by two independent prop-

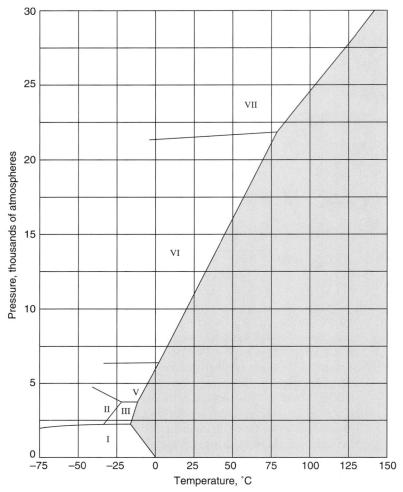

FIGURE 3.6 Phase diagram of water (Adapted from the *American Institute of Physics Handbook,* 2nd ed., 1963, McGraw-Hill).

erties. For example, if the specific volume and temperature of superheated steam are specified, the state of the steam is determined.

To understand the significance of the term independent property, consider the saturated-liquid and saturated-vapor states of a pure substance. These two states have the same pressure and the same temperature, but they are definitely not the same state. In a saturation state, therefore, pressure and temperature are not independent properties. Two independent properties such as pressure and specific volume or pressure and quality are required to specify a saturation state of a pure substance.

The reason for mentioning previously that a mixture of gases, such as air, has the same characteristics as a pure substance as long as only one phase is present, concerns precisely this point. The state of air, which is a mixture of gases of definite composition, is determined by specifying two properties as long as it remains in the gaseous phase. Air then can be treated as a pure substance.

3.4 EQUATIONS OF STATE FOR THE VAPOR PHASE OF A SIMPLE COMPRESSIBLE SUBSTANCE

From experimental observations it has been established that the P–v–T behavior of gases at low density is closely given by the following equation of state,

$$P\bar{v} = \bar{R}T \qquad (3.1)$$

where $\bar{R}$ is the universal gas constant. The value of $\bar{R}$ is

$$\bar{R} = 8.3145\frac{\text{kN m}}{\text{kmol K}} = 8.3145\frac{\text{kJ}}{\text{kmol } K}$$

In the English Engineering system,

$$\bar{R} = 1545\frac{\text{ft lbf}}{\text{lb mol } R}$$

Dividing Eq. 3.1 by M, the molecular weight, we have the equation of state on a unit mass basis,

$$\frac{P\bar{v}}{M} = \frac{\bar{R}T}{M}$$

or

$$Pv = RT \qquad (3.2)$$

where

$$R = \frac{\bar{R}}{M} \qquad (3.3)$$

Here R is a constant for a particular gas. The value of R for a number of substances is given in Table A.10 of the Appendix. It follows from Eqs. 3.1 and 3.2 that this equation of state can be written in terms of the total volume:

$$PV = n\bar{R}T$$
$$PV = mRT \qquad (3.4)$$

It should also be noted that Eq. 3.4 can be written alternately in the form

$$\frac{P_1V_1}{T_1} = \frac{P_2V_2}{T_2} \qquad (3.5)$$

That is, gases at low density closely follow the well-known Boyle's and Charles' laws. Boyle and Charles, of course, based their statements on experimental observations. (Strictly speaking, neither of these statements should be called a law, since they are only approximately true and even then only under conditions of low density.)

The equation of state given by Eq. 3.1 (or Eq. 3.2) is referred to as the ideal-gas equation of state. At very low density, all gases and vapors approach ideal-gas behavior, with the P–v–T relationship being given by the ideal-gas equation of state. At higher densities the behavior may deviate substantially from the ideal-gas equation of state.

Because of its simplicity, the ideal-gas equation of state is very convenient to use in thermodynamic calculations. However, two questions are now appropriate. The ideal-gas equation of state is a good approximation at low density. But what constitutes low density? Or, expressed in other words, over what range of density will the ideal-gas equation of state hold with accuracy? The second question is, how much does an actual gas at a given pressure and temperature deviate from ideal-gas behavior?

To answer both questions, we introduce the concept of the compressibility factor, Z, which is defined by the relation

$$Z = \frac{P\bar{v}}{\bar{R}T}$$

or

$$P\bar{v} = Z\bar{R}T \tag{3.6}$$

Note that for an ideal gas $Z = 1$, and the deviation of Z from unity is a measure of the deviation of the actual relation from the ideal-gas equation of state.

Figure 3.7 shows a skeleton compressibility chart for nitrogen. From this chart we make three observations. The first is that at all temperatures $Z \to 1$ as $P \to 0$. That is, as the pressure approaches zero, the P–v–T behavior closely approaches that predicted by the ideal-gas equation of state. Note also that at temperatures of 300 K and above (that is, room temperature and above) the compressibility factor is near unity up to pressure of about 10 MPa. This means that the ideal-gas equation of state can be used for nitrogen (and, as it happens, air) over this range with considerable accuracy.

Now suppose we reduce the temperature from 300 K but keep the pressure constant at 4 MPa. The density will increase, and we note a sharp decrease below unity in the value of the compressibility factor. Values of $Z < 1$ mean that the actual density is greater than would be predicted by ideal-gas behavior. The physical explanation of this is as follows. As the temperature is reduced from 300 K and the pressure remains constant at 4 MPa, the molecules are brought closer together. In this range of intermolecular distances, and at this pressure and temperature, there is an attractive force between the

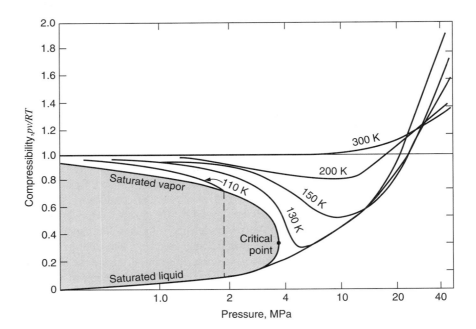

FIGURE 3.7
Compressibility of
nitrogen.

molecules. The lower the temperature the greater is this intermolecular attractive force. This attractive force between the molecules means that the density is greater than would be predicted by the ideal-gas behavior, which assumes no intermolecular forces. Note also from the compressibility chart that at very high densities, for pressures above 30 MPa, the compressibility factor is always greater than unity. In this range the intermolecular distances are very small, and there is a repulsive force between the molecules. This factor tends to make the density less than would otherwise be expected.

The precise nature of intermolecular forces is a rather complex matter. These forces are a function of the temperature as well as the density. The preceding discussion should be considered a qualitative analysis to assist in gaining some understanding of the ideal-gas equation of state and how the P–v–T behavior of actual gases deviates from this equation.

For other gases, the behavior of Z with respect to temperature and pressure is very similar to that of nitrogen, at least in a qualitative sense. To quantify this relation, we divide the temperature by the critical temperature of the substance, calling the result the reduced temperature, T_r. In the same manner, the pressure divided by the critical pressure is termed reduced pressure, P_r. A plot of Z versus P_r at various T_r is in reasonably close agreement quantitatively for many different gases. This plot is called a generalized compressibility chart. The basic chart for simple spherical molecules is included in the Appendix as Fig. A.7. Corrections to this basic generalized chart value will be discussed in detail in Chapter 10. For the present, the chart will help us decide whether, in a given circumstance, it is reasonable to assume ideal-gas behavior as a model. For example, we note from the chart that if the pressure is very low (that is, $\ll P_c$), the ideal-gas model can be assumed with good accuracy, regardless of the temperature. Furthermore, at high temperatures (that is, greater than about twice T_c), the ideal-gas model can be assumed with good accuracy to pressures as high as four or five times P_c. When the temperature is less than about twice the critical temperature and the pressure is not extremely low, we are in a region, commonly termed superheated vapor, in which the deviation from ideal-gas behavior may be considerable. In this region it is preferable to use tables of thermodynamic properties or charts for a particular substance. These tables are considered in the following section.

Instead of the ideal-gas model to represent gas behavior, or even the generalized compressibility chart, which is approximate, it is desirable to have an equation of state that accurately represents the P–v–T behavior for a particular gas over the entire superheated vapor region. Such an equation is necessarily more complicated and consequently more difficult to use. Many such equations have been proposed and used to correlate the observed behavior of gases. To illustrate the nature and complexity of these equations, we present one of the best known, the Benedict–Webb–Rubin equation of state:

$$P = \frac{RT}{v} + \frac{RTB_0 - A_0 - C_0/T^2}{v^2} + \frac{RTb - a}{v^3} + \frac{a\alpha}{v^6} + \frac{c}{v^3 T^2}\left(1 + \frac{\gamma}{v^2}\right)e^{-\gamma/v^2} \qquad (3.7)$$

This equation contains eight empirical constants and is accurate to densities of about twice the critical density. The empirical constants for the Benedict–Webb–Rubin equation for a number of substances are given in Table 3.3.

The matter of equations of state will be discussed further in Chapter 10. Note that an equation of state that accurately describes the relation between pressure, temperature, and specific volume is rather cumbersome and that the solution requires considerable time. When we use a digital computer, it is often most convenient to determine the ther-

TABLE 3.3
Empirical Constants for Benedict-Webb-Rubin Equation

UNITS: ATMOSPHERES, LITERS, MOLES, K. *GAS CONSTANTS:* $R = 0.082\ 06$; $T = 273.15 + T(°C)$

Gas	A_0	B_0	$C_0 \cdot 10^{-6}$	a	b	$c \cdot 10^{-6}$	$\alpha \cdot 10^3$	$\gamma \cdot 10^2$
Methane	1.855 00	0.042 600	0.022 570	0.494 00	0.003 380 04	0.002 545	0.124 359	0.600
Ethylene	3.339 58	0.055 683 3	0.131 140	0.259 00	0.008 600	0.021 120	0.178 00	0.923
Ethane	4.155 56	0.062 772 4	0.179 592	0.345 16	0.011 122	0.032 767	0.243 389	1.180
Propylene	6.112 20	0.085 064 7	0.439 182	0.774 056	0.018 705 9	0.102 611	0.455 696	1.829
Propane	6.872 25	0.097 313	0.508 256	0.947 70	0.022 500	0.129 00	0.607 175	2.200
n-Butane	10.084 7	0.124 361	0.992 830	1.882 31	0.039 998 3	0.316 400	1.101 32	3.400
n-Pentane	12.179 4	0.156 751	2.121 21	4.074 80	0.066 812	0.824 17	1.810 00	4.750
n-Hexane	14.437 3	0.177 813	3.319 35	7.116 71	0.109 131	1.512 76	2.810 86	6.668 49
n-Heptane	17.520 6	0.199 005	4.745 74	10.364 75	0.151 954	2.470 00	4.356 11	9.000
Nitrogen	1.192 50	0.045 80	0.005 889 1	0.014 90	0.001 981 54	0.000 548 064	0.291 545	0.750
Oxygen	1.498 80	0.046 524	0.003 861 7	-0.040 507	-0.000 279 63	-0.000 203 76	0.008 641	0.359
Ammonia	3.789 28	0.051 646 1	0.178 567	0.103 54	0.000 719 561	0.000 157 536	0.004 651 89	1.980
Carbon dioxide	2.673 40	0.045 628	0.113 33	0.051 689	0.003 081 9	0.007 067 2	0.112 71	0.494

Sources: M. Benedict, G. Webb, and L. Rubin, *Chem. Eng. Progress*, **47**, 419 (1951) (first nine entries).
S. M. Wales, *Phase Equilibrium in Chemical Engineering*, Butterworth Publishing (1985) (last four entries).

modynamic properties in a given state from such equations. However, in hand calculations, it is much more convenient to tabulate values of pressure, temperature, specific volume, and other thermodynamic properties for various substances. The Appendix includes summary tables and graphs of the thermodynamic properties of water, ammonia, chlorofluorocarbon refrigerants R-12 and R-22, the new refrigerant R-134a, nitrogen, and methane. The tables of the properties of water are usually referred to as the "steam tables," where the first set was presented by Keenan, Keyes, Hill, and Moore. The method for compiling the P–v–T data for such a table is to find an equation of state that accurately fits the experimental data and then to solve the equation of state for the values listed in the table.

EXAMPLE 3.1

What is the mass of air contained in a room 6 m × 10 m × 4 m if the pressure is 100 kPa and the temperature is 25°C? Assume air to be an ideal gas.

By using Eq. 3.4 and the value of R from Table A.10, we have

$$m = \frac{PV}{RT} = \frac{100 \text{ kN} / \text{m}^2 \times 240 \text{ m}^3}{0.287 \text{ kN m} / \text{kg K} \times 298.2 \text{ K}} = 280.5 \text{ kg}$$

EXAMPLE 3.2

A tank has a volume of 0.5 m³ and contains 10 kg of an ideal gas having a molecular weight of 24. The temperature is 25°C. What is the pressure?

The gas constant is determined first:

$$R = \frac{\overline{R}}{M} = \frac{8.3145 \text{ kN m} / \text{kmol K}}{24 \text{ kg} / \text{kmol}}$$

$$= 0.346 \ 44 \text{ kN m} / \text{kg K}$$

We now solve for P:

$$P = \frac{mRT}{V} = \frac{10 \text{ kg} \times 0.346 \ 44 \text{ kN m} / \text{kg K} \times 298.2 \text{ K}}{0.5 \text{ m}^3}$$

$$= 2066 \text{ kPa}$$

EXAMPLE 3.2E

A tank has a volume of 15 ft³ and contains 20 lbm of an ideal gas having a molecular weight of 24. The temperature is 80 F. What is the pressure?

The gas constant is determined first:

$$R = \frac{\overline{R}}{M} = \frac{1545 \text{ ft lbf} / \text{lb mole R}}{24 \text{ lbm} / \text{lb mole}} = 64.4 \text{ ft lbf} / \text{lbm R}$$

We now solve for P.

$$P = \frac{mRT}{V} = \frac{20 \text{ lbm} \times 64.4 \text{ ft lbf} / \text{lbm R} \times 540 \text{R}}{144 \text{ in.}^2 / \text{ft}^2 \times 15 \text{ ft}^3} = 321 \text{ lbf} / \text{in.}^2$$

3.5 TABLES OF THERMODYNAMIC PROPERTIES

Tables of thermodynamic properties of many substances are available, and in general, all these tables have the same form. In this section we will refer to the steam tables. The steam tables are selected both because they are a vehicle for presenting thermodynamic tables and because steam is used extensively in power plants and industrial processes. Once the steam tables are understood, other thermodynamic tables can be readily used.

Before we introduce and discuss the actual steam tables, it is worthwhile to examine the ideas behind thermodynamic tables and to explore some of the initial difficulties that students frequently encounter. We therefore introduce these concepts on an artificial, oversimplified, and somewhat abstract basis, after which we will proceed to the actual tables. These "steam tables" actually consist of four separate tables, say A, B, C, and D, each of which is associated with a different region (phase) of the possible combinations of the two variables, T and P, as shown in Fig. 3.8. Each table is listed according to values of T and P. For each set of T and P (a state), the table contains values for four other variables, v, u, h, and s.

If the values of T and P are given, then by comparing them with the boundary values of T' and P', we are able to choose which of the four tables (A, B, C, and D) has the desired values of v, u, h, and s. Table A is the correct choice, for example, only if $T < T'$ and $P > P'$ are both true.

The principal difficulty in first learning how to use the tables is that the state may actually be specified by any two of the variables (assuming that T, P, v, u, h, and s are all independent). We then wish to find the other four variables from the appropriate table. If the two given values are not T and P, it may not be obvious which table to choose to find the other four variables. Even when we have had experience comparing the appropriate values at the table boundaries and understand through practice in which directions the changes go, we must go through the process of finding the appropriate table each and every time.

In addition to finding the correct table, the other nuisance of everyday use of the tables is interpolation, that is, when one (or both) of the stated values is not exactly equal to a value listed in the table.

Computerized tables eliminate both of these problems, but the student nevertheless must learn to understand the significance, construction, and limitations of the tables because there will be occasions when it will be necessary to use printed tables.

Let us use this background, to introduce the actual steam tables, as presented in the Appendix, Table A.1. Table A.1 is a summary table based on a complicated curve fit to the behavior of water. It is very similar to the *Steam Tables,* by Keenan, Keyes, Hill, and Moore, published in 1969 and 1978, revisions of the original tables published by Keenan and Keyes in 1936. We will concentrate on the three properties already discussed in Chapter 2, namely T, P, and v, and note that the other properties listed in Table A.1, u, h,

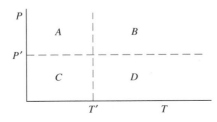

FIGURE 3.8 Schematic diagram of "steam tables" principle.

and s, will be discussed later. We further note that the separation of phases in terms of values of T and P is actually described by the relations illustrated in Fig. 3.4, not Fig. 3.8. The region of superheated vapor in Fig. 3.4 is given in Table A.1.3, and that of compressed liquid is given in Table A.1.4. The compressed-solid region shown in Fig. 3.4 is not listed in the Appendix. The saturated-liquid and saturated-vapor region, as seen in the T and v diagram of Fig. 3.3 (and as the vaporization line in Fig. 3.4), is listed according to the values of T in Table A.1.1 and according to the values of P (T and P are not independent in the two-phase regions) in Table A.1.2. Similarly, the saturated-solid and saturated-vapor region is listed according to T in Table A.1.5, but the saturated-solid and saturated-liquid region, the third phase boundary line shown in Fig. 3.4, is not listed in the Appendix.

In Table A.1.1, the first column after the temperature gives the corresponding saturation pressure in kilopascals or megapascals. The next two columns give specific volume in cubic meters per kilogram. The first of these columns gives the specific volume of the saturated liquid, v_f; the second column gives the specific volume of saturated vapor, v_g. The difference between these two, $v_g - v_f$, represents the increase in specific volume when the state changes from saturated liquid to saturated vapor and is designated v_{fg}.

The specific volume of a substance having a given quality can be found by using the definition of quality. Quality has already been defined as the ratio of the mass of vapor to the total mass of liquid plus vapor when a substance is in a saturation state. Let us consider a mass m having a quality x. The volume is the sum of the volume of the liquid and the volume of the vapor:

$$V = V_{liq} + V_{vap} \tag{3.8}$$

In terms of the masses, Eq. 3.8 can be written in the form

$$mv = m_{liq} v_f + m_{vap} v_g$$

Dividing by the total mass and introducing the quality x, we have

$$v = (1-x)v_f + xv_g \tag{3.9}$$

Using the definition

$$v_{fg} = v_g - v_f$$

Equation 3.9 can then be written in the form

$$v = v_f + xv_{fg} \tag{3.10}$$

As an example, let us calculate the specific volume of saturated steam at 200°C having a quality of 70%. Using Eq. 3.9 gives

$$v = 0.3(0.001\ 157) + 0.7(0.127\ 36)$$

$$= 0.0895 \text{ m}^3 / \text{kg}$$

In Table A.1.2, the first column after the pressure lists the saturation temperature for each pressure. The next column lists specific volume in a manner similar to Table A.1.1. When necessary, v_{fg} can readily be found by subtracting v_f from v_g.

Table 3 of the steam tables, which is summarized in Table A.1.3, gives the properties of superheated vapor. In the superheated region, pressure and temperature are independent properties and, therefore, for each pressure a large number of temperatures is given, and for each temperature four thermodynamic properties are listed, the first one

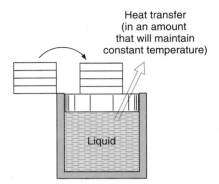

FIGURE 3.9 Illustration of compressed-liquid state.

being specific volume. Thus, the specific volume of steam at a pressure of 0.5 MPa and 200°C is 0.4249 m³/kg.

Table 4 of the steam tables, summarized in Table A.1.4, gives the properties of the compressed liquid. To demonstrate the use of this table, consider a piston and a cylinder (as shown in Fig. 3.9) that contains 1 kg of saturated-liquid water at 100°C. Its properties are given in Table A.1.1, and we note that the pressure is 0.1013 MPa and the specific volume is 0.001 044 m³/kg. Suppose the pressure is increased to 10 MPa while the temperature is held constant at 100°C by the necessary transfer of heat, Q. Since water is slightly compressible, we would expect a slight decrease in specific volume during this process. Table A.1.4 gives this specific volume as 0.001 039 m³/kg. This is only a slight decrease, and only a small error would be made if one assumed that the volume of a compressed liquid was equal to the specific volume of the saturated liquid at the same temperature. In many situations this is the most convenient procedure, particularly when compressed-liquid data are not available.

Furthermore, because specific volume does change rapidly with temperature, care should be exercised in interpolation over the wide temperature ranges in Table A.1.4. (Sometimes it may be more accurate to use the saturated-liquid data from Table A.1.1 and interpolate differences between Table A.1.1, the saturated-liquid data, and Table A.1.4, the compressed-liquid data.)

Table 6 of the steam tables, which is summarized in Table A.1.5, gives the properties of saturated solid and saturated vapor that are in equilibrium. The first column gives the temperature and the second column gives the corresponding saturation pressure. As would be expected, all these pressures are less than the triple-point pressure. The next two columns give the specific volume of the saturated solid and saturated vapor (note that the tabulated value is $v_i \times 10^3$).

3.6 COMPUTERIZED TABLES

The tables in the Appendix are supplied as a set of computer programs on the disk in the book. The programs come in two versions, one is a menu-driven executable version, and the other is a collection of callable subroutines in FORTRAN arranged in a library. The subroutines are identical for the two versions, only the main program is different. It is envisioned that the menu-driven program should be used for the solution of problems by hand, avoiding the table look up and iterpolations for most problems. The callable subroutines can be linked with a FORTRAN main program and are ideally suited for more complicated problems that requires repetitive calculations for parametric studies and optimization.

The menu-driven program is self-explanatory and allows a selection of the substance for which the properties should be evaluated. It also allows a choice of units so metric (SI) or English units for the properties can be used regardless of the substance. These then provide English-unit properties for the cases where only SI-unit tables are printed in Appendix A.

Details for the interface and call of the subroutines are presented in Appendix B. The package also includes sample programs and utility routines in FORTRAN source code to help in the understanding and implementation of user programs.

3.7 THERMODYNAMIC SURFACES

The matter discussed in this chapter can be well summarized by a consideration of a pressure-specific volume–temperature surface. Two such surfaces are shown in Figs. 3.10 and 3.11. Figure 3.10 shows a substance such as water in which the specific volume

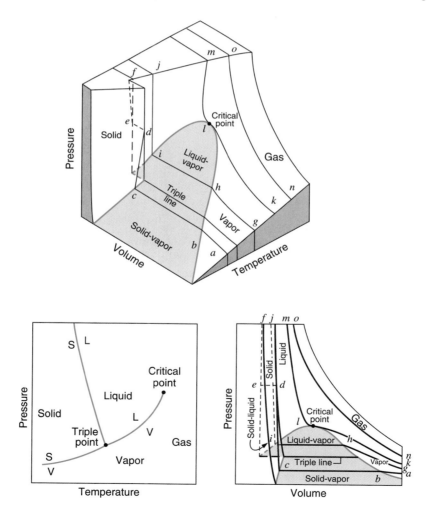

FIGURE 3.10
Pressure–volume–temperature surface for a substance that expands on freezing.

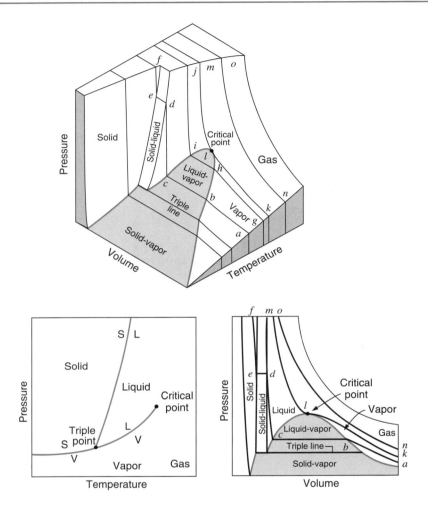

FIGURE 3.11
Pressure–volume–temperature surface for a substance that contracts on freezing.

increases during freezing. Figure 3.11 shows a substance in which the specific volume decreases during freezing.

In these diagrams the pressure, specific volume, and temperature are plotted on mutually perpendicular coordinates, and each possible equilibrium state is thus represented by a point on the surface. This follows directly from the fact that a pure substance has only two independent intensive properties. All points along a quasi-equilibrium process lie on the P–v–T surface, since such a process always passes through equilibrium states.

The regions of the surface that represent a single phase—the solid, liquid, and vapor phases—are indicated. These surfaces are curved. The two-phase regions—the solid–liquid, solid–vapor, and liquid–vapor regions—are ruled surfaces. By this we understand that they are made up of straight lines parallel to the specific-volume axis. This, of course, follows from the fact that in the two-phase region, lines of constant pressure are also lines of constant temperature, although the specific volume may change. The triple point actually appears as the triple line on the P–v–T surface, since the pressure and temperature of the triple point are fixed, but the specific volume may vary, depending on the proportion of each phase.

It is also of interest to note the pressure–temperature and pressure–volume projections of these surfaces. We have already considered the pressure–temperature diagram for a substance such as water. It is on this diagram that we observe the triple point. Various lines of constant temperature are shown on the pressure–volume diagram, and the corresponding constant-temperature sections are lettered identically on the P–v–T surface. The critical isotherm has a point of inflection at the critical point.

One notices that for a substance such as water, which expands on freezing, the freezing temperature decreases with an increase in pressure. For a substance that contracts on freezing, the freezing temperature increases as the pressure increases. Thus, as the pressure of vapor is increased along the constant-temperature line *abcdef* in Fig. 3.10, a substance that expands on freezing first becomes solid and then liquid. For the substance that contracts on freezing, the corresponding constant-temperature line, Fig. 3.11 indicates that as the pressure on the vapor is increased, it first becomes liquid and then solid.

EXAMPLE 3.3   A vessel having a volume of 0.4 m³ contains 2.0 kg of a liquid water and water vapor mixture in equilibrium at a pressure of 600 kPa. Calculate

1. The volume and mass of liquid.
2. The volume and mass of vapor.

The specific volume is calculated first:

$$v = \frac{0.4}{2.0} = 0.20 \text{ m}^3 / \text{kg}$$

From the steam tables (Table A.1.2),

$$v_{fg} = 0.3157 - 0.001\ 101 = 0.3146$$

The quality can now be calculated using Eq. 3.10:

$$0.20 = 0.001\ 101 + x0.3146$$

$$x = 0.6322$$

Therefore, the mass of liquid is

$$2.0(0.3678) = 0.7356 \text{ kg}$$

The mass of vapor is

$$2.0(0.6322) = 1.2644 \text{ kg}$$

The volume of liquid is

$$V_{liq} = m_{liq} v_f = 0.7356(0.001\ 101) = 0.0008 \text{ m}^3$$

The volume of vapor is

$$V_{vap} = m_{vap} v_g = 1.2644(0.3157) = 0.3992 \text{ m}^3$$

EXAMPLE 3.4   A rigid vessel contains saturated ammonia vapor at 20°C. Heat is transferred to the system until the temperature reaches 40°C. What is the final pressure?

Since the volume does not change during this process, the specific volume also remains constant. From the ammonia tables, Table A.2,

$$v_1 = v_2 = 0.149 \ 28 \ m^3 / kg$$

Since v_g at 40°C is less than 0.149 28 m³–kg, it is evident that in the final state the ammonia is superheated vapor. By interpolating between the 900- and 1000-kPa columns of Table A.2.2, we find that

$$P_2 = 936 \ kPa$$

EXAMPLE 3.4E   A rigid vessel contains saturated ammonia vapor at 70F. Heat is transferred to the system until the temperature reaches 120F. What is the final pressure?

Since the volume does not change during this process, the specific volume also remains constant. From the ammonia tables, Table A.2E,

$$v_1 = v_2 = 2.311 \ ft^3 / lbm$$

Since v_g at 120F is less than 2.311 ft³/lbm, it is evident that in the final state the ammonia is superheated vapor. By interpolating between the 140- and 180- lbf/in.² columns of Table A.2.2E, we find that

$$P_2 = 145 \ lbf / in.^2$$

PROBLEMS

3.1 Neon gas at 50°C in a vertical cylinder of diameter 150 mm has a 6-kg frictionless piston mounted. The outside atmospheric pressure is 98 kPa and the neon volume is 4000 cm³. Find the mass of neon.

3.2 A 1-m³ rigid tank with air at 1 MPa, 400 K is connected to an air line as shown in Fig. P3.2. The valve is opened and air flows into the tank until the pressure reaches 5 MPa, at which point the valve is closed and the temperature inside is 450 K.

    a. What is the mass of air in the tank before and after the process?

    b. The tank eventually cools to room temperature, 300 K. What is the pressure inside the tank then?

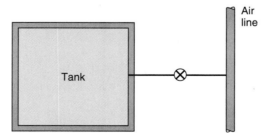

FIGURE P3.2

3.3 A cylindrical gas cylinder is 1 m long, inside diameter of 20 cm, is evacuated and then filled with carbon dioxide gas at 25°C. To what pressure should it be charged if there should be 1.2 kg of carbon dioxide?

3.4 A hollow metal sphere of 150-mm inside diameter is weighed on a precision beam balance when evacuated and again after being filled to 875 kPa with an unknown gas. The difference in mass is 0.0025 kg, and the temperature is 25°C. What is the gas, assuming it is a pure substance?

3.5 A piston/cylinder arrangement, shown in Fig. P3.5, contains air at 250 kPa, 300°C. The 50-kg piston has a diameter of 0.1 m and initially pushes against the stops. The atmosphere is at 100 kPa and 20°C. The cylinder now cools as heat is transferred to the ambient.

   a. At what temperature does the piston begin to move down?

   b. How far has the piston dropped when the temperature reaches ambient?

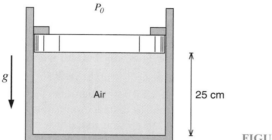

FIGURE P3.5

3.6 A vacuum pump is used to evacuate a chamber where some specimens are dried at 50°C. The pump rate of volume displacement is 0.5 m^3/s with an inlet pressure of 0.1 kPa and temperature 50°C. How much water vapor has been removed over a 30-min period?

3.7 A cylinder has a thick piston initially held by a pin as shown in Fig. P3.7. The cylinder contains carbon dioxide at 150 kPa and ambient temperature of 290 K. The metal piston has a density of 8000 kg/m^3 and the atmospheric pressure is 101 kPa. The pin is now removed, allowing the piston to move and after a while the gas returns to ambient temperature. Is the piston against the stops?

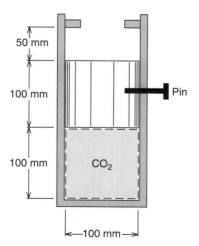

FIGURE P3.7

3.8 An initially deflated and flat balloon is connected by a valve to a storage tank containing helium gas at 1 MPa at ambient temperature of 20°C. The valve is opened and the balloon is inflated at constant pressure, 100 kPa, equal to ambient pressure until it becomes spherical at $D_1 = 1$ m. If the balloon is larger than this, the balloon material is stretched giving a pressure inside as

$$P = P_0 + C\left(1 - \frac{D_1}{D}\right)\frac{D_1}{D}$$

The balloon is slowly inflated to a final diameter of 4 m, at which point the pressure inside is 400 kPa. The temperature remains constant at 20°C. Determine the minimum volume required of the helium tank to inflate the balloon.

3.9 Consider the process of inflating a helium balloon as described in Problem 3.8. What is the maximum pressure inside the balloon at any time during this inflation process? What is the pressure inside the helium storage tank at this time?

3.10 The helium balloon described in Problems 3.8 and 3.9 is released into the atmosphere and rises to an elevation of 5000 m, where the local ambient pressure is 50 kPa and the temperature is −20°C. What is then the diameter of the balloon?

3.11 A cylinder is fitted with a 10-cm-diameter piston that is restrained by a linear spring (force proportional to distance) as shown in Fig. P3.11. The spring force constant is 80 kN/m and the piston initially rests on the stops, with a cylinder volume of 1 L. The valve to the air line is opened and the piston begins to rise when the cylinder pressure is 150 kPa. When the valve is closed, the cylinder volume is 1.5 L and the temperature is 80°C. What mass of air is inside the cylinder?

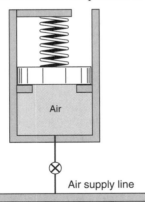

Air

Air supply line

FIGURE P3.11

3.12 Air in a tire is initially at −10°C, 190 kPa. After driving awhile, the temperature goes up to 10°C. Find the new pressure. You must make one assumption on your own.

3.13 Is it reasonable to assume that at the given states the substance behaves as an ideal gas?
   a. Oxygen at 30°C, 3 MPa
   b. Methane at 30°C, 3 MPa
   c. Water at 1000°C, 3 MPa
   d. R-134a at 20°C, 100 kPa
   e. R-134a at −30°C, 100 kPa

3.14 Methane gas is stored in a 2 m³ tank at −30°C, 3 MPa.
   a. Determine the mass inside the tank.
   b. Estimate the percent error in (a) if the ideal gas model is used.

3.15 Repeat Problem 3.14 for argon gas instead of methane.

3.16 Determine whether water at each of the following states is a compressed liquid, a superheated vapor, or a mixture of saturated liquid and vapor.

a. 18 MPa, 0.003 m³/kg

b. 1 MPa, 150°C

c. 200°C, 0.2 m³/kg

d. 10 kPa, 10°C

e. 130°C, 200 kPa

f. 70°C, 1 m³/kg

3.17 Determine whether refrigerant R-22 in each of the following states is a compressed liquid, a superheated vapor, or a mixture of saturated liquid and vapor.

a. 50°C, 0.5 m³/kg

b. 1.0 MPa, 20°C

c. 0.1 MPa, 0.1 m³/kg

d. 50°C, 0.3 m³/kg

e. −20°C, 200 kPa

f. 2 MPa, 0.012 m³/kg

3.18 Determine the qualify (if saturated) or temperature (if superheated) of the following substances at the given two states:

a. Water at 1: 120°C, 1 m³/kg; 2: 10 MPa, 0.02 m³/kg

b. Nitrogen at 1: 1 MPa, 0.03 m³/kg; 2: 100 K, 0.03 m³/kg

c. Ammonia at 1: 0°C, 0.1 m³/kg; 2: 1000 kPa, 0.145 m³/kg

d. R-22 at 1: 130 kPa, 0.1 m³/kg; 2: 150 kPa, 0.17 m³/kg

3.19 Calculate the following specific volumes

a. R-134a 50°C, 80% quality

b. Water 4 MPa, 90% quality

c. Methane 140 K, 60% quality

d. Ammonia 10°C, 25% quality

3.20 Give the phase and the specific volume.

a.	H_2O	$T = 275°C$	$P = 5$ MPa
b.	H_2O	$T = -2°C$	$P = 100$ kPa
c.	CO_2	$T = 267°C$	$P = 0.5$ MPa
d.	Air	$T = 20°C$	$P = 200$ kPa
e.	NH_3	$T = 65°C$	$P = 600$ kPa

3.21 Give the phase and the specific volume.

a.	R-12	$T = -5°C$	$P = 200$ kPa
b.	R-12	$T = -5°C$	$P = 300$ kPa
c.	R-22	$T = 5°C$	$P = 500$ kPa
d.	Ar	$T = 200°C$	$P = 200$ kPa
e.	NH_3	$T = 20°C$	$P = 100$ kPa

3.22 Find the phase, quality x if applicable and the missing property P or T.

a. H_2O $T = 120°C$ $v = 0.5$ m³/kg
b. H_2O $P = 100$ kPa $v = 1.695$ m³/kg
c. H_2O $T = 263$ K $v = 200$ m³/kg
d. Ne $P = 750$ kPa $v = 0.2$ m³/kg
e. NH_3 $T = 20°C$ $v = 0.1$ m³/kg

3.23 Give the phase and the missing properties of P, T, v and x.

a. R-22 $T = 10°C$ $v = 0.01$ m³/kg
b. H_2O $T = 350°C$ $v = 0.2$ m³/kg
c. CO_2 $T = 800$ K $P = 200$ kPa
d. N_2 $T = 200$ K $P = 100$ kPa
e. CH_4 $T = 190$ K $x = 0.75$

3.24 Give the phase and the missing properties of P, T, v and x. These may be a little more difficult if the appendix tables are used instead of the software.

a. R-22 $T = 10°C$ $v = 0.036$ m³/kg
b. H_2O $v = 0.2$m³/kg $x = 0.5$
c. H_2O $T = 60°C$ $v = 0.001016$ m³/kg
d. NH_3 $T = 30°C$ $P = 60$ kPa
e. R-134a $v = 0.005$m³/kg $x = 0.5$

3.25 Plot a P–v diagram on log-log paper (3×5 cycles) for water. Show the following lines

a. Saturated liquid
b. Saturated vapor
c. The constant quality lines for $x = 0.1$ and $x = 0.5$

3.26 What is the percent error in pressure if the ideal gas model is used to represent the behavior of superheated ammonia at 40°C, 500 kPa? What if the generalized compressibility chart, Fig. A.7, is used instead?

3.27 What is the percent error in pressure if the ideal gas model is used to represent the behavior of superheated vapor R-22 at 50°C, 0.03 m³/kg? What if the generalized compressibility chart, Fig. A.7, is used instead?

3.28 A water storage tank contains liquid and vapor in equilibrium at 110°C. The distance from the bottom of the tank to the liquid level is 8 m. What is the absolute pressure at the bottom of the tank?

3.29 A sealed rigid vessel has volume of 1 m³ and contains 1 kg of water at 100°C. The vessel is now heated. If a safety pressure valve is installed, at what pressure should the valve be set to have a maximum temperature of 200°C?

3.30 A 500-L tank stores 100 kg of nitrogen gas at 150 K. To design the tank the pressure must be estimated and four different methods are suggested. Which is the most accurate, and how different in percent are the other three?

a. Benedict–Webb–Rubin equation of state
b. Nitrogen tables, Table A.6
c. Ideal gas
d. Generalized compressibility chart, Fig. A.7

3.31 A 400-m³ storage tank is being constructed to hold LNG, liquified natural gas, which may be assumed to be essentially pure methane. If the tank is to contain 90% liquid and 10% vapor, by volume, at 100 kPa, what mass of LNG (kg) will the tank hold? What is the quality in the tank?

3.32 The storage tank in the previous problem warms up by 5°C per hour due to a failure in the refrigeration system. The tank is designed to hold a pressure of 600 kPa. How much time is available to repair the system before it reaches the design pressure?

3.33 Saturated liquid water at 60°C is put under pressure to decrease the volume by 1% keeping the temperature constant. To what pressure should it be compressed?

3.34 Saturated water vapor at 60°C has its pressure decreased to increase the volume by 10% keeping the temperature constant. To what pressure should it be expanded?

3.35 A boiler feed pump delivers 0.05 m³/s of water at 240°C, 20 MPa. What is the mass flowrate (kg/s)? What would be the percent error if the properties of saturated liquid at 240°C were used in the calculation? What if the properties of saturated liquid at 20 MPa were used?

3.36 A glass jar is filled with saturated water at 100 kPa, quality 25%, and a tight lid is put on. Now it is cooled to −10°C. What is the mass fraction of solid at this temperature?

3.37 A cylinder/piston arrangement contains water at 105°C, 85% quality with a volume of 1 L. The system is heated, causing the piston to rise and encounter a linear spring as shown in Fig. P3.37. At this point the volume is 1.5 L, piston diameter is 150 mm, and the spring constant is 100 N/mm. The heating continues, so the piston compresses the spring. What is the cylinder pressure when the temperature reaches 600°C?

FIGURE P3.37

3.38 Saturated (liquid + vapor) refrigerant R-134a at 0°C is contained in a rigid steel tank. It is used in an experiment, where it should pass through the critical point when the system is heated. What should the initial mass fraction of liquid be?

3.39 For a certain experiment, R-22 vapor is contained in a sealed glass tube at 20°C. It is desired to know the pressure at this condition, but there is no means of measuring it, since the tube is sealed. However, if the tube is cooled to −20°C small droplets of liquid are observed on the glass walls. What is the initial pressure?

3.40 A steel tank contains 6 kg of propane (liquid + vapor) at 20°C with a volume of 0.015 m³. The tank is now slowly heated. Will the liquid level inside eventually rise to the top or drop to the bottom of the tank? What if the initial mass is 1 kg instead of 6 kg?

3.41 A cylinder containing ammonia is fitted with a piston restrained by an external force that is proportional to cylinder volume squared. Initial conditions are 10°C, 90% quality and a volume of 5 L. A valve on the cylinder is opened and additional ammonia flows into the cylinder until the mass inside has doubled. If at this point the pressure is 1.2 MPa, what is the final temperature?

3.42 A container with liquid nitrogen at 500 kPa has a cross sectional area of 0.5 m². Due to heat transfer, some of the liquid evaporates and in one hour the liquid level drops 30 mm. The vapor leaving the container passes through a heater and exits at 500 kPa, 275 K. Calculate the volume rate of flow of nitrogen gas exiting the heater.

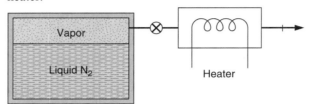

FIGURE P3.42

3.43 A pressure cooker (closed tank) contains water at 100°C with the liquid volume being 1/10 of the vapor volume. It is heated until the pressure reaches 2.0 MPa. Find the final temperature. Has the final state more or less vapor than the initial state?

3.44 Ammonia in a piston/cylinder arrangement is at 700 kPa, 80°C. It is now cooled at constant pressure to saturated vapor (state 2) at which point the piston is locked with a pin. The cooling continues to −10°C (state 3). Show the processes 1 to 2 and 2 to 3 on both a P–v and T–v diagram.

3.45 A piston/cylinder arrangement is loaded with a linear spring and the outside atmosphere. It contains water at 5 MPa, 400°C with the volume being 0.1 m³. If the piston is at the bottom, the spring exerts a force such that $P_{\text{lift}} = 200$ kPa. The system now cools until the pressure reaches 1200 kPa. Find the mass of water and v_1. Find the final state (T_2, v_2) and plot the P–v diagram for the process.

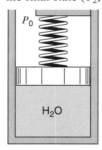

FIGURE P3.45

3.46 A spring-loaded piston/cylinder contains water at 90°C, 100 kPa. The setup is such that pressure is proportional to volume, $P = CV$. Heat is now added until the temperature reaches 200°C. Find the final pressure and also the qualify if in the two-phase region.

3.47 A spring-loaded piston/cylinder contains water at 500°C, 3 MPa. The setup is such that pressure is proportional to volume, $P = CV$. It is now cooled until the water becomes saturated vapor. Find the final pressure.

3.48 Refrigerant-12 in a piston/cylinder arrangement is initially at 50°C, $x = 1$. It is then expanded in a process so that $P = Cv^{-1}$ to a pressure of 100 kPa. Find the final temperature and specific volume.

3.49 A sealed rigid vessel of 2 m³ contains a saturated mixture of liquid and vapor R-134a at 10°C. If it is heated to 50°C, the liquid phase disappears. Find the pressure at 50°C and the initial mass of the liquid.

3.50 Two tanks are connected as shown in Fig. P3.50, both containing water. Tank A is at 200 kPa, $v = 0.5$ m³–kg, $V_A = 1$ m³ and tank B contains 3.5 kg at 0.5 MPa, 400°C. The valve is now opened and the two come to a uniform state. Find the final specific volume.

FIGURE P3.50

3.51 A tank contains 2 kg of nitrogen at 100 K with a quality of 50%. Through a volume flowmeter and valve, 0.5 kg is now removed while the temperature remains constant. Find the final state inside the tank and the volume of nitrogen removed if the valve/meter is located at

a. The top of the tank

b. The bottom of the tank

3.52 Consider two tanks, A and B, connected by a valve, as shown in Fig. P3.52. Each has a volume of 200 L and tank A has R-12 at 25°C, 10% liquid and 90% vapor by volume, while tank B is evacuated. The valve is now opened and saturated vapor flows from A to B until the pressure in B has reached that in A, at which point the valve is closed. This process occurs slowly such that all temperatures stay at 25°C throughout the process. How much has the quality changed in tank A during the process?

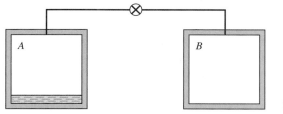

FIGURE P3.52

3.53E A cylindrical gas cylinder 3 ft long, inside diameter of 8 in., is evacuated and then filled with carbon dioxide gas at 77 F. To what pressure should it be charged if there should be 2.6 lbm of carbon dioxide?

3.54E A hollow metal sphere of 6 in. inside diameter is weighed on a precision beam balance when evacuated and again after being filled to 110 lbf/in.² with an unknown gas. The difference in mass is 0.005 lbm, and the temperature is 80 F. What is the gas, assuming it is a pure substance?

3.55E A vacuum pump is used to evacuate a chamber where some specimens are dried at 120 F. The pump rate of volume displacement is 900 ft³/min with an inlet pressure of 1 mm Hg and temperature 120 F. How much water vapor has been removed over a 30-min period?

3.56E A cylinder is fitted with a 4-in.-diameter piston that is restrained by a linear spring (force proportional to distance) as shown in Fig. P3.11. The spring force constant is 400 lbf–in. and the piston initially rests on the stops, with a cylinder volume of 60 in.^{3}. The valve to the air line is opened and the piston begins to rise when the cylinder pressure is 22 lbf/in.^{2}. When the valve is closed, the cylinder volume is 90 in.^{3} and the temperature is 180 F. What mass of air is inside the cylinder?

3.57E Air in an automobile tire is initially at 15 F, 1.9 atm. After the car is driven awhile, the temperature goes up to 40 F. Find the new pressure. You must make one assumption on your own.

3.58E Methane gas is stored in a 70 ft^3 tank at −25 F, 440 lbf/in.^{2}
  a. Find the mass of methane using the generalized compressibility chart.
  b. Estimate the percent error in (a) if the ideal gas model is used.

3.59E Repeat Problem 3.58 for argon gas instead of methane.

3.60E Determine whether water at each of the following states is a compressed liquid, a superheated vapor, or a mixture of saturated liquid and vapor.
  a. 1800 lbf/in.^{2}, 0.03 ft^3/lbm
  b. 150 lbf/in.^{2}, 320 F
  c. 380 F, 3 ft^3/lbm
  d. 2 lbf/in.^{2}, 50 F
  e. 270 F, 30 lbf/in.^{2}
  f. 160 F, 10 ft^3/lbm

3.61E Give the phase and the specific volume.
  a. H$_2$O $T = 520$F $P = 700$ lbf/in.^{2}
  b. H$_2$O $T = 30$ F $P = 15$ lbf/in.^{2}
  c. CO$_2$ $T = 510$ F $P = 75$ lbf/in.^{2}
  d. Air $T = 68$ F $P = 2$ atm
  e. NH$_3$ $T = 150$ F $P = 90$ lbf/in.^{2}

3.62E Give the phase and the specific volume.
  a. R-12 $T = 20$ F $P = 30$ lbf/in.^{2}
  b. R-12 $T = 20$ F $P = 40$ lbf/in.^{2}
  c. R-22 $T = 40$ F $P = 70$ lbf/in.^{2}
  d. Ar $T = 300$ F $P = 30$ lbf/in.^{2}
  e. NH$_3$ $T = 60$ F $P = 15$ lbf/in.^{2}

3.63E Give the phase and the missing properties of P, T, v and x. These may be a little more difficult if the appendix tables are used instead of the software.
  a. R-22 $T = 50$ F $v = 0.6$ ft^3/lbm
  b. H$_2$O $v = 2$ ft^3/lbm $x = 0.5$
  c. H$_2$O $T = 150$ F $v = 0.01632$ ft^3/lbm
  d. NH$_3$ $T = 80$ F $P = 13$ lbf/in.^{2}
  e. R-134a $v = 0.08$ ft^3/lbm $x = 0.5$

3.64E What is the percent error in specific volume if the ideal gas model is used to represent the behavior of superheated ammonia at 100 F, 80 lbf/in.^{2}? What if the generalized compressibility chart, Fig. A.7, is used instead?

3.65E A water storage tank contains liquid and vapor in equilibrium at 220 F. The distance from the bottom of the tank to the liquid level is 25 ft. What is the absolute pressure at the bottom of the tank?

3.66E A sealed rigid vessel has volume of 35 ft³ and contains 2 lbm of water at 200 F. The vessel is now heated. If a safety pressure valve is installed, at what pressure should the valve be set to have a maximum temperature of 400 F?

3.67E Saturated liquid water at 200 F is put under pressure to decrease the volume by 1%, keeping the temperature constant. To what pressure should it be compressed?

3.68E Saturated water vapor at 200 F has its pressure decreased to increase the volume by 10%, keeping the temperature constant. To what pressure should it be expanded?

3.69E A boiler feed pump delivers 100 ft³/min of water at 400 F, 3000 lbf/in.². What is the mass flowrate (lbm/s)? What would be the percent error if the properties of saturated liquid at 400 F were used in the calculation? What if the properties of saturated liquid at 3000 lbf/in.² were used?

3.70E Saturated (liquid + vapor) refrigerant R-134a at 30 F is contained in a rigid steel tank. It is used in an experiment, where it should pass through the critical point when the system is heated. What should the initial mass fraction of liquid be?

3.71E A steel tank contains 14 lbm of propane (liquid + vapor) at 70 F with a volume of 0.25 ft³. The tank is now slowly heated. Will the liquid level inside eventually rise to the top or drop to the bottom of the tank? What if the initial mass is 2 lbm instead of 14 lbm?

3.72E A pressure cooker (closed tank) contains water at 200 F with the liquid volume being 1/10 of the vapor volume. It is heated until the pressure reaches 300 lbf/in.². Find the final temperature. Has the final state more or less vapor than the initial state?

3.73E A spring-loaded piston/cylinder contains water at 200 F, 14.7 lbf/in.². The setup is such that pressure is proportional to volume, $P = CV$. Heat is now added until the temperature reaches 390 F. Find the final pressure and also the quality if in the two-phase region.

3.74E A spring-loaded piston/cylinder contains water at 900 F, 450 lbf/in.². The setup is such that pressure is proportional to volume, $P = CV$. It is now cooled until the water becomes saturated vapor. Find the final pressure.

3.75E Refrigerant-12 in a piston/cylinder arrangement is initially at 120 F, $x = 1$. It is then expanded in a process so that $P = Cv^{-1}$ to a pressure of 15 lbf/in.². Find the final temperature and specific volume.

COMPUTER, DESIGN, AND OPEN ENDED PROBLEMS

3.76 Write a program that lists the saturated P, T for ammonia with pressure as an entry, beginning with $P = 100$ kPa and ending with $P = 500$ kPa, in steps of 25 kPa.

3.77 Write a program that tabulates values of P and T along a constant specific volume line for water. The starting state is 100 kPa, quality of $x = 0.5$, and the ending state is 1 MPa.

3.78 Write a computer program that lists the states P, T, v along the process curves in Problems 3.46 and 3.47.

3.79 Write a computer program that lists the states P, T, v along the process curves in Problem 3.48.

3.80 We wish to be able to calculate "standard" atmospheric pressure at any given elevation up to 10 km. The "standard" atmosphere assumes that gravitational acceleration drops by 0.0031 m–s^2 from the sea level value of 9.8066 m–s^2 and that the ambient temperature drops by 6.5°C from the sea level standard of 15°C, both for each kilometer of elevation.

3.81 In Problem 3.37, we wish to follow the path of the process for the steam for any state between the initial and final states inside the cylinder.

3.82 In Problem 3.41, we wish to follow the path of the process for the ammonia for any state between the initial and final states inside the cylinder.

3.83 In Problem 3.52, we wish to follow the path of the process for the R~12 for any state between the initial and final states inside the two tanks.

3.84 For any specified substance, calculate the pressure according to the Benedict–Webb–Rubin equation of state for any given set(s) of temperature and specific volume. Compare the result with the ideal gas equation of state.

3.85 For any specified substance, solve the Benedict–Webb–Rubin equation of state for specific volume at any given set(s) of pressure and temperature. Compare the result with the ideal gas equation of state.

3.86 For any specified substance in Tables A.1–A.7, fit a polynomial equation of degree n to tabular data for pressure as a function of density along any given isotherm in the superheated vapor region.

3.87 The refrigerant fluid in a household refrigerator changes phase from liquid to vapor at the low temperature in the refrigerator. It changes phase from vapor to liquid at the higher temperature in the heat exchanger that gives the energy to the room air. Measure or otherwise estimate these temperatures. Based on these temperatures make a table with the refrigerant pressures for the refrigerants for which tables are available in Appendix A. Discuss the results and the requirements for a substance to be a potential refrigerant.

3.88 Repeat the previous problem for refrigerants that are listed in Table A.8 and use the compressibility chart Fig. A.7 or Table A.15 to estimate the pressures.

3.89 The saturated pressure as a function of temperature follows the correlation developed by Wagner as

$$\ln P_r = \left[w_1 \tau + w_2 \tau^{1.5} + w_3 \tau^3 + w_4 \tau^6\right]/T_r$$

where the reduced pressure and temperature are $P_r = P/P_c$ and $T_r = T/T_c$. The temperature variable is $\tau = 1 - T_r$. The parameters are found for R-12 and R-134a as

	w_1	w_2	w_3	w_4
R-12	-6.91826	1.49560	-2.65015	-0.63170
R-134a	-7.59884	1.48886	-3.79873	1.81379

Compare these correlations to the tables in Appendix A.

3.90 Find the constants in the curve fit for the saturation pressure using Wagner's correlation as shown in the previous problem for water and methane. Find other correlations in the literature and compare them to the tables and give the maximum deviation.

3.91 The specific volume of saturated liquid can be approximated by the Rackett equation as

$$v_f = \frac{\overline{R}T_c}{MP_c} Z_c^{1+(1-T_r)^{2/7}}$$

with the reduced temperature, $T_r = T/T_c$, and the compressibility factor, $\overline{Z}_c = P_c\overline{v}_c/\overline{R}T_c$. Using values from Table A.8 with the critical constants, compare the formula to the tables for substances where the saturated specific volume is available.

4 WORK AND HEAT

In this chapter we consider work and heat. It is essential for the student of thermodynamics to understand clearly the definitions of both work and heat, because the correct analysis of many thermodynamic problems depends on distinguishing between them.

4.1 DEFINITION OF WORK

Work is usually defined as a force F acting through a displacement x, the displacement being in the direction of the force. That is,

$$W = \int_1^2 F\,dx \qquad (4.1)$$

This is a very useful relationship because it enables us to find the work required to raise a weight, to stretch a wire, or to move a charged particle through a magnetic field.

However, when treating thermodynamics from a macroscopic point of view, it is advantageous to tie in the definition of work with the concepts of systems, properties, and processes. We therefore define work as follows: work is done by a system if the sole effect on the surroundings (everything external to the system) could be the raising of a weight. Notice that the raising of a weight is in effect a force acting through a distance. Notice also that our definition does not state that a weight was actually raised or that a force actually acted through a given distance, but that the sole effect external to the system could be the raising of a weight. Work done *by* a system is considered positive and work done *on* a system is considered negative. The symbol W designates the work done by a system.

In general, we will speak of work as a form of energy. No attempt will be made to give a rigorous definition of energy. Since the concept is familiar, the term energy will be used as appropriate, and various forms of energy will be identified. Work is the form of energy that fulfills the definition.

Let us illustrate this definition of work with a few examples. Consider as a system the battery and motor of Fig. 4.1a and let the motor drive a fan. Does work cross the boundary of the system? To answer this question using the definition of work given earlier, replace the fan with the pulley and weight arrangement shown in Fig. 4.1b. As the motor turns, the weight is raised, and the sole effect external to the system is the raising of a weight. Thus, for our original system of Fig. 4.1a, we conclude that work is crossing the boundary of the system since the sole effect external to the system could be the raising of a weight.

Let the boundaries of the system be changed now to include only the battery shown in Fig. 4.2. Again we ask the question, does work cross the boundary of the system? To

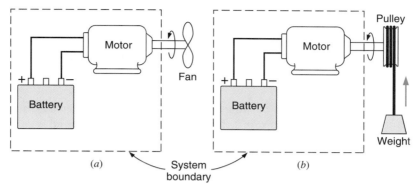

FIGURE 4.1 Example of work crossing the boundary of a system.

answer this question, we need to ask a more general question. Does the flow of electrical energy across the boundary of a system constitute work?

The only limiting factor in having the sole external effect be the raising of a weight is the inefficiency of the motor. However, as we design a more efficient motor, with lower bearing and electrical losses, we recognize that we can approach a certain limit that meets the requirement of having the only external effect be the raising of a weight. Therefore, we can conclude that when there is a flow of electricity across the boundary of a system, as in Fig. 4.2, it is work.

4.2 UNITS FOR WORK

As already noted, work done *by* a system, such as that done by a gas expanding against a piston, is positive, and work done *on* a system, such as that done by a piston compressing a gas, is negative. Thus, positive work means that energy leaves the system, and negative work means that energy is added to the system.

Our definition of work involves the raising of a weight, that is, the product of a unit force (one newton) acting through a unit distance (one meter). This unit for work in SI units is called the joule (J).

$$1\,J = 1\,N\,m$$

Power is the time rate of doing work and is designated by the symbol $\dot{W}$:

$$\dot{W} \equiv \frac{\delta W}{dt}$$

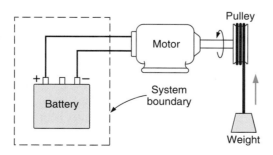

FIGURE 4.2 Example of work crossing the boundary of a system because of a flow of an electric current across the system boundary.

The unit for power is a rate of work of one joule per second, which is a watt (W):

$$1 \text{ W} = 1 \text{ J} / \text{s}$$

A familiar unit for power in English units is the horsepower (hp), where

$$1 \text{ hp} = 550 \text{ ft lbf} / \text{s}$$

It is often convenient to speak of the work per unit mass of the system, often termed "specific work." This quantity is designated w and is defined

$$w \equiv \frac{W}{m}$$

4.3 WORK DONE AT THE MOVING BOUNDARY OF A SIMPLE COMPRESSIBLE SYSTEM IN A QUASI-EQUILIBRIUM PROCESS

We have already noted that there are a variety of ways in which work can be done on or by a system. These include work done by a rotating shaft, electrical work, and the work done by the movement of the system boundary, such as the work done in moving the piston in a cylinder. In this section we will consider in some detail the work done at the moving boundary of a simple compressible system during a quasi-equilibrium process.

Consider as a system the gas contained in a cylinder and piston, as in Fig. 4.3. Let one of the small weights be removed from the piston, which will cause the piston to move upward a distance dL. We can consider this quasi-equilibrium process and calculate the amount of work W done by the system during this process. The total force on the piston is $P\mathcal{A}$, where P is the pressure of the gas and $\mathcal{A}$ is the area of the piston. Therefore, the work δW is

$$\delta W = P\mathcal{A} \ dL$$

But $\mathcal{A} \ dL = dV$, the change in volume of the gas. Therefore,

$$\delta W = P \ dV \qquad (4.2)$$

The work done at the moving boundary during a given quasi-equilibrium process can be found by integrating Eq. 4.2. However, this integration can be performed only if we know the relationship between P and V during this process. This relationship may be expressed in the form of an equation, or it may be shown in the form of a graph.

Let us consider a graphical solution first. We use as an example a compression process such as occurs during the compression of air in a cylinder, Fig. 4.4. At the beginning of the process the piston is at position 1, and the pressure is relatively low. This state is represented on a pressure–volume diagram (usually referred to as a P–V diagram). At the conclusion of the process the piston is in position 2, and the corresponding state of the gas is shown at point 2 on the P–V diagram. Let us assume that this compression was a quasi-equilibrium process, and that during the process the system passed through the states shown by the line connecting states 1 and 2 on the P–V diagram. The assumption of a quasi-equilibrium process is essential here because each point on line 1–2 represents a definite state, and these states will correspond to the actual state of the system only if the deviation from equilibrium is infinitesimal. The work done on the air

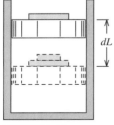

FIGURE 4.3
Example of work done at the moving boundary of a system in a quasi-equilibrium process.

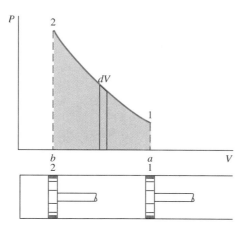

FIGURE 4.4 Use of pressure–volume diagram to show work done at the moving boundary of a system in a quasi-equilibrium process.

during this compression process can be found by integrating Eq. 4.2:

$$_1W_2 = \int_1^2 \delta W = \int_1^2 P\,dV \tag{4.3}$$

The symbol $_1W_2$ is to be interpreted as the work done during the process form state 1 to state 2. It is clear from examining the P–V diagram that the work done during this process,

$$\int_1^2 P\,dV$$

is represented by the area under the curve 1–2, area a–1–2–b–a. In this example the volume decreased, and the area a–1–2–b–a represents work done on the system. If the process had proceeded from state 2 to state 1 along the same path, the same area would represent work done by the system.

Further consideration of a P–V diagram, such as Fig. 4.5, leads to another important conclusion. It is possible to go from state 1 to state 2 along many different quasi-equilibrium paths, such as A, B, or C. Since the area underneath each curve represents the work for each process, it is evident that the amount of work done during each process not only is a function of the end states of the process, but depends on the path that is followed in going from one state to another. For this reason work is called a path function or, in mathematical parlance, δW is an inexact differential.

This concept leads to a brief consideration of point and path functions or, to use another term, exact and inexact differentials. Thermodynamic properties are point functions, a name that comes from the fact that for a given point on a diagram (such as Fig. 4.5) or surface (such as Fig. 3.10), the state is fixed, and thus there is a definite value of each property corresponding to this point. The differentials of point functions are exact differentials, and the integration is simply

$$\int_1^2 dV = V_2 - V_1$$

Thus, we can speak of the volume in state 2 and the volume in state 1, and the change in volume depends only on the initial and final states.

Work, on the other hand, is a path function, for, as has been indicated, the work done in a quasi-equilibrium process between two given states depends on the path fol-

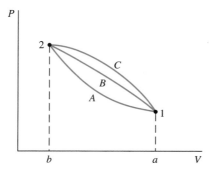

FIGURE 4.5 Various quasi-equilibrium processes between two given states, indicating that work is a path function.

lowed. The differentials of path functions are inexact differentials, and the symbol δ will be used in this text to designate inexact differentials (in contrast to d for exact differentials). Thus, for work, we write

$$\int_1^2 \delta W = {}_1W_2$$

It would be more precise to use the notation, ${}_1W_{2A}$, which would indicate the work done during the change from state 1 to state 2 along path A. However, it is implied in the notation ${}_1W_2$ that the process between states 1 and 2 has been specified. It should be noted that we never speak about the work in the system in state 1 or state 2, and thus we would never write $W_2 - W_1$.

In evaluating the integral of Eq. 4.3, we should always keep in mind that we wish to determine the area under the curve on Fig. 4.5. In connection with this point, we identify the following two classes of problems:

1. The relationship between P and V is given in terms of experimental data or in graphical form (as, for example, the trace on an oscilloscope). Therefore, we may evaluate the integral, Eq. 4.3, by graphical or numerical integration.

2. The relationship between P and V makes it possible to fit an analytical relationship between them. We may then integrate directly.

One common example of this second type of functional relationship is a process called a polytropic process, one in which

$$PV^n = \text{constant}$$

throughout the process. The exponent n may possibly be any value from $-\infty$ to $+\infty$, depending on the particular process. For this type of process, we can integrate Eq. 4.3 as follows:

$$PV^n = \text{constant} = P_1V_1^n = P_2V_2^n$$

$$P = \frac{\text{constant}}{V^n} = \frac{P_1V_1^n}{V^n} = \frac{P_2V_2^n}{V^n}$$

$$\int_1^2 P\,dV = \text{constant} \int_1^2 \frac{dV}{V^n} = \text{constant} \left(\frac{V^{-n+1}}{-n+1} \right) \Bigg|_1^2$$

$$\int_1^2 P\,dV = \frac{\text{constant}}{1-n}(V_2^{1-n} - V_1^{1-n}) = \frac{P_2 V_2^n V_2^{1-n} - P_1 V_1^n V_1^{1-n}}{1-n}$$

$$= \frac{P_2 V_2 - P_1 V_1}{1-n} \tag{4.4}$$

Note that the resulting Eq. 4.4 is valid for any exponent n, except $n = 1$. Where $n = 1$,

$$PV = \text{constant} = P_1 V_1 = P_2 V_2$$

and

$$\int_1^2 P\,dV = P_1 V_1 \int_1^2 \frac{dV}{V} = P_1 V_1 \ln \frac{V_2}{V_1} \tag{4.5}$$

Note that in Eqs. 4.4 and 4.5 we did not say that the work is equal to the expressions given in these equations. These expressions give us the value of a certain integral, that is, a mathematical result. Whether or not that integral equals the work in a particular process depends on the result of a thermodynamic analysis of that process. It is important to keep the mathematical result separate from the thermodynamic analysis, for there are many situations in which work is not given by Eq. 4.3.

The polytropic process as described demonstrates one special functional relationship between P and V during a process. There are many other possible relations, some of which will be examined in the problems at the end of this chapter.

EXAMPLE 4.1   Consider as a system the gas in the cylinder shown in Fig. 4.6; the cylinder is fitted with a piston on which a number of small weights are placed. The initial pressure is 200 kPa, and the initial volume of the gas is 0.04 m³.

1. Let a Bunsen burner be placed under the cylinder, and let the volume of the gas increase to 0.1 m³ while the pressure remains constant. Calculate the work done by the system during this process.

$$_1W_2 = \int_1^2 P\,dV$$

Since the pressure is constant, we conclude from Eq. 4.3 that

$$_1W_2 = P\int_1^2 dV = P(V_2 - V_1)$$

$$_1W_2 = 200\ \text{kPa} \times (0.1 - 0.04)\text{m}^3 = 12.0\ \text{kJ}$$

2. Consider the same system and initial conditions, but at the same time the Bunsen burner is under the cylinder and the piston is rising, let weights be removed from the piston at such a rate that, during the process, the temperature of the gas remains constant.

    If we assume that the ideal-gas model is valid, then, from Eq. 3.4,

$$PV = mRT$$

We note that this is a polytropic process with exponent $n = 1$. From our analysis,

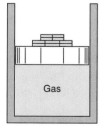

FIGURE 4.6
Sketch for Example 4.1.

Gas

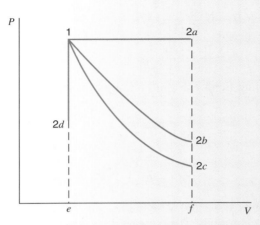

FIGURE 4.7 Pressure–volume diagram showing work done in the various processes of Example 4.1.

we conclude that the work is given by Eq. 4.3 and that the integral in this equation is given by Eq. 4.5. Therefore,

$$_1W_2 = \int_1^2 P\,dV = P_1V_1 \ln\frac{V_2}{V_1}$$

$$= 200 \text{ kPa} \times 0.04 \text{ m}^3 \times \ln\frac{0.10}{0.04} = 7.33 \text{ kJ}$$

3. Consider the same system, but during the heat transfer let the weights be removed at such a rate that the expression $PV^{1.3} = $ constant describes the relation between pressure and volume during the process. Again the final volume is 0.1 m³. Calculate the work.

This is a polytropic process in which $n = 1.3$. Analyzing the process, we conclude again that the work is given by Eq. 4.3, and that the integral is given by Eq. 4.4. Therefore,

$$P_2 = 200\left(\frac{0.04}{0.10}\right)^{1.3} = 60.77 \text{ kPa}$$

$$_1W_2 = \int_1^2 P\,dV = \frac{P_2V_2 - P_1V_1}{1 - 1.3} = \frac{60.77 \times 0.1 - 200 \times 0.04}{1 - 1.3}$$

$$= 6.41 \text{ kJ}$$

4. Consider the system and initial state given in the first three examples, but let the piston be held by a pin so that the volume remains constant. In addition, let heat be transferred from the system until the pressure drops to 100 kPa. Calculate the work.

Since $\delta W = P\,dV$ for a quasi-equilibrium process, the work is zero, because there is no change in volume.

The process for each of the four examples is shown on the P–V diagram of Fig. 4.7. Process 1–2a is a constant-pressure process, and area 1–2a–f–e–1 represents the work. Similarly, line 1–2b represents the process in which $PV = $ constant, line 1–2c

the process in which $PV^{1.3}$ = constant, and line 1–2d the constant-volume process. The student should compare the relative areas under each curve with the numerical results obtained for the amounts of work done.

EXAMPLE 4.2   Consider a slightly different piston cylinder arrangement as shown in Fig. 4.8. In this example the piston is loaded with a mass, m_p, the outside atmosphere P_0, a linear spring and a single point force F_1. The piston is restricted in its motion by lower and upper stops trapping the gas with a pressure P. A force balance on the piston in the direction of motion yields

$$m_p a \simeq 0 = \sum F_\uparrow - \sum F_\downarrow$$

with a zero acceleration in a quasi-equilibrium process. The forces, when the piston is between the stops, are

$$\sum F_\uparrow = P\mathcal{A} \qquad \sum F_\downarrow = m_p g + P_0 \mathcal{A} + k_s(x - x_0) + F_1$$

with the linear spring constant, k_s. The piston position for a relaxed spring is x_0, which depends on how the spring is installed. The force balance then gives the gas pressure by division with the area, $\mathcal{A}$ as

$$P = P_0 + \left[m_p g + F_1 + k_s(x - x_0) \right] / \mathcal{A}$$

To illustrate the process in a P–V diagram, the distance x is converted to volume by division and multiplication with $\mathcal{A}$:

$$P = P_0 + \frac{m_p g}{\mathcal{A}} + \frac{F_1}{\mathcal{A}} + \frac{k_s}{\mathcal{A}^2}(V - V_0) = C_1 + C_2 V$$

This relation gives the pressure as a linear function of the volume, with the line having a slope of $C_2 = k_s / \mathcal{A}^2$. With the stops installed, the minimum and maximum volumes limit the possible states of the system to the combination of P and V as shown in Fig. 4.9. Regardless of what substance is inside, any process must proceed along the lines in the P–V diagram. The work term in a quasi-equilibrium process then follows as

$$_1W_2 = \int_1^2 P \, dV = \text{ AREA under the process curve}$$

$$_1W_2 = \frac{1}{2}\left(P_1' + P_2'\right)\left(V_2 - V_1\right)$$

with $P_1' = P_1$ and $P_2' = P_2$, subject to the constraint that

$$P_{\min} \le P_1', \; P_2' \le P_{\max}$$

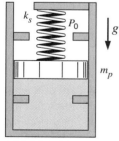

FIGURE 4.8

Sketch of physical system for Example 4.2.

These limits show that only the part of the process that follows the sloped line, when the piston moves, contributes to the work. Any part of the process with a pressure smaller than $P_{\min}$ or larger than $P_{\max}$ does not involve work as the piston is

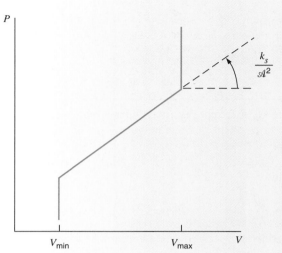

FIGURE 4.9 The process curve showing possible P–V combinations for Example 4.2.

held in a fixed position by one of the stops. The maximum work then arises, when the piston travels the total distance between the two stops.

In this section we have discussed boundary movement work in a quasi-equilibrium process. We should also realize that there may very well be boundary movement work in a nonequilibrium process. Then the total force exerted on the piston by the gas inside the cylinder, $P\mathscr{A}$, does not equal the external force, F_{ext}, and the work is not given by Eq. 4.2. The work can, however, be evaluated in terms of F_{ext} or, dividing by area, an equivalent external pressure, P_{ext}. The work done at the moving boundary in this case is

$$\delta W = F_{ext}\, dL = P_{ext}\, dV \tag{4.6}$$

Evaluation of Eq. 4.6 in any particular instance requires a knowledge of how the external force or pressure changes during the process.

4.4 SOME OTHER SYSTEMS IN WHICH WORK IS DONE AT A MOVING BOUNDARY

In the preceding section we considered the work done at the moving boundary of a simple compressible system during a quasi-equilibrium process and also during a nonequilibrium process. There are other types of systems in which work is done at a moving boundary. In this section we briefly consider two such systems, a stretched wire and a surface film.

Consider as a system a stretched wire that is under a given tension $\mathscr{T}$. When the length of the wire changes by the amount dL, the work done by the system is

$$\delta W = -\mathscr{T}\, dL \tag{4.7}$$

The minus sign is necessary because work is done by the system when dL is negative. This equation can be integrated to give

$$_1W_2 = -\int_1^2 \mathcal{T}\, dL \tag{4.8}$$

The integration can be performed either graphically or analytically if the relation between $\mathcal{T}$ and L is known. The stretched wire is a simple example of the type of problem in solid-body mechanics that involves the calculation of work.

EXAMPLE 4.3   A metallic wire of initial length L_0 is stretched. Assuming elastic behavior, determine the work done in terms of the modulus of elasticity and the strain.
Let $\sigma =$ stress, $e =$ strain, and $E =$ the modulus of elasticity.

$$\sigma = \frac{\mathcal{T}}{\mathcal{A}} = Ee$$

Therefore,

$$\mathcal{T} = \mathcal{A}Ee$$

From the definition of strain,

$$de = \frac{dL}{L_0}$$

Therefore,

$$\delta W = -\mathcal{T}\, dL = -\mathcal{A}\, EeL_0\, de$$

$$W = -\mathcal{A}\, EL_0 \int_{e=0}^{e} e\, de = -\frac{\mathcal{A}EL_0}{2}(e)^2$$

Now consider a system that consists of a liquid film having a surface tension $\mathcal{S}$. A schematic arrangement of such a film, maintained on a wire frame, one side of which can be moved, is shown in Fig. 4.10. When the area of the film is changed, for example, by sliding the movable wire along the frame, work is done on or by the film. When the area changes by an amount $d\mathcal{A}$, the work done by the system is

$$\delta W = -\mathcal{S}\, d\mathcal{A} \tag{4.9}$$

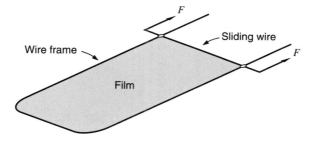

FIGURE 4.10 Schematic arrangement showing work done on a surface film.

For finite changes,

$$_1W_2 = -\int_1^2 \mathscr{S} \, d\mathscr{A} \tag{4.10}$$

4.5 SYSTEMS THAT HAVE OTHER MODES OF WORK

There are systems that have other modes of work. In this section we consider two of these systems—those for which the mode of work is magnetic and those for which the mode of work is electric. We will consider a quasi-equilibrium process for these systems and present expressions for the work done during such a process.

   To visualize how work can be accomplished by magnetic effects, let us briefly describe magnetic cooling, or adiabatic demagnetization, which is a process used to produce temperatures well below 1 K. A temperature of 1.0 K can be produced by pumping a vacuum over a bath of liquid helium (helium has the lowest normal boiling point of any substance, namely 4.2 K at one atmosphere pressure). An apparatus in which the magnetic cooling is accomplished is shown schematically in Fig. 4.11. The paramagnetic salt is the magnetic substance in which temperatures well below 1 K are achieved. When the magnetic field is slowly increased, work is done on the paramagnetic salt. From a microscopic point of view, this work is associated with the fact that in the presence of the magnetic field the ions in the salt tend to align themselves with their magnetic axes in the di-

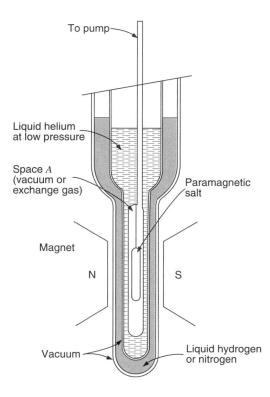

FIGURE 4.11 Schematic arrangement for magnetic cooling.

rection of the field. Through this work done on the salt, the temperature of the salt tends to increase. However, at this point in the experiment, space A is filled with low-pressure helium gas, and heat is transferred from the paramagnetic salt to the liquid helium, which is maintained at about 1 K. When the magnetic field is at full strength and the paramagnetic salt is at the temperature of the liquid helium, space A is evacuated, thus insulating the paramagnetic salt. The magnetic field is now reduced to zero. In this process work is done by the paramagnetic salt, and its temperature drops sharply. This entire process may be compared to the compression of a gas that is initially at ambient pressure and temperature. As a result of the compression, the temperature of the gas tends to increase. However, the high-pressure gas can be cooled to the ambient temperature. If this gas is then isolated from the surroundings and allowed to expand and do work (against a piston, for example), the temperature of the gas will decrease below the ambient temperature during the expansion process.

The basic parameters in this process are the intensity of the magnetic field and the magnetization. It may be shown that in a reversible quasi-equilibrium process, the work done on a simple magnetic substance is

$$\delta W = -\mu_0 \, \mathscr{H} \, d\left(V\mathfrak{M}\right) \tag{4.11}$$

where

$$\mu_0 = \text{permeability of free space}$$

$$V = \text{volume}$$

$$\mathscr{H} = \text{intensity of the magnetic field}$$

$$\mathfrak{M} = \text{magnetization}$$

The minus sign indicates that as the magnetization increases, work is done on the simple magnetic substance.

We have already noted that electrical energy flowing across the boundary of a system is work. We can gain further insight into such a process by considering a system in which the only work mode is electrical. As examples of such a system, we can think of a charged condenser, an electrolytic cell, and the type of fuel cell described in Chapter 1. Consider a quasi-equilibrium process for such a system, and during this process let the potential difference be $\mathscr{E}$ and the amount of electrical charge that flows into the system be dZ. For this quasi-equilibrium process the work is given by the relation

$$\delta W = -\mathscr{E} \, dZ \tag{4.12}$$

Since the current, i, equals dZ/dt (where $t =$ time), we can also write

$$\delta W = -\mathscr{E} i \, dt$$

$$_1W_2 = -\int_1^2 \mathscr{E} i \, dt \tag{4.13}$$

Equation 4.13 may also be written as a rate equation for work (the power).

$$\frac{\delta W}{dt} = -\mathscr{E} i \tag{4.14}$$

Since the ampere (electric current) is one of the fundamental units in the International System, and the watt has been defined previously, this relation serves as the definition of the unit for electric potential, the volt (V), which is one watt divided by one ampere.

4.6 SOME CONCLUDING REMARKS REGARDING WORK

The similarity of the expressions for work in the two processes discussed in Section 4.5 and in the three processes in which work is done at a moving boundary should be noted. In each of these quasi-equilibrium processes, the work is given by the integral of the product of an intensive property and the change of an extensive property. The following is a summary list of these processes and their work expressions

$$
\begin{aligned}
\text{Simple compressible system} \qquad & {}_1W_2 = \int_1^2 P\,dV \\[1em]
\text{Stretched wire} \qquad & {}_1W_2 = -\int_1^2 \mathcal{T}\,dL \\[1em]
\text{Surface film} \qquad & {}_1W_2 = -\int_1^2 \mathcal{S}\,d\mathcal{A} \qquad\qquad (4.15) \\[1em]
\text{System in which the work is} \atop \text{completely magnetic} \qquad & {}_1W_2 = -\int_1^2 \mu_0\,\mathcal{H}\,d(V\mathcal{A}) \\[1em]
\text{System in which the work is} \atop \text{completely electrical} \qquad & {}_1W_2 = -\int_1^2 \mathcal{E}\,dZ
\end{aligned}
$$

Although we will deal primarily with systems in which there is only one mode of work, it is quite possible to have more than one work mode in a given process. Thus, we could write

$$
\delta W = P\,dV - \mathcal{T}\,dL - \mathcal{S}\,d\mathcal{A} - \mu_0\,\mathcal{H}\,d(V\mathfrak{M}) - \mathcal{E}\,dZ + \cdots \qquad (4.16)
$$

where the dots represent other products of an intensive property and the derivative of a related extensive property. In each term the intensive property can be viewed as the driving force that causes a change to occur in the related extensive property, which is often termed the displacement.

It should also be noted that many other forms of work can be identified in processes that are not quasi-equilibrium processes. For example, there is the work done by shearing forces in the friction in a viscous fluid or the work done by a rotating shaft that crosses the system boundary.

The identification of work is an important aspect of many thermodynamic problems. We have already noted that work can be identified only at the boundaries of the system. For example, consider Fig. 4.12, which shows a gas separated from the vacuum by a membrane. Let the membrane rupture and the gas fill the entire volume. Neglecting any work associated with the rupturing of the membrane, we can ask whether work is done in the process. If we take as our system the gas and the vacuum space, we readily conclude that no work is done because no work can be identified at the system boundary. If we take the gas as a system, we do have a change of volume, and we might be

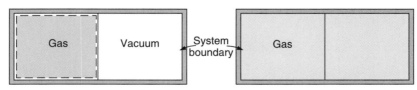

FIGURE 4.12 Example of process involving a change of volume for which the work is zero.

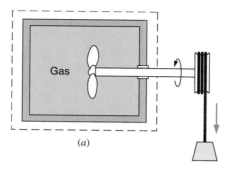

(a)

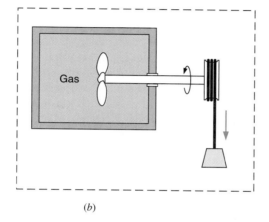

(b)

FIGURE 4.13 Example showing how selection of the system determines whether work is involved in a process.

tempted to calculate the work from the integral

$$\int_1^2 P\,dV$$

However, this is not a quasi-equilibrium process, and therefore the work cannot be calculated from this relation. Because there is no resistance at the system boundary as the volume increases, we conclude that for this system no work is done in this process of filling the vacuum.

Another example can be cited with the aid of Fig. 4.13. In Fig. 4.13*a* the system consists of the container plus the gas. Work crosses the boundary of the system at the point where the system boundary intersects the shaft, and can be associated with the shearing forces in the rotating shaft. In Fig. 4.13*b* the system includes shaft and weight as well as the gas and the container. Therefore, no work crosses the system boundary as the weight moves downward. As we will see in the next chapter, we can identify a change of potential energy within the system, but this should not be confused with work crossing the system boundary.

4.7 DEFINITION OF HEAT

The thermodynamic definition of heat is somewhat different from the everyday understanding of the word. It is essential to understand clearly the definition of heat given here, because it plays a part in so many thermodynamic problems.

If a block of hot copper is placed in a beaker of cold water, we know from experience that the block of copper cools down and the water warms up until the copper and water reach the same temperature. What causes this decrease in the temperature of the copper and the increase in the temperature of the water? We say that it is the result of the transfer of energy from the copper block to the water. It is out of such a transfer of energy that we arrive at a definition of heat.

Heat is defined as the form of energy that is transferred across the boundary of a system at a given temperature to another system (or the surroundings) at a lower temperature by virtue of the temperature difference between the two systems. That is, heat is transferred from the system at the higher temperature to the system at the lower temperature, and the heat transfer occurs solely because of the temperature difference between the two systems. Another aspect of this definition of heat is that a body never contains heat. Rather, heat can be identified only as it crosses the boundary. Thus, heat is a transient phenomenon. If we consider the hot block of copper as one system and the cold water in the beaker as another system, we recognize that originally neither system contains any heat (they do contain energy, of course). When the copper block is placed in the water and the two are in thermal communication, heat is transferred from the copper to the water until equilibrium of temperature is established. At this point we no longer have heat transfer, because there is no temperature difference. Neither of the systems contains heat at the conclusion of the process. It also follows that heat is identified at the boundary of the system, for heat is defined as energy being transferred across the system boundary.

4.8 UNITS OF HEAT

Heat, like work, is a form of energy transfer to or from a system. Therefore, the units for heat, and to be more general, for any other form of energy as well, are the same as the units for work, or are at least directly proportional to them. In the International System the unit for heat (energy) is the joule. Similarly, in the English System, the foot pound force is an appropriate unit for heat. However, another unit came to be used naturally over the years, the result of an association with the process of heating water, such as that used in connection with defining heat in the previous section. Consider as a system 1 lbm of water at 59.5F. Let a block of hot copper of appropriate mass and temperature be placed in the water, so that when thermal equilibrium is established the temperature of the water is 60.5F. This unit amount of heat transferred from the copper to the water in this process is called the British thermal unit (Btu). More specifically, it is called the 60-degree Btu, defined as the amount of heat required to raise 1 lbm of water from 59.5F to 60.5F. (The Btu as used today is actually defined in terms of the standard SI units.) It is worth noting here that a unit of heat in metric units, the calorie, originated naturally in a manner similar to the origin of the Btu in the English system. The calorie is defined as the amount of heat required to raise 1 gram of water from 14.5°C to 15.5°C.

Heat transferred *to* a system is considered positive, and heat transferred *from* a system is negative. Thus, positive heat represents energy transferred to a system, and negative heat represents energy transferred from a system. The symbol Q represents heat. A process in which there is no heat transfer ($Q = 0$) is called an adiabatic process.

From a mathematical perspective, heat, like work, is a path function and is recognized as an inexact differential. That is, the amount of heat transferred when a system undergoes a change from state 1 to state 2 depends on the path that the system follows during the change of state. Since heat is an inexact differential, the differential is written δQ.

On integrating, we write

$$\int_1^2 \delta Q =\, _1Q_2$$

In words, $_1Q_2$ is the heat transferred during the given process between states 1 and 2.
The rate at which heat is transferred to a system is designated by symbol $\dot{Q}$.

$$\dot{Q} \equiv \frac{\delta Q}{dt}$$

It is also convenient to speak of the heat transfer per unit mass of the system, q, often termed "specific heat transfer," which is defined as

$$q \equiv \frac{Q}{m}$$

4.9 COMPARISON OF HEAT AND WORK

At this point it is evident that there are many similarities between heat and work.

1. Heat and work are both transient phenomena. Systems never possess heat or work, but either or both cross the system boundary when a system undergoes a change of state.
2. Both heat and work are boundary phenomena. Both are observed only at the boundaries of the system, and both represent energy crossing the boundary of the system.
3. Both heat and work are path functions and inexact differentials.

It should also be noted that in our sign convention, $+Q$ represents heat transferred *to* the system and thus is energy added to the system, and $+W$ represents work done *by* the system and thus represents energy leaving the system.

A final illustration may help explain the difference between heat and work. Figure 4.14 shows a gas contained in a rigid vessel. Resistance coils are wound around the outside of the vessel. When current flows through the resistance coils, the temperature of the gas increases. Which crosses the boundary of the system, heat or work?

In Fig. 4.14a we consider only the gas as the system. The energy crosses the boundary of the system because the temperature of the walls is higher than the temperature of the gas. Therefore, we recognize that heat crosses the boundary of the system.

In Fig. 4.14b the system includes the vessel and the resistance heater. Electricity crosses the boundary of the system, and as indicated earlier, this is work.

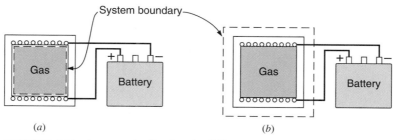

(a)   (b)

FIGURE 4.14 An example showing the difference between heat and work.

PROBLEMS

4.1 A cylinder fitted with a frictionless piston contains 5 kg of superheated refrigerant R-134a vapor at 1000 kPa, 140°C. The setup is cooled at constant pressure until the R-134a reaches a quality of 25%. Calculate the work done in the process.

4.2 A piston/cylinder arrangement shown in Fig. P4.2 initially contains air at 150 kPa, 400°C. The setup is allowed to cool to the ambient temperature of 20°C.

a. Is the piston resting on the stops in the final state? What is the final pressure in the cylinder?

b. What is the work done by the air during this process?

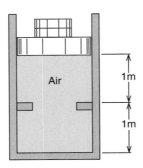

FIGURE P4.2

4.3 Consider the process described in Problem 2.14. For a control volume that consists of the space inside the cylinder, determine the boundary movement work done from the point at which the valve is first opened (piston resting at the bottom of the cylinder) to the final state given in Problem 2.14.

4.4 A linear spring, $F = k_s(x - x_o)$, with spring constant $k_s = 500$ N/m, is stretched until it is 100 mm longer. Find the required force and work input.

4.5 A nonlinear spring has the force versus displacement relation of $F = k_{ns}(x - x_o)^n$. If the spring end is moved to x_1 from the relaxed state, determine the formula for the required work.

4.6 Repeat the previous problem, but let the spring end be moved from x_1 to x_2, both different from x_o so the spring is not initially relaxed.

4.7 The refrigerant R-22 is contained in a piston/cylinder as shown in Fig. P4.7, where the volume is 11 L when the piston hits the stops. The initial state is −30°C, 150 kPa with a volume of 10 L. This system is brought indoors and warms up to 15°C.

a. Is the piston at the stops in the final state?

b. Find the work done by the R-22 during this process.

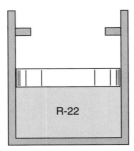

FIGURE P4.7

4.8 A piston/cylinder contains butane, C_4H_{10}, at 300°C, 100 kPa with a volume of 0.02 m³. The gas is now compressed slowly in an isothermal process to 300 kPa.

    a. Show that it is reasonable to assume that butane behaves as an ideal gas during this process.

    b. Determine the work done by the butane during the process.

4.9 The piston/cylinder shown in Fig. P4.9 contains carbon dioxide at 300 kPa, 100°C with a volume of 0.2 m³. Mass is added at such a rate that the gas compresses according to the relation $PV^{1.2} = $ constant to a final temperature of 200°C. Determine the work done during the process.

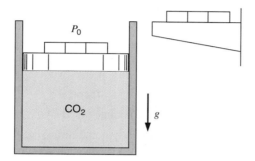

FIGURE P4.9

4.10 Determine the work done by the air inside the cylinder for the process described in Problem 3.11.

4.11 A gas initially at 1 MPa, 500°C is contained in a piston and cylinder arrangement with an initial volume of 0.1 m³. The gas is then slowly expanded according to the relation $PV = $ constant until a final pressure of 100 kPa is reached. Determine the work for this process.

4.12 Consider the two-part process described in Problem 3.37 and determine the work done by the water inside the cylinder.

4.13 A cylinder fitted with a piston contains propane gas at 100 kPa, 300 K with a volume of 0.2 m³. The gas is now slowly compressed according to the relation $PV^{1.1} = $ constant to a final temperature of 340 K.

    a. What is the final pressure?

    b. Justify the use of the ideal gas model.

    c. How much work is done during the process?

4.14 Consider the process described in Problem 3.41. With the ammonia as a control mass, determine the boundary work during the process.

4.15 The gas space above the water in a closed storage tank contains nitrogen at 25°C, 100 kPa. Total tank volume is 4 m³, and there is 500 kg of water at 25°C. An additional 500 kg water is now forced into the tank. Assuming constant temperature throughout, find the final pressure of the nitrogen and the work done on the nitrogen in this process.

4.16 Consider the process of inflating a helium balloon, as described in Problem 3.8. For a control volume that consists of the space inside the balloon, determine the work done during the overall process.

4.17 Ammonia vapor is compressed inside a cylinder by an external force acting on the piston. The ammonia is initially at 30°C, 500 kPa, and the final pressure is 1400 kPa. The following data have been measured for the process:

Pressure, kPa	500	653	802	945	1100	1248	1400
Volume, L	1.25	1.08	0.96	0.84	0.72	0.60	0.50

   a. Determine the work done by the ammonia in the process.

   b. What is the final temperature of the ammonia?

4.18 A balloon behaves such that the pressure inside is proportional to the diameter squared. It contains 2 kg of ammonia at 0°C, 60% quality. The balloon and ammonia are now heated so that a final pressure of 600 kPa is reached. Considering the ammonia as a control mass, find the amount of work done in the process.

4.19 A piston/cylinder arrangement of initial volume 0.025 m^3 contains saturated water vapor at 200°C. The steam now expands in a quasi-equilibrium isothermal process to a final pressure of 200 kPa, while it does work against the piston.

   a. Determine the work done in this process.

   b. How different is the work calculated if ideal gas is assumed?

4.20 A spherical elastic balloon initially containing 5 kg ammonia as saturated vapor at 20°C is connected by a valve to a 3-m^3 evacuated tank. The balloon is made such that the pressure inside is proportional to the diameter. The valve is now opened, allowing ammonia to flow into the tank until the pressure in the balloon has dropped to 600 kPa, at which point the valve is closed. The final temperature in both the balloon and the tank is 20°C. Determine

   a. The final pressure in the tank

   b. The work done by the ammonia in the process

4.21 A cylinder having an initial volume of 3 m^3 contains 0.1 kg of water at 40°C. The water is then compressed in an isothermal quasi-equilibrium process until it has a quality of 50%. Calculate the work done in the process.

4.22 Consider the nonequilibrium process described in Problem 3.7. Determine the work done by the carbon dioxide in the cylinder during the process.

4.23 Two kilograms of water is contained in a piston/cylinder (Fig. P4.23) with a massless piston loaded with a linear spring and the outside atmosphere. Initially the spring force is zero and $P_1 = P_o = 100$ kPa with a volume of 0.2 m^3. If the piston just hits the upper stops the volume is 0.8 m^3 and $T = 600$°C. Heat is now added until the pressure reaches 1.2 MPa. Find the final temperature, show the P–V diagram and find the work done during the process.

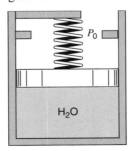

FIGURE P4.23

4.24 A piston/cylinder (Fig. P4.24) with $A_{cyl} = 0.01$ m^2 and $m_p = 101$ kg contains 1 kg of water at 20°C with a volume of 0.1 m^3. Initially the piston rests on some stops with the top surface open to the atmosphere, P_o. To what temperature should the water be heated to lift the piston? If it is heated to saturated vapor find the final temperature, volume and the work, $_1W_2$.

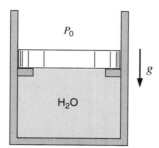

FIGURE P4.24

4.25 Assume the same system as in the previous problem, but let the piston be locked with a pin. If the water is heated to saturated vapor find the final temperature, volume and the work, $_1W_2$.

4.26 Air at 200 kPa, 30°C is contained in a cylinder/piston arrangement with initial volume 0.1 m^3. The inside pressure balances ambient pressure of 100 kPa plus an externally imposed force that is proportional to $V^{0.5}$. Now heat is transferred to the system to a final temperature of 200°C. Find the final pressure and the work done in the process.

4.27 A 400-L tank, A (see Fig. P4.27) contains argon gas at 250 kPa, 30°C. Cylinder B, having a frictionless piston of such mass that a pressure of 150 kPa will float it, is initially empty. The valve is opened and argon flows into B and eventually reaches a uniform state of 150 kPa, 30°C throughout. What is the work done by the argon?

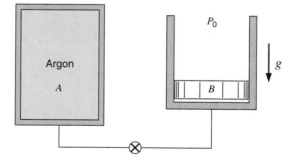

FIGURE P4.27

4.28 A spring-loaded piston/cylinder arrangement contains R-134a at 20°C, 24% quality with a volume 50 L. The setup is heated and thus expands, moving the piston. It is noted that when the last drop of liquid disappears the temperature is 40°C. The heating is stopped when $T = 130$°C. Find the final pressure and the work done in the process.

4.29 A cylinder containing 1 kg of ammonia has an externally loaded piston. Initially the ammonia is at 2 MPa, 180°C and is now cooled to saturated vapor at 40°C, and then further cooled to 20°C, at which point the quality is 50%. Find the total work for the process, assuming a piecewise linear variation of P versus V.

4.30 A vertical cylinder (Fig. P4.30) has a 90-kg piston locked with a pin trapping 10 L of R-22 at 10°C, 90% quality inside. Atmospheric pressure is 100 kPa, and the cylinder cross-sectional area is 0.006 m². The pin is removed, allowing the piston to move and come to rest with a final temperature of 10°C for the R-22.

a. Find the final pressure and volume.

b. Find the work done by the R-22.

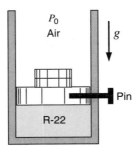

FIGURE P4.30

4.31 A steam radiator in a room at 25°C has saturated water vapor at 110 kPa flowing through it, when the inlet and exit valves are closed. What is the pressure and the quality of the water, when it has cooled to 25°C? How much work is done?

4.32 Two springs with same spring constant are installed in a massless piston/cylinder with the outside air at 100 kPa. If the piston is at the bottom, both springs are relaxed and the second spring comes in contact with the piston at $V = 2$ m³. The cylinder (Fig. P4.32) contains ammonia initially at -2°C, $x = 0.13$, $V = 1$ m³, which is then heated until the pressure finally reaches 1200 kPa. At what pressure will the piston touch the second spring? Find the final temperature and the total work done by the ammonia.

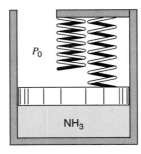

FIGURE P4.32

4.33 A cylinder (Fig. P4.33), $A_{cyl} = 7.012$ cm², has two pistons mounted, the upper one, $m_{p1} = 100$ kg, initially resting on the stops. The lower piston, $m_{p2} = 0$ kg, has 2 kg water below it, with a spring in vacuum connecting the two pistons. The spring force is zero when the lower piston stands at the bottom, and when the lower piston hits the stops the volume is 0.3 m³. The water, initially at 50 kPa, $V = 0.00206$ m³, is then heated to saturated vapor.

a. Find the initial temperature and the pressure that will lift the upper piston.

b. Find the final T, P, v and the work done by the water.

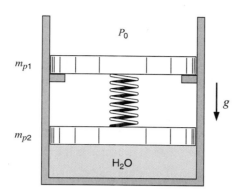

FIGURE P4.33

4.34 Find the work for Problem 3.44.

4.35 Find the work for Problem 3.45.

4.36 Find the work for Problem 3.46.

4.37 Find the work for Problem 3.47.

4.38 Find the work for Problem 3.48.

4.39 A 0.5-m-long steel rod with a 1-cm diameter is stretched in a tensile test. What is the required work to obtain a relative strain of 0.1%? The modulus of elasticity of steel is 2×10^8 kPa.

4.40 A film of ethanol at 20°C has a surface tension of 22.3 mN/m and is maintained on a wire frame as shown in Fig. P4.40. Consider the film with two surfaces as a control mass and find the work done when the wire is moved 10 mm to make the film 20×40 mm.

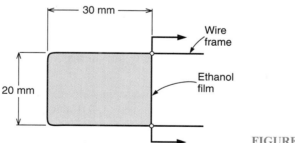

FIGURE P4.40

4.41 Consider a paramagnetic salt such as that described in Section 4.5. Assume that the magnetization of the substance is proportional to the magnetic field intensity divided by the temperature. Derive an expression for the work done in changing the magnetization from M_1 to M_2 in an isothermal process.

4.42 Repeat the previous problem for a process with constant magnetic field intensity.

4.43 A battery is well insulated while being charged by 12.3 V at a current of 6 A during 4 h. Take the battery as a control mass and find the work.

4.44 Consider a window-mounted air conditioning unit used in the summer to cool incoming air. Examine the system boundaries for rates of work and heat transfer, including signs.

4.45 Consider a hot-air heating system for a home. Examine the following systems for heat transfer.

    a. The combustion chamber and combustion gas side of the heat transfer area

    b. The furnace as a whole, including the hot- and cold-air ducts and chimney

4.46 Consider a household refrigerator that has just been filled up with room-temperature food. Define a control volume (mass) and examine its boundaries for rates of work and heat transfer, including sign.

    a. Immediately after the food is placed in the refrigerator

    b. After a long period of time has elapsed and the food is cold

4.47 A room is heated with steam radiators on a winter day. Examine the following control volumes, regarding heat transfer, including sign.

    a. Radiator

    b. Room

    c. Radiator and room together

ENGLISH UNIT PROBLEMS

4.48E A cylinder fitted with a frictionless piston contains 10 lbm of superheated refrigerant R-134a vapor at 100 lbf/in.^{2}, 300 F. The setup is cooled at constant pressure until the water reaches a quality of 25%. Calculate the work done in the process.

4.49E A linear spring, $F = k_s(x - x_o)$, with spring constant $k_s = 35$ lbf/ft, is stretched until it is 2.5 in. longer. Find the required force and work input.

4.50E The piston/cylinder shown in Fig. P4.9 contains carbon dioxide at 50 lbf/in.^{2}, 200 F with a volume of 5 ft^3. Mass is added at such a rate that the gas compresses according to the relation $PV^{1.2}$ = constant to a final temperature of 350 F. Determine the work done during the process.

4.51E Determine the work done by the air inside the cylinder for the process described in Problem 3.56.

4.52E The gas space above the water in a closed storage tank contains nitrogen at 80 F, 15 lbf/in.^{2}. Total tank volume is 150 ft^3 and there is 1000 lbm of water at 80 F. An additional 1000 lbm water is now forced into the tank. Assuming constant temperature throughout, find the final pressure of the nitrogen and the work done on the nitrogen in this process.

4.53E A piston/cylinder arrangement of initial volume 0.3 ft^3 contains saturated water vapor at 360 F. The steam now expands in a quasi-equilibrium isothermal process to a final pressure of 30 lbf/in.^{2}, while it does work against the piston.

    a. Determine the work done in this process.

    b. How different is the work calculated if ideal gas is assumed?

4.54E A cylinder having an initial volume of 100 ft^3 contains 0.2 lbm of water at 100 F. The water is then compressed in an isothermal quasi-equilibrium process until it has a quality of 50%. Calculate the work done in the process.

4.55E Air at 30 lbf/in.^{2}, 85 F is contained in a cylinder/piston arrangement with initial volume 3.5 ft^3. The inside pressure balances ambient pressure of 14.7 lbf/in.^{2}

plus an externally imposed force that is proportional to $V^{0.5}$. Now heat is transferred to the system to a final temperature of 400 F. Find the final pressure and the work done in the process.

4.56E A cylinder containing 2 lbm of ammonia has an externally loaded piston. Initially the ammonia is at 280 lbf/in.^{2}, 360 F and is now cooled to saturated vapor at 105 F, and then further cooled to 65 F, at which point the quality is 50%. Find the total work for the process, assuming a piecewise linear variation of P versus V.

4.57E A steam radiator in a room at 75 F has saturated water vapor at 16 lbf/in.^{2} flowing through it, when the inlet and exit valves are closed. What is the pressure and the quality of the water, when it has cooled to 75F? How much work is done?

4.58E Find the work for Problem 3.73.

4.59E Find the work for Problem 3.74.

4.60E Find the work for Problem 3.75.

4.61E A 1-ft-long steel rod with a 0.5-in. diameter is stretched in a tensile test. What is the required work to obtain a relative strain of 0.1%? The modulus of elasticity of steel is 30×10^6 lbf/in.^{2}.

COMPUTER, DESIGN, AND OPEN-ENDED PROBLEMS

4.62 In Problem 3.37, determine the work done by the water at any point during the process.

4.63 In Problem 3.41, determine the work done by the ammonia at any point during the process.

4.64 Consider the process described in Problem 4.20, in which ammonia is transferred from a balloon to a rigid tank. Follow the process from the initial state, point by point, until the pressure in the balloon and tank eventually equalize.

4.65 Reconsider the process in Problem 4.29 in which three states were specified. Solve the problem by fitting a single smooth curve (P versus v) through the three points. Map out the path followed (including temperature and quality) during the process.

4.66 The elastic balloon described in Problem 3.8 is to be filled with one of the real working fluids listed in Tables A.1–A.6. Determine the state inside the balloon at any intermediate point and the boundary movement work done to this point during the inflation process to a given final pressure.

4.67 Write a computer program to determine the boundary movement work for a specified substance undergoing a process for a given set of data (values of pressure and corresponding volume during the process).

4.68 Write a computer program to determine the boundary movement work for a specified gas undergoing an isothermal process at temperature T from v_1 to v_2. Use the Benedict–Webb–Rubin equation of state. Compare the result with that found assuming ideal gas behavior.

4.69 A substance is brought from a state of P_1, v_1 to a state P_2, v_2 in a piston cylinder arrangement. Assume the process can be approximated as a polytropic process. Write a program that will find the polytropic exponent, n, and the boundary work per unit mass. The four state properties are input variables. Check the program with cases that you can easily hand calculate.

4.70 Heat transfer by conduction through a media gives a heat transfer rate proportional to the temperature difference, conductivity and area. It is inversely proportional to the thickness and expressed as

$$\dot{Q} = \frac{\kappa A}{L} \Delta T$$

Assume you have a plate of $A = 1$ m^2 with thickness $L = 0.02$ m over which there is a temperature difference of 20°C. Find the conductivity, κ, from the literature and compare the heat transfer rates if the plate substance is a metal like aluminum or steel, wood, foam insulation, air, argon or liquid water. Assume the average substance temperature is 25°C.

4.71 Make a list of household appliances such as refrigerators, electric heaters, vacuum cleaner, hair dryer, TV, stereo set, and any others you may think of. For each, list its energy consumption and explain where you have energy transfer as work and where there is heat transfer.

4.72 A gasoline car engine burns fuel with air to release energy. The energy is converted so that approximately one third goes out as shaft work, one third is carried away by the coolant to the radiator, and the rest goes out through the exhaust manifold. Make a list of the energy used to drive at 50 km/h (30 mi/h) and at 100 km/h (60 mi/h) with typical cars (small, mid-sized, full-sized, minivan). How much energy have each of these cars given off to the atmosphere over 50 000 km (30 000 mi) if a suitable average speed is assumed?

4.73 A water storage tank of 1 m^3 is used to store water from a well so it is under pressure. This is accomplished by air trapped in a bag inside the tank, which is compressed as the pump adds liquid water. Assume the air is at constant temperature equal to the water temperature of 10°C. When water is used in the house the air bag acts as a cushion and expands so the pressure drops. At a preset low pressure the pump adds more water and stops at a preset high pressure. Design the system by selection of the amount of air in the bag and the preset high and low pressures. For an estimated standard household water use, also determine the number of times the pump cycles on and off.

THE FIRST LAW OF
THERMODYNAMICS 5

Having completed our consideration of basic definitions and concepts, we are ready to proceed to a discussion of the first law of thermodynamics. This law is often called the law of the conservation of energy and, as we will see later, this is essentially true. Our procedure will be to state this law for a system (control mass) undergoing a cycle, and then for a change of state of a system. The law of the conservation of matter will also be considered in this chapter, as will the first law for a control volume.

5.1 THE FIRST LAW OF THERMODYNAMICS FOR A CONTROL MASS UNDERGOING A CYCLE

The first law of thermodynamics states that during any cycle a system (control mass) undergoes, the cyclic integral of the heat is proportional to the cyclic integral of the work.

To illustrate this law, consider as a control mass the gas in the container shown in Fig. 5.1. Let this system go through a cycle that is made up of two processes. In the first process work is done on the system by the paddle that turns as the weight is lowered. Let the system then return to its initial state by transferring heat from the system until the cycle has been completed.

Historically, work was measured in mechanical units of force times distance, such as, foot pounds force or joules, and heat was measured in thermal units, such as the British thermal unit or the calorie. Measurements of work and heat were made during a cycle for a wide variety of systems and for various amounts of work and heat. When the amounts of work and heat were compared, it was found that they were always proportional. Such observations led to the formulation of the first law of thermodynamics, which in equation form is written

$$J \oint \delta Q = \oint \delta W \qquad (5.1)$$

The symbol $\oint \delta Q$, which is called the cyclic integral of the heat transfer, represents the net heat transfer during the cycle, and $\oint \delta W$, the cyclic integral of the work, represents the net work during the cycle. Here, J is a proportionality factor that depends on the units used for work and heat.

The basis of every law of nature is experimental evidence, and this is true also of the first law of thermodynamics. Many different experiments have been conducted on the first law, and every one thus far has verified it either directly or indirectly. The first law has never been disproved.

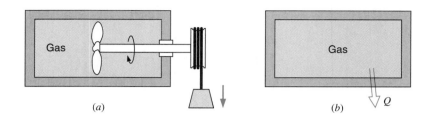

FIGURE 5.1
Example of a control mass undergoing a cycle.

(a)

(b)

As was discussed in Chapter 4, the units for work and heat or for any other form of energy either are the same or are directly proportional. In SI units, the joule is used as the unit for both work and heat and for any other energy unit. In English units, the basic unit for work is the foot pound force, and the basic unit for heat is the British thermal unit (Btu). James P. Joule (1818–1889) did the first accurate work in the 1840s on measurement of the proportionality factor J, which relates these units. Today, the Btu is defined in terms of the basic SI metric units,

$$1 \text{ Btu} = 778.17 \text{ ft lbf}$$

This unit is termed the International British thermal unit. For much engineering work, the accuracy of other data does not warrant more accuracy than the relation 1 Btu = 778 ft lbf, which is the value used with English units in the problems in this text. Because these units are equivalent, it is not necessary to include the factor J explicitly in Eq. 5.1, but simply to recognize that for any system of units, each equation must have consistent units throughout. Therefore, we may write Eq. 5.1 as

$$\oint \delta Q = \oint \delta W \qquad (5.2)$$

which can be considered the basic statement of the first law of thermodynamics.

5.2 THE FIRST LAW OF THERMODYNAMICS FOR A CHANGE IN STATE OF A CONTROL MASS

Equation 5.2 states the first law of thermodynamics for a control mass during a cycle. Many times, however, we are concerned with a process rather than a cycle. We now consider the first law of thermodynamics for a control mass that undergoes a change of state. We begin by introducing a new property, the energy, which is given the symbol E. Consider a system that undergoes a cycle, in which it changes from state 1 to state 2 by process A, and returns from state 2 to state 1 by process B. This cycle is shown in Fig. 5.2 on a pressure (or other intensive property)–volume (or other extensive property) diagram. From the first law of thermodynamics, Eq. 5.2,

$$\oint \delta Q = \oint \delta W$$

Considering the two separate processes, we have

$$\int_1^2 \delta Q_A + \int_2^1 \delta Q_B = \int_1^2 \delta W_A + \int_2^1 \delta W_B$$

Now consider another cycle in which the control mass changes from state 1 to state 2 by

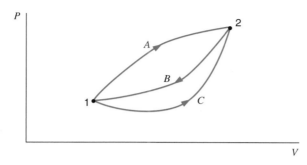

FIGURE 5.2 Demonstration of the existence of thermodynamic property E.

process C and returns to state 1 by process B, as before. For this cycle we can write

$$\int_1^2 \delta Q_C + \int_2^1 \delta Q_B = \int_1^2 \delta W_C + \int_2^1 \delta W_B$$

Subtracting the second of these equations from the first, we have

$$\int_1^2 \delta Q_A - \int_1^2 \delta Q_C = \int_1^2 \delta W_A - \int_1^2 \delta W_C$$

or, by rearranging,

$$\int_1^2 (\delta Q - \delta W)_A = \int_1^2 (\delta Q - \delta W)_C \tag{5.3}$$

Since A and C represents arbitrary processes between states 1 and 2, the quantity $\delta Q - \delta W$ is the same for all processes between states 1 and 2. Therefore, $\delta Q - \delta W$ depends only on the initial and final states and not on the path followed between the two states. We conclude that this is a point function, and therefore it is the differential of a property of the mass. This property is the energy of the mass and is given the symbol E. Thus we can write

$$dE = \delta Q - \delta W$$

$$\delta Q = dE + \delta W \tag{5.4}$$

Because E is a property, its derivative is written dE. When Eq. 5.4 is integrated from an initial state 1 to a final state 2, we have

$$_1Q_2 = E_2 - E_1 + {}_1W_2 \tag{5.5}$$

where E_1 and E_2 are the initial and final values of the energy E of the control mass, $_1Q_2$ is the heat transferred to the control mass during the process from state 1 to state 2 and $_1W_2$ is the work done by the control mass during the process.

The physical significance of the property E is that it represents all the energy of the system in the given state. This energy might be present in a variety of forms, such as the kinetic or potential energy of the system as a whole with respect to the chosen coordinate frame, energy associated with the motion and position of the molecules, energy associated with the structure of the atom, chemical energy present in a storage battery, energy present in a charged condenser, or in any of a number of other forms.

In the study of thermodynamics, it is convenient to consider the bulk kinetic and potential energy separately and then to consider all the other energy of the control mass in a single property that we call the internal energy, and to which we give the symbol U.

Thus, we would write

$$E = \text{Internal energy} + \text{Kinetic energy} + \text{Potential energy}$$

or

$$E = U + \text{KE} + \text{PE}$$

The kinetic and potential energy of the control mass are associated with the coordinate frame that we select and can be specified by the macroscopic parameters of mass, velocity, and elevation. The internal energy U includes all other forms of energy of the control mass and is associated with the thermodynamic state of the system.

Since the terms comprising E are point functions, we can write

$$dE = dU + d(\text{KE}) + d(\text{PE}) \tag{5.6}$$

The first law of thermodynamics for a change of state may therefore be written

$$\delta Q = dU + d(\text{KE}) + d(\text{PE}) + \delta W \tag{5.7}$$

In words this equation states that as a control mass undergoes a change of state, energy may cross the boundary as either heat or work, and each may be positive or negative. The net change in the energy of the system will be exactly equal to the net energy that crosses the boundary of the system. The energy of the system may change in any of three ways—by a change in internal energy, in kinetic energy, or in potential energy.

This section concludes by deriving an expression for the kinetic and potential energy of a control mass. Consider a mass that is initially at rest relative to the earth, which is taken as the coordinate frame. Let this system be acted on by an external horizontal force F that moves the mass a distance dx in the direction of the force. Thus, there is no change in potential energy. Let there be no heat transfer and no change in internal energy. Then from the first law, Eq. 5.7,

$$\delta W = -F\,dx = -d\text{KE}$$

But

$$F = ma = m\frac{d\mathbf{V}}{dt} = m\frac{dx}{dt}\frac{d\mathbf{V}}{dx} = m\mathbf{V}\frac{d\mathbf{V}}{dx}$$

Then

$$d\text{KE} = F\,dx = m\mathbf{V}\,d\mathbf{V}$$

Integrating, we obtain

$$\int_{\text{KE}=0}^{\text{KE}} d\text{KE} = \int_{\mathbf{V}=0}^{\mathbf{V}} m\mathbf{V}\,d\mathbf{V}$$

$$\text{KE} = \frac{1}{2}m\mathbf{V}^2 \tag{5.8}$$

A similar expression for potential energy can be found. Consider a control mass that is initially at rest and at the elevation of some reference level. Let this mass be acted on by a vertical force F of such magnitude that it raises (in elevation) the mass with constant velocity an amount dZ. Let the acceleration due to gravity at this point be g. From the first law, Eq. 5.7,

$$\delta W = -F\,dZ = -d\text{PE}$$

$$F = ma = mg$$

Then

$$dPE = F\,dZ = mg\,dZ$$

Integrating gives

$$\int_{PE_1}^{PE_2} d\,PE = m\int_{Z_1}^{Z_2} g\,dZ$$

Assuming that g does not vary with Z (which is a very reasonable assumption for moderate changes in elevation),

$$PE_2 - PE_1 = mg(Z_2 - Z_1) \tag{5.9}$$

Substituting the expressions for kinetic and potential energy into Eq. 5.6, we have

$$dE = dU + m\mathbf{V}\,d\mathbf{V} + mg\,dZ$$

Integrating for a change of state from state 1 to state 2 with constant g, we have

$$E_2 - E_1 = U_2 - U_1 + \frac{m\mathbf{V}_2^2}{2} - \frac{m\mathbf{V}_1^2}{2} + mgZ_2 - mgZ_1$$

Similarly, substituting these expressions for kinetic and potential energy into Eq. 5.7, we have

$$\delta Q = dU + \frac{d\left(m\mathbf{V}^2\right)}{2} + d\left(mgZ\right) + \delta W \tag{5.10}$$

Assuming g is a constant, in the integrated form of this equation,

$$_1Q_2 = U_2 - U_1 + \frac{m\left(\mathbf{V}_2^2 - \mathbf{V}_1^2\right)}{2} + mg(Z_2 - Z_1) + {_1}W_2 \tag{5.11}$$

Three observations should be made regarding this equation. The first observation is that the property E, the energy of the control mass, was found to exist, and we were able to write the first law for a change of state using Eq. 5.5. However, rather than deal with this property E, we find it more convenient to consider the internal energy and the kinetic and potential energies of the mass. In general, this procedure will be followed in the rest of this book.

The second observation is that Eqs. 5.10 and 5.11 are in effect a statement of the conservation of energy. The net change of the energy of the control mass is always equal to the net transfer of energy across the boundary as heat and work. This is somewhat analogous to a joint checking account shared by a husband and wife. There are two ways in which deposits and withdrawals can be made—either by the husband or by the wife—and the balance will always reflect the net amount of the transaction. Similarly, there are two ways in which energy can cross the boundary of a control mass—either as heat or as work—and the energy of the mass will change by the exact amount of the net energy crossing the boundary. The concept of energy and the law of the conservation of energy are basic to thermodynamics.

The third observation is that Eqs. 5.10 and 5.11 can give only changes in internal energy, kinetic energy, and potential energy. We can learn nothing about absolute values of these quantities from these equations. If we wish to assign values to internal energy, kinetic energy, and potential energy, we must assume reference states and assign a value to the quantity in this reference state. The kinetic energy of a body with zero velocity relative to the earth is assumed to be zero. Similarly, the value of the potential energy is assumed to be zero when the body is at some reference elevation. With internal energy, therefore, we must also have a reference state if we wish to assign values of this property. This matter is considered in the following section.

5.3 INTERNAL ENERGY—A THERMODYNAMIC PROPERTY

Internal energy is an extensive property, because it depends on the mass of the system. Similarly, the kinetic and potential energies are extensive properties.

The symbol U designates the internal energy of a given mass of a substance. Following the convention used with other extensive properties, the symbol u designates the internal energy per unit mass. We could speak of u as the specific internal energy, as we do with specific volume. However, because the context will usually make it clear whether u or U is referred to, we will simply use the term internal energy to refer to both internal energy per unit mass and the total internal energy.

In Chapter 3 it was noted that in the absence of the effects of motion, gravity, surface effects, electricity, and magnetism, the state of a pure substance is specified by two independent properties. It is very significant that with these restrictions, the internal energy may be one of the independent properties of a pure substance. This means, for example, that if we specify the pressure and internal energy (with reference to an arbitrary base) of superheated steam, the temperature is also specified.

Thus, in a table of thermodynamic properties such as the steam tables, the value of internal energy can be tabulated along with other thermodynamic properties. Tables 1 and 2 of the steam tables by Keenan and his colleagues (Tables A.1.1 and A.1.2) list the internal energy for saturated states. Included are the internal energy of saturated liquid u_f, the internal energy of saturated vapor u_g, and the difference between the internal energy of saturated liquid and saturated vapor u_{fg}. The values are given in relation to an arbitrarily assumed reference state, which will be discussed later. The internal energy of saturated steam of a given quality is calculated in the same way as specific volume. The relations are

$$U = U_{\text{liq}} + U_{\text{vap}}$$

or

$$mu = m_{\text{liq}} u_f + m_{\text{vap}} u_g$$

Dividing by m and introducing the quality x, gives

$$u = (1-x)u_f + xu_g$$

$$u = u_f + xu_{fg}$$

As an example, the specific internal energy of saturated steam having a pressure of 0.6 MPa and a quality of 95% can be calculated as

$$u = u_f + xu_{fg} = 669.9 + 0.95(1897.5) = 2472.5 \text{ kJ / kg}$$

EXAMPLE 5.1 A tank containing a fluid is stirred by a paddle wheel. The work input to the paddle wheel is 5090 kJ. The heat transfer from the tank is 1500 kJ. Consider the tank and the fluid inside a control surface and determine the change in internal energy of this control mass.

The first law of thermodynamics is (Eq. 5.11)

$$_1Q_2 = U_2 - U_1 + \frac{1}{2}m\left(\mathbf{V}_2^2 - \mathbf{V}_1^2\right) + mg\left(Z_2 - Z_1\right) + _1W_2$$

Since there is no change in kinetic and potential energy, this reduces to

$$_1Q_2 = U_2 - U_1 + _1W_2$$
$$U_2 - U_1 = -1500 - \left(-5090\right) = 3590 \text{ kJ}$$

EXAMPLE 5.2 Consider a stone having a mass of 10 kg and a bucket containing 100 kg of liquid water. Initially the stone is 10.2 m above the water, and the stone and the water are at the same temperature, state 1. The stone then falls into the water.

Determine $\Delta U, \Delta KE, \Delta PE, Q$, and W for the following changes of state, assuming standard gravitational acceleration of 9.806 65 m/s².

a. The stone is about to enter the water, state 2.

b. The stone has just come to rest in the bucket, state 3.

c. Heat has been transferred to the surroundings in such an amount that the stone and water are at the temperature, T_1, state 4.

Analysis and Solution

The first law for any of the steps is

$$Q = \Delta U + \Delta KE + \Delta PE + W$$

and each term can be identified for each of the changes of state.

1. The stone has fallen from Z_1 to Z_2, and we assume no heat transfer as it falls. The water has not changed state, thus

$$\Delta U = 0; \quad _1Q_2 = 0; \quad _1W_2 = 0$$

and the first law reduces to

$$\Delta KE + \Delta PE = 0$$
$$\Delta KE = -\Delta PE = -mg\left(Z_2 - Z_1\right)$$
$$= -10 \text{ kg} \times 9.806\ 65 \text{ m/s}^2 \times \left(-10.2 \text{ m}\right)$$
$$= 1000 \text{ J} = 1 \text{ kJ}$$

That is, for the process from state 1 to state 2,

$$\Delta KE = 1 \text{ kJ} \quad \text{and} \quad \Delta PE = -1\text{kJ}$$

2. For the process from state 2 to state 3 with zero kinetic energy, we have

$$\Delta PE = 0 \qquad {}_2Q_3 = 0 \qquad {}_2W_3 = 0$$

Then

$$\Delta U + \Delta KE = 0$$

$$\Delta U = -\Delta KE = 1 \text{ kJ}$$

3. In the final state, there is no kinetic, nor potential energy and the internal energy is the same as in state 1.

$$\Delta U = -1 \text{kJ} \qquad \Delta KE = 0 \qquad \Delta PE = 0 \qquad {}_3W_4 = 0$$

$$_3Q_4 = \Delta U = -1 \text{kJ}$$

5.4 PROBLEM ANALYSIS AND SOLUTION TECHNIQUE

At this point in our study of thermodynamics, we have progressed sufficiently far (that is, we have accumulated sufficient tools to work with) that it is worthwhile to develop a somewhat formal technique or procedure for analyzing and solving thermodynamic problems. For the time being it may not seem entirely necessary to use such a rigorous procedure for many of our problems, but we should keep in mind that as we acquire more analytical tools the problems that we are capable of dealing with will become much more complicated. Thus, it is appropriate that we begin to practice this technique now in anticipation of these future problems.

Our problem analysis and solution technique is contained within the framework of the following set of questions that must be answered in the process of an orderly solution of a thermodynamic problem.

1. What is the control mass or control volume? Is it useful, or necessary, to choose more than one? It may be helpful to draw a sketch of the system at this point, illustrating all heat and work flows, and indicating forces such as external pressures and gravitation.

2. What do we know about the initial state (which properties)?

3. What do we know about the final state?

4. What do we know about the process that takes place? Is anything constant or zero? Is there some known functional relation between two properties?

5. Is it helpful to draw a diagram of the information in steps 2 to 4 (for example, a T–v or P–v diagram)?

6. What is our thermodynamic model for the behavior of the substance (for example, steam tables, ideal gas, and so on)?

7. What is our analysis of the problem (examine control surfaces for various work modes, first law, conservation of mass)?

8. What is our solution technique (in other words, from what we have done so far in steps 1–7, how do we proceed to find whatever it is that is desired)? Is a trial-and-error solution necessary?

It is not always necessary to write out all these steps, and in the majority of the examples throughout this text we will not do so. However, when faced with a new and unfamiliar problem, the student should always at least think through this set of questions to develop the ability to solve increasingly more challenging problems. In solving the following example, we will use this technique in detail.

EXAMPLE 5.3   A vessel having a volume of 5 m³ contains 0.05 m³ of saturated liquid water and 4.95 m³ of saturated water vapor at 0.1 MPa. Heat is transferred until the vessel is filled with saturated vapor. Determine the heat transfer for this process.

Control mass: All the water inside the vessel.

Sketch: Fig. 5.3.

Initial state: Pressure, volume of liquid, volume of vapor; therefore, state 1 is fixed.

Final state: Somewhere along the saturated-vapor curve; the water was heated, so $P_2 > P_1$.

Process: Constant volume and mass; therefore, constant specific volume.

Diagram: Fig. 5.4.

Model: Steam tables.

Analysis:

First law:
$$_1Q_2 = U_2 - U_1 + m\frac{\mathbf{V}_2^2 - \mathbf{V}_1^2}{2} + mg\left(Z_2 - Z_1\right) + {}_1W_2$$

From examining the control surface for various work modes, we conclude that the work for this process is zero. Furthermore, the system is not moving, so there is no change in kinetic energy. There is a small change in the center of mass of the system but we will assume that the corresponding change in potential energy is negligible (in kilojoules). Therefore,

$$_1Q_2 = U_2 - U_1$$

Solution

The heat transfer will be found from the first law. State 1 is known, so U_1 can be calculated. The specific volume at state 2 is also known (from state 1 and the process). Since

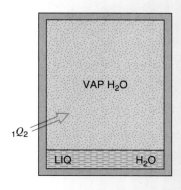

FIGURE 5.3   Sketch for Example 5.3.

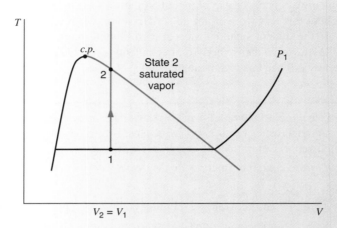

FIGURE 5.4 Diagram for Example 5.3.

state 2 is saturated vapor, state 2 is fixed, as is seen from Fig. 5.4. Therefore, U_2 can also be found.

The solution proceeds as follows:

$$m_{1\ \text{liq}} = \frac{V_{\text{liq}}}{v_f} = \frac{0.05}{0.001\ 043} = 47.94 \text{ kg}$$

$$m_{1\ \text{vap}} = \frac{V_{\text{vap}}}{v_g} = \frac{4.95}{1.6940} = 2.92 \text{ kg}$$

Then

$$U_1 = m_{1\ \text{liq}} u_{1\ \text{liq}} + m_{1\ \text{vap}} u_{1\ \text{vap}}$$

$$= 47.94(417.36) + 2.92(2506.1) = 27\ 326 \text{ kJ}$$

To determine u_2 we need to know two thermodynamic properties, since this determines the final state. The properties we know are the quality, $x = 100$ percent, and v_2, the final specific volume, which can readily be determined.

$$m = m_{1\ \text{liq}} + m_{1\ \text{vap}} = 47.94 + 2.92 = 50.86 \text{ kg}$$

$$v_2 = \frac{V}{m} = \frac{5.0}{50.86} = 0.098\ 31 \text{ m}^3/\text{kg}$$

In Table 2 of the steam tables we find, by interpolation, that at a pressure of 2.03 MPa, $v_g = 0.098\ 31$ m³/kg. The final pressure of the steam is therefore 2.03 MPa. Then

$$u_2 = 2600.5 \text{ kJ}/\text{kg}$$

$$U_2 = mu_2 = 50.86(2600.5) = 132\ 261 \text{ kJ}$$

$${}_1Q_2 = U_2 - U_1 = 132\ 261 - 27\ 326 = 104\ 935 \text{ kJ}$$

EXAMPLE 5.3E A vessel having a volume of 100 ft³ contains 1 ft³ of saturated liquid water and 99 ft³ of saturated water vapor at 14.7 lbf/in². Heat is transferred until the vessel is filled with saturated vapor. Determine the heat transfer for this process.

Control mass: All the water inside the vessel.

Sketch: Fig. 5.3.

Initial state: Pressure, volume of liquid, volume of vapor; therefore, state 1 is fixed.

Final state: Somewhere along the saturated-vapor curve; the water was heated, so $P_2 > P_1$.

Process: Constant volume and mass; therefore, constant specific volume.

Diagram: Fig. 5.4.

Model: Steam tables.

Analysis:

First law:
$$_1Q_2 = U_2 - U_1 + m\frac{\left(\mathbf{V}_2^2 - \mathbf{V}_1^2\right)}{2} + mg(Z_2 - Z_1) + {}_1W_2$$

From examining the control surface for various work modes, we conclude that the work for this process is zero. Furthermore, the system is not moving, so there is no change in kinetic energy. There is a small change in center of mass of the system but we will assume that the corresponding change in potential energy is negligible (compared to other terms). Therefore,

$$_1Q_2 = U_2 - U_1$$

Solution

The heat transfer will be found from the first law above. State 1 is known, so U_1 can be calculated. Also, the specific volume at state 2 is known (from state 1 and the process). Since state 2 is saturated vapor, state 2 is fixed, as is seen from Fig. 5.4. Therefore, U_2 can also be found.

The solution proceeds as follows:

$$m_{1\,\text{liq}} = \frac{V_{\text{liq}}}{v_f} = \frac{1}{0.01672} = 59.81\ \text{lbm}$$

$$m_{1\,\text{vap}} = \frac{V_{\text{vap}}}{v_g} = \frac{99}{26.80} = 3.69\ \text{lbm}$$

Then,

$$U_1 = m_{1\,\text{liq}}u_{1\,\text{liq}} + m_{1\,\text{vap}}u_{1\,\text{vap}}$$

$$= 59.81(180.1) + 3.69(1077.6) = 14\,748\ \text{Btu}$$

To determine u_2 we need to know two thermodynamic properties, since this determines the final state. The properties we know are the quality, $x = 100\%$, and v_2, the final specific volume, which can readily be determined.

$$m = m_{1\,liq} + m_{1\,vap} = 59.81 + 3.69 = 63.50 \text{ lbm}$$

$$v_2 = \frac{V}{m} = \frac{100}{63.50} = 1.575 \text{ ft}^3 / \text{lbm}$$

In Table 2 of the steam tables we find, by interpolation, that at a pressure of 294 lbf/in.^{2}, $v_g = 1.575$ ft^3/plbm. The final pressure of the steam is therefore 294 lbf/in.^{2}.

$$u_2 = 1117.0 \text{ Btu} / \text{lbm}$$

$$U_2 = mu_2 = 63.50(1117.0) = 70\,930 \text{ Btu}$$

$$_1Q_2 = U_2 - U_1 = 70\,930 - 14\,748 = 56\,182 \text{ Btu}$$

Up to this point, we have discussed how internal energies are listed in the saturation tables of the steam tables. The internal energy of superheated vapor steam is listed as a function of temperature and pressure in Table A.1.3. Similarly, Tables A.1.4 and A.1.5 list values in the compressed-liquid region and in the saturated-solid–saturated-vapor region, respectively. In summary, all the values of internal energy are listed in the tables in the same manner as are values of specific volume.

5.5 THE THERMODYNAMIC PROPERTY ENTHALPY

In analyzing specific types of processes, we frequently encounter certain combinations of thermodynamic properties, which are therefore also properties of the substance undergoing the change of state. To demonstrate one such situation, let us consider a control mass undergoing a quasi-equilibrium constant-pressure process, as shown in Fig. 5.5. Assume that there are no changes in kinetic or potential energy and that the only work done during the process is that associated with the boundary movement. Taking the gas as our control mass and applying the first law, Eq. 5.11, we have

$$_1Q_2 = U_2 - U_1 + {}_1W_2$$

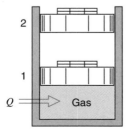

FIGURE 5.5 The constant-pressure quasi-equilibrium process.

The work can be calculated from the relation

$$_1W_2 = \int_1^2 P\,dV$$

Since the pressure is constant,

$$_1W_2 = P\int_1^2 dV = P(V_2 - V_1)$$

Therefore,

$$_1Q_2 = U_2 - U_1 + P_2V_2 - P_1V_1$$
$$= (U_2 + P_2V_2) - (U_1 + P_1V_1)$$

We find that in this very restricted case, the heat transfer during the process is given in terms of the change in the quantity $U + PV$ between the initial and final states. Because all these quantities are thermodynamic properties, that is, functions only of the state of the system, their combination must also have these same characteristics. Therefore, we find it convenient to define a new extensive property, the enthalpy,

$$H = U + PV \tag{5.12}$$

or, per unit mass,

$$h = u + Pv \tag{5.13}$$

As for internal energy, we could speak of specific enthalpy, h, and total enthalpy, H. However, we will refer to both as enthalpy, since the context will make it clear which is being discussed.

The heat transfer in a constant-pressure quasi-equilibrium process is equal to the change in enthalpy, which includes both the change in internal energy and the work for this particular process. This is by no means a general result. It is valid for this special case only because the work done during the process is equal to the difference in the PV product for the final and initial states. This would not be true if the pressure had not remained constant during the process.

The significance and use of enthalpy is not restricted to the special process just described. Other cases in which this same combination of properties $u + Pv$ appear will be developed later, notably in Section 5.11 in which we discuss control volume analyses. Our reason for introducing enthalpy at this time is that although the steam tables list values for internal energy, many other tables and charts of thermodynamic properties give values for enthalpy but not for the internal energy. Therefore, it is necessary to calculate the internal energy at a state using the tabulated values and Eq. 5.13:

$$u = h - Pv$$

Students often become confused about the validity of this calculation when analyzing system processes that do not occur at constant pressure, for which enthalpy has no physical significance. We must keep in mind that enthalpy, being a property, is a state or point function, and its use in calculating internal energy at the same state is not related to, or dependent on, any process that may be taking place.

Tabular values of enthalpy, such as those included in Tables A.1 through A.7, are all relative to some arbitrarily selected base. In the steam tables, the internal energy of saturated liquid at $0.01°C$ is the reference state and is given a value of zero. For refrigerants, such as ammonia and chlorofluorocarbons R-12 and R-22, the reference state is arbitrarily taken as saturated liquid at −40°C. The enthalpy in this reference state is assigned the value of zero. Cryogenic fluids, such as nitrogen or methane, have other arbitrary reference states chosen for enthalpy values listed in their tables. Because each of these reference states is arbitrarily selected, it is always possible to have negative values for enthalpy, as for saturated-solid water in Table A.1.5. When enthalpy and internal energy are given values relative to the same reference state, as they are in essentially all thermodynamic tables, the difference between internal energy and enthalpy at the reference state is equal to Pv. Since the specific volume of the liquid is very small, this product is negligible as far as the significant figures of the tables are concerned, but the principle should be kept in mind, for in certain cases it is significant.

In the superheat region of many thermodynamic tables, values of the specific internal energy u are not given. As mentioned earlier, these values can be readily calculated from the relation $u = h − Pv$, though it is important to keep the units in mind. As an example, let us calculate the internal energy u of superheated R-134a at 0.4 MPa, 70°C.

$$u = h - Pv$$
$$= 460.545 - 400 \times 0.066\,484$$
$$= 433.951 \text{ kJ / kg}$$

The enthalpy of a substance in a saturation state and with a given quality is found in the same way as the specific volume and internal energy. The enthalpy of saturated liquid has the symbol h_f, saturated vapor h_g, and the increase in enthalpy during vaporization h_{fg}. For a saturation state, the enthalpy can be calculated by one of the following relations:

$$h = (1 - x)h_f + xh_g$$
$$h = h_f + xh_{fg}$$

The enthalpy of compressed liquid water may be found from Table A.1.4. For substances for which compressed-liquid tables are not available, the enthalpy is taken as that of saturated liquid at the same temperature.

EXAMPLE 5.4   A cylinder fitted with a piston has a volume of 0.1 m³ and contains 0.5 kg of steam at 0.4 MPa. Heat is transferred to the steam until the temperature is 300°C, while the pressure remains constant.

Determine the heat transfer and the work for this process.

Control mass: Water inside cylinder.

Initial state: P_1, V_1, m; therefore v_1 is known, state 1 is fixed (at P_1, v_1, check steam tables—two-phase region).

Final state: P_2, T_2; therefore state 2 is fixed (superheated).

Process: Constant pressure.

Diagram: Fig. 5.6.

Model: Steam tables.

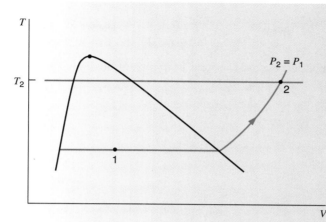

FIGURE 5.6 Diagram for Example 5.4.

Analysis:

No change in kinetic energy, no change in potential energy. Work is done by movement at the boundary. Assume the process to be quasi-equilibrium. Since the pressure is constant,

$$_1W_2 = \int_1^2 P\,dV = P\int_1^2 dV = P(V_2 - V_1) = m(P_2 v_2 - P_1 v_1)$$

Therefore, the first law is

$$_1Q_2 = m(u_2 - u_1) + {_1W_2}$$

$$= m(u_2 - u_1) + m(P_2 v_2 - P_1 v_1) = m(h_2 - h_1)$$

Solution

There is a choice of procedures to follow. State 1 is known, so v_1 and h_1 (or u_1) can be found. State 2 is also known, so v_2 and h_2 (or u_2) can be found. Using the first law and the work equation, we can calculate the heat transfer and work. Using the enthalpies, we have

$$v_1 = \frac{V_1}{m} = \frac{0.1}{0.5} = 0.2 = 0.001\,084 + x_1\,0.4614$$

$$x_1 = \frac{0.1989}{0.4614} = 0.4311$$

$$h_1 = h_f + x_1 h_{fg}$$

$$= 604.74 + 0.4311 \times 2133.8 = 1524.7$$

$$h_2 = 3066.8$$

$$_1Q_2 = 0.5(3066.8 - 1524.7) = 771.1 \text{ kJ}$$

$$_1W_2 = mP(v_2 - v_1) = 0.5 \times 400(0.6548 - 0.2) = 91.0 \text{ kJ}$$

Therefore,

$$U_2 - U_1 = {}_1Q_2 - {}_1W_2 = 771.1 - 91.0 = 680.1 \text{ kJ}$$

The heat transfer could also have been found from u_1 and u_2,

$$u_1 = u_f + x_1 u_{fg}$$

$$= 604.31 + 0.4311 \times 1949.3 = 1444.7$$

$$u_2 = 2804.8$$

and

$${}_1Q_2 = U_2 - U_1 + {}_1W_2$$

$$= 0.5(2804.8 - 1444.7) + 91.0 = 771.1 \text{ kJ}$$

5.6 THE CONSTANT-VOLUME AND CONSTANT-PRESSURE SPECIFIC HEATS

In this section we will consider a homogeneous phase of a substance of constant composition. This phase may be a solid, a liquid, or a gas, but no change of phase will occur. We will then define a variable termed the specific heat, the amount of heat required per unit mass to raise the temperature by one degree. Since it would be of interest to examine the relation between the specific heat and other thermodynamic variables, we note first that the heat transfer is given by Eq. 5.10. Neglecting changes in kinetic and potential energies, and assuming a simple compressible substance and quasi-equilibrium process, for which the work in Eq. 5.10 is given by Eq. 4.2, we have

$$\delta Q = dU + \delta W = dU + P \, dV$$

We find that this expression can be evaluated for two separate special cases.

1. Constant volume, for which the work term ($P \, dV$) is zero, so that the specific heat (at constant volume) is

$$C_v = \frac{1}{m}\left(\frac{\delta Q}{\delta T}\right)_v = \frac{1}{m}\left(\frac{\partial U}{\partial T}\right)_v = \left(\frac{\partial h}{\partial T}\right)_v \qquad (5.14)$$

2. Constant pressure, for which the work term can be integrated and the resulting PV terms at the initial and final states be associated with the internal energy terms, as in Section 5.5, thereby leading to the conclusion that the heat transfer can be expressed in terms of the enthalpy change. The corresponding specific heat (at constant pressure) is

$$C_p = \frac{1}{m}\left(\frac{\delta Q}{\delta T}\right)_p = \frac{1}{m}\left(\frac{\partial H}{\partial T}\right)_p = \left(\frac{\partial h}{\partial T}\right)_p \qquad (5.15)$$

Note that in each of these special cases, the resulting expression, Eq. 5.14 or 5.15, contains only thermodynamic properties, from which we conclude that the constant-volume and constant-pressure specific heats must themselves be thermodynamic properties.

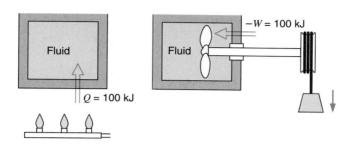

FIGURE 5.7 Sketch showing two ways in which a given ΔU may be achieved.

This means that, although we began this discussion by considering the amount of heat transfer required to cause a unit temperature change and then proceeded through a very specific development leading to Eq. 5.14 (or 5.15), the result ultimately expresses a relation among a set of thermodynamic properties and therefore constitutes a definition that is independent of the particular process leading to it (in the same sense that the definition of enthalpy in the previous section is independent of the process used to illustrate one situation in which the property is useful in a thermodynamic analysis). As an example, consider the two identical fluid masses shown in Fig. 5.7. In the first system 100 kJ of heat is transferred to it, and in the second system 100 kJ of work is done on it. Thus, the change of internal energy is the same for each, and therefore the final state and the final temperature are the same in each. In accordance with Eq. 5.14, therefore, exactly the same value for the average constant-volume specific heat would be found for this substance for the two processes, even though the two processes are very different as far as heat transfer is concerned.

It should also be noted that this development was carried out for a simple compressible substance, for which the work was given by Eq. 4.2. Where there are different work modes, such as those described in Sections 4.4 through 4.6, different specific heats become appropriate. For any particular quasi-equilibrium work mode, there is one specific heat at constant displacement and one at constant driving force, as these terms were discussed in connection with Eq. 4.16. For example, in a system involving magnetic effects, it is convenient to utilize a specific heat at constant magnetization and one at constant magnetic field intensity.

EXAMPLE 5.5   Estimate the constant-pressure specific heat of steam at 0.5 MPa, 375°C.

Solution

If we consider a change of state at constant pressure, Eq. 5.15 may be written

$$C_p \approx \left(\frac{\Delta h}{\Delta T} \right)_p$$

From the steam tables,

$$\text{at } 0.5 \text{ MPa, } 350°\text{C, } \quad h = 3167.7$$

$$\text{at } 0.5 \text{ MPa, } 400°\text{C, } \quad h = 3271.8$$

Since we are interested in C_p at 0.5 MPa, 375°C,

$$C_p \simeq \frac{104.1}{50} = 2.082 \ \text{kJ}/\text{kg K}$$

As a special case, consider either a solid or a liquid. Since both of these phases are nearly incompressible,

$$dh = du + d(Pv) \approx du + vdP \tag{5.16}$$

Also, for both of these phases, the specific volume is very small, such that in many cases

$$dh \approx du \approx C \ dT \tag{5.17}$$

where C is either the constant-volume or the constant-pressure specific heat, as the two would be nearly the same. In many processes involving a solid or a liquid, we might further assume that the specific heat in Eq. 5.17 is constant (unless the process is at low temperature or over a wide range of temperature). Equation 5.17 can then be integrated to

$$h_2 - h_1 \simeq u_2 - u_1 \simeq C(T_2 - T_1) \tag{5.18}$$

Specific heats for various solids and liquids are listed in Table A.9.

In other processes for which it is not possible to assume constant specific heat, there may be a known relation for C as a function of temperature. Equation 5.17 could then also be integrated.

5.7 THE INTERNAL ENERGY, ENTHALPY, AND SPECIFIC HEAT OF IDEAL GASES

At this point certain comments about the internal energy, enthalpy, and the constant-pressure and constant-volume specific heats of an ideal gas should be made. An ideal gas has been defined in Chapter 3 as a gas at sufficiently low density so that intermolecular forces and the associated energy are negligibly small. Therefore, an ideal gas has the equation of state

$$Pv = RT$$

It can be shown that for an ideal gas, the internal energy is a function of the temperature only. That is, for an ideal gas,

$$u = f(T) \tag{5.19}$$

This means that an ideal gas at a given temperature has a certain definite specific internal energy u, regardless of the pressure. This will be demonstrated mathematically using the methods of classical thermodynamics in Chapter 10.

In 1843, Joule demonstrated this fact when he conducted the following classical experiment in thermodynamics. Two pressure vessels (Fig. 5.8), connected by a pipe and valve, were immersed in a bath of water. Initially vessel A contained air at 22-atm pressure and vessel B was highly evacuated. When thermal equilibrium was attained, the valve was opened, allowing the pressures in A and B to equalize. No change in the

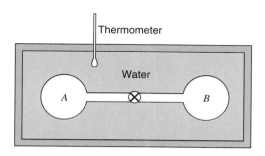

FIGURE 5.8 Apparatus for conducting Joule's experiment.

temperature of the bath was detected during or after this process. Because there was no change in the temperature of the bath, Joule concluded that no heat had been transferred to the air. Because the work was also zero, he concluded from the first law of thermodynamics that there was no change in the internal energy of the gas. Because the pressure and volume changed during this process, one concludes that internal energy is not a function of pressure and volume. Because air does not conform exactly to the definition of an ideal gas, a small change in temperature will be detected when very accurate measurements are made in Joule's experiment.

The relation between the internal energy u and the temperature can be established by using the definition of constant-volume specific heat given by Eq. 5.14.

$$C_v = \left(\frac{\partial u}{\partial T}\right)_v$$

Since the internal energy of an ideal gas is not a function of volume, for an ideal gas we can write

$$C_{v0} = \frac{du}{dT}$$

$$du = C_{v0}\, dT \qquad (5.20)$$

where the subscript 0 denotes the specific heat of an ideal gas. For a given mass m,

$$dU = m C_{v0}\, dT \qquad (5.21)$$

From the definition of enthalpy and the equation of state of an ideal gas, it follows that

$$h = u + Pv = u + RT \qquad (5.22)$$

Since R is a constant and u is a function of temperature only, it follows that the enthalpy, h, of an ideal gas is also a function of temperature only. That is,

$$h = f(T) \qquad (5.23)$$

The relation between enthalpy and temperature is found from the constant-pressure specific heat as defined by Eq. 5.15.

$$C_p = \left(\frac{\partial h}{\partial T}\right)_p$$

Since the enthalpy of an ideal gas is a function of the temperature only and is independent of the pressure, it follows that

$$C_{p0} = \frac{dh}{dT}$$

$$dh = C_{p0}\,dT \tag{5.24}$$

For a given mass m,

$$dH = mC_{p0}\,dT \tag{5.25}$$

The consequences of Eqs. 5.20 and 5.24 are demonstrated in Fig. 5.9, which shows two lines of constant temperature. Since internal energy and enthalpy are functions of temperature only, these lines of constant temperature are also lines of constant internal energy and constant enthalpy. From state 1 the high temperature can be reached by a variety of paths, and in each case the final state is different. However, regardless of the path, the change in internal energy is the same, as is the change in enthalpy, for lines of constant temperature are also lines of constant u and constant h.

Because the internal energy and enthalpy of an ideal gas are functions of temperature only, it also follows that the constant-volume and constant-pressure specific heats are also functions of temperature only. That is,

$$C_{v0} = f(T) \qquad C_{p0} = f(T) \tag{5.26}$$

Because all gases approach ideal-gas behavior as the pressure approaches zero, the ideal-gas specific heat for a given substance is often called the zero-pressure specific heat, and the zero-pressure, constant-pressure specific heat is given the symbol C_{p0}. The zero-pressure, constant-volume specific heat is given the symbol C_{p0}. Figure 5.10 shows $\overline{C}_{p0}$ as a function of temperature for a number of different substances. These values are determined by the techniques of statistical thermodynamics and will not be discussed in this text. However, a brief qualitative discussion at this point should provide some insight into this behavior and will be beneficial in determining under what conditions an assumption of constant specific heat is justified.

As was discussed in Section 2.6, the energy possessed by molecules may be stored in several forms. The translational and rotational energies increase linearly with temperature, which means that these contributions to the specific heat are not temperature-dependent. Contributions from vibrational and electronic modes, on the other hand, are temperature-dependent (the electronic usually being very small). From Fig. 5.10, it is evident that the specific heat of a diatomic gas (such as hydrogen or oxygen) increases with an increase in temperature, primarily because of the vibration. A polyatomic gas (such as

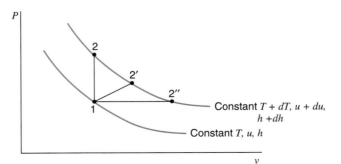

Constant $T + dT$, $u + du$,
$h + dh$

Constant T, u, h

FIGURE 5.9 Pressure–volume diagram for an ideal gas.

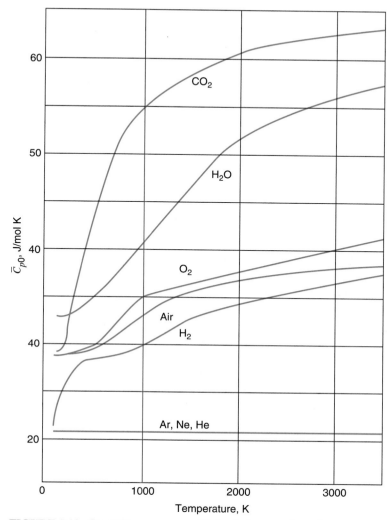

FIGURE 5.10 Constant-pressure specific heats for a number of gases at zero pressure.

carbon dioxide or water) shows a much greater increase in specific heat as the temperature increases, and this is due to the additional vibrational modes of a polyatomic molecule. A monatomic gas (such as helium, argon, or neon), possessing only translational and electronic energies, shows little or no variation of specific heat over a wide range of temperatures.

A very important relation between the constant-pressure and constant-volume specific heats of an ideal gas may be developed from the definition of enthalpy.

$$h = u + Pv = u + RT$$

Differentiating and substituting Eqs. 5.20 and 5.24, we have

$$dh = du + R \ dT$$

$$C_{p0} \ dT = C_{v0} \ dT + R \ dT$$

Therefore,

$$C_{p0} - C_{v0} = R \tag{5.27}$$

On a mole basis this equation is written

$$\overline{C}_{p0} - \overline{C}_{v0} = \overline{R} \tag{5.28}$$

This tells us that the difference between the constant-pressure and constant-volume specific heats of an ideal gas is always constant, though both are functions of temperature. Thus, we need examine only the temperature dependency of one, and the other is given by Eq. 5.27.

Let us consider the specific heat C_{p0}. There are three possibilities to examine. The situation is simplest if we assume constant specific heat, that is, no temperature dependence. Then it is possible to integrate Eq. 5.24 directly to

$$h_2 - h_1 = C_{p0}(T_2 - T_1) \tag{5.29}$$

We note from Fig. 5.10 the circumstances under which this will be an accurate model. It should be added, however, that it may be a reasonable approximation under other conditions, especially if an average specific heat in the particular temperature range is used in Eq. 5.29. Values of specific heat at room temperature and gas constants for various gases are given in Table A.10.

The second possibility for the specific heat is to use an analytical equation for C_{p0} as a function of temperature. Because the results of specific-heat calculations from statistical thermodynamics do not lend themselves to convenient mathematical forms, these results have been approximated empirically. The equations for C_{p0} as a function of temperature are listed in Table A.11 for a number of gases.

The third possibility is to integrate the results of the calculations of statistical thermodynamics from an arbitrary reference temperature to any other temperature T, and to define a function

$$h_T = \int_{T_0}^{T} C_{p0} \, dT$$

This function can then be tabulated in a single-entry (temperature) table. Then, between any two states 1 and 2,

$$h_2 - h_1 = \int_{T_0}^{T_2} C_{p0} \, dT - \int_{T_0}^{T_1} C_{p0} \, dT = h_{T_2} - h_{T_1} \tag{5.30}$$

and it is seen that the reference temperature cancels out. This function h_T (and a similar function $u_T = h_T - RT$) is listed for air in Table A.12. The enthalpy function (with respect to a reference temperature of 25°C) is listed for other gases in Table A.13.

To summarize the three possibilities, we note that using the ideal-gas tables, Tables A.12 and A.13, gives us the most accurate answer, but that the equations in Table A.11 would give a close empirical approximation. Constant specific heat would be less accurate, except for monatomic gases and gases below room temperature. It should be remembered that all these results are a part of the ideal-gas model, which in many of our problems is not a valid assumption for the behavior of the substance.

EXAMPLE 5.6   Calculate the change of enthalpy as 1 kg of oxygen is heated from 300 to 1500 K. Assume ideal-gas behavior.

Solution

For an ideal gas, the enthalpy change is given by Eq. 5.24. However, we also need to make an assumption about the dependence of specific heat on temperature. Let us solve this problem in several ways and compare the answers.

Our most accurate answer for the ideal-gas enthalpy change for oxygen between 300 and 1500 K would be from the ideal-gas tables, Table A.13. This result is, using Eq. 5.30,

$$h_2 - h_1 = \frac{\bar{h}_{1500} - \bar{h}_{300}}{M} = \frac{40\,600 - 54}{32} = 1267.0 \text{ kJ} / \text{kg}$$

The empirical equation from Table A.11 should give a good approximation to this result. Integrating Eq. 5.24, we have

$$\bar{h}_2 - \bar{h}_1 = \int_{T_1}^{T_2} \bar{C}_{p0} \, dT = \int_{\theta_1}^{\theta_2} \bar{C}_{p0}(\theta) \times 100 \, d\theta$$

$$\bar{h}_{1500} - \bar{h}_{300} = 100 \left(37.432\theta + \frac{0.020\,102}{2.5} \theta^{2.5} + \frac{178.57}{0.5} \theta^{-0.5} - 236.88\theta^{-1} \right) \Bigg|_{\theta_1 = 3}^{\theta_2 = 15}$$

$$= 40\,525 \text{ kJ} / \text{kmol}$$

$$h_2 - h_1 = \frac{\bar{h}_{1500} - \bar{h}_{300}}{M} = \frac{40\,525}{32}$$

$$= 1266.4 \text{ kJ} / \text{kg}$$

which is different than the first result by less than 0.1%.

If we assume constant specific heat, we must be concerned about what value we are going to use. If we use the value at 300 K from Table A.10, we find, from Eq. 5.29, that

$$h_2 - h_1 = C_{p0}(T_2 - T_1) = 0.9216 \times 1200 = 1105.9 \text{ kJ} / \text{kg}$$

which is low by 12.7%. On the other hand, suppose we assume that the specific heat is constant at its value at 900 K, the average temperature. Substituting 900 K into the equation for specific heat from Table A.11, we have

$$\bar{C}_{p0} = 37.432 + 0.020\,102(9)^{1.5} - 178.57(9)^{-1.5} + 236.88(9)^{-2}$$

$$= 34.2855 \text{ kJ} / \text{kmol K}$$

or

$$C_{p0} = \frac{34.2855}{32} = 1.0714 \text{ kJ} / \text{kg K}$$

Substituting this value into Eq. 5.29 gives the result

$$h_2 - h_1 = 1.0714 \times 1200 = 1285.7 \text{ kJ} / \text{kg}$$

which is high by about 1.5%, a much closer result than the one using the room temperature specific heat. It should be kept in mind that part of the model involving ideal gas with constant specific heat is the choice of what value is to be used.

EXAMPLE 5.7   A cylinder fitted with a piston has an initial volume of 0.1 m³ and contains nitrogen at 150 kPa, 25°C. The piston is moved, compressing the nitrogen until the pressure is 1 MPa and the temperature is 150°C. During this compression process heat is transferred from the nitrogen, and the work done on the nitrogen is 20 kJ. Determine the amount of this heat transfer.

Control mass: Nitrogen.

Initial state: P_1, T_1, V_1; state 1 fixed.

Final state: P_2, T_2; state 2 fixed.

Process: Work input known.

Model: Ideal gas, constant specific heat with value at 300 K, Table A.10.

Analysis:

First law:
$$_1Q_2 = m(u_2 - u_1) + {}_1W_2$$

Solution

The mass of nitrogen is found from the equation of state with the value of R from Table A.10.

$$m = \frac{PV}{RT} = \frac{150 \times 0.1}{0.2968 \times 298.15} = 0.1695 \text{ kg}$$

Assuming constant specific heat as given in Table A.10, we have

$$_1Q_2 = mC_{v0}(T_2 - T_1) + {}_1W_2$$
$$= 0.1695(0.7448)(150 - 25) - 20.0$$
$$= 15.8 - 20.0 = -4.2 \text{ kJ}$$

It would, of course, be somewhat more accurate to use Table A.13 than to assume constant specific heat (room temperature value), but often the slight increase in accuracy does not warrant the added difficulties of manually interpolating the tables.

EXAMPLE 5.7E   A cylinder fitted with a piston has an initial volume of 2 ft³ and contains nitrogen at 20 lbf/in.², 80 F. The piston is moved, compressing the nitrogen until the pressure is 160 lbf/in.² and the temperature is 300F. During this compression process heat is transferred from the nitrogen, and the work done on the nitrogen is 9.15 Btu. Determine the amount of this heat transfer.

Control mass: Nitrogen.

Initial state: P_1, T_1, V_1; state 1 fixed.

Final state: P_2, T_2; state 2 fixed.

Process: Work input known.

Model: Ideal gas, constant specific heat with value at 540 R, Table A.10.

Analysis:

First law:
$$_1Q_2 = m(u_2 - u_1) + {_1}W_2$$

Solution

The mass of nitrogen is found from the equation of state with the value of R from Table A.10.

$$m = \frac{PV}{RT} = \frac{20 \times 144 \times 2}{55.15 \times 540} = 0.1934 \text{ lbm}$$

Assuming constant specific heat as given in Table A.10,

$$_1Q_2 = mC_{v0}(T_2 - T_1) + {_1}W_2$$
$$= 0.1934(0.177)(300 - 80) - 9.15$$
$$= 7.53 - 9.15 = -1.62 \text{ Btu}$$

It would, of course, be somewhat more accurate to use Table A.13 than to assume constant specific heat (room temperature value), but often the slight increase in accuracy does not warrant the added difficulties of manually interpolating the tables.

5.8 THE FIRST LAW AS A RATE EQUATION

We frequently find it desirable to use the first law as a rate equation that expresses either the instantaneous or average rate at which energy crosses the control surface as heat and work and the rate at which the energy of the control mass changes. In so doing we are departing from a strictly classical point of view, because basically classical thermodynamics deals with systems that are in equilibrium, and time is not a relevant parameter for systems that are in equilibrium. However, since these rate equations are developed from the concepts of classical thermodynamics and are used in many applications of thermodynamics, they are included in this book. This rate form of the first law will be used in the development of the first law for the control volume in Section 5.11, and in this form the first law finds extensive applications in thermodynamics, fluid mechanics, and heat transfer.

Consider a time internal δt during which an amount of heat δQ crosses the control surface, an amount of work δW is done by the control mass, the internal energy change is ΔU, the kinetic energy change is ΔKE, and the potential energy change is ΔPE. From the first law we can write

$$\delta Q = \Delta U + \Delta KE + \Delta PE + \delta W$$

Dividing by δt we have the average rate of energy transfer as heat and work and increase of the energy of the control mass.

$$\frac{\delta Q}{\delta t} = \frac{\Delta U}{\delta t} + \frac{\Delta KE}{\delta t} + \frac{\Delta PE}{\delta t} + \frac{\delta W}{\delta t}$$

Taking the limit for each of these quantities as δt approaches zero, we have

$$\lim_{\delta t \to 0} \frac{\delta Q}{\delta t} = \dot{Q}, \quad \text{the heat transfer rate}$$

$$\lim_{\delta t \to 0} \frac{\delta W}{\delta t} = \dot{W}, \quad \text{the power}$$

$$\lim_{\delta t \to 0} \frac{\Delta U}{\delta t} = \frac{dU}{dt} \qquad \lim_{\delta t \to 0} \frac{\Delta (KE)}{\delta t} = \frac{d(KE)}{dt} \qquad \lim_{\delta t \to 0} \frac{\Delta (PE)}{\delta t} = \frac{d(PE)}{dt}$$

Therefore, the rate equation form of the first law is

$$\dot{Q} = \frac{dU}{dt} + \frac{d(KE)}{dt} + \frac{d(PE)}{dt} + \dot{W} \tag{5.31}$$

We could also write this in the form

$$\dot{Q} = \frac{dE}{dt} + \dot{W} \tag{5.32}$$

EXAMPLE 5.8 During the charging of a storage battery, the current is 20 A and the voltage is 12.8 V. The rate of heat transfer from the battery is 10 W. At what rate is the internal energy increasing?

Solution

Since changes in kinetic and potential energy are insignificant, the first law can be written as a rate equation in the form, Eq. 5.31,

$$\dot{Q} = \frac{dU}{dt} + \dot{W}$$

$$\dot{W} = -\mathscr{E}i = -20 \times 12.8 = -256 \text{ W}$$

Therefore,

$$\frac{dU}{dt} = \dot{Q} - \dot{W} = -10 \text{ W} - (-256) = 246 \text{ J/s}$$

5.9 CONSERVATION OF MASS

In the previous sections we considered the first law of thermodynamics for a control mass undergoing a change of state. A control mass is defined as a fixed quantity of mass. The question now is whether the mass of such a system changes when its energy changes? If it does, our definition of a control mass as a fixed quantity of mass is no longer valid when the energy changes.

We know from relativistic considerations that mass and energy are related by the well-known equation

$$E = mc^2 \tag{5.33}$$

where c = velocity of light and E = energy. We conclude from this equation that the mass of a control mass does change when its energy changes. Let us calculate the magnitude of this change of mass for a typical problem and determine whether this change in mass is significant.

Consider a rigid vessel that contains a 1-kg stoichiometric mixture of a hydrocarbon fuel (such as gasoline) and air. From our knowledge of combustion, we know that after combustion takes place it will be necessary to transfer about 2900 kJ from the system to restore it to its initial temperature. From the first law

$$_1Q_2 = U_2 - U_1 + \,_1W_2$$

we conclude that since $_1W_2 = 0$ and $_1Q_2 = -2900$ kJ, the internal energy of this system decreases by 2900 kJ during the heat transfer process. Let us now calculate the decrease in mass during this process using Eq. 5.33.

The velocity of light, c, is 2.9979×10^8 m/s. Therefore,

$$2900 \text{ kJ} = 2\,900\,000 \text{ J} = m \,(\text{kg}) \times \left(2.9979 \times 10^8 \text{ m/s}\right)^2$$

$$m = 3.23 \times 10^{-11} \text{ kg}$$

Thus, when the energy of the control mass decreases by 2900 kJ, the decrease in mass is 3.23×10^{-11} kg.

A change in mass of this magnitude cannot be detected by even our most accurate chemical balance. Certainly, a fractional change in mass of this magnitude is beyond the accuracy required in essentially all engineering calculations. Therefore, if we use the laws of conservation of mass and conservation of energy as separate laws, we will not introduce significant error into most thermodynamic problems, and our definition of a control mass as having a fixed mass can be used even though the energy changes.

5.10 CONSERVATION OF MASS AND THE CONTROL VOLUME

A control volume is a volume in space in which one has interest for a particular study or analysis. The surface of this control volume is referred to as a control surface and always consists of a closed surface. The size and shape of the control volume are completely arbitrary, and are so defined as to best suit the analysis that is to be made. The surface may be fixed, or it may move or expand. However, the surface must be defined relative to some coordinate system. In some analyses it may be desirable to consider a rotating or moving coordinate system, and to describe the position of the control surface relative to such a coordinate system.

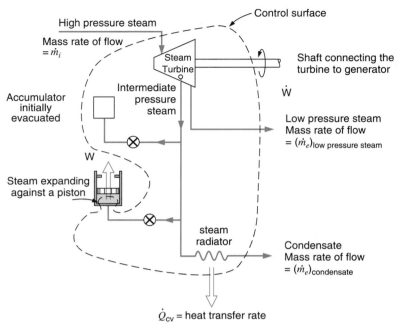

FIGURE 5.11 Schematic diagram of a control volume showing mass and energy transfers and accumulation.

Mass as well as heat and work can cross the control surface and the mass in the control volume, as well as the properties of this mass, can change with time. Figure 5.11 shows a schematic diagram of a control volume, with heat transfer, shaft work, accumulation of mass within the control volume, and a moving boundary.

Let us consider the law of the conservation of mass as it relates to the control volume. We first consider the mass flow into and out of the control volume and the net increase of mass within the control volume. During a time interval δt let the mass δm_i, as shown in Fig. 5.12, enter the control volume, and let the mass δm_e leave the control volume. Further, let us designate the mass within the control volume at the beginning of this time interval as m_t and the mass within the control volume after this interval as $m_{t+\delta t}$. From the law of the conservation of mass we can write

$$m_t + \delta m_i = m_{t+dt} + \delta m_e$$

We can also look at this from the point of view of the net flow across the control surface and the change of mass within the control volume.

Net flow into the control volume during δt = increase of mass within the control volume during δt:

$$\left(\delta m_i - \delta m_e\right) = m_{t+\delta t} - m_t$$

or

$$\left(m_{t+dt} - m_t\right) + \left(\delta m_e - \delta m_i\right) = 0 \tag{5.34}$$

As written, this equation simply states that the change of mass within the control volume during δt, $(m_{t+\delta t} - m_t)$, and the net mass flow into the control volume during δt, $(\delta m_i - \delta m_e)$, are equal. However, in many problems requiring a thermodynamic analysis, we find it very convenient to have the law of the conservation of mass (as well as the

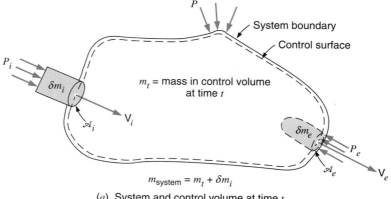

(a) System and control volume at time t

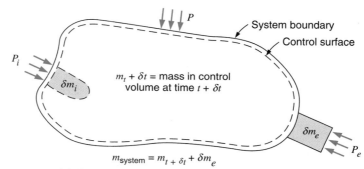

(b) System and control volume at time $t + \delta t$

FIGURE 5.12 Schematic diagram of a control volume for analysis of the continuity equation as applied to a control volume. (a) System and control volume at time t. (b) System and control volume at time $t + \delta t$.

first and second laws of thermodynamics and the momentum equation) expressed as a rate equation for the control volume. This involves the instantaneous rate of mass flow across the control surface and the instantaneous rate of change of mass within the control volume. Let us therefore write an expression for the average rate of change of mass within the control volume during δt and the average mass rates of flow across the control surface during δt by dividing Eq. 5.34 by δt.

$$\left(\frac{m_{t+dt} - m_t}{\delta t} \right) + \frac{\delta m_e}{\delta t} - \frac{\delta m_i}{\delta t} = 0 \tag{5.35}$$

To obtain the rate equation for the control volume, we find the limit of each term as δt is made to approach zero, at which point the system and control volume coincide.

$$\lim_{\delta t \to 0} \left(\frac{m_{t+\delta t} - m_t}{\delta t} \right) = \frac{dm_{c.v.}}{dt}$$

$$\lim_{\delta t \to 0} \left(\frac{\delta m_e}{\delta t} \right) = \dot{m}_e$$

$$\lim_{\delta t \to 0} \left(\frac{\delta m_i}{\delta t} \right) = \dot{m}_i$$

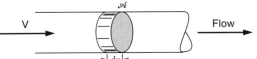

FIGURE 5.13 Flow across a control surface.

The symbol $m_{c.v.}$ denotes the instantaneous mass inside the control volume; $\dot{m}_i$ is the instantaneous rate of flow entering the control volume across area $\mathscr{A}_i$; and $\dot{m}_e$ is that leaving across area $\mathscr{A}_e$. We recognize that, in practice, there may well be several areas on the control surface across which flow occurs (in mixing processes or processes involving chemical reactions, for example). We account for this possibility by including summation signs, implying a summation of flows over the various discrete flow areas. Thus the instantaneous rate form becomes

$$\frac{dm_{c.v.}}{dt} + \sum \dot{m}_e - \sum \dot{m}_i = 0 \tag{5.36}$$

Equation 5.36 for the conservation of mass is commonly termed the continuity equation. While this form of the equation is sufficient for the majority of applications in thermodynamics, it is frequently rewritten in terms of local fluid properties in the study of fluid mechanics and heat transfer. In this text, we consider Eq. 5.36 as the general expression of the continuity equation.

Let us consider one further aspect of mass flow across a control surface. For simplicity, we assume that a fluid is flowing uniformly inside a pipe or duct, as shown in Fig. 5.13.

We wish to examine the flow in terms of the amount of mass crossing the surface $\mathscr{A}$ during the time interval δt. As seen from Fig. 5.13, the fluid moves a distance dx during this time, and therefore the volume of fluid crossing surface $\mathscr{A}$ is $\mathscr{A}\, dx$. Consequently, the mass crossing surface $\mathscr{A}$ is given by

$$\delta m = \frac{\mathscr{A}\, dx}{v}$$

Now, if we divide both sides of this expression by δt and take the limit as $\delta t \to 0$, the result is

$$\dot{m} = \frac{\mathscr{A}\, \mathbf{V}}{v} \tag{5.37}$$

It should be noted that this result, Eq. 5.37, has been developed for a stationary control surface $\mathscr{A}$, and that we tacitly assumed that the flow was normal to that surface and was uniform across the surface. It should also be pointed out that Eq. 5.37 applies to any of the various flow streams entering or leaving the control volume, subject to the assumptions mentioned.

EXAMPLE 5.9   Air is flowing in a 0.2-m-diameter pipe at a uniform velocity of 0.1 m/s. The temperature and pressure are 25°C and 150 kPa. Determine the mass flow rate.

Solution

From Eq. 5.37,

$$\dot{m} = \frac{\mathcal{A} \mathbf{V}}{v}$$

For air, using R from Table A.10,

$$v = \frac{RT}{P} = \frac{0.287 \times 298.2}{150} = 0.5705 \text{ m}^3 / \text{kg}$$

The cross-sectional area is

$$\mathcal{A} = \frac{\pi}{4}(0.2)^2 = 0.0314 \text{ m}^2$$

Therefore,

$$\dot{m} = \frac{0.0314 \times 0.1}{0.5705} = 0.0055 \text{ kg} / \text{s}$$

5.11 THE FIRST LAW OF THERMODYNAMICS FOR A CONTROL VOLUME

We have already considered the first law of thermodynamics for a control mass, which consists of a fixed quantity of mass, and noted, Eq. 5.5, that it may be written

$$_1Q_2 = E_2 - E_1 + {_1}W_2$$

We have also noted that this may be written as an average rate equation over the time interval δt by dividing by δt:

$$\frac{\delta Q}{\delta t} = \frac{E_2 - E_1}{\delta t} + \frac{\delta W}{\delta t} \tag{5.38}$$

To write the first law as a rate equation for a control volume, we proceed in a manner analogous to that used in developing a rate equation for the law of the conservation of mass. A control mass and control volume are shown in Fig. 5.14. The system consists of all the mass initially in the control volume plus the mass δm_i.

Consider the changes that take place in the control mass and control volume during the time interval δt. During this time δt, the mass δm_i enters the control volume through the discrete area $\mathcal{A}_i$, and the mass δm_e leaves the control volume through the area $\mathcal{A}_e$.

In our analysis we assume that the increment of mass, δm_i, has uniform properties, and similarly that δm_e has uniform properties. The total work done by the control mass during this process, δW, is that associated with the masses δm_i and δm_e crossing the control surface, usually referred to as flow work, and the work $\delta W_{c.v.}$, which includes all other forms of work, such as work associated with a rotating shaft that crosses the control mass boundary, shear forces, electrical, magnetic, or surface effects, or expansion or contraction of the control volume. An amount of heat δQ crosses the boundary during δt.

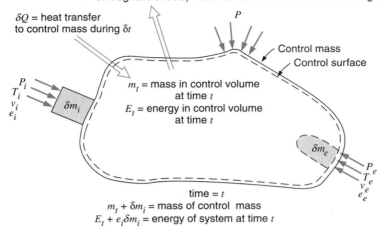

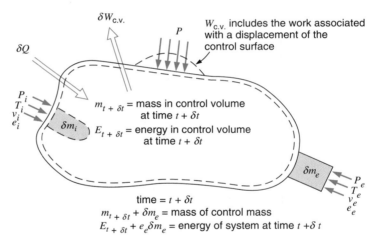

FIGURE 5.14 Schematic diagram for a first law analysis of control volume, showing heat and work as well as mass crossing the control surface.

Let us now consider each of the terms of the first law as it is written for the control mass and transform each term into an equivalent form that applies to the control volume. Consider first the term $E_2 - E_1$.

Let

$$E_t = \text{the energy in the control volume at time } t$$

$$E_{t+\delta t} = \text{the energy in the control volume at time } t + \delta t$$

Then

$$E_1 = E_t + e_i \delta m_i = \text{the energy of the control mass at time } t$$

$$E_2 = E_{t+\delta t} + e_e \delta m_e = \text{the energy of the control mass at time } t + \delta t \qquad (5.39)$$

Therefore,

$$E_2 - E_1 = E_{t+\delta t} + e_e \delta m_e - E_t - e_i \delta m_i$$

$$= \left(E_{t+\delta t} - E_t \right) + \left(e_e \delta m_e - e_i \delta m_i \right)$$

The term $(e_e \delta m_e - e_i \delta m_i)$ represents the net flow of energy that crosses the control surface during δt as the result of the masses δm_e and δm_i crossing the control surface.

Let us consider in more detail the work associated with the masses δm_i and δm_e crossing the control surface. Work is done by the normal force (normal to area $\mathcal{A}$) acting on δm_i and δm_e as these masses cross the control surface. This normal force is equal to the product of the normal tensile stress, $-\sigma_n$, and the area, $\mathcal{A}$. The work done is

$$-\sigma_n \mathcal{A}\, dl = -\sigma_n \delta V = -\sigma_n v \delta m \tag{5.40}$$

A complete analysis of the nature of the normal stress, σ_n, for an actual fluid involves both static pressure and viscous effects, and is beyond the scope of this book. We will assume in this book that the normal stress, σ_n, at a point is equal to the static pressure at this point, P, and simply note that in many applications this is a very reasonable assumption and yields results of good accuracy.

With this assumption the work done on mass δm_i as it enters the control volume is $P_i v_i \delta m_i$, and the work done by mass δm_e as it leaves the control volume is $P_e v_e \delta m_e$. We refer to these terms as the flow work. Various other terms are encountered in the literature such as flow energy and work of introduction and work of expulsion.

Thus the total work done by the system during time δt is

$$\delta W = \delta W_{\text{c.v.}} + \left(P_e v_e \delta m_e - P_i v_i \delta m_i \right) \tag{5.41}$$

Let us now divide Eqs. 5.39 and 5.41 by δt and substitute into the first law, Eq. 5.38. Combining terms and rearranging,

$$\frac{\delta Q}{\delta t} + \frac{\delta m_i}{\delta t}\left(e_i + P_i v_i \right) = \left(\frac{E_{t+\delta t} - E_t}{\delta t} \right) + \frac{\delta m_e}{\delta t}\left(e_e + P_e v_e \right) + \frac{\delta W_{\text{c.v.}}}{\delta t} \tag{5.42}$$

We recognize that each of the flow terms in this expression can be rewritten in the form

$$e + Pv = u + Pv + \frac{\mathbf{V}^2}{2} + gZ = h + \frac{\mathbf{V}^2}{2} + gZ \tag{5.43}$$

using the definition of the thermodynamic property enthalpy, Eq. 5.13. It should be apparent that the appearance of the combination $u + Pv$ whenever mass flows across a control surface is the principal reason for defining the property enthalpy. Its introduction earlier in conjunction with the constant-pressure process was to facilitate use of the tables of thermodynamic properties at that time.

Utilizing Eq. 5.43 for the masses entering and leaving the control volume, Eq. 5.42 becomes

$$\frac{\delta Q}{\delta t} + \frac{\delta m_i}{\delta t}\left(h_i + \frac{\mathbf{V}_i^2}{2} + gZ_i \right) = \left(\frac{E_{t+\delta t} - E_t}{\delta t} \right) + \frac{\delta m_e}{\delta t}\left(h_e + \frac{\mathbf{V}_e^2}{2} + gZ_e \right) + \frac{\delta W_{\text{c.v.}}}{\delta t}$$

$$\tag{5.44}$$

To reduce this expression to a rate equation, we consider what happens to each of the terms in Eq. 5.44 as δt is made to approach zero. The heat transfer and work terms become rate transfer quantities, as was the case in Section 5.8. Similarly, the two mass quantities become mass flow rates, as in Section 5.10, and the energy term becomes the time rate of change of energy in the control volume, in a manner analogous to the mass term in the conservation of mass. In addition, we originally assumed uniform properties for the incremental mass δm_i entering the control volume across area $\mathscr{A}_i$, and made a similar assumption for δm_e leaving across area $\mathscr{A}_e$. Consequently, in taking the limits above, this reduces to the restriction of uniform properties across area $\mathscr{A}_i$ and also across $\mathscr{A}_e$ at an instant of time. These may, of course, be time dependent.

In utilizing the limiting values to express the rate equation of the first law for a control volume, we again include summation signs on the flow terms to account for the possibility of additional flow streams entering or leaving the control volume. Therefore, the result is

$$\dot{Q}_{c.v.} + \sum \dot{m}_i \left(h_i + \frac{\mathbf{V}_i^2}{2} + gZ_i \right) = \frac{dE_{c.v.}}{dt} + \sum \dot{m}_e \left(h_e + \frac{\mathbf{V}_e^2}{2} + gZ_e \right) + \dot{W}_{c.v.} \qquad (5.45)$$

which is, for our purposes, the general expression of the first law of thermodynamics. In words, this equation says that the rate of heat transfer into the control volume plus rate of energy flowing in as a result of mass transfer is equal to the rate of change of energy inside the control volume plus rate of energy flowing out as a result of mass transfer plus the power output associated with shaft, shear, and electrical effects and other factors that have already been mentioned.

Equation 5.45 can be integrated over the total time of a process to give the total energy changes that occur during that period. However, to do so requires knowledge of the time dependency of the various mass flow rates and the states at which mass enters and leaves the control volume. An example of this type of process will be considered in Section 5.14.

Another point to be noted is that if there is no mass flow into or out of the control volume, those terms simply drop out of Eq. 5.45, which then reduces to the rate equation form of the first law for a control mass:

$$\dot{Q} = \frac{dE}{dt} + \dot{W}$$

discussed in Section 5.8.

Since the control volume approach is more general, and reduces to the usual statement of the first law for a control mass when there is no mass flow across the control surface, we will, when making a general statement of the first law, utilize Eq. 5.45, the rate form of the first law for a control volume.

The term flux should also be introduced at this point. Flux is defined as a rate of flow of any quantity per unit area across a control surface. Thus the mass flux is the rate of mass flow per unit area and heat flux is the heat flow rate per unit area across the control surface.

Finally, it should be pointed out that in fluid mechanics and heat transfer the control volume expression for the first law is commonly written in terms of local properties, as was the case with the conservation of mass. That form will not be introduced in this text, and for our purposes we will consider Eq. 5.45 as the general expression of the first law for a control volume analysis.

5.12 THE STEADY-STATE, STEADY-FLOW PROCESS

Our first application of the control volume equations will be to develop a suitable analytical model for the long-term steady operation of devices such as turbines, compressors, nozzles, boilers, condensers—a very large class of problems of interest in thermodynamic analysis. This model will not include the short-term transient start-up or shutdown of such devices, but only the steady operating period of time.

Let us consider a certain set of assumptions (beyond those leading to Eqs. 5.36 and 5.45) that lead to a reasonable model for this type of process, which we refer to as the steady-state, steady-flow process. For convenience, we often refer to this process as the SSSF process.

1. The control volume does not move relative to the coordinate frame.
2. The state of the mass at each point in the control volume does not vary with time.
3. As for the mass that flows across the control surface, the mass flux and the state of this mass at each discrete area of flow on the control surface do not vary with time. The rates at which heat and work cross the control surface remain constant.

As an example of a steady-state, steady-flow process consider a centrifugal air compressor that operates with constant mass rate of flow into and out of the compressor, constant properties at each point across the inlet and exit ducts, a constant rate of heat transfer to the surroundings, and a constant power input. At each point in the compressor the properties are constant with time, even though the properties of a given elemental mass of air vary as it flows through the compressor. Usually, such a process is referred to simply as a steady-flow process, since we are concerned primarily with the properties of the fluid entering and leaving the control volume. On the other hand, in the analysis of certain heat transfer problems in which the same assumptions apply, we are primarily interested in the spatial distribution of properties, particularly temperature, and such a process is often referred to as a steady-state process. Since this is an introductory book we use the term steady-state, steady-flow process to emphasize the basic assumptions involved. The student should realize that the terms steady-state process and steady-flow process are both used extensively in the literature.

Let us now consider the significance of each of these assumptions for the SSSF process.

1. The assumption that the control volume does not move relative to the coordinate frame means that all velocities measured relative to the coordinate frame are also velocities relative to the control surface, and there is no work associated with the acceleration of the control volume.
2. The assumption that the state of the mass at each point in the control volume does not vary with time requires that

$$\frac{dm_{c.v.}}{dt} = 0$$

and also

$$\frac{dE_{c.v.}}{dt} = 0$$

Therefore, we conclude that for the SSSF process we can write, from Eqs. 5.36 and 5.45,

Continuity equation:

$$\sum \dot{m}_i = \sum \dot{m}_e \qquad (5.46)$$

First law:

$$\dot{Q}_{c.v.} + \sum \dot{m}_i \left(h_i + \frac{\mathbf{V}_i^2}{2} + gZ_i \right) = \sum \dot{m}_e \left(h_e + \frac{\mathbf{V}_e^2}{2} + gZ_e \right) + \dot{W}_{c.v.} \qquad (5.47)$$

3. The assumption that the various mass flows, states, and rates at which heat and work cross the control surface remain constant requires that every quantity in Eqs. 5.46 and 5.47 is steady with time. This means that application of Eqs. 5.46 and 5.47 to the operation of some device is independent of time.

Many of the applications of the SSSF model are such that there is only one flow stream entering and one leaving the control volume. For this type of process, we can write

Continuity equation:

$$\dot{m}_i = \dot{m}_e = \dot{m} \qquad (5.48)$$

First law:

$$\dot{Q}_{c.v.} + \dot{m} \left(h_i + \frac{\mathbf{V}_i^2}{2} + gZ_i \right) = \dot{m} \left(h_e + \frac{\mathbf{V}_e^2}{2} + gZ_e \right) + \dot{W}_{c.v.} \qquad (5.49)$$

Rearranging this equation, we have

$$q + h_i + \frac{\mathbf{V}_i^2}{2} + gZ_i = h_e + \frac{\mathbf{V}_e^2}{2} + gZ_e + w \qquad (5.50)$$

where, by definition,

$$q = \frac{\dot{Q}_{c.v.}}{\dot{m}} \qquad \text{and} \qquad w = \frac{\dot{W}_{c.v.}}{\dot{m}} \qquad (5.51)$$

Note that the units for q and w are kJ/kg. From their definition, q and w can be thought of as the heat transfer and work (other than flow work) per unit mass flowing into and out of the control volume for this particular SSSF process.

The symbols q and w are also used for the heat transfer and work per unit mass of a control mass. However, since it is always evident from the context whether it is a control mass (fixed mass) or control volume (involving a flow of mass) with which we are concerned, the significance of the symbols q and w will also be readily evident in each situation.

The SSSF process is often used in the analysis of reciprocating machines, such as reciprocating compressors or engines. In this case the rate of flow, which may actually be pulsating, is considered to be the average rate of flow for an integral number of cycles. A similar assumption is made regarding the properties of the fluid flowing across the control surface, and the heat transfer and work crossing the control surface. It is also assumed that for an integral number of cycles the reciprocating device undergoes, the energy and mass within the control volume do not change.

A number of examples are now given to illustrate the analysis of SSSF processes.

EXAMPLE 5.10   The mass rate of flow into a steam turbine is 1.5 kg/s, and the heat transfer from the turbine is 8.5 kW. The following data are known for the steam entering and leaving the turbine.

	Inlet Conditions	Exit Conditions
Pressure	2.0 MPa	0.1 MPa
Temperature	350°C	
Quality		100%
Velocity	50 m/s	200 m/s
Elevation above reference plane	6 m	3 m
g = 9.8066 m/s²		

Determine the power output of the turbine.

Control volume: Turbine (Fig. 5.15).

Inlet state: Fixed (above).

Exit state: Fixed (above).

Process: SSSF.

Model: Steam tables.

Analysis:

First law (Eq. 5.49):

$$\dot{Q}_{c.v.} + \dot{m}\left(h_i + \frac{\mathbf{V}_i^2}{2} + gZ_i \right) = \dot{m}\left(h_e + \frac{\mathbf{V}_e^2}{2} + gZ_e \right) + \dot{W}_{c.v.}$$

with

$$\dot{Q}_{c.v.} = -8.5 \text{ kW}$$

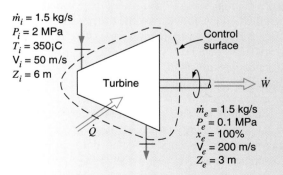

$\dot{m}_i = 1.5$ kg/s
$P_i = 2$ MPa
$T_i = 350$¡C
$V_i = 50$ m/s
$Z_i = 6$ m

Control surface

Turbine

$\dot{W}$

$\dot{Q}$

$\dot{m}_e = 1.5$ kg/s
$P_e = 0.1$ MPa
$x_e = 100\%$
$V_e = 200$ m/s
$Z_e = 3$ m

FIGURE 5.15 Illustration for Example 5.10.

Solution

$h_i = 3137.0$ kJ/kg (from the steam tables)

$$\frac{\mathbf{V}_i^2}{2} = \frac{50 \times 50}{2 \times 1000} = 1.25 \text{ kJ/kg}$$

$$gZ_i = \frac{6 \times 9.8066}{1000} = 0.059 \text{ kJ/kg}$$

Similarly, $h_e = 2675.5$ kJ/kg

$$\frac{\mathbf{V}_e^2}{2} = \frac{200 \times 200}{2 \times 1000} = 20.0 \text{ kJ/kg}$$

$$gZ_e = \frac{3 \times 9.8066}{1000} = 0.029 \text{ kJ/kg}$$

Therefore, substituting into Eq. 5.49,

$$-8.5 + 1.5(3137 + 1.25 + 0.059) = 1.5(2675.5 + 20.0 + 0.029) + \dot{W}_{c.v.}$$

$$\dot{W}_{c.v.} = -8.5 + 4707.5 - 4043.3 = 655.7 \text{ kW}$$

If Eq. 5.50 is used, the work per kilogram of fluid flowing is found first.

$$q + h_i + \frac{\mathbf{V}_i^2}{2} + gZ_i = h_e + \frac{\mathbf{V}_e^2}{2} + gZ_e + w$$

$$q = \frac{-8.5}{1.5} = -5.667 \text{ kJ/kg}$$

Therefore, substituting into Eq. 5.50,

$$-5.667 + 3137 + 1.25 + 0.059 = 2675.5 + 20.0 + 0.029 + w$$

$$w = 437.11 \text{ kJ/kg}$$

$$\dot{W}_{c.v.} = 1.5 \text{ kg/s} \times 437.11 \text{ kJ/kg} = 655.7 \text{ kW}$$

Two further observations can be made by referring to this example. First, in many engineering problems, potential energy changes are insignificant when compared with the other energy quantities. In the above example the potential energy change did not affect any of the significant figures. In most problems where the change in elevation is small the potential energy terms may be neglected.

Second, if velocities are small, say, under 20 m/s, in many cases the kinetic energy is insignificant compared with other energy quantities. Furthermore, when the velocities entering and leaving the system are essentially the same, the change in kinetic energy is small. Since it is the change in kinetic energy that is important in the SSSF energy equation, the kinetic energy terms can usually be neglected when there is no significant difference between the velocity of the fluid entering and leaving the control volume. Thus, in many thermodynamic problems, one must make judgments as to which quantities may be negligible for a given analysis.

EXAMPLE 5.11

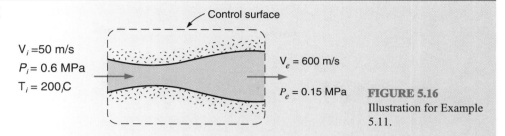

FIGURE 5.16
Illustration for Example 5.11.

Steam at 0.6 MPa, 200°C enters an insulated nozzle with a velocity of 50 m/s. It leaves at a pressure of 0.15 MPa and a velocity of 600 m/s. Determine the final temperature if the steam is superheated in the final state, and the quality if it is saturated.

Control volume: Nozzle.

Inlet state: Fixed (see Fig. 5.16).

Exit state: P_e known.

Process: SSSF.

Model: Steam tables.

Analysis:

$$\dot{Q}_{c.v.} \approx 0 \ (\text{nozzle insulated})$$

$$\dot{W}_{c.v.} = 0 \qquad PE_i \approx PE_e$$

First law (Eq. 5.50):

$$h_i + \frac{\mathbf{V}_i^2}{2} = h_e + \frac{\mathbf{V}_e^2}{2}$$

Solution

$$h_e = 2850.1 + \frac{(50)^2}{2 \times 1000} - \frac{(600)^2}{2 \times 1000} = 2671.4 \ \text{kJ/kg}$$

The two properties of the fluid leaving that we now know are pressure and enthalpy, and therefore the state of this fluid is determined. Since h_e is less than h_g at 0.15 MPa, the quality is calculated.

$$h = h_f + x h_{fg}$$

$$2671.4 = 467.1 + x_e 2226.5$$

$$x_e = 0.99$$

EXAMPLE 5.11E   Steam at 100 lbf/in.², 400 F, enters an insulated nozzle with a velocity of 200 ft/s. It leaves at a pressure of 20 lbf/in.² and a velocity of 2000 ft/s. Determine the final temperature if the steam is superheated in the final state, and the quality if it is saturated.

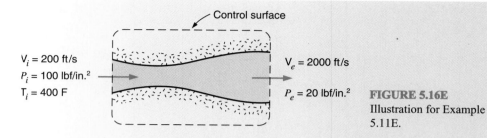

FIGURE 5.16E
Illustration for Example 5.11E.

Control volume: Nozzle.

Inlet state: Fixed (see Fig. 5.16E).

Exit state: P_e known.

Process: SSSF.

Model: Steam tables.

Analysis:

$$\dot{Q}_{c.v.} \approx 0 \ (\text{nozzle insulated})$$

$$\dot{W}_{c.v.} = 0 \qquad \text{PE}_i \approx \text{PE}_e$$

First law (Eq. 5.50):

$$h_i + \frac{\mathbf{V}_i^2}{2} = h_e + \frac{\mathbf{V}_e^2}{2}$$

Solution

$$h_e = 1227.5 + \frac{(200)^2}{2 \times 32.17 \times 778} - \frac{(2000)^2}{2 \times 32.17 \times 778} = 1148.3 \ \text{Btu} / \text{lbm}$$

The two properties of the fluid leaving that we now know are pressure and enthalpy, and therefore the state of this fluid is determined. Since h_e is less than h_g at 20 lbf/in.², the quality is calculated.

$$h = h_f + x h_{fg}$$

$$1148.3 = 196.26 + x_e \, 960.1$$

$$x_e = 0.992$$

EXAMPLE 5.12   In a refrigeration system, in which R-134a is the refrigerant, the R-134a enters the compressor at 200 kPa, −10°C and leaves at 1.0 MPa, 70°C. The mass rate of flow is 0.015 kg/s, and the power input to the compressor is 1 kW.

On leaving the compressor the refrigerant enters a water-cooled condenser at 1.0

MPa, 60°C and leaves as a liquid at 0.95 MPa, 35°C. Water enters the condenser at 10°C and leaves at 20°C. Determine:

1. The heat transfer rate from the compressor.

2. The rate at which cooling water flows through the condenser.

First control volume: Compressor.

Inlet state: P_i, T_i; state fixed.

Exit state: P_e, T_e; state fixed.

Process: SSSF.

Model: R-134a tables.

Analysis:

Since the velocity of the vapor entering a refrigeration compressor is low and not greatly different from the velocity leaving, kinetic energy changes, as well as potential energy changes, can be neglected. Therefore, the first law, Eq. 5.50, is

$$q + h_i = h_e + w$$

Solution

$$w = \frac{-1}{0.015} = 66.67 \text{ kJ / kg}$$

From the R-134a tables,

$$h_i = 392.34 \text{ kJ / kg} \qquad h_e = 452.35 \text{ kJ / kg}$$

Therefore,

$$q = 452.35 - 66.67 - 392.34 = -6.66 \text{ kJ / kg}$$

$$\dot{Q}_{c.v.} = 0.015 \text{ kg / s} \times (-6.66 \text{ kJ / kg}) = -0.10 \text{ kW}$$

Second control volume: Condenser.

Sketch: Fig. 5.17

Inlet states: R-134a—fixed; water—fixed.

Exit states: R-134a—fixed; water—fixed.

Process: SSSF.

Model: R-134a tables; steam tables.

Analysis:

With this control volume we have two fluid streams, the R-134a and the water, entering and leaving the control volume. It is reasonable to assume that the kinetic and potential energy changes are negligible. We note that the work is zero, and we make the

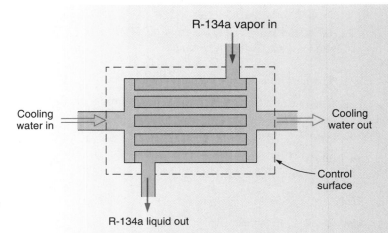

FIGURE 5.17
Schematic diagram of a R-134a condenser.

other reasonable assumption that there is no heat transfer across the control surface. Therefore, the first law, Eq. 5.47, reduces to

$$\sum \dot{m}_i h_i = \sum \dot{m}_e h_e$$

Using the subscript r for refrigerant and w for water,

$$\dot{m}_r \left(h_i \right)_r + \dot{m}_w \left(h_i \right)_w = \dot{m}_r \left(h_e \right)_r + \dot{m}_w \left(h_e \right)_w$$

Solution

From the R-134a and steam tables,

$$\left(h_i \right)_r = 441.89 \text{ kJ / kg} \qquad \left(h_i \right)_w = 42.00 \text{ kJ / kg}$$

$$\left(h_e \right)_r = 249.10 \text{ kJ / kg} \qquad \left(h_e \right)_w = 83.95 \text{ kJ / kg}$$

Solving the above equation for $\dot{m}_w$, the rate of flow of water,

$$\dot{m}_w = \dot{m}_r \frac{\left(h_i - h_e \right)_r}{\left(h_e - h_i \right)_w} = 0.015 \text{ kg / s} \frac{\left(441.89 - 249.10 \right) \text{ kJ / kg}}{\left(83.95 - 42.00 \right) \text{ kJ / kg}} = 0.0689 \text{ kg / s}$$

This problem can also be solved by considering two separate control volumes, one of which has the flow of R-134a across its control surface, and the other having the flow of water across its control surface. Further there is heat transfer from one control volume to the other. This is shown schematically in Fig. 5.18.

The heat transfer for the control volume involving R-134a is calculated first. In this case the SSSF energy equation, Eq. 5.49, reduces to

$$\dot{Q}_{c.v.} = \dot{m}_r \left(h_e - h_i \right)_r$$

$$\dot{Q}_{c.v.} = 0.015 \text{ kg / s} \times \left(249.10 - 441.89 \right) \text{ kJ / kg} = -2.892 \text{ kW}$$

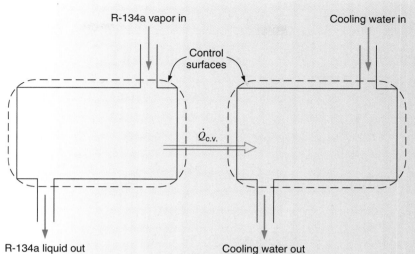

R-134a vapor in   Cooling water in

R-134a liquid out   Cooling water out

FIGURE 5.18 Schematic diagram of R-134a condenser considering two control surfaces.

This is also the heat transfer to the other control volume, for which $\dot{Q}_{c.v.} = +2.892$ kW.

$$\dot{Q}_{c.v.} = \dot{m}_w \left(h_e - h_i \right)_w$$

$$\dot{m}_w = \frac{2.892 \text{ kW}}{(83.95 - 42.00)\text{kJ}/\text{kg}} = 0.0689 \text{ kg}/\text{s}$$

EXAMPLE 5.13   Consider the simple steam power plant, as shown in Fig. 5.19. The following data are for such a power plant.

Location	Pressure	Temperature or Quality
Leaving boiler	2.0 MPa	300°C
Entering turbine	1.9 MPa	290°C
Leaving turbine, entering condenser	15 kPa	90%
Leaving condenser, entering pump	14 kPa	45°C
Pump work = 4 kJ/kg		

Determine the following quantities per kilogram flowing through the unit.

1. Heat transfer in line between boiler and turbine.

2. Turbine work.

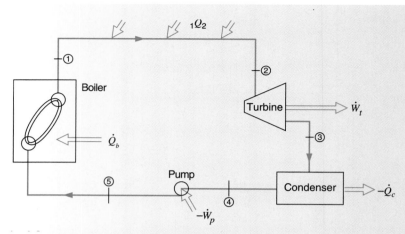

FIGURE 5.19 Simple steam power plant.

3. Heat transfer in condenser.

4. Heat transfer in boiler.

There is a certain advantage in assigning a number to various points in the cycle. For this reason the subscripts i and e in the steady-state, steady-flow energy equation are often replaced by appropriate numbers.

Since there are several control volumes to be considered in the solution to this problem, let us consolidate our solution procedure somewhat in this example. Using the notation of Fig. 5.19:

All processes: SSSF.

Model: Steam tables.

From the steam tables:

$$h_1 = 3023.5 \text{ kJ / kg}$$

$$h_2 = 3002.5 \text{ kJ / kg}$$

$$h_3 = 226.0 + 0.9(2373.1) = 2361.8 \text{ kJ / kg}$$

$$h_4 = 188.5 \text{ kJ / kg}$$

All analyses: No changes in kinetic or potential energy will be considered in the solution. In each case, the first law is given by Eq. 5.50.

Now, we proceed to answer the specific questions raised in the problem statement.

1. *Control volume* Pipe line between the boiler and the turbine.

First law and solution:

$$_1q_2 + h_1 = h_2$$

$$_1q_2 = h_2 - h_1 = 3002.5 - 3023.5 = -21.0 \text{ kJ / kg}$$

2. *Control volume*   Turbine.

First law and solution: A turbine is essentially an adiabatic machine. Therefore, it is reasonable to neglect heat transfer, so that

$$h_2 = h_3 + {}_2w_3$$

$${}_2w_3 = 3002.5 - 2361.8 = 640.7 \text{ kJ} / \text{kg}$$

3. *Control volume*   Condenser.

First law and solution: There is no work for this control volume. Therefore,

$${}_3q_4 + h_3 = h_4$$

$${}_3q_4 = 188.5 - 2361.8 = -2173.3 \text{ kJ} / \text{kg}$$

4. *Control volume*   Boiler.

First law: The work is equal to zero, so that

$${}_5q_1 + h_5 = h_1$$

A solution requires a value for h_5, which can be found by taking a control volume around the pump:

$$h_4 = h_5 + {}_4w_5$$

$$h_5 = 188.5 - (-4) = 192.5 \text{ kJ} / \text{kg}$$

Therefore, for the boiler,

$${}_5q_1 + h_5 = h_1$$

$${}_5q_1 = 3023.5 - 192.5 = 2831 \text{ kJ} / \text{kg}$$

EXAMPLE 5.14   The centrifugal air compressor of a gas turbine receives air from the ambient atmosphere where the pressure is 1 bar and the temperature is 300 K. At the discharge of the compressor the pressure is 4 bar, the temperature is 480 K, and the velocity is 100 m/s. The mass rate of flow into the compressor is 15 kg/s. Determine the power required to drive the compressor.

Control volume: We consider a control volume around the compressor, and locate the control volume at some distance from the compressor so that the air crossing the control surface has a very low velocity and is essentially at ambient conditions. If we located our control volume directly across the inlet section, it would be necessary to know the temperature and velocity at the compressor inlet.

Sketch: Fig. 5.20.

Inlet and exit states: Both states fixed.

Process: SSSF.

Model: Ideal gas with constant specific heat, value from Table A.10 (300 K).

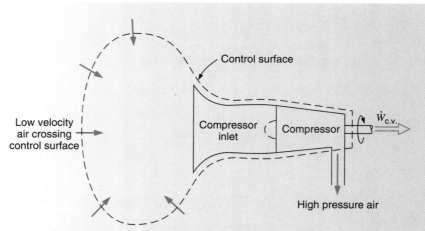

FIGURE 5.20 Sketch for Example 5.14.

Analysis:

A large, high-flow rate compressor such as this is essentially an adiabatic machine, inasmuch as the air passes through the compressor very quickly and there is no particular effort made to promote any heat transfer. Therefore, we assume the process to be adiabatic. We also neglect any change in potential energy as well as the inlet kinetic energy. The first law, Eq. 5.50, reduces to

$$h_i = h_e + \frac{\mathbf{V}_e^2}{2} + w$$

Solution

$$-w = h_e - h_i + \frac{\mathbf{V}_e^2}{2} = C_{po}\left(T_e - T_i\right) + \frac{\mathbf{V}_e^2}{2}$$

$$= 1.0035(480 - 300) + \frac{100 \times 100}{2 \times 1000}$$

$$= 180.6 + 5.0 = 185.6 \text{ kJ} / \text{kg}$$

$$-\dot{W}_{c.v.} = 15 \times 185.6 = 2784 \text{ kW}$$

A more accurate model for the behavior of the air in this process would be the ideal gas and air tables, A.12. For this model, the solution is

$$h_i = 300.19 \text{ kJ} / \text{kg} \qquad h_e = 482.48 \text{ kJ} / \text{kg}$$

$$-w = h_e - h_i + \frac{\mathbf{V}_e^2}{2} = 482.48 - 300.19 + \frac{100 \times 100}{2 \times 1000}$$

$$= 182.3 + 5.0 = 187.3 \text{ kJ} / \text{kg}$$

$$-\dot{W}_{c.v.} = 15 \times 187.3 = 2810 \text{ kW}$$

EXAMPLE 5.14E   The centrifugal air compressor of a gas turbine receives air from the ambient atmosphere where the pressure is 14.5 lbf/in.^{2} and the temperature is 80F. At the discharge of the compressor the pressure is 54 lbf/in.^{2}, the temperature is 400F, and the velocity is 300 ft/s. The mass rate of flow into the compressor is 200 lbm/min. Determine the power required to drive the compressor.

Control volume: We consider a control volume around the compressor, and locate the control volume at some distance from the compressor so that the air crossing the control surface has a very low velocity and is essentially at ambient conditions. If we located our control volume directly across the inlet section, it would be necessary to know the temperature and velocity at the compressor inlet.

Sketch: Fig. 5.20.

Inlet and exit states: Both states fixed.

Process: SSSF.

Model: Ideal gas with constant specific heat, value from Table A.10 (540 R).

Analysis:

A large, high-flow rate compressor such as this is essentially an adiabatic machine, inasmuch as the air passes through the compressor very quickly and there is no particular effort made to promote any heat transfer. Therefore, we assume the process to be adiabatic. We also neglect any change in potential energy as well as the inlet kinetic energy. The first law, Eq. 5.50, reduces to

$$h_i = h_e + \frac{\mathbf{V}_e^2}{2} + w$$

Solution

$$-w = h_e - h_i + \frac{\mathbf{V}_e^2}{2} = C_{po}\left(T_e - T_i\right) + \frac{\mathbf{V}_e^2}{2}$$

$$= 0.24\left(400 - 80\right) + \frac{\left(300\right)^2}{2 \times 32.17 \times 778}$$

$$= 76.8 + 1.8 = 78.6 \text{ Btu / lbm}$$

$$-\dot{W}_{c.v.} = \frac{78.6\left(2000\right)}{42.4} = 3708 \text{ hp}$$

A more accurate model for the behavior of the air in this process would be the ideal gas and air tables, A.12. For this model, the solution is

$$h_i = 129.06 \text{ Btu / lbm} \qquad h_e = 206.46 \text{ Btu / lbm}$$

$$-w = h_e - h_i + \frac{\mathbf{V}_e^2}{2}$$

$$= 206.46 - 129.06 + \frac{\left(300\right)^2}{2 \times 32.17 \times 778}$$

$$-\dot{W}_{c.v.} = \frac{79.2\left(2000\right)}{42.4} = 3736 \text{ hp}$$

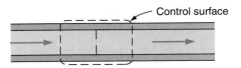

FIGURE 5.21 The throttling process.

5.13 THE JOULE–THOMSON COEFFICIENT AND THE THROTTLING PROCESS

The Joule–Thomson coefficient, μ_J, is defined by the relation

$$\mu_J \equiv \left(\frac{\partial T}{\partial P}\right)_h \tag{5.52}$$

As in the definition of specific heats in Section 5.6, this quantity is defined in terms of thermodynamic properties, and therefore is itself a property of a substance.

The significance of the Joule–Thomson coefficient may be demonstrated by considering a throttling process that is a steady-state, steady-flow process across a restriction, with a resulting drop in pressure. A typical example is the flow through a partially opened valve or a restriction in the line. In most cases this occurs so rapidly and in such a small space that there is neither sufficient time nor a large enough area for much heat transfer. Therefore, we usually may assume such processes to be adiabatic.

If we consider the control surface shown in Fig. 5.21 we can write the SSSF energy equation for this process. There is no work, no change in potential energy, and we make the reasonable assumption that there is no heat transfer. The SSSF energy equation, Eq. 5.50, reduces to

$$h_i + \frac{\mathbf{V}_i^2}{2} = h_e + \frac{\mathbf{V}_e^2}{2}$$

If the fluid is a gas, the specific volume always increases in such a process and, therefore, if the pipe is of constant diameter, the kinetic energy of the fluid increases. In many cases, however, this increase in kinetic energy is small (or perhaps the diameter of the exit pipe is larger than that of the inlet pipe) and we can say with a high degree of accuracy that in this process the final and initial enthalpies are equal. Unless otherwise specified, we will assume this to be the case for throttling processes throughout the remainder of this text.

It is for such a process that the Joule–Thomson coefficient, μ_J, is significant. A positive Joule–Thomson coefficient means that the temperature drops during throttling, and when the Joule–Thomson coefficient is negative the temperature rises during throttling.

EXAMPLE 5.15   Steam at 800 kPa, 300°C is throttled to 200 kPa. Changes in kinetic energy are negligible for this process. Determine the final temperature of the steam, and the average Joule–Thomson coefficient.

Control volume: Throttle valve (or other restriction).

Inlet state: P_i, T_i known; state fixed.

Exit state: P_e known.

Process: SSSF.

Model: Steam tables.

Analysis:

First law, Eq. 5.50. Work equals zero. Neglect heat transfer and changes in kinetic and potential energies. Therefore, the first law reduces to

$$h_i = h_e$$

Solution

Since

$$h_e = h_i = 3056.5 \text{ kJ} / \text{kg}$$

and

$$P_e = 200 \text{ kPa}$$

these two properties determine the final state. From the superheat table for steam,

$$T_e = 292.4°\text{C}$$

$$\mu_{J(av)} = \left(\frac{\Delta T}{\Delta P} \right)_h = \frac{-7.6}{-600} = 0.0127 \text{ K} / \text{kPa}$$

Frequently, a throttling process involves a change in the phase of the fluid. A typical example is the flow through the expansion valve of a vapor-compression refrigeration system. The following example deals with this problem.

EXAMPLE 5.16 Consider the throttling process across the expansion valve or through the capillary tube in a vapor-compression refrigeration cycle. In this process the pressure of the refrigerant drops from the high pressure in the condenser to the low pressure in the evaporator, and during this process some of the liquid flashes into vapor. If we consider this process to be adiabatic, the quality of the refrigerant entering the evaporator can be calculated.

Consider the following process, in which ammonia is the refrigerant. The ammonia enters the expansion valve at a pressure of 1.50 MPa and a temperature of 32°C. Its pressure on leaving the expansion valve is 268 kPa. Calculate the quality of the ammonia leaving the expansion valve.

Control volume: Expansion valve or capillary tube.

Inlet state: P_i, T_i known; state fixed.

Exit state: P_e known.

Process: SSSF.

Model: Ammonia tables.

Analysis:

Standard throttling process analysis and assumptions, as in Example 5.15. The first law reduces to

$$h_i = h_e$$

Solution

From the ammonia tables

$$h_i = 332.6 \text{ kJ / kg}$$

(The enthalpy of a slightly compressed liquid is essentially equal to the enthalpy of saturated liquid at the same temperature.)

$$h_e = h_i = 332.6 = 126.0 + x_e \left(1303.5\right)$$

$$x_e = 0.1585 = 15.85\%$$

5.14 THE UNIFORM-STATE, UNIFORM-FLOW PROCESS

In Section 5.12 we considered the steady-state, steady-flow process and several examples of its application. Many processes of interest in thermodynamics involve unsteady flow and do not fit into this category. A certain group of these—for example, filling closed tanks with a gas or liquid, or discharge from closed vessels—can be reasonably represented to a first approximation by another simplified model. We call this process the uniform-state, uniform-flow process, or for convenience, the USUF process. The basic assumptions are as follows:

1. The control volume remains constant relative to the coordinate frame.

2. The state of the mass within the control volume may change with time, but at any instant of time the state is uniform throughout the entire control volume (or over several identifiable regions that make up the entire control volume).

3. The state of the mass crossing each of the areas of flow on the control surface is constant with time although the mass flow rates may be time varying.

Let us examine the consequence of these assumptions and derive an expression for the first law that applies to this process. The assumption that the control volume remains stationary relative to the coordinate frame has already been discussed in Section 5.12. The remaining assumptions lead to the following simplifications for the continuity equation and the first law.

The overall process occurs during time t. At any instant of time during the process, the continuity equation is

$$\frac{dm_{c.v.}}{dt} + \sum \dot{m}_e - \sum \dot{m}_i = 0$$

where the summation is over all areas on the control surface through which flow occurs. Integrating over time t gives the change of mass in the control volume during the overall process.

$$\int_0^t \left(\frac{dm_{c.v.}}{dt}\right) dt = \left(m_2 - m_1\right)_{c.v.}$$

The total mass leaving the control volume during time t is

$$\int_0^t \left(\Sigma \dot{m}_e\right) dt = \Sigma m_e$$

and the total mass entering the control volume during time t is

$$\int_0^t \left(\Sigma \dot{m}_i\right) dt = \Sigma m_i$$

Therefore, for this period of time t, we can write the continuity equation for the USUF process as

$$\left(m_2 - m_1\right)_{c.v.} + \Sigma m_e - \Sigma m_i = 0 \qquad (5.53)$$

In writing the first law for the USUF process we consider Eq. 5.45, which applies at any instant of time during the process.

$$\dot{Q}_{c.v.} + \Sigma \dot{m}_i \left(h_i + \frac{\mathbf{V}_i^2}{2} + gZ_i \right) = \frac{dE_{c.v.}}{dt} + \Sigma \dot{m}_e \left(h_e + \frac{\mathbf{V}_e^2}{2} + gZ_e \right) + \dot{W}_{c.v.}$$

Since at any instant of time the state within the control volume is uniform, the first law for the USUF process becomes

$$\dot{Q}_{c.v.} + \Sigma \dot{m}_i \left(h_i + \frac{\mathbf{V}_i^2}{2} + gZ_i \right) = \Sigma \dot{m}_e \left(h_e + \frac{\mathbf{V}_e^2}{2} + gZ_e \right)$$

$$+ \frac{d}{dt} \left[m \left(u + \frac{\mathbf{V}^2}{2} + gZ \right) \right]_{c.v.} + \dot{W}_{c.v.}$$

Let us now integrate this equation over time t, during which time we have

$$\int_0^t \dot{Q}_{c.v.} dt = Q_{c.v.}$$

$$\int_0^t \left[\Sigma \dot{m}_i \left(h_i + \frac{\mathbf{V}_i^2}{2} + gZ_i \right) \right] dt = \Sigma m_i \left(h_i + \frac{\mathbf{V}_i^2}{2} + gZ_i \right)$$

$$\int_0^t \left[\Sigma \dot{m}_e \left(h_e + \frac{\mathbf{V}_e^2}{2} + gZ_e \right) \right] dt = \Sigma m_e \left(h_e + \frac{\mathbf{V}_e^2}{2} + gZ_e \right)$$

$$\int_0^t \dot{W}_{c.v.} dt = W_{c.v.}$$

$$\int_0^t \frac{d}{dt} \left[m \left(u + \frac{\mathbf{V}^2}{2} + gZ \right) \right]_{c.v.} dt = \left[m_2 \left(u_2 + \frac{\mathbf{V}_2^2}{2} + gZ_2 \right) - m_1 \left(u_1 + \frac{\mathbf{V}_1^2}{2} + gZ_1 \right) \right]_{c.v.}$$

Therefore, for this period of time t, we can write the first law for the uniform-state, uniform-flow process as

$$Q_{\text{c.v.}} + \sum m_i \left(h_i + \frac{\mathbf{V}_i^2}{2} + gZ_i \right)$$

$$= \sum m_e \left(h_e + \frac{\mathbf{V}_e^2}{2} + gZ_e \right)$$

$$+ \left[m_2 \left(u_2 + \frac{\mathbf{V}_2^2}{2} + gZ_2 \right) - m_1 \left(u_1 + \frac{\mathbf{V}_1^2}{2} + gZ_1 \right) \right]_{\text{c.v.}} + W_{\text{c.v.}} \qquad (5.54)$$

As an example of the type of problem for which these assumptions are valid and Eq. 5.54 is appropriate, let us consider the classic problem of flow into an evacuated vessel. This is the subject of Example 5.17.

EXAMPLE 5.17   Steam at a pressure of 1.4 MPa, 300°C is flowing in a pipe, Fig. 5.22. Connected to this pipe through a valve is an evacuated tank. The valve is opened and the tank fills with steam until the pressure is 1.4 MPa, and then the valve is closed. The process takes place adiabatically and kinetic energies and potential energies are negligible. Determine the final temperature of the steam.

Control volume: Tank, as shown in Fig. 5.22.

Initial state (in tank):   Evacuated, mass $m_1 = 0$.

Final state: P_2 known.

Inlet state: P_i, T_i (in line) known.

Process: USUF.

Model: Steam tables.

Analysis:

First law, Eq. 5.54:

$$Q_{\text{c.v.}} + \sum m_i \left(h_i + \frac{\mathbf{V}_i^2}{2} + gZ_i \right)$$

$$= \sum m_e \left(h_e + \frac{\mathbf{V}_e^2}{2} + gZ_e \right)$$

$$+ \left[m_2 \left(u_2 + \frac{\mathbf{V}_2^2}{2} + gZ_2 \right) - m_1 \left(u_1 + \frac{\mathbf{V}_1^2}{2} + gZ_1 \right) \right]_{\text{c.v.}} + W_{\text{c.v.}}$$

We note that $Q_{\text{c.v.}} = 0$, $W_{\text{c.v.}} = 0$, $m_e = 0$, and $(m_1)_{\text{c.v.}} = 0$. We further assume that changes in kinetic and potential energy are negligible. Therefore, the statement of the first law for this process reduces to

$$m_i h_i = m_2 u_2$$

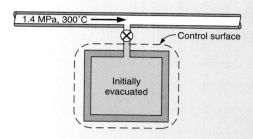

FIGURE 5.22 Flow into an evacuated vessel—control volume analysis.

From the continuity equation for this process, Eq. 5.53, we conclude that

$$m_2 = m_i$$

Therefore, combining the continuity equation with the first law, we have

$$h_i = u_2$$

That is, the final internal energy of the steam in the tank is equal to the enthalpy of the steam entering the tank.

Solution

From the steam tables

$$h_i = u_2 = 3040.4 \text{ kJ} / \text{kg}$$

Since the final pressure is given as 1.4 MPa, we know two properties at the final state and therefore the final state is determined. The temperature corresponding to a pressure of 1.4 MPa and an internal energy of 3040.4 kJ/kg is found to be 452°C. Had this problem involved a substance for which internal energies are not listed in the thermodynamic tables, it would have been necessary to calculate a few values for u before an interpolation could be made for the final temperature.

   This problem can also be solved by considering the steam that enters the tank and the evacuated space as a control mass, as indicated in Fig. 5.23.

   The process is adiabatic, but we must examine the boundaries for work. If we visualize a piston between the steam that is included in the control mass and the steam that flows behind, we readily recognize that the boundaries move and that the steam in the pipe does work on the steam that comprises the control mass. The amount of this work is

$$-W = P_1 V_1 = m P_1 v_1$$

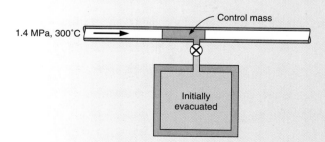

FIGURE 5.23 Flow into an evacuated vessel—control mass.

Writing the first law for the control mass, Eq. 5.11, and noting that kinetic and potential energies can be neglected, we have

$$_1Q_2 = U_2 - U_1 + {_1W_2}$$

$$0 = U_2 - U_1 - P_1V_1$$

$$0 = mu_2 - mu_1 - mP_1v_1 = mu_2 - mh_1$$

Therefore,

$$u_2 = h_1$$

which is the same conclusion that was reached using a control volume analysis.

The two other examples that follow illustrate further the uniform-state, uniform-flow process.

EXAMPLE 5.18 Let the tank of the previous example have a volume of 0.4 m³ and initially contain saturated vapor at 350 kPa. The valve is then opened and steam from the line at 1.4 MPa, 300°C flows into the tank until the pressure is 1.4 MPa.

Calculate the mass of steam that flows into the tank.

Control volume: Tank, as in Fig. 5.22.

Initial state: P_1, saturated vapor; state fixed.

Final state: P_2.

Inlet state: P_i, T_i; state fixed.

Process: USUF.

Model: Steam tables.

Analysis:

Same as in Example 5.17, except that the tank is not evacuated initially. Again we note that $Q_{c.v.} = 0$, $W_{c.v.} = 0$, $m_e = 0$, and we assume that changes in kinetic energy and potential energy are zero. The statement of the first law for this process, Eq. 5.54, reduces to

$$m_i h_i = m_2 u_2 - m_1 u_1$$

The continuity equation, Eq. 5.53, reduces to

$$m_2 - m_1 = m_i$$

Therefore, combining the continuity equation and the first law, we have

$$(m_2 - m_1)h_i = m_2 u_2 - m_1 u_1$$

$$m_2(h_i - u_2) = m_1(h_i - u_1) \tag{a}$$

There are two unknowns in this equation—m_2 and u_2. However, we have one additional equation:

$$m_2 v_2 = V = 0.4\,\text{m}^3 \tag{b}$$

Substituting (b) into (a) and rearranging, we have

$$\frac{V}{v_2}(h_i - u_2) - m_1(h_i - u_1) = 0 \tag{c}$$

in which the only unknowns are v_2 and u_2, both functions of T_2 and P_2. Since T_2 is unknown, it means that there is only one value of T_2 for which Eq. (c) will be satisfied, and we must find it by trial and error.

Solution

$$v_1 = 0.5243 \text{ m}^3 / \text{kg} \qquad m_1 = \frac{0.4}{0.5243} = 0.763 \text{ kg}$$

$$h_i = 3040.4 \text{ kJ} / \text{kg} \qquad u_1 = 2548.9 \text{ kJ} / \text{kg}$$

Assume

$$T_2 = 342°C$$

For this temperature and the known value of P_2,

$$v_2 = 0.1974 \text{ m}^3 / \text{kg} \qquad u_2 = 2855.8 \text{ kJ} / \text{kg}$$

Substituting into (c),

$$\frac{0.4}{0.1974}(3040.4 - 2855.8) - 0.763(3040.4 - 2548.9) \simeq 0$$

and we conclude that the assumed value for $T_2 = 342°C$ is correct. The final mass inside the tank is

$$m_2 = \frac{0.4}{0.1974} = 2.026 \text{ kg}$$

and the mass of steam that flows into the tank is

$$m_i = m_2 - m_1 = 2.026 - 0.763 = 1.263 \text{ kg}$$

EXAMPLE 5.19

A tank of 2 m³ volume contains saturated ammonia at a temperature of 40°C. Initially the tank contains 50% liquid and 50% vapor by volume. Vapor is withdrawn from the top of the tank until the temperature is 10°C. Assuming that only vapor (i.e., no liquid) leaves and that the process is adiabatic, calculate the mass of ammonia that is withdrawn.

Control volume: Tank.

Initial state: T_1, V_{liq}, V_{vap}; state fixed.

Final state: T_2.

Exit state: Saturated vapor (temperature changing).

Process: USUF.

Model: Ammonia tables.

Analysis:

In the first law, Eq. 5.54, we note that $Q_{c.v.} = 0$, $W_{c.v.} = 0$, $m_i = 0$, and we assume that changes in kinetic and potential energy are negligible. However, the enthalpy of saturated vapor varies with temperature, and therefore we cannot simply assume that the enthalpy of the vapor leaving the tank remains constant. However, we note that at 40°C, $h_g = 1472.2$ kJ/kg and at 10°C, $h_g = 1453.3$ kJ/kg. Since the change in h_g during this process is small, we may accurately assume that h_e is the average of the two values given above. Therefore,

$$\left(h_e\right)_{av} = 1462.8 \text{ kJ}/\text{kg}$$

and the first law reduces to

$$m_e h_e + m_2 u_2 - m_1 u_1 = 0$$

and the continuity equation (from Eq. 5.53)

$$\left(m_2 - m_1\right)_{c.v.} + m_e = 0$$

Combining these two equations we have

$$m_2\left(h_e - u_2\right) = m_1 h_e - m_1 u_1$$

Solution

The following values are from the ammonia tables:

$$v_{f1} = 0.001\ 726 \text{ m}^3/\text{kg} \qquad v_{g1} = 0.0833 \text{ m}^3/\text{kg}$$

$$v_{f2} = 0.001\ 601 \qquad v_{fg2} = 0.2040$$

$$u_{f1} = 371.7 - 1554.33 \times 0.001\ 726 = 369.0 \text{ kJ}/\text{kg}$$

$$u_{g1} = 1472.2 - 1554.33 \times 0.0833 = 1342.7$$

$$u_{f2} = 227.6 - 614.95 \times 0.001\ 601 = 226.6$$

$$u_{g2} = 1453.3 - 614.95 \times 0.2056 = 1326.9$$

$$u_{fg2} = 1326.9 - 226.6 = 1100.3$$

Calculating first the initial mass, m_1, in the tank, the mass of the liquid initially present, m_{f1}, is

$$m_{f1} = \frac{1.0}{0.001\ 726} = 579.4 \text{ kg}$$

Similarly, the initial mass of vapor, m_{g1}, is

$$m_{g1} = \frac{1.0}{0.0833} = 12.0 \text{ kg}$$

$$m_1 = m_{f1} + m_{g1} = 579.4 + 12.0 = 591.4 \text{ kg}$$

$$m_1 h_e = 591.4 \times 1462.8 = 865\ 100 \text{ kJ}$$

$$m_1 u_1 = (mu)_{f1} + (mu)_{g1} = 579.4 \times 369.0 + 12.0 \times 1342.7$$
$$= 229\,910 \text{ kJ}$$

Substituting these into the first law,

$$m_2(h_e - u_2) = m_1 h_e - m_1 u_1 = 865\,100 - 229\,910 = 635\,190 \text{ kJ}$$

There are two unknowns, m_2 and u_2, in this equation. However,

$$m_2 = \frac{V}{v_2} = \frac{2.0}{0.001\,601 + x_2(0.2040)}$$

and

$$u_2 = 226.6 + x_2(1100.3)$$

both are functions only of x_2, the quality at the final state. Consequently,

$$\frac{2.0(1462.8 - 226.6 - 1100.3 x_2)}{0.001\,601 + 0.204 x_2} = 635\,190$$

Solving,

$$x_2 = 0.011\,04$$

Therefore,

$$v_2 = 0.001\,601 + 0.011\,04 \times 0.2040 = 0.003\,854 \text{ m}^3/\text{kg}$$

$$m_2 = \frac{2}{0.003\,854} = 518.9 \text{ kg}$$

and the mass of ammonia withdrawn, m_e, is

$$m_e = m_1 - m_2 = 591.4 - 518.9 = 72.5 \text{ kg}$$

EXAMPLE 5.19E A tank of 50 ft^3 volume contains saturated ammonia at a pressure of 200 lbf/in.^{2}. Initially the tank contains 50% liquid and 50% vapor by volume. Vapor is withdrawn from the top of the tank until the pressure is 100 lbf/in.^{2}. Assuming that only vapor (i.e., no liquid) leaves and that the process is adiabatic, calculate the mass of ammonia that is withdrawn.

Control volume: Tank.

Initial state: T_1, V_{liq}, V_{vap}; state fixed.

Final state: T_2.

Exit state: Saturated vapor (temperature changing).

Process: USUF.

Model: Ammonia tables.

Analysis:

In the first law, Eq. 5.54, we note that $Q_{c.v.} = 0$, $W_{c.v.} = 0$, $m_i = 0$, and we assume that changes in kinetic and potential energy are negligible. However, the enthalpy of saturated vapor varies with temperature, and therefore we cannot simply assume that the enthalpy of the vapor leaving the tank remains constant. We note that at 200 lbf/in.^{2}, $h_g = 632.7$ Btu/lbm and at 100 lbf/in.^{2}, $h_g = 626.5$ Btu/lbm. Since the change in h_g during this process is small, we may accurately assume that h_e is the average of the two values given above. Therefore

$$\left(h_e\right)_{av} = 629.6 \text{ Btu}/\text{lbm}$$

and the first law reduces to

$$m_e h_e + m_2 u_2 - m_1 u_1 = 0$$

and the continuity equation (from Eq. 5.53)

$$\left(m_2 - m_1\right)_{c.v.} + m_e = 0$$

Combining these two equations we have

$$m_2\left(h_e - u_2\right) = m_1 h_e - m_1 u_1$$

The following values are from the ammonia tables:

$$v_{f1} = 0.02732 \text{ ft}^3/\text{lbm} \qquad v_{g1} = 1.502 \text{ ft}^3/\text{lbm}$$

$$v_{f2} = 0.02584 \qquad v_{g2} = 2.952$$

$$u_{f1} = 150.9 - \frac{200 \times 144 \times 0.0273}{778} = 149.9 \text{ Btu}/\text{lbm}$$

$$u_{f2} = 104.7 - \frac{100 \times 144 \times 0.0258}{778} = 104.2$$

$$u_{g1} = 632.7 - \frac{200 \times 144 \times 1.502}{778} = 577.1$$

$$u_{g2} = 626.5 - \frac{100 \times 144 \times 2.952}{778} = 571.9$$

$$u_{fg2} = 571.9 - 104.2 = 467.7$$

Calculating first the initial mass, m_1, in the tank, the mass of the liquid initially present, m_{f1}, is

$$m_{f1} = \frac{25}{0.02732} = 915 \text{ lbm}$$

Similarly, the initial mass of vapor, m_{g1}, is

$$m_{g1} = \frac{25}{1.502} = 16.6 \text{ lbm}$$

$$m_1 = m_{f1} + m_{g1} = 915 + 17 = 932 \text{ lbm}$$

$$m_1 h_e = 932 \times 629.6 = 586\,800 \text{ Btu}$$

$$m_1 u_1 = (mu)_{f1} + (mu)_{g1} = 915 \times 149.9 + 16.6 \times 577.0 = 146\,700 \text{ Btu}$$

Substituting these into the first law,

$$m_2(h_e - u_2) = m_1 h_e - m_1 u_1 = 586\,800 - 146\,700 = 440\,100 \text{ Btu}$$

There are two unknowns, m_2 and u_2, in this equation. However,

$$m_2 = \frac{V}{v_2} = \frac{50}{0.0258 + x_2(2.952 - 0.0258)}$$

and

$$u_2 = 104.2 + x_2(467.7)$$

both functions only of x_2, the quality at the final state. Consequently,

$$\frac{50(629.6 - 104.2 - 467.7 x_2)}{0.0258 + 2.926 x_2} = 440\,100$$

Solving, $x_2 = 0.01137$
Therefore,

$$v_2 = 0.0258 + 0.01137(2.926) = 0.0590 \text{ ft}^3 / \text{lbm}$$

$$m_2 = \frac{50}{0.0590} = 847 \text{ lbm}$$

and the mass of ammonia withdrawn, m_e, is

$$m_e = m_1 - m_2 = 932 - 847 = 85 \text{ lbm}$$

PROBLEMS

5.1 A small elevator is being designed for a construction site. It is expected to carry four 75-kg workers to the top of a 100-m tall building in less than 2 min. The elevator cage will have a counterweight to balance its mass. What is the smallest size (power) electric motor that can drive this unit?

5.2 The rate of heat transfer to the surroundings from a person at rest is about 400 kJ/h. Suppose that the ventilation system fails in an auditorium containing 100 people.

a. How much will the internal energy of the air in the auditorium increase during the first 10 min after the ventilation system fails?

b. Considering the auditorium and all the people in it as a control mass, how much does the internal energy of this control mass change? How do you explain the fact that the temperature of the air increases?

5.3 A "bomb calorimeter," a closed rigid vessel, will be used to measure the energy released by a certain chemical reaction. This calorimeter initially contains the appropriate chemicals and is located in a large tank of water. When the chemicals react, heat is transferred from the bomb to the water, causing its temperature to rise. The electric power input to a stirrer used to circulate the water is 0.05 kW.

During a 25-min period, the heat transfer from the bomb to the water is 1400 kJ and the heat transfer from the water to the surrounding air is 70 kJ. Assuming that no water evaporates, what is the increase in the internal energy of the water during this period of time?

5.4 Find the missing properties
  a. H_2O $T = 300°C$, $u = 1700$ kJ/kg $P = ?$ $v = ?$
  b. CO_2 $T = 267°C$, $P = 0.5$ MPa $x = ?$ $h = ?$
  c. H_2O $T = -2°C$, $P = 100$ kPa $u = ?$ $v = ?$
  d. Air $P = 200$ kPa, $u = 210$ kJ/kg $v = ?$ $T = ?$
  e. NH_3 $T = 65°C$, $P = 600$ kPa $u = ?$ $v = ?$

5.5 Find the missing properties and give the phase of the substance
  a. H_2O $u = 2390$ kJ/kg, $v = 0.46$ m³/kg $h = ?$ $T = ?$ $x = ?$
  b. H_2O $u = 1200$ kJ/kg, $P = 10$ MPa $T = ?$ $x = ?$ $v = ?$
  c. R-12 $T = -5°C$, $P = 300$ kPa $h = ?$ $x = ?$
  d. R-134a $T = 60°C$, $h = 430$ kJ/kg $v = ?$ $x = ?$
  e. NH_3 $T = 20°C$, $P = 100$ kPa $u = ?$ $v = ?$ $x = ?$

5.6 Find the missing properties and give the phase of the substance
  a. H_2O $T = 120°C$, $v = 0.5$ m³/kg $u = ?$ $P = ?$ $x = ?$
  b. H_2O $T = 100°C$, $P = 10$ MPa $u = ?$ $x = ?$ $v = ?$
  c. CO_2 $T = 800$ K, $P = 200$ kPa $v = ?$ $u = ?$
  d. Ne $T = 100°C$, $v = 0.1$ m³/kg $P = ?$ $x = ?$
  e. CH_4 $T = 190$ K, $x = 0.75$ $v = ?$ $u = ?$

5.7 Find the missing properties among (P, T, v, u, h) together with x if applicable and give the phase of the substance.
  a. R-22 $T = 10°C$, $u = 200$ kJ/kg
  b. H_2O $T = 350°C$, $h = 3150$ kJ/kg
  c. R-12 $P = 600$ kPa, $h = 230$ kJ/kg
  d. R-134a $T = 40°C$, $u = 407$ kJ/kg
  e. NH_3 $T = 20°C$, $v = 0.1$ m³/kg

5.8 A 100-L rigid tank contains nitrogen (N_2) at 900 K, 12 MPa. The tank is now cooled to 100 K. What are the work and heat transfer for this process?

5.9 Water in a 150-L closed, rigid tank is at 100°C, 90% quality. The tank is then cooled to −10°C. Calculate the heat transfer during the process.

5.10 A cylinder fitted with a frictionless piston contains 2 kg of superheated refrigerant R-134a vapor at 1 MPa, 100°C. The cylinder is now cooled so the R-134a remains at constant pressure until it reaches a quality of 75%. Calculate the heat transfer in the process.

5.11 A test cylinder with constant volume of 0.1 L contains water at the critical point. It now cools down to room temperature of 20°C. Calculate the heat transfer from the water.

5.12 Ammonia at 0°C, quality 60% is contained in a rigid 200-L tank. The tank and ammonia is now heated to a final pressure of 1 MPa. Determine the heat transfer for the process.

5.13 A 10-L rigid tank contains R-22 at −10°C, 80% quality. A 10-A electric current (from a 6-V battery) is passed through a resistor inside the tank for 10 min, after which the R-22 temperature is 40°C. What was the heat transfer to or from the tank during this process?

5.14 Consider the 100-L Dewar (a rigid double-walled vessel for storing cryogenic liquids) shown in Fig. P5.14. The Dewar contains nitrogen at 1 atm, 90% liquid and 10% vapor by volume. The insulation holds heat transfer into the Dewar from the ambient to a very low rate, 5 J/s. The vent valve is accidentally closed so that the pressure inside slowly rises. It is estimated that the Dewar will rupture when the pressure reaches 500 kPa. How long will it take to reach this pressure?

FIGURE P5.14

5.15 A piston/cylinder arrangement contains 1 kg of water, shown in Fig. P5.15. The piston is spring loaded and initially rests on some stops. A pressure of 300 kPa will just float the piston and, at a volume of 1.5 m³, a pressure of 500 kPa will balance the piston. The initial state of the water is 100 kPa with a volume of 0.5 m³. Heat is now added until a pressure of 400 kPa is reached.

a. Find the initial temperature and the final volume.

b. Find the work and heat transfer in the process and plot the *P–V* diagram.

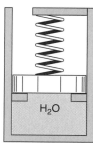

H_2O

FIGURE P5.15

5.16 A closed steel bottle contains ammonia at −20°C, $x = 20\%$ and the volume is 0.05 m³. It has a safety valve that opens at a pressure of 1.4 MPa. By accident, the bottle is heated until the safety valve opens. Find the temperature and heat transfer when the valve first opens.

5.17 A piston/cylinder arrangement B is connected to a 1-m³ tank A by a line and valve, shown in Fig. P5.17. Initially both contain water, with A at 100 kPa, saturated vapor and B at 400°C, 300 kPa, 1 m³. The valve is now opened and, the water in both A and B comes to a uniform state.

  a. Find the initial mass in A and B.

  b. If the process results in $T_2 = 200°C$, find the heat transfer and work.

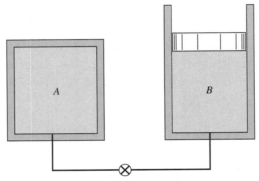

FIGURE P5.17

5.18 Consider the same setup and initial conditions as in the previous problem. Assuming that the process is adiabatic, find the final temperature and work.

5.19 A vertical cylinder fitted with a piston contains 5 kg of R-22 at 10°C, shown in Fig. P5.19. Heat is transferred to the system, causing the piston to rise until it reaches a set of stops at which point the volume has doubled. Additional heat is transferred until the temperature inside reaches 50°C, at which point the pressure inside the cylinder is 1.3 MPa.

  a. What is the quality at the initial state?

  b. Calculate the heat transfer for the overall process.

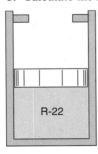

FIGURE P5.19

5.20 Ten kilograms of water in a piston/cylinder with constant pressure is at 450°C and a volume of 0.633 m³. It is now cooled to 20°C. Show the P–v diagram and find the work and heat transfer for the process.

5.21 A cylinder fitted with a piston restrained by a linear spring has a cross-sectional area of 0.05 m² and initial volume of 20 L, shown in Fig. P5.21. The cylinder contains ammonia at 1 MPa, 60°C. The spring constant is 150 kN/m. Heat is rejected from the system, and the piston moves until 6.25 kJ of work has been done on the ammonia.

a. Find the final temperature of the ammonia.

b. Calculate the heat transfer for the process.

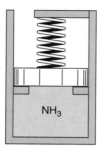

FIGURE P5.21

5.22 An insulated cylinder fitted with a piston contains R-12 at 25°C with a quality of 90% and a volume of 45 L. The piston is allowed to move, and the R-12 expands until it exists as saturated vapor. During this process the R-12 does 7.0 kJ of work against the piston. Determine the final temperature, assuming the process is adiabatic.

5.23 Two kilograms of nitrogen at 100 K, $x = 0.5$ is heated in a constant pressure process to 300 K in a piston/cylinder arrangement. Find the initial and final volumes and the total heat transfer required.

5.24 A spherical balloon contains 2 kg of R-22 at 0°C, 30% quality. This system is heated until the pressure in the balloon reaches 600 kPa. In the range of variables of this process, it can be assumed that the pressure in the balloon is directly proportional to the balloon diameter. What is the heat transfer for the process?

5.25 A piston/cylinder arrangement has the piston loaded with outside atmospheric pressure and the piston mass to a pressure of 150 kPa, shown in Fig. P5.25. It contains water at −2°C, which is then heated until the water becomes saturated vapor. Find the final temperature and specific work and heat transfer for the process.

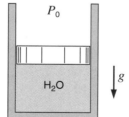

FIGURE P5.25

5.26 Consider the system shown in Fig. P5.26. Tank A has a volume of 100 L and contains saturated vapor R-134a at 30°C. When the valve is cracked open, R-134a flows slowly into cylinder B. The piston mass requires a pressure of 300 kPa in cylinder B to raise the piston. The process ends when the pressure in tank A has fallen to 300 kPa. During this process heat is exchanged with the surroundings such that the R-134a always remains at 30°C. Calculate the heat transfer for the process.

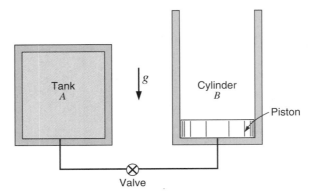

FIGURE P5.26

5.27 A piston held by a pin in an insulated cylinder, shown in Fig. P5.27, contains 2 kg water at 100°C, quality 98%. The piston and weights have a total mass of 102 kg, with cross-sectional area of 100 cm^2, and the ambient pressure is 100 kPa. The pin is released, which allows the piston to move. Determine the final state of the water, assuming the process to be adiabatic.

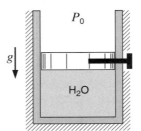

FIGURE P5.27

5.28 A piston/cylinder arrangement has a linear spring and the outside atmosphere acting on the piston, shown in Fig. P5.28. It contains water at 3 MPa, 400°C with the volume being 0.1 m^3. If the piston is at the bottom, the spring exerts a force such that a pressure of 200 kPa inside is required to balance the forces. The system now cools until the pressure reaches 1 MPa. Find the heat transfer for the process.

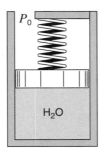

FIGURE P5.28

5.29 Two heavily insulated tanks are connected by a valve, as shown in Fig. P5.29. Tank A contains 0.6 kg of water at 300 kPa, 300°C. Tank B has a volume of 300 L and contains water at 600 kPa, 80% quality. The valve is opened, and the two tanks eventually come to a uniform state. Assuming the process to be adiabatic, what is the final pressure?

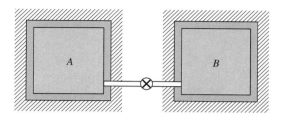

FIGURE P5.29

5.30 A cylinder/piston arrangement contains 5 kg of water at 100°C with $x = 20\%$ and the piston, $m_P = 75$ kg, resting on some stops, shown in Fig. P.5.30. The outside pressure is 100 kPa, and the cylinder area is $A = 24.5$ cm². Heat is now added until the water reaches a saturated vapor state. Find the initial volume, final pressure, work, and heat transfer terms and show the P–v diagram.

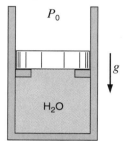

P_0

g

H_2O

FIGURE P5.30

5.31 A rigid tank is divided into two rooms by a membrane, both containing water, shown in Fig. P5.31. Room A is at 200 kPa, $v = 0.5$ m³/kg, $V_A = 1$ m³, and room B contains 3.5 kg at 0.5 MPa, 400°C. The membrane now ruptures and heat transfer takes place so the water comes to a uniform state at 100°C. Find the heat transfer during the process.

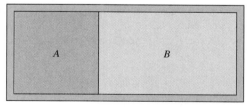

A   B

FIGURE P5.31

5.32 Two tanks are connected by a valve and line as shown in Fig. P5.32. The volumes are both 1 m³ with R-134a at 20°C, quality 25% in A and tank B is evacuated. The valve is opened and saturated vapor flows from A into B until the pressures become equal. The process occurs slowly enough that all temperatures stay at 20°C during the process. Find the total heat transfer to the R-134a during the process.

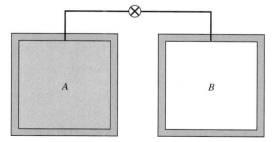

A   B

FIGURE P5.32

5.33 Consider the same system as in the previous problem. Let the valve be opened and transfer enough heat to both tanks so all the liquid disappears. Find the necessary heat transfer.

5.34 A cylinder having a piston restrained by a linear spring contains 0.5 kg of saturated vapor water at 120°C, as shown in Fig. P5.34. Heat is transferred to the water, causing the piston to rise, and during the process the spring resisting force is proportional to the distance moved. The spring constant is 15 kN/m. The piston cross-sectional area is 0.05 m^2.

a. What is the pressure in the cylinder when the temperature inside reaches 600°C?

b. Calculate the heat transfer for the process.

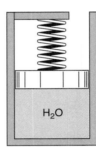

FIGURE P5.34

5.35 A water-filled reactor with volume of 1 m^3 is at 20 MPa, 360°C and placed inside a containment room as shown in Fig. P5.35. The room has a volume of 100 m^3, well insulated and initially evacuated. Due to a failure, the reactor ruptures and the water fills the containment room. Find the final pressure.

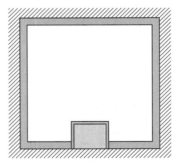

FIGURE P5.35

5.36 Refrigerant-12 is contained in a piston/cylinder arrangement at 2 MPa, 150°C with a massless piston against the stops, at which point $V = 0.5$ m^3. The side above the piston is connected by an open valve to an air line at 10°C, 450 kPa, shown in Fig. P5.36. The whole setup now cools to the surrounding temperature of 10°C. Find the heat transfer and show the process in a P–v diagram.

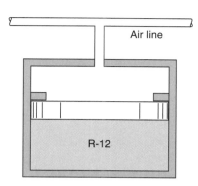

FIGURE P5.36

5.37 A 10-m high open cylinder, $A_{cyl} = 0.1 \text{ m}^2$, contains 20°C water above and 2 kg of 20°C water below a 198.5-kg thin insulated floating piston, shown in Fig. P5.37. Assume standard g, P_o. Now heat is added to the water below the piston so that it expands, pushing the piston up, causing the water on top to spill over the edge. This process continues until the piston reaches the top of the cylinder. Find the final state of the water below the piston (T, P, v) and the heat added during the process.

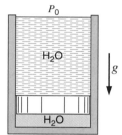

FIGURE P5.37

5.38 A cylinder fitted with a piston contains 2 kg of R-12 at 10°C, 90% quality. The system undergoes a quasi-equilibrium polytropic expansion to 100 kPa, during which the system receives a heat transfer of 52.5 kJ. What is the final temperature of the R-12?

5.39 A piston/cylinder arrangement of initial volume 0.025 m³ contains saturated water vapor at 180°C. The steam now expands in a polytropic process with exponent $n = 1$ to a final pressure of 200 kPa, while it does work against the piston. Determine the heat transfer in this process.

5.40 Calculate the heat transfer for the process described in Problem 4.23.

5.41 A piston/cylinder contains 1 kg of water at 20°C with a volume of 0.1 m³, shown in Fig. P5.41. Initially the piston rests on some stops with the top surface open to the atmosphere, P_o, so a pressure of 300 kPa is required to lift it. To what temperature should the water be heated to lift the piston? If it is heated to saturated vapor find the final temperature, volume, and the heat transfer, $_1Q_2$.

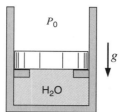

FIGURE P5.41

5.42 Consider the piston/cylinder arrangement shown in Fig. P5.42, in which a frictionless piston is free to move between two sets of stops. When the piston rests on the lower stops, the enclosed volume is 400 L. When the piston reaches the upper stops, the volume is 600 L. The cylinder initially contains water at 100 kPa, 20% quality. The system is heated until the water eventually exists as saturated vapor. If the mass of the piston requires 300 kPa pressure to move it against the outside ambient pressure, determine

a. The final pressure in the cylinder

b. The heat transfer and the work for the overall process

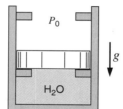

FIGURE P5.42

5.43 Calculate the heat transfer for the process described in Problem 4.29.

5.44 A cylinder fitted with a frictionless piston that is restrained by a linear spring contains R-22 at 20°C, quality 60% with a volume of 8 L, shown in Fig. P5.44. The piston cross-sectional area is 0.04 m², and the spring constant is 500 kN/m. A total of 60 kJ of heat is now added to the R-22. What are the final pressure and temperature of the R-22?

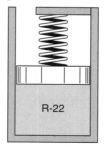

FIGURE P5.44

5.45 A 1-L capsule of water at 700 kPa, 150°C is placed in a larger insulated and otherwise evacuated vessel. The capsule breaks and its contents fill the entire volume. If the final pressure should not exceed 200 kPa, what should the vessel volume be?

5.46 A cylinder with a frictionless piston contains steam at 2 MPa, 500°C with a volume of 5 L, shown in Fig. P5.46. The external piston force is proportional to cylinder volume cubed. Heat is transferred out of the cylinder, reducing the volume and thus the force until the cylinder pressure has dropped to 500 kPa. Find the work and heat transfer for this process.

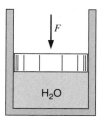

FIGURE P5.46

5.47 A spherical balloon initially 150 mm in diameter and containing R-12 at 100 kPa is connected to a 30-L uninsulated, rigid tank containing R-12 at 500 kPa. Everything is at the ambient temperature of 20°C. A valve connecting the tank and balloon is opened slightly and remains so until the pressures equalize. During this process heat is exchanged so the temperature remains constant at 20°C. For this range of variables the pressure inside the balloon is proportional to the diameter at any time. Calculate

a. The final pressure

b. Work done by the R-12 during the process

c. Heat transferred to the R-12 during the process

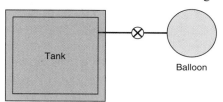

FIGURE P5.47

5.48 Superheated refrigerant R-134a at 20°C, 0.5 MPa is cooled in a piston/cylinder arrangement at constant temperature to a final two-phase state with quality of 50%. The refrigerant mass is 5 kg, and during this process 500 kJ of heat is removed. Find the initial and final volumes and the necessary work.

5.49 Calculate the heat transfer for the process described in Problem 4.32.

5.50 Calculate the heat transfer for the process described in Problem 4.33.

5.51 A piston/cylinder, shown in Fig. P5.51, contains R-12 at $-30°C$, $x = 20\%$. The volume is 0.2 m³. It is known that $V_{stop} = 0.4$ m³, and if the piston sits at the bottom, the spring force balances the other loads on the piston. It is now heated up to 20°C. Find the mass of the fluid and show the P–v diagram. Find the work and heat transfer.

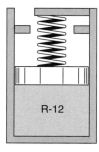

FIGURE P5.51

5.52 A cylinder, $A_{cyl} = 0.1$ m^2, has two pistons as shown in Fig. P5.52, each side A and B contains 1 kg of water at 20°C, with $P_{A1} = 150$ kPa and $P_{B1} = 500$ kPa. The sides are insulated, and heat is added through the bottom to B until it reaches 200°C. The piston separating A and B conducts heat to the extent that $T_{A2} = 50$°C. Neglecting the potential energy of the water in A and B, find the final volumes, the heat transfer to A, and the total work done by the water in B.

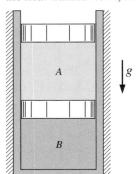

FIGURE P5.52

5.53 Ammonia, NH_3, is contained in a sealed rigid tank at 0°C, $x = 75\%$ and is then heated to 100°C. Find the final state P_2, u_2 and the specific work and heat transfer.

5.54 A house is being designed to use a thick concrete floor mass as thermal storage material for solar energy heating. The concrete is 30 cm thick and the area exposed to the sun during the day time is 4 m × 6 m. It is expected that this mass will undergo an average temperature rise of about 3°C during the day. How much energy will be available for heating during the nighttime hours?

5.55 A copper block of volume 1 L is heat treated at 500°C and now cooled in a 100-L oil bath initially at 20°C, shown in Fig. P5.55. Assuming no heat transfer with the surroundings, what is the final temperature?

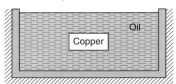

FIGURE P5.55

5.56 Saturated, $x = 1\%$, water at 25°C is contained in a hollow spherical aluminum vessel with inside diameter of 0.5 m and a 1-cm thick wall. The vessel is heated until the water inside is saturated vapor. Considering the vessel and water together as a control mass, calculate the heat transfer for the process.

5.57 An ideal gas is heated from 500 to 1500 K. Find the change in enthalpy using constant specific heat from Table A.10 (room temperature value) and discuss the accuracy of the result if the gas is

a. Argon

b. Oxygen

c. Carbon dioxide

5.58 A computer in a closed room of volume 150 m^3 dissipates energy at a rate of 10 kW. The air in the room was at 300 K, 100 kPa, when the air conditioner suddenly stops. What is the air temperature after 15 min?

5.59 The heaters in a spacecraft suddenly fail. Heat is lost by radiation at the rate of 50 kJ/h, and the electric instruments generate 25 kJ/h. Initially, the air is at 100 kPa, 25°C with a volume of 10 m³. How long will it take to reach an air temperature of −40°C?

5.60 An insulated cylinder is divided into two parts of 1 m³ each by an initially locked piston, as shown in Fig. P5.60. Side A has air at 200 kPa, 300 K, and side B has air at 1.0 MPa, 1400 K. The piston is now unlocked so it is free to move, and it conducts heat so the air comes to a uniform temperature $T_A = T_B$. Find the mass in both A and B, and the final T and P.

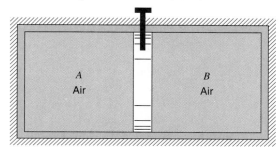

FIGURE P5.60

5.61 A cylinder with a piston restrained by a linear spring contains 2 kg of carbon dioxide at 500 kPa, 400°C. It is cooled to 40°C, at which point the pressure is 300 kPa. Calculate the heat transfer for the process.

5.62 A piston/cylinder in a car contains 0.2 L of air at 90 kPa, 20°C, shown in Fig. P5.62. The air is compressed in a quasi-equilibrium polytropic process with polytropic exponent $n = 1.25$ to a final volume seven times smaller. Determine the final pressure, temperature, and the heat transfer for the process.

FIGURE P5.62

5.63 Water at 20°C, 100 kPa, is brought to 200 kPa, 1500°C. Find the change in the specific internal energy, using the tables and an appropriate model(s).

5.64 For an application the change in enthalpy of carbon dioxide from 30 to 1500°C at 100 kPa is needed. Consider the following methods and indicate the most accurate one.

a. Constant specific heat, value from Table A.10.

b. Constant specific heat, value at average temperature from the equation in Table A.11.

c. Variable specific heat, integrating the equation in Table A.11.

d. Enthalpy from ideal gas tables in Table A.13.

5.65 Air in a piston/cylinder at 200 kPa, 600 K, is expanded in a constant-pressure process to twice the initial volume (state 2), shown in Fig. P5.65. The piston is then locked with a pin and heat is transferred to a final temperature of 600 K. Find P, T, and h for states 2 and 3, and find the work and heat transfer in both processes.

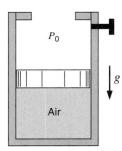

FIGURE P5.65

5.66 An insulated floating piston divides a cylinder into two volumes each of 1 m³, as shown in Fig. P5.66. One contains water at 100°C and the other air at −3°C and both pressures are 200 kPa. A line with a safety valve that opens at 400 kPa is attached to the water side of the cylinder. Assume no heat transfer to the water. Show possible air states in a P–v diagram, and find the air temperature when the safety valve opens. How much heat transfer is needed to bring the air to 1300 K?

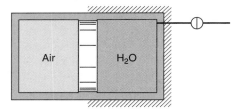

FIGURE P5.66

5.67 Two containers are filled with air, one a rigid tank A, and the other a piston/cylinder B that is connected to A by a line and valve, as shown in Fig. P5.67. The initial conditions are: $m_A = 2$ kg, $T_A = 600$ K, $P_A = 500$ kPa and $V_B = 0.5$ m³, $T_B = 27$°C, $P_B = 200$ kPa. The piston in B is loaded with the outside atmosphere and the piston mass in the standard gravitational field. The valve is now opened, and the air comes to a uniform condition in both volumes. Assuming no heat transfer, find the initial mass in B, the volume of tank A, the final pressure and temperature and the work, $_1W_2$.

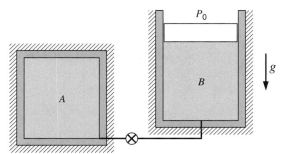

FIGURE P5.67

5.68 A 250-L rigid tank contains methane gas at 500°C, 600 kPa. The tank is cooled to 300 K.

    a. Find the final pressure and the heat transfer for the process.

    b. What is the percent error in the heat transfer if the specific heat is assumed constant at the room temperature value?

5.69 A piston/cylinder arrangement, shown in Fig. P5.69, contains 5 g of air at 250 kPa, 300°C. The 50-kg piston has a diameter of 0.1 m and initially pushes against the stops. The atmosphere is at 100 kPa and 20°C. The cylinder now cools to 20°C as heat is transferred to the ambient. Calculate the heat transfer.

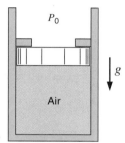

FIGURE P5.69

5.70 Oxygen at 300 kPa, 100°C is in a piston/cylinder arrangement with a volume of 0.1 m^3. It is now compressed in a polytropic process with exponent, $n = 1.2$, to a final temperature of 200°C. Calculate the heat transfer for the process.

5.71 A piston/cylinder contains 2 kg of air at 27°C, 200 kPa, shown in Fig. P5.71. The piston is loaded with a linear spring, mass and the atmosphere. Stops are mounted so that $V_{stop} = 3$ m^3, at which point $P = 600$ kPa is required to balance the piston forces. The air is now heated to 1500 K. Find the final pressure and volume and the work and heat transfer. Find the work done on the spring.

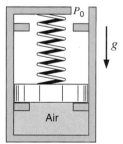

FIGURE P5.71

5.72 An air pistol contains compressed air in a small cylinder, shown in Fig. P5.72. Assume that the volume is 1 cm^3, pressure is 1 MPa, and the temperature is 27°C when armed. A bullet, $m = 15$ g, acts as a piston initially held by a pin (trigger); when released, the air expands in an isothermal process ($T =$ constant). If the air pressure is 0.1 MPa in the cylinder as the bullet leaves the gun, find

    a. The final volume and the mass of air.

    b. The work done by the air and work done on the atmosphere.

    c. The work to the bullet and the bullet exit velocity.

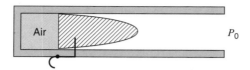

FIGURE P5.72

5.73 A certain elastic balloon will support an internal pressure equal to $P_0 = 100$ kPa until the balloon becomes spherical at a diameter of $D_0 = 1$ m, beyond which

$$P = P_0 + C\left[1 - \left(\frac{D_0}{D}\right)^6\right]\frac{D_0}{D}$$

because of the offsetting effects of balloon curvature and elasticity. This balloon contains helium gas at 250 K, 100 kPa, with a 0.4 m³ volume. The balloon is heated until the volume reaches 2 m³. During the process the maximum pressure inside the balloon is 200 kPa.

a. What is the temperature inside the balloon when pressure is maximum?

b. What are the final pressure and temperature inside the balloon?

c. Determine the work and heat transfer for the overall process.

5.74 A 10-m high cylinder, cross-sectional area 0.1 m², has a massless piston at the bottom with water at 20°C on top of it, shown in Fig. P5.74. Air at 300 K, volume 0.3 m³, under the piston is heated so that the piston moves up, spilling the water out over the side. Find the total heat transfer to the air when all the water has been pushed out.

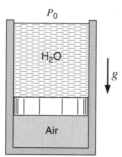

FIGURE P5.74

5.75 A rigid 50-L tank A and cylinder are connected as shown in Fig. P5.75. A thin frictionless piston separates B and C, each with initial volume 100 L. A and B contain ammonia and C contains air. Initially, the quality in A is 40%, the pressures in B and C are 100 kPa. The valve is opened slowly, and the system reaches a common pressure. All temperatures are ambient, 20°C, throughout the process.

a. Determine the final pressure.

b. Calculate the work done on the air.

c. Calculate the heat transfer to the combined system.

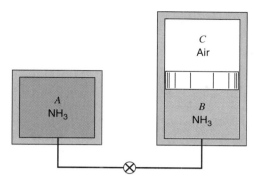

FIGURE P5.75

5.76 A piston/cylinder contains argon gas at 140 kPa, 10°C, and the volume is 100 L. The gas is compressed in a polytropic process to 700 kPa, 280°C. Calculate the heat transfer during the process.

5.77 Water at 150°C, quality 50% is contained in a cylinder/piston arrangement with initial volume 0.05 m³. The loading of the piston is such that the inside pressure is linear with the square root of volume as $P = 100 + CV^{0.5}$ kPa. Now heat is transferred to the cylinder to a final pressure of 600 kPa. Find the heat transfer in the process.

5.78 A piston/cylinder has 1 kg propane gas at 700 kPa, 40°C. The piston cross-sectional area is 0.5 m², and the total external force restraining the piston is directly proportional to the cylinder volume squared. Heat is transferred to the propane until its temperature reaches 1100°C. Determine the final pressure inside the cylinder, the work done by the propane, and the heat transfer during the process.

5.79 A closed cylinder is divided into two rooms by a frictionless piston held in place by a pin, as shown in Fig. P5.79. Room A has 10 L air at 100 kPa, 30°C, and room B has 300 L saturated water vapor at 30°C. The pin is pulled, releasing the piston, and both rooms come to equilibrium at 30°C. Considering a control mass of the air and water, determine the work done by the system and the heat transfer to the cylinder.

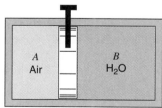

FIGURE P5.79

5.80 Repeat Problem 5.79, with the initial air volume in room A as 40 L instead of 10 L.

5.81 Heat is transferred at a given rate to a mixture of liquid and vapor in equilibrium in a closed container, shown in Fig. P5.81. Determine the rate of change of temperature as a function of the thermodynamic properties of the liquid and vapor and the masses of liquid and vapor.

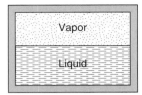

FIGURE P5.81

5.82 A frictionless, thermally conducting piston separates the air and water in the cylinder shown in Fig. P5.82. The initial volumes of A and B are each 500 L, and the initial pressure on each side is 700 kPa. The volume of the liquid in B is 2% of the volume of B at this state. Heat is transferred to both A and B until all the liquid in B evaporates. Determine

   a. The total heat transfer during the process.

   b. The work done by the piston on the air and the heat transferred to the air during the process.

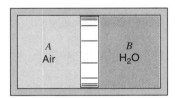

FIGURE P5.82

5.83 Air at 35°C, 105 kPa, flows in a 100 mm × 150 mm rectangular duct in a heating system. The volumetric flow rate is 0.015 m^3/s. What is the velocity of the air flowing in the duct?

5.84 Nitrogen gas flowing in a 50-mm diameter pipe at 15°C, 200 kPa, at the rate of 0.05 kg/s, encounters a partially closed valve. If there is a pressure drop of 30 kPa across the valve and essentially no temperature change, what are the velocities upstream and downstream of the valve?

5.85 Saturated vapor R-134a leaves the evaporator in a heat pump system at 10°C, with a steady mass flow rate of 0.1 kg/s. What is the smallest diameter tubing that can be used at this location if the velocity of the refrigerant is not to exceed 7 m/s?

5.86 Carbon dioxide enters a steady-state, steady-flow heater at 300 kPa, 15°C, and exits at 275 kPa, 1200°C, as shown in Fig. P5.86. Changes in kinetic and potential energies are negligible. Calculate the required heat transfer per kilogram of carbon dioxide flowing through the heater.

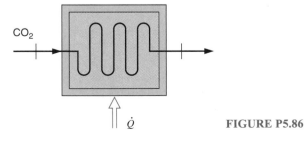

FIGURE P5.86

5.87 The compressor of a large gas turbine receives air from the ambient at 95 kPa, 20°C, with a low velocity. At the compressor discharge, air exits at 1.52 MPa, 430°C, with velocity of 90 m/s. The power input to the compressor is 5000 kW. Determine the mass flow rate of air through the unit.

5.88 Hoover Dam across the Colorado River dams up Lake Mead 200 m higher than the river downstream. The electric generators driven by water-powered turbines deliver 1300 MW of power. If the water is 17.5°C, find the minimum amount of water running through the turbines.

5.89 Calculate the heat transfer for the process described in Problem 3.42.

5.90 The purpose of a nozzle is to produce a high velocity stream of fluid at the expense of its pressure, shown in Fig. P5.90. Superheated vapor ammonia enters an insulated nozzle at 20°C, 800 kPa, with a low velocity and at the steady rate of 0.01 kg/s. The ammonia exits at 300 kPa with a velocity of 450 m/s. Determine the temperature (or quality, if saturated) and the exit area of the nozzle.

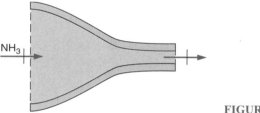

FIGURE P5.90

5.91 A diffuser is a steady-state, steady-flow device, shown in Fig. P5.91, used to decelerate a high-velocity fluid in a manner designed to increase the fluid pressure (essentially the opposite process of a nozzle expansion). Consider a diffuser in which air enters at 100 kPa, 300 K, with a velocity of 200 m/s. The inlet cross-sectional area of the diffuser is 100 mm². At the exit, the area is 860 mm², and the exit velocity is 20 m/s. Determine the exit pressure and temperature of the air.

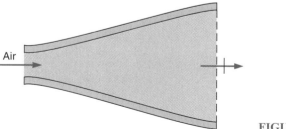

FIGURE P5.91

5.92 A steam turbine receives water at 15 MPa, 600°C at a rate of 100 kg/s, shown in Fig. P5.92. In the middle section 20 kg/s is withdrawn at 2 MPa, 350°C, and the rest exits the turbine at 75 kPa, and 95% quality. Assuming no heat transfer and no changes in kinetic energy, find the total turbine work.

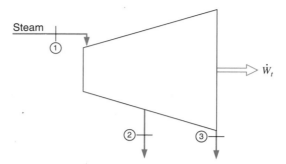

FIGURE P5.92

5.93 A small water pump is used in an irrigation system. The pump takes water in from a river at 10°C, 100 kPa at a rate of 5 kg/s. The exit line enters a pipe that goes up to an elevation 20 m above the pump and river, where the water runs into an open channel. Assume the process is adiabatic and that the water stays at 10°C. Find the required pump work.

5.94 A condenser (heat exchanger) brings 1 kg/s water flow at 10 kPa from 300°C to saturated liquid at 10 kPa, as shown in Fig. P5.94. The cooling is done by lake water at 20°C that returns to the lake at 30°C. For an insulated condenser, find the flow rate of cooling water.

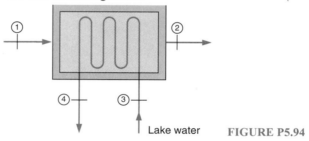

Lake water   FIGURE P5.94

5.95 Two kg of water at 500 kPa, 20°C is heated in a constant pressure process (SSSF) to 1700°C. Find the best estimate for the heat transfer.

5.96 A small, high-speed turbine operating on compressed air produces a power output of 100 W. The inlet state is 400 kPa, 50°C, and the exit state is 150 kPa, −30°C. Assuming the velocities to be low and the process to be adiabatic, find the required mass flow rate of air through the turbine.

5.97 A steam pipe for a 1500-m tall building receives superheated steam at 200 kPa at ground level. At the top floor the pressure is 125 kPa and the heat loss in the pipe is 110 kJ/kg. What should the inlet temperature be so that no water will condense inside the pipe?

5.98 In a steam generator, compressed liquid water at 10 MPa, 30°C, enters a 30-mm diameter tube at the rate of 3 L/s. Steam at 9 MPa, 400°C exits the tube. Find the rate of heat transfer to the water.

5.99 A heat exchanger, shown in Fig. P5.99, is used to cool an air flow from 800 to 360 K, both states at 1 MPa. The coolant is a water flow at 15°C, 0.1 MPa. If the water leaves as saturated vapor, find the ratio of the flow rates $\dot{m}_{water}/\dot{m}_{air}$.

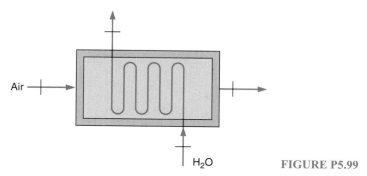

Air →
H_2O

FIGURE P5.99

5.100 Compressed liquid R-22 at 1.5 MPa, 10°C is mixed in a steady-state, steady-flow process with saturated vapor R-22 at 1.5 MPa. Both flow rates are 0.1 kg/s, and the exiting flow is at 1.2 MPa and a quality of 85%. Find the rate of heat transfer to the mixing chamber.

5.101 Two steady flows of air enters a control volume, shown in Fig. P5.101. One is 0.025 kg/s flow at 350 kPa, 150°C, state 1, and the other enters at 350 kPa, 15°C, both flows with low velocity. A single flow of air exits at 100 kPa, −40°C through a 25-mm diameter pipe, state 3. The control volume rejects 1.2 kW heat to the surroundings and produces 4.5 kW of power. Determine the flow rate of air at the inlet at state 2.

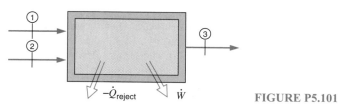

$-\dot{Q}_{reject}$ $\dot{W}$ FIGURE P5.101

5.102 An air compressor takes in air at 100 kPa, 17°C and delivers it at 1 MPa, 600 K to a constant-pressure cooler, which it exits at 300 K. Find the specific compressor work and the specific heat transfer.

5.103 A desuperheater mixes superheated water vapor with liquid water in a ratio that produces saturated water vapor as output without any external heat transfer. A flow of 0.5 kg/s superheated vapor at 5 MPa, 400°C and a flow of liquid water at 5 MPa, 40°C enter a desuperheater. If saturated water vapor at 4.5 MPa is produced, determine the flow rate of the liquid water.

5.104 The following data are for a simple steam power plant as shown in Fig. P5.104.

State	1	2	3	4	5	6	7
P MPa	6.2	6.1	5.9	5.7	5.5	0.01	0.009
T°C		45	175	500	490		40

State 6 has $x_6 = 0.92$, and velocity of 200 m/s. The rate of steam flow is 25 kg/s, with 300 kW power input to the pump. Piping diameters are 200 mm from steam generator to the turbine and 75 mm from the condenser to the steam generator. The economizer is a low temperature heat exchanger. Determine

a. The power output of the turbine

b. Heat transfer rates in the condenser, economizer, and steam generator

c. The flow rate of cooling water through the condenser, if the cooling water increases from 15° to 25°C in the condenser

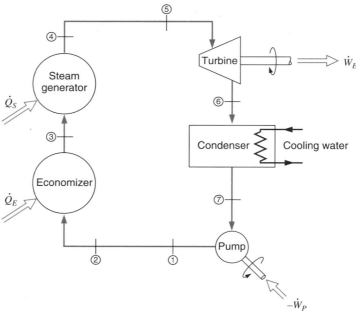

FIGURE P5.104

5.105 Cogeneration is often used where a steam supply is needed for industrial process energy. Assume a supply of 5 kg/s steam at 0.5 MPa is needed. Rather than generating this from a pump and boiler, the setup in Fig. P5.105 is used so the supply is extracted from the high-pressure turbine. Find the power the turbine now cogenerates in this process.

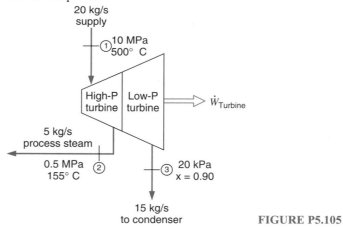

FIGURE P5.105

5.106 A somewhat simplified flow diagram for a nuclear power plant shown in Fig. 1.4 is given in Fig. P5.106. Mass flow rates and the various states in the cycle are shown in the accompanying table.

The cycle includes a number of heaters in which heat is transferred from steam, taken out of the turbine at some intermediate pressure, to liquid water pumped from the condenser on its way to the steam drum. The heat exchanger in the reactor supplies 157 MW, and it may be assumed that there is no heat transfer in the turbines.

Point	$\dot{m}$, kg/s	P, kPa	T, °C	h, kJ/kg
1	75.6	7240	sat vap	
2	75.6	6900		2765
3	62.874	345		2517
4		310		
5		7		2279
6	75.6	7	33	
7		415		140
8	2.772	35		2459
9	4.662	310		558
10		35	34	
11	75.6	380	68	
12	8.064	345		2517
13	75.6	330		
14				349
15	4.662	965	139	584
16	75.6	7930		565
17	4.662	965		2593
18	75.6	7580		688
19	1386	7240	277	
20	1386	7410		1221
21	1386	7310		

a. Assume the moisture separator has no heat transfer between the two turbine sections, determine the enthalpy and quality (h_4, x_4).

b. Determine the power output of the low-pressure turbine.

c. Determine the power output of the high-pressure turbine.

d. Find the ratio of the total power output of the two turbines to the total power delivered by the reactor.

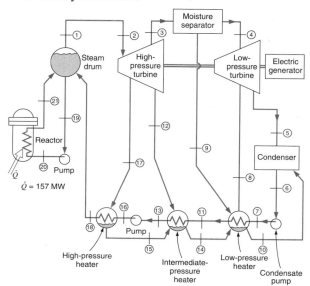

FIGURE P5.106

5.107 Consider the powerplant as described in the previous problem.

    a. Determine the quality of the steam leaving the reactor.

    b. What is the power to the pump that feeds water to the reactor?

5.108 Consider the powerplant as described in Problem 5.106.

    a. Determine the temperature of the water leaving the intermediate pressure heater, T_{13}, assuming no heat transfer to the surroundings.

    b. Determine the pump work, between states 13 and 16.

5.109 Consider the powerplant as described in Problem 5.106.

    a. Find the power removed in the condenser by the cooling water (not shown).

    b. Find the power to the condensate pump.

    c. Do the energy terms balance for the low pressure heater or is there a heat transfer not shown?

5.110 Helium is throttled from 1.2 MPa, 20°C, to a pressure of 100 kPa. The diameter of the exit pipe is so much larger than the inlet pipe that the inlet and exit velocities are equal. Find the exit temperature of the helium and the ratio of the pipe diameters.

5.111 Water flowing in a line at 400 kPa, saturated vapor, is taken out through a valve to 100 kPa. What is the temperature as it leaves the valve assuming no changes in kinetic energy and no heat transfer?

5.112 Methane at 3 MPa, 300 K, is throttled to 100 kPa. Calculate the exit temperature assuming no changes in the kinetic energy and

    a. Ideal-gas behavior

    b. Real-gas behavior

5.113 Water at 1.5 MPa, 150°C, is throttled adiabatically through a valve to 200 kPa. The inlet velocity is 5 m/s, and the inlet and exit pipe diameters are the same. Determine the state and the velocity of the water at the exit.

5.114 A proposal is made to use a geothermal supply of hot water to operate a steam turbine, as shown in Fig. P5.114. The high-pressure water at 1.5 MPa, 180°C, is

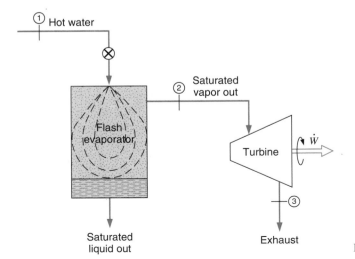

FIGURE P5.114

throttled into a flash evaporator chamber, which forms liquid and vapor at a lower pressure of 400 kPa. The liquid is discarded while the saturated vapor feeds the turbine and exits at 10 kPa, 90% quality. If the turbine should produce 1 MW, find the required mass flow rate of hot geothermal water in kilograms per hour.

5.115 A small turbine, shown in Fig. P5.115, is operated at part load by throttling a 0.25-kg/s steam supply at 1.4 MPa, 250°C down to 1.1 MPa before it enters the turbine and the exhaust is at 10 kPa. If the turbine produces 110 kW, find the exhaust temperature (and quality if saturated).

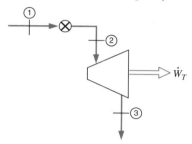

FIGURE P5.115

5.116 A R-12 heat pump cycle shown in Fig. P5.116 has a R-12 flow rate of 0.05 kg/s with 4 kW into the compressor. The following data are given

State	1	2	3	4	5	6
P kPa	1250	1230	1200	320	300	290
T °C	120	110	45		0	5

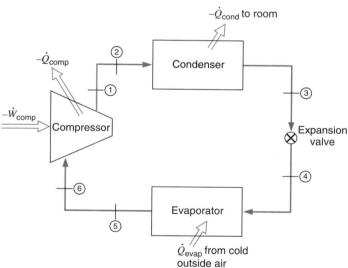

FIGURE P5.116

Calculate

a. The heat transfer from the compressor

b. The heat transfer from the R-12 in the condenser

c. The heat transfer to the R-12 in the evaporator.

5.117 A 25-L tank, shown in Fig. P5.117, that is initially evacuated is connected by a valve to an air supply line flowing air at 20°C, 800 kPa. The valve is opened, and air flows into the tank until the pressure reaches 600 kPa.

a. Determine the final temperature and mass inside the tank, assuming the process is adiabatic.

b. Develop an expression for the relation between the line temperature and the final temperature using constant specific heats.

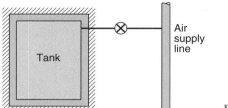

FIGURE P5.117

5.118 A 2-m³ insulated tank containing ammonia at −20°C, 80% quality, is connected by a valve to a line flowing ammonia at 2 MPa, 60°C. The valve is opened, allowing ammonia to flow into the tank. At what pressure should the valve be closed if the manufacturer wishes to have 15 kg of ammonia inside at the final state?

5.119 A 100-L rigid tank contains carbon dioxide gas at 1 MPa, 300 K. A valve is cracked open, and carbon dioxide escapes slowly until the tank pressure has dropped to 500 kPa. At this point the valve is closed. The gas remaining inside the tank may be assumed to have undergone a polytropic expansion, with polytropic exponent $n = 1.15$. Find the final mass inside and the heat transferred to the tank during the process.

5.120 A 1-m³ tank contains ammonia at 150 kPa, 25°C. The tank is attached to a line flowing ammonia at 1200 kPa, 60°C. The valve is opened, and mass flows in until the tank is half full of liquid, by volume at 25°C. Calculate the heat transferred from the tank during this process.

5.121 An evacuated 150-L tank is connected to a line flowing air at room temperature, 25°C, and 8 MPa pressure. The valve is opened allowing air to flow into the tank until the pressure inside is 6 MPa. At this point the valve is closed. This filling process occurs rapidly and is essentially adiabatic. The tank is then placed in storage where it eventually returns to room temperature. What is the final pressure?

5.122 A 0.5-m diameter balloon containing air at 200 kPa, 300 K, is attached by a valve to an air line flowing air at 400 kPa, 400 K. The valve is now opened, allowing air to flow into the balloon until the pressure inside reaches 300 kPa, at which point the valve is closed. The final temperature inside the balloon is 350 K. The pressure is directly proportional to the diameter of the balloon. Find the work and heat transfer during the process.

5.123 A 500-L insulated tank contains air at 40°C, 2 MPa. A valve on the tank is opened, and air escapes until half the original mass is gone, at which point the valve is closed. What is the pressure inside then?

5.124 A steam engine based on a turbine is shown in Fig. P5.124. The boiler tank has a volume of 100 L and initially contains saturated liquid with a very small amount of vapor at 100 kPa. Heat is now added by the burner, and the pressure regulator does not open before the boiler pressure reaches 700 kPa, which it keeps constant. The saturated vapor enters the turbine at 700 kPa and is discharged to the atmosphere as saturated vapor at 100 kPa. The burner is turned off when no more liquid is present in the as boiler. Find the total turbine work and the total heat transfer to the boiler for this process.

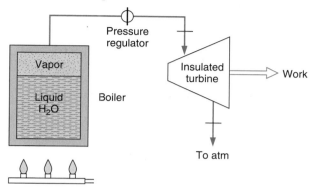

5.125 Air is contained in the insulated cylinder shown in Fig. P5.125. At this point the air is at 140 kPa, 25°C, and the cylinder volume is 15 L. The piston cross-sectional area is 0.045 m², and the spring is linear with spring constant 35 kN/m. The valve is opened, and air from the line at 700 kPa, 25°C, flows into the cylinder until the pressure reaches 700 kPa, and then the valve is closed. Find the final temperature.

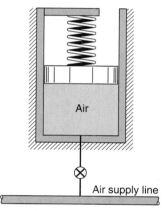

FIGURE P5.125

5.126 A 2-m³ insulated vessel, shown in Fig. P5.126, contains saturated vapor steam at 4 MPa. A valve on the top of the tank is opened, and steam is allowed to escape. During the process any liquid formed collects at the bottom of the vessel, so that only saturated vapor exits. Calculate the total mass that has escaped when the pressure inside reaches 1 MPa.

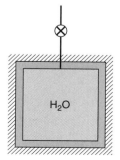

FIGURE P5.126

5.127 A 1-m³ insulated, 40-kg rigid steel tank contains air at 500 kPa, and both tank and air are at 20°C. The tank is connected to a line flowing air at 2 MPa, 20°C. The valve is opened, allowing air to flow into the tank until the pressure reaches 1.5 MPa and is then closed. Assume the air and tank are always at the same temperature and find the final temperature.

5.128 An inflatable bag, initially flat and empty, is connected to a supply line of saturated vapor R-22 at ambient temperature of 10°C. The valve is opened, and the bag slowly inflates at constant temperature to a final diameter of 2 m. The bag is inflated at constant pressure, $P_o = 100$ kPa, until it becomes spherical at $D_o = 1$ m. After this the pressure and diameter are related according to

$$P = P_o + C\left[1-\left(\frac{D_o}{D}\right)^6\right]\frac{D_o}{D}$$

A maximum pressure of 500 kPa is recorded for the whole process. Find the heat transfer to the bag during the inflation process.

5.129 A 750-L rigid tank, shown in Fig. P5.129, initially contains water at 250°C, 50% liquid and 50% vapor, by volume. A valve at the bottom of the tank is opened, and liquid is slowly withdrawn. Heat transfer takes place such that the temperature remains constant. Find the amount of heat transfer required to the state where half the initial mass is withdrawn.

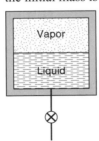

FIGURE P5.129

5.130 A cylinder, shown in Fig. P5.130, fitted with a piston restrained by a linear spring contains 1 kg of R-12 at 100°C, 800 kPa. The spring constant is 50 kN/m, and the piston cross-sectional area is 0.05 m². A valve on the cylinder is opened and R-12 flows out until half the initial mass is left. Heat is transferred so the final temperature of the R-12 is 10°C. Find the final state of the R-12, (P_2, x_2), and the heat transfer to the cylinder.

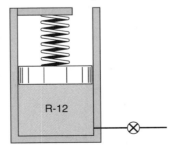

FIGURE P5.130

5.131 An initially empty bottle, $V = 0.25$ m³, is filled with water from a line at 0.8 MPa, 350°C. Assume no heat transfer and that the bottle is closed when the pressure reaches line pressure. Find the final temperature and mass in the bottle.

5.132 A supply line of ammonia at 0°C, 450 kPa is used to fill a 0.05-m³ container initially storing ammonia at 20°C, 100 kPa. The supply line valve is closed when the pressure inside reaches 290.9 kPa. Find the final mass and temperature in the container.

5.133 An example of transfer of liquids by a pressurized gas is shown in Fig. P5.133, in which both tanks A and B contains R-134a. The R-134a in A is initially saturated vapor at 20°C with a volume of 90 L, and in the insulated tank B it is at −30°C, quality of 0.01 with a volume of 75 L. The valve is opened slightly, allowing the gas to flow from A to B. The pressure regulator allows liquid to flow out when the pressure inside B reaches 150 kPa and this continues until the pressure in A has dropped to 150 kPa. Enough heat is transferred to tank A, so its and the R-134a temperature stays at 20°C. Determine

a. The quality in tank B at the end of the process.

b. The mass of liquid transferred from tank B.

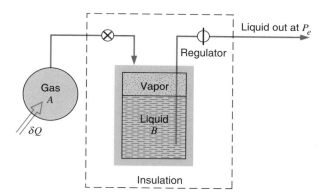

FIGURE P5.133

5.134 An insulated spring-loaded piston/cylinder, shown in Fig. P5.134, is connected to an air line flowing air at 600 kPa, 700 K by a valve. Initially the cylinder is empty and the spring force is zero. The valve is then opened until the cylinder pressure reaches 300 kPa. By noting that $u_2 = u_{line} + C_v(T_2 - T_{line})$ and $h_{line} - u_{line} = RT_{line}$ find an expression for T_2 as a function of P_2, P_o, T_{line}. With $P_o = 100$ kPa, find T_2.

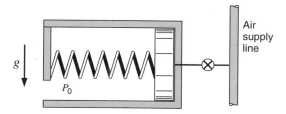

FIGURE P5.134

5.135 A mass-loaded piston/cylinder, shown in Fig. P5.135, containing air is at 300 kPa, 17°C with a volume of 0.25 m³, while at the stops $V = 1$ m³. An air line, 500 kPa, 600 K, is connected by a valve that is then opened until a final inside pressure of 400 kPa is reached, at which point $T = 350$ K. Find the air mass that enters, the work, and heat transfer.

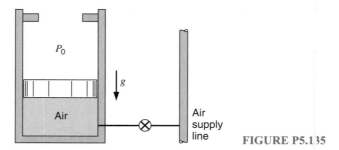

FIGURE P5.135

5.136 An elastic balloon behaves such that pressure is proportional to diameter and the balloon contains 0.5 kg air at 200 kPa, 30°C. The balloon is momentarily connected to an air line at 400 kPa, 100°C. Air is let in until the volume doubles, during which process there is a heat transfer of 50 kJ out of the balloon. Find the final temperature and the mass of air that enters the balloon.

5.137 A 2-m³ storage tank contains 95% liquid and 5% vapor by volume of liquified natural gas (LNG) at 160 K, as shown in Fig. P5.137. It may be assumed that LNG has the same properties as pure methane. Heat is transferred to the tank and saturated vapor at 160 K flows into the a steady flow heater which it leaves at 300 K. The process continues until all the liquid in the storage tank is gone. Calculate the total amount of heat transfer to the tank and the total amount of heat transferred to the heater.

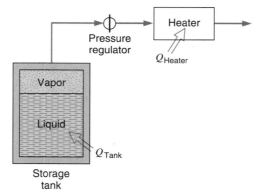

FIGURE P5.137

5.138 An uninsulated cylinder is fitted with a frictionless piston that is restrained by a variable external force. The cylinder initially contains 0.5 kg of water at 100 kPa, 30% quality. A valve is opened to a line flowing steam at 2 MPa, 400°C, and steam flows into the cylinder. It is noted that when the cylinder volume is twice its initial value, the pressure inside the cylinder is 400 kPa. When the cylinder pressure reaches 1 MPa, the valve is closed, at which point the cylinder contains 3 kg of water. During the process there is a heat transfer of 75 kJ from the cylinder to the ambient. Determine the final temperature inside the cylinder, assuming that P versus V is piecewise linear.

5.139 A nitrogen line, 300 K and 0.5 MPa, shown in Fig. P5.139, is connected to a turbine that exhausts to a closed initially empty tank of 50 m³. The turbine operates to a tank pressure of 0.5 MPa, at which point the temperature is 250 K. Assuming the entire process is adiabatic, determine the turbine work.

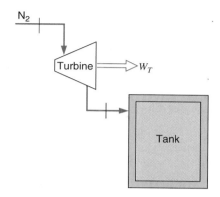

FIGURE P5.139

ENGLISH UNIT PROBLEMS

5.140E A small elevator is being designed for a construction site. It is expected to carry four 150 lbm workers to the top of a 300-ft-tall building in less than 2 min. The elevator cage will have a counterweight to balance its mass. What is the smallest size (power) electric motor that can drive this unit?

5.141E Find the missing properties and give the phase of the substance.

a. H_2O	$u = 1000$ Btu/lbm, $v = 7$ ft³/lbm	$h = ?$ $T = ?$ $x = ?$
b. H_2O	$u = 450$ Btu/lbm, $P = 1500$ lbf/in.²	$T = ?$ $x = ?$ $v = ?$
c. R-12	$T = 30$ F, $P = 45$ lbf/in.²	$h = ?$ $x = ?$
d. R-134a	$T = 140$ F, $h = 185$ Btu/lbm	$v = ?$ $x = ?$
e. NH_3	$T = 60$ F, $P = 15$ lbf/in.² $u = ?$	$v = ?$ $x = ?$

5.142E Find the missing properties among (P, T, v, u, h) together with x, if applicable, and give the phase of the substance.

a. R-22	$T = 50$ F,	$u = 85$ Btu/lbm
b. H_2O	$T = 600$ F,	$h = 1322$ Btu/lbm
c. R-12	$P = 150$ lbf/in.²,	$h = 100$ Btu/lbm
d. R-134a	$T = 100$ F,	$u = 175$ Btu/lbm
e. NH_3	$T = 70$ F,	$v = 2$ ft³/lbm

5.143E Water in a 6-ft³ closed, rigid tank is at 220 F, 90% quality. The tank is then cooled to 10 F. Calculate the heat transfer during the process.

5.144E A cylinder fitted with a frictionless piston contains 4 lbm of superheated refrigerant R-134a vapor at 100 lbf/in.², 200 F. The cylinder is now cooled so the R-134a remains at constant pressure until it reaches a quality of 75%. Calculate the heat transfer in the process.

5.145E A test cylinder with constant volume of 5 in.³ contains water at the critical point. It now cools down to room temperature of 70 F. Calculate the heat transfer from the water.

5.146E Ammonia at 30 F, quality 60% is contained in a rigid 8-ft³ tank. The tank and ammonia are now heated to a final pressure of 140 lbf/in.². Determine the heat transfer for the process.

5.147E A piston/cylinder arrangement B is connected to a 10-ft³ tank A by a line and valve, as shown in Fig. P5.17. Initially, both contain water, with A at 15 lbf/in.², saturated vapor, and B at 700 F, 60 lbf/in.², 10 ft³. The valve is now opened and the water in both A and B comes to a uniform state.

a. Find the initial mass in A and B.

b. If the process results in $T_2 = 400$ F, find the heat transfer and work.

5.148E A vertical cylinder fitted with a piston contains 10 lbm of R-22 at 50 F. Heat is transferred to the system causing the piston to rise until it reaches a set of stops at which point the volume has doubled. Additional heat is transferred until the temperature inside reaches 120 F, at which point the pressure inside the cylinder is 200 lbf/in.².

a. What is the quality at the initial state?

b. Calculate the heat transfer for the overall process.

5.149E Twenty pound-mass of water in a piston/cylinder with constant pressure is at 1100 F and a volume of 22.6 ft³. It is now cooled to 100 F. Show the $P-v$ diagram and find the work and heat transfer for the process.

5.150E A spherical balloon contains 1 lbm of R-22 at 30 F, 30% quality. This system is heated until the pressure in the balloon reaches 90 lbf/in.². In the range of variables of this process, it can be assumed that the pressure in the balloon is directly proportional to the balloon diameter. What is the heat transfer for the process?

5.151E A piston/cylinder arrangement has the piston loaded with outside atmospheric pressure and the piston mass to a pressure of 20 lbf/in.². It contains water at 25 F, which is then heated until the water becomes saturated vapor. Find the final temperature and specific work and heat transfer for the process.

5.152E Two heavily insulated tanks are connected by a valve, as shown in Fig. P5.29. Tank A contains 1.3 lbm of water at 40 lbf/in.², 600 F. Tank B has a volume of 15 ft³ and contains water at 90 lbf/in.², 80% quality. The valve is opened, and the two tanks eventually come to a uniform state. Assuming the process to be adiabatic, what is the final pressure?

5.153E Two tanks are connected by a valve and line as shown in Fig. P5.32. The volumes are both 35 ft³ with R-134a at 70 F, quality 25% in A and tank B is evacuated. The valve is opened and saturated vapor flows from A into B until the pressures become equal. The process occurs slowly enough that all temperatures stay at 70 F during the process. Find the total heat transfer to the R-134a during the process.

5.154E A water-filled reactor with volume of 50 ft³ is at 2000 lbf/in.², 560 F and placed inside a containment room, as shown in Fig. P5.35. The room has a volume of 5000 ft³, well insulated and initially evacuated. Due to a failure, the reactor ruptures and the water fills the containment room. Find the final pressure.

5.155E A cylinder fitted with a piston contains 4 lbm of R-12 at 50 F, 90% quality. The system undergoes a quasi-equilibrium polytropic expansion to 15 lbf/in.², during which the system receives a heat transfer of 50 Btu. What is the final temperature of the R-12?

5.156E A piston/cylinder arrangement of initial volume 0.3 ft³ contains saturated water vapor at 360 F. The steam now expands in a polytropic process with exponent n = 1 to a final pressure of 30 lbf/in.², while it does work against the piston. Determine the heat transfer in this process.

5.157E Calculate the heat transfer for the process described in Problem 4.56.

5.158E A 1-ft³ capsule of water at 100 lbf/in.², 300 F is placed in a larger insulated and otherwise evacuated vessel. The capsule breaks, and its contents fill the entire volume. If the final pressure should not exceed 30 lbf/in.², what should the vessel volume be?

5.159E A spherical balloon initially 6 in. in diameter and containing R-12 at 15 lbf/in.² is connected to a 1-ft³ uninsulated, rigid tank containing R-12 at 80 lbf/in.². Everything is at the ambient temperature of 80 F and is shown in Fig. P5.47. A valve connecting the tank and balloon is opened slightly and remains so until the pressures equalize. During this process heat is exchanged so the temperature remains constant at 80 F. For this range of variables the pressure inside the balloon is proportional to the diameter at any time. Calculate

a. The final pressure

b. Work done by the R-12 during the process

c. Heat transferred to the R-12 during the process

5.160E Ammonia, NH_3, is contained in a sealed rigid tank at 30 F, $x = 75\%$ and is then heated to 200 F. Find the final state P_2, u_2 and the specific work and heat transfer.

5.161E A copper block of volume 60 in.³ is heat treated at 900 F and now cooled in a 3.5-ft³ oil bath initially at 70 F. Assuming no heat transfer with the surroundings, what is the final temperature?

5.162E A computer in a closed room of volume 5000 ft³ dissipates energy at a rate of 10 hp. The air in the room was at 540 R, 1 atm, when the air conditioner suddenly stops. What is the air temperature after 15 min?

5.163E An insulated cylinder is divided into two parts of 10 ft³ each by an initially locked piston. Side A has air at 2 atm, 600 R and side B has air at 10 atm, 2500 R as shown in Fig. P5.60. The piston is now unlocked so it is free to move, and it conducts heat so the air comes to a uniform temperature $T_A = T_B$. Find the mass in both A and B and also the final T and P.

5.164E A piston/cylinder in a car contains 12 in.³ of air at 13 lbf/in.², 68 F, shown in Fig. P5.62. The air is compressed in a quasi-equilibrium polytropic process with polytropic exponent $n = 1.25$ to a final volume seven times smaller. Determine the final pressure, temperature, and the heat transfer for the process.

5.165E Water at 70 F, 15 lbf/in.², is brought to 30 lbf/in.², 2700 F. Find the change in the specific internal energy, using the tables and an appropriate model(s).

5.166E Air in a piston/cylinder at 30 lbf/in.², 1080 R, is shown in Fig. P5.65. It is expanded in a constant-pressure process to twice the initial volume (state 2). The piston is then locked with a pin, and heat is transferred to a final temperature of 1080 R. Find P, T, and h for states 2 and 3, and find the work and heat transfer in both processes.

5.167E Two containers are filled with air, one a rigid tank A, and the other a piston/cylinder B that is connected to A by a line and valve, as shown in Fig. P5.67. The initial conditions are: $m_A = 4$ lbm, $T_A = 1080$ R, $P_A = 75$ lbf/in.^{2} and $V_B = 17$ ft^3, $T_B = 80$ F, $P_B = 30$ lbf/in.^{2}. The piston in B is loaded with the outside atmosphere and the piston mass in the standard gravitational field. The valve is now opened, and the air comes to a uniform condition in both volumes. Assuming no heat transfer, find the initial mass in B, the volume of tank A, the final pressure and temperature and the work, $_1W_2$.

5.168E A 2-ft^3 rigid tank contains methane gas at 1000 F, 100 lbf/in.^{2}. The tank is cooled to 540 R.

  a. Find the final pressure and the heat transfer for the process.

  b. What is the percent error in the heat transfer if the specific heat is assumed constant at the room temperature value?

5.169E Oxygen at 50 lbf/in.^{2}, 200 F is in a piston/cylinder arrangement with a volume of 3 ft^3. It is now compressed in a polytropic process with exponent, $n = 1.2$, to a final temperature of 350 F. Calculate the heat transfer for the process.

5.170E A piston/cylinder contains 4 lbm of air at 100 F, 2 atm, as shown in Fig. P5.71. The piston is loaded with a linear spring, mass, and the atmosphere. Stops are mounted so that $V_{stop} = 100$ ft^3, at which point $P = 6$ atm is required to balance the piston forces. The air is now heated to 2600 R. Find the final pressure and volume, and the work and heat transfer. Find the work done on the spring.

5.171E An air pistol contains compressed air in a small cylinder, as shown in Fig. P5.72. Assume that the volume is 1 in.^{3}, pressure is 10 atm, and the temperature is 80 F when armed. A bullet, $m = 0.04$ lbm, acts as a piston initially held by a pin (trigger); when released, the air expands in an isothermal process (T = constant). If the air pressure is 1 atm in the cylinder as the bullet leaves the gun, find

  a. The final volume and the mass of air.

  b. The work done by the air and work done on the atmosphere.

  c. The work to the bullet and the bullet exit velocity.

5.172E A certain elastic balloon will support an internal pressure equal to $P_o = 14.7$ lbf/in.^{2} until the balloon becomes spherical at a diameter of $D_o = 3$ ft, beyond-which

$$P = P_o + C\left[1 - \left(\frac{D_o}{D}\right)^6\right]\frac{D_o}{D}$$

because of the offsetting effects of balloon curvature and elasticity. This balloon contains helium gas at 450 R, 14.7 lbf/in.^{2}, with a 10 ft^3 volume. The balloon is heated until the volume reaches 50 ft^3. During the process the maximum pressure inside the balloon is 30 lbf/in.^{2}.

a. What is the pressure inside the balloon when pressure is maximum?

b. What are the final pressure and temperature inside the balloon?

c. Determine the work and heat transfer for the overall process.

5.173E A 30-ft high cylinder, cross-sectional area 1 ft^2, has a massless piston at the bottom with water at 70 F on top of it, as shown in Fig. P5.74. Air at 540 R, volume 10 ft^3 under the piston is heated so that the piston moves up, spilling the water out over the side. Find the total heat transfer to the air when all the water has been pushed out.

5.174E Water at 300 F, quality 50% is contained in a cylinder/piston arrangement with initial volume 2 ft^3. The loading of the piston is such that the inside pressure is linear with the square root of volume as $P = 14.7 + CV^{0.5}$ lbf/in.^{2}. Now heat is transferred to the cylinder to a final pressure of 90 lbf/in.^{2}. Find the heat transfer in the process.

5.175E A closed cylinder is divided into two rooms by a frictionless piston held in place by a pin, as shown in Fig. P5.79. Room A has 0.3 ft^3 air at 14.7 lbf/in.^{2}, 90 F, and room B has 10 ft^3 saturated water vapor at 90 F. The pin is pulled, releasing the piston and both rooms come to equilibrium at 90 F. Considering a control mass of the air and water, determine the work done by the system and the heat transfer to the cylinder.

5.176E Repeat Problem 5.175, with the initial air volume in room A as 1.2 ft^3 instead of 0.3 ft^3.

5.177E Air at 95 F, 16 lbf/in.^{2}, flows in a 4 in. × 6 in. rectangular duct in a heating system. The volumetric flow rate is 30 cfm (ft^3/min). What is the velocity of the air flowing in the duct?

5.178E Saturated vapor R-134a leaves the evaporator in a heat pump at 50 F, with a steady mass flow rate of 0.2 lbm/s. What is the smallest diameter tubing that can be used at this location if the velocity of the refrigerant is not to exceed 20 ft/s?

5.179E Carbon dioxide gas enters a steady-state, steady-flow heater at 45 lbf/in.^{2} 60 F, and exits at 40 lbf/in.^{2}, 1800 F. It is shown in Fig. P5.86, where changes in kinetic and potential energies are negligible. Calculate the required heat transfer per lbm of carbon dioxide flowing through the heater.

5.180E Hoover Dam across the Colorado River dams up Lake Mead 600 ft higher than the river downstream. The electric generators driven by water-powered turbines deliver 1.2×10^6 Btu/s. If the water is 65 F, find the minimum amount of water running through the turbines.

5.181E A diffuser is a steady-state, steady-flow device used to decelerate a high-velocity fluid in a manner designed to increase the fluid pressure (essentially the opposite process of a nozzle expansion). Consider a diffuser in which air enters at 14.7 lbf/in.^{2}, 540 R, with a velocity of 600 ft/s, as shown in Fig. P5.91. The inlet cross-sectional area of the diffuser is 0.2 in.^{2}. At the exit, the area is 1.75 in.^{2}, and the exit velocity is 60 ft/s. Determine the exit pressure and temperature of the air.

5.182E A steam turbine receives water at 2000 lbf/in.^{2}, 1200 F at a rate of 200 lbm/s as shown in Fig. P5.92. In the middle section 40 lbm/s is withdrawn at 300 lbf/in.^{2}, 650 F and the rest exits the turbine at 10 lbf/in.^{2}, 95% quality. Assuming no heat transfer and no changes in kinetic energy, find the total turbine work.

5.183E A small water pump is used in an irrigation system. The pump takes water in from a river at 50 F, 1 atm at a rate of 10 lbm/s. The exit line enters a pipe that goes up to an elevation 60 ft above the pump and river, where the water runs into an open channel. Assume the process is adiabatic and that the water stays at 50 F. Find the required pump work.

5.184E A condenser, as the heat exchanger shown in Fig. P5.94, brings 1 lbm/s water flow at 1 lbf/in.^{2} from 500 F to saturated liquid at 1 lbf/in.^{2}. The cooling is done by lake water at 70 F that returns to the lake at 90 F. For an insulated condenser, find the flow rate of cooling water.

5.185E Four pound-mass of water at 80 lbf/in.^{2}, 70 F is heated in a constant pressure process (SSSF) to 2600 F. Find the best estimate for the heat transfer.

5.186E A small, high-speed turbine operating on compressed air produces a power output of 0.1 hp. The inlet state is 60 lbf/in.^{2}, 120 F, and the exit state is 14.7 lbf/in.^{2}, −20 F. Assuming the velocities to be low and the process to be adiabatic, find the required mass flow rate of air through the turbine.

5.187E In a steam generator, compressed liquid water at 1500 lbf/in.^{2}, 100 F, enters a 1-in. diameter tube at the rate of 5 ft^3/min. Steam at 1250 lbf/in.^{2}, 750 F exits the tube. Find the rate of heat transfer to the water.

5.188E A heat exchanger is used to cool an air flow from 1400 to 680 R, both states at 150 lbf/in.^{2}. The coolant is a water flow at 60 F, 15 lbf/in.^{2} and it is shown in Fig. P5.99. If the water leaves as saturated vapor, find the ratio of the flow rates $\dot{m}_{water}/\dot{m}_{air}$.

5.18E9 An air compressor takes in air at 14 lbf/in.^{2}, 60 F and delivers it at 140 lbf/in.^{2}, 1080 R to a constant-pressure cooler, which it exits at 560 R. Find the specific compressor work and the specific heat transfer.

5.190E The following data are for a simple steam power plant as shown in Fig. P5.104.

State	1	2	3	4	5	6	7
P 1bf/in.^{2}	900	890	860	830	800	1.5	1.4
T F		115	350	920	900		110

State 6 has $x_6 = 0.92$, and velocity of 600 ft/s. The rate of steam flow is 200 000 lbm/h, with 400 hp input to the pump. Piping diameters are 8 in. from steam generator to the turbine and 3 in. from the condenser to the steam generator. The economizer is a low temperature heat exchanger. Determine

a. The power output of the turbine.

b. Heat transfer rates in the condenser, economizer, and steam generator.

c. The flow rate of cooling water through the condenser, if the cooling water increases from 55 to 75 F in the condenser.

5.191E Helium is throttled from 175 lbf/in.^{2}, 70 F, to a pressure of 15 lbf/in.^{2}. The diameter of the exit pipe is so much larger than the inlet pipe that the inlet and exit velocities are equal. Find the exit temperature of the helium and the ratio of the pipe diameters.

5.192E Water flowing in a line at 60 lbf/in.^{2}, saturated vapor, is taken out through a valve to 14.7 lbf/in.^{2}. What is the temperature as it leaves the valve assuming no changes in kinetic energy and no heat transfer?

5.193E A proposal is made to use a geothermal supply of hot water to operate a steam turbine, as shown in Fig. P5.114. The high pressure water at 200 lbf/in.^{2}, 350 F, is throttled into a flash evaporator chamber, which forms liquid and vapor at a lower pressure of 60 lbf/in.^{2}. The liquid is discarded while the saturated vapor feeds the turbine and exits at 1 lbf/in.^{2}, 90% quality. If the turbine should produce 1000 hp, find the required mass flow rate of hot geothermal water in pound-mass per hour.

5.194E A 1-ft^3 tank, shown in Fig. P5.117, that is initially evacuated is connected by a valve to an air supply line flowing air at 70 F, 120 lbf/in.^{2}. The valve is opened, and air flows into the tank until the pressure reaches 90 lbf/in.^{2}.

a. Determine the final temperature and mass inside the tank, assuming the process is adiabatic.

b. Develop an expression for the relation between the line temperature and the final temperature using constant specific heats.

5.195E A 20-ft^3 tank contains ammonia at 20 lbf/in.^{2}, 80 F. The tank is attached to a line flowing ammonia at 180 lbf/in.^{2}, 140 F. The valve is opened, and mass flows in until the tank is half full of liquid, by volume at 80 F. Calculate the heat transferred from the tank during this process.

5.196E A 18-ft^3 insulated tank contains air at 100 F, 300 lbf/in.^{2}. A valve on the tank is opened, and air escapes until half the original mass is gone, at which point the valve is closed. What is the pressure inside then?

5.197E Air is contained in the insulated cylinder shown in Fig. P5.125. At this point the air is at 20 lbf/in.^{2}, 80 F, and the cylinder volume is 0.5 ft^3. The piston cross-sectional area is 0.5 ft^2, and the spring is linear with spring constant 200 lbf/in. The valve is opened, and air from the line at 100 lbf/in.^{2}, 80 F, flows into the cylinder until the pressure reaches 100 lbf/in.^{2}, and then the valve is closed. Find the final temperature.

5.198E A 35-ft^3 insulated, 90-lbm rigid steel tank contains air at 75 lbf/in.^{2}, and both tank and air are at 70 F. The tank is connected to a line flowing air at 300 lbf/in.^{2}, 70 F. The valve is opened, allowing air to flow into the tank until the pressure reaches 250 lbf/in.^{2} and is then closed. Assume the air and tank are always at the same temperature and find the final temperature.

5.199E A cylinder fitted with a piston restrained by a linear spring contains 2 lbm of R-12 at 220 F, 125 lbf/in.^{2}. The system is shown in Fig. P5.130 where the spring constant is 285 lbf/in., and the piston cross-sectional area is 75 in.^{2}. A valve on the cylinder is opened and R-12 flows out until half the initial mass is left. Heat is transferred so the final temperature of the R-12 is 50 F. Find the final state of the R-12, (P_2, x_2), and the heat transfer to the cylinder.

5.200E An initially empty bottle, $V = 10$ ft^3, is filled with water from a line at 120 lbf/in.^{2}, 500 F. Assume no heat transfer and that the bottle is closed when the pressure reaches line pressure. Find the final temperature and mass in the bottle.

5.201E A mass-loaded piston/cylinder containing air is at 45 lbf/in.^{2}, 60 F with a volume of 9 ft^3, while at the stops $V = 36$ ft^3. An air line, 75 lbf/in.^{2}, 1100 R, is connected by a valve, as shown in Fig. P5.135. The valve is then opened until a final inside pressure of 60 lbf/in.^{2} is reached, at which point T = 630R. Find the air mass that enters, the work, and heat transfer.

5.202E A nitrogen line, 540 R, and 75 lbf/in.^{2}, is connected to a turbine that exhausts to a closed initially empty tank of 2000 ft^3, as shown in Fig. P5.139. The turbine operates to a tank pressure of 75 lbf/in.^{2}, at which point the temperature is 450 R. Assuming the entire process is adiabatic, determine the turbine work.

COMPUTER, DESIGN, AND OPEN-ENDED PROBLEMS

5.203 Write a program that tracks the process in Problem 5.9 in steps of 10°C. At each T write out P, T and the heat transfer to reach that T from the initial state.

5.204 Write a program that tracks the process in Problem 5.10 in steps of 5°C until the two-phase region is reached after that step with jumps of 5% in the quality. At each step write out T, x, and the heat transfer to reach that state from the initial state.

5.205 For one of the substances in Table A.11, compare the enthalpy change between any two temperatures, T_1 and T_2, as calculated by integrating the specific heat equation; by assuming constant specific heat at the average temperature; and by assuming constant specific heat at temperature T_1.

5.206 Write a program to solve Problem 5.27, in which the initial state of the water and the piston mass and area are input variables.

5.207 Consider a general version of Problem 5.34. Write a program in which the initial state and the final temperature of the water and, in addition, the spring constant are input variables.

5.208 Write a program to solve Problem 5.45, for the general case in which the two volumes and the initial state of the water are input variables and the final pressure and temperature are output variables.

5.209 Write a program to solve Problem 5.56, for the case in which the vessel wall thickness, initial temperature, and water quality are input variables.

5.210 Write a program for Problem 5.62, where the initial state, the volume ratio, and the polytropic exponent are input variables.

5.211 Consider a general version of Problem 5.68, in which a substance listed in Table A.11, the initial temperature and pressure, and the final temperature are program inputs.

5.212 Write a program to follow the state path of the helium gas in the process of Problem 5.73. The initial state, the initial and final volumes, and the maximum pressure should be input variables in the program.

5.213 Consider the general case of Problem 5.82. Write a program in which the initial volumes and the volume percent of liquid water are all input variables.

5.214 (Adv.) Fit a polynomial expression of degree n in the temperature for ideal gas specific heat. Use the ideal gas enthalpy values for one of the substances listed in Table A.13 as data. The accuracy of the correlation should be studied as a function of the temperature range of the fit, as well as of the polynomial degree n.

5.215 An insulated tank of volume V contains a specified ideal gas (with constant specific heat) at P_1, T_1. A valve is opened, allowing the gas to flow out until the pressure inside drops to P_2. Determine T_2 and m_2 using a stepwise solution in increments of pressure between P_1 and P_2; the number of increments is variable.

5.216 Write a program to solve the general case of Problem 5.97, in which all the parameters given in the problem are input variables.

5.217 Write a program to solve the general case of Problem 5.118, in which the final mass inside the tank is specified as a program parameter.

5.218 We wish to solve Problem 5.126, using a stepwise solution, whereby the process is subdivided into several parts to minimize the effects of a linear average enthalpy approximation. Write a program to solve this problem, making the number of steps in the solution an input variable.

5.219 Write a program to solve the general case of the balloon inflation of Problem 5.128 using ammonia as the working fluid. Follow the path of the ammonia inside the balloon at all diameters during the process, and determine the pressure, the work, and the heat transfer up to each point.

5.220 Repeat the previous problem, assuming that the balloon inflation process is adiabatic instead of isothermal.

5.221 Write a program to solve Problem 5.138, incorporating an empirical smooth curve for the relation between cylinder pressure and volume throughout the process.

5.222 Examine a process where air at 300 K, 100 kPa is compressed in a piston cylinder arrangement to 600 kPa. Assume the process is polytropic with exponents in the range 1.2–1.6. Find the work and heat transfer per unit mass of air. Discuss the different cases and how they may be accomplished by insulating the cylinder or by providing heating or cooling.

5.223 A cylindrical tank of height 2 m with a cross-sectional area of 0.5 m^2 contains hot water at 80°C, 125 kPa. It is in a room with temperature $T_o = 20°C$ so it slowly loses energy to the room air proportional to the temperature difference as

$$\dot{Q}_{loss} = CA(T - T_o)$$

with the tank surface area, A, and C is a constant. For different values of the constant C estimate the time it takes to bring the water to 50°C. Make enough simplifying assumptions so you can solve the problem mathematically (i.e., find a formula for $T(t)$).

5.224 The air–water counterflowing heat exchanger given in Problem 5.99 has an air exit temperature of 360 K. Suppose the air exit temperature is listed as 300 K, then a ratio of the mass flow rates is found from the energy equation to be 5.

Show that this is an impossible process by looking at air and water temperatures at several locations inside the heat exchanger. Discuss how this puts a limit on the energy that can be extracted from the air.

5.225 A coflowing heat exchanger receives air at 800 K, 1 MPa and liquid water at 15°C, 100 kPa, shown in Fig. P5.225. The air line heats the water so at the exit the air temperature is 20°C above the water temperature. Investigate the limits for the air and water exit temperatures as a function of the ratio of the two mass flow rates. Plot the temperatures of the air and water inside the heat exchanger along the flow path.

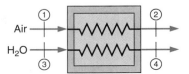

FIGURE P5.225

5.226 Consider the geothermal supply used to feed a turbine in Problem 5.114. Investigate the possibility for two flash evaporators, as shown in Fig. P5.226. Assume the turbine exit state is as listed and examine if there is an optimal choice for P_2 that will give the most turbine work per unit mass of the hot water supply.

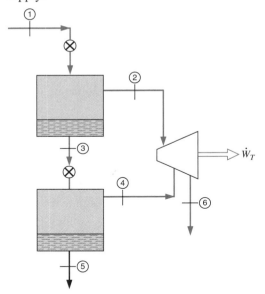

FIGURE P5.226

5.227 Consider the R-12 heat pump described in Problem 5.116. The compressor had a heat loss since the temperature is significantly higher than the surrounding temperature, 20°C. Investigate the effect of the heat loss on the work required to drive the compressor. Assume the same inlet state and exit pressure as given and vary the exit temperature. To calculate the specific work term assume the process is polytropic and make a listing of corresponding work, heat transfer, and exit temperatures. Should the compressor be insulated?

THE SECOND LAW OF THERMODYNAMICS 6

The first law of thermodynamics states that during any cycle that a system undergoes, the cyclic integral of the heat is equal to the cyclic integral of the work. The first law, however, places no restrictions on the direction of flow of heat and work. A cycle in which a given amount of heat is transferred from the system and an equal amount of work is done on the system satisfies the first law just as well as a cycle in which the flows of heat and work are reversed. However, we know from our experience that because a proposed cycle does not violate the first law does not ensure that the cycle will actually occur. It is this kind of experimental evidence that led to the formulation of the second law of thermodynamics. Thus, a cycle will occur only if both the first and second laws of thermodynamics are satisfied.

In its broader significance the second law acknowledges that processes proceed in a certain direction but not in the opposite direction. A hot cup of coffee cools by virtue of heat transfer to the surroundings, but heat will not flow from the cooler surroundings to the hotter cup of coffee. Gasoline is used as a car drives up a hill, but the fuel level in the gasoline tank cannot be restored to its original level when the car coasts down the hill. Such familiar observations as these, and a host of others, are evidence of the validity of the second law of thermodynamics.

In this chapter, we consider the second law for a system undergoing a cycle, and in the next two chapters we extend the principles to a system undergoing a change of state and then to a control volume.

6.1 HEAT ENGINES AND REFRIGERATORS

Consider the system and the surroundings previously cited in the development of the first law, as shown in Fig. 6.1. Let the gas constitute the system and, as in our discussion of the first law, let this system undergo a cycle in which work is first done on the system by the paddle wheel as the weight is lowered. Then let the cycle be completed by transferring heat to the surroundings.

We know from our experience that we cannot reverse this cycle. That is, if we transfer heat to the gas, as shown by the dotted arrow, the temperature of the gas will increase, but the paddle wheel will not turn and raise the weight. With the given surroundings (the container, the paddle wheel, and the weight) this system can operate in a cycle in which the heat transfer and work are both negative, but it cannot operate in a cycle in which both the heat transfer and work are positive, even though this would not violate the first law.

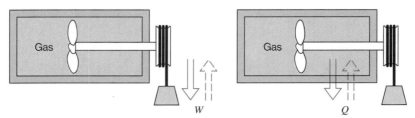

FIGURE 6.1 A system that undergoes a cycle involving work and heat.

Consider another cycle, which we know from our experience is impossible actually to complete. Let two systems, one at a high temperature and the other at a low temperature, undergo a process in which a quantity of heat is transferred from the high-temperature system to the low-temperature system. We know that this process can take place. We also know that the reverse process, in which heat is transferred from the low-temperature system to the high-temperature system, does not occur, and that it is impossible to complete the cycle by heat transfer only. This impossibility is illustrated in Fig. 6.2.

These two examples lead us to a consideration of the heat engine and the refrigerator, which is also referred to as a heat pump. With the heat engine we can have a system that operates in a cycle and performs a net positive work and a net positive heat transfer. With the heat pump we can have a system that operates in a cycle and has heat transferred to it from a low-temperature body and heat transferred from it to a high-temperature body, though work is required to do this. Three simple heat engines and two simple refrigerators will be considered.

The first heat engine is shown in Fig. 6.3. It consists of a cylinder fitted with appropriate stops and a piston. Let the gas in the cylinder constitute the system. Initially the piston rests on the lower stops, with a weight on the platform. Let the system now undergo a process in which heat is transferred from some high-temperature body to the gas, causing it to expand and raise the piston to the upper stops. At this point the weight is removed. Now let the system be restored to its initial state by transferring heat from the gas to a low-temperature body, thus completing the cycle. Since the weight was raised during the cycle, it is evident that work was done by the gas during the cycle. From the first law we conclude that the net heat transfer was positive and equal to the work done during the cycle.

Such a device is called a heat engine, and the substance to which and from which heat is transferred is called the working substance or working fluid. A heat engine may be defined as a device that operates in a thermodynamic cycle and does a certain amount of net positive work through the transfer of heat from a high-temperature body and to a low-temperature body. Often the term heat engine is used in a broader sense to include all devices that produce work, either through heat transfer or through combustion, even though the device does not operate in a thermodynamic cycle. The internal combustion

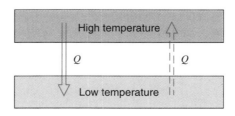

FIGURE 6.2 An example showing the impossibility of completing a cycle by transferring heat from a low-temperature body to a high-temperature body.

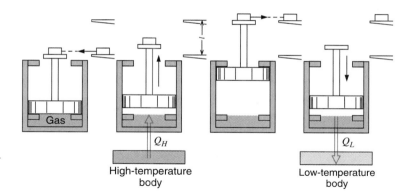

FIGURE 6.3 A simple heat engine.

engine and the gas turbine are examples of such devices, and calling them heat engines is an acceptable use of the term. In this chapter, however, we are concerned with the more restricted form of heat engine, as just defined, one which operates on a thermodynamic cycle.

A simple steam power plant is an example of a heat engine in this restricted sense. Each component in this plant may be analyzed individually as a steady-state, steady-flow process, but as a whole it may be considered a heat engine (Fig. 6.4) in which water (steam) is the working fluid. An amount of heat, Q_H, is transferred from a high-temperature body, which may be the products of combustion in a furnace, a reactor, or a secondary fluid that in turn has been heated in a reactor. In Fig. 6.4 the turbine is shown schematically as driving the pump. What is significant, however, is the net work that is delivered during the cycle. The quantity of heat Q_L is rejected to a low-temperature body, which is usually the cooling water in a condenser. Thus, the simple steam power plant is a heat engine in the restricted sense, for it has a working fluid, to which and from which heat is transferred, and which does a certain amount of work as it undergoes a cycle.

Another example of a heat engine is the thermoelectric power generation device that was discussed in Chapter 1 and shown schematically in Fig. 1.10. Heat is transferred from a high-temperature body to the hot junction (Q_H), and heat is transferred from the cold junction to the surroundings (Q_L). Work is done in the form of electrical energy.

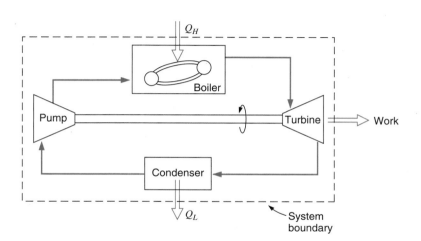

FIGURE 6.4 A heat engine involving steady-state, steady-flow processes.

Since there is no working fluid, we do not usually think of this as a device that operates in a cycle. However, if we adopt a microscopic point of view, we could regard a cycle as the flow of electrons. Furthermore, as with the steam power plant, the state at each point in the thermoelectric power generator does not change with time under steady-state conditions.

Thus, by means of a heat engine, we are able to have a system operate in a cycle and have both the net work and the net heat transfer positive, which we were not able to do with the system and surroundings of Fig. 6.1.

We note that in using the symbols Q_H and Q_L, we have departed from our sign connotation for heat, because for a heat engine Q_L is negative when the working fluid is considered as the system. In this chapter it will be advantageous to use the symbol Q_H to represent the heat transfer to or from the high-temperature body, and Q_L to represent the heat transfer to or from the low-temperature body. The direction of the heat transfer will be evident from the context.

At this point it is appropriate to introduce the concept of thermal efficiency of a heat engine. In general, we say that efficiency is the ratio of output, the energy sought, to input, the energy that costs, but these output and input must be clearly defined. At the risk of oversimplification, we may say that in a heat engine the energy sought is the work, and the energy that costs money is the heat from the high-temperature source (indirectly, the cost of the fuel). Thermal efficiency is defined as

$$\eta_{\text{thermal}} = \frac{W\,(\text{energy sought})}{Q_H\,(\text{energy that costs})} = \frac{Q_H - Q_L}{Q_H} = 1 - \frac{Q_L}{Q_H} \tag{6.1}$$

The second cycle that we were not able to complete was the one indicating the impossibility of transferring heat directly from a low-temperature body to a high-temperature body. This can of course be done with a refrigerator or heat pump. A vapor-compression refrigerator cycle, which was introduced in Chapter 1 and shown in Fig. 1.7, is shown again in Fig. 6.5. The working fluid is the refrigerant, such as R-134a or ammonia, which goes through a thermodynamic cycle. Heat is transferred to the refrigerant in the evaporator, where its pressure and temperature are low. Work is done on the refrigerant in the compressor and heat is transferred from it in the condenser, where its pressure and temperature are high. The pressure drops as the refrigerant flows through the throttle valve or capillary tube.

Thus, in a refrigerator or heat pump, we have a device that operates in a cycle, that requires work, and that accomplishes the objective of transferring heat from a low-tem-

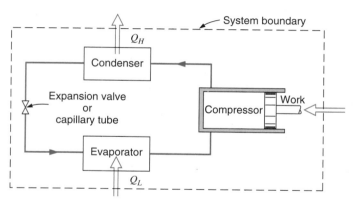

FIGURE 6.5 A simple refrigeration cycle.

perature body to a high-temperature body.

The thermoelectric refrigerator, which was discussed in Chapter 1 and is shown schematically in Fig. 1.9, is another example of a device that meets our definition of a refrigerator. The work input to the thermoelectric refrigerator is in the form of electrical energy, and heat is transferred from the refrigerated space to the cold junction (Q_L) and from the hot junction to the surroundings (Q_H).

The "efficiency" of a refrigerator is expressed in terms of the coefficient of performance, which we designate with the symbol β. For a refrigerator the objective, that is, the energy sought, is Q_L, the heat transferred from the refrigerated space. The energy that costs is the work W. Thus, the coefficient of performance, β,[1] is

$$\beta = \frac{Q_L(\text{energy sought})}{W(\text{energy that costs})} = \frac{Q_L}{Q_H - Q_L} = \frac{1}{Q_H/Q_L - 1} \qquad (6.2)$$

Before we state the second law, the concept of a thermal reservoir should be introduced. A thermal reservoir is a body to which and from which heat can be transferred indefinitely without change in the temperature of the reservoir. Thus, a thermal reservoir always remains at constant temperature. The ocean and the atmosphere approach this definition very closely. Frequently it will be useful to designate a high-temperature reservoir and a low-temperature reservoir. Sometimes a reservoir from which heat is transferred is called a source, and a reservoir to which heat is transferred is called a sink.

6.2 SECOND LAW OF THERMODYNAMICS

On the basis of the matter considered in the previous section, we are now ready to state the second law of thermodynamics. There are two classical statements of the second law, known as the Kelvin–Planck statement and the Clausius statement.

> ***The Kelvin–Planck statement:*** It is impossible to construct a device that will operate in a cycle and produce no effect other than the raising of a weight and the exchange of heat with a single reservoir.

This statement ties in with our discussion of the heat engine. In effect, it states that it is impossible to construct a heat engine that operates in a cycle, receives a given amount of heat from a high-temperature body, and does an equal amount of work. The

---

[1]It should be noted that a refrigeration or heat pump cycle can be used with either of two objectives. It can be used as a refrigerator, in which case the primary objective is Q_L, the heat transferred to the refrigerant from the refrigerated space. It can also be used as a heating system (in which case it is usually referred to as a heat pump), the objective being Q_H, the heat transferred from the refrigerant to the high-temperature body, which is the space to be heated. Q_L is transferred to the refrigerant from the ground, the atmospheric air, or well water. The coefficient of performance for this case, β' is

$$\beta' = \frac{Q_H(\text{energy sought})}{W(\text{energy that costs})} = \frac{Q_H}{Q_H - Q_L} = \frac{1}{1 - Q_L/Q_H}$$

It also follows that for a given cycle,

$$\beta' - \beta = 1$$

Unless otherwise specified, the term coefficient of performance will always refer to a refrigerator as defined by Eq. 6.2.

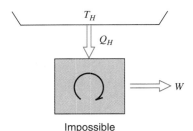

FIGURE 6.6 The Kelvin-Planck statement.

only alternative is that some heat must be transferred from the working fluid at a lower temperature to a low-temperature body. Thus, work can be done by the transfer of heat only if there are two temperature levels, and heat is transferred from the high-temperature body to the heat engine and also from the heat engine to the low-temperature body. This implies that it is impossible to build a heat engine that has a thermal efficiency of 100%.

> *The Clausius statement:* It is impossible to construct a device that operates in a cycle and produces no effect other than the transfer of heat from a cooler body to a hotter body.

This statement is related to the refrigerator or heat pump. In effect, it states that it is impossible to construct a refrigerator that operates without an input of work. This also implies that the coefficient of performance is always less than infinity.

Three observations should be made about these two statements. The first observation is that both are negative statements. It is of course impossible to "prove" a negative statement. However, we can say that the second law of thermodynamics (like every other law of nature) rests on experimental evidence. Every relevant experiment that has been conducted either directly or indirectly verifies the second law, and no experiment has ever been conducted that contradicts the second law. The basis of the second law is therefore experimental evidence.

A second observation is that these two statements of the second law are equivalent. Two statements are equivalent if the truth of each statement implies the truth of the other, or if the violation of each statement implies the violation of the other. That a violation of the Clausius statement implies a violation of the Kelvin–Planck statement may be shown. The device at the left in Fig. 6.8 is a refrigerator that requires no work and thus violates the Clausius statement. Let an amount of heat Q_L be transferred from the low-

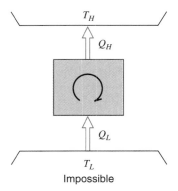

FIGURE 6.7 The Clausius statement.

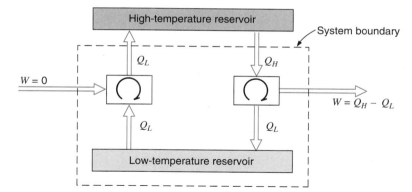

FIGURE 6.8
Demonstration of the equivalence of the two statements of the second law.

temperature reservoir to this refrigerator, and let the same amount of heat Q_L be transferred to the high-temperature reservoir. Let an amount of heat Q_H that is greater than Q_L be transferred from the high-temperature reservoir to the heat engine, and let the engine reject the amount of heat Q_L as it does an amount of work W, which equals $Q_H - Q_L$. Because there is no net heat transfer to the low-temperature reservoir, the low-temperature reservoir, the heat engine, and the refrigerator can be considered together as a device that operates in a cycle and produces no effect other than the raising of a weight (work) and the exchange of heat with a single reservoir. Thus, a violation of the Clausius statement implies a violation of the Kelvin–Planck statement. The complete equivalence of these two statements is established when it is also shown that a violation of the Kelvin–Planck statement implies a violation of the Clausius statement. This is left as an exercise for the student.

The third observation is that frequently the second law of thermodynamics has been stated as the impossibility of constructing a perpetual-motion machine of the second kind. A perpetual-motion machine of the first kind would create work from nothing or create mass or energy, thus violating the first law. A perpetual-motion machine of the second kind would extract heat from a source and then convert this heat completely into other forms of energy, thus violating the second law. A perpetual-motion machine of the third kind would have no friction, and thus would run indefinitely but produce no work.

A heat engine that violated the second law could be made into a perpetual-motion machine of the second kind by taking the following steps. Consider Fig. 6.9, which might be the power plant of a ship. An amount of heat Q_L is transferred from the ocean to a

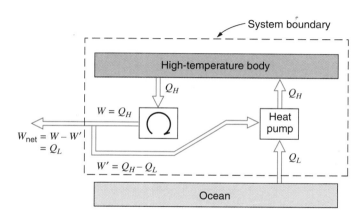

FIGURE 6.9 A perpetual-motion machine of the second kind.

high-temperature body by means of a heat pump. The work required is W', and the heat transferred to the high-temperature body is Q_H. Let the same amount of heat be transferred to a heat engine that violates the Kelvin–Planck statement of the second law and does an amount of work $W = Q_H$. Of this work an amount $Q_H - Q_L$ is required to drive the heat pump, leaving the net work ($W_{net} = Q_L$) available for driving the ship. Thus, we have a perpetual-motion machine in the sense that work is done by utilizing freely available sources of energy such as the ocean or atmosphere.

6.3 THE REVERSIBLE PROCESS

The question that can now logically be posed is this. If it is impossible to have a heat engine of 100% efficiency, what is the maximum efficiency one can have? The first step in the answer to this question is to define an ideal process, which is called a reversible process.

A reversible process for a system is defined as a process that once having taken place can be reversed and in so doing leave no change in either system or surroundings.

Let us illustrate the significance of this definition for a gas contained in a cylinder that is fitted with a piston. Consider first Fig. 6.10, in which a gas, which we define as the system, is restrained at high pressure by a piston that is secured by a pin. When the pin is removed, the piston is raised and forced abruptly against the stops. Some work is done by the system, since the piston has been raised a certain amount. Suppose we wish to restore the system to its initial state. One way of doing this would be to exert a force on the piston and thus compress the gas until the pin can be reinserted in the piston. Since the pressure on the face of the piston is greater on the return stroke than on the initial stroke, the work done on the gas in this reverse process is greater than the work done by the gas in the initial process. An amount of heat must be transferred from the gas during the reverse stroke so that the system has the same internal energy as it had originally. Thus, the system is restored to its initial state, but the surroundings have changed by virtue of the fact that work was required to force the piston down and heat was transferred to the surroundings. The initial process therefore is an irreversible one because it could not be reversed without leaving a change in the surroundings.

In Fig. 6.11 let the gas in the cylinder comprise the system, and let the piston be loaded with a number of weights. Let the weights be slid off horizontally one at a time, allowing the gas to expand and do work in raising the weights that remain on the piston.

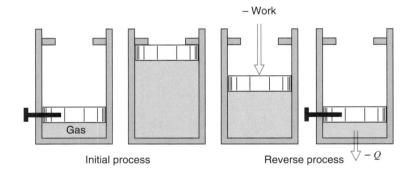

FIGURE 6.10
An example of an irreversible process.

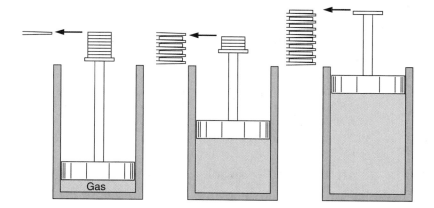

FIGURE 6.11 An example of a process that approaches being reversible.

As the size of the weights is made smaller and their number is increased, we approach a process that can be reversed, for at each level of the piston during the reverse process there will be a small weight that is exactly at the level of the platform and thus can be placed on the platform without requiring work. In the limit, therefore, as the weights become very small, the reverse process can be accomplished in such a manner that both the system and surroundings are in exactly the same state they were initially. Such a process is a reversible process.

6.4 FACTORS THAT RENDER PROCESSES IRREVERSIBLE

There are many factors that make processes irreversible. Four—friction, unrestrained expansion, heat transfer through a finite temperature difference, and mixing of two different substances—are considered in this section.

Friction

It is readily evident that friction makes a process irreversible, but a brief illustration may amplify the point. Let a block and an inclined plane make up a system, as in Fig. 6.12, and let the block be pulled up the inclined plane by weights that are lowered. A certain

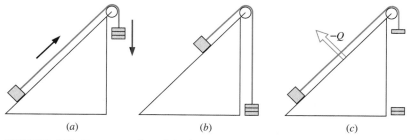

(a)   (b)   (c)

FIGURE 6.12 Demonstration of the fact that friction makes processes irreversible.

amount of work is needed to do this. Some of this work is required to overcome the friction between the block and the plane, and some is required to increase the potential energy of the block. The block can be restored to its initial position by removing some of the weights and thus allowing the block to slide back down the plane. Some heat transfer from the system to the surroundings will no doubt be required to restore the block to its initial temperature. Since the surroundings are not restored to their initial state at the conclusion of the reverse process, we conclude that friction has rendered the process irreversible. Another type of frictional effect is that associated with the flow of viscous fluids in pipes and passages and in the movement of bodies through viscous fluids.

Unrestrained Expansion

The classic example of an unrestrained expansion, as shown in Fig. 6.13, is a gas separated from a vacuum by a membrane. Consider what happens when the membrane breaks and the gas fills the entire vessel. It can be shown that this is an irreversible process by considering what would be necessary to restore the system to its original state. The gas would have to be compressed and heat transferred from the gas until its initial state is reached. Since the work and heat transfer involve a change in the surroundings, the surroundings are not restored to their initial state, indicating that the unrestrained expansion was an irreversible process. The process described in Fig. 6.10 is also an example of an unrestrained expansion.

In the reversible expansion of a gas there must be only an infinitesimal difference between the force exerted by the gas and the restraining force, so that the rate at which the boundary moves will be infinitesimal. In accordance with our previous definition, this is a quasi-equilibrium process. However, actual systems have a finite difference in forces, which causes a finite rate of movement of the boundary, and thus the processes are irreversible in some degree.

Heat Transfer Through a Finite Temperature Difference

Consider as a system a high-temperature body and a low-temperature body, and let heat be transferred from the high-temperature body to the low-temperature body. The only way in which the system can be restored to its initial state is to provide refrigeration, which requires work from the surroundings, and some heat transfer to the surroundings will also be necessary. Because of the heat transfer and the work, the surroundings are not restored to their original state, indicating that the process was irreversible.

An interesting question is now posed. Heat is defined as energy that is transferred through a temperature difference. We have just shown that heat transfer through a temperature difference is an irreversible process. Therefore, how can we have a reversible heat-transfer process? A heat-transfer process approaches a reversible process as the tem-

FIGURE 6.13
Demonstration of the fact that unrestrained expansion makes processes irreversible.

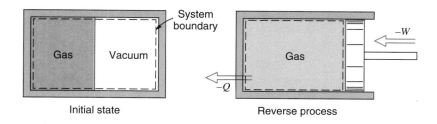

perature difference between the two bodies approaches zero. Therefore, we define a reversible heat-transfer process as one in which the heat is transferred through an infinitesimal temperature difference. We realize of course that to transfer a finite amount of heat through an infinitesimal temperature difference would require an infinite amount of time or infinite area. Therefore, all actual heat transfers are through a finite temperature difference and hence are irreversible, the greater the temperature difference, the greater the irreversibility. We will find, however, that the concept of reversible heat transfer is very useful in describing ideal processes.

Mixing of Two Different Substances

Figure 6.14 illustrates the process of mixing two different gases separated by a membrane. When the membrane is broken, a homogeneous mixture of oxygen and nitrogen fills the entire volume. This process will be considered in some detail in Chapter 11. We can say here that this may be considered a special case of an unrestrained expansion, for each gas undergoes an unrestrained expansion as it fills the entire volume. A certain amount of work is necessary to separate these gases. Thus, an air separation plant such as described in Chapter 1 requires an input of work to accomplish the separation.

Other Factors

There are a number of other factors that make processes irreversible, but they will not be considered in detail here. Hysteresis effects and the i^2R loss encountered in electrical circuits are both factors that make processes irreversible. Ordinary combustion is also an irreversible process.

It is frequently advantageous to distinguish between internal and external irreversibility. Figure 6.15 shows two identical systems to which heat is transferred. Assuming each system to be a pure substance, the temperature remains constant during the heat-transfer process. In one system the heat is transferred from a reservoir at a temperature $T + dT$, and in the other the reservoir is at a much higher temperature, $T + \Delta T$, than the system. The first is a reversible heat-transfer process and the second is an irreversible heat-transfer process. However, as far as the system itself is concerned, it passes through exactly the same states in both processes, which we assume are reversible. Thus, we can say for the second system that the process is internally reversible but externally irreversible because the irreversibility occurs outside the system.

We should also note the general interrelation of reversibility, equilibrium, and time. In a reversible process, the deviation from equilibrium is infinitesimal, and therefore it occurs at an infinitesimal rate. Since it is desirable that actual processes proceed at a finite rate, the deviation from equilibrium must be finite, and therefore the actual

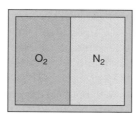
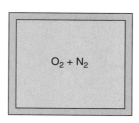

FIGURE 6.14 Demonstration of the fact that the mixing of two different substances is an irreversible process.

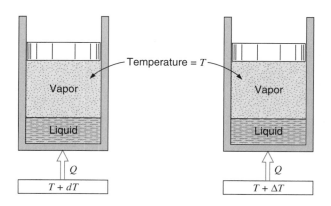

FIGURE 6.15 Illustration of the difference between an internally and externally reversible process.

process is irreversible in some degree. The greater the deviation from equilibrium, the greater the irreversibility, and the more rapidly the process will occur. It should also be noted that the quasi-equilibrium process, which was described in Chapter 2, is a reversible process, and hereafter the term reversible process will be used.

6.5 THE CARNOT CYCLE

Having defined the reversible process and considered some factors that make processes irreversible, let us again pose the question raised in Section 6.3. If the efficiency of all heat engines is less than 100%, what is the most efficient cycle we can have? Let us answer this question for a heat engine that receives heat from a high-temperature reservoir and rejects heat to a low-temperature reservoir. Since we are dealing with reservoirs, we recognize that both the high temperature and the low temperature of the reservoirs are constant and remain constant regardless of the amount of heat transferred.

Let us assume that this heat engine, which operates between the given high-temperature and low-temperature reservoirs, does so in a cycle in which every process is reversible. If every process is reversible, the cycle is also reversible; and if the cycle is reversed, the heat engine becomes a refrigerator. In the next section we will show that this is the most efficient cycle that can operate between two constant-temperature reservoirs. It is called the Carnot cycle and is named after a French engineer, Nicolas Leonard Sadi Carnot (1796–1832), who expressed the foundations of the second law of thermodynamics in 1824.

We now turn our attention to the Carnot cycle. Figure 6.16 shows a power plant that is similar in many respects to a simple steam power plant and, we assume, operates on the Carnot cycle. Consider the working fluid to be a pure substance, such as steam. Heat is transferred from the high-temperature reservoir to the water (steam) in the boiler. For this process to be a reversible heat transfer, the temperature of the water (steam) must be only infinitesimally lower than the temperature of the reservoir. This result also implies, since the temperature of the reservoir remains constant, that the temperature of the water must remain constant. Therefore, the first process in the Carnot cycle is a reversible isothermal process in which heat is transferred from the high-temperature reservoir to the working fluid. A change of phase from liquid to vapor at constant pressure is of course an isothermal process for a pure substance.

The next process occurs in the turbine without heat transfer and is therefore adiabatic. Since all processes in the Carnot cycle are reversible, this must be a reversible adi-

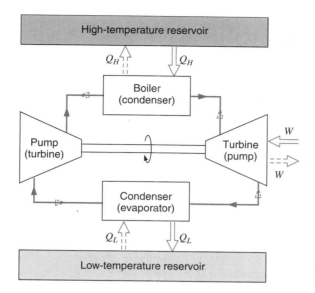

FIGURE 6.16 Example of a heat engine that operates on a Carnot cycle.

abatic process, during which the temperature of the working fluid decreases from the temperature of the high-temperature reservoir to the temperature of the low-temperature reservoir.

In the next process heat is rejected from the working fluid to the low-temperature reservoir. This must be a reversible isothermal process in which the temperature of the working fluid is infinitesimally higher than that of the low-temperature reservoir. During this isothermal process some of the steam is condensed.

The final process, which completes the cycle, is a reversible adiabatic process in which the temperature of the working fluid increases from the low temperature to the high temperature. If this were to be done with water (steam) as the working fluid, a mixture of liquid and vapor would have to be taken from the condenser and compressed. (This would be very inconvenient in practice, and therefore in all power plants the working fluid is completely condensed in the condenser. The pump handles only the liquid phase.)

Since the Carnot heat engine cycle is reversible, every process could be reversed, in which case it would become a refrigerator. The refrigerator is shown by the dotted lines and parentheses in Fig. 6.16. The temperature of the working fluid in the evaporator would be infinitesimally less that the temperature of the low-temperature reservoir, and in the condenser it is infinitesimally higher than that of the high-temperature reservoir.

It should be emphasized that the Carnot cycle can be executed in many different ways. Many different working substances can be used, such as a gas, a thermoelectric device, or a paramagnetic substance in a magnetic field such as described in Chapter 4. There are also various possible arrangements of machinery. For example, a Carnot cycle can be devised that takes place entirely within a cylinder, using a gas as a working substance, as shown in Fig. 6.17.

The important point to be made here is that the Carnot cycle, regardless of what the working substance may be, always has the same four basic processes. These processes are

1. A reversible isothermal process in which heat is transferred to or from the high-temperature reservoir.

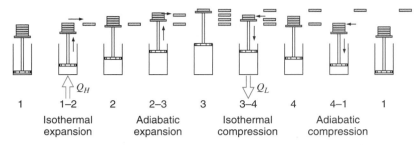

FIGURE 6.17 Example of a gaseous system operating on a Carnot cycle.

2. A reversible adiabatic process in which the temperature of the working fluid decreases from the high temperature to the low temperature.

3. A reversible isothermal process in which heat is transferred to or from the low-temperature reservoir.

4. A reversible adiabatic process in which the temperature of the working fluid increases from the low temperature to the high temperature.

6.6 TWO PROPOSITIONS REGARDING THE EFFICIENCY OF A CARNOT CYCLE

There are two important propositions regarding the efficiency of a Carnot cycle.

First Proposition

It is impossible to construct an engine that operates between two given reservoirs and is more efficient than a reversible engine operating between the same two reservoirs.

The proof of this statement is accomplished through a "thought experiment." An initial assumption is made, and it is then shown that this assumption leads to impossible conclusions. The only possible conclusion is that the initial assumption was incorrect.

Let us assume that there is an irreversible engine operating between two given reservoirs that has a greater efficiency than a reversible engine operating between the same two reservoirs. Let the heat transfer to the irreversible engine be Q_H, the heat rejected be Q'_L, and the work be W_{IE} (which equals $Q_H - Q'_L$) as shown in Fig. 6.18. Let the reversible engine operate as a refrigerator (this is possible since it is reversible). Finally, let the heat transfer with the low-temperature reservoir be Q_L, the heat transfer with the high-temperature reservoir be Q_H, and the work required be W_{RE} (which equals $Q_H - Q_L$).

Since the initial assumption was that the irreversible engine is more efficient, it follows (because Q_H is the same for both engines) that $Q'_L < Q_L$ and $W_{IE} > W_{RE}$. Now the irreversible engine can drive the reversible engine and still deliver the net work W_{net}, which equals $W_{IE} - W_{RE} = Q_L - Q'_L$. If we consider the two engines and the high-temperature reservoir as a system, as indicated in Fig. 6.18 it is a system that operates in a cycle, exchanges heat with a single reservoir, and does a certain amount of work. However, this would constitute a violation of the second law, and we conclude that our initial as-

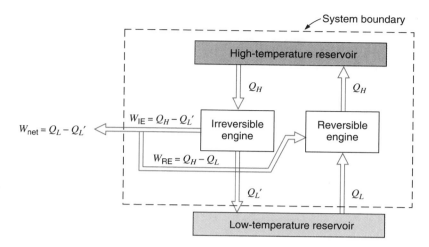

FIGURE 6.18
Demonstration of the fact that the Carnot cycle is the most efficient cycle operating between two fixed-temperature reservoirs.

sumption (that the irreversible engine is more efficient than a reversible engine) is incorrect. Therefore, we cannot have an irreversible engine that is more efficient than a reversible engine operating between the same two reservoirs.

Second Proposition

All engines that operate on the Carnot cycle between two given constant-temperature reservoirs have the same efficiency. The proof of this proposition is similar to the proof just outlined, which assumes that there is one Carnot cycle that is more efficient than another Carnot cycle operating between the same temperature reservoirs. Let the Carnot cycle with the higher efficiency replace the irreversible cycle of the previous argument, and the Carnot cycle with the lower efficiency operate as the refrigerator. The proof proceeds with the same line of reasoning as in the first proposition. The details are left as an exercise for the student.

6.7 THE THERMODYNAMIC TEMPERATURE SCALE

In discussing the matter of temperature in Chapter 2, we pointed out that the zeroth law of thermodynamics provides a basis for temperature measurement, but that a temperature scale must be defined in terms of a particular thermometer substance and device. A temperature scale that is independent of any particular substance, which might be called an absolute temperature scale, would be most desirable. In the last paragraph we noted that the efficiency of a Carnot cycle is independent of the working substance and depends only on the temperature. This fact provides the basis for such an absolute temperature scale, which we call the thermodynamic scale.

The concept of this temperature scale may be developed with the help of Fig. 6.19, which shows three reservoirs and three engines that operate on the Carnot cycle. T_1 is the highest temperature, T_3 is the lowest temperature, and T_2 is an intermediate temperature, and the engines operate between the various reservoirs as indicated. Q_1 is the same for both A and C and, since we are dealing with reversible cycles, Q_3 is the same for B and C.

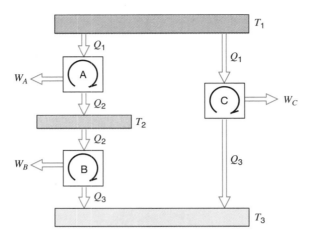

FIGURE 6.19 Arrangement of heat engines to demonstrate the thermodynamic temperature scale.

Since the efficiency of a Carnot cycle is a function only of the temperature, we can write

$$\eta_{\text{thermal}} = 1 - \frac{Q_L}{Q_H} = \psi(T_L, T_H)$$

(6.3)

where ψ designates a functional relation.

Let us apply this functional relation to the three Carnot cycles of Fig. 6.19.

$$\frac{Q_1}{Q_2} = \psi(T_1, T_2)$$

$$\frac{Q_2}{Q_3} = \psi(T_2, T_3)$$

$$\frac{Q_1}{Q_3} = \psi(T_1, T_3)$$

Since

$$\frac{Q_1}{Q_3} = \frac{Q_1 Q_2}{Q_2 Q_3}$$

it follows that

$$\psi(T_1, T_3) = \psi(T_1, T_2) \times \psi(T_2, T_3)$$

(6.4)

Note that the left side is a function of T_1 and T_3 (and not of T_2), and therefore the right side of this equation must also be a function of T_1 and T_3 (and not of T_2). From this fact we can conclude that the form of the function ψ must be such that

$$\psi(T_1, T_2) = \frac{f(T_1)}{f(T_2)}$$

$$\psi(T_2, T_3) = \frac{f(T_2)}{f(T_3)}$$

for in this way $f(T_2)$ will cancel from the product of $\psi(T_1, T_2) \times \psi(T_2, T_3)$. Therefore, we conclude that

$$\frac{Q_1}{Q_3} = \psi(T_1, T_3) = \frac{f(T_1)}{f(T_3)} \tag{6.5}$$

In general terms,

$$\frac{Q_H}{Q_L} = \frac{f(T_H)}{f(T_L)} \tag{6.6}$$

Now there are several functional relations that will satisfy this equation. For the thermodynamic scale of temperature, which was originally proposed by Lord Kelvin, the selected relation is

$$\frac{Q_H}{Q_L} = \frac{(T_H)}{(T_L)} \tag{6.7}$$

With absolute temperatures so defined, the efficiency of a Carnot cycle may be expressed in terms of the absolute temperatures.[2]

$$\eta_{\text{thermal}} = 1 - \frac{Q_L}{Q_H} = 1 - \frac{T_L}{T_H} \tag{6.8}$$

This means that if the thermal efficiency of a Carnot cycle operating between two given constant-temperature reservoirs is known, the ratio of the two absolute temperatures is also known.

It should be noted that Eq. 6.7 gives us a ratio of absolute temperatures, but it does not give us information about the magnitude of the degree. Let us first consider a qualitative approach to this matter and then a more rigorous statement.

Suppose we had a heat engine operating on the Carnot cycle that received heat at the temperature of the steam point and rejected heat at the temperature of the ice point. (Because a Carnot cycle involves only reversible processes, it is impossible to construct such a heat engine and perform the proposed experiment. However, we can follow the

---

[2]Lord Kelvin also proposed a logarithmic scale of the form

$$\frac{Q_H}{Q_L} = \frac{e^{"T_H"}}{e^{"T_L"}}$$

where $"T_H"$ and $"T_L"$ designate the absolute temperatures on this proposed logarithmic scale. This relation can also be written

$$\ln\frac{Q_H}{Q_L} = "T_H" - "T_L"$$

The form Kelvin actually proposed was

$$\log_{10}\frac{Q_H}{Q_L} = "T_H" - "T_L"$$

Thus, the relation between the scale in use and the proposed logarithmic scale is

$$"T" = \log_{10} T + L$$

where L is a constant that determines the level of temperature that corresponds to zero on the logarithmic scale. On this logarithmic scale temperatures range from $-\infty$ to $+\infty$, whereas on the thermodynamic scale in use they vary from 0 to $+\infty$ for ordinary systems.

reasoning as a "thought experiment" and gain additional understanding of the thermodynamic temperature scale.) If the efficiency of such an engine could be measured, we would find it to be 26.80%. Therefore, from Eq. 6.8,

$$\eta_{th} = 1 - \frac{T_L}{T_H} = 1 - \frac{T_{\text{ice point}}}{T_{\text{steam point}}} = 0.2680$$

$$\frac{T_{\text{ice point}}}{T_{\text{steam point}}} = 0.7320$$

This gives us one equation concerning the two unknowns T_H and T_L. The second equation comes from an arbitrary decision regarding the magnitude of the degree on the thermodynamic temperature scale. If we wish to have the magnitude of the degree on the absolute scale correspond to the magnitude of the degree on the Celsius scale, we can write

$$T_{\text{steam point}} - T_{\text{ice point}} = 100$$

Solving these two equations simultaneously, we find

$$T_{\text{steam point}} = 373.15 \text{ K}, \qquad T_{\text{ice point}} = 273.15 \text{ K}$$

It follows that

$$T(\text{°C}) + 273.15 = T(\text{K})$$

The absolute scale related to the Fahrenheit scale is the Rankine scale, designated by R. On both these scales there are 180 degrees between the ice point and the steam point. Therefore, for a Carnot cycle heat engine operating between the steam point and the ice point, we would have the two relations

$$T_{\text{steam point}} - T_{\text{ice point}} = 180$$

$$\frac{T_{\text{ice point}}}{T_{\text{steampoint}}} = 0.7320$$

Solving these two equations simultaneously, we find

$$T_{\text{steam point}} = 671.67 \text{ R}, \qquad T_{\text{ice point}} = 491.67 \text{ R}$$

It follows that temperatures on the Fahrenheit and Rankine scales are related as follows:

$$T(\text{F}) + 459.67 = T(\text{R})$$

As already noted, the measurement of efficiencies of Carnot cycles is, however, not a practical way to approach the problem of temperature measurement on the thermodynamic scale of temperature. The actual approach is based on the ideal-gas thermometer and an assigned value for the triple point of water. At the Tenth Conference on Weights and Measures, which was held in 1954, the temperature of the triple point of water was assigned the value 273.16 K. [The triple point of water is approximately 0.01°C above the ice point. The ice point is defined as the temperature of a mixture of ice and water at a pressure of 1 atm (101.3 kPa) of air that is saturated with water vapor.] The ideal-gas thermometer is discussed in the following section.

6.8 THE IDEAL-GAS TEMPERATURE SCALE

In this section we consider the ideal-gas scale of temperature. This scale is based on the observation that as the pressure of a gas approaches zero, its equation of state approaches that of an ideal gas:

$$Pv = RT$$

Consider how an ideal gas might be used to measure temperature in a constant-volume gas thermometer, which is shown schematically in Fig. 6.20. Let the gas bulb be placed in the location where the temperature is to be measured, and let the mercury column be adjusted so that the level of mercury stands at the reference mark A. Thus, the volume of the gas remains constant. Assume that the gas in the capillary tube is at the same temperature as the gas in the bulb. Then the pressure of the gas, which is indicated by the height L of the mercury column, is a measure of the temperature.

Let the pressure that is associated with the temperature of the triple point of water (273.16 K) also be measured, and let us designate this pressure $P_{t.p.}$. Then, from the definition of an ideal gas, any other temperature T could be determined from a pressure measurement P by the relation

$$T = 273.16\left(\frac{P}{P_{t.p.}}\right)$$

The temperature so measured is referred to as the ideal-gas temperature, and it can be shown that the temperature so measured is exactly equal to the thermodynamic temperature.

From a practical point of view, we have the problem that no gas behaves exactly like an ideal gas. However, we do know that as the pressure approaches zero, the behavior of all gases approaches that of an ideal gas. Suppose then that a series of measurements is made with varying amounts of gas in the gas bulb. This means that the pressure measured at the triple point, and also the pressure at any other temperature, will vary. If the indicated temperature T_i (obtained by assuming that the gas is ideal) is plotted against the pressure of gas with the bulb at the triple point of water, a curve like the one shown

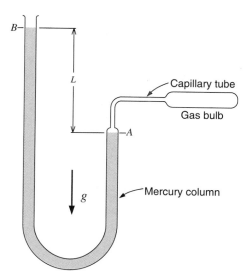

FIGURE 6.20 Schematic diagram of a constant-volume gas thermometer.

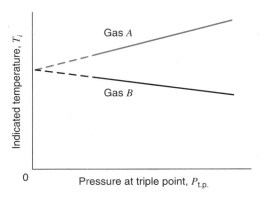

FIGURE 6.21 Sketch showing how the ideal-gas temperature is determined.

in Fig. 6.21 is obtained. When this curve is extrapolated to zero pressure, the correct ideal-gas temperature is obtained. Different curves might result from different gases, but they would all indicate the same temperature at zero pressure.

We have outlined only the general features and principles for measuring temperature on the ideal-gas scale of temperatures. Precision work in this field is difficult and laborious, and there are only a few laboratories in the world where this precision work is carried on. The International Temperature Scale, which was presented in Chapter 2, closely approximates the thermodynamic temperature scale and is much easier to work with in actual temperature measurement.

6.9 EQUIVALENCE OF IDEAL-GAS AND THERMODYNAMIC TEMPERATURE SCALES

In this section we demonstrate that the ideal-gas temperature scale discussed in the previous section is, in fact, identical to the thermodynamic temperature scale, which was defined in the discussion of the Carnot cycle and the second law. Our objective can be achieved by using an ideal gas as the working fluid for a Carnot-cycle heat engine and analyzing the four processes that make up the cycle. The four state points, 1, 2, 3, 4, and the four processes are as shown in Fig. 6.17. For convenience, let us consider a unit mass of gas inside the cylinder. Now for each of the four processes, the reversible work done at the moving boundary is given by Eq. 4.2.

$$\delta w = P \, dv$$

Similarly, for each process the gas behavior is, from the ideal-gas relation, Eq. 3.2,

$$Pv = RT$$

and the internal energy change, from Eq. 5.20, is

$$du = C_{v0} dT$$

Assuming no changes in kinetic or potential energies, the first law is, from Eq. 5.7 at unit mass,

$$\delta q = du + \delta w$$

Substituting the three previous expressions into this equation, we have for each of the four processes

$$\delta q = C_{v0}\, dT + \frac{RT}{v}\, dv \tag{6.9}$$

We now proceed to integrate Eq. 6.9 for each of the four processes that make up the Carnot cycle. For the isothermal heat addition process 1–2, we have

$$q_H = {}_1q_2 = 0 + RT_H \ln\frac{v_2}{v_1} \tag{6.10}$$

For the adiabatic expansion process 2–3,

$$0 = \int_{T_H}^{T_L} \frac{C_{v0}}{T}\, dT + R \ln\frac{v_3}{v_2} \tag{6.11}$$

For the isothermal heat rejection process 3–4,

$$q_L = -{}_3q_4 = -0 - RT_L \ln\frac{v_4}{v_3}$$

$$= +RT_L \ln\frac{v_3}{v_4} \tag{6.12}$$

and for the adiabatic compression process 4–1,

$$0 = \int_{T_L}^{T_H} \frac{C_{v0}}{T}\, dT + R \ln\frac{v_1}{v_4} \tag{6.13}$$

From Eqs. 6.11 and 6.13,

$$\int_{T_L}^{T_H} \frac{C_{v0}}{T}\, dT = R \ln\frac{v_3}{v_2} = -R\ln\frac{v_1}{v_4}$$

Therefore,

$$\frac{v_3}{v_2} = \frac{v_4}{v_1}, \quad \text{or} \quad \frac{v_3}{v_4} = \frac{v_2}{v_1} \tag{6.14}$$

Thus, from Eqs. 6.10 and 6.12 and substituting Eq. 6.14, we find that

$$\frac{q_H}{q_L} = \frac{R\,T_H \ln\dfrac{v_2}{v_1}}{R\,T_L \ln\dfrac{v_3}{v_4}} = \frac{T_H}{T_L}$$

which is Eq. 6.7, the definition of the thermodynamic temperature scale in connection with the second law.

A final point needs to be made about the significance of absolute-zero temperature in connection with the second law and the thermodynamic temperature scale. Consider a Carnot-cycle heat engine that receives a given amount of heat from a given high-temperature reservoir. As the temperature at which heat is rejected from the cycle is lowered, the net work output increases and the amount of heat rejected decreases. In the limit, the heat rejected is zero, and the temperature of the reservoir corresponding to this limit is absolute zero.

Similarly, for a Carnot-cycle refrigerator, the amount of work required to produce a given amount of refrigeration increases as the temperature of the refrigerated space decreases. Absolute zero represents the limiting temperature that can be achieved, and the amount of work required to produce a finite amount of refrigeration approaches infinity as the temperature at which refrigeration is provided approaches zero.

6.10 POWER OUTPUT AND THE CARNOT CYCLE

Consider an internally reversible (i.e., Carnot) cyclic heat engine operating between two fixed-temperature reservoirs, with the condition that the system is not externally reversible because the two heat transfers between the heat engine and the reservoirs both take place over finite temperature differences. The reservoirs are at temperatures T_H, T_L and the heat engine has high and low temperatures of T_a and T_b, respectively, as shown in Fig. 6.22.

Assume that the heat transfer rates can be expressed as

$$\dot{Q}_H = C_H\left(T_H - T_a\right) \tag{6.15}$$

$$\dot{Q}_L = C_L\left(T_b - T_L\right) \tag{6.16}$$

such that the net power (rate of work) output is, from the first law

$$\dot{W} = \dot{Q}_H - \dot{Q}_L = \eta_{th}\dot{Q}_H \tag{6.17}$$

It is noted that if the temperatures in the heat engine cycle are equal to the reservoir temperatures, then this system would be externally as well as internally reversible, and the cycle would have its maximum thermal efficiency. However, since there are no temperature differences between the reservoirs and the heat engine, the rate of heat transfer is zero, and consequently the net power output is also zero. On the other hand, if T_a is much lower than T_H, and T_b much higher than T_L, the heat transfer rates are high, but the thermal efficiency of the engine becomes low, in the limit approaching zero, and again the power output would be zero. Thus, there must be an optimal set of the cycle temperatures

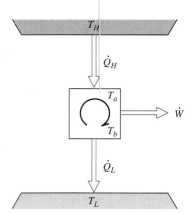

FIGURE 6.22 The externally irreversible and internally reversible heat engine.

in between these two extremes that produces the maximum power output for this situation. Applying the second law to the reversible heat engine we have

$$\frac{\dot{Q}_H}{T_a} = \frac{\dot{Q}_L}{T_b} \tag{6.18}$$

into which we substitute Eqs. 6.15 and 6.16 and thus can express T_b as a function of T_a as

$$C_H\left(\frac{T_H}{T_a} - 1\right) = C_L\left(1 - \frac{T_L}{T_b}\right)$$

$$T_b = T_L\left[1 - \frac{C_H}{C_L}\left(\frac{T_H}{T_a} - 1\right)\right]^{-1} \tag{6.19}$$

The power output, Eq. 6.17, is then rewritten, using Eqs. 6.15 and 6.18,

$$\dot{W} = \eta_{th}\dot{Q}_H = \left(1 - \frac{T_b}{T_a}\right)C_H(T_H - T_a) \tag{6.20}$$

and substitution of Eq. 6.19 for T_b gives the net power as a function of T_a and the other parameters. To find the maximum power output, we differentiate this expression with respect to T_a and set the resulting derivative equal to zero.

$$\frac{\partial\dot{W}}{\partial T_a} = \left(\frac{T_b}{T_a} - \frac{\partial T_b}{\partial T_a}\right)C_H\left(\frac{T_H}{T_a} - 1\right) - C_H\left(1 - \frac{T_b}{T_a}\right) = 0 \tag{6.21}$$

Here the derivative of T_b is

$$\frac{\partial T_b}{\partial T_a} = -T_L\left[1 - \frac{C_H}{C_L}\left(\frac{T_H}{T_a} - 1\right)\right]^{-2}\frac{C_H T_H}{C_L T_a^2} \tag{6.22}$$

After substituting Eqs. 6.19 and 6.22 into Eq. 6.21, the result becomes somewhat lengthy, but it can be simplified to give

$$T_a = \frac{C_H}{C_H + C_L}T_H + \frac{C_L}{C_H + C_L}\sqrt{T_H T_L} \tag{6.23}$$

$$T_b = \frac{C_L}{C_H + C_L}T_L + \frac{C_H}{C_H + C_L}\sqrt{T_H T_L} \tag{6.24}$$

The cycle temperatures depend on the heat transfer coefficients and the reservoir temperatures. If one of the coefficients, C_H or C_L, is significantly larger than the other, the corresponding temperature approaches that of the reservoir.

In general, using Eqs. 6.23 and 6.24 the thermal efficiency becomes

$$\eta_{th} = 1 - \frac{T_b}{T_a} = 1 - \frac{C_L T_L + C_H\sqrt{T_H T_L}}{C_H T_H + C_L\sqrt{T_H T_L}} \tag{6.25}$$

which reduces to

$$\eta_{th} = 1 - \sqrt{\frac{T_L}{T_H}} \tag{6.26}$$

and is independent of the heat transfer coefficients, C_H, C_L. The heat engine cycle that produces the maximum power thus has an efficiency given only by the reservoir temperatures and it is less than the corresponding Carnot cycle efficiency.

PROBLEMS

6.1 Calculate the thermal efficiency of the steam power plant cycle described in Problem 5.104.

6.2 Calculate the coefficient of performance of the R-12 heat pump cycle described in Problem 5.116.

6.3 Prove that a cyclic device that violates the Kelvin–Planck statement of the second law also violates the Clausius statement of the second law.

6.4 Discuss the factors that would make the power plant cycle described in Problem 5.104 an irreversible cycle.

6.5 Discuss the factors that would make the heat pump described in Problem 5.116 an irreversible cycle.

6.6 Calculate the thermal efficiency of a Carnot-cycle heat engine operating between reservoirs at 500°C and 40°C. Compare the result with that of Problem 6.1.

6.7 Calculate the thermal efficiency of a Carnot-cycle heat pump operating between reservoirs at 0°C and 45°C. Compare the result with that of Problem 6.2.

6.8 In a steam power plant 1 MW is added at 700°C in the boiler, 0.58 MW is taken out at 40°C in the condenser and the pump work is 0.02 MW. Find the plant thermal efficiency. Assuming the same pump work and heat transfer to the boiler is given, how much turbine power could be produced if the plant were running in a Carnot cycle?

6.9 At certain locations geothermal energy in undergound water is available and used as the energy source for a power plant. Consider a supply of saturated liquid water at 150°C. What is the maximum possible thermal efficiency of a cyclic heat engine using this source of energy with the ambient at 20°C? Would it be better to locate a source of saturated vapor at 150°C than use the saturated liquid at 150°C?

6.10 Find the maximum coefficient of performance for the refrigerator in your kitchen, assuming it runs in a Carnot cycle.

6.11 A sales person selling refrigerators and deep freezers will guarantee a minimum coefficient of performance of 4.5 year round. How would you evaluate that? Are they all the same?

6.12 We propose to heat a house in the winter with a heat pump. The house is to be maintained at 20°C at all times. When the ambient temperature outside drops to −10°C, the rate at which heat is lost from the house is estimated to be 25 kW. What is the minimum electrical power required to drive the heat pump?

6.13 A cyclic machine, shown in Fig. P6.13, receives 325 kJ from a 1000 K energy reservoir. It rejects 125 kJ to a 400 K energy reservoir and the cycle produces 200 kJ of work as output. Is this cycle reversible, irreversible, or impossible?

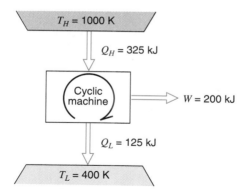

6.14 A household freezer operates in a room at 20°C. Heat must be transferred from the cold space at a rate of 2 kW to maintain its temperature at −30°C. What is the theoretically smallest (power) motor required to operate this freezer?

6.15 Differences in surface water and deep water temperature can be utilized for power generation. It is proposed to construct a cyclic heat engine that will operate near Hawaii, where the ocean temperature is 20°C near the surface and 5°C at some depth. What is the possible thermal efficiency of such a heat engine?

6.16 An inventor has developed a refrigeration unit that maintains the cold space at −10°C, while operating in a 25°C room. A coefficient of performance of 8.5 is claimed. How do you evaluate this?

6.17 A certain solar-energy collector produces a maximum temperature of 100°C. The energy is used in a cyclic heat engine that operates in a 10°C environment. What is the maximum thermal efficiency? What is it, if the collector is redesigned to focus the incoming light to produce a maximum temperature of 300°C?

6.18 Liquid sodium leaves a nuclear reactor at 800°C and is used as the energy souce in a steam power plant. The condenser cooling water comes from a cooling tower at 15°C. Determine the maximum thermal efficiency of the power plant. Is it misleading to use the temperatures given to calculate this value?

6.19 A house is heated by a heat pump driven by an electric motor using the outside as the low-temperature reservoir. The house loses energy directly proportional to the temperature difference as $Q_{loss} = K(T_H - T_L)$. Determine the minimum electric power to drive the heat pump as a function of the two temperatures.

6.20 A house is heated by an electric heat pump using the outside as the low-temperature reservoir. For several different winter outdoor temperatures, estimate the percent savings in electricity if the house is kept at 20°C instead of 24°C. Assume that the house is losing energy to the outside as described in the previous problem.

6.21 A house is cooled by an electric heat pump using the outside as the high-temperature reservoir. For several different summer outdoor temperatures, estimate the percent savings in electricity if the house is kept at 25°C instead of 20°C. Assume that the house is gaining energy from the outside directly proportional to the temperature difference.

6.22 Helium has the lowest normal boiling point of any of the elements at 4.2 K. At this temperature the enthalpy of evaporation is 83.3 kJ/kmol. A Carnot refrigeration cycle is analyzed for the production of 1 kmol of liquid helium at 4.2 K from saturated vapor at the same temperature. What is the work input to the refrigerator and the coefficient of performance for the cycle with an ambient at 300 K?

6.23 A temperature of about 0.01 K can be achieved by magnetic cooling, which was described in Section 4.5. In this process a strong magnetic field is imposed on a paramagnetic salt, maintained at 1 K by transfer of energy to liquid helium boiling at low pressure. The salt is then thermally isolated from the helium, the magnetic field is removed, and the salt temperature drops. Assume that 1 mJ is removed at an average temperature of 0.1 K to the helium by a Carnot-cycle heat pump. Find the work input to the heat pump and the coefficient of performance with an ambient at 300 K.

6.24 The lowest temperature that has been achieved is about 1×10^{-6} K. To achieve this an additional stage of cooling is required beyond that described in the previous problem, namely nuclear cooling. This process is similar to magnetic cooling, but it involves the magnetic moment associated with the nucleus rather than that associated with certain ions in the paramagnetic salt. Suppose that 10 μJ is to be removed from a specimen at an average temperature of 10^{-5} K (ten microjoules is about the potential energy loss of a pin dropping 3 mm). Find the work input to a Carnot heat pump and its coefficient of performance to do this assuming the ambient is at 300 K.

6.25 We wish to produce refrigeration at −30°C. A reservoir, shown in Fig. P6.25, is available at 200°C and the ambient temperature is 30°C. Thus, work can be done by a cyclic heat engine operating between the 200°C reservoir and the ambient. This work is used to drive the refrigerator. Determine the ratio of the heat transferred from the 200°C reservoir to the heat transferred from the −30°C reservoir, assuming all processes are reversible.

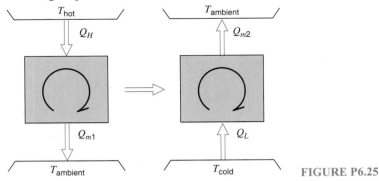

FIGURE P6.25

6.26 Consider a Carnot cycle heat engine operating in outer space. Heat can be rejected from this engine only by thermal radiation, which is proportional to the radiator area and the fourth power of absolute temperature, $Q_{rad} \sim KAT^4$. Show that for a given engine work output and given T_H, the radiator area will be minimum when the ratio $T_L/T_H = \frac{3}{4}$.

6.27 A heat pump heats a house in the winter and then reverses to cool it in the summer. The interior temperature should be 20°C in the winter and 25°C in the summer. Heat transfer through the walls and ceilings is estimated to be 2400 kJ per hour per degree temperature difference between the inside and outside.

a. If the winter outside temperature is 0°C, what is the minimum power required to drive the heat pump?

b. For the same power as in part (a), what is the maximum outside summer temperature for which the house can be maintained at 25°C?

6.28 It is proposed to build a 1000-MW electric power plant with steam as the working fluid. The condensers are to be cooled with river water (see Fig. P6.28). The maximum steam temperature is 550°C, and the pressure in the condensers will be 10 kPa. Estimate the temperature rise of the river downstream from the power plant.

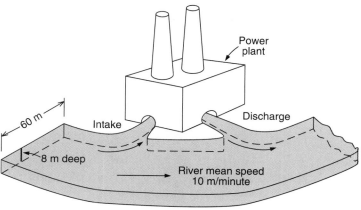

FIGURE P6.28

6.29 Two different fuels can be used in a heat engine, operating between the fuel-burning temperature and a low temperature of 350 K. Fuel A burns at 2500 K delivering 52 000 kJ/kg and costs $1.75 per kilogram. Fuel B burns at 1700 K, delivering 40 000 kJ/kg and costs $1.50 per kilogram. Which fuel would you buy and why?

6.30 Refrigerant-12 at 95°C, $x = 0.1$ flowing at 2 kg/s is brought to saturated vapor in a constant-pressure heat exchanger. The energy is supplied by a heat pump with a low temperature of 10°C. Find the required power input to the heat pump.

6.31 A furnace, shown in Fig. P6.31, can deliver heat, Q_{H1} at T_{H1} and it is proposed to use this to drive a heat engine with a rejection at T_{atm} instead of direct room heating. The heat engine drives a heat pump that delivers Q_{H2} at T_{room} using the atmosphere as the cold reservoir. Find the ratio Q_{H2}/Q_{H1} as a function of the temperatures. Is this a better set-up than direct room heating from the furnace?

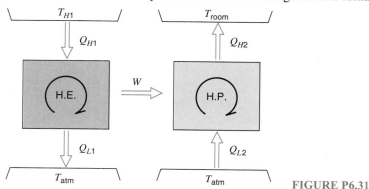

FIGURE P6.31

6.32 In a cryogenic experiment you need to keep a container at −125°C although it gains 100 W due to heat tranfer. What is the smallest motor you would need for a heat pump absorbing heat from the container and rejecting heat to the room at 20°C?

6.33 Sixty kilograms per hour of water runs through a heat exchanger, entering as saturated liquid at 200 kPa and leaving as saturated vapor. The heat is supplied by a Carnot heat pump operating from a low-temperature reservoir at 16°C. Find the rate of work into the heat pump.

6.34 A Carnot heat engine, shown in Fig. P6.34, receives energy from a reservoir at T_{res} through a heat exchanger where the heat transferred is proportional to the temperature difference as $Q_H = K(T_{res} − T_H)$. It rejects heat at a given low temperature T_L. To design the heat engine for maximum work output show that the high temperature, T_H, in the cycle should be selected as $T_H = \sqrt{(T_L T_{res})}$.

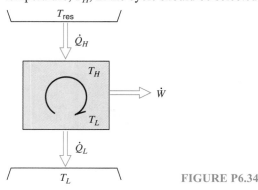

FIGURE P6.34

6.35 A heat engine has equal sized heat exchangers to the two reservoirs at 700°C and 40°C, respectively. What are the high- and low-cycle temperatures if the overall heat engine is designed for maximum power? Find the cycle efficiency and compare it to a Carnot cycle operating between the reservoir temperatures.

6.36 A 10-m³ tank of air at 500 kPa, 600 K acts as the high-temperature reservoir for a Carnot heat engine that rejects heat at 300 K. A temperature difference of 25°C between the air tank and the Carnot cycle high temperature is needed to transfer the heat. The heat engine runs until the air temperature has dropped to 400 K and then stops. Assume constant specific heat capacities for air and find how much work is given out by the heat engine.

6.37 Obtain information from manufacturers of heat pumps for domestic use. Make a listing of the coefficient of performance and compare those to corresponding Carnot cycle devices operating between the same temperature reservoirs.

ENGLISH UNIT
PROBLEMS

6.38E Calculate the thermal efficiency of the steam power plant cycle described in Problem 5.190.

6.39E Calculate the thermal efficiency of a Carnot-cycle heat engine operating between reservoirs at 920 F and 110 F. Compare the result with that of Problem 6.38.

6.40E In a steam power plant 1000 Btu/s is added at 1200 F in the boiler, 580 Btu/s is taken out at 100 F in the condenser and the pump work is 20 Btu/s. Find the plant thermal efficiency. Assume the same pump work and heat transfer to the boiler as given, how much turbine power could be produced if the plant were running in a Carnot cycle?

6.41E We propose to heat a house in the winter with a heat pump. The house is to be maintained at 68 F at all times. When the ambient temperature outside drops to 15 F, the rate at which heat is lost from the house is estimated to be 80000 Btu/h. What is the minimum electrical power required to drive the heat pump?

6.42E An inventor has developed a refrigeration unit that maintains the cold space at 14 F, while operating in a 77 F room. A coefficient of performance of 8.5 is claimed. How do you evaluate this?

6.43E Liquid sodium leaves a nuclear reactor at 1500 F and is used as the energy source in a steam power plant. The condenser cooling water comes from a cooling tower at 60 F. Determine the maximum thermal efficiency of the power plant. Is it misleading to use the temperatures given to calculate this value?

6.44E A house is heated by an electric heat pump using the outside as the low-temperature reservoir. For several different winter outdoor temperatures, estimate the percent savings in electricity if the house is kept at 68 F instead of 75 F. Assume that the house is losing energy to the outside directly proportional to the temperature difference as $\dot{Q}_{loss} = K(T_H - T_L)$.

6.45E A house is cooled by an electric heat pump using the outside as the high-temperature reservoir. For several different summer outdoor temperatures estimate the percent savings in electricity if the house is kept at 77 F instead of 68 F. Assume that the house is gaining energy from the outside directly proportional to the temperature difference.

6.46E We wish to produce refrigeration at −20 F. A reservoir is available at 400 F and the ambient temperature is 80 F, as shown in Fig. P6.25. Thus, work can be done by a cyclic heat engine operating between the 400 F reservoir and the ambient. This work is used to drive the refrigerator. Determine the ratio of the heat transferred from the 400 F reservoir to the heat transferred from the −20 F reservoir, assuming all processes are reversible.

6.47E Refrigerant-12 at 180 F, $x = 0.1$ flowing at 4 lbm/s is brought to saturated vapor in a constant-pressure heat exchanger. The energy is supplied by a heat pump with a low temperature of 50 F. Find the required power input to the heat pump.

6.48E Six-hundred pound-mass per hour of water runs through a heat exchanger, entering as saturated liquid at 30 lbf/in.² and leaving as saturated vapor. The heat is supplied by a Carnot heat pump operating from a low-temperature reservoir at 60 F. Find the rate of work into the heat pump.

6.49E A heat engine has equal sized heat exchangers to the two reservoirs at 1300 F and 100 F, respectively. What are the high- and low-cycle temperatures if the overall heat engine is designed for maximum power? Find the cycle efficiency and compare it to a Carnot cycle operating between the reservoir temperatures.

6.50E A 350-ft^3 tank of air at 80 lbf/in.^{2}, 1080 R acts as the high-temperature reservoir for a Carnot heat engine that rejects heat at 540 R. A temperature difference of 45 F between the air tank and the Carnot cycle high temperature is needed to transfer the heat. The heat engine runs until the air temperature has dropped to 700 R and then stops. Assume constant specific heat capacities for air and find how much work is given out by the heat engine.

ENTROPY 7

Up to this point in our consideration of the second law of thermodynamics, we have dealt only with thermodynamic cycles. Although this is a very important and useful approach, we are often concerned with processes rather than cycles. Thus, we might be interested in the second-law analysis of processes we encounter daily, such as the combustion process in an automobile engine, the cooling of a cup of coffee, or the chemical processes that take place in our bodies. It would also be beneficial to be able to deal with the second law quantitatively as well as qualitatively.

In our consideration of the first law, we initially stated the law in terms of a cycle, but then defined a property, the internal energy, which enabled us to use the first law quantitatively for processes. Similarly, we have stated the second law for a cycle, and we now find that the second law leads to a property, entropy, which enables us to treat the second law quantitatively for processes. Energy and entropy are both abstract concepts that help to describe certain observations. As we noted in Chapter 2, thermodynamics can be described as the science of energy and entropy. The significance of this statement will become increasingly evident.

7.1 INEQUALITY OF CLAUSIUS

The first step in our consideration of the property we call entropy is to establish the inequality of Clausius, which is

$$\oint \frac{\delta Q}{T} \leq 0$$

The inequality of Clausius is a corollary or a consequence of the second law of thermodynamics. It will be demonstrated to be valid for all possible cycles, including both reversible and irreversible heat engines and refrigerators. Since any reversible cycle can be represented by a series of Carnot cycles, in this analysis we need consider only a Carnot cycle that leads to the inequality of Clausius.

Consider first a reversible (Carnot) heat engine cycle operating between reservoirs at temperatures T_H and T_L, as shown in Fig. 7.1. For this cycle, the cyclic integral of the heat transfer, $\oint \delta Q$, is greater than zero.

$$\oint \delta Q = Q_H - Q_L > 0$$

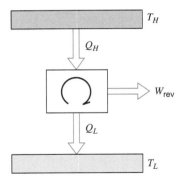

FIGURE 7.1 Reversible heat engine cycle for demonstration of the inequality of Clausius.

Since T_H and T_L are constant, from the definition of the absolute temperature scale and from the fact this is a reversible cycle, it follows that

$$\oint \frac{\delta Q}{T} = \frac{Q_H}{T_H} - \frac{Q_L}{T_L} = 0$$

If $\oint \delta Q$, the cyclic integral of δQ, approaches zero (by making T_H approach T_L) and the cycle remains reversible, the cyclic integral of $\delta Q/T$ remains zero. Thus, we conclude that for all reversible heat engine cycles

$$\oint \delta Q \geq 0$$

and

$$\oint \frac{\delta Q}{T} = 0$$

Now consider an irreversible cyclic heat engine operating between the same T_H and T_L as the reversible engine of Fig. 7.1 and receiving the same quantity of heat Q_H. Comparing the irreversible cycle with the reversible one, we conclude from the second law that

$$W_{irr} < W_{rev}$$

Since $Q_H - Q_L = W$ for both the reversible and irreversible cycles, we conclude that

$$Q_H - Q_{L\,irr} < Q_H - Q_{L\,rev}$$

and therefore

$$Q_{L\,irr} > Q_{L\,rev}$$

Consequently, for the irreversible cyclic engine,

$$\oint \delta Q = Q_H - Q_{L\,irr} > 0$$

$$\oint \frac{\delta Q}{T} = \frac{Q_H}{T_H} - \frac{Q_{L\,irr}}{T_L} < 0$$

Suppose that we cause the engine to become more and more irreversible, but keep Q_H, T_H, and T_L fixed. The cyclic integral of δQ then approaches zero, and that for $\delta Q/T$

becomes a progressively larger negative value. In the limit, as the work output goes to zero,

$$\oint \delta Q = 0$$

$$\oint \frac{\delta Q}{T} < 0$$

Thus, we conclude that for all irreversible heat engine cycles

$$\oint \delta Q \geq 0$$

$$\oint \frac{\delta Q}{T} < 0$$

To complete the demonstration of the inequality of Clausius, we must perform similar analyses for both reversible and irreversible refrigeration cycles. For the reversible refrigeration cycle shown in Fig. 7.2,

$$\oint \delta Q = -Q_H + Q_L < 0$$

and

$$\oint \frac{\delta Q}{T} = -\frac{Q_H}{T_H} + \frac{Q_L}{T_L} = 0$$

As the cyclic integral of δQ approaches zero reversibly (T_H approaches T_L), the cyclic integral of $\delta Q/T$ remains at zero. In the limit,

$$\oint \delta Q = 0$$

$$\oint \frac{\delta Q}{T} = 0$$

Thus, for all reversible refrigeration cycles,

$$\oint \delta Q \leq 0$$

$$\oint \frac{\delta Q}{T} = 0$$

Finally, let an irreversible cyclic refrigerator operate between temperatures T_H and T_L and receive the same amount of heat Q_L as the reversible refrigerator of Fig. 7.2. From the second law, we conclude that the work input required will be greater for the irreversible refrigerator, or

$$W_{irr} > W_{rev}$$

Since $Q_H - Q_L = W$ for each cycle, it follows that

$$Q_{H\,irr} - Q_L > Q_{H\,rev} - Q_L$$

and therefore,

$$Q_{H\,irr} > Q_{H\,rev}$$

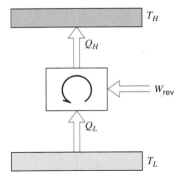

FIGURE 7.2 Reversible refrigeration cycle for demonstration of the inequality of Clausius.

That is, the heat rejected by the irreversible refrigerator to the high-temperature reservoir is greater than the heat rejected by the reversible refrigerator. Therefore, for the irreversible refrigerator,

$$\oint \delta Q = -Q_{H\,\mathrm{irr}} + Q_L < 0$$

$$\oint \frac{\delta Q}{T} = -\frac{Q_{H\,\mathrm{irr}}}{T_H} + \frac{Q_L}{T_L} < 0$$

As we make this machine progressively more irreversible, but keep Q_L, T_H, and T_L constant, the cyclic integrals of δQ and $\delta Q/T$ both become larger in the negative direction. Consequently, a limiting case as the cyclic integral of δQ approaches zero does not exit for the irreversible refrigerator.

Thus, for all irreversible refrigeration cycles,

$$\oint \delta Q < 0$$

$$\oint \frac{\delta Q}{T} < 0$$

Summarizing, we note that, in regard to the sign of $\oint \delta Q$, we have considered all possible reversible cycles (i.e., $\oint \delta Q \gtrless 0$), and for each of these reversible cycles

$$\oint \frac{\delta Q}{T} = 0$$

We have also considered all possible irreversible cycles for the sign of $\oint \delta Q$ (that is, $\oint \delta Q \gtrless 0$), and for all these irreversible cycles

$$\oint \frac{\delta Q}{T} < 0$$

Thus, for all cycles we can write

$$\oint \frac{\delta Q}{T} \le 0 \tag{7.1}$$

where the equality holds for reversible cycles and the inequality for irreversible cycles. This relation, Eq. 7.1, is known as the inequality of Clausius.

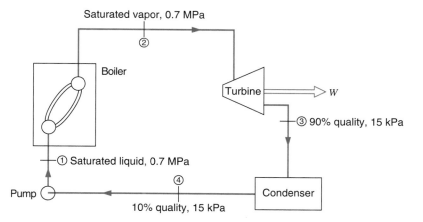

FIGURE 7.3 A simple steam power plant that demonstrates the inequality of Clausius.

The significance of the inequality of Clausius may be illustrated by considering the simple steam power plant cycle shown in Fig. 7.3. This cycle is slightly different from the usual cycle for steam power plants in that the pump handles a mixture of liquid and vapor in such proportions that saturated liquid leaves the pump and enters the boiler. Suppose that someone reports that the pressure and quality at various points in the cycle are as given in Fig. 7.3. Does this cycle satisfy the inequality of Clausius?

Heat is transferred in two places, the boiler and the condenser. Therefore,

$$\oint \frac{\delta Q}{T} = \int \left(\frac{\delta Q}{T}\right)_{boiler} + \int \left(\frac{\delta Q}{T}\right)_{condenser}$$

Since the temperature remains constant in both the boiler and condenser, this may be integrated as follows:

$$\oint \frac{\delta Q}{T} = \frac{1}{T_1}\int_1^2 \delta Q + \frac{1}{T_3}\int_3^4 \delta Q = \frac{{}_1Q_2}{T_1} + \frac{{}_3Q_4}{T_3}$$

Let us consider a 1-kg mass as the working fluid.

$${}_1q_2 = h_2 - h_1 = 2066.3 \text{ kJ / kg}, \qquad T_1 = 164.97°C$$

$${}_3q_4 = h_4 - h_3 = 463.4 - 2361.8 = -1898.4 \text{ kJ / kg} \qquad T_3 = 53.97°C$$

Therefore,

$$\oint \frac{\delta Q}{T} = \frac{2066.3}{164.97 + 273.15} - \frac{1898.4}{53.97 + 273.15} = -1.087 \text{ kJ / kg K}$$

Thus, this cycle satisfies the inequality of Clausius, which is equivalent to saying that it does not violate the second law of thermodynamics.

7.2 ENTROPY—A PROPERTY OF A SYSTEM

By applying Eq. 7.1 and Fig. 7.4 we can demonstrate that the second law of thermodynamics leads to a property of a system that we call entropy. Let a system (control mass) undergo a reversible process from state 1 to state 2 along a path A, and let the cycle be completed along path B, which is also reversible.

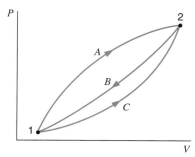

FIGURE 7.4 Two reversible cycles demonstrating the fact that entropy is a property of a substance.

Because this is a reversible cycle, we can write

$$\oint \frac{\delta Q}{T} = 0 = \int_1^2 \left(\frac{\delta Q}{T}\right)_A + \int_2^1 \left(\frac{\delta Q}{T}\right)_B$$

Now consider another reversible cycle, which proceeds first along path C and is then completed along path B. For this cycle we can write

$$\oint \frac{\delta Q}{T} = 0 = \int_1^2 \left(\frac{\delta Q}{T}\right)_C + \int_2^1 \left(\frac{\delta Q}{T}\right)_B$$

Subtracting the second equation from the first, we have

$$\int_1^2 \left(\frac{\delta Q}{T}\right)_A = \int_1^2 \left(\frac{\delta Q}{T}\right)_C$$

Since the $\int \delta Q/T$ is the same for all reversible paths between states 1 and 2, we conclude that this quantity is independent of the path and it is a function of the end states only; it is therefore a property. This property is called entropy and is designated S. It follows that entropy may be defined as a property of a substance in accordance with the relation

$$dS \equiv \left(\frac{\delta Q}{T}\right)_{rev} \tag{7.2}$$

Entropy is an extensive property, and the entropy per unit mass is designated s. It is important to note that entropy is defined here in terms of a reversible process.

The change in the entropy of a system as it undergoes a change of state may be found by integrating Eq. 7.2. Thus,

$$S_2 - S_1 = \int_1^2 \left(\frac{\delta Q}{T}\right)_{rev} \tag{7.3}$$

To perform this integration, we must know the relation between T and Q, and illustrations will be given subsequently. The important point is that since entropy is a property, the change in the entropy of a substance in going from one state to another is the same for all processes, both reversible and irreversible, between these two states. Equation 7.3 enables us to find the change in entropy only along a reversible path. However, once the change has been evaluated, this value is the magnitude of the entropy change for all processes between these two states.

Equation 7.3 enables us to determine changes of entropy, but it tells us nothing about absolute values of entropy. However, from the third law of thermodynamics, which is discussed in Chapter 12, it follows that the entropy of all pure substances can be assigned the value of zero at the absolute zero of temperature. This value allows in turn the assignment of absolute values of entropy and is particularly important when chemical reactions are involved.

However, when there is no change of composition, it is quite adequate to give values of entropy relative to some arbitrarily selected reference state. This is the procedure followed in most tables of thermodynamic properties, such as the steam tables and ammonia tables. Therefore, until absolute entropy is introduced in Chapter 12, values of entropy will always be given relative to some arbitrary reference state.

A word should be added here regarding the role of T as an integrating factor. We noted in Chapter 4 that Q is a path function, and therefore δQ is an inexact differential. However, since $(\delta Q/T)_{\text{rev}}$ is a thermodynamic property, it is an exact differential. From a mathematical perspective, we note that an inexact differential may be converted to an exact differential by the introduction of an integrating factor. Therefore, $1/T$ serves as the integrating factor in converting the inexact differential δQ to the exact differential $\delta Q/T$ for a reversible process.

7.3 THE ENTROPY OF A PURE SUBSTANCE

Entropy is an extensive property of a system. Values of specific entropy (entropy per unit mass) are given in tables of thermodynamic properties in the same manner as specific volume and specific enthalpy. The units of specific entropy in the steam tables, refrigerant tables, and ammonia tables are kJ/kg K, and the values are given relative to an arbitrary reference state. In the steam tables the entropy of saturated liquid at 0.01°C is given the value of zero. For many refrigerants, the entropy of saturated liquid at −40°C is assigned the value of zero.

In general, we use the term "entropy" to refer to both total entropy and entropy per unit mass, since the context or appropriate symbol will clearly indicate the precise meaning of the term.

In the saturation region the entropy may be calculated using the quality. The relations are similar to those for specific volume and enthalpy.

$$s = (1+x)s_f + xs_g$$

$$s = s_f + xs_{fg} \tag{7.4}$$

The entropy of a compressed liquid is tabulated in the same manner as the other properties. These properties are primarily a function of the temperature and are not greatly different from those for saturated liquid at the same temperature. Table 4 of the steam tables, which is summarized in Table A.1.4, gives the entropy of compressed-liquid water in the same manner as for other properties.

The thermodynamic properties of a substance are often shown on a temperature–entropy diagram and on an enthalpy–entropy diagram, which is also called a Mollier diagram, after Richard Mollier (1863–1935) of Germany. Figures 7.5 and 7.6 show the essential elements of temperature–entropy and enthalpy–entropy diagrams for steam. The

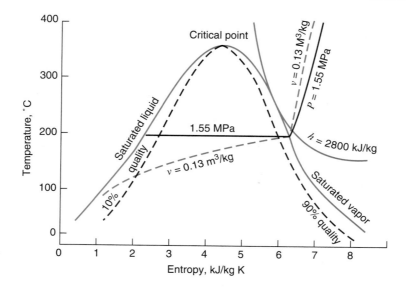

FIGURE 7.5
Temperature–entropy
diagram for steam.

general features of such diagrams are the same for all pure substances. A more complete temperature–entropy diagram for steam is shown in Fig A.1.

These diagrams are valuable both because they present thermodynamic data and because they enable us to visualize the changes of state that occur in various processes. As our study progresses, the student should acquire facility in visualizing thermodynamic processes on these diagrams. The temperature–entropy diagram is particularly useful for this purpose.

One more observation should be made here concerning the compressed-liquid lines on the temperature–entropy diagram for water. Reference to Table 4 of the steam tables (Table A.1.4) indicates that the entropy of the compressed liquid is less than that of saturated liquid at the same temperature for all the temperatures listed except 0°C. It can be shown that each constant-pressure line crosses the saturated-liquid line at the point of maximum density, which is about 4°C for water (the exact temperature at which maximum density occurs varies with pressure). For temperatures less than the temperature at

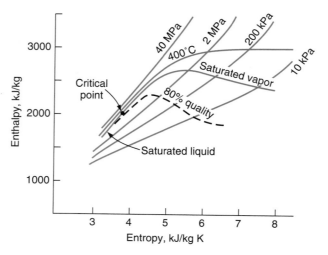

FIGURE 7.6 Enthalpy–entropy diagram for steam.

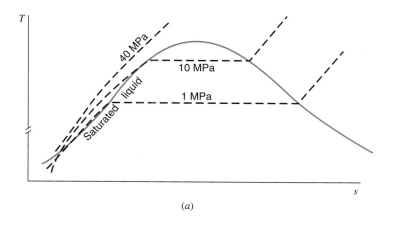

(a)

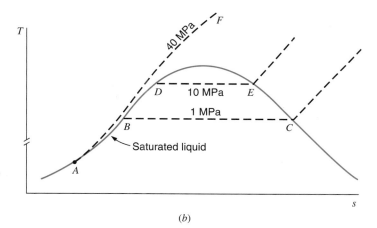

FIGURE 7.7
Temperature–entropy diagram to show properties of a compressed liquid, water.

(b)

which the density is a maximum, the entropy of compressed liquid is greater than that of the saturated liquid. Thus, lines of constant pressure in the compressed-liquid region appear (on a magnified scale) as shown in Fig. 7.7a. It is important to understand the general shape of these lines when showing the pumping process for liquids.

Having made this observation for water, we should state that for most substances the difference in the entropy of a compressed liquid and a saturated liquid at the same temperature is so small that a process in which liquid is heated at constant pressure nearly coincides with the saturated-liquid line until the saturation temperature is reached (Fig. 7.7b). Thus, if water at 10 MPa is heated from 0°C to the saturation temperature, it would be shown by line *ABD*, which coincides with the saturated-liquid line.

7.4 ENTROPY CHANGE IN REVERSIBLE PROCESSES

Having established that entropy is a thermodynamic property of a system, we now consider its significance in various processes. In this section we will limit ourselves to systems that undergo reversible processes and consider the Carnot cycle, reversible heat-transfer processes, and reversible adiabatic processes.

Let the working fluid of a heat engine operating on the Carnot cycle make up the system. The first process is the isothermal transfer of heat to the working fluid from the high-temperature reservoir. For this process we can write

$$S_2 - S_1 = \int_1^2 \left(\frac{\delta Q}{T}\right)_{rev}$$

Since this is a reversible process in which the temperature of the working fluid remains constant, the equation can be integrated to give

$$S_2 - S_1 = \frac{1}{T_H}\int_1^2 \delta Q = \frac{{}_1 Q_2}{T_H}$$

This process is shown in Fig. 7.8a, and the area under line 1–2, area 1–2–b–a–1, represents the heat transferred to the working fluid during the process.

The second process of a Carnot cycle is a reversible adiabatic one. From the definition of entropy,

$$dS = \left(\frac{\delta Q}{T}\right)_{rev}$$

it is evident that the entropy remains constant in a reversible adiabatic process. A constant-entropy process is called an isentropic process. Line 2–3 represents this process, and this process is concluded at state 3 when the temperature of the working fluid reaches T_L.

The third process is the reversible isothermal process in which heat is transferred from the working fluid to the low-temperature reservoir. For this process we can write

$$S_4 - S_3 = \int_3^4 \left(\frac{\delta Q}{T}\right)_{rev} = \frac{{}_3 Q_4}{T_L}$$

Because during this process the heat transfer is negative (in regard to the working fluid), the entropy of the working fluid decreases. Moreover, because the final process 4–1, which completes the cycle, is a reversible adiabatic process (and therefore isentropic), it is evident that the entropy decrease in process 3–4 must exactly equal the entropy increase in process 1–2. The area under line 3–4, area 3–4–a–b–3, represents the heat transferred from the working fluid to the low-temperature reservoir.

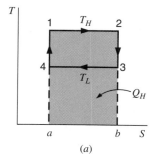

(a)

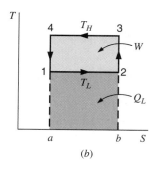

(b)

FIGURE 7.8 The Carnot cycle on the temperature–entropy diagram.

Since the net work of the cycle is equal to the net heat transfer, it is evident that area 1–2–3–4–1 represents the net work of the cycle. The efficiency of the cycle may also be expressed in terms of areas.

$$\eta_{th} = \frac{W_{net}}{Q_H} = \frac{\text{area } 1\text{-}2\text{-}3\text{-}4\text{-}1}{\text{area } 1\text{-}2\text{-}b\text{-}a\text{-}1}$$

Some statements made earlier about efficiencies may now be understood graphically. For example, increasing T_H while T_L remains constant increases the efficiency. Decreasing T_L as T_H remains constant increases the efficiency. It is also evident that the efficiency approaches 100% as the absolute temperature at which heat is rejected approaches zero.

If the cycle is reversed, we have a refrigerator or heat pump. The Carnot cycle for a refrigerator is shown in Fig. 7.8b. Notice that the entropy of the working fluid increases at T_L, since heat is transferred to the working fluid at T_L. The entropy decreases at T_H because of heat transfer from the working fluid.

Let us next consider reversible heat-transfer processes. Actually, we are concerned here with processes that are internally reversible, that is, processes that have no irreversibilities within the boundary of the system. For such processes the heat transfer to or from a system can be shown as an area on a temperature–entropy diagram. For example, consider the change of state from saturated liquid to saturated vapor at constant pressure. This process would correspond to the process 1–2 on the T–s diagram of Fig. 7.9 (note that absolute temperature is required here), and the area 1–2–b–a–1 represents the heat transfer. Since this is a constant-pressure process, the heat transfer per unit mass is equal to h_{fg}. Thus,

$$s_2 - s_1 = s_{fg} = \frac{1}{m} \int_1^2 \left(\frac{\delta Q}{T} \right)_{rev} = \frac{1}{mT} \int_1^2 \delta Q = \frac{_1 q_2}{T} = \frac{h_{fg}}{T}$$

This relation gives a clue about how s_{fg} is calculated for tabulation in tables of thermodynamic properties. For example, consider steam at 10 MPa. From the steam tables we have

$$h_{fg} = 1317.1 \text{ kJ} / \text{kg}$$

$$T = 311.06 + 273.15 = 584.21 \text{ K}$$

Therefore,

$$s_{fg} = \frac{h_{fg}}{T} = \frac{1317.1}{584.21} = 2.2544 \text{ kJ} / \text{kg K}$$

This is the value listed for s_{fg} in the steam tables.

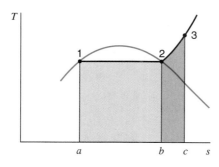

FIGURE 7.9 A temperature–entropy diagram to show areas that represent heat transfer for an internally reversible process.

If heat is transferred to the saturated vapor at constant pressure, the steam is superheated along line 2–3. For this process we can write

$$_2q_3 = \frac{1}{m}\int_2^3 \delta Q = \int_2^3 T\,ds$$

Since T is not constant, this equation cannot be integrated unless we know a relation between temperature and entropy. However, we do realize that the area under line 2–3, area 2–3–c–b–2, represents $\int_2^3 T\,ds$, and therefore represents the heat transferred during this reversible process.

The important conclusion to draw here is that for processes that are internally reversible, the area underneath the process line on a temperature–entropy diagram represents the quantity of heat transferred. This is not true for irreversible processes, as will be demonstrated later.

There are many situations in which essentially adiabatic processes take place. We have already noted that in such cases the ideal process, which is a reversible adiabatic process, is isentropic. We will consider here an example of this ideal process for a control mass and then consider later in the chapter the reversible adiabatic process for a control volume. We will also note in a later section of this chapter that by comparing an actual process with the ideal or isentropic process, we have a basis for defining the efficiency of certain classes of machines.

EXAMPLE 7.1   Consider a cylinder fitted with a piston that contains saturated R-134a vapor at −5°C. Let this vapor be compressed in a reversible adiabatic process until the pressure is 1.0 MPa. Determine the work per kilogram of R-134a for this process.

Control mass: R-134a

Initial state: T_1, saturated vapor; state fixed.

Final state: P_2 known.

Process: Reversible and adiabatic.

Model: R-134a tables.

Analysis:

First law, adiabatic:

$$_1q_2 = u_2 - u_1 + {_1w_2} = 0$$

$$_1w_2 = u_1 - u_2$$

Second law, reversible and adiabatic:

$$s_1 = s_2$$

Therefore, we know entropy and pressure in the final state, which is sufficient to specify the final state, since we are dealing with a pure substance.

Solution

From the R-134a tables,

$$h_1 = 395.34 \text{ kJ} / \text{kg} \qquad s_1 = 1.7288 \text{ kJ} / \text{kg K}$$
$$v_1 = 0.082\ 58 \text{ m}^3 / \text{kg} \qquad P_1 = P_g = 245 \text{ kPa}$$

Then,

$$u_1 = 395.34 - 245 \times 0.082\ 58 = 375.11 \text{ kJ} / \text{kg}$$

Since

$$P_2 = 1.0 \text{ MPa}, \ s_2 = 1.7288 \text{ kJ} / \text{kg K},$$

from the superheat tables for R-134a, we find

$$T_2 = 44.1°C$$

$$h_2 = 424.70 \text{ kJ} / \text{kg}, \qquad v_2 = 0.021\ 03 \text{ m}^3 / \text{kg}$$

Therefore,

$$u_2 = 424.70 - 1000 \times 0.021\ 03 = 403.67 \text{ kJ} / \text{kg}$$

$$_1w_2 = u_1 - u_2 = 375.11 - 403.67 = -28.56 \text{ kJ} / \text{kg}$$

7.5 THE THERMODYNAMIC PROPERTY RELATION

At this point we derive two important thermodynamic relations for a simple compressible substance. These relations are

$$T \ dS = dU + P \ dV$$

$$T \ dS = dH - V \ dP$$

The first of these relations can be derived by considering a simple compressible substance in the absence of motion or gravitational effects. The first law for a change of state under these conditions can be written

$$\delta Q = dU + \delta W$$

The equations we are deriving here deal first with the changes of state in which the state of the substance can be identified at all times. Thus, we must consider a quasi-equilibrium process or, to use the term introduced in the last chapter, a reversible process. For a reversible process of a simple compressible substance, we can write

$$\delta Q = T \ dS \qquad \text{and} \qquad \delta W = P \ dV$$

Substituting these relations into the first-law equation, we have

$$T \ dS = dU + P \ dV \tag{7.5}$$

which is the first equation we set out to derive. Note that this equation was derived by assuming a reversible process, and this equation can therefore be integrated for any reversible process, for during such a process the state of the substance can be identified at any point during the process. We also note that Eq. 7.5 deals only with properties. Suppose we have an irreversible process taking place between the given initial and final states. The properties of a substance depend only on the state, and therefore the change in the properties during a given change of state are the same for an irreversible process as for a reversible process. Therefore, Eq. 7.5 is often applied to an irreversible process between two given states, but the integration of Eq. 7.5 is performed along a reversible path between the same two states.

Since enthalpy is defined as

$$H = U + PV$$

it follows that

$$dH = dU + P\,dV + V\,dP$$

Substituting this relation into Eq. 7.5, we have

$$T\,dS = dH - V\,dP \tag{7.6}$$

which is the second relation that we set out to derive. These two expressions, Eqs. 7.5 and 7.6, are two forms of the thermodynamic property relation, and are frequently called Gibbs equations.

These equations can also be written for a unit mass,

$$T\,ds = du + P\,dv$$
$$T\,ds = dh - v\,dP \tag{7.7}$$

or on a mole basis,

$$T\,d\overline{s} = d\overline{u} + P\,d\overline{v}$$
$$T\,d\overline{s} = d\overline{h} - \overline{v}\,dP \tag{7.8}$$

The Gibbs equations will be used extensively in certain subsequent sections of this book.

If we consider substances of fixed composition other than a simple compressible substance, we can write "$T\,dS$" equations other than those just given for a simple compressible substance. In Eq. 4.16 we noted that for a reversible process we can write the following expression for work:

$$\delta W = P\,dV - \mathcal{T}\,dL - \mathcal{S}\,d\mathcal{A} - \mu_0\,\mathcal{H}\,d(V\mathcal{M}) - \mathcal{E}\,dZ + \cdots$$

It follows that a more general expression for the thermodynamic property relation would be

$$T\,dS = dU + P\,dV - \mathcal{T}\,dL - \mathcal{S}\,d\mathcal{A} - \mu_0\,\mathcal{H}\,d(V\mathcal{M}) - \mathcal{E}\,dZ + \cdots \tag{7.9}$$

7.6 ENTROPY CHANGE OF A CONTROL MASS DURING AN IRREVERSIBLE PROCESS

Consider a control mass that undergoes the cycles shown in Fig. 7.10. The cycle made up of the reversible processes A and B is a reversible cycle. Therefore we can write

$$\oint \frac{\delta Q}{T} = \int_1^2 \left(\frac{\delta Q}{T}\right)_A + \int_2^1 \left(\frac{\delta Q}{T}\right)_B = 0$$

The cycle made of the irreversible process C and the reversible process B is an irreversible cycle. Therefore, for this cycle the inequality of Clausius may be applied, giving the result

$$\oint \frac{\delta Q}{T} = \int_1^2 \left(\frac{\delta Q}{T}\right)_C + \int_2^1 \left(\frac{\delta Q}{T}\right)_B < 0$$

Subtracting the second equation from the first and rearranging, we have

$$\int_1^2 \left(\frac{\delta Q}{T}\right)_A > \int_1^2 \left(\frac{\delta Q}{T}\right)_C$$

Since path A is reversible, and since entropy is a property

$$\int_1^2 \left(\frac{\delta Q}{T}\right)_A = \int_1^2 dS_A = \int_1^2 dS_C$$

Therefore,

$$\int_1^2 dS_C > \int_1^2 \left(\frac{\delta Q}{T}\right)_C$$

As path C was arbitrary, the general result is

$$dS \geq \frac{\delta Q}{T}$$

$$S_2 - S_1 \geq \int_1^2 \frac{\delta Q}{T} \tag{7.10}$$

In these equations the equality holds for a reversible process and the inequality for an irreversible process.

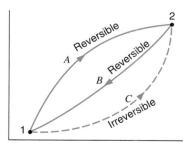

FIGURE 7.10 Entropy change of a control mass during an irreversible process.

This is one of the most important equations of thermodynamics. It is used to develop a number of concepts and definitions. In essence, this equation states the influence of irreversibility on the entropy of a control mass. Thus, if an amount of heat δQ is transferred to a control mass at temperature T in a reversible process, the change of entropy is given by the relation

$$dS = \left(\frac{\delta Q}{T} \right)_{rev}$$

If any irreversible effects occur while the amount of heat δQ is transferred to the control mass at temperature T, however, the change of entropy will be greater than for the reversible process. We would then write

$$dS > \left(\frac{\delta Q}{T} \right)_{irr}$$

Equation 7.10 holds when $\delta Q = 0$, when $\delta Q < 0$ and when $\delta Q > 0$. If δQ is negative, the entropy will tend to decrease as a result of the heat transfer. However, the influence of irreversibilities is still to increase the entropy of the mass and from the absolute numerical perspective we can still write for δQ:

$$dS \geq \frac{\delta Q}{T}$$

7.7 Entropy Generation

The conclusion from the previous considerations is that the entropy change for an irreversible process is larger than the change in a reversible process for the same δQ and T. This can be written out in a common form as an equality

$$dS = \frac{\delta Q}{T} + \delta S_{gen} \tag{7.11}$$

provided the last term is positive

$$\delta S_{gen} \geq 0 \tag{7.12}$$

The amount of entropy, δS_{gen}, is the entropy generation in the process due to irreversibilities occuring inside the system, a control mass for now, but later extended to the more general control volume. This internal generation can be caused by the processes mentioned in Section 6.4 such as friction, unrestrained expansions, internal transfer of energy (redistribution) over a finite temperature difference. In addition to this internal entropy generation, external irreversibilities are possible by heat transfer over finite temperature differences as the δQ is transferred from a reservoir or by the mechanical transfer of work.

Equation 7.12 is then valid with the equal sign for a reversible process and the greater than sign for an irreversible process. Since the entropy generation is always positive and the smallest in a reversible process, namely zero, we may deduce some limits for the heat transfer and work terms.

Consider a reversible process, for which the entropy generation is zero, and the heat transfer and work terms therefore are

$$\delta Q = T \, dS \qquad \text{and} \qquad \delta W = P \, dV$$

For an irreversible process with a nonzero entropy generation, the heat transfer from Eq. 7.11 becomes

$$\delta Q_{irr} = T \, dS - T\delta S_{gen}$$

and thus smaller than the reversible case for the same change of state, dS. We also note that for the irreversible process, the work is no longer equal to $P \, dV$, but is smaller. Further, since the first law is

$$\delta Q_{irr} = dU + \delta W_{irr}$$

and the property relation is valid,

$$T \, dS = dU + P \, dV$$

it is found that

$$\delta W_{irr} = P \, dV - T\delta S_{gen} \tag{7.13}$$

showing that the work is reduced by an amount proportional to the entropy generation. For this reason the term $T \, \delta S_{gen}$ is often called "lost work," although it is not a real work or energy quantity lost, but a lost opportunity to extract work.

Equation 7.11 can be integrated between initial and final states, to

$$S_2 - S_1 = \int_1^2 dS = \int_1^2 \frac{\delta Q}{T} + {}_1 S_{2 \, gen} \tag{7.14}$$

Thus we have an expression for the change of entropy for an irreversible process as an equality, whereas in the last section we had an inequality. In the limit of a reversible process, with a zero entropy generation, the change in S expressed in Eq. 7.14 becomes identical to Eq. 7.10 as the equal sign applies and the work term becomes $\int P \, dV$.

Some important conclusions can now be drawn from Eqs. 7.11, 7.12, and 7.13. First, there are two ways in which the entropy of a system can be increased—by transferring heat to it and by having an irreversible process. Since the entropy generation cannot be less than zero, there is only way in which the entropy of a system can be decreased, and that is to transfer heat from the system.

Second, as we have already noted for an adiabatic process, $\delta Q = 0$, and therefore the increase in entropy is always associated with the irreversibilities.

Finally, the presence of irreversibilities will cause the work to be smaller than the reversible work. This means less work out in an expansion process and more work into the control mass ($\delta W < 0$) in a compression process.

One other point, concerning the representation of irreversible processes on P–V and T–S diagrams should be made. The work for an irreversible process is not equal to $\int P \, dV$, and the heat transfer is not equal to $\int T \, dS$. Therefore, the area underneath the path does not represent work and heat on the P–V and T–S diagrams, respectively. In fact, in many situations we are not certain of the exact state through which a system passes when it undergoes an irreversible process. For this reason it is advantageous to show irreversible processes as dashed lines and reversible processes as solid lines. Thus, the area underneath the dashed line will never represent work or heat. Figure 7.11a shows an irreversible process and, because the heat transfer and work for this process are

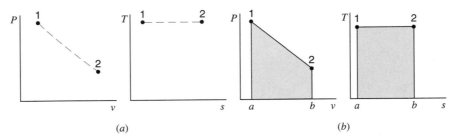

FIGURE 7.11 Reversible and irreversible processes on pressure-volume and temperature–entropy diagrams.

zero, the area underneath the dashed line has no significance. Figure 7.11*b* shows the reversible process, and area 1–2–*b*–*a*–1 represents the work on the *P–V* diagram and the heat transfer on the *T–S* diagram.

7.8 PRINCIPLE OF THE INCREASE OF ENTROPY

In the previous section, we considered irreversible processes in which the irreversibilities occurred inside the system or control mass. We also found that the entropy change of a control mass could be either positive or negative, since entropy can be increased by internal entropy generation and either increased or decreased by heat transfer, depending on the direction of that transfer. In this section, we examine the effect of heat transfer on the change of state in the surroundings, as well as the control mass itself.

Consider the process shown in Fig. 7.12 in which a quantity of heat δQ is transferred from the surroundings at temperature T_0 to the control mass at temperature T. Let the work done during this process be δW. For this process we can apply Eq. 7.10 to the control mass and write

$$dS_{\text{c.m.}} \geq \frac{\delta Q}{T}$$

$$dS_{\text{surr}} = \frac{-\delta Q}{T_0}$$

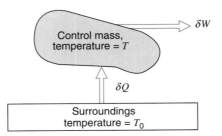

FIGURE 7.12 Entropy change for the control mass plus surroundings.

The total net change of entropy is therefore

$$dS_{net} = dS_{c.m.} + dS_{surr} \geq \frac{\delta Q}{T} - \frac{\delta Q}{T_0}$$

$$\geq \delta Q \left(\frac{1}{T} - \frac{1}{T_0} \right) \tag{7.15}$$

Since $T_0 > T$, the quantity $[(1/T) - (1/T_0)]$ is positive and we conclude that

$$dS_{net} = dS_{c.m.} + dS_{surr} \geq 0$$

If $T > T_0$, the heat transfer is from the control mass to the surroundings, and both δQ and the quantity $[(1/T) - (1/T_0)]$ are negative, thus yielding the same result.

It should be noted that the right-hand side of Eq. 7.15 represents an external entropy generation, that due to heat transfer through a finite temperature difference. To amplify this point take as a control mass the system that connects the surroundings at T_0 with the previous control mass at T, which typically is the walls. This mass does not experience any change of state, yet it has fluxes of entropy flowing in and out due to the heat transfer and it is an irreversible process. For this mass, Eq. 7.11 gives

$$dS_{c.m.2} = 0 = \frac{\delta Q}{T_0} - \frac{\delta Q}{T} + \delta S_{gen,2}$$

and it is realized that the difference in the two $\delta Q/T$ terms (fluxes of S) is the entropy generated in this control mass

$$\delta S_{gen,2} = \delta Q \left(\frac{1}{T} - \frac{1}{T_0} \right)$$

This is also precisely the location in space, where the heat transfer takes place over the finite temperature difference $T_0 - T$. This term is always positive (or zero for an adiabatic process), but as the temperature difference is made to approach zero, this term also approaches zero.

There could also be additional entropy generation terms in the surroundings of the types discussed in the previous section, if those factors were also present in the surroundings, and those will be positive, as well. Thus we conclude that the net entropy change is the sum of a number of terms, each of which is positive, due to a specific cause of irreversible entropy generation, such that the net entropy change could also be termed the total entropy generation.

$$dS_{net} = dS_{c.m.} + dS_{surr} = \sum \delta S_{gen} \geq 0 \tag{7.16}$$

where the equality holds for reversible processes and the inequality for irreversible processes. This is a very important equation, not only for thermodynamics but also for philosophical thought. This equation is referred to as the principle of the increase of entropy. The great significance is that the only processses that can take place are those in which the net change in entropy of the control mass plus its surroundings increase (or in the limit, remains constant). The reverse process, in which both the control mass and surroundings are returned to their original state, can never be made to occur. In other words, Eq. 7.16 dictates the single direction in which any process can proceed. Thus, the principle of the increase of entropy can be considered a quantitative general statement of the second law from the macroscopic point of view and applies to the combustion of fuel in our automobile engines, the cooling of our coffee, and the processes that take place in our body.

Sometimes this principle of the increase of entropy is stated in terms of an isolated system, one in which there is no interaction between the the system and its surroundings. Then there is no change in the surroundings, and we then conclude that

$$dS_{\text{isolated system}} = \delta S_{\text{gen, system}} \geq 0 \qquad (7.17)$$

That is, in an isolated system, the only processes that can occur are those that have an associated increase in entropy.

EXAMPLE 7.2 Suppose that 1 kg of saturated water vapor at 100°C is condensed to a saturated liquid at 100°C in a constant-pressure process by heat transfer to the surrounding air, which is at 25°C. What is the net increase in entropy of the water plus surroundings?

Solution

For the control mass (water), from the steam tables,

$$\Delta S_{\text{c.m.}} = -ms_{fg} = -1 \times 6.0480 = -6.0480 \text{ kJ / K}$$

Concerning the surroundings, we have

$$Q_{\text{to surroundings}} = mh_{fg} = 1 \times 2257.0 = 2257 \text{ kJ} \backslash \text{K}$$

$$\Delta S_{\text{surr}} = \frac{Q}{T_0} = \frac{2257}{298.15} = 7.5700 \text{ kJ / K}$$

$$\Delta S_{\text{net}} = \Delta S_{\text{c.m.}} + \Delta S_{\text{surr}} = -6.0480 + 7.5700 = 1.5220 \text{ kJ / K}$$

This increase in entropy is in accordance with the principle of the increase of entropy, and tells us, as does our experience, that this process can take place.

It is interesting to note how this heat transfer from the water to the surroundings might have taken place reversibly. Suppose that an engine operating on the Carnot cycle received heat from the water and rejected heat to the surroundings, as shown in Fig.

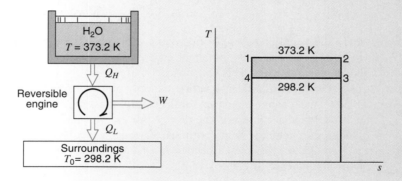

FIGURE 7.13
Reversible heat transfer with the surroundings.

7.13. The decrease in the entropy of the water is equal to the increase in the entropy of the surroundings.

$$\Delta S_{\text{c.m.}} = -6.0480 \text{ kJ / K}$$

$$\Delta S_{\text{surr}} = 6.0480 \text{ kJ / K}$$

$$Q_{\text{to surroundings}} = T_0 \Delta S = 298.15(6.0480) = 1803.2 \text{ kJ}$$

$$W = Q_H - Q_L = 2257 - 1803.2 = 453.8 \text{ kJ}$$

Since this is a reversible cycle, the engine could be reversed and operated as a heat pump. For this cycle the work input to the heat pump would be 453.8 kJ.

EXAMPLE 7.2E Suppose that 1 lbm of saturated water vapor at 212 F is condensed to a saturated liquid at 212 F in a constant-pressure process by heat transfer to the surrounding air, which is at 80 F. What is the net increase in entropy of the water plus surroundings?

Solution

For the control mass (water), from the steam tables,

$$\Delta S_{\text{system}} = -s_{fg} = -1.4446 \text{ Btu / 1bm R}$$

Considering the surroundings, we have

$$Q_{\text{to surroundings}} = h_{fg} = 970.3 \text{ Btu / 1bm}$$

$$\Delta S_{\text{surr}} = \frac{Q}{T_0} = \frac{970.3}{540} = 1.7980 \text{ Btu / 1bm R}$$

$$\Delta S_{\text{system}} + \Delta S_{\text{surr}} = -1.4446 + 1.7980 = 0.3534 \text{ Btu / 1bm R}$$

This increase in entropy is in accordance with the principle of the increase of entropy, and tells us, as does our experience, that this process can take place.

It is interesting to note how this heat transfer from the water to the surroundings might have taken place reversibly. Suppose that an engine operating on the Carnot cycle received heat from the water and rejected heat to the surroundings, as shown in Fig.

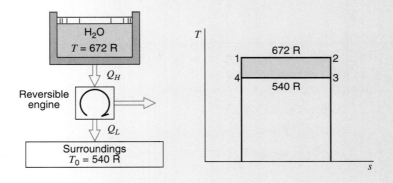

FIGURE 7.13E
Reversible heat transfer with the surroundings.

7.13E. The decrease in the entropy of the water is equal to the increase in the entropy of the surroundings.

$$\Delta S_{\text{c.m.}} = -14446.6 \text{ Btu} / 1\text{bm}$$

$$\Delta S_{\text{surr}} = 1.4446.6 \text{ Btu} / 1\text{bm}$$

$$Q_{\text{to surroundings}} = T_0 \Delta S = 540(1.4446) = 780.1 \text{ Btu} / 1\text{bm}$$

$$W = Q_H - Q_L = 970.3 - 780.1 = 190.2 \text{ Btu} / 1\text{bm}$$

Since this is a reversible cycle, the engine could be reversed and operated as a heat pump. For this cycle the work input to the heat pump would be 190.2 Btu/lbm.

7.9 ENTROPY CHANGE OF A SOLID OR LIQUID

In Section 5.6 we considered the calculation of the internal energy and enthalpy changes with temperature for solids and liquids and found that, in general, it is possible to express both in terms of the specific heat, in the simple manner of Eq. 5.17, and in most instances in the integrated form of Eq. 5.18. We can now use this result and the thermodynamic property relation, Eq. 7.7, to calculate the entropy change for a solid or liquid. Note that for such a phase the specific volume term in Eq. 7.7 is very small, so that substituting Eq. 5.17 yields

$$ds \simeq \frac{du}{T} \simeq \frac{C}{T} dT \tag{7.18}$$

Now, as was mentioned in Section 5.6, for many processes involving a solid or liquid, we may assume that the specific heat remains constant, in which case Eq. 7.18 can be integrated. The result is

$$s_2 - s_1 \simeq C \ln \frac{T_2}{T_1} \tag{7.19}$$

If the specific heat is not constant, then commonly C is known as a function of T, in which case Eq. 7.18 can also be integrated to find the entropy change.

EXAMPLE 7.3   One kilogram of liquid water is heated from 20°C to 90°C. Calculate the entropy change, assuming constant specific heat, and compare the result with that found when using the steam tables.

Control mass: Water.

Initial and final states: Known.

Model: Constant specific heat, value at room temperature.

Solution

For constant specific heat, from Eq. 7.19,

$$s_2 - s_1 = 4.184 \ \ln\left(\frac{363.2}{293.2}\right) = 0.8958 \ \text{kJ} / \text{kg K}$$

Comparing this result with that obtained by using the steam tables, we have

$$s_2 - s_1 = s_{f\,90°C} - s_{f\,20°C} = 1.1925 - 0.2966$$

$$= 0.8959 \ \text{kJ} / \text{kg K}$$

7.10 ENTROPY CHANGE OF AN IDEAL GAS

Two very useful equations for computing the entropy change of an ideal gas can be developed from Eq. 7.7 by substituting Eqs. 5.20 and 5.24:

$$T \ ds = du + P \ dv$$

For an ideal gas

$$du = c_{v0}dT \qquad \text{and} \qquad \frac{P}{T} = \frac{R}{v}$$

Therefore,

$$ds = C_{v0}\frac{dT}{T} + \frac{R \ dv}{v} \tag{7.20}$$

$$s_2 - s_1 = \int_1^2 C_{v0}\frac{dT}{T} + R \ln\frac{v_2}{v_1} \tag{7.21}$$

Similarly,

$$T \ ds = dh - v \ dP$$

For an ideal gas

$$dh = C_{p0} \ dT \qquad \text{and} \qquad \frac{v}{T} = \frac{R}{P}$$

Therefore,

$$ds = C_{p0}\frac{dT}{T} - R\frac{dp}{P} \tag{7.22}$$

$$s_2 - s_1 = \int_1^2 C_{p0}\frac{dT}{T} - R \ \ln\frac{P_2}{P_1} \tag{7.23}$$

To integrate Eqs. 7.21 and 7.23, we must know the temperature dependence of the specific heats. However, if we recall that their difference is always constant as expressed by Eq. 5.27, we realize that we need to examine the temperature dependence of only one of the specific heats.

As in Section 5.7, let us consider the specific heat C_{p0}. Again, there are three possibilities to examine, the simplest of which is the assumption of constant specific heat. In this instance it is possible to integrate Eq. 7.23 directly, to

$$s_2 - s_1 = C_{p0} \ln \frac{T_2}{T_1} - R \ln \frac{P_2}{P_1} \tag{7.24}$$

Similarly, integrating Eq. 7.21 for constant specific heat, we have

$$s_2 - s_1 = C_{v0} \ln \frac{T_2}{T_1} + R \ln \frac{v_2}{v_1} \tag{7.25}$$

The second possibility for the specific heat is to use an analytical equation for C_{p0} as a function of temperature, for example, one of those listed in Table A.11. The third possibility is to integrate the results of the calculations of statistical thermodynamics from reference temperature T_0 to any other temperature T and define a function

$$s_T^0 = \int_{T_0}^{T} \frac{C_{pO}}{T} dT \tag{7.26}$$

This function can then be tabulated in the single-entry (temperature) ideal-gas table, as for air in Table A.12 or for other gases in Table A.13. The entropy change between any two states 1 and 2 is then given by

$$s_2 - s_1 = \left(s_{T2}^0 - s_{T1}^0\right) - R \ln \frac{P_2}{P_1} \tag{7.27}$$

As with the energy functions discussed in Section 5.7, the ideal-gas tables, Tables A.12 and A.13, would give the most accurate results, and the equations listed in Table A.11 would give a close empirical approximation. Constant specific heat would be less accurate, except for monoatomic gases and for other gases below room temperature. Again, it should be remembered that all these results are part of the ideal-gas model, which may or may not be appropriate in any particular problem.

EXAMPLE 7.4 Consider Example 5.6, in which oxygen is heated from 300 to 1500 K. Assume that during this process the pressure dropped from 200 to 150 kPa. Calculate the change in entropy per kilogram.

Solution

The most accurate answer for the entropy change, assuming ideal-gas behavior, would be from the ideal-gas tables, Table A.13. This result is, using Eq. 7.27,

$$\bar{s}_2 - \bar{s}_1 = \left(258.068 - 205.329\right) - 8.3145 \ln\left(\frac{150}{200}\right)$$

$$= 52.739 + 2.392 = 55.131 \text{ kJ / kmol K}$$

$$s_2 - s_1 = \frac{55.131}{32} = 1.7228 \text{ kJ / kg K}$$

The empirical equation from Table A.11 should give a good approximation to this result. Integrating Eq. 7.23, we have

$$\bar{s}_2 - \bar{s}_1 = \int_{T_1}^{T_2} \bar{C}_{p0} \frac{dT}{T} - \bar{R} \ln \frac{P_2}{P_1}$$

$$\bar{s}_2 - \bar{s}_1 = \left(37.432 \ln \theta + \frac{0.020\,102}{1.5} \theta^{1.5} \right.$$

$$\left. + \frac{178.57}{1.5} \theta^{-1.5} - \frac{236.88}{2} \theta^{-2} \right) \Bigg|_{\theta_1=3}^{\theta_2=1.5} - 8.3145 \ln \left(\frac{150}{200} \right)$$

$$= 52.726 = 2.392 = 55.118 \text{ kJ / kmol K}$$

$$s_2 - s_1 = \frac{\bar{s}_2 - \bar{s}_1}{M} = \frac{55.118}{32} = 1.7224 \text{ kJ / kg K}$$

which is within 0.1% of the previous value. For constant specific heat, using the value at 300 K from Table A.10, we have

$$s_2 - s_1 = 0.9216 \ln \left(\frac{1500}{300} \right) - 0.259\,83 \ln \left(\frac{150}{200} \right)$$

$$= 1.4833 + 0.0747 = 1.558 \text{ kJ / kg K}$$

which is too low by 9.6%. If, on the other hand, we assume that the specific heat is constant at its value at 900 K, the average temperature, as in Example 5.6, then

$$s_2 - s_1 = 1.0714 \ln \left(\frac{1500}{300} \right) + 0.0747 = 1.7991 \text{ kJ / kg K}$$

which is high by 4.4%.

EXAMPLE 7.5   Calculate the change in entropy per kilogram as air is heated from 300 to 600 K while pressure drops from 400 to 300 kPa. Assume:

1. Constant specific heat.
2. Variable specific heat.

Solution

1. From Table A.10 for air at 300 K,

$$C_{p0} = 1.0035 \text{ kJ / kg K}$$

Therefore, using Eq. 7.24, we have

$$s_2 - s_1 = 1.0035 \ln \left(\frac{600}{300} \right) - 0.287 \ln \left(\frac{300}{400} \right) = 0.7781 \text{ kJ / kg K}$$

2. From Table A.12,

$$s_{T1}^0 = 6.8693 \text{ kJ / kg K}, \qquad s_{T2}^0 = 7.5764 \text{ kJ / kg K}$$

Using Eq. 7.27 gives

$$s_2 - s_1 = 7.5764 - 6.8693 - 0.287 \ln\left(\frac{300}{400}\right) = 0.7897 \text{ kJ / kg K}$$

EXAMPLE 7.5E Calculate the change in entropy per pound as air is heated from 540 R to 1200 R while pressure drops from 50 lbf/in.^{2} to 40 lbf/in.^{2}. Assume:

1. Constant specific heat.
2. Variable specific heat.

Solution

1. From Table A.10E for air at 80 F,

$$C_{p0} = 0.24 \text{ Btu / lbm R}$$

Therefore, using Eq. 7.24, we have

$$s_2 - s_1 = 0.24 \ln\left(\frac{1200}{540}\right) - \frac{53.34}{778} \ln\left(\frac{40}{50}\right) = 0.2068 \text{ Btu / lbm R}$$

2. From Table A.12E

$$s_{T_1}^0 = 0.6008 \text{ Btu / lbm R} \qquad s_{T_2}^0 = 0.7963 \text{ Btu / lbm R}$$

Using Eq. 7.27 gives

$$s_2 - s_1 = 0.7963 - 0.6008 - \frac{53.34}{778} \ln\frac{40}{50} = 0.2108 \text{ Btu / lbm R}$$

The air tables can be used for reversible adiabatic processes by employing the relative pressure P_r and relative specific volume v_r. The definition of these terms and the derivation follow.

For the reversible adiabatic process

$$T \, ds = dh - v \, dP = 0$$

Therefore,

$$dh = C_{p0} dT = v \, dP = RT \frac{dP}{P}$$

$$\frac{dP}{P} = \frac{C_{p0}}{R} \frac{dT}{T}$$

Let this equation be integrated between a reference state having a temperature T_0 and a pressure P_0, and a given arbitrary state having a temperature T and a pressure P.

Then

$$\ln \frac{P}{P_0} = \frac{1}{R} \int_{T_0}^{T} C_{p0} \frac{dT}{T}$$

The right side of this equation is a function of temperature only. The relative pressure P_r is defined as

$$\ln P_r \equiv \ln \frac{P}{P_0} = \frac{1}{R} \int_{T_0}^{T} C_{p0} \frac{dT}{T} = \frac{s_T^0}{R} \qquad (7.28)$$

Thus, a value of P_r can be tabulated as a function of temperature.

If we consider two states, 1 and 2, along a constant-entropy line, it follows from Eq. 7.28 that

$$\frac{P_1}{P_2} = \left(\frac{P_{r1}}{P_{r2}} \right)_{s=\text{constant}} \qquad (7.29)$$

This equation states that the ratio of the relative pressures for two states having the same entropy is equal to the ratio of the absolute pressures.

The development of the relative specific volume is similar, and the ratio of the relative specific volumes v_r, in an isentropic process is equal to the ratio of the specific volumes. That is,

$$\frac{v_1}{v_2} = \left(\frac{v_{r1}}{v_{r2}} \right)_{s=\text{constant}} \qquad (7.30)$$

The isentropic functions P_r and v_r are tabulated for air, Table A.12, but are not listed in the other gas tables in the Appendix, Table A.13. In analyzing an isentropic process for these gases, we need to use Eq. 7.27 with the left side of the equation equal to zero and the standard-state entropies from Table A.13.

EXAMPLE 7.6 One kilogram of air is contained in a cylinder fitted with a piston at a pressure of 400 kPa and a temperature of 600 K. The air is expanded to 150 kPa in a reversible, adiabatic process. Calculate the work done by the air.

    Control mass: Air.

    Initial state: P_1, T_1; state 1 fixed.

    Final state: P_2.

    Process: Reversible and adiabatic.

    Model: Ideal-gas and air tables, Table A.11.

Analysis:

First law:

$$0 = u_2 - u_1 + w$$

Second law:

$$s_2 = s_1$$

Solution

From Table A.12,

$$u_1 = 435.10 \text{ kJ / kg} \qquad P_{r1} = 13.0923$$

From Eq. 7.29,

$$P_{r2} = P_{r1} \times \frac{P_2}{P_1} = 13.0923 \times \frac{150}{400} = 4.9096$$

From Table A.12,

$$T_2 = 457 \text{ K} \qquad u_2 = 328.14$$

Therefore,

$$w = 435.10 - 328.14 = 106.96 \text{ kJ / kg}$$

At this point, it is advantageous to introduce the specific heat ratio k, which is defined as the ratio of the constant-pressure specific heat to the constant-volume specific heat at zero pressure.

$$k = \frac{C_{p0}}{C_{v0}} \tag{7.31}$$

Because the difference between C_{p0} and C_{v0} is a constant, Eq. 5.27, and because C_{p0} and C_{v0} are functions of temperature, it follows that k is also a function of temperature. However, when we consider the specific heat to be constant, k is also constant.

From the definition of k and Eq. 5.27, it follows that

$$C_{v0} = \frac{R}{k-1} \qquad C_{p0} = \frac{kR}{k-1} \tag{7.32}$$

Some very useful and simple relations for the reversible adiabatic process can be developed when the specific heats are assumed to be constant.

For the reversible adiabatic process, $ds = 0$. Therefore,

$$T \, ds = du + P \, dv = C_{v0} dT + P \, dv = 0$$

From the equation of state for an ideal gas,

$$dT = \frac{1}{R} \left(P \, dv + v \, dP \right)$$

Therefore,

$$\frac{C_{v0}}{R} \left(P \, dv + v \, dP \right) + P \, dv = 0$$

Substituting Eq. 7.32 into the expression and rearranging gives

$$\frac{1}{k-1} \left(P \, dv + v \, dP \right) + P \, dv = 0$$

$$v\, dP + kP\, dv = 0$$

$$\frac{dP}{P} + k\frac{dv}{v} = 0$$

Because k is constant when the specific heat is constant, this equation can be integrated under these conditions to give

$$Pv^k = \text{constant} \tag{7.33}$$

Equation 7.33 holds for all reversible adiabatic processes that involve an ideal gas with constant specific heat. It is usually advantageous to express this constant in terms of the initial and final states.

$$Pv^k = P_1 v_1^k = P_2 v_2^k = \text{constant} \tag{7.34}$$

From this equation and the ideal-gas equation of state, the following expressions relating the initial and final states of an isentropic process can be derived.

$$\frac{P_2}{P_1} = \left(\frac{v_1}{v_2}\right)^k = \left(\frac{V_1}{V_2}\right)^k \tag{7.35}$$

$$\frac{T_2}{T_1} = \left(\frac{P_2}{P_1}\right)^{(k-1)/k} = \left(\frac{v_1}{v_2}\right)^{k-1} \tag{7.36}$$

With the assumption of constant specific heat, some convenient equations can be derived for the work done by an ideal gas during an adiabatic process. Consider a control mass consisting of an ideal gas that undergoes a process in which work is done only at the moving boundary.

$$_1Q_2 = m\left(u_2 - u_1\right) + {}_1W_2 = 0$$

$$_1W_2 = -m\left(u_2 - u_1\right) = -mC_{v0}\left(T_2 - T_1\right)$$

$$= \frac{mR}{1-k}\left(T_2 - T_1\right) = \frac{P_2 V_2 - P_1 V_1}{1-k} \tag{7.37}$$

It should be noted that Eq. 7.37 applies to adiabatic processes only. Since no assumption was made regarding reversibility, it applies to both reversible and irreversible processes. Equation 7.37 is frequently derived for reversible processes by starting with the relation

$$_1W_2 = \int_1^2 P\, dV$$

for the control mass.

7.11 THE REVERSIBLE POLYTROPIC PROCESS FOR AN IDEAL GAS

When a gas undergoes a reversible process in which there is heat transfer, the process frequently takes place in such a manner that a plot of log P versus log V is a straight line, as shown in Fig. 7.14. For such a process PV^n is a constant.

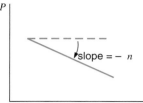

log V   **FIGURE 7.14** Example of a polytropic process.

A process having this relation between pressure and volume is called a polytropic process. An example is the expansion of the combustion gases in the cylinder of a water-cooled reciprocating engine. If the pressure and volume are measured during the expansion stroke of a polytropic process, as might be done with an engine indicator, and the logarithms of the pressure and volume are plotted, the result would be similar to the straight line in Fig. 7.14. From this figure it follows that

$$\frac{d \ln P}{d \ln V} = -n$$

$$d \ln P + n \, d \ln V = 0$$

If n is a constant (which implies a straight line on the log P versus log V plot), this equation can be integrated to give the following relation:

$$PV^n = \text{constant} = P_1 V_1^n = P_2 V_2^n \tag{7.38}$$

From this equation it is evident that the following relations can be written for a polytropic process.

$$\frac{P_2}{P_1} = \left(\frac{V_1}{V_2}\right)^n$$

$$\frac{T_2}{T_1} = \left(\frac{P_2}{P_1}\right)^{(n-1)/n} = \left(\frac{V_1}{V_2}\right)^{n-1} \tag{7.39}$$

For a control mass consisting of an ideal gas, the work done at the moving boundary during a reversible polytropic process can be derived from the relations

$$_1W_2 = \int_1^2 P \, dV \qquad \text{and} \qquad PV^n = \text{constant}$$

$$_1W_2 = \int_1^2 P \, dV = \text{constant} \int_1^2 \frac{dV}{V^n}$$

$$= \frac{P_2 V_2 - P_1 V_1}{1-n} = \frac{mR(T_2 - T_1)}{1-n} \tag{7.40}$$

for any value of n except $n = 1$.

The polytropic processes for various values of n are shown in Fig. 7.15 on P–V and T–S diagrams. The values of n for some familiar processes are

Isobaric process	$n = 0$,	$P = \text{constant}$
Isothermal process	$n = 1$,	$T = \text{constant}$
Isentropic process	$n = k$,	$S = \text{constant}$
Isochoric process	$n = \infty$,	$V = \text{constant}$

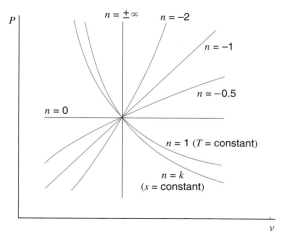 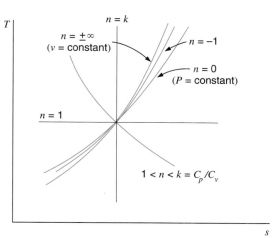

FIGURE 7.15 Polytropic processes on P–v T–s diagrams.

EXAMPLE 7.7

In a reversible process, nitrogen is compressed in a cylinder from 100 kPa, 20°C to 500 kPa. During this compression process the relation between pressure and volume is $PV^{1.3}$ = constant. Calculate the work and heat transfer per kilogram, and show this process on P–V and T–S diagrams.

Control mass: Nitrogen.

Initial state: P_1, T_1; state 1 known.

Final state: P_2.

Process: Reversible, polytropic with exponent $n < k$

Diagram: Fig. 7.16.

Model: Ideal gas, constant specific heat—value at 300 K.

Analysis:

Boundary movement work. From Eq. 7.40,

$$_1W_2 = \int_1^2 P\, dV = \frac{P_2V_2 - P_1V_1}{1-n} = \frac{mR(T_2 - T_1)}{1-n}$$

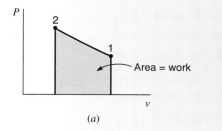 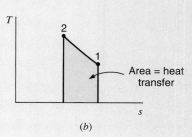

FIGURE 7.16
Diagram for Example
7.7.

(a)   (b)

First law:

$$_1 q_2 = u_2 - u_1 + {_1w_2} = C_{v0}\left(T_2 - T_1\right) + {_1w_2}$$

Solution

From Eq. 7.39,

$$\frac{T_2}{T_1} = \left(\frac{P_2}{P_1}\right)^{(n-1)/n} = \left(\frac{500}{100}\right)^{(1.3-1)/1.3} = 1.4498$$

$$T_2 = 293.2 \times 1.4498 = 425 \text{ K}$$

Then

$$_1w_2 = \frac{R\left(T_2 - T_1\right)}{1-n} = \frac{0.2968\left(425 - 293.2\right)}{\left(1 - 1.3\right)} = -130.4 \text{ kJ} / \text{kg}$$

and from the first law,

$$_1q_2 = C_{v0}\left(T_2 - T_1\right) + {_1w_2}$$

$$= 0.7448\left(425 - 293.2\right) - 130.4 = -32.2 \text{ kJ} / \text{kg}$$

The reversible isothermal process for an ideal gas is of particular interest. In this process

$$PV = \text{constant} = P_1 V_1 = P_2 V_2 \tag{7.41}$$

The work done at the boundary of a simple compressible mass during a reversible isothermal process can be found by integrating the equation

$$_1W_2 = \int_1^2 P \, dV$$

The integration is

$$_1W_2 = \int_1^2 P \, dV = \text{constant} \int_1^2 \frac{dV}{V} = P_1 V_1 \ln \frac{V_2}{V_1} = P_1 V_1 \ln \frac{P_1}{P_2} \tag{7.42}$$

or

$$_1W_2 = mRT \ln \frac{V_2}{V_1} = mRT \ln \frac{P_1}{P_2} \tag{7.43}$$

Because there is no change in internal energy or enthalpy in an isothermal process, the heat transfer is equal to the work (neglecting changes in kinetic and potential energy). Therefore, we could have derived Eq. 7.42 by calculating the heat transfer.

For example, using Eq. 7.7, we have

$$\int_1^2 T \, ds = {_1q_2} = \int_1^2 du + \int_1^2 P \, dv$$

But $du = 0$ and $Pv = \text{constant} = P_1v_1 = P_2v_2$, such that

$$_1q_2 = \int_1^2 P \, dv = P_1v_1 \ln\frac{v_2}{v_1}$$

which yields the same result as Eq. 7.42.

7.12 THE SECOND LAW OF THERMODYNAMICS FOR A CONTROL VOLUME

The second law of thermodynamics can be applied to a control volume by a procedure similar to that used in Section 5.11 for writing the first law for a control volume. Equation 7.11 states the second law for a control mass analysis in the form

$$dS = \frac{\delta Q}{T} + \delta S_{gen}$$

For a change in entropy, $S_2 - S_1$, that occurs during a time interval δt, we can write

$$\frac{S_2 - S_1}{\delta t} = \frac{1}{\delta t}\left(\frac{\delta Q}{T}\right) + \frac{1}{\delta t}\left(\delta S_{gen}\right) \tag{7.44}$$

Consider the control mass and control volume shown in Fig. 7.17. During time δt the mass δm_i enters the control volume across the discrete area $\mathcal{A}_i$, the mass δm_e leaves across the discrete area $\mathcal{A}_e$, an amount of heat δQ is transferred to the control mass across an element of area where the surface temperature is T, and work δW is done by the control mass. As in the analysis of Section 5.11, we assume that the increment of mass, δm_i, has uniform properties, and similarly that δm_e has uniform properties.

Let

$$S_t = \text{the entropy in the control volume at time } t$$

$$S_{t+\delta t} = \text{the entropy in the control volume at time } t + \delta t$$

Then

$$S_1 = S_t + s_t\delta m_i = \text{entropy of the control mass at time } t$$

$$S_2 = S_{t+\delta t} + s_e\delta m_e = \text{entropy of the control mass at time } t + \delta t$$

Therefore,

$$S_2 - S_1 = \left(S_{t+\delta t} - S_t\right) + \left(s_e\delta m_e - s_i\delta m_i\right) \tag{7.45}$$

The term $s_e\delta m_e - s_i\delta m_i$ represents the net flow of entropy out of the control volume during δt as the result of the masses δm_e and δm_i crossing the control surface.

The significance of the remaining terms of Eq. 7.45 must be carefully considered when applied to a control volume. When we write the terms $\delta Q/T$ and δS_{gen} for a system, we normally consider a quantity of mass at uniform temperature at any instant of time. As noted before, for a control volume analysis, we deviate from a strictly classical concept of thermodynamics and are prepared to consider a variation in temperature throughout the control volume, the assumption of "local" equilibrium.

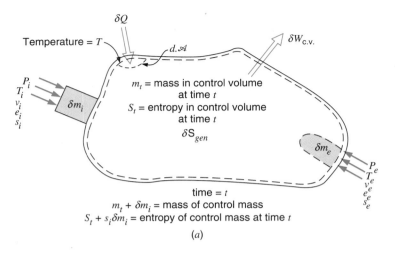

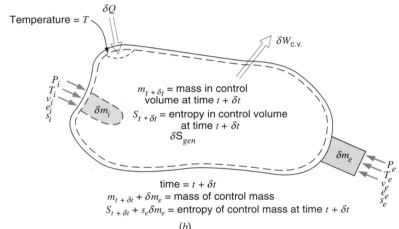

FIGURE 7.17
Schematic diagram for
a second-law analysis
of a control volume.

As a consequence, for the heat-transfer term, we must consider each element of area for the surface across which heat flows and the surface temperature of this local area. Therefore, this term in Eq. 7.44 must be written as the summation of all local $\delta Q/T$ terms over the control surface, or

$$\frac{1}{\delta t}\left(\frac{\delta Q}{T}\right) = \frac{1}{\delta t}\sum_{\text{c.v.}}\left(\frac{\delta Q}{T}\right)_{\text{c.v.}} \tag{7.46}$$

In considering the entropy generation term, we realize that this term in Eq. 7.44 represents the summation of all the local irreversible entropy generations throughout the interior of the control mass, at whatever the local interior states are.

Now, substituting Eqs. 7.45 and 7.46 into Eq. 7.44, we have

$$\frac{S_{t+\delta t} - S_t}{\delta t} + \frac{s_e \delta m_e}{\delta t} - \frac{s_i \delta m_i}{\delta t} = \frac{1}{\delta t}\sum_{\text{c.v.}}\left(\frac{\delta Q}{T}\right)_{\text{c.v.}} + \frac{1}{\delta t}\left(\delta S_{gen}\right) \tag{7.47}$$

To reduce this expression to a rate equation, we consider what happens to each of the terms in Eq. 7.47 as δt approaches zero. As in Section 5.11, the two mass quantities become mass flow rates, and the entropy term becomes the time rate of change of entropy in the control volume. The manner is analogous to the energy term in the first-law equation becoming the time rate of change of energy in the control volume. Similarly, the heat-transfer and generation terms become rate quantities, the first of which is divided by the appropriate local temperature.

In utilizing these limiting values to express the rate form of the entropy equation for a control volume, we again include summation signs on the flow terms to account for the possibility of additional flow streams entering or leaving the control volume. The resulting expression is

$$\frac{dS_{c.v.}}{dt} + \sum \dot{m}_e s_e - \sum \dot{m}_i s_i = \sum_{c.v.} \frac{\dot{Q}_{c.v.}}{T} + \dot{S}_{gen} \qquad (7.48)$$

which, for our purposes, is the general expression of the second-law entropy equation. In words, this expression states that the rate of change of entropy inside the control volume plus the net rate of entropy flow out is equal to the sum of two terms, the overall heat-transfer surface term and the positive internal-irreversibility entropy generation term. Since this second term is necessarily positive (or zero) and is difficult if not impossible to evaluate quantitatively, Eq. 7.48 is frequently written in the form

$$\frac{dS_{c.v.}}{dt} + \sum \dot{m}_e s_e - \sum \dot{m}_i s_i \geq \sum_{c.v.} \frac{\dot{Q}_{c.v.}}{T} \qquad (7.49)$$

where the equality applies to internally reversible processes and the inequality to internally irreversible processes.

If there is no mass flow into or out of the control volume and if temperature is considered uniform at any instant of time, Eq. 7.48 reduces to a rate form of the control mass entropy equation, Eq. 7.11. In this sense, Eq. 7.48 (or 7.49) can be considered a general expression, as was pointed out for the control volume form of the first law.

7.13 THE STEADY-STATE, STEADY-FLOW PROCESS AND THE UNIFORM-STATE, UNIFORM-FLOW PROCESS

We now consider the application of the second-law control volume equation, Eq. 7.48 or 7.49, to the two control volume model processes developed in Chapter 5.

For the steady-state, steady-flow process, which has been defined in Section 5.12, we conclude that there is no change with time of the entropy per unit mass at any point within the control volume, and therefore the first term of Eq. 7.48 equals zero. That is,

$$\frac{dS_{c.v.}}{dt} = 0 \qquad (7.50)$$

so that, for the SSSF process,

$$\sum \dot{m}_e s_e - \sum \dot{m}_i s_i \geq \sum_{c.v.} \frac{\dot{Q}_{c.v.}}{T} + \dot{S}_{gen} \qquad (7.51)$$

in which the various mass flows, heat transfer and entropy generation, rates and states are all constant with time.

If in a steady-state, steady-flow process there is only one area over which mass enters the control volume at a uniform rate and only one area over which mass leaves the control volume at a uniform rate, we can write

$$\dot{m}(s_e - s_i) = \sum_{\text{c.v.}} \frac{\dot{Q}_{\text{c.v.}}}{T} + \dot{S}_{\text{gen}} \tag{7.52}$$

For an adiabatic process with these assumptions, it follows that

$$s_e \geq s_i \tag{7.53}$$

where the equality holds for a reversible adiabatic process.

EXAMPLE 7.8

Steam enters a steam turbine at a pressure of 1 MPa, a temperature of 300°C, and a velocity of 50 m/s. The steam leaves the turbine at a pressure of 150 kPa and a velocity of 200 m/s. Determine the work per kilogram of steam flowing through the turbine, assuming the process to be reversible and adiabatic.

Control volume: Turbine.

Sketch: Fig. 7.18.

Inlet state: Fixed (Fig. 7.18).

Exit state: P_e, $\mathbf{V}_e$ known.

Process: SSSF.

Model: Steam tables.

Analysis:

Continuity equation:

$$\dot{m}_e = \dot{m}_i = \dot{m}$$

First law:

$$h_i + \frac{\mathbf{V}_i^2}{2} = h_e + \frac{\mathbf{V}_e^2}{2} + w$$

Second law:

$$s_e = s_i$$

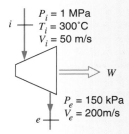

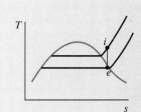

FIGURE 7.18 Sketch for Example 7.8.

Solution

From the steam tables,

$$h_i = 3051.2 \text{ kJ / kg} \qquad s_i = 7.1229 \text{ kJ / kg K}$$

The two properties known in the final state are pressure and entropy.

$$P_e = 0.15 \text{ MPa} \qquad s_e = s_i = 7.1229 \text{ kJ / kg K}$$

The quality and enthalpy of the steam leaving the turbine can be determined.

$$s_e = 7.1229 = s_f + x_e s_{fg} = 1.4336 + x_e 5.7897$$

$$x_e = 0.9827$$

$$h_e = h_f + x_e h_{fg} = 467.1 + 0.9827(2226.5)$$

$$= 2655.0 \text{ kJ / kg}$$

Therefore, the work per kilogram of steam for this isentropic process may be found using the equation for the first law.

$$w = 3051.2 + \frac{50 \times 50}{2 \times 1000} - 2655.0 - \frac{200 \times 200}{2 \times 1000} = 377.5 \text{ kJ / kg}$$

EXAMPLE 7.9 Consider the reversible adiabatic flow of steam through a nozzle. Steam enters the nozzle at 1 MPa, 300°C, with a velocity of 30 m/s. The pressure of the steam at the nozzle exit is 0.3 MPa. Determine the exit velocity of the steam from the nozzle, assuming a reversible, adiabatic, steady-state, steady-flow process.

Control volume: Nozzle.

Sketch: Fig. 7.19.

Inlet state: Fixed (Fig. 7.19).

Exit state: P_e known.

Process: SSSF.

Model: Steam tables.

Analysis:

Because this is a steady-state, steady-flow process in which the work, the heat transfer, and the changes in potential energy are zero, we can write

FIGURE 7.19
Sketch for Example 7.9.

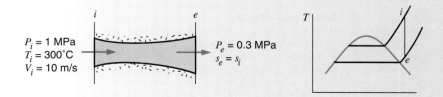

Continuity equation:

$$\dot{m}_e = \dot{m}_i = \dot{m}$$

First law:

$$h_i + \frac{\mathbf{V}_i^2}{2} = h_e + \frac{\mathbf{V}_e^2}{2}$$

Second law:

$$s_e = s_i$$

Solution

From the steam tables,

$$h_i = 3051.2 \text{ kJ / kg} \qquad s_i = 7.1229 \text{ kJ / kg K}$$

The two properties that we know in the final state are entropy and pressure.

$$s_e = s_i = 7.1229 \text{ kJ / kg K} \qquad P_e = 0.3 \text{ MPa}$$

Therefore,

$$T_e = 159.1^\circ\text{C} \qquad h_e = 2780.2 \text{ kJ / kg}$$

Substituting into the equation for the first law, we have

$$\frac{\mathbf{V}_e^2}{2} = h_i - h_e + \frac{\mathbf{V}_i^2}{2}$$

$$= 3051.2 - 2780.2 + \frac{30 \times 30}{2 \times 1000} = 271.5 \text{ kJ / kg}$$

$$\mathbf{V}_e = 737 \text{ m/s}$$

EXAMPLE 7.9E Consider the reversible adiabatic flow of steam through a nozzle. Steam enters the nozzle at 100 lbf/in.², 500 F, with a velocity of 100 ft/s. The pressure of the steam at the nozzle exit is 40 lbf/in.². Determine the exit velocity of the steam from the nozzle, assuming a reversible adiabatic, steady-state, steady-flow process.

Control volume: Nozzle.

Sketch: Fig. 7.19E.

Inlet state: Fixed (Fig. 7.19E).

Exit state: P_e known.

Process: SSSF.

Model: Steam tables.

Analysis:

Because this is a steady-state, steady-flow process in which the work, the heat transfer, and the changes in potential energy are zero, we can write

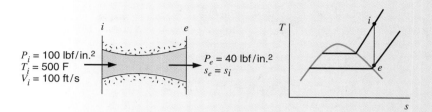

FIGURE 7.19E
Sketch for Example
7.9E.

Continuity equation:

$$\dot{m}_e = \dot{m}_i = \dot{m}$$

First law:

$$h_i + \frac{\mathbf{V}_i^2}{2} = h_e + \frac{\mathbf{V}_e^2}{2}$$

Second law:

$$s_e = s_i$$

Solution

From the steam tables,

$$h_i = 1279.1 \text{ Btu / lbm} \qquad s_i = 1.7085 \text{ Btu / lbm R}$$

The two properties that we know in the final state are entropy and pressure.

$$s_e = s_i = 1.7085 \text{ Btu / lbm R}, \; P_e = 40 \text{ lbf / in.}^2$$

Therefore,

$$T_e = 314.2 \text{ F} \qquad h_e = 1193.9 \text{ Btu / lbm}$$

Substituting into the equation for the first law, we have

$$\frac{\mathbf{V}_e^2}{2} = h_i - h_e + \frac{\mathbf{V}_i^2}{2}$$

$$= 1279.1 - 1193.9 + \frac{100 \times 100}{2 \times 32.17 \times 778} = 85.4 \text{ Btu / lbm}$$

$$\mathbf{V}_e = \sqrt{2 \times 32.17 \times 778 \times 85.4} = 2070 \text{ ft / s}$$

EXAMPLE 7.10   An inventor reports that she has a refrigeration compressor that receives saturated R-134a vapor at −20°C and delivers the vapor at 1 MPa, 40°C. The compression process is adiabatic. Does the process described violate the second law?

Control volume: Compressor.

Inlet state: Fixed (saturated vapor at T_i).

Exit state: Fixed (P_e, T_e known).

Process: SSSF, adiabatic.

Model: R-134a tables.

Analysis:

Because this is a steady-state, steady-flow, adiabatic process, we can write
Second law:

$$s_e \geq s_i$$

Solution

From the R-134a tables,

$$s_e = 1.71479 \text{ kJ / kg K} \qquad s_i = 1.7395 \text{ kJ / kg K}$$

Therefore, $s_e < s_i$, whereas for this process the second law requires that $s_e \geq s_i$. The process described involves a violation of the second law and thus would be impossible.

EXAMPLE 7.11 Air is compressed in a centrifugal compressor from ambient conditions, 100 kPa and 300 K, to a pressure of 450 kPa. Assume the process to be reversible and adiabatic with negligible changes in kinetic and potential energy. Calculate the work input per kilogram of air flowing through the compressor.

Control volume: Compressor.

Inlet state: P_i, T_i known; state fixed.

Exit state: P_e known.

Process: SSSF.

Model: Ideal-gas and air tables, A.12.

Analysis:

Because this is a steady-state, steady-flow reversible process, we can write
Continuity equation:

$$\dot{m}_e = \dot{m}_i = \dot{m}$$

First law:

$$h_i = h_e + w$$

Second law:

$$s_e = s_i$$

Solution

From Table A.12,

$$h_i = 300.47 \text{ kJ / kg} \qquad P_{ri} = 1.1146$$

From Eq. 7.29,

$$P_{re} = P_{ri} \frac{P_e}{P_i} = 1.1146 \frac{450}{100} = 5.0157$$

Therefore,

$$T_e = 460 \text{ K} \qquad h_e = 462.14$$

and from the first law,

$$w = h_i - h_e = 300.47 - 462.14 = -161.67 \text{ kJ/kg}$$

EXAMPLE 7.11E Air is compressed in a centrifugal compressor from ambient conditions, 14.7 lbf/in.^{2} and 520 R, to a pressure of 65 lbf/in.^{2}. Assume the process to be reversible and adiabatic with negligible changes in kinetic and potential energy. Calculate the work input per poundmeter of air flowing through the compressor.

  Control volume: Compressor.

  Inlet state: P_i, T_i known; state fixed.

  Exit state: P_e known.

  Process: SSSF.

  Model: Ideal-gas and air tables, A.12.

Analysis:

Since this is a steady-state, steady-flow reversible process we can write
Continuity equation:

$$\dot{m}_e = \dot{m}_i = \dot{m}$$

First law:

$$h_i = h_e + w$$

Second law:

$$s_e = s_i$$

Solution

From Table A.12,

$$h_i = 124.38 \text{ Btu/lbm} \qquad P_{ri} = 0.9767$$

From Eq. 7.29,

$$P_{re} = P_{ri} \left(\frac{P_e}{P_i} \right) = 0.9767 \frac{65}{14.7} = 4.3187$$

Therefore,

$$T_e = 794 \text{ R} \qquad h_e = 190.37$$

and from the first law,

$$w = h_i - h_e = 124.38 - 190.37 = -66.0 \text{ kJ/kg}$$

For the uniform-state, uniform-flow process, which was described in Section 5.14, the second law for a control volume, Eq. 7.49, can be written in the following form.

$$\frac{d}{dt}(ms)_{\text{c.v.}} + \sum \dot{m}_e s_e - \sum \dot{m}_i s_i \geq \sum_{\text{c.v.}} \frac{\dot{Q}_{\text{c.v.}}}{T} \tag{7.54}$$

If this is integrated over the time interval t, we have

$$\int_0^t \frac{d}{dt}(ms)_{\text{c.v.}} \, dt = (m_2 s_2 - m_1 s_1)_{\text{c.v.}}$$

$$\int_0^t \left(\sum \dot{m}_e s_e\right) dt = \sum m_e s_e \qquad \int_0^t \left(\sum \dot{m}_i s_i\right) dt = \sum m_i s_i$$

Therefore, for this period of time t, we can write the second law for the uniform-state, uniform-flow process as

$$(m_2 s_2 - m_1 s_1)_{\text{c.v.}} + \sum m_e s_e - \sum m_i s_i \geq \int_0^t \sum_{\text{c.v.}} \frac{\dot{Q}_{\text{c.v.}}}{T} \, dt \tag{7.55}$$

Since in this process the temperature is uniform throughout the control volume at any instant of time, the integral on the right reduces to

$$\int_0^t \sum_{\text{c.v.}} \frac{\dot{Q}_{\text{c.v.}}}{T} \, dt = \int_0^t \frac{1}{T} \sum_{\text{c.v.}} \dot{Q}_{\text{c.v.}} \, dt = \int_0^t \frac{\dot{Q}_{\text{c.v.}}}{T} \, dt$$

and therefore the second law for the uniform-state, uniform-flow process can be written

$$(m_2 s_2 - m_1 s_1)_{\text{c.v.}} + \sum m_e s_e - \sum m_i s_i \geq \int_0^t \frac{\dot{Q}_{\text{c.v.}}}{T} \, dt \tag{7.56}$$

By introducing the rate of internal entropy generation, we can write this as an equality. Integrating over the time interval t, we have the total amount of internal entropy generation during the process, $_1 S_{2\text{gen}}$. Therefore,

$$(m_2 s_2 - m_1 s_1)_{\text{c.v.}} + \sum m_e s_e - \sum m_i s_i = \int_0^t \frac{\dot{Q}_{\text{c.v.}}}{T} \, dt +_1 S_{2\text{gen}} \tag{7.57}$$

7.14 THE REVERSIBLE STEADY-STATE, STEADY-FLOW PROCESS

An expression can be derived for the work in a reversible adiabatic, steady-state, steady-flow process that is of great help in understanding its significant variables. We have noted that when a steady-state, steady-flow process involves a single flow of fluid into and out of the control volume, the first law, Eq. 5.50, can be written,

$$q + h_i + \frac{\mathbf{v}_i^2}{2} + gZ_i = h_e + \frac{\mathbf{v}_e^2}{2} + gZ_e + w$$

and the second law, Eq. 7.52 is

$$\dot{m}(s_e - s_i) = \sum_{c.v.} \frac{\dot{Q}_{c.v.}}{T} + \dot{S}_{gen}$$

Let us now consider two types of flow, a reversible adiabatic process and a reversible isothermal process.

If the process is reversible and adiabatic, the second-law equation reduces to

$$s_e = s_i$$

It follows from the property relation

$$T\,ds = dh - v\,dP$$

that

$$h_e - h_i = \int_i^e v\,dP \tag{7.58}$$

Substituting these relations into Eq. 5.50 and noting that $q = 0$, we have for the reversible, adiabatic process

$$w = (h_i - h_e) + \frac{\mathbf{v}_i^2 - \mathbf{v}_e^2}{2} + g(Z_i - Z_e)$$

$$= -\int_i^e v\,dP + \frac{\mathbf{v}_i^2 - \mathbf{v}_e^2}{2} + g(Z_i - Z_e) \tag{7.59}$$

If, instead, the process is reversible and isothermal, the second law reduces to

$$\dot{m}(s_e - s_i) = \frac{1}{T}\sum_{c.v.}\dot{Q}_{c.v.} = \frac{\dot{Q}_{c.v.}}{T} \tag{7.60}$$

or

$$T(s_e - s_i) = \frac{\dot{Q}_{c.v.}}{\dot{m}} = q$$

and the property relation can be integrated to give

$$T(s_e - s_i) = (h_e - h_i) - \int_i^e v\,dP \tag{7.61}$$

Substituting Eqs. 7.60 and 7.61 into the first law, Eq. 5.50, gives us the same expression as for the reversible adiabatic process, Eq. 7.59. We further note that any other reversible process can be constructed, in the limit, from a series of alternate adiabatic and isothermal processes. Thus, we may conclude that Eq. 7.59 is valid for any reversible, steady-state, steady-flow process without the restriction that it be either adiabatic or isothermal.

This expression has a wide range of application. If we consider a reversible steady-state, steady-flow process in which the work is zero (such as flow through a nozzle) and the fluid is incompressible ($v = $ constant), Eq. 7.59 can be integrated to give

$$v(P_e - P_i) + \frac{\mathbf{v}_e^2 - \mathbf{v}_i^2}{2} + g(Z_e - Z_i) = 0 \tag{7.62}$$

Known as the Bernoulli equation (after Daniel Bernoulli), this is a very important equation in fluid mechanics.

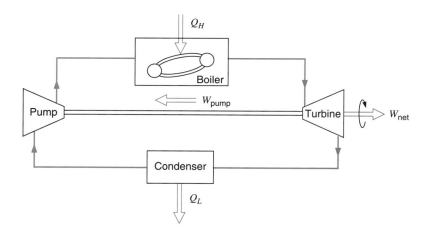

FIGURE 7.20
Simple steam power
plant.

Equation 7.59 is also frequently applied to the large class of flow processes involving work (such as turbines and compressors) in which changes in kinetic and potential energies of the working fluid are small. The model process for these machines is then a reversible, SSSF process with no change in kinetic or potential energy (and commonly, although not necessarily, adiabatic as well). For this process Eq. 7.59 reduces to the form

$$w = -\int_i^e v \, dP \tag{7.63}$$

From this result, we conclude that the shaft work associated with this type of process is closely related to the specific volume of the fluid during the process. To amplify this point further, consider the simple steam power plant shown in Fig. 7.20. Suppose this is an ideal power plant with no pressure drop in the piping, the boiler, or the condenser. Thus, the pressure increase in the pump is equal to the pressure decrease in the turbine. Neglecting kinetic and potential energy changes, the work done in each of these processes is given by Eq. 7.63. Since the pump handles liquid, which has a very small specific volume compared to that of the vapor that flows through the turbine, the power input to the pump is much less than the power output of the turbine. The difference is the net power output of the power plant.

This same line of reasoning can be qualitatively applied to actual devices that involve steady-state, steady-flow processes, even though the processes are not exactly reversible and adiabatic.

EXAMPLE 7.12   Calculate the work per kilogram to pump water isentropically from 100 kPa, 30°C, to 5 MPa.

       Control volume: Pump.

       Inlet state: P_i, T_i known; state fixed.

       Exit state: P_e known.

       Process: SSSF.

       Model: Steam tables.

Analysis:

Since the process is SSSF, reversible, and adiabatic, and changes in kinetic and potential energies can be neglected,

First law:

$$h_i = h_e + w$$

Second law:

$$s_e - s_i = 0$$

Solution

Since P_e and s_e are known, state e is fixed and, therefore h_e is known and w can be found from the first law. However, the process is reversible, SSSF, with negligible changes in kinetic and potential energies, so that Eq. 7.63 is also valid. Furthermore, since a liquid is being pumped, the specific volume will change very little during the process.

From the steam tables, $v_i = 0.001\,004$ m³/kg. Assuming that the specific volume remains constant and using Eq. 7.63, we have

$$-w = \int_1^2 v\, dP = v\left(P_2 - P_1\right) = 0.001\,004\left(5000 - 100\right) = 4.92 \text{ kJ/kg}$$

As a final application of Eq. 7.59, we recall the reversible polytropic process for an ideal gas, discussed in Section 7.11 for a control mass process. For the SSSF process with no change in kinetic and potential energies, from the relations,

$$w = -\int_i^e v\, dP \qquad \text{and} \qquad Pv^n = \text{constant} = C^n$$

$$w = -\int_i^e v\, dP = -C\int_i^e \frac{dP}{P^{1/n}} \tag{7.64}$$

$$= -\frac{n}{n-1}\left(P_e v_e - P_i v_i\right) = -\frac{nR}{n-1}\left(T_e - T_i\right)$$

If the process is isothermal, then $n = 1$ and the integral becomes

$$w = -\int_i^e v\, dP = -\text{constant}\int_i^e \frac{dP}{P} = -P_i v_i \ln\frac{P_e}{P_i} \tag{7.65}$$

These evaluations of the integral

$$\int_i^e v\, dP$$

may also be used in conjunction with Eq. 7.59 for instances in which kinetic and potential energy changes are not negligibly small.

7.15 PRINCIPLE OF THE INCREASE OF ENTROPY

The principle of the increase of entropy for a control mass analysis was discussed in Section 7.8. The same general conclusion is reached for a control volume analysis. To demonstrate this, consider a control volume, Fig. 7.21, that exchanges both mass and heat with its surroundings. At the point in the surroundings where the heat transfer occurs, the temperature is T_0. From Eq. 7.49, the second law for this process is

$$\frac{dS_{c.v.}}{dt} + \sum \dot{m}_e s_e - \sum \dot{m}_i s_i \geq \sum_{c.v.} \frac{\dot{Q}_{c.v.}}{T}$$

We recall that the first term represents the rate of change of entropy within the control volume, and the next terms the net entropy flow out of the control volume resulting from the mass flow. Therefore, for the surroundings, we can write

$$\frac{dS_{surr}}{dt} = \sum \dot{m}_e s_e - \sum \dot{m}_i s_i - \frac{\dot{Q}_{c.v.}}{T_0} \tag{7.66}$$

Adding Eqs. 7.49 and 7.66, we have

$$\frac{dS_{net}}{dt} = \frac{dS_{c.v.}}{dt} + \frac{dS_{surr}}{dt} \geq \sum_{c.v.} \frac{\dot{Q}_{c.v.}}{T} - \frac{\dot{Q}_{c.v.}}{T_0} \tag{7.67}$$

Because $\dot{Q}_{c.v.} > 0$ when $T_0 > T$ and $\dot{Q}_{c.v.} < 0$ when $T_0 < T$, it follows that

$$\frac{dS_{net}}{dt} = \frac{dS_{c.v.}}{dt} + \frac{dS_{surr}}{dt} = \sum \dot{S}_{gen} \geq 0 \tag{7.68}$$

which can be termed the general statement of the principle of the increase of entropy.

When we use Eq. 7.68 to check any particular process for a possible violation of the second law, it will be in connection with one of our model processes. For example, in a steady-state, steady-flow process, as we consider the two terms in Eq. 7.68, we realize that, in accordance with Eq. 7.50, the first term is zero. As a result, all the entropy change that is due to irreversibilities in this type of process is observed in the surroundings. This term may then be evaluated using Eq. 7.66. On the other hand, for the uniform-state, uniform-flow process, there are both control volume and surroundings terms to evaluate. Each term is integrated over the time t of the process, as was done in Section 7.13. Thus, Eq. 7.68 is integrated to

$$\Delta S_{net} = \Delta S_{c.v.} + \Delta S_{surr} \tag{7.69}$$

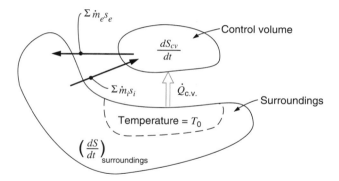

FIGURE 7.21 Entropy change for a control volume plus surroundings.

in which the control volume term is

$$\Delta S_{\text{c.v.}} = \left(m_2 s_2 - m_1 s_1 \right)_{\text{c.v.}} \tag{7.70}$$

The term for the surroundings is, after applying Eq. 7.57 to the surroundings and integrating,

$$\Delta S_{\text{surr}} = \frac{-Q_{\text{c.v.}}}{T_0} + \sum m_e s_e - \sum m_i s_i \tag{7.71}$$

7.16 EFFICIENCY

In Chapter 6 we noted that the second law of thermodynamics led to the concept of thermal efficiency for a heat engine cycle, namely

$$\eta_{\text{th}} = \frac{W_{\text{net}}}{Q_H}$$

where W_{net} is the net work of the cycle and Q_H is the heat transfer from the high-temperature body.

In this chapter we have extended our consideration of the second law to control volume processes, which leads us now to consider the efficiency of a process. For example, we might be interested in the efficiency of a turbine in a steam power plant or the compressor in a gas turbine engine.

In general, we can say that to determine the efficiency of a machine in which a process takes place, we compare the actual performance of the machine under given conditions and the performance that would have been achieved in an ideal process. It is in the definition of this ideal process that the second law becomes a major consideration. For example, a steam turbine is intended to be an adiabatic machine. The only heat transfer is the unavoidable heat transfer that takes place between the given turbine and the surroundings. We also note that for a given steam turbine operating in a steady-state, steady-flow manner, the state of the steam entering the turbine and the exhaust pressure are fixed. Therefore, the ideal process is a reversible adiabatic process, which is an isentropic process, between the inlet state and the turbine exhaust pressure. If we denote the actual work done per unit mass of steam flow through the turbine as w_a and the work that would have been done in a reversible adiabatic process between the inlet state and the turbine exhaust pressure as w_s, the efficiency of the turbine is defined as

$$\eta_{\text{turbine}} = \frac{w_a}{w_s} \tag{7.72}$$

The same relation holds for a gas turbine.

A few other examples may help to clarify this point. For a nozzle, the objective is to have the maximum kinetic energy leaving the nozzle for the given inlet conditions and exhaust pressure. The nozzle is also an adiabatic device, and therefore the ideal process is a reversible adiabatic or isentropic process. The efficiency of a nozzle is the ratio of the actual kinetic energy leaving the nozzle, $\mathbf{V}_a^2/2$, to the kinetic energy for an isentropic process between the same inlet conditions and exhaust pressure—$\mathbf{V}_s^2/2$. Thus we have

$$\eta_{\text{nozzle}} = \frac{\mathbf{V}_a^2/2}{\mathbf{V}_s^2/2} \tag{7.73}$$

For compressors of air or other gases, there are two ideal processes to which their actual performance can be compared. If no effort is made to cool the gas during compression (i.e., when the process is adiabatic), the ideal process is a reversible adiabatic or isentropic process between the given inlet state and exhaust pressure. If we denote the work per unit mass of gas flow through the compressor for this isentropic process as w_s and the actual work as w_a (the actual work input will be greater than the work input for an isentropic process), the efficiency is defined by the relation

$$\eta_{\text{adiabatic compressor}} = \frac{w_s}{w_a} \tag{7.74}$$

If an attempt is made to cool the air during compression by using a water jacket or fins, the ideal process is considered a reversible isothermal process. If w_t is the work for the reversible isothermal process between the given inlet state and exhaust pressure, and w_a the actual work, the efficiency is defined by the relation

$$\eta_{\text{cooled compressor}} = \frac{w_t}{w_a} \tag{7.75}$$

Thus, to determine the efficiency of a device that carries out a process (rather than a cycle), we compare the actual performance to what would be achieved in a related, but well-defined ideal process.

EXAMPLE 7.13   A steam turbine receives steam at a pressure of 1 MPa, 300°C. The steam leaves the turbine at a pressure of 15 kPa. The work output of the turbine is measured and is found to be 600 kJ/kg of steam flowing through the turbine. Determine the efficiency of the turbine.

    Control volume: Turbine.

    Inlet state: P_i, T_i known; state fixed.

    Exit state: P_e known.

    Process: SSSF.

    Model: Steam tables.

Analysis:

The efficiency of the turbine is given by Eq. 7.72.

$$\eta_{\text{turbine}} = \frac{w_a}{w_s}$$

Thus, to determine the turbine efficiency, we calculate the work that would be done in an isentropic process between the given inlet state and final pressure. For this isentropic process,

Continuity equation:

$$\dot{m}_i = \dot{m}_e = \dot{m}$$

First law:

$$h_i = h_{es} + w_s$$

Second law:

$$s_i = s_{es}$$

Solution

From the steam tables,

$$h_i = 3051.2 \text{ kJ} / \text{kg} \qquad s_i = 7.1229 \text{ kJ} / \text{kg K}$$

Therefore, at $P_e = 15$ kPa,

$$s_{es} = s_i = 7.1229 = 0.7549 + x_{es} 7.2536$$

$$x_{es} = 0.8779$$

$$h_{es} = 225.9 + 0.8779(2373.1) = 2309.3 \text{ kJ} / \text{kg}$$

From the first law for the isentropic process,

$$w_s = h_i - h_{es} = 3051.2 - 2309.3 = 741.9 \text{ kJ} / \text{kg}$$

But, since

$$w_a = 600 \text{ kJ} / \text{kg}$$

we find that

$$\eta_{\text{turbine}} = \frac{w_a}{w_s} = \frac{600}{741.9} = 0.809 = 80.9\%$$

In connection with this example, it should be noted that to find the actual state e of the steam exiting the turbine, we need to analyze the real process taking place. For the real process

$$\dot{m}_i = \dot{m}_e = \dot{m}$$

$$h_i = h_e + w_a$$

$$s_e > s_i$$

Therefore, from the first law for the real process, we have

$$h_e = 3051.2 - 600 = 2451.2 \text{ kJ} / \text{kg}$$

$$2451.2 = 225.9 + x_e 2373.1$$

$$x_e = 0.9377$$

7.17 SOME GENERAL COMMENTS REGARDING ENTROPY

It is quite possible at this point that a student may have a good grasp of the material that has been covered, and yet may have only a vague understanding of the significance of entropy. In fact, the question "What is entropy?" is frequently raised by students with the implication that no one really knows! This section has been included in an attempt to give insight into the qualitative and philosophical aspects of the concept of entropy, and to illustrate the broad application of entropy to many different disciplines.

First, we recall that the concept of energy rises from the first law of thermodynamics and the concept of entropy from the second law of thermodynamics. Actually it is just as difficult to answer the question "What is energy?" as it is to answer the question "What is entropy?" However, since we regularly use the term energy and are able to relate this term to phenomena that we observe every day, the word energy has a definite meaning to us and thus serves as an effective vehicle for thought and communication. The word entropy could serve in the same capacity. If, when we observed a highly irreversible process (such as cooling coffee by placing an ice cube in it), we said, "That surely increases the entropy," we would soon be as familiar with the word *entropy* as we are with the word *energy*. In many cases when we speak about a higher efficiency we are actually speaking about accomplishing a given objective with a smaller total increase in entropy.

A second point to be made regarding entropy is that in statistical thermodynamics, the property entropy is defined in terms of probability. Although this topic will not be examined in detail in this text, a few brief remarks regarding entropy and probability may prove helpful. From this point of view the net increase in entropy that occurs during an irreversible process can be associated with a change of state from a less probable state to a more probable state. For instance, to use a previous example, one is more likely to find gas on both sides of ruptured membrane in Fig. 6.11 than to find a gas on one side and a vacuum on the other. Thus, when the membrane ruptures, the direction of the process is from a less probable state to a more probable state and associated with this process is an increase in entropy. Similarly, the more probable state is that a cup of coffee will be at the same temperature as its surroundings than at a higher (or lower) temperature. Therefore, as the coffee cools as the result of a transferring of heat to the surroundings, there is a change from a less probable to a more probable state, and associated with this is an increase in entropy.

The final point to be made is that the second law of thermodynamics and the principle of the increase of entropy have philosophical implications. Does the second law of thermodynamics apply to the universe as a whole? Are there processes unknown to us that occur somewhere in the universe, such as "continual creation," that have a decrease in entropy associated with them, and thus offset the continual increase in entropy that is associated with the natural processes that are known to us? If the second law is valid for the universe (we of course do not know if the universe can be considered as an isolated system), how did it get in the state of low entropy? On the other end of the scale, if all processes known to us have an increase in entropy associated with them, what is the future of the natural world as we know it?

Quite obviously it is impossible to give conclusive answers to these questions on the basis of the second law of thermodynamics alone. However, we see the second law of thermodynamics as a description of the prior and continuing work of a creator, who also holds the answer to our future destiny and that of the universe.

PROBLEMS

7.1 Find the missing properties and give the phase of the substance

a. H_2O $s = 7.70$, kJ/kg K, $P = 25$ kPa $h = ?\ T = ?\ x = ?$

b. H_2O $u = 3400$ kJ/kg, $P = 10$ MPa $T = ?\ x = ?\ s = ?$

    c. R-12 $T = 0°C$, $P = 250$ kPa $s = ?$ $x = ?$

    d. R-134a $T = -10°C$, $x = 0.45$ $v = ?$ $s = ?$

    e. NH_3 $T = 20°C$, $s = 5.50$ kJ/kg K $u = ?$ $x = ?$

7.2 Consider a Carnot-cycle heat engine with water as the working fluid. The heat transfer to the water occurs at 300°C, during which process the water changes from saturated liquid to saturated vapor. The heat is rejected from the water at 40°C.

    a. Show the cycle on a T–s diagram.

    b. Find the quality of the water at the beginning and end of the heat rejection process.

    c. Determine the net work output per kilogram of water and the cycle thermal efficiency.

7.3 In a Carnot engine with water as the working fluid, the high temperature is 250°C and as Q_H is received, the water changes from saturated liquid to saturated vapor. The water pressure at the low temperature is 100 kPa. Find T_L, the cycle thermal efficiency, the heat added per kilogram, and the entropy, s, at the beginning of the heat rejection process.

7.4 Consider a Carnot-cycle heat pump with R-22 as the working fluid. Heat is rejected from the R-22 at 40°C, during which process the R-22 changes from saturated vapor to saturated liquid. The heat is transferred to the R-22 at 0°C.

    a. Show the cycle on a T–s diagram.

    b. Find the quality of the R-22 at the beginning and end of the isothermal heat addition process at 0°C.

    c. Determine the coefficient of performance for the cycle.

7.5 Do the Problem 7.4 using refrigerant R-134a instead of R-22.

7.6 A closed tank, $V = 10$ L, containing 5 kg of water initially at 25°C, is heated to 175°C by a heat pump that is receiving heat from the surroundings at 25°C. Assume that this process is reversible. Find the heat transfer to the water and the work input to the heat pump.

7.7 One kilogram of ammonia in a piston/cylinder at 50°C, 1000 kPa is expanded in a reversible isothermal process to 100 kPa. Find the work and heat transfer for this process.

7.8 One kilogram of ammonia in a piston/cylinder at 50°C, 1000 kPa is expanded in a reversible isobaric process to 140°C. Find the work and heat transfer for this process.

7.9 One kilogram of ammonia in a piston/cylinder at 50°C, 1000 kPa is expanded in a reversible adiabatic process to 100 kPa. Find the work and heat transfer for this process.

7.10 A cylinder fitted with a piston contains ammonia at 50°C, 20% quality with a volume of 1 L. The ammonia expands slowly, and during this process heat is transferred to maintain a constant temperature. The process continues until all the liquid is gone. Determine the work and heat transfer for this process.

7.11 An insulated cylinder fitted with a piston contains 0.1 kg of water at 100°C, 90% quality. The piston is moved, compressing the water until it reaches a pressure of 1.2 MPa. How much work is required in the process?

7.12 A cylinder containing R-134a at 10°C, 150 kPa, has an initial volume of 20 L. A piston compresses the R-134a in a reversible, isothermal process until it reaches the saturated vapor state. Calculate the required work and heat transfer to accomplish this process.

7.13 One kilogram of water at 300°C expands against a piston in a cylinder until it reaches ambient pressure, 100 kPa, at which point the water has a quality of 90%. It may be assumed that the expansion is reversible and adiabatic.

a. What was the initial pressure in the cylinder?

b. How much work is done by the water?

7.14 A spring-loaded piston/cylinder, shown in Fig. P7.14, contains water at 100 kPa with $v = 0.07237$ m³/kg. The water is now heated to a pressure of 3 MPa by a reversible heat pump extracting Q from a reservoir at 300 K. It is known that the water will pass through saturated vapor at 1.5 MPa. Find the final temperature, the heat transfer to the water and the work input to the heat pump.

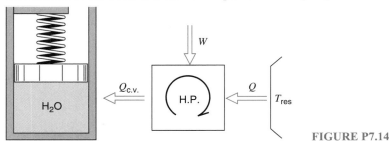

FIGURE P7.14

7.15 A heavily insulated cylinder/piston contains ammonia at 1200 kPa, 60°C. The piston is moved, expanding the ammonia in a reversible process until the temperature is −20°C. During the process 600 kJ of work is given out by the ammonia. What was the initial volume of the cylinder?

7.16 A rigid, insulated vessel contains superheated vapor steam at 3 MPa, 400°C. A valve on the vessel is opened, allowing steam to escape. It may be assumed that the steam remaining inside the vessel goes through a reversible adiabatic expansion. Determine the fraction of steam that has escaped, when the final state inside is saturated vapor.

7.17 An insulated cylinder fitted with a piston contains 0.1 kg of superheated vapor steam. The steam expands to ambient pressure, 100 kPa, at which point the steam inside the cylinder is at 150°C. The steam does 50 kJ of work against the piston during the expansion. What were the initial pressure and temperature?

7.18 A certain elastic balloon supports an internal pressure of $P_0 = 100$ kPa until it is spherical at a diameter of $D_0 = 0.5$ m, above which

$$P = P_o + C\left(D^{*-1} - D^{*-7}\right) \qquad D^* = D / D_o$$

The balloon, which is initially flat and empty, is connected by a valve to a 130-L insulated, rigid tank with ammonia at 1 MPa, 50°C. The valve is opened, and ammonia flows from the tank into the balloon, inflating it until the balloon goes through a maximum point in pressure of 300 kPa. During this process the ammonia remaining in the tank has gone through a reversible, adiabatic expansion and the temperature inside the balloon stays constant at the ambient temperature, 20°C. What is the pressure inside the tank, when the balloon reaches the maximum pressure of 300 kPa?

7.19 Consider the previous problem and suppose the valve is left open and the process continues until the pressure in the tank and the balloon equalizes. What is the pressure? What is the temperature inside the tank at this point?

7.20 Water in a piston/cylinder is at 1 MPa, 500°C. There are two stops, a lower one at which $V_{min} = 1$ m³ and an upper one at $V_{max} = 3$ m³. The piston is loaded with a mass and outside atmosphere such that it floats when the pressure is 500 kPa. This setup is now cooled to 100°C by rejecting heat to the surroundings at 20°C. Find the total entropy generated in the process.

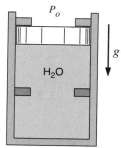

FIGURE P7.20

7.21 Two tanks contain steam, and they are both connected to a piston/cylinder as shown in Fig. P7.21. Initially the piston is at the bottom and the mass of the piston is such that a pressure of 1.4 MPa below it will be able to lift it. Steam in A is 4 kg at 7 MPa, 700°C and B has 2 kg at 3 MPa, 350°C. The two valves are opened, and the water comes to a uniform state. Find the final temperature and the total entropy generation, assuming no heat transfer.

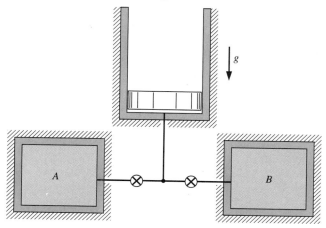

FIGURE P7.21

7.22 A cylinder/piston contains 3 kg of water at 500 kPa, 600°C. The piston has a cross-sectional area of 0.1 m² and is restrained by a linear spring with spring constant 10 kN/m. The setup is allowed to cool down to room temperature due to heat transfer to the room at 20°C. Calculate the total (water and surroundings) change in entropy for the process.

7.23 An insulated cylinder with a frictionless piston, shown in Fig. P7.23, contains water at ambient pressure, 100 kPa, a quality of 0.8 and the volume is 8 L. A force is now applied, slowly compressing the water until it reaches a set of stops, at which point the cylinder volume is 1 L. The insulation is then removed from the cylinder walls, and the water cools to ambient temperature, 20°C. Calculate the work and the heat transfer for the overall process.

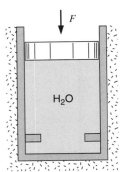

FIGURE P7.23

7.24 Calculate the net change in entropy for the overall process described in Problems 7.18 and 7.19, assuming that the heat transfer is with the ambient.

7.25 An insulated cylinder/piston has an initial volume of 0.15 m³ and contains steam at 400 kPa, 200°C. The steam is expanded adiabatically, and the work output is measured very carefully to be 30 kJ. It is claimed that the final state of the water is in the two-phase (liquid and vapor) region. What is your evaluation of the claim?

7.26 (Adv.) Consider the process shown in Fig. P7.26. The insulated tank A has a volume of 600 L, and contains steam at 1.4 MPa, 300°C. The uninsulated tank B has a volume of 300 L and contains steam at 200 kPa, 200°C. A valve connecting the two tanks is opened, and steam flows from A to B until the temperature in A reaches 250°C. The valve is closed. During the process heat is transferred from B to the surroundings at 25°C, such that the temperature in B remains at 200°C. It may be assumed that the steam remaining in A has undergone a reversible adiabatic expansion. Determine

a. The final pressure in each tank

b. The final mass in tank B

c. The net entropy change, system plus surroundings, for the process

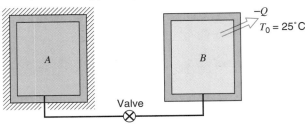

FIGURE P7.26

7.27 A mass and atmosphere loaded piston/cylinder contains 2 kg of water at 5 MPa, 100°C. Heat is added from a reservoir at 700°C to the water until it reaches 700°C. Find the work, heat transfer, and total entropy production for the system and surroundings.

7.28 A piece of hot metal should be cooled rapidly (quenched) to 25°C, which requires removal of 1000 kJ from the metal. The cold space that absorbs the energy could be one of three possibilities: (1) Submerge the metal into a bath of liquid water and ice, thus melting the ice. (2) Let saturated liquid R-22 at −20°C absorb the energy so that it becomes saturated vapor. (3) Absorb the energy by vaporizing liquid nitrogen at 101.3 kPa pressure.

    a. Calculate the change of entropy of the cooling media for each of the three cases.

    b. Discuss the significance of the results.

7.29 An insulated cylinder/piston contains R-134a at 1 MPa, 50°C, with a volume of 100 L. The R-134a expands, moving the piston until the pressure in the cylinder has dropped to 100 kPa. It is claimed that the R-134a does 190 kJ of work against the piston during the process. Is that possible?

7.30 A cylinder/piston contains water at 200 kPa, 200°C with a volume of 20 L. The piston is moved slowly, compressing the water to a pressure of 800 kPa. The loading on the piston is such that the product PV is a constant. Assuming that the room temperature is 20°C, show that this process does not violate the second law.

7.31 One kilogram of ammonia (NH_3) is contained in a spring-loaded piston/cylinder as saturated liquid at −20°C. Heat is added from a reservoir at 100°C until a final condition of 800 kPa, 70°C is reached. Find the work, heat transfer, and entropy generation, assuming the process is internally reversible.

7.32 (Adv.) A vertical cylinder/piston contains R–22 at −20°C, 70% quality, and the volume is 50 L, shown in Fig. P7.32. This cylinder is brought into a 20°C room, and an electric current of 10 A is passed through a resistor inside the cylinder. The voltage drop across the resistor is 12 V. It is claimed that after 30 min the temperature inside the cylinder is 40°C. Is this possible?

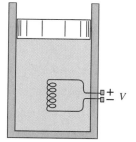

FIGURE P7.32

7.33 A foundry form box with 25 kg of 200°C hot sand is dumped into a bucket with 50 L water at 15°C. Assuming no heat transfer with the surroundings and no boiling away of liquid water, calculate the net entropy change for the process.

7.34 A large slab of concrete, $5 \times 8 \times 0.3$ m, is used as a thermal storage mass in a solar-heated house. If the slab cools overnight from 23°C to 18°C in an 18°C house, what is the net entropy change associated with this process?

7.35 Liquid lead initially at 500°C is poured into a form so that it holds 2 kg. It then cools at constant pressure down to room temperature of 20°C as heat is transferred to the room. The melting point of lead is 327°C and the enthalpy change between the phases, h_{if}, is 24.6 kJ/kg. The specific heat is 0.138 kJ/kg K for the solid and 0.155 kJ/kg K for the liquid. Calculate the net entropy change for this process.

7.36 A hollow steel sphere with a 0.5-m inside diameter and a 2-mm thick wall contains water at 2 MPa, 250°C. The system (steel plus water) cools to the ambient temperature, 30°C. Calculate the net entropy change of the system and surroundings for this process.

7.37 A mass of 1 kg of air contained in a cylinder at 1.5 MPa, 1000 K, expands in a reversible isothermal process to a volume 10 times larger. Calculate

a. The heat transfer during the process

b. The change of entropy of the air

7.38 A mass of 1 kg of air contained in a cylinder at 1.5 MPa, 1000 K, expands in a reversible adiabatic process to 100 kPa. Calcuate the final temperature and the work done during the process, using

a. Constant specific heat, value from Table A.10

b. The ideal gas tables, Table A.12

7.39 Consider a Carnot-cycle heat pump having 1 kg of nitrogen gas in a cylinder/piston arrangement. This heat pump operates between reservoirs at 300 K and 400 K. At the beginning of the low-temperature heat addition, the pressure is 1 MPa. During this processs the volume triples. Analyze each of the four processes in the cycle and determine

a. The pressure, volume, and temperature at each point

b. The work and heat transfer for each process

7.40 A handheld pump for a bicycle has a volume of 25 cm^3 when fully extended. You now press the plunger (piston) in while holding your thumb over the exit hole so that an air pressure of 300 kPa is obtained. The outside atmosphere is at P_0, T_0. Consider two cases: (1) it is done quickly (~1 s), and (2) it is done very slowly (~1 h).

a. State assumptions about the process for each case.

b. Find the final volume and temperature for both cases.

7.41 An insulated cylinder/piston contains carbon dioxide gas at 120 kPa, 400 K. The gas is compressed to 2.5 MPa in a reversible adiabatic process. Calculate the final temperature and the work per unit mass, assuming

a. Variable specific heat, Table A.13

b. Constant specific heat, value from Table A.10

c. Constant specific heat, value at an intermediate temperature from Table A.11

7.42 A piston/cylinder, shown in Fig. P7.42, contains air at 1380 K, 15 MPa, with $V_1 =$ 10 cm^3, $A_{cyl} = 5$ cm^2. The piston is released, and just before the piston exits the end of the cylinder the pressure inside is 200 kPa. If the cylinder is insulated, what is its length? How much work is done by the air inside?

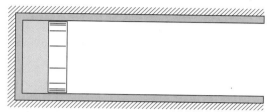

FIGURE P7.42

7.43 Two rigid tanks each contain 10 kg N$_2$ gas at 1000 K, 500 kPa. They are now thermally connected to a reversible heat pump, which heats one and cools the other with no heat transfer to the surroundings. When one tank is heated to 1500 K the process stops. Find the final (P, T) in both tanks and the work input to the heat pump, assuming constant heat capacities.

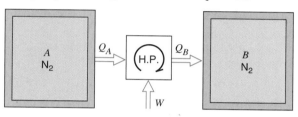

FIGURE P7.43

7.44 Repeat the previous problem, but with variable heat capacities.

7.45 We wish to obtain a supply of cold helium gas by applying the following technique. Helium contained in a cylinder at ambient conditions, 100 kPa, 20°C, is compressed in a reversible isothermal process to 600 kPa, after which the gas is expanded back to 100 kPa in a reversible adiabatic process.

a. Show the process on a T–s diagram.

b. Calculate the final temperature and the net work per kilogram of helium.

c. If a diatomic gas, such as nitrogen or oxygen, is used instead, would the final temperature be higher, lower, or the same?

7.46 A 1-m^3 insulated, rigid tank contains air at 800 kPa, 25°C. A valve on the tank is opened, and the pressure inside quickly drops to 150 kPa, at which point the valve is closed. Assuming that the air remaining inside has undergone a reversible adiabatic expansion, calculate the mass withdrawn during the process.

7.47 (Adv.) Redo the previous Problem 7.46, but calculate the mass withdrawn by a first-law, control-volume analysis. Compare the result to that obtained in Problem 7.46. Show from a differential step of mass out that the first law leads to the same result. (Find the relation between dP and dT.)

7.48 A rigid container with volume 200 L is divided into two equal volumes by a partition. Both sides contain nitrogen, one side is at 2 MPa, 200°C, and the other at 200 kPa, 100°C. The partition ruptures, and the nitrogen comes to a uniform state at 70°C. Assume the temperature of the surroundings is 20°C, determine the work done and the net entropy change for the process.

7.49 Neon at 400 kPa, 20°C is brought to 100°C in a polytropic process with $n = 1.4$. Give the sign for the heat transfer and work terms and explain.

7.50 A gas in a rigid vessel is at ambient temperature and at a pressure, P_1, slightly higher than ambient pressure, P_0. A valve on the vessel is opened, so gas escapes and the pressure drops quickly to ambient pressure. The valve is closed and after a long time the remaining gas returns to ambient temperature at which point the pressure is P_2. Develop an expression that allows a determination of the ratio of specific heats, k, in terms of the pressures.

7.51 A cylinder/piston contains carbon dioxide at 1 MPa, 300°C with a volume of 200 L. The total external force acting on the piston is proportional to V^3. This system is allowed to cool to room temperature, 20°C. What is the total entropy generation for the process?

7.52 A cylinder with a spring-loaded piston contains carbon dioxide gas at 2 MPa with a volume of 50 L. The device is of aluminum and has a mass of 4 kg. Everything (Al and gas) is initially at 200°C. By heat transfer the whole system cools to the ambient temperature of 25°C, at which point the gas pressure is 1.5 MPa. Find the total entropy generation for the process.

7.53 A cylinder/piston contains 1 kg methane gas at 100 kPa, 20°C. The gas is compressed reversibly to a pressure of 800 kPa. Calculate the work required if the process is
   a. Adiabatic
   b. Isothermal
   c. Polytropic, with exponent $n = 1.15$

7.54 The power stroke in an internal combustion engine can be approximated with a polytropic expansion. Consider air in a cylinder volume of 0.2 L at 7 MPa, 1800 K. It now expands in a reversible polytropic process with exponent, $n = 1.5$, through a volume ratio of $8 : 1$. Show this process on P–v and T–s diagrams, and calculate the work and heat transfer for the process.

7.55 A mass of 2 kg ethane gas at 500 kPa, 100°C, undergoes a reversible polytropic expansion with exponent, $n = 1.3$, to a final temperature of the ambient, 20°C. Calculate the total entropy generation for the process if the heat is exchanged with the ambient.

7.56 A cylinder/piston contains air at ambient conditions, 100 kPa and 20°C with a volume of 0.3 m^3. The air is compressed to 800 kPa in a reversible polytropic process with exponent, $n = 1.2$, after which it is expanded back to 100 kPa in a reversible adiabatic process.
   a. Show the two processes in P–v and T–s diagrams.
   b. Determine the final temperature and the net work.
   c. What is the potential refrigeration capacity (in kilojoules) of the air at the final state?

7.57 An ideal gas having a constant specific heat undergoes a reversible polytropic expansion with exponent, $n = 1.4$. If the gas is carbon dioxide will the heat transfer for this process be positive, negative, or zero?

7.58 A cylinder/piston contains 100 L of air at 110 kPa, 25°C. The air is compressed in a reversible polytropic process to a final state of 800 kPa, 200°C. Assume the heat transfer is with the ambient at 25°C and determine

a. The polytropic exponent n

b. The final volume of the air

c. The work done by the air and the heat transfer for the process

d. The total entropy generation

7.59 (Adv.) A closed, partly insulated cylinder divided by an insulated piston contains air in one side and water on the other, as shown in Fig. P7.59. There is no insulation on the end containing water. Each volume is initially 100 L, with the air at 40°C and the water at 90°C, quality 10%. Heat is slowly transferred to the water, until a final state of saturated vapor. Calculate the final pressure and the amount of heat transferred.

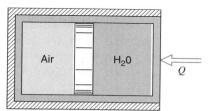

FIGURE P7.59

7.60 A cylinder/piston contains saturated vapor R-22 at 10°C; the volume is 10 L. The R-22 is compressed to 2 MPa, 60°C in a reversible (internally) polytropic process. If all the heat transfer during the process is with the ambient at 10°C, calculate the net entropy change.

7.61 Consider a small air pistol with a cylinder volume of 1 cm^3 at 250 kPa, 27°C. The bullet acts as a piston initially held by a trigger. The bullet is released so the air expands in an adiabatic process. If the pressure should be 100 kPa as the bullet leaves the cylinder find the final volume and the work done by the air.

7.62 (Adv.) Consider the system shown in Fig. P7.62. Tank A has a volume of 300 L and initially contains air at 700 kPa, 40°C. Cylinder B has a piston resting on the bottom at which point the linear spring is fully extended. The piston has a cross-sectional area of 0.065 m^2, a mass of 40 kg, and the spring constant is 17.5 kN/m. Atmospheric pressure is 100 kPa. The valve is opened and air flows into the cylinder until the pressures in A and B becomes equal and the valve is closed. The entire process is adiabatic, and the air remaining in A has undergone a reversible adiabatic expansion. Determine the final pressure in the system and the final temperature in cylinder B.

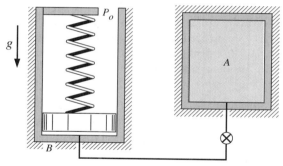

FIGURE P7.62

7.63 Steam enters a turbine at 3 MPa, 450°C, expands in a reversible adiabatic process and exhausts at 10 kPa. Changes in kinetic and potential energies between the inlet and the exit of the turbine are small. The power output of the turbine is 800 kW. What is the mass flow rate of steam through the turbine?

7.64 (Adv.) In a heat pump that uses R-134a as the working fluid, the R-134a enters the compressor at 150 kPa, −10°C at a rate of 0.1 kg/s. In the compressor the R-134a is compressed so that when it subsequently cools at constant pressure in the condenser condensation will occur at 40°C. Calculate the power input required to the compressor, assuming the process to be reversible and adiabatic.

7.65 Two flowstreams of water, one at 0.6 MPa, saturated vapor, and the other at 0.6 MPa, 600°C, mix adiabatically in a SSSF process to produce a single flow out at 0.6 MPa, 400°C. Find the total entropy generation for this process.

7.66 Consider the design of a nozzle in which nitrogen gas flowing in a pipe at 500 kPa, 200°C, and at a velocity of 10 m/s, is to be expanded to produce a velocity of 300 m/s. Determine the exit pressure and cross-sectional area of the nozzle if the mass flow rate is 0.15 kg/s, and the expansion is reversible and adiabatic.

7.67 A counterflowing heat exchanger, shown in Fig. P7.67, is used to cool air at 540 K, 400 kPa to 360 K by using a 0.05 kg/s supply of water at 20°C, 200 kPa. The air flow is 0.5 kg/s in a 10-cm diameter pipe. Find the air inlet velocity, the water exit temperature, and total entropy generation in the process.

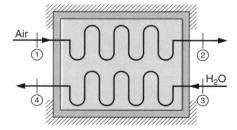

FIGURE P7.67

7.68 A small pump takes in water at 20°C, 100 kPa and pumps it to 2.5 MPa at a flow rate of 100 kg/min. Find the required pump power input.

7.69 In a heat-driven refrigerator with ammonia as the working fluid, a turbine with inlet conditions of 2.0 MPa, 70°C is used to drive a compressor with inlet saturated vapor at −20°C. The exhausts, both at 1.2 MPa, are then mixed together. The ratio of the mass flow rate to the turbine to the total exit flow was measured to be 0.62. Can this be true?

7.70 A diffuser is a steady-state, steady-flow device in which a fluid flowing at high velocity is decelerated such that the pressure increases in the process. Air at 120 kPa, 30°C enters a diffuser with velocity 200 m/s and exits with a velocity of 20 m/s. Assuming the process is reversible and adiabatic what are the exit pressure and temperature of the air?

7.71 A mixing chamber receives 5 kg/min ammonia as saturated liquid at −20°C from one line and ammonia at 40°C, 250 kPa from another line through a valve. The chamber also receives 325 kJ/min energy as heat transferred from a 40°C reservoir. This should produce saturated ammonia vapor at −20°C in the exit line.

What is the mass flow rate at state 2 and what is the total entropy generation in the process?

7.72 One technique for operating a steam turbine in part-load power output is to throttle the steam to a lower pressure before it enters the turbine, as shown in Fig. P7.72. The steamline conditions are 2 MPa, 400°C, and the turbine exhaust pressure is fixed at 10 kPa. Assuming the expansion inside the turbine to be reversible and adiabatic, determine

  a. The full-load specific work output of the turbine

  b. The pressure the steam must be throttled to for 80% of full-load output

  c. Show both processes in a T–s diagram.

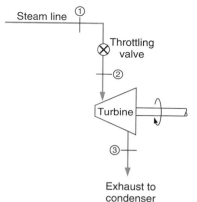

FIGURE P7.72

7.73 An air turbine with inlet conditions 1200 K, 1 MPa and exhaust pressure of 100 kPa pulls a sledge over a leveled plane surface, $T = 20°C$. The turbine work overcomes the friction between the sledge and the surface. Find the total entropy generation per kilogram of air through the turbine.

7.74 A certain industrial process requires a steady supply of saturated vapor steam at 200 kPa, at a rate of 0.5 kg/s. Also required is a steady supply of compressed air at 500 kPa, at a rate of 0.1 kg/s. Both are to be supplied by the process shown in Fig. P7.74. Steam is expanded in a turbine to supply the power needed to drive the air compressor, and the exhaust steam exits the turbine at the desired state. Air into the compressor is at the ambient conditions, 100 kPa, 20°C. Give the required steam inlet pressure and temperature, assuming that both the turbine and the compressor are reversible and adiabatic.

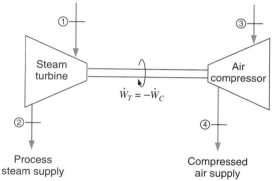

FIGURE P7.74

7.75 Air enters a turbine at 800 kPa, 1200 K, and expands in a reversible adiabatic process to 100 kPa. Calculate the exit temperature and the work output per kilogram of air, using

a. The ideal gas tables, Table A.12

b. Constant specific heat, value at 300 K from table A.10

c. Constant specific heat, value at an intermediate temperature from Fig. 5.10

Discuss why the method of part (b) gives a poor value for the exit temperature and yet a relatively good value for the work output.

7.76 (Adv.) Consider a geothermal supply of hot water available as saturated liquid at $P_1 = 1.5$ MPa. The liquid is to be flashed (throttled) to some lower pressure, P_2. The saturated liquid and saturated vapor at this pressure are separated and the vapor is expanded through a reversible adiabatic turbine to the exhaust pressure, $P_3 = 10$ kPa. Study the turbine power output per unit initial mass, m_1 as a function of the pressure, P_2.

7.77 Liquid water at ambient conditions, 100 kPa, 25°C, enters a pump at the rate of 0.5 kg/s. Power input to the pump is 3 kW. Assuming the pump process to be reversible, determine the pump exit pressure and temperature.

7.78 A large storage tank contains liquefied natural gas (LNG), which may be assumed to be pure methane. The tank contains saturated liquid at ambient pressure, 100 kPa; it is to be pumped to 500 kPa and fed to a pipeline at the rate of 0.5 kg/s. How much power input is required for the pump, assuming it to be reversible?

7.79 Consider a steam turbine power plant operating at supercritical pressure, as shown in Fig. P7.79. As a first approximation, it may be assumed that the turbine and the pump processes are reversible and adiabatic. Neglecting any changes in kinetic and potential energies, calculate

a. The specific turbine work output and the turbine exit state

b. The pump work input and enthalpy at the pump exit state

c. The thermal efficient of the cycle

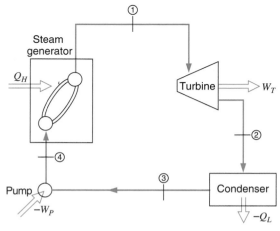

$P_4 = P_1 = 25$ MPa $T_1 = 700°$C
$P_2 = P_3 = 20$ kPa $T_3 = 40°$C

FIGURE P7.79

7.80 Helium gas enters a steady-flow expander at 800 kPa, 300°C, and exits at 120 kPa. The mass flow rate is 0.2 kg/s, and the expansion process can be considered as a reversible polytropic process with exponent, $n = 1.3$. Calculate the power output of the expander.

7.81 (Adv.) Ammonia enters a nozzle at 800 kPa, 50°C, at a velocity of 10 m/s and at the rate of 0.1 kg/s. The nozzle expansion is assumed to be a reversible, polytropic SSSF process. Ammonia exits the nozzle at 200 kPa; the rate of heat transfer to the nozzle is 8.2 kW. What is the velocity of the ammonia exiting the nozzle?

7.82 One type of feedwater heater for preheating the water before entering a boiler operates on the principle of mixing the water with steam that has been bled from the turbine. For the states as shown in Fig. P7.82, calculate the rate of net entropy increase for the process, assuming the process to be steady flow and adiabatic.

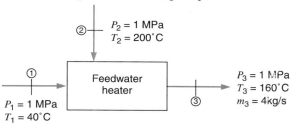

FIGURE P7.82

7.83 (Adv.) Consider the scheme shown in Fig. P7.83 for producing fresh water from salt water. The conditions are as shown in the figure. Assume that the properties of salt water are the same as for pure water, and that the pump is reversible and adiabatic.

a. Determine the ratio $(\dot{m}_7/\dot{m}_1)$, the fraction of salt water purified.

b. Determine the input quantities, w_p and q_H.

c. Make a second law analysis of the overall system.

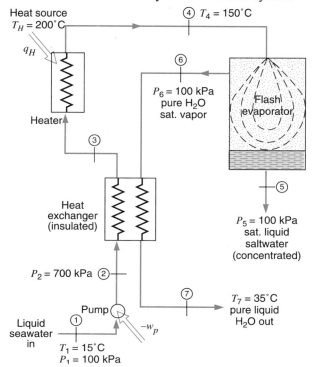

7.84 A pump/compressor pumps a substance from 100 kPa, 10°C to 1 MPa in a reversible adiabatic SSSF process. The exit pipe has a small crack, so that a small amount leaks to the atmosphere at 100 kPa. If the substance is (a) water, (b) R-12, find the temperature after compression and the temperature of the leak flow as it enters the atmosphere neglecting kinetic energies.

7.85 A stream of ammonia enters a steady flow device at 100 kPa, 50°C, at the rate of 1 kg/s. Two streams exit the device at equal mass flow rates; one is at 200 kPa, 50°C, and the other as saturated liquid at 10°C. It is claimed that the device operates in a room at 25°C on an electrical power input of 250 kW. Is this possible?

7.86 Air from a line at 12 MPa, 15°C, flows into a 500-L rigid tank that initially contained air at ambient conditions, 100 kPa, 15°C. The process occurs rapidly and is essentially adiabatic. The valve is closed when the pressure inside reaches some value, P_2. The tank eventually cools to room temperature, at which time the pressure inside is 5 MPa. What is the pressure P_2? What is the net entropy change for the overall process?

7.87 An initially empty spring-loaded piston/cylinder requires 100 kPa to float the piston. A compressor with a line and valve now charges the cylinder with water to a final pressure of 1.4 MPa at which point the volume is 0.6 m³, state 2. The inlet condition to the reversible adiabatic compressor is saturated vapor at 100 kPa. After charging the valve is closed and the water eventually cools to room temperature, 20°C, state 3. Find the final mass of water, the piston work from 1 to 2, the required compressor work, and the final pressure, P_3.

7.88 A 1-m³ rigid tank contains 100 kg R-22 at ambient temperature, 15°C. A valve on top of the tank is opened, and saturated vapor is throttled to ambient pressure, 100 kPa, and flows to a collector system. During the process the temperature inside the tank remains at 15°C. The valve is closed when no more liquid remains inside. Calculate

a. The heat transfer to the tank

b. The total entropy generation in the process

7.89 (Adv.) A cylinder with a piston restrained by an external force contains R-12 at 50°C, 90% quality with a volume of 100 L. The cylinder is attached to a line flowing R-12 at 3 MPa, 150°C. The valve is opened and mass flows into the cylinder until it reaches a pressure of 3 MPa, at which point the temperature is 100°C. It is claimed that during the process the R-12 does 150 kJ of work against the external force and any heat transfer takes place with the ambient at 20°C. Assume that the R-12 has followed a polytropic process. Does this process violate the second law?

7.90 An old abandoned saltmine, 100 000 m³ in volume, contains air at 290 K, 100 kPa. The mine is used for energy storage so the local power plant pumps it up to 2.1 MPa using outside air at 290 K, 100 kPa. Assume the pump is ideal and the process is adiabatic. Find the final mass and temperature of the air and the required pump work. Overnight, the air in the mine cools down to 400 K. Find the final pressure and heat transfer.

7.91 An initially empty 0.1 m³ cannister is filled with R-12 from a line flowing saturated liquid at −5°C. This is done quickly such that the process is adiabatic. Find the final mass, liquid and vapor volumes, if any, in the cannister. Is the process reversible?

7.92 A rigid steel bottle, $V = 0.25$ m^3, contains air at 100 kPa, 300 K. The bottle is now charged with air from a line at 260 K, 6 MPa to a bottle pressure of 5 MPa, state 2, and the valve is closed. Assume that the process is adiabatic, and the charge always is uniform. In storage, the bottle slowly returns to room temperature at 300 K, state 3. Find the final mass, the temperature T_2, the final pressure P_3, the heat transfer $_1Q_3$ and the total entropy generation.

7.93 (Adv.) A spherical aluminum tank having an inside diameter of 1 m and a wall thickness of 5 mm contains water at 2 MPa, 300°C. A valve on the tank is opened, allowing steam to flow out until the tank and steam remaining inside both reach 150°C, at which point the valve is closed. The steam inside now has a quality of 90%.

  a. How much steam was lost from the tank?

  b. Take a control volume around the tank and its content, calculate the entropy change in the control volume and that of its surroundings for this process.

7.94 Liquid water enters a pump at 15°C, 100 kPa, and exits at a pressure of 5 MPa. If the isentropic efficiency of the pump is 75%, determine the enthalpy (steam table reference) of the water at the pump exit.

7.95 Repeat Problem 7.74 assuming the steam turbine and the air compressor each have an isentropic efficiency of 80%.

7.96 A small air turbine with an isentropic efficiency of 80% should produce 270 kJ/kg of work. The inlet temperature is 1000 K and it exhausts to the atmosphere. Find the required inlet pressure and the exhaust temperature.

7.97 Repeat Problem 7.79 assuming the turbine and the pump each have an isentropic efficiency of 85%.

7.98 In a heat-powered refrigerator, a turbine is used to drive the compressor using the same working fluid. Consider the combination shown in Fig. P7.98 where the turbine produces just enough power to drive the compressor and the two exit flows are mixed together. List any assumptions made and find the ratio of mass flow rates $\dot{m}_3/\dot{m}_1$ and T_5 (x_5 if in two-phase region) if

  a. The turbine and the compressor are reversible and adiabatic

  b. The turbine and the compressor both have an isentropic efficiency of 70%

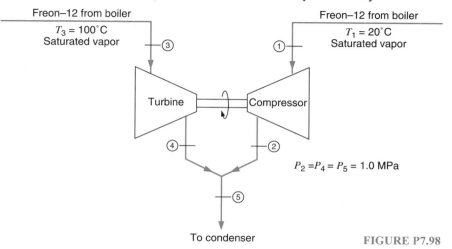

FIGURE P7.98

7.99 Air enters an insulated compressor at ambient conditions, 100 kPa, 20°C, at the rate of 0.1 kg/s and exits at 200°C. The isentropic efficiency of the compressor is 70%. What is the exit pressure? How much power is required to drive the compressor?

7.100 Supercharging of an engine is used to increase the inlet air density so that more fuel can be added, the result of which is an increased power output. Assume that ambient air, 100 kPa and 27°C, enters the supercharger at a rate of 250 L/s. The supercharger (compressor) has an isentropic efficiency of 75%, and uses 20 kW of power input. Assume that the ideal and actual compressor have the same exit pressure. Find the ideal specific work and verify that the exit pressure is 175 kPa. Find the percent increase in air density entering the engine due to the supercharger and the reversible work.

7.101 Steam enters a turbine at 300°C and exhausts at 20 kPa. It is estimated that the isentropic efficiency of the turbine is 70%. What is the maximum turbine inlet pressure if the exhaust is not to be in the two-phase region?

7.102 A nozzle is required to produce a stream of air at 200 m/s at 20°C, 100 kPa. It is estimated that the nozzle has an isentropic efficiency of 92%. What nozzle inlet pressure and temperature is required?

7.103 A compressor is used to bring saturated water vapor at 1 MPa up to 17.5 MPa, where the actual exit temperature is 650°C. Find the isentropic compressor efficiency and the entropy generation.

7.104 Air at 100 kPa, 17°C is compressed to 400 kPa after which it is expanded through a nozzle back to the atmosphere. The compressor and the nozzle both have efficiency of 90% and kinetic energy in/out of the compressor can be neglected. Find the actual compressor work and its exit temperature and find the actual nozzle exit velocity.

7.105 A small turbine delivers 150 kW and is supplied with steam at 700°C, 2 MPa. The exhaust passes through a heat exchanger where the pressure is 10 kPa and exits as saturated liquid. The isentropic turbine efficiency is 88%. Find the actual specific turbine work, the entropy generated in the turbine and the heat transfer in the heat exchanger.

7.106 A centrifugal compressor takes in ambient air at 100 kPa, 15°C, and discharges it at 450 kPa. The compressor has an isentropic efficiency of 80%. What is your best estimate for the discharge temperature?

7.107 A jet-ejector pump, shown schematically in Fig. P7.107, is a device in which a low-pressure (secondary) fluid is compressed by entrainment in a high-velocity (primary) fluid stream. The compression results from the deceleration in a diffuser. For purposes of analysis this can be considered as equivalent to the turbine-compressor unit shown in Fig. P7.98 with the states 1, 3, and 5 corresponding to those in Fig. P7.107. Consider a steam jet-pump with state 1 as saturated vapor at 35 kPa; state 3 is 300 kPa, 150°C; and the discharge pressure, P_5, is 100 kPa.

a. Calculate the ideal mass flow ratio, $\dot{m}_1/\dot{m}_3$.

b. The efficiency of a jet pump is defined as

$$\eta_{\text{jet pump}} = \frac{\left(\dot{m}_1 / \dot{m}_3\right)_{\text{actual}}}{\left(\dot{m}_1 / \dot{m}_3\right)_{\text{ideal}}}$$

for the same inlet conditions and discharge pressure. Determine the discharge temperature of the jet pump if its efficiency is 10%.

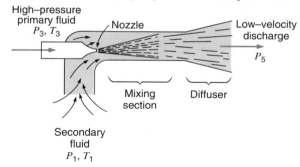

FIGURE P7.107

7.108 A geothermal supply of hot water at 500 kPa, 150°C is fed to an insulated flash evaporator at the rate of 1.5 kg/s. A stream of saturated liquid at 200 kPa is drained from the bottom of the chamber, and a stream of saturated vapor at 200 kPa is drawn from the top and fed to a turbine. The turbine has an isentropic efficiency of 70% and an exit pressure of 15 kPa. Evaluate the second law for a control volume that includes the flash evaporator and the turbine.

7.109 (Adv.) A turbo charger boosts the inlet air pressure to an automobile engine. It consists of an exhaust gas driven turbine directly connected to an air compressor, as shown in Fig. P7.109. For a certain engine load the conditions are given in the figure. If it is assumed that both the turbine and the compressor are reversible and adiabatic, calculate

a. The turbine exit temperature and power output

b. The compressor exit pressure and temperature

c. Repeat parts (a) and (b), but assume the isentropic efficiency of the turbine is 85% and that of the compressor is 80%. Assume the actual and ideal compressor have the same exit pressure

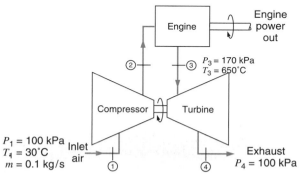

FIGURE P7.109

7.110 A firefighter on a ladder 40 m above ground should be able to spray water an additional 10 m up with the hose nozzle of exit diameter 2.5 cm. For a water pump on the ground with an efficiency of 85% (hose, nozzle included) find the minimum required power.

7.111 A flow of 20 kg/s steam at 10 MPa, 550°C enters a two-stage turbine. The exit of the first stage is at 2 MPa where 4 kg/s is taken out for process steam and the rest continues through the second stage, which has an exit at 50 kPa. Assume both stages have an isentropic efficiency of 85% find the total actual turbine work and the entropy generation.

7.112 (Adv.) A nozzle is required to produce a steady stream of R–134a at 240 m/s at ambient conditions, 100 kPa, 20°C. The isentropic efficiency may be assumed to be 90%. What pressure and temperature are required in the line upstream of the nozzle?

7.113 Calculate the isentropic efficiency for each of the stages in the stream turbine shown in Problem 5.105. Find also the total entropy generated in the turbine.

7.114 (Adv.) A certain industrial process requires a steady stream of saturated vapor water at 200 kPa at a rate of 2 kg/s. There are two alternatives for supplying this steam from ambient liquid water at 20°C, 100 kPa.

1. Pump the water to 200 kPa and feed it to a steam generator (heater).

2. Pump the water to 5 MPa, feed it to a steam generator and then expand it through a turbine from which the steam exhausts at the desired state.

a. Compare these two alternatives, making reasonable assumptions about the various processes including efficiencies.

b. What is the total entropy generation for each alternative?

7.115 A two-stage compressor having an interstage cooler takes in air, 300 K, 100 kPa, and compresses it to 2 MPa, as shown in Fig. P7.115. The cooler then cools the air to 340 K, after which it enters the second stage, which has an exit pressure of 15.74 MPa. The isentropic efficiency of stage one is 90% and the air exits the second stage at 630 K. Both stages are adiabatic, and the cooler dumps Q to reservoir at T_0. Find Q in the cooler, the efficiency of the second stage, and the total entropy generated in this process.

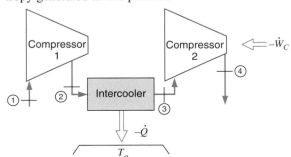

FIGURE P7.115

7.116 (Adv.) A cylinder/piston containing 2 kg of ammonia at −10°C, 90% quality is brought into a 20°C room and attached to a line flowing ammonia at 800 kPa, 40°C. The total restraining force on the piston is proportional to the cylinder volume squared. The valve is opened and ammonia flows into the cylinder until the mass inside is twice the initial mass and the valve is closed. An electrical current of 15 A is passed through a 2-Ω resistor inside the cylinder for 20 min. It is claimed that the final pressure in the cylinder is 600 kPa. Is this possible?

7.117 Carbon dioxide, CO_2, enters an adiabatic compressor at 100 kPa, 300 K, and exits at 1000 kPa, 520 K. Find the compressor efficiency and the entropy generation for the process.

7.118 A paper mill, shown in Fig. P7.118, has two steam generators, one at 4.5 MPa, 300°C and one at 8 MPa, 500°C. Each generator feeds a turbine, both of which have an exhaust pressure of 1.2 MPa and isentropic efficiency of 87%, such that their combined power output is 20 MW. The two exhaust flows are mixed adiabatically to produce saturated vapor at 1.2 MPa. Find the two mass flow rates and the entropy produced in each turbine and in the mixing chamber.

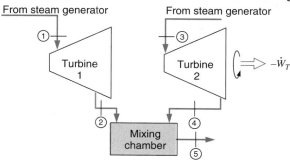

FIGURE P7.118

7.119 (Adv.) A rigid 1-m^3 storage tank of nitrogen at 600 kPa, room temperature of 300 K. The tank is connected by an adjustable pressure regulator to an insulated turbine that exhausts to atmospheric pressure, 100 kPa. The regulator maintains a fixed turbine-inlet pressure by throttling gas from the storage tank until the storage tank pressure drops to that value of pressure, at which point the nitrogen flow stops and the process ends. During the process enough heat is transferred to the storage tank that the temperature inside remains constant at 300 K. We wish to obtain the maximum total work output of the turbine. Assuming a turbine isentropic efficiency of 80%, determine

a. The optimum regulator pressure setting

b. The total work output

c. The net entropy change for the process

7.120 A two-stage turbine receives air at 1160 K, 5.0 MPa. The first stage exit at 1 MPa then enters stage 2, which has an exit pressure of 200 kPa. Each stage has an isentropic efficiency of 85%. Find the specific work in each stage, the overall isentropic efficiency, and the total entropy generation.

7.121 A heat-powered portable air compressor consists of three components: (a) an adiabatic compressor; (b) a constant pressure heater (heat supplied from an outside source); and (c) an adiabatic turbine. The compressor and the turbine each have an isentropic efficiency of 85%. Ambient air enters the compressor at 100 kPa, 300 K, and is compressed to 600 kPa. All of the power from the turbine goes into the compressor, and the turbine exhaust is the supply of compressed air. If this pressure is required to be 200 kPa, what must the temperature be at the exit of the heater?

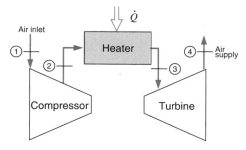

FIGURE P7.121

7.122E Find the missing properties and give the phase of the substance

    a. H_2O $s = 1.75$, Btu/lbm R, $P = 4$ lbf/in.^{2} $h = ?\ T = ?\ x = ?$

    b. H_2O $u = 1350$ Btu/lbm, $P = 1500$ lbf/in.^{2} $T = ?\ x = ?\ s = ?$

    c. R-12 $T = 30$ F, $P = 35$ lbf/in.^{2} $s = ?\ x = ?$

    d. R-134a $T = 10$ F, $x = 0.45$ $v = ?\ s = ?$

    e. NH_3 $T = 60$ F, $s = 1.35$ Btu/lbm R $u = ?\ x = ?$

7.123E In a Carnot engine with water as the working fluid, the high temperature is 450 F and as Q_H is received, the water changes from saturated liquid to saturated vapor. The water pressure at the low temperature is 14.7 lbf/in.^{2}. Find T_L, cycle thermal efficiency, heat added per pound-mass, and entropy, s, at the beginning of the heat rejection process.

7.124E Consider a Carnot-cycle heat pump with R-22 as the working fluid. Heat is rejected from the R-22 at 100 F, during which process the R-22 changes from saturated vapor to saturated liquid. The heat is transferred to the R-22 at 30 F.

    a. Show the cycle on a T–s diagram.

    b. Find the quality of the R-22 at the beginning and end of the isothermal heat addition process at 30 F.

    c. Determine the coefficient of performance for the cycle.

7.125E Do the Problem 7.124 using refrigerant R-134a instead of R-22.

7.126E A closed tank, $V = 0.35$ ft^3, containing 10 lbm of water initially at 77 F is heated to 350 F by a heat pump that is receiving heat from the surroundings at 77 F. Assume that this process is reversible. Find the heat transfer to the water and the work input to the heat pump.

7.127E Two pound-mass of ammonia in a piston/cylinder at 120 F, 150 lbf/in.^{2} is expanded in a reversible adiabatic process to 15 lbf/in.^{2}. Find the work and heat transfer for this process.

7.128E A cylinder fitted with a piston contains ammonia at 120 F, 20% quality with a volume of 60 in.^{3}. The ammonia expands slowly, and during this process heat is transferred to maintain a constant temperature. The process continues until all the liquid is gone. Determine the work and heat transfer for this process.

7.129E A cylinder containing R-134a at 50 F, 20 lbf/in.^{2}, has an initial volume of 1 ft^3. A piston compresses the R-134a in a reversible, isothermal process until it reaches the saturated vapor state. Calculate the required work and heat transfer to accomplish this process.

7.130E One pound-mass of water at 600 F expands against a piston in a cylinder until it reaches ambient pressure, 14.7 lbf/in.^{2}, at which point the water has a quality of 90%. It may be assumed that the expansion is reversible and adiabatic.

    a. What was the initial pressure in the cylinder?

    b. How much work is done by the water?

7.131E A rigid, insulated vessel contains superheated vapor steam at 450 lbf/in.^{2}, 700 F. A valve on the vessel is opened, allowing steam to escape. It may be assumed that the steam remaining inside the vessel goes through a reversible adiabatic expansion. Determine the fraction of steam that has escaped, when the final state inside is saturated vapor.

7.132E A certain elastic balloon supports an internal pressure of $P_0 = 14.7$ lbf/in.^{2} until it is spherical at a diameter of $D_o = 1.5$ ft, above which

$$P = P_o + C\left(D^{*-1} - D^{*-7}\right) \qquad D^* = D/D_o$$

The balloon, which is initially flat and empty, is connected by a valve to a 5-ft^3 insulated, rigid tank with ammonia at 140 lbf/in.^{2}, 120 F. The valve is opened, and ammonia flows from the tank into the balloon, inflating it until the balloon goes through a maximum point in pressure of 45 lbf/in.^{2}. During this process the ammonia remaining in the tank has gone through a reversible, adiabatic expansion and the temperature inside the balloon stays constant at the ambient temperature, 60 F. What is the pressure inside the tank, when the balloon reaches the maximum pressure of 45 lbf/in.^{2}?

7.133E Water in a piston/cylinder is at 150 lbf/in.^{2}, 900 F, as shown in Fig. P7.20. There are two stops, a lower one at which $V_{min} = 35$ ft^3 and an upper one at $V_{max} = 105$ ft^3. The piston is loaded with a mass and outside atmosphere such that it floats when the pressure is 75 lbf/in.^{2}. This setup is now cooled to 210 F by rejecting heat to the surroundings at 70 F. Find the total entropy generated in the process.

7.134E A cylinder/piston contains 5 lbm of water at 80 lbf/in.^{2}, 1000 F. The piston has cross-sectional area of 1 ft^2 and is restrained by a linear spring with spring constant 60 lbf/in. The setup is allowed to cool down to room temperature due to heat transfer to the room at 70 F. Calculate the total (water and surroundings) change in entropy for the process.

7.135E A mass and atmosphere loaded piston/cylinder contains 4 lbm of water at 500 lbf/in.^{2}, 200 F. Heat is added from a reservoir at 1200 F to the water until it reaches 1200 F. Find the work, heat transfer, and total entropy production for the system and surroundings.

7.136E An insulated cylinder/piston contains R-134a at 150 lbf/in.^{2}, 120 F, with a volume of 3.5 ft^3. The R-134a expands, moving the piston until the pressure in the cylinder has dropped to 15 lbf/in.^{2}. It is claimed that the R-134a does 180 Btu of work against the piston during the process. Is that possible?

7.137E A cylinder/piston contains water at 30 lbf/in.^{2}, 400 F with a volume of 1 ft^3. The piston is moved slowly, compressing the water to a pressure of 120 lbf/in.^{2}. The loading on the piston is such that the product PV is a constant. Assuming that the room temperature is 70 F, show that this process does not violate the second law.

7.138E A foundry form box with 50 lbm of 400 F hot sand is dumped into a bucket with 2 ft^3 water at 60 F. Assuming no heat transfer with the surroundings and no boiling away of liquid water, calculate the net entropy change for the process.

7.139E A hollow steel sphere with a 2-ft inside diameter and a 0.1-in. thick wall contains water at 300 lbf/in.^{2}, 500 F. The system (steel plus water) cools to the ambient temperature, 90 F. Calculate the net entropy change of the system and surroundings for this process.

7.140E A handheld pump for a bicycle has a volume of 2 in.^{3} when fully extended. You now press the plunger (piston) in while holding your thumb over the exit hole so an air pressure of 45 lbf/in.^{2} is obtained. The outside atmosphere is at P_0, T_0. Consider two cases: (1) it is done quickly (~1 s), and (2) it is done very slowly (~1 h).

    a. State assumptions about the process for each case.

    b. Find the final volume and temperature for both cases.

7.141E A piston/cylinder contains air at 2500 R, 2200 lbf/in.^{2}, with $V_1 = 1$ in.^{3}, $A_{cyl} = 1$ in.^{2} as shown in Fig. P7.42. The piston is released and just before the piston exits the end of the cylinder the pressure inside is 30 lbf/in.^{2}. If the cylinder is insulated, what is its length? How much work is done by the air inside?

7.142E We wish to obtain a supply of cold helium gas by applying the following technique. Helium contained in a cylinder at ambient conditions, 14.7 lbf/in.^{2}, 70 F, is compressed in a reversible isothermal process to 90 lbf/in.^{2}, after which the gas is expanded back to 14.7 lbf/in.^{2} in a reversible adiabatic process.

    a. Show the process on a T–s diagram.

    b. Calculate the final temperature and the net work per kilogram of helium.

    c. If a diatomic gas, such as nitrogen or oxygen, is used instead, would the final temperature be higher, lower, or the same?

7.143E A 25-ft^3 insulated, rigid tank contains air at 110 lbf/in.^{2}, 75 F. A valve on the tank is opened, and the pressure inside quickly drops to 15 lbf/in.^{2}, at which point the valve is closed. Assuming that the air remaining inside has undergone a reversible adiabatic expansion, calculate the mass withdrawn during the process.

7.144E (Adv.) Redo the previous Problem 7.143, but calculate the mass withdrawn by a first-law, control-volume analysis. Compare the result to that obtained in Problem 7.143. Show from a differential step of mass out that the first law leads to the same result. (Find the relation between dP and dT.)

7.145E A cylinder/piston contains carbon dioxide at 150 lbf/in.^{2}, 600 F with a volume of 7 ft^3. The total external force acting on the piston is proportional to V^3. This system is allowed to cool to room temperature, 70 F. What is the total entropy generation for the process?

7.146E A cylinder/piston contains 1 lbm methane gas at 15 lbf/in.^{2}, 70 F. The gas is compressed reversibly to a pressure of 120 lbf/in.^{2}. Calculate the work required if the process is

    a. Adiabatic

    b. Isothermal

    c. Polytropic, with exponent $n = 1.15$

7.147E A cylinder/piston contains air at ambient conditions, 14.7 lbf/in.^{2} and 70 F with a volume of 10 ft^3. The air is compressed to 100 lbf/in.^{2} in a reversible polytropic process with exponent, $n = 1.2$, after which it is expanded back to 14.7 lbf/in.^{2} in a reversible adiabatic process.

    a. Show the two processes in P–v and T–s diagrams.

b. Determine the final temperature and the net work.

c. What is the potential refrigeration capacity (in British thermal units) of the air at the final state?

7.148E (Adv.) A closed, partly insulated cylinder divided by an insulated piston contains air in one side and water on the other, as shown in Fig. P7.59. There is no insulation on the end containing water. Each volume is initially 3 ft^3, with the air at 100 F and the water at 190 F, quality 10%. Heat is slowly transferred to the water, until a final state of saturated vapor. Calculate the final pressure and the amount of heat transferred.

7.149E Steam enters a turbine at 450 lbf/in.^{2}, 900 F, expands in a reversible adiabatic process and exhausts at 2 lbf/in.^{2}. Changes in kinetic and potential energies between the inlet and the exit of the turbine are small. The power output of the turbine is 800 Btu/s. What is the mass flow rate of steam through the turbine?

7.150E (Adv.) In a heat pump that uses R-134a as the working fluid, the R-134a enters the compressor at 20 lbf/in.^{2}, 10 F at a rate of 0.1 lbm/s. In the compressor the R-134a is compressed so that when it subsequently cools at constant pressure in the condenser condensation will occur at 100 F. Calculate the power input required to the compressor, assuming the process to be reversible and adiabatic.

7.151E Two flowstreams of water, one at 100 lbf/in.^{2}, saturated vapor, and the other at 100 lbf/in.^{2}, 1000 F, mix adiabatically in a SSSF process to produce a single flow out at 100 lbf/in.^{2}, 600 F. Find the total entropy generation for this process.

7.152E A small pump takes in water at 70 F, 14.7 lbf/in.^{2} and pumps it to 250 lbf/in.^{2} at a flow rate of 200 lbm/min. Find the required pump power input.

7.153E A diffuser is a steady-state, steady-flow device in which a fluid flowing at high velocity is decelerated such that the pressure increases in the process. Air at 18 lbf/in.^{2}, 90 F enters a diffuser with velocity 600 ft/s and exits with a velocity of 60 ft/s. Assuming the process is reversible and adiabatic what are the exit pressure and temperature of the air?

7.154E A mixing chamber receives 10 lbm/min ammonia as saturated liquid at 0 F from one line and ammonia at 100 F, 40 lbf/in.^{2} from another line through a valve. The chamber also receives 340 Btu/min energy as heat transferred from a 100-F reservoir. This should produce saturated ammonia vapor at 0 F in the exit line. What is the mass flow rate at state 2 and what is the total entropy generation in the process?

7.155E One technique for operating a steam turbine in part-load power output is to throttle the steam to a lower pressure before it enters the turbine, as shown in Fig. P7.72. The steamline conditions are 200 lbf/in.^{2}, 600 F, and the turbine exhaust pressure is fixed at 1 lbf/in.^{2}. Assuming the expansion inside the turbine to be reversible and adiabatic, determine

a. The full-load specific work output of the turbine

b. The pressure the steam must be throttled to for 80% of full-load output

c. Show both processes in a T–s diagram

7.156E Liquid water at ambient conditions, 14.7 lbf/in.^{2}, 75 F, enters a pump at the rate of 1 lbm/s. Power input to the pump is 3 Btu/s. Assuming the pump process to be reversible, determine the pump exit pressure and temperature.

7.157E Helium gas enters a steady-flow expander at 120 lbf/in.^{2}, 500 F, and exits at 18 lbf/in.^{2}. The mass flow rate is 0.4 lbm/s, and the expansion process can be considered as a reversible polytropic process with exponent, $n = 1.3$. Calculate the power output of the expander.

7.158E Air from a line at 1800 lbf/in.^{2}, 60 F, flows into a 20-ft^3 rigid tank that initially contained air at ambient conditions, 14.7 lbf/in.^{2}, 60 F. The process occurs rapidly and is essentially adiabatic. The valve is closed when the pressure inside reaches some value, P_2. The tank eventually cools to room temperature, at which time the pressure inside is 750 lbf/in.^{2}. What is the pressure P_2? What is the net entropy change for the overall process?

7.159E An old abandoned saltmine, 3.5×10^6 ft^3 in volume, contains air at 520 R, 14.7 lbf/in.^{2}. The mine is used for energy storage so the local power plant pumps it up to 310 lbf/in.^{2} using outside air at 520 R, 14.7 lbf/in.^{2}. Assume the pump is ideal and the process is adiabatic. Find the final mass and temperature of the air and the required pump work. Overnight, the air in the mine cools down to 720 R. Find the final pressure and heat transfer.

7.160E (Adv.) A spherical aluminum tank having an inside diameter of 3 ft and a wall thickness of 0.2 in. contains water at 300 lbf/in.^{2}, 550 F. A valve on the tank is opened, allowing steam to flow out until the tank and steam remaining inside both reach 300 F, at which point the valve is closed. The steam inside now has a quality of 90%.

a. How much steam was lost from the tank?

b. Take a control volume around the tank and its content, calculate the entropy change in the control volume and that of its surroundings for this process.

7.161E Liquid water enters a pump at 60 F, 14.7 lbf/in.^{2}, and exits at a pressure of 1000 lbf/in.^{2}. If the isentropic efficiency of the pump is 75%, determine the enthalpy (steam table reference) of the water at the pump exit.

7.162E A small air turbine with an isentropic efficiency of 80% should produce 120 Btu/lbm of work. The inlet temperature is 1800 R and it exhausts to the atmosphere. Find the required inlet pressure and the exhaust temperature.

7.163E Air enters an insulated compressor at ambient conditions, 14.7 lbf/in.^{2}, 70 F, at the rate of 0.1 lbm/s and exits at 400 F. The isentropic efficiency of the compressor is 70%. What is the exit pressure? How much power is required to drive the compressor?

7.164E Steam enters a turbine at 500 F and exhausts at 2 lbf/in.^{2}. It is estimated that the isentropic efficiency of the turbine is 70%. What is the maximum turbine inlet pressure if the exhaust is not to be in the two-phase region?

7.165E A compressor is used to bring saturated water vapor at 150 lbf/in.^{2} up to 2500 lbf/in.^{2}, where the actual exit temperature is 1200 F. Find the isentropic compressor efficiency and the entropy generation.

7.166E Air at 1 atm, 60 F is compressed to 4 atm, after which it is expanded through a nozzle back to the atmosphere. The compressor and the nozzle both have efficiency of 90% and kinetic energy in/out of the compressor can be neglected.

Find the actual compressor work and its exit temperature and find the actual nozzle exit velocity.

7.167 E A geothermal supply of hot water at 80 lbf/in.^{2}, 300 F is fed to an insulated flash evaporator at the rate of 10,000 lbm/h. A stream of saturated liquid at 30 lbf/in.^{2} is drained from the bottom of the chamber and a stream of saturated vapor at 30 lbf/in.^{2} is drawn from the top and fed to a turbine. The turbine has an isentropic efficiency of 70% and an exit pressure of 2 lbf/in.^{2}. Evaluate the second law for a control volume that includes the flash evaporator and the turbine.

7.168 E A fireman on a ladder 100 ft above ground should be able to spray water an additional 30 ft up with the hose nozzle of exit diameter 1 in. For a water pump on the ground with an efficiency of 85% (hose, nozzle included) find the minimum required power.

7.169 E (Adv.) A nozzle is required to produce a steady stream of R-134a at 790 ft/s at ambient conditions, 14.7 lbf/in.^{2}, 70 F. The isentropic efficiency may be assumed to be 90%. What pressure and temperature are required in the line upstream of the nozzle?

7.170 E A two-stage compressor having an interstage cooler as shown in Fig. P7.115. It takes in air at 540 R, 14 lbf/in.^{2}, and compresses it to 280 lbf/in.^{2}. The cooler then cools the air to 610 R, after which it enters the second stage, which has an exit pressure of 2200 lbf/in.^{2}. The isentropic efficiency of stage one is 90% and the air exits the second stage at 1140 R. Both stages are adiabatic, and the cooler dumps Q to reservoir at T_0. Find Q in the cooler, the efficiency of the second-stage, and the total entropy generated in this process.

7.171 E A paper mill has two steam generators, one at 600 lbf/in.^{2}, 550 F and one at 1250 lbf/in.^{2}, 900 F. The setup is shown in Fig. P7.118. Each generator feeds a turbine, both of which have an exhaust pressure of 160 lbf/in.^{2} and isentropic efficiency of 87%, such that their combined power output is 20 000 Btu/s. The two exhaust flows are mixed adiabatically to produce saturated vapor at 160 lbf/in.^{2}. Find the two mass flow rates and the entropy produced in each turbine and in the mixing chamber.

7.172 E A two-stage turbine receives air at 2100 R, 750 lbf/in.^{2}. The first stage exit at 150 lbf in.^{2} then enters stage 2, which has an exit pressure of 30 lbf/in.^{2}. Each stage has an isentropic efficiency of 85%. Find the specific work in each stage, the overall isentropic efficiency, and the total entropy generation.

COMPUTER,
DESIGN, AND
OPEN-ENDED
PROBLEMS

7.173 Use the menu-driven software to get the properties for the calculation of the isentropic efficiency of the pump in the steam power plant of Problem 5.104.

7.174 Write a program to solve Problem 7.3. Let the high temperature and the low pressure be program input variables.

7.175 Saturated water vapor at 700 kPa is expanded in a reversible adiabatic process down to 100 kPa in a piston/cylinder arrangement. Write a program that tracks the process in steps of 25 kPa writing out the temperature, quality and the total work out to that state.

7.176 Write a program to track the process in Problem 7.7. Step with 50 kPa and print out at each step the total heat transfer and work terms to reach that state from the beginning.

7.177 Write a program to solve the following problem. One of the gases listed in Table A.11 undergoes a reversible adiabatic process in a cylinder from P_1, T_1 to P_2. We wish to calculate the final temperature and the work for the process by three methods

a. Integrating the specific heat equation

b. Assuming constant specific heat at temperature, T_1

c. Assuming constant specific heat at the average temperature (by iteration)

7.178 Write a program to follow the path of the ammonia in the tank and balloon for Problems 7.18, 7.19, and 7.24.

7.179 Write a program to solve the general case of Problem 7.22, in which the initial state, spring constant, and final pressure all are variables.

7.180 (Adv.) Write a program to follow the path of the processes in the two tanks of Problem 7.26. The initial states and volumes and the final temperature in tank A are all input variables.

7.181 Write a program to solve the general case of Problem 7.36, in which the sphere dimensions and the initial state of the water are input variables.

7.182 (Adv.) Write a program to solve the general case of Problem 7.59, in which both initial volumes and states are program input variables.

7.183 Write a program to solve the general case of Problems 7.74 and 7.95, in which the states, flow rates, and isentropic efficiencies of both the steam turbine and the air compressor are to be input variables.

7.184 Write a program to solve Problem 7.72 in which the supply line pressure and temperature and the percent full-load power output are variables. Also include turbine isentropic efficiency as a variable in this study.

7.185 Write a program to solve the general case of the power plant described in Problems 7.79 and 7.97. The states given in Problem 7.79 and the isentropic efficiencies are to be program input variables.

7.186 Write a program to solve the general storage tank turbine expansion optimization described in Problem 7.119, using real ammonia as the working fluid. The storage tank initial state and the turbine efficiency are to be input variables.

7.187 Write a program to solve the general version of Problem 7.68. Initial state, flow rate, and final pressure are input variables. Compute the required pump power from the property program values of h and compute it also with the assumption of constant specific volume equal to the inlet state.

7.188 Write a program to solve Problem 7.65 with the pressure as an input variable. Print out the ratio of the mass flow rates and the entropy generation per unit mass flowing out.

7.189 (Adv.) Write a program to solve the following problem. An ideal gas with constant specific heat enters a nozzle at a known inlet state P_1, T_1 with velocity $\mathbf{V}_1$

and area A_1, and expands in a reversible adiabatic process. We wish to map out the nozzle contour in terms of the cross-sectional area A as a function of the local pressure P, where P is any value less than the inlet pressure. This problem should be studied over a range of conditions for given values of the constants R and k.

7.190 (Adv.) Write a program to solve the following problem. In Problem 5.214, the enthalpy values in the ideal gas tables, Table A.13, were used to curve-fit polynomial equations for ideal-gas specific heat as functions of temperature. We wish to use these equations to calculate ideal-gas entropy functions, and to compare the results with the values listed in Table A.13 and with polynomial equations that are curve fitted directly to the entropy functions as listed in Table A.13. Why are the results different?

7.191 A piston/cylinder maintaining constant pressure contains 0.5 kg of water at room temperature 20°C and 100 kPa. An electric heater of 500 W heats the water up to 500°C. Assume no heat losses to the ambient and plot the temperature and total accumulated entropy production as a function of time. Investigate the first part of the process, namely bringing the water to the boiling point, by measuring it in your kitchen and knowing the rate of power added.

7.192 Air in a piston/cylinder is used as a small air-spring that should support a steady load of 200 N. Assume that the load can vary with ±10% over a period of 1 s and that the displacement should be limited to ±0.01 m. For some choice of sizes show the spring displacement, x, as a function of load and compare that to an elastic linear coil spring designed for the same conditions.

7.193 Consider a small air compressor taking atmospheric air in and compressing it to 1 MPa in a SSSF process. For a maximum flowrate of 0.1 kg/s, discuss the necessary sizes for the piping and the motor to drive the unit.

7.194 Small gasoline engine or electric motor driven air compressors are used to supply compressed air to power tools, machine shops, etc. The compressor charges air into a tank that acts as a storage buffer. Find examples of these and discuss their sizes in terms of tank volume, charging pressure, engine, or motor power. Also find the time it will take to charge the system from start and its continuous supply capacity.

7.195 Consider a piston/cylinder arrangement with ammonia at −10°C, 50 kPa that is compressed to 200 kPa. Examine the effect of heat transfer to/from the ambient at 15°C on the process and the required work. Some limiting processes are a reversible adiabatic compression giving an exit temperature of about 90°C and as mentioned in the text an isothermal compression. Evaluate the work and heat transfer for both cases and for cases inbetween assuming a polytropic process. Which processes are actually possible and how would they proceed?

7.196 (Adv.) Consider a compression or expansion process for a control mass that is an ideal gas. Assume the heat transfer rate is proportional to the temperature difference, $T - T_0$, so the energy equation on a rate form becomes

$$\frac{dE}{dt} = \dot{Q} - \dot{W}$$

$$mC_v \dot{T} = -C_H \left(T - T_o\right) - Pm\dot{v}$$

This can be written as a rate equation for the temperature, T, as

$$\dot{T}/T = -C_1\left(1 - T_o/T\right) - \frac{R}{C_v}\frac{\dot{v}}{v}$$

with the factor, $C_1 = C_H/mC_v$, units 1/s. Investigate different processes (vary C_1) with expansion or compression at a constant relative volume change, $\dot{v}/v$ and different initial conditions. Plot various properties along the process path and include total entropy generation.

7.197 A coflowing heat exchanger receives air at 800 K, 15 MPa and water at 15°C, 100 kPa. The two flows exchange energy as they flow alongside each other to the exit, where the air should be cooled to 350 K. Investigate the range of water flows necessary per kilogram per second air flow and the possible water exit temperatures with the restriction that the minimum temperature difference between the water and air should be 25°C. Include an estimation for the overall entropy generation in the process per kilogram of air flow.

7.198 A reversible adiabatic compressor receives air at the state of the surroundings, 20°C, 100 kPa. It should compress the air to a pressure of 1.2 MPa in two stages with a constant pressure intercooler between the two stages. Investigate the work input as a function of the pressure between the two stages assuming the intercooler brings the air down to 50°C.

7.199 (Adv.) Investigate the optimal pressure, P_2, for a constant pressure intercooler between two stages in a compressor. Assume the compression process in each stage follows a polytropic process and that the intercooler brings the substance to the original inlet temperature, T_1. Show that the minimal work for the combined stages arises when

$$P_2 = \sqrt{P_1 P_3}$$

where P_3 is the final exit pressure.

7.200 (Adv.) Reexamine the previous problem when the intercooler cools the substance to a temperature, $T_2 > T_1$, due to finite heat transfer rates. What is the effect of having isentropic efficiencies for the compressor stages less than 100% on the total work and selection of P_2?

7.201 Investigate the sizes of turbochargers and superchargers available for automobiles. Look at their boost pressures and check if they have intercoolers mounted also. Analyze an example with respect to the power input and the air it can deliver to the engine and estimate its isentropic efficiency if enough data are found.

IRREVERSIBILITY AND AVAILABILITY 8

We now turn our attention to irreversibility and availability, two additional concepts that have found increasing use in recent years. These concepts are particularly applicable in the analysis of complex thermodynamic systems, for with the aid of a digital computer, irreversibility and availability are very powerful tools in design and optimization studies of such systems.

8.1 AVAILABLE ENERGY, REVERSIBLE WORK, AND IRREVERSIBILITY

In the previous chapter, we introduced the concept of the efficiency of a device, such as a turbine, nozzle, or compressor (perhaps more correctly termed a first-law efficiency, since it is given as the ratio of two energy terms). We proceed now to develop concepts that include more meaningful second-law analysis. Our ultimate goal is to use this analysis to manage our natural resources and environment better.

We first focus our attention on the potential for producing useful work from some source or supply of energy. Consider the simple situation shown in Fig. 8.1a, in which there is an energy source Q in the form of heat transfer from a very large and, therefore, constant-temperature reservoir at temperature T. What is the ultimate potential for producing work?

To answer this question, we imagine that a cyclic heat engine is available, as shown in Fig. 8.1b. To convert the maximum fraction of Q to work requires that the engine be completely reversible, that is, a Carnot cycle, and that the lower-temperature reservoir be at the lowest temperature possible, often, but not necessarily, at the ambient temperature. From the first and second laws for the Carnot cycle and the usual consideration of all the Q's as positive quantities, we find

$$W_{\text{rev H.E.}} = Q - Q_0$$

$$\frac{Q}{T} = \frac{Q_0}{T_0}$$

so that

$$W_{\text{rev H.E.}} = Q\left(1 - \frac{T_0}{T}\right) \tag{8.1}$$

We might say that the fraction of Q given by the right side of Eq. 8.1 is the available portion of the total energy quantity Q. To carry this thought one step farther, consider the

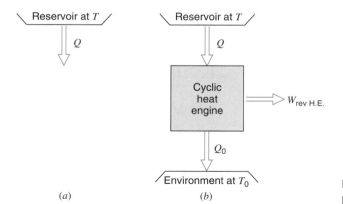

(a)　　　　　　　　　(b)

FIGURE 8.1 Constant-temperature energy source.

situation shown on the *T–S* diagram in Fig. 8.2. The total shaded area is *Q*. The portion of *Q* that is below T_0, the environment temperature, cannot be converted into work by the heat engine and must instead be thrown away. This portion is therefore the unavailable portion of total energy *Q*, and the portion lying between the two temperatures *T* and T_0 is the available energy.

　　　Let us next consider the same situation, except that the heat transfer *Q* is available from a constant-pressure source, for example, a simple heat exchanger as shown in Fig. 8.3*a*. The Carnot cycle must now be replaced by a sequence of such engines, with the result shown in Fig. 8.3*b*. The only difference between the first and second laws is that the second law includes an integral, which corresponds to ΔS.

$$\Delta S = \int \frac{\delta Q_{\mathrm{rev}}}{T} = \frac{Q_0}{T_0}$$

Substituting into the first law, we have

$$W_{\mathrm{rev\ H.E.}} = Q - T_0\,\Delta S \qquad (8.2)$$

Note that this ΔS quantity does not include the standard sign convention. It corresponds to the amount of change of entropy shown in Fig. 8.3*b*. Equation 8.2 specifies the available portion of the quantity *Q*. The portion unavailable for producing work in this circumstance lies below T_0 in Fig. 8.3*b*.

　　　We now proceed along a similar but slightly different line. Consider the general control volume shown in Fig. 8.4, containing an actual system or devices for which the

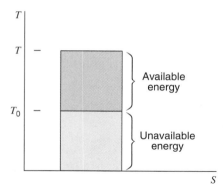

FIGURE 8.2 *T–S* diagram for constant-temperature energy source.

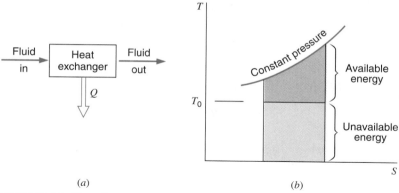

(a)   (b)

FIGURE 8.3   Changing-temperature energy source.

balance equations for mass, energy and entropy are

$$\frac{dm_{\text{c.v.}}}{dt} = \sum \dot{m}_i - \sum \dot{m}_e \tag{8.3}$$

$$\frac{dE_{\text{c.v.}}}{dt} = \sum \dot{m}_i h_{\text{tot}, i} - \sum \dot{m}_e h_{\text{tot}, e} + \sum \dot{Q}_{\text{c.v.}, j} - \dot{W}_{\text{c.v.}} \tag{8.4}$$

$$\frac{dS_{\text{c.v.}}}{dt} = \sum \dot{m}_i s_i - \sum \dot{m}_e s_e + \sum \frac{\dot{Q}_{\text{c.v.}, j}}{T_j} + \dot{S}_{\text{gen, c.v.}} \tag{8.5}$$

where the total heat transfer rate may be the sum over several quantities exchanged with different reservoirs. The temperature T_j, at which $\dot{Q}_{\text{c.v.}, j}$ is added, is the temperature at the control volume surface where the heat transfer crosses and therefore not necessarily equal to the reservoir temperature. Also to simplify notation, we have defined the total enthalpy as

$$h_{\text{tot}} \equiv h + \frac{1}{2}\mathbf{V}^2 + gZ \tag{8.6}$$

Let us now compare this actual system to an ideal one where all processes are reversible. The ideal system should have the same mass flow rates entering and leaving at the same states as the actual system, and have the same change of state for the mass inside the control volume as the actual one. The heat transfer rates should also be the same except for one, namely the reversible heat transfer from the surroundings at T_0. We can now solve

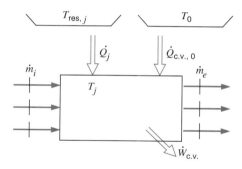

FIGURE 8.4   General control volume.

for this reversible heat transfer rate from the second law written for the ideal system with a zero entropy generation by multiplication with T_0 as

$$\dot{Q}^{\text{rev}}_{\text{c.v.},0} = T_0 \left[\frac{dS_{\text{c.v.}}}{dt} + \sum \dot{m}_e s_e - \sum \dot{m}_i s_i \right] - \sum_{j \neq 0} \frac{T_0}{T_j} \dot{Q}_{\text{c.v.}j} \tag{8.7}$$

The energy equation is solved for the reversible rate of work,

$$\dot{W}^{\text{rev}}_{\text{c.v.}} = \sum \dot{m}_i h_{\text{tot},i} - \sum \dot{m}_e h_{\text{tot},e} + \dot{Q}^{\text{rev}}_{\text{c.v.},0} + \sum_{j \neq 0} \dot{Q}_{\text{c.v.},j} - \frac{dE_{\text{c.v.}}}{dt}$$

into which the reversible heat transfer from the ambient is substituted. After a rearrangement of the terms the reversible rate of work becomes

$$\dot{W}^{\text{rev}}_{\text{c.v.}} = \sum \dot{m}_i \left(h_{\text{tot},i} - T_0 s_i \right) - \sum \dot{m}_e \left(h_{\text{tot},e} - T_0 s_e \right)$$

$$+ \sum_{j \neq 0} \left(1 - \frac{T_0}{T_j} \right) \dot{Q}_{\text{c.v.},j} + T_0 \frac{dS_{\text{c.v.}}}{dt} - \frac{dE_{\text{c.v.}}}{dt} \tag{8.9}$$

Notice how the heat transfer rates from the reservoirs other than the ambient contribute with a rate of work as if these were used in heat engines operating between the temperatures T_j and the ambient with the Carnot heat engine efficiency. This reversible work is the maximum possible work that can be achieved with the given system interacting with the environment at T_0. We recognize that the real system with the same change of state and heat transfer rates would not produce this amount of work, since the real system is not internally reversible. In addition, whatever heat is transferred in the real process does not interact with the energy reservoirs and the environment through reversible heat engines. Instead, the quantity $\dot{W}^{\text{rev}}_{\text{c.v.}}$ establishes a theoretical upper limit for what might be accomplished in a totally idealized situation; the quantity may therefore be useful to us in evaluating real processes.

As the real process produces an amount of work that is less than the ideal reversible work, the difference between these is defined as the irreversibility for the control volume.

$$\dot{I}_{\text{c.v.}} = \dot{W}^{\text{rev}}_{\text{c.v.}} - \dot{W}^{\text{ac}}_{\text{c.v.}} \tag{8.10}$$

Irreversibility is plainly a measure of the "inefficiency" of an actual process, since the less the actual work done for a given change of state, inlet and exit states, the greater the irreversibility. The irreversibility would be zero for a completely reversible process, and otherwise is always greater than zero.

To amplify the last point let us relate the irreversibility to the entropy generation for the control volume by taking the difference between the reversible work from Eq. 8.8 and the actual work in Eq. 8.4:

$$\dot{I}_{\text{c.v.}} = \dot{W}^{\text{rev}}_{\text{c.v.}} - \dot{W}^{\text{ac}}_{\text{c.v.}} = \dot{Q}^{\text{rev}}_{\text{c.v.},0} - \dot{Q}^{\text{ac}}_{\text{c.v.},0} \tag{8.11}$$

Now take the difference between the actual and the reversible heat transfer from the entropy balance equations, multiplied by T_0, for the actual Eq. 8.5 and the ideal systems, Eq. 8.7, where we find

$$\dot{I}_{\text{c.v.}} = \dot{Q}^{\text{rev}}_{\text{c.v.},0} - \dot{Q}^{\text{ac}}_{\text{c.v.},0} = T_0 \dot{S}_{\text{gen,c.v.}} \tag{8.12}$$

As the entropy generation is positive it follows that the irreversibility is positive. The irreversibility thus expresses the amount of energy that could be converted from heat transfer to work if the system has a change in entropy as $\dot{S}_{\text{gen,c.v.}}$ take place at temperature T_0.

The irreversibility expressed in Eq. 8.12 is directly proportional to the entropy generation inside the control volume and only includes irreversibilities inside the chosen control volume. The actual process may result in external irreversibilities if for instance the temperatures T_j are different from the reservoir temperatures and there is heat transfer over a finite temperature difference. To include these in the irreversibility, choose the control volume to extend out to the reservoirs so the temperatures at the control volume surface is equal to the reservoir temperatures. When this is the case, the control volume includes all parts of space where a heat transfer may occur over a finite temperature difference. The entropy generation is then larger and the irreversibility is then the total irreversibility as

$$\dot{I}_{\text{c.v.,tot}} = T_0\,\dot{S}_{\text{gen,c.v.,tot}} = T_0\,\frac{\partial S_{\text{net}}}{\partial t} \tag{8.13}$$

expressing the total irreversibility generated due to the processes that occur in the system including the surroundings.

The preceding development of the reversible work and irreversibility was done for the general unsteady process on a rate form. The cases of a control mass with no flows into or out of the control volume, the steady-state steady-flow situation and the uniform-state uniform-flow situation are all special cases that follow from the foregoing formulation.

The concepts of reversible work and irreversibility are illustrated by examples in the following sections. Unless otherwise stated, we will use 25°C for the temperature of the surroundings. This temperature has been selected because thermochemical data are frequently given relative to this base and because it is a reasonable temperature to assume for the surroundings.

The Control Mass

For a control mass there are no flows in or out, and a process can take place in a steady state with constant rates of heat and work to or from the control mass or in an unsteady process from a state 1 to a state 2. The steady state process is the situation where the working substance inside the control mass goes through a cycle like in a heat engine or heat pump. With no net accumulation of energy or entropy for the control mass the rate of the reversible heat transfer from the ambient is from Eq. 8.7

$$\dot{Q}_{\text{c.v.,0}}^{\text{rev}} = -\sum_{j\neq0}\frac{T_0}{T_j}\dot{Q}_{\text{c.v.,}j} \tag{8.14}$$

The rate of reversible work is from Eq. 8.9:

$$\dot{W}_{\text{c.v.}}^{\text{rev}} = \sum_{j\neq0}\left(1 - \frac{T_0}{T_j}\right)\dot{Q}_{\text{c.v.,}j} \tag{8.15}$$

so the expression for the rate of irreversibility becomes identical to the previously given forms in Eqs. 8.10–8.12.

For the unsteady process the rate equations are integrated over time from state 1 to state 2, and, since we do have the changes in energy and entropy during the process, the original form must be used rather than the foregoing rate equations. The expression for the reversible heat transfer is from Eq. 8.7 integrated over time:

$$_1Q_{\text{c.v.,0}}^{\text{rev}} = mT_0\left(s_2 - s_1\right) - \sum_{j\neq0}\frac{T_0}{T_j}\,_1Q_{2.\text{c.v.,}j} \tag{8.16}$$

and the work term is from Eq. 8.9

$$_1W_{2\text{c.v.}}^{\text{rev}} = \sum_{j\neq0}\left(1-\frac{T_0}{T_j}\right){}_1Q_{2\text{c.v.},j} - m\left[(e_2-T_0s_2)-(e_1-T_0s_1)\right] \tag{8.17}$$

Finally, the irreversibility from Eqs. 8.10–8.12:

$$_1I_2 = mT_0(s_2-s_1) - \sum\frac{T_0}{T_j}{}_1Q_{2\text{c.v.},j}^{\text{ac}}$$

$$=_1W_{2,\text{c.v.}}^{\text{rev}} -_1W_{2,\text{c.v.}}^{\text{ac}} = T_0\,_1S_{2,\text{gen,c.v.}} \tag{8.18}$$

EXAMPLE 8.1

The insulated cylinder with a frictionless piston as shown in Fig. 8.5 is divided into two rooms A and B by a membrane where each room has a volume of 1 m³. Initially, A contains water at room temperature 20°C, quality 50% and room B is evacuated. The membrane then ruptures and the water fills out the total volume without any piston motion and no heat transfer. Determine the reversible work and the irreversibility in the process from state 1 to state 2.

From state 2 the piston now compresses the water back until it has the original temperature 20°C in a reversible process with no heat transfer. Find the final volume and the necessary work.

Part 1

Control mass: Room A and room B.

Initial state: T_1, x_1 known; state fixed.

Final state: V_2 known.

Process: Adiabatic, no work, no change in kinetic or potential energy.

Model: Steam tables.

Analysis:

The continuity equation for a control mass says mass is constant

$$m_2 = m_1 = m \Rightarrow v_2 = V_2/m$$

and the energy equation has zero heat transfer and work so

$$m(e_2-e_1) = m(u_2-u_1) = 0$$

With no heat transfer the entropy increase is only due to entropy generation

$$m(s_2-s_1) = _1S_{2,\text{gen}}$$

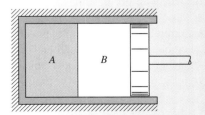

FIGURE 8.5   Illustration for Example 8.1.

Solution

From the steam tables we have

$$u_1 = 1243.45 \qquad v_1 = 28.895 \qquad s_1 = 4.48186$$
$$v_2 = V_2 / m = 2 \times v_1 = 57.79 \qquad u_2 = u_1 = 1243.45$$

We must now iterate in the steam tables and find it by guessing T:

$$T_2 = 5°C \text{ and } v_2 \Rightarrow u = 948.5 \text{ kJ} / \text{kg}$$
$$T_2 = 10°C \text{ and } v_2 \Rightarrow u = 1317.09 \text{ kJ} / \text{kg}$$

so the final interpolation in u gives a temperature of 9°C. If the menu-driven software is used the final state is interpolated to be

$$T_2 = 9.1°C \qquad x_2 = 0.513 \qquad s_2 = 4.644$$

with the given u and v. Since the actual work is zero we have

$$_1W_{2,c.v.}^{rev} = mT_0 \left(s_2 - s_1 \right) = {_1}I_2$$
$$= \left(1/28.895 \right) 293.15 \left(4.644 - 4.48186 \right) = 1.645 \text{ kJ}$$

Part 2

Control mass: The cylinder volume.

Initial state: u_2, v_2 known; state fixed.

Final state: T_3 known.

Process: Reversible adiabatic, no change in kinetic or potential energy.

Model: Steam tables.

Analysis:

The energy equation has zero heat transfer so

$$m\left(e_3 - e_2 \right) = m\left(u_3 - u_2 \right) = -{_2}W_3$$

and in the second law there is no entropy generation

$$m\left(s_3 - s_2 \right) = 0 \Rightarrow s_3 = s_2$$

so this determines the second thermodynamic property at state 3.

Solution

From the steam tables we have with $T_3 = 20°C$ and $s_3 = 4.644$ kJ/kg K

$$x_3 = 0.5194 \qquad v_3 = 30.015 \qquad u_3 = 1288.36 \text{ kJ} / \text{kg}$$

The final volume and the work thus become

$$V_3 = mv_3 = 1.04 \text{ m}^3$$
$$_2W_3 = -m\left(u_3 - u_2 \right) = -\left(1/28.895 \right) \left(1288.36 - 1243.45 \right) = -1.554 \text{ kJ}$$

Since the final volume is slightly larger than the initial volume, the work input is also slightly smaller than the reversible work in the first process.

The Steady-State Steady-Flow Process

The steady-state and steady-flow form has no accumulation of mass, energy, or entropy inside the control volume and the equations become

$$\dot{Q}_{c.v.,0}^{rev} = \sum \dot{m}_e T_0 s_e - \sum \dot{m}_i T_0 s_i - \sum_{j \neq 0} \frac{T_0}{T_j} \dot{Q}_{c.v.,j} \tag{8.19}$$

$$\dot{W}_{c.v.}^{rev} = \sum \dot{m}_i (h_{tot,i} - T_0 s_i) - \sum \dot{m}_e (h_{tot,e} - T_0 s_e) + \sum_{j \neq 0} \left(1 - \frac{T_0}{T_j}\right) \dot{Q}_{c.v.,j} \tag{8.20}$$

$$\dot{I}_{c.v.} = \sum \dot{m}_e T_0 s_e - \sum \dot{m}_i T_0 s_i - \sum \frac{T_0}{T_j} \dot{Q}_{c.v.,j}$$

$$= \dot{W}_{c.v.}^{rev} - \dot{W}_{c.v.}^{ac} = T_0 \dot{S}_{gen,c.v.} \tag{8.21}$$

Frequently these quantities are divided with a mass flow rate and given as specific quantities as will be illustrated in a following example. For devices that have a single inlet and exit, the mass flow rate is divided out to give

$$q_o^{rev} = T_0 (s_e - s_i) - \sum_{j \neq 0} \frac{T_o}{T_j} q_{c.v.,j} \tag{8.22}$$

$$w^{rev} = (h_{tot,i} - T_0 s_i) - (h_{tot,e} - T_0 s_e) + \sum_{j \neq 0} \left(1 - \frac{T_o}{T_j}\right) q_{c.v.,j} \tag{8.23}$$

$$i = \dot{I}_{c.v.}/\dot{m} = T_0 (s_e - s_i) - \sum_{j \neq 0} \frac{T_o}{T_j} q_{c.v.,j}$$

$$= w^{rev} - w^{ac} = T_0 s_{gen,c.v.} \tag{8.24}$$

EXAMPLE 8.2   Consider an air compressor that receives ambient air at 100 kPa, 25°C. It compresses the air to a pressure of 1 MPa, where it exits at a temperature of 540 K. Since the air and compressor housing are hotter than the ambient it loses 50 kJ per kilogram air flowing through the compressor. Find the reversible work, reversible heat transfer, and irreversibility in the process.

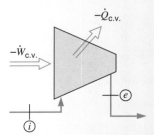

FIGURE 8.6 Illustration for Example 8.2.

Control volume: The air compressor.

Inlet state: P_i, T_i known; state fixed.

Exit state: P_e, T_e known; state fixed.

Process: Nonadiabatic compression with no change in kinetic or potential energy.

Model: Ideal gas.

Analysis:

This SSSF process has a single inlet and exit flow so all quantities are done on a mass basis as specific quantities. From the ideal gas air tables.

$$h_i = 298.615 \text{ kJ} / \text{kg} \qquad s_{T_i}^0 = 6.86285 \text{ kJ} / \text{kg K}$$

$$h_e = 544.686 \text{ kJ} / \text{kg} \qquad s_{T_e}^0 = 7.46642 \text{ kJ} / \text{kg K}$$

so the energy equation for the actual compressor gives the work as

$$q^{ac} = -50 \text{ kJ} / \text{kg}$$

$$w^{ac} = h_i - h_e + q^{ac} = 298.615 - 544.686 - 50 = -296.07 \text{ kJ} / \text{kg}$$

The reversible heat transfer from Eq. 8.22:

$$q^{rev} = T_0 \left(s_e - s_i \right) = T_0 \left(s_{T_e}^0 - s_{T_i}^0 - R \ln \frac{P_e}{P_i} \right)$$

$$= 298.15 \left(7.46642 - 6.86285 - 0.287 \ln 10 \right) = -17.08 \text{ kJ} / \text{kg}$$

and the reversible work

$$w^{rev} = \left(h_i - T_0 s_i \right) - \left(h_e - T_0 s_e \right) = h_i - h_e + q^{rev}$$

$$= 298.615 - 544.686 - 17.08 = -263.15 \text{ kJ} / \text{kg}$$

Finally, from the definition of the irreversibility,

$$i = w^{rev} - w^{ac} = -263.15 + 296.07 = 32.92 \text{ kJ} / \text{kg}$$

$$= q^{rev} - q^{ac} = -17.08 + 50 = 32.92 \text{ kJ} / \text{kg}$$

EXAMPLE 8.2E   Consider an air compressor that receives ambient air at 14.7 lbf/in.^{2}, 80 F. It compresses the air to a pressure of 150 lbf/in.^{2}, where it exits at a temperature of 960 R. Since the air and compressor housing are hotter than the ambient it loses 22 Btu/lbm air flowing through the compressor. Find the reversible work, reversible heat transfer, and irreversibility in the process.

Control volume: The air compressor.

Inlet state: P_i, T_i known; state fixed.

Exit state: P_e, T_e known; state fixed.

Process: Nonadiabatic compression with no change in kinetic or potential energy.

Model: Ideal gas.

Analysis:

This SSSF process has a single inlet and exit flow so all quantities are done on a mass basis as specific quantities. From the ideal gas air tables

$$h_i = 129.182 \text{ Btu} / \text{lbm} \qquad s^0_{T_i} = 1.64053 \text{ Btu} / \text{lbm R}$$

$$h_e = 231.202 \text{ Btu} / \text{lbm} \qquad s^0_{T_e} = 1.78025 \text{ Btu} / \text{lbm R}$$

so the energy equation for the actual compressor gives the work as

$$q^{ac} = -22 \text{ Btu} / \text{lbm}$$

$$w^{ac} = h_i + h_e + q^{ac} = 129.182 - 231.202 - 22 = -124.02 \text{ Btu} / \text{lbm}$$

The reversible heat transfer from Eq. 8.22

$$q^{rev} = T_0(s_e - s_i) = T_0\left(s^0_{T_e} - s^0_{T_i} - R \ln \frac{P_e}{P_i}\right)$$

$$= 539.67(1.78025 - 1.64053 - 0.06855 \ln 10) = -9.785 \text{ Btu} / \text{lbm}$$

and the reversible work

$$w^{rev} = (h_i - T_0 s_i) - (h_e - T_0 s_e) = h_i - h_e + q^{rev}$$
$$= 129.182 - 231.202 - 9.785 = -111.805 \text{ Btu} / \text{lbm}$$

Finally from the definition of the irreversibility

$$i = w^{rev} - w^{ac} = -111.805 + 124.02 = 12.215 \text{ Btu} / \text{lbm}$$

$$= q^{rev} - q^{ac} = -9.785 + 22 = 12.215 \text{ Btu} / \text{lbm}$$

EXAMPLE 8.3   A feedwater heater has 5 kg/s water at 5 MPa, 40°C flowing through it, being heated from two sources as shown in Fig. 8.6. One source adds 900 kW from a 100°C reservoir and the other source adds heat transfer from a 200°C reservoir such that the water exit condition is 5 MPa, 180°C. Find the reversible heat transfer, the reversible work and the irreversibility.

Control volume: Feedwater heater extending out to the two reservoirs.

Inlet state: P_i, T_i known; state fixed.

Exit state: P_e, T_e known; state fixed.

Process: Constant pressure heat addition with no change in kinetic or potential energy.

Model: Steam tables.

Analysis:

This control volume has a single inlet and exit flow with two heat transfer rates coming from reservoirs different from the ambient. There is no actual work or actual heat transfer with the surroundings at 25°C. For the actual feedwater heater, the energy equation becomes

$$h_i + q_1 + q_2 = h_e$$

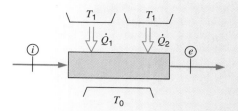

FIGURE 8.7 The feedwater heater for Example 8.3.

which determines the heat transfer q_2 since the rest of the terms are known. The reversible heat transfer is then found from Eq. 8.22, the reversible work from the energy equation and the irreversibility from its definition.

Solution

From the steam tables the inlet and exit state properties are

$$h_i = 171.97 \qquad s_i = 0.5705$$
$$h_e = 765.25 \qquad s_e = 2.1341$$

The second heat transfer is found from the energy equation as

$$q_2 = h_e - h_i - q_1 = 765.25 - 171.97 - 900/5 = 413.28 \text{ kJ/kg}$$

The reversible heat transfer from the surroundings is then

$$q_0^{rev} = T_0(s_e - s_i) - \frac{T_0}{T_1}q_1 - \frac{T_0}{T_2}q_2$$

$$= 298.15(2.1341 - 0.5705) - \frac{298.15}{373.15}180 - \frac{298.15}{473.15}413.28$$

$$= 466.19 - 143.82 - 260.42 = 61.94 \text{ kJ/kg}$$

and the reversible work becomes

$$w^{rev} = h_i - h_e + q_0^{rev} + q_1 + q_2$$

$$= 171.97 - 765.25 + 61.94 + 900/5 + 413.28 = 61.94 \text{ kJ/kg}$$

identical to the reversible heat transfer since the actual work is zero and there is no actual heat transfer from the surroundings.

The Uniform-State Uniform-Flow Process

The uniform state and uniform flow has a control volume change from state 1 to state 2 with mass flows in or out. Again integrating the rate equations in time from the initial to the final state gives the expressions for the reversible heat transfer, work and the irreversibility.

$$_1Q_{2,c.v.,0}^{rev} = T_0\left[m_2s_2 - m_1s_1 + \sum m_e s_e - \sum m_i s_i\right] - \sum_{j\neq 0}\frac{T_0}{T_j}\,_1Q_{2,c.v.,j} \tag{8.25}$$

$$_1W_{2,\text{c.v.}}^{\text{rev}} = \sum m_i \left(h_{\text{tot},i} - T_0 s_i \right) - \sum m_e \left(h_{\text{tot},e} - T_0 s_e \right)$$

$$+ \sum_{j \neq 0} \left(1 - \frac{T_0}{T_j} \right) {}_1Q_{2,\text{c.v.},j} - \left[m_2 \left(e_2 - T_0 s_2 \right) - m_1 \left(e_1 - T_0 s_1 \right) \right] \qquad (8.26)$$

$$_1I_2 = T_0 \left(\sum m_e s_e - \sum m_i s_i + m_2 s_2 - m_1 s_1 \right) - \sum \frac{T_0}{T_j} {}_1Q_{2,\text{c.v.},j}^{\text{ac}}$$

$$= {}_1W_{2,\text{c.v.}}^{\text{rev}} - {}_1W_{2,\text{c.v.}}^{\text{ac}} = T_0 {}_1S_{2,\text{gen,c.v.}} \qquad (8.27)$$

EXAMPLE 8.4   A 1-m^3 rigid tank contains ammonia at 200 kPa and the ambient temperature 20°C. The tank is connected with a valve to a line flowing saturated liquid ammonia at −10°C. The valve is opened and the tank is charged quickly until the flow stops and the valve is closed. As the process happens very quickly there is no heat transfer. Determine the final mass in the tank and the irreversibility in the process.

Control volume: The tank and the valve.

Initial state: T_1, P_1 known; state fixed.

Inlet state: T_i, x_i known; state fixed.

Final state: $P_2 = P_{\text{line}}$ known.

Process: Adiabatic, no kinetic or potential energy change.

Model: Ammonia tables.

Analysis:

Since the line pressure is higher than the initial pressure inside the tank flow is going into the tank and the flow stops when the tank pressure has increased to the line pressure. The continuity, energy, and entropy equations are

$$m_2 - m_1 = m_i$$

$$m_2 u_2 - m_1 u_1 = m_i h_i = \left(m_2 - m_1 \right) h_i$$

$$m_2 s_2 - m_1 s_1 = m_i s_i + {}_1S_{2,\text{gen}}$$

where kinetic and potential energies are zero for the initial and final states and neglected for the inlet flow.

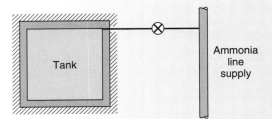

FIGURE 8.8 Ammonia tank and line for Example 8.4.

Solution

From the ammonia tables the initial and line state properties are

$$v_1 = 0.6995 \text{ m}^3 \qquad u_1 = h_1 - P_1v_1 = 1369.5 \text{ kJ} / \text{kg} \qquad s_1 = 5.927 \text{ kJ} / \text{kg K}$$

$$h_i = 134.41 \text{ kJ} / \text{kg} \qquad s_i = 0.5408 \text{ kJ} / \text{kg K}$$

The initial mass is therefore

$$m_1 = V / v_1 = 1 / 0.6995 = 1.4296 \text{ kg}$$

It is observed that only the final pressure is known so one property is needed. The unknowns are the final mass and final internal energy in the energy equation. Since only one property is unknown, the two quantities are not independent. From the energy equation we have

$$m_2(u_2 - h_i) = m_1(u_1 - h_i)$$

from which it is seen that $u_2 > h_i$ and the state therefore is two-phase or superheated vapor. Assume that the state is two phase, then

$$m_2 = V / v_2 = 1 / (0.001534 + x_2 \times 0.41684)$$

$$u_2 = 133.964 + x_2 \times 1175.257$$

so the energy equation is

$$\frac{133.964 + x_2 \times 1175.257 - 134.41}{0.001534 + x_2 \times 0.041684} = 1.4296(1369.5 - 134.41) = 1765.67 \text{ kJ}$$

This equation is solved for the quality and the rest of the properties to give

$$x_2 = 0.007182 \qquad v_2 = 0.0045276 \text{ m}^3 / \text{kg} \qquad s_2 = 0.5762 \text{ kJ} / \text{kg}$$

Now the final mass and the irreversibility are found

$$m_2 = V / v_2 = 1 / 0.0045276 = 220.87 \text{ kg}$$

$$_1S_{2,\text{gen}} = m_2 s_2 - m_1 s_1 - m_i s_i = 127.265 - 8.473 - 118.673 = 0.119 \text{ kJ} / \text{K}$$

$$_1I_2 = T_0 \ _1S_{2,\text{gen}} = 293.15 \times 0.119 = 34.885 \text{ kJ}$$

8.2 AVAILABILITY AND SECOND-LAW EFFICIENCY

What is the maximum reversible work that can be done by a given mass in a given state? In the previous section, we developed expressions for the reversible work for a given change of state for a control mass and control volume undergoing specific types of processes. For any given case, what final state will give the maximum reversible work?

The answer to this question is that, for any type of process, when the mass comes into equilibrium with the environment, no spontaneous change of state will occur and the mass will be incapable of doing any work. Therefore, if a mass in a given state undergoes a completely reversible process until it reaches a state in which it is in equilibrium with the environment, the maximum reversible work will have been done by the mass. In this sense, we refer to the availability at the original state in terms of the potential for achieving the maximum possible work by the mass.

If a control mass is in equilibrium with the surroundings, it must certainly be in pressure and temperature equilibrium with the surroundings, that is, at pressure P_0 and temperature T_0. It must also be in chemical equilibrium with the surroundings, which implies that no further chemical reaction will take place. Equilibrium with the surroundings also requires that the system have zero velocity and minimum potential energy. Similar requirements could be set forth regarding magnetic, electrical, and surface effects if these are relevant to a given problem.

The same general remarks can be made about a quantity of mass that undergoes a steady-state steady-flow process. With a given state for the mass entering the control volume, the reversible work will be a maximum when this mass leaves the control volume in equilibrium with the surroundings. This means that as the mass leaves the control volume, it must be at the pressure and temperature of the surroundings, in chemical equilibrium with the surroundings, and have minimum potential energy and zero velocity. (The mass leaving the control volume must of necessity have some velocity but it can be made to approach zero.)

Let us first consider the availability associated with a steady-state steady-flow process. Consider a control volume with a single flow for which the specific reversible work is given by Eq. 8.23 as

$$w^{\mathrm{rev}} = \left(h_{\mathrm{tot},i} - T_0 s_i\right) - \left(h_{\mathrm{tot},e} - T_0 s_e\right) + \sum_{j \neq 0}\left(1 - \frac{T_o}{T_j}\right) q_{\mathrm{c.v.},j}$$

From the discussion of the heat engine that lead to Eq. 8.1, it is clear that the last term in the expression for the reversible work is the contribution to the net reversible work from the heat transfers. These can be viewed as transfer of availability associated with q_j, which gives a potential to do work as in a heat engine. Such contributions are separate from the availability in the flow itself. The SSSF flow reversible work will be maximum, relative to the surroundings, when the mass leaving the control volume is in equilibrium with the surroundings. The state in which the fluid is in equilibrium with the surroundings is designated with subscript 0, and the reversible work will be maximum when $h_e = h_0$, $s_e = s_0$, $\mathbf{V}_e = 0$, and $Z_e = Z_0$. This maximum reversible work per unit mass flow without the additional heat transfers is the flow availability or exergy and assigned the symbol ψ

$$\psi = \left(h - T_0 s + \frac{1}{2}\mathbf{V}^2 + gZ\right) - \left(h_0 - T_0 s_0 + gZ_0\right) \tag{8.28}$$

It is written without subscript for the inlet state to indicate that this is the flow availability associated with a substance in any state as it enters the control volume in a steady-state steady-flow process. The reversible work is therefore seen to be equal to the decrease in flow availability plus the reversible work that can be extracted from heat engines operating with the heat transfers at T_j and the ambient.

The irreversibility can then be related to the changes in the availability through the reversible work from Eq. 8.20 and the irreversibility from Eq. 8.21 as

$$\dot{I}_{\mathrm{c.v.}} = \left(\sum \dot{m}_i \psi_i - \sum \dot{m}_e \psi_e\right) + \sum\left(1 - \frac{T_0}{T_j}\right)\dot{Q}_{\mathrm{c.v.},j} - \dot{W}_{\mathrm{c.v.}}^{\mathrm{ac}} \tag{8.29}$$

In this form the irreversibility is equal to the decrease in the availability of the mass flows plus the decrease of availability of the heat transfer at T_j minus the increase in availability of the surroundings that receive the actual work. The rate of irreversibility is thus seen to be the rate of destruction of availability which is also directly proportional to the entropy generation, recall Eq. 8.21. This rate of destruction of availability is only due to processes that occur inside the control volume and it does not include possible irreversibilities destroying availability outside the control volume.

For a control mass, a similar consideration of the maximum reversible work will lead to a nonflow availability concept. In this case the volume may change and some work is exchanged with the ambient, which is not available as useful work. The reversible work per unit mass from Eq. 8.17 is

$$_1w_2^{\text{rev}} = \left(e_1 - T_0 s_1\right) - \left(e_2 - T_0 s_2\right) + \sum\left(1 - \frac{T_0}{T_j}\right){}_1q_{2,\text{c.v.},j}$$

which is the maximum between the two given states. This work is available if the final state is in equilibrium with the surroundings, for which we must have $e_2 = e_0 = u_0 + gZ_0$, the kinetic energy being zero, and $s_2 = s_0$. The work done against the surroundings, w_{surr}, is

$$w_{\text{surr}} = P_0\left(v_0 - v_1\right) = -P_0\left(v_1 - v_0\right)$$

such that the maximum available work is

$$w_{\text{avail}}^{\text{max}} = w_{\text{max}}^{\text{rev}} - w_{\text{surr}}$$

$$= \left(e - T_0 s\right) - \left(e_0 - T_0 s_0\right) + P_0\left(v - v_0\right) + \sum\left(1 - \frac{T_0}{T_j}\right)q_j \qquad (8.30)$$

The subscript is again dropped to indicate that this is the maximum available work at a given state having also q_j available from a source at T_j. The nonflow availability is defined as the maximum available work from a state without the heat transfers included as

$$\phi = \left(e - T_0 s\right) - \left(e_o - T_o s_o\right) + P_o\left(v - v_0\right)$$
$$\phi = \left(e + P_0 v - T_0 s\right) - \left(e_0 + P_0 v_0 - T_0 s_0\right) \qquad (8.31)$$

Sometimes the definition excludes the kinetic and potential energies, in which case u should be used instead of e.

The irreversibility may again be related to the changes in availability through the difference between the reversible work and the actual work. The reversible work from above is expressed with the availability from which the actual work is subtracted to give

$$_1I_2 = m\left(\phi_1 - \phi_2\right) + \sum\left(1 - \frac{T_0}{T_j}\right){}_1Q_{2,\text{c.v.},j} - \left({}_1W_2^{\text{ac}} - P_0\left(V_2 - V_1\right)\right) \qquad (8.32)$$

The irreversibility is then equal to the decrease in availability of the control mass plus the decrease in availability of the heat transfers at T_j minus the increase in availability of the surroundings that received the actual work. It is noted again that the irreversibility expresses the net destruction of availability of the control mass and its influence on the surroundings which is proportional to the entropy generation inside the control mass.

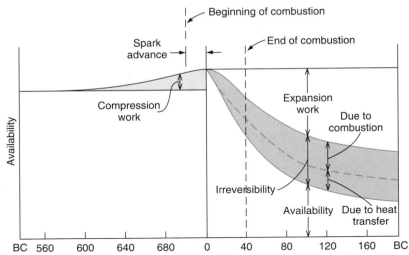

FIGURE 8.9 Availability versus crank angle of the charge in a spark-ignited internal combustion engine. (From D. J. Patterson and G. J. Van Wylen, "A Digitial Computer Simulation for Spark Ignited Engine Cycles," *SAE Progress in Technology Series,* 7, p. 88. Published by SAE Inc., New York, 1964.)

The use of availability and irreversibility in an actual thermodynamic problem is shown in Fig. 8.9. A theoretical analysis was made of a reciprocating internal-combustion automotive engine to see what happened to the availability of the air–fuel mixture that entered the engine and where the irreversibility occurred during the process. The abscissa on Fig. 8.9 is the crank angle, the left side representing bottom dead center when the cylinder is assumed to be filled with an air–fuel mixture having the availability indicated on the ordinate. As the compression process takes place, this mixture becomes more available through the work done in compressing the mixture. When the piston passes top dead center, the expansion process takes place. The beginning and end of combustion are also indicated on the diagram. During the combustion and expansion process irreversibilities occur. Those associated with the combustion process itself and those associated with heat transfer to the cooling water or surroundings are both indicated. The work done during the expansion process and the availability at the end of the expansion are indicated on the right ordinate. The availability that remains in the cylinder at the end of the expansion stroke is exhausted to the atmosphere.

The less the irreversibility associated with a given change of state, the greater the amount of work that will be done (or the smaller the amount of work that will be required). This relation is significant in at least two regards. The first is that availability is one of our natural resources. This availability is found in such forms as oil reserves, coal reserves, and uranium reserves. Suppose we wish to accomplish a given objective that requires a certain amount of work. If this work is produced reversibly while drawing on one of the availability reserves, the decrease in availability is exactly equal to the reversible work. However, since there are irreversibilities in producing this required amount of work, the actual work will be less than the reversible work, and the decrease in availability will be greater (by the amount of the irreversibility) than if this work had been produced reversibly. Thus the more irreversibilities we have in all our processes,

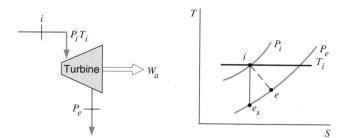

FIGURE 8.10 Irreversible turbine.

the greater will be the decrease in our availability reserves.[1] The conservation and effective use of these availability reserves is an important responsibility for all of us.

The second reason that it is desirable to accomplish a given objective with the smallest irreversibility is an economic one. Work costs money, and in many cases a given objective can be accomplished at less cost when the irreversibility is less. It should be noted, however, that many factors enter into the total cost of accomplishing a given objective, and an optimization process that considers many factors is often necessary to arrive at the most economical design. For example, in a heat-transfer process, the smaller the temperature difference across which the heat is transferred, the less the irreversibility. However, for a given rate of heat transfer, a smaller temperature difference will require a larger (and therefore more expensive) heat exchanger. These various factors must all be considered in developing the optimum and most economical design.

In many engineering decisions other factors, such as the impact on the environment (for example, air pollution and water pollution) and the impact on society must be considered in developing the optimum design.

Along with the increased use of availability analysis in recent years, a term called the "second-law efficiency" has come into more common use. This term refers to comparison of the desired output of a process with the cost, or input, in terms of the thermodynamic availability. Thus, the isentropic turbine efficiency defined by Eq. 7.72 as the actual work output divided by the work for a hypothetical isentropic expansion from the same inlet state to the same exit pressure might well be called a "first-law efficiency," in that it is a comparison of two energy quantities. The "second-law efficiency," as just described, would be the actual work output of the turbine divided by the decrease in availability from the same inlet state to the same exit state. For the turbine shown in Fig. 8.10, the second-law efficiency is

$$n_{\text{2nd law}} = \frac{w_a}{\psi_i - \psi_e} \tag{8.33}$$

In this sense, this concept provides a rating or measure of the real process in terms of the actual change of state, and is simply another convenient way of utilizing the concept of thermodynamic availability. In a similar manner, the second-law efficiency of a pump or compressor is the ratio of the increase in availability to the work input to the device.

---

[1]In many popular talks reference is made to our energy reserves. From a thermodynamic point of view, availability reserves would be a much more acceptable term. There is much energy in the atmosphere and the ocean, but relatively little availability.

EXAMPLE 8.5   An insulated steam turbine, Fig. 8.11, receives 30 kg of steam per second at 3 MPa, 350°C. At the point in the turbine where the pressure is 0.5 MPa, steam is bled off for processing equipment at the rate of 5 kg/s. The temperature of this steam is 200°C. The balance of the steam leaves the turbine at 15 kPa, 90% quality. Determine the availability per kilogram of the steam entering and at both points at which steam leaves the turbine, the isentropic efficiency and the second-law efficiency for this process.

> *Control volume:*   Turbine.
>
> *Inlet state:* P_1, T_1 known; state fixed.
>
> *Exit state:* P_2, T_2 known; P_3, x_3 known; both states fixed.
>
> *Process:*   SSSF.
>
> *Model:*   Steam tables.

Analysis:

The availability at any point for the steam entering or leaving the turbine is given by Eq. 8.28,

$$\psi = (h - h_0) - T_0(s - s_0) + \frac{\mathbf{V}^2}{2} + g(Z - Z_0)$$

Since there are no changes in kinetic and potential energy in this problem, this equation reduces to

$$\psi = (h - h_o) - T_0(s - s_0)$$

For the ideal isentropic turbine,

$$s_{e_s} - s_i = 0 \qquad s_{3_s} = s_1$$

$$\dot{W}_s = \dot{m}(h_i - h_{es}) = \dot{m}_1(h_1 - h_{es})$$

For the actual turbine,

$$\dot{W} = \dot{m}_1 h_1 - \dot{m}_2 h_2 - \dot{m}_3 h_3$$

Solution

At the pressure and temperature of the surroundings, 0.1 MPa, 25°C, the water is a slightly compressed liquid, and the properties of the water are essentially equal to those for saturated liquid at 25°C.

$$h_0 = 104.9 \text{ kJ / kg} \qquad s_0 = 0.3674 \text{ kJ / kg K}$$

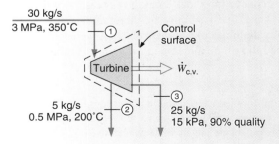

FIGURE 8.11   Sketch for Example 8.5.

From Eq. 8.28

$$\psi_1 = (3115.3 - 104.9) - 298.15(6.7428 - 0.3674) = 1109.6 \text{ kJ/kg}$$

$$\psi_2 = (2855.4 - 104.9) - 298.15(7.0592 - 0.3674) = 755.3 \text{ kJ/kg}$$

$$\psi_3 = (2361.8 - 104.9) - 298.15(7.2831 - 0.3674) = 195.0 \text{ kJ/kg}$$

$$\dot{m}_1 \psi_1 - \dot{m}_2 \psi_2 - \dot{m}_3 \psi_3 = 30(1109.6) - 5(755.3) - 25(195.0) = 24\ 637 \text{ kW}$$

For the ideal isentropic turbine,

$$s_{3_s} = 6.7428 = 0.7549 + x_{3_s} \times 7.2536$$

$$x_{3_s} = 0.8255$$

$$h_{3_s} = 225.9 + 0.8255 \times 2373.1 = 2184.9$$

$$\dot{W}_s = 30(3115.3 - 2184.9) = 27912 \text{ kW}$$

For the actual turbine,

$$\dot{W}_s = 30(3115.3) - 5(2855.4) - 25(2361.8) = 20\ 137 \text{ kW}$$

The isentropic efficiency is

$$\eta_s = \frac{20\ 137}{27\ 912} = 0.721$$

and the second-law efficiency is

$$\eta_{2\text{nd law}} = \frac{20\ 137}{24\ 637} = 0.817$$

For a device that does not involve the production or the input of work, the definition of second-law efficiency refers to the accomplishment of the goal of the process relative to the process input, in terms of availability changes or transfers. For example, in a heat exchanger, where energy is transferred from a high-temperature fluid stream to a low-temperature fluid stream, as shown in Fig. 8.12, in which case the second-law efficiency is defined as

$$\eta_{2\text{nd law}} = \frac{\dot{m}_1(\psi_2 - \psi_1)}{\dot{m}_3(\psi_3 - \psi_4)} \tag{8.34}$$

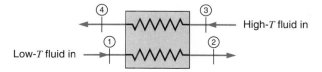

FIGURE 8.12 A two-fluid heat exchanger.

EXAMPLE 8.6   In a boiler, heat is transferred from the products of combustion to the steam. The temperature of the products of combustion decreases from 1100°C to 550°C while the pressure remains constant at 0.1 MPa. The average constant-pressure specific heat of the products of combustion is 1.09 kJ/kg K. The water enters at 0.8 MPa, 150°C and leaves at 0.8 MPa, 250°C. Determine the second-law efficiency for this process and the irreversibility per kilogram of water evaporated.

   Control volume: Overall heat exchanger.

   Sketch: Fig. 8.13.

   Inlet states: Both known, given in Fig. 8.13.

   Exit states: Both known, given in Fig. 8.13.

   Process: Overall, adiabatic.

   Diagram: Fig. 8.14.

   Model: Products—ideal gas, constant specific heat. Water—steam tables.

Analysis:

For the products, the entropy change for this constant-pressure process is

$$\left(s_e - s_i\right)_{\text{prod}} = C_{po} \ln \frac{T_e}{T_i}$$

For this control volume we can write the following governing equations:
Continuity equation:

$$\left(\dot{m}_i\right)_{H_2O} = \left(\dot{m}_e\right)_{H_2O} \tag{a}$$

$$\left(\dot{m}_i\right)_{\text{prod}} = \left(\dot{m}_e\right)_{\text{prod}} \tag{b}$$

First law (a steady-state, steady-flow process):

$$\left(\dot{m}_i h_i\right)_{H_2O} + \left(\dot{m}_i h_i\right)_{\text{prod}} = \left(\dot{m}_e h_e\right)_{H_2O} + \left(\dot{m}_e h_e\right)_{\text{prod}} \tag{c}$$

Second law (the process is adiabatic for the control volume shown):

$$\left(\dot{m}_e s_e\right)_{H_2O} + \left(\dot{m}_e s_e\right)_{\text{prod}} \geq \left(\dot{m}_i s_i\right)_{H_2O} + \left(\dot{m}_i s_i\right)_{\text{prod}}$$

Solution

From Eqs. a, b, and c, we can calculate the ratio of the mass flow of products to the mass flow of water.

$$\dot{m}_{\text{prod}}\left(h_i - h_e\right)_{\text{prod}} = \dot{m}_{H_2O}\left(h_e - h_i\right)_{H_2O}$$

$$\frac{\dot{m}_{\text{prod}}}{\dot{m}_{H_2O}} = \frac{\left(h_e - h_i\right)_{H_2O}}{\left(h_i - h_e\right)_{\text{prod}}} = \frac{2950 - 632.2}{1.09\left(1100 - 550\right)} = 3.866$$

The increase in availability of the water is, per kilogram of water,

$$\psi_2 - \psi_1 = \left(h_2 - h_1\right) T_0 \left(s_2 - s_1\right)$$

$$= \left(2950 - 632.2\right) - 298.15\left(7.0384 - 1.8418\right)$$

$$= 768.4 \text{ kJ / kg } H_2O$$

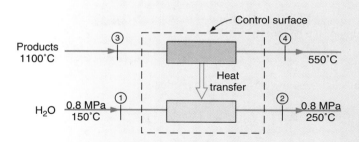

FIGURE 8.13 Sketch for Example 8.6.

The decrease in availability of the products, per kilogram of water, is

$$\frac{\dot{m}_{prod}}{\dot{m}_{H_2O}}\left(\psi_3 - \psi_4\right) = \frac{\dot{m}_{prod}}{\dot{m}_{H_2O}}\left[\left(h_3 - h_4\right) - T_0\left(s_3 - s_4\right)\right]$$

$$= 3.866\left[1.09\left(1100 - 550\right) - 298.15\left(1.09\ln\frac{1373.15}{823.15}\right)\right]$$

$$= 1674.7 \text{ kJ / kg } H_2O$$

Therefore, the second-law efficiency is, from Eq. 8.34,

$$\eta_{2nd\ law} = \frac{768.4}{1674.7} = 0.459$$

From Eq. 8.29, the process irreversibility per kilogram of water is

$$\frac{I}{\dot{m}_{H_2O}} = \sum_i \frac{\dot{m}_i}{\dot{m}_{H_2O}}\psi_i - \sum_2 \frac{\dot{m}_e}{\dot{m}_{H_2O}}\psi_e$$

$$= \left(\psi_1 - \psi_2\right) + \frac{\dot{m}_{prod}}{\dot{m}_{H_2O}}\left(\psi_3 - \psi_4\right)$$

$$= \left(-768.4 + 1674.7\right) = 906.3 \text{ kJ / kg } H_2O$$

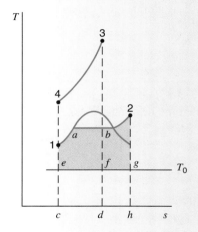

FIGURE 8.14 Temperature-entropy diagram for Example 8.6.

It is also of interest to determine the net change of entropy. The change in the entropy of the water is

$$(s_2 - s_1)_{H_2O} = 7.0384 - 1.8418 = 5.1966 \text{ kJ / kg } H_2O \text{ K}$$

The change in the entropy of the products is

$$\frac{\dot{m}_{prod}}{\dot{m}_{H_2O}}(s_4 - s_3)_{prod} = -3.866\left(1.09\ln\frac{1373.15}{823.15}\right) = -2.1564 \text{ kJ / kg } H_2O \text{ K}$$

Thus there is a net increase in entropy during the process. The irreversibility could also have been calculated from Eq. 8.21:

$$\dot{I} = \sum \dot{m}_e T_0 s_e - \sum \dot{m}_i T_0 s_i - \dot{Q}_{c.v.}$$

For the control volume selected, $\dot{Q}_{c.v.} = 0$, and therefore

$$\frac{\dot{I}}{\dot{m}_{H_2O}} = T_0(s_2 - s_1)_{H_2O} + \frac{\dot{m}_{prod}}{\dot{m}_{H_2O}}T_0(s_4 - s_3)_{prod}$$

$$= 298.15(5.1966) + 298.15(-2.1564)$$

$$= 906.3 \text{ kJ / kg } H_2O$$

These two processes are shown on the T–s diagram of Fig. 8.14. Line 3–4 represents the process for the 3.866 kg of products. Area 3–4–c–d–3 represents the heat transferred from the 3.866 kg of products of combustion, and area 3–4–e–f–3 represents the decrease in availability of these products. Area 1–a–b–2–h–c–1 represents the heat transferred to the water, and this is equal to area 3–4–c–d–3, which represents the heat transferred from the products of combustion. Area 1–a–b–2–g–e–1 represents the increase in availability of the water. The difference between area 3–4–e–f–3 and area 1–a–b–2–g–e–1 represents the net decrease in availability. It is readily shown that this net change is equal to area f–g–h–d–f, or $T_0(\Delta s)_{net}$. Since the actual work is zero, this area also represents the irreversibility, which agrees with our calculation above.

It is essential to note that when the change of state involving heat transfer takes place reversibly, the net change in entropy is zero, and therefore the decrease in the entropy of the body from which heat is transferred must be equal to the increase in entropy of the body to which heat is transferred. This is best demonstrated by considering an example similar to Example 8.5.

EXAMPLE 8.7   Consider the process in Example 8.6, but assume that the heat transfer could be made to be entirely reversible, that is, that the heat transfer takes place through a reversible engine.

Sketch: Fig. 8.15.

Analysis:

Schematically this would involve heat transfer from the products of combustion to reversible engines that reject heat to the water, as shown in Fig. 8.15.

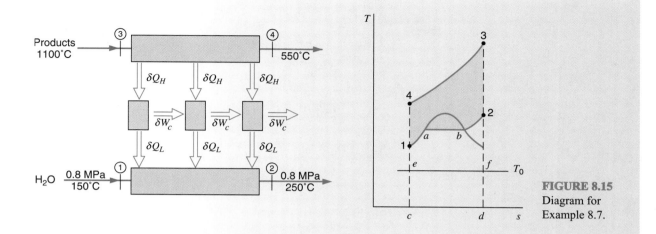

FIGURE 8.15
Diagram for
Example 8.7.

We again write the governing equations:

Continuity equation:

$$\left(\dot{m}_i\right)_{H_2O} = \left(\dot{m}_e\right)_{H_2O}$$

$$\left(\dot{m}_i\right)_{prod} = \left(\dot{m}_e\right)_{prod}$$

First law:

$$\left(\dot{m}_i h_i\right)_{H_2O} + \left(\dot{m}_i h_i\right)_{prod} = \left(\dot{m}_e h_e\right)_{H_2O} + \left(\dot{m}_e h_e\right)_{prod} + \dot{W}_{c.v.}$$

Second law: Since this is a reversible adiabatic process,

$$\left(\dot{m}_e s_e\right)_{H_2O} + \left(\dot{m}_e s_e\right)_{prod} = \left(\dot{m}_i s_i\right)_{H_2O} + \left(\dot{m}_i s_i\right)_{prod}$$

Solution

From the continuity equation and second law we can determine flow of products per unit flow of water.

$$\left(s_2 - s_1\right)_{H_2O} + \frac{\dot{m}_{prod}}{\dot{m}_{H_2O}}\left(s_4 - s_3\right)_{prod} = 0$$

$$\frac{\dot{m}_{prod}}{\dot{m}_{H_2O}}\left(s_3 - s_4\right)_{prod} = \left(7.0384 - 1.8418\right)$$

$$\frac{\dot{m}_{prod}}{\dot{m}_{H_2O}}\left(1.09\ln\frac{1373.15}{823.15}\right) = 5.1966$$

$$\frac{\dot{m}_{prod}}{\dot{m}_{H_2O}} = 9.317$$

We now calculate the decrease in availability of the water and the products:

$$\left(\psi_1 - \psi_2\right) = \left(h_1 - h_2\right) - T_0\left(s_1 - s_2\right)$$

$$\frac{\dot{m}_{prod}}{\dot{m}_{H_2O}}\left(\psi_3 - \psi_4\right) = \left[\left(h_3 - h_4\right) - T_0\left(s_3 - s_4\right)\right]\frac{\dot{m}_{prod}}{\dot{m}_{H_2O}}$$

But, since

$$\left(s_2 - s_1\right)_{H_2O} = \frac{\dot{m}_{prod}}{\dot{m}_{H_2O}}\left(s_3 - s_4\right)_{prod}$$

it follows that

$$\left(\psi_1 - \psi_2\right) + \frac{\dot{m}_{prod}}{\dot{m}_{H_2O}}\left(\psi_3 - \psi_4\right) = \left(h_1 - h_2\right) + \frac{\dot{m}_{prod}}{\dot{m}_{H_2O}}\left(h_3 - h_4\right)$$

We note that this net decrease in availability is exactly equal to the work that would be determined from the first law. We would expect this, of course, since the process is completely reversible.

$$w = w_{rev} = \left(632.2 - 2950\right) + 9.317 \times 1.09\left(1100 - 550\right)$$

$$= 3267.7 \text{ kJ} / \text{kg H}_2\text{O}$$

Since the net change in entropy is zero, the T–s diagram is as shown in Fig. 8.15. Area 3–4–c–d–3 represents the heat transferred from the products to the engines, and area 1–a–b–2–d–c–1 represents the heat received by the water. Area 3–4–1–a–b–2–3 represents the work done by the heat engines, which is equal to the reversible work for this process.

8.3 PROCESSES INVOLVING CHEMICAL REACTIONS

Although chemical reaction will not be considered in detail until Chapter 12 some preliminary remarks regarding availability in such processes can be made here. In the first place, we observe that the reactants are often in pressure and temperature equilibrium with the surroundings before the reaction takes place, and the same is true of the products after the reaction. An automobile engine would be an example of such a process if we visualized the products being cooled to atmospheric temperature before being discharged from the engine.

Let us first consider for a control mass the implications of temperature equilibrium with the surroundings during a chemical reaction in a system. The temperature of the system T is equal to T_0, the temperature of the surroundings. Therefore, from Eq. 8.17 we can write (noting that in this case $T_0 = T$)

$$_1w_2^{rev} = \left(u_1 - T_1s_1 \frac{\mathbf{V}_1^2}{2} + gZ_1\right) - \left(u_2 - T_2s_2 + \frac{\mathbf{V}_2^2}{2} + gZ_2\right)$$

The quantity $(U - TS)$ is a thermodynamic property of a substance and is called the

Helmholtz function. It is an extensive property and we designate it by the symbol A. Thus

$$A = U - TS$$

$$a = u - Ts \tag{8.35}$$

Therefore, when a system undergoes a change of state while in temperature equilibrium with the surroundings, the reversible work is given by the relation

$$_1(w_{rev})_2 = \left(a_1 + \frac{\mathbf{V}_1^2}{2} + gZ_1\right) - \left(a_2 + \frac{\mathbf{V}_2^2}{2} + gZ_2\right)$$

In those cases where the kinetic and potential energy changes are not significant, this reduces to

$$_1(W_{rev})_2 = A_1 - A_2 = m(a_1 - a_2) \tag{8.36}$$

Let us now consider a control mass that undergoes a chemical reaction while in both pressure and temperature equilibrium with the surroundings. The availability function ϕ for a control mass has been defined as

$$\phi = (u + P_0 v - T_0 s) - (u + P_0 v_0 - T_0 s_0)$$

If $P = P_0$ and $T = T_0$, then

$$\phi = (u + Pv - Ts) - (u + P_0 v_0 - T_0 s_0)$$

$$= (h - Ts) - (h_0 - T_0 s_0) \tag{8.37}$$

The quantity $h - Ts$ is a thermodynamic property and is termed the Gibbs function, designated by the symbol G.

$$G = H - TS$$

$$g = h - Ts \tag{8.38}$$

Introducing the Gibbs function into Eq. 8.37 we have, when a system is in pressure and temperature equilibrium with the surroundings,

$$\phi = g - g_0$$

It follows that under these same conditions

$$_1w_2^{rev} = (g_1 - g_2) - P_0(v_1 - v_2) \frac{\mathbf{V}_1^2 - \mathbf{V}_2^2}{2} + g(Z_1 - Z_2) \tag{8.39}$$

The Gibbs function is also of significance in a steady-state, steady-flow process that takes place in temperature equilibrium with the surroundings. Since in this case $T_i = T_e = T_0$, Eq. 8.20 reduces to

$$\dot{W}_{c.v.}^{rev} = \dot{m}_i \left(h_i - T_i s_i + \frac{\mathbf{V}_i^2}{2} + gZ_i \right) - \dot{m}_e \left(h_e - T_e s_e + \frac{\mathbf{V}_e^2}{2} + gZ_e \right)$$

Introducing the Gibbs function, we have

$$\dot{W}_{c.v.}^{rev} = \dot{m}_i \left(g_i + \frac{\mathbf{V}_i^2}{2} + gZ_i \right) - \dot{m}_e \left(g_e + \frac{\mathbf{V}_e^2}{2} + gZ_e \right) \tag{8.40}$$

Thus we have introduced two new properties, the Helmholtz function, A, and the Gibbs function, G. Both these functions are very important in the thermodynamics of chemical reactions and will be used extensively in later chapters of this book.

PROBLEMS

8.1 The compressor in a refrigerator takes refrigerant R-134a in at 100 kPa, −20°C and compresses it to 1 MPa, 40°C. With the room at 20°C find the reversible heat transfer and the minimum compressor work.

8.2 Calculate the reversible work out of the two-stage turbine shown in Problem 5.105, assuming the ambient is at 25°C. Compare this to the actual work which was found to be 18.08 MW.

8.3 A household refrigerator has a freezer at T_F and a cold space at T_C from which energy is removed and rejected to the ambient at T_A as shown in Fig. P8.3. Assume that the rate of heat transfer from the cold space, $\dot{Q}_C$, is the same as from the freezer, $\dot{Q}_F$, find an expression for the minimum power into the heat pump. Evaluate this power when $T_A = 20°C$, $T_C = 5°C$, $T_F = -10°C$, and $\dot{Q}_F = 3$ kW.

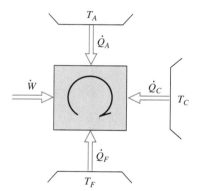

FIGURE P8.3

8.4 Consider an internally reversible Carnot heat engine with heat exchangers designed so that $C_H = 5$ kW/K and $C_L = 10$ kW/K, see Section 6.10. It operates between a high-temperature reservoir at 700°C and the ambient at 25°C. The temperatures in the Carnot cycle are selected to give maximum work output. Find the actual power produced and the rate of irreversibility.

8.5 An air compressor takes air in at the state of the surroundings 100 kPa, 300 K. The air exits at 400 kPa, 200°C at the rate of 2 kg/s. Determine the minimum compressor work input.

8.6 A supply of steam at 100 kPa, 150°C is needed in a hospital for cleaning purposes at a rate of 15 kg/s. A supply of steam at 150 kPa, 250°C is available from a boiler and tap water at 100 kPa, 15°C is also available. The two sources are then mixed in an SSSF mixing chamber to generate the desired state as output. Determine the rate of irreversibility of the mixing process.

8.7 Two flows of air both at 200 kPa of equal flow rates mixes in an insulated mixing chamber. One flow is at 1500 K and the other is at 300 K. Find the irreversibility in the process per kilogram of air flowing out.

8.8 (Adv.) Water as saturated liquid at 200 kPa goes through a constant pressure heat exchanger as shown in Fig. P8.8. The heat input is supplied from a reversible heat pump extracting heat from the surroundings at 17°C. The water flow rate is 2 kg/min and the whole process is reversible. If the heat pump receives 40 kW of work find the water exit state and the increase in availability of the water.

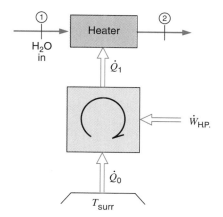

FIGURE P8.8

8.9 A steam turbine receives steam at 6 MPa, 800°C. It has a heat loss of 49.7 kJ/kg and an isentropic efficiency of 90%. For an exit pressure of 15 kPa and surroundings at 20°C, find the actual work and the reversible work between the inlet and the exit.

8.10 Fresh water can be produced from saltwater by evaporation and subsequent condensation. An example is shown in Fig. P8.10 where 150-kg/s saltwater, state 1, comes from the condenser in a large power plant. The water is throttled to the saturated pressure in the flash evaporator and the vapor, state 2, is then condensed by cooling with sea water. As the evaporation takes place below atmospheric pressure, pumps must bring the liquid water flows back up to P_0. Assume that the saltwater has the same properties as pure water, the ambient is at 20°C and that there are no external heat transfers. With the states as shown in the table below find the irreversibility in the throttling valve and in the condenser.

State	1	2	3	4	5	6	7	8
$T°C$	30	25	25	–	23	–	17	20

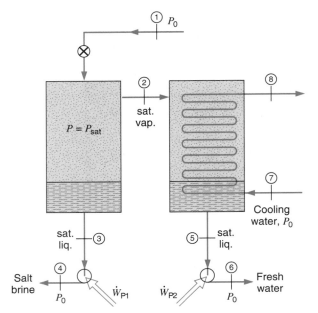

FIGURE P8.10

8.11 An air compressor receives atmospheric air at $T_0 = 17°C$, 100 kPa, and compresses it up to 1400 kPa. The compressor has an isentropic efficiency of 88% and it loses energy by heat transfer to the atmosphere as 10% of the isentropic work. Find the actual exit temperature, the reversible work, and reversible heat transfer.

8.12 Calculate the irreversibility for the process described in Problem 5.138.

8.13 Air enters the turbocharger compressor (see Fig. P8.13), of an automotive engine at 100 kPa, 30°C, and exits at 170 kPa. The air is cooled by 50°C in an intercooler before entering the engine. The isentropic efficiency of the compressor is 75%. Determine the temperature of the air entering the engine and the irreversibility of the compression-cooling process.

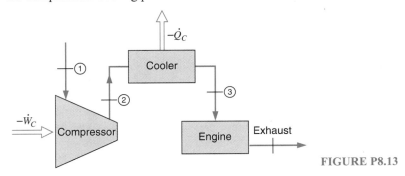

FIGURE P8.13

8.14 A car air-conditioning unit has a 0.5-kg aluminum storage cylinder that is sealed with a valve and it contains 2 L of refrigerant R-134a at 500 kPa and both are at room temperature 20°C. It is now installed in a car sitting outside where the whole system cools down to ambient temperature at −10°C. What is the irreversibility of this process?

8.15 A 2-kg piece of iron is heated from room temperature 25°C to 400°C by a heat source at 600°C. What is the irreversibility in the process?

8.16 A 2-kg/s flow of steam at 1 MPa, 700°C should be brought to 500°C by spraying in liquid water at 1 MPa, 20°C in an SSSF setup. Find the rate of irreversibility, assuming that surroundings are at 20°C.

8.17 (Adv.) Refrigerant-22 is flowing in a pipeline at 10°C, 600 kPa, with a velocity of 200 m/s, at a steady flowrate of 0.1 kg/s. It is desired to decelerate the fluid and increase its pressure by installing a diffuser in the line (a diffuser is basically the opposite of a nozzle in this respect). The R-22 exits the diffuser at 30°C, with a velocity of 100 m/s. It may be assumed that the diffuser process is SSSF, polytropic, and internally reversible. Determine the diffuser exit pressure and the rate of irreversibility for the process.

8.18 A control mass gives out 10 kJ of energy in the form of

a. Electrical work from a battery

b. Mechanical work from a spring

c. Heat transfer at 500°C

Find the change in availability of the control mass for each of the three cases.

8.19 A steady stream of R-22 at ambient temperature, 10°C, and at 750 kPa enters a solar collector. The stream exits at 80°C, 700 kPa. Calculate the change in availability of the R-22 between these two states.

8.20 Nitrogen flows in a pipe with velocity 300 m/s at 500 kPa, 300°C. What is its availability with respect to an ambient at 100 kPa, 20°C?

8.21 A 10-kg iron disk brake on a car is initially at 10°C. Suddenly the brake pad hangs up, increasing the brake temperature by friction to 110°C while the car maintains constant speed. Find the change in availability of the disk and the energy depletion of the car's gas tank due to this process alone. Assume that the engine has a thermal efficiency of 35%.

8.22 Consider the following device proposed as an outdoor portable power supply for winter use. A 100-L rigid tank contains 90% liquid water and 10% nitrogen gas, by volume, at 200 kPa, 0°C. The tank is left outdoors overnight, and the water freezes, compressing the nitrogen. A regulating valve is opened, allowing nitrogen at 200 kPa, 0°C, to flow to an expansion engine to produce power, as desired. This device will operate until the tank pressure drops back to 200 kPa. What is the availability of the nitrogen?

8.23 Consider the springtime melting of ice in the mountains, which gives cold water running in a river at 2°C while the air temperature is 20°C. What is the availability of the water (SSSF) relative to the temperature of the ambient?

8.24 Refrigerant R-12 at 30°C, 0.75 MPa enters a SSSF device and exits at 30°C, 100 kPa. Assume the process is isothermal and reversible. Find the change in availability of the refrigerant.

8.25 A wooden bucket (2 kg) with 10 kg hot liquid water, both at 85°C, is lowered 400 m down into a mineshaft. What is the availability of the bucket and water with respect to the surface ambient at 20°C?

8.26 A rigid container with volume 200 L is divided into two equal volumes by a partition. Both sides contains nitrogen, one side is at 2 MPa, 300°C, and the other at 1 MPa, 50°C. The partition ruptures, and the nitrogen comes to a uniform state at 100°C. Assuming the surroundings are at 25°C find the actual heat transfer and the irreversibility in the process.

8.27 An air compressor is used to charge an initially empty 200-L tank with air up to 5 MPa. The air inlet to the compressor is at 100 kPa, 17°C and the compressor isentropic efficiency is 80%. Find the total compressor work and the change in energy of the air.

8.28 Steam enters a turbine at 25 MPa, 550°C and exits at 5 MPa, 325°C at a flow rate of 70 kg/s. Determine the total power output of the turbine, its isentropic efficiency and the second law efficiency.

8.29 A piston/cylinder contains ammonia at −20°C, quality 80%, and a volume of 10 L. A force is now applied to the piston so it compresses the ammonia in an adiabatic process to a volume of 5 L, where the piston is locked. Now heat transfer with the ambient takes place so the ammonia reaches the temperature of the ambient at 20°C. Find the work and heat transfer. If it is done in a reversible process, how much work and heat transfer would be involved?

8.30 Consider the irreversible process in Problem 7.21. Assume that the process could be done reversibly by adding heat engines/pumps between tanks A and B and the cylinder. The total system is insulated, so there is no heat transfer to or from the ambient. Find the final state, the work given out to the piston and the total work to or from the heat engines/pumps.

8.31 (Adv.) Water in a piston/cylinder is at 100 kPa, 34°C, shown in Fig. P8.31. The cylinder has stops mounted so $V_{min} = 0.01$ m^3 and $V_{max} = 0.5$ m^3. The piston is loaded with a mass and outside P_0, so a pressure inside of 5 MPa will float it. Heat of 15 000 kJ from a 400°C source is added. Find the total change in availability of the water and the total irreversibility.

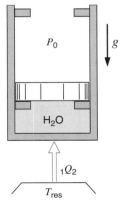

FIGURE P8.31

8.32 Air enters a compressor at ambient conditions, 100 kPa, 300 K, and exits at 800 kPa. If the isentropic compressor efficiency is 85%, what is the second-law efficiency of the compressor process?

8.33 A compressor is used to bring saturated water vapor at 1 MPa up to 17.5 MPa, where the actual exit temperature is 650°C. Find the irreversibility and the second-law efficiency.

8.34 A flow of steam at 10 MPa, 550°C goes through a two-stage turbine. The pressure between the stages is 2 MPa and the second stage has an exit at 50 kPa. Assume both stages have an isentropic efficiency of 85%. Find the second law efficiencies for both stages of the turbine.

8.35 Consider the two-stage turbine in the previous problem as a single turbine from inlet to final actual exit and find its second-law efficiency.

8.36 A rock bed consists of 6000 kg granite and is at 70°C. A small house with lumped mass of 12 000 kg wood and 1000 kg iron is at 15°C. They are now brought to a uniform final temperature with no external heat transfer.

a. For a reversible process, find the final temperature and the work done in the process.

b. If they are connected thermally by circulating water between the rock bed and the house, find the final temperature and the irreversibility of the process, assuming that surroundings are at 15°C.

8.37 A compressor brings R-12 from 150 kPa, 10°C to 800 kPa, 60°C. Assume that the process follows a polytropic process and that any heat transfer is exchanged with the ambient at 20°C. Find the work and heat transfer for the process and the second-law efficiency.

8.38 A steady combustion of natural gas yields 0.15 kg/s of products (having approximately the same properties as air) at 1100°C, 100 kPa. The products are passed through a heat exchanger and exit at 550°C. What is the maximum theoretical power output from a cyclic heat engine operating on the heat rejected from the combustion products, assuming that the ambient temperature is 20°C?

8.39 Air flows into a heat engine at ambient conditions 100 kPa, 300 K, as shown in Fig. P8.39. Energy is supplied as 1200 kJ per kg air from a 1500 K source and in some part of the process a heat transfer loss of 300 kJ/kg air happens at 750 K. The air leaves the engine at 100 kPa, 800 K. Find the first and the second law efficiencies.

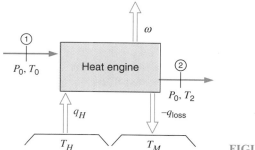

FIGURE P8.39

8.40 (Adv.) Consider the heat engine in Problem 8.39. The exit temperature was given as 800 K, but what are the theoretical limits for this temperature? Find the lowest and the highest, assuming that the heat transfers are as given. For an exit temperature that is the average of the highest and lowest possible, find the first- and second-law efficiencies for the heat engine.

8.41 The simple steam power plant shown in Problem 5.104 has a turbine with given inlet and exit states. Find the availability at the turbine exit, state 6. Find the second law efficiency for the turbine, neglecting kinetic energy at state 5.

8.42 A compressor takes in saturated vapor R-134a at −20°C and delivers it at 30°C, 0.4 MPa. Assuming that the compression is adiabatic, find the isentropic efficiency and the second law efficiency.

8.43 Consider the two turbines in Problem 7.118. What is the second-law efficiency for the combined system?

8.44 Air in a piston/cylinder arrangement, shown in Fig. P8.44, is at 200 kPa, 300 K with a volume of 0.5 m³. If the piston is at the stops, the volume is 1 m³ and a pressure of 400 kPa is required to raise the piston. The air is then heated from the initial state to 1500 K by a 1900 K reservoir. Find the total irreversibility in the process assuming surroundings are at 20°C.

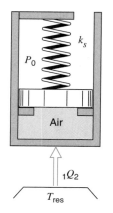

FIGURE P8.44

8.45 Carbon dust of 1 kmol is burned with 1 kmol of oxygen to form 1 kmol of carbon dioxide in a steady-flow process. The temperature of the carbon and the oxygen before combustion is 25°C, and the temperature of the carbon dioxide after combustion is also 25°C. The heat transferred during the process is −393 522 kJ and the entropy of the carbon dioxide is 2.908 kJ/kmol K higher than the entropy of the carbon and oxygen before combustion. Calculate the reversible work and the irreversibility for this process.

8.46 Steam is supplied in a line at 3 MPa, 700°C. A turbine with an isentropic efficiency of 85% is connected to the line by a valve and it exhausts to the atmosphere at 100 kPa. If the steam is throttled down to 2 MPa before entering the turbine find the

a. Actual turbine specific work

b. Change in availability through the valve

c. Second law efficiency of the turbine

8.47 Air flows through a constant pressure heating device, shown in Fig. P8.47. It is heated up in a reversible process with a work input of 200 kJ/kg air flowing. The device exchanges heat with the ambient at 300 K. The air enters at 300 K, 400 kPa. Assuming constant specific heat develop an expression for the exit temperature and solve for it.

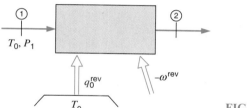

FIGURE P8.47

8.48 (Adv.) A cylinder/piston shown in Fig. P8.48 contains 0.1 kg of air at room temperature 300 K and a pressure of 200 kPa. The piston mass and spring is such that pressure is proportional to volume, $P = CV$. The air is now heated by a reversible heat pump/engine exchanging energy with a 500 K reservoir to a final temperature of 1200 K. Find the change in availability of the air and the net work out of the heat pump/engine.

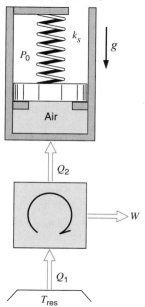

FIGURE P8.48

8.49 A piston/cylinder has forces on the piston so it keeps constant pressure. It contains 2 kg of ammonia at 1 MPa, 40°C and is now heated to 100°C by a reversible heat engine that receives heat from a 200°C source. Find the work out of the heat engine.

8.50 A spring-loaded piston/cylinder, shown in Fig. P8.50, contains 3 kg of water at 1 MPa, 700°C. The linear spring exerts zero force when the piston is at the bottom of the cylinder. Heat is going out to the ambient to a final pressure of 500 kPa is reached. The ambient is at 300 K, 100 kPa. Find the heat transfer, the work, and the total irreversibility.

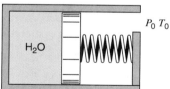

FIGURE P8.50

8.51 Air in a piston/cylinder arrangement is at 110 kPa, 25°C, with a volume of 50 L. It goes through a reversible polytropic process to a final state of 700 kPa, 500 K, and exchanges heat with the ambient at 25°C through a reversible device. Find the total work (including the external device) and the heat transfer from the ambient.

8.52 A counterflowing heat exchanger cools air at 600 K, 400 kPa to 320 K using a supply of water at 20°C, 200 kPa. The water flow rate is 0.1 kg/s and the air flow rate is 1 kg/s. Assume this can be done in a reversible process by the use of heat engines and neglect kinetic energy changes. Find the water exit temperature and the power out of the heat engine(s).

8.53 Consider a gasoline engine for a car as an SSSF device where air and fuel enters at the surrounding conditions 25°C, 100 kPa and leaves the engine exhaust manifold at 1000 K, 100 kPa as products assumed to be air. The engine cooling system removes 750 kJ/kg air through the engine to the ambient. For the analysis take the fuel as air where the extra energy of 2200 kJ/kg of air released in the combustion process, is added as heat transfer from a 1800 K reservoir. Find the work out of the engine, the irreversibility per kilogram of air, and the first- and second-law efficiencies.

8.54 The condenser in a refrigerator receives R-134a at 700 kPa, 50°C and it exits as saturated liquid at 25°C. The flowrate is 0.1 kg/s and the condenser has air flowing in at ambient 15°C and leaving at 35°C. Find the minimum flow rate of air and the heat exchanger second-law efficiency.

8.55 Consider the high-pressure closed feedwater heater in the nuclear power plant described in Problem 5.106. Determine its second-law efficiency.

8.56 A piston/cylinder arrangement has a load on the piston so it maintains constant pressure. It contains 1 kg of steam at 500 kPa, 50% quality. Heat from a reservoir at 700°C brings the steam to 600°C. Find the second-law efficiency for this process. Note that no formula is given for this particular case so determine a reasonable expression for it.

8.57 (Adv.) Consider two rigid containers each of volume $1m^3$ containing air at 100 kPa, 400 K. An internally reversible Carnot heat pump is then thermally connected between them so it heats one up and cools the other down. In order to transfer heat at a reasonable rate, the temperature difference between the working substance inside the heat pump and the air in the containers is set to 20°C. The process stops when the air in the coldest tank reaches 300 K. Find the final temperature of the air that is heated up, the work input to the heat pump, and the overall second-law efficiency.

8.58E The compressor in a refrigerator takes refrigerant R-134a in at 15 lbf/in.², 0 F and compresses it to 125 lbf/in.², 100 F. With the room at 70 F find the reversible heat transfer and the minimum compressor work.

8.59E Consider an internally reversible Carnot heat engine with heat exchangers designed so that $C_H = 3$ Btu/sR and $C_L = 6$ Btu/sR; see Section 6.10. It operates between a high-temperature reservoir at 1300 F and the ambient at 77 F. The temperatures in the Carnot cycle are selected to give maximum work output. Find the actual power produced and the rate of irreversibility.

8.60E A supply of steam at 14.7 lbf/in.^{2}, 320 F is needed in a hospital for cleaning purposes at a rate of 30 lbm/s. A supply of steam at 20 lbf/in.^{2}, 500 F is available from a boiler and tap water at 14.7 lbf/in.^{2}, 60 F is also available. The two sources are then mixed in an SSSF mixing chamber to generate the desired state as output. Determine the rate of irreversibility of the mixing process.

8.61E Fresh water can be produced from saltwater by evaporation and subsequent condensation. An example is shown in Fig. P8.10 where 300-lbm/s saltwater, state 1, comes from the condenser in a large power plant. The water is throttled to the saturated pressure in the flash evaporator and the vapor, state 2, is then condensed by cooling with sea water. As the evaporation takes place below atmospheric pressure, pumps must bring the liquid water flows back up to P_0. Assume that the saltwater has the same properties as pure water, the ambient is at 68 F, and that there are no external heat transfers. With the states as shown in the table below find the irreversibility in the throttling valve and in the condenser.

State	1	2	3	4	5	6	7	8
T F	86	77	77	-	74	-	63	68

8.62E Air enters the turbocharger compressor of an automotive engine at 14.7 lbf/in.^{2}, 90 F, and exits at 25 lbf/in.^{2}, as shown in Fig. P8.13. The air is cooled by 90 F in an intercooler before entering the engine. The isentropic efficiency of the compressor is 75%. Determine the temperature of the air entering the engine and the irreversibility of the compression-cooling process.

8.63E A 4-lbm piece of iron is heated from room temperature 77 F to 750 F by a heat source at 1100 F. What is the irreversibility in the process?

8.64E (Adv.) Refrigerant-22 is flowing in a pipeline at 40 F, 80 lbf/in.^{2}, with a velocity of 650 ft/s, at a steady flowrate of 0.2 lbm/s. It is desired to decelerate the fluid and increase its pressure by installing a diffuser in the line (a diffuser is basically the opposite of a nozzle in this respect). The R-22 exits the diffuser at 80 F, with a velocity of 160 ft/s. It may be assumed that the diffuser process is SSSF, polytropic, and internally reversible. Determine the diffuser exit pressure and the rate of irreversibility for the process.

8.65E A control mass gives out 1000 Btu of energy in the form of

a. Electrical work from a battery

b. Mechanical work from a spring

c. Heat transfer at 700 F

Find the change in availability of the control mass for each of the three cases.

8.66E A steady stream of R-22 at ambient temperature, 50 F, and at 110 lbf/in.^{2} enters a solar collector. The stream exits at 180 F, 100 lbf/in.^{2}. Calculate the change in availability of the R-22 between these two states.

8.67 E A 20-lbm iron disk brake on a car is at 50 F. Suddenly the brake pad hangs up, increasing the brake temperature by friction to 230 F while the car maintains constant speed. Find the change in availability of the disk and the energy depletion of the car's gas tank due to this process alone. Assume that the engine has a thermal efficiency of 35%.

8.68 E Consider the springtime melting of ice in the mountains, which gives cold water running in a river at 34 F while the air temperature is 68 F. What is the availability of the water (SSSF) relative to the temperature of the ambient?

8.69 E A wood bucket (4 lbm) with 20 lbm hot liquid water, both at 180 F, is lowered 1300 ft down into a mineshaft. What is the availability of the bucket and water with respect to the surface ambient at 70 F?

8.70 E An air compressor is used to charge an initially empty 7-ft^3 tank with air up to 750 lbf/in.^{2}. The air inlet to the compressor is at 14.7 lbf/in.^{2}, 60 F and the compressor isentropic efficiency is 80%. Find the total compressor work and the change in energy of the air.

8.71 E (Adv.) Water in a piston/cylinder is at 14.7 lbf/in.^{2}, 90 F, as shown in Fig. P8.31. The cylinder has stops mounted so that $V_{min} = 0.36$ ft^3 and $V_{max} = 18$ ft^3. The piston is loaded with a mass and outside P_0, so a pressure inside of 700 lbf/in.^{2} will float it. Heat of 14 000 Btu from a 750 F source is added. Find the total change in availability of the water and the total irreversibility.

8.72 E A compressor is used to bring saturated water vapor at 150 lbf/in.^{2} up to 2500 lbf/in.^{2}, where the actual exit temperature is 1200 F. Find the irreversibility and the second law efficiency.

8.73 E A rock bed consists of 12 000 lbm granite and is at 160 F. A small house with lumped mass of 24 000 lbm wood and 2000 lbm iron is at 60 F. They are now brought to a uniform final temperature with no external heat transfer.

a. For a reversible process, find the final temperature and the work done in the process.

b. If they are connected thermally by circulating water between the rock bed and the house, find the final temperature and the irreversibility of the process assuming that surroundings are at 60 F.

8.74 E A compressor brings R-12 from 20 lbf/in.^{2}, 20 F to 125 lbf/in.^{2}, 100 F. Assume that the process follows a polytropic process and that any heat transfer is exchanged with the ambient at 68 F. Find the work and heat transfer for the process and the second-law efficiency.

8.75 E Air flows into a heat engine at ambient conditions 14.7 lbf/in.^{2}, 540 R, as shown in Fig. P8.39. Energy is supplied as 540 Btu per lbm air from a 2700 R source and in some part of the process a heat transfer loss of 135 Btu per lbm air happens at 1350 R. The air leaves the engine at 14.7 lbf/in.^{2}, 1440 R. Find the first- and the second-law efficiencies.

8.76 E (Adv.) Consider the heat engine in Problem 8.75. The exit temperature was given as 1350 R, but what are the theoretical limits for this temperature? Find the lowest and the highest, assuming that the heat transfers are as given. For an exit temperature that is the average of the highest and lowest possible, find the first- and second-law efficiencies for the heat engine?

8.77E Consider the two turbines in Problem 7.171, shown in Fig. P7.118. What is the second-law efficiency of the combined system?

8.78E Air in a piston/cylinder arrangement, shown in Fig. P8.44, is at 30 lbf/in.², 540 R with a volume of 20 ft³. If the piston is at the stops the volume is 40 ft³ and a pressure of 60 lbf/in.² is required. The air is then heated from the initial state to 2700 R by a 3400 R reservoir. Find the total irreversibility in the process assuming surroundings are at 70 F.

8.79E Steam is supplied in a line at 450 lbf/in.², 1200 F. A turbine with an isentropic efficiency of 85% is connected to the line by a valve and it exhausts to the atmosphere at 14.7 lbf/in.². If the steam is throttled down to 300 lbf/in.² before entering the turbine find the:

    a. Actual turbine specific work

    b. Change in availability through the valve

    c. Second law efficiency of the turbine

8.80E Air flows through a constant pressure heating device as shown in Fig. P8.47. It is heated up in a reversible process with a work input of 85 Btu/lbm air flowing. The device exchanges heat with the ambient at 540 R. The air enters at 540 R, 60 lbf/in.². Assuming constant specific heat develop an expression for the exit temperature and solve for it.

8.81E (Adv.) A cylinder/piston shown in Fig. P8.48 contains 0.2 lbm of air at room temperature 540 R and a pressure of 30 lbf/in.². The piston mass and spring is such that pressure is proportional to volume, $P = CV$. The air is now heated by a reversible heat pump/engine exchanging energy with a 900 R reservoir to a final temperature of 2160 R. Find the change in availability of the air and the net work out of the heat pump/engine.

8.82E A spring-loaded piston/cylinder contains 6 lbm of water at 140 lbf/in.², 1200 F, shown in Fig. P8.50. The linear spring exerts zero force when the piston is at the bottom of the cylinder. Heat is going out to the ambient to a final pressure of 75 lbf/in.² is reached. The ambient is at 540 R, 15 lbf/in.². Find the heat transfer, the work and the total irreversibility.

8.83E Consider a gasoline engine for a car as an SSSF device where air and fuel enters at the surrounding conditions 77 F, 14.7 lbf/in.² and leaves the engine exhaust manifold at 1800 R, 14.7 lbf/in.² as products assumed to be air. The engine cooling system removes 320 Btu/lbm air through the engine to the ambient. For the analysis take the fuel as air where the extra energy of 950 Btu/lbm of air released in the combustion process, is added as heat transfer from a 3240 R reservoir. Find the work out of the engine, the irreversibility per pound-mass of air, and the first- and second-law efficiencies.

8.84E A piston/cylinder arrangement has a load on the piston so it maintains constant pressure. It contains 1 lbm of steam at 80 lbf/in.², 50% quality. Heat from a reservoir at 1300 F brings the steam to 1000 F. Find the second-law efficiency for this process. Note that no formula is given for this particular case, so determine a reasonable expression for it.

COMPUTER, DESIGN, AND OPEN-ENDED PROBLEMS

8.85 Use the menu-driven software to get the properties of water as needed for consideration of the moisture separator in Problem 5.106. Steam comes in at state 3 and leaves as liquid, state 9, with the rest at state 4 going to the low pressure turbine. Assume no heat transfer to the surroundings at 20°C and find the total entropy generation and irreversibility in the process.

8.86 Use the menu-driven software to get the properties of water as needed and calculate the second law efficiency of the low pressure turbine in Problem 5.106.

8.87 Write a program to solve the general case of Problem 8.14, using ammonia as the working fluid. The initial state is to be the program input variable.

8.88 Consider Problem 8.22 and pose the question: Would it be better to fill the tank with a different proportion of water and nitrogen? Write a program to test for a possible optimal ratio of the proportion that will give the maximum availability.

8.89 (Adv.) Write a program to trace the process described in Problem 8.31. Let the balancing pressure that will float the piston and the total heat added be program input variables. Step the process through increments of 1000 kJ (1000 Btu) heat added but limit it so states where the piston starts/stops moving falls on a step. Print out the state, the availability, and the irreversibility to reach that state at each step.

8.90 Write a program to follow the process through the compressor as described in Problem 8.37. Step through the process with steps of 50 kPa in pressure and for each step write out the work, heat transfer and the change in availability to reach that state from the initial state.

8.91 Consider the heat engine in Problem 8.39. The exit temperature was given as 800 K, but what are the theoretical limits for this temperature? Write a program to find the lowest and the highest limits, assuming that the heat transfers are as given. Make a table of the first- and second-law efficiencies as a function of the exit temperature in steps of 50 K (100 R).

8.92 Write a program to solve Problem 8.47. Use variable heat capacities and let the work input be a program input variable.

8.93 The maximum power a windmill can possibly extract from the wind is

$$\dot{W} = \frac{16}{27}\rho A \mathbf{V} \frac{1}{2}\mathbf{V}^2 = \frac{16}{27}\dot{m}_{air} \times KE$$

Water flowing through Hoover Dam, see Problem 5.88, produces $W = 0.8\dot{m}_{water}\, gh$. Burning 1 kg of coal gives 24 000 kJ delivered at 900 K to a heat engine. Find other examples in the literature and from problems in the previous chapters with steam and gases into turbines. Make a list of the availability (exergy) for a flow of 1 kg/s of substance with the above examples. Use a reasonable choice for the values of the parameters and do the necessary analysis.

8.94 Consider the condenser in the simple steam power plant described in Problem 5.104. The cooling water is lake water at 20°C and it should not be heated more than 5°C as it goes back to the lake. Assume the heat transfer rate inside the condenser is 350 W/m²K so $Q = 350 \times A\Delta T$ in watts. Estimate the flow rate of the cooling water and the needed interface area inside the condenser, A. Find the change in the availability of the cooling water and the steam inside the condenser and compare. Discuss your estimates and the size of the pump for the cooling water.

8.95 Consider the nuclear power plant shown in Problem 5.106. Select one feedwater heater and one pump and make an analysis of their performance. Check the energy balances and do the second-law analysis. Determine the change of availability in all the flows and discuss measures of performance for both the pump and the feedwater heater.

8.96 Reconsider the use of the geothermal energy as discussed in Problem 5.114. The analysis that was done and the original problem statement specified the turbine exit state as 10 kPa, 90% quality. Reconsider this problem with an adiabatic turbine having an isentropic efficiency of 85% and an exit pressure of 10 kPa. Include a second-law analysis and discuss the changes in availability. Describe another way of using the geothermal energy and make appropriate calculations.

8.97 Reconsider the dual flash chamber version of the use of geothermal hot water as described in Problem 5.226. With your knowledge of the second law, repeat the problem assuming that the turbine exit pressure is 10 kPa and disregard the listed exit quality. Include a comparison of the change in availability of the geothermal water with the turbine work.

8.98 An air gun should shoot a harpoon of mass 5 kg out so that it has a velocity of 75 m/s as it leaves the gun. The harpoon acts as the piston in a cylinder and air is trapped below the piston (end of harpoon) that can be initially locked. The air is charged so the initial state is at high pressure and temperature. Determine sizes for the cylinder diameter, cylinder length, air mass, and initial (P,T) of the air. Make reasonable assumptions about the process and include a determination of the state of the air during the process.

8.99 Energy can be stored in many different forms. Thermal energy can be stored as internal energy in a mass like a rock bed, water, metals, etc. Mechanical energy (potential or kinetic) can be stored in springs, rotating flywheels, elevated masses, etc. A tank with a compressed gas is used that can drive a turbine. Batteries are used in cars. Make a list with at least 5 different ways of storing 1000 MJ of energy and size the systems. Note how the energy is taken out and find the availability for each case. Discuss the various alternatives.

8.100 Find from the literature the amount of energy that must be stored in a car to start the engine. Size three different systems to provide that energy and compare those to an ordinary car battery. Discuss the feasibility and cost.

8.101 Reconsider Problem 8.56 where a reversible heat engine is placed between the reservoir and the piston/cylinder. What is the second-law efficiency of this new configuration? Find the thermal efficiency of the heat engine at the beginning and at the end of the process. Show how the heat engine might be done and discuss the practical implementation.

9 POWER AND REFRIGERATION SYSTEMS

Some power plants, such as the simple steam power plant, which we have considered several times, operate in a cycle. That is, the working fluid undergoes a series of processes and finally returns to the initial state. In other power plants, such as the internal-combustion engine and the gas turbine, the working fluid does not go through a thermodynamic cycle, even though the engine itself may operate in a mechanical cycle. In this instance the working fluid has a different composition or is in a different state at the conclusion of the process than it had or was at the beginning. Such equipment is sometimes said to operate on the open cycle (the word cycle is really a misnomer), whereas the steam power plant operates on a closed cycle. The same distinction between open and closed cycles can be made regarding refrigeration devices. For both the open- and closed-cycle apparatus, however, it is advantageous to analyze the performance of an idealized closed cycle similar to the actual cycle. Such a procedure is particularly advantageous for determining the influence of certain variables on performance. For example, the spark-ignition internal-combustion engine is usually approximated by the Otto cycle. From an analysis of the Otto cycle we conclude that increasing the compression ratio increases the efficiency. This is also true for the actual engine, even though the Otto-cycle efficiencies may deviate significantly from the actual efficiencies.

This chapter is concerned with these idealized cycles for both power and refrigeration apparatus. Both vapors and ideal gases are considered as working fluids. An attempt will be made to point out how the processes in actual apparatus deviate from the ideal. Consideration is also given to certain modifications of the basic cycles that are intended to improve performance. These modifications include the use of devices such as regenerators, multistage compressors and expanders, and intercoolers. Various combinations of these types of systems and also special applications, such as cogeneration of electrical power and energy, combined cycles, topping and bottoming cycles, and binary cycle systems, are also discussed in this chapter.

9.1 INTRODUCTION TO POWER SYSTEMS

In introducing the second law of thermodynamics in Chapter 6, we considered cyclic heat engines consisting of four separate processes. We noted there that it is possible to have these engines operate as steady-state, steady-flow devices involving shaft work, as shown in Fig. 6.14, or instead as cylinder/piston devices involving boundary-movement work, as shown in Fig. 6.15. The former may have a working fluid that changes phase during

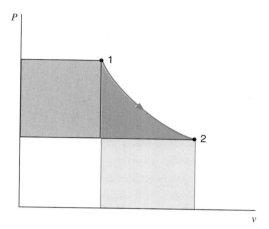

FIGURE 9.1 Comparison of shaft work and boundary-movement work.

the processes in the cycle, or may have a single-phase (usually gaseous) working fluid throughout. The latter type would normally have a gaseous working fluid throughout the cycle.

For a reversible SSSF process involving negligible kinetic and potential energy changes, the shaft work per unit mass is given by Eq. 7.63,

$$w = -\int v\, dP$$

For a reversible process involving a simple compressible substance, the boundary movement work per unit mass is given by Eq. 4.3,

$$w = \int P\, dv$$

The areas represented by these two integrals are shown in Fig. 9.1. It is of interest to note that, in the former case, there is no work involved in a constant-pressure process, while in the latter case, there is no work involved in a constant-volume process.

Let us now consider a power system consisting of four SSSF processes, as in Fig. 6.14. We assume that each process is internally reversible and has negligible changes in kinetic and potential energies, which results in the work for each process being given by Eq. 7.63. For convenience of operation, we will make the two heat-transfer processes (boiler and condenser) constant-pressure processes, such that those are simple heat exchangers involving no work. Let us also assume that the turbine and pump processes are both adiabatic, such that they are therefore isentropic processes. Thus, the four processes comprising the cycle are as shown in Fig. 9.2. Note that if the entire cycle takes place inside the two-phase liquid-vapor dome, the resulting cycle is the Carnot cycle, since the two constant-pressure processes are also isothermal. Otherwise, this cycle is not a Carnot cycle. In either case, we find that the net work output for this power system is given by

$$w_{\text{net}} = -\int_1^2 v\, dP + 0 - \int_3^4 v\, dP + 0 = -\int_1^2 v\, dP + \int_4^3 v\, dP$$

and, since $P_2 = P_3$ and also $P_1 = P_4$, we find that the system produces a net work output because the specific volume is larger during the expansion from 3 to 4 than it is during the compression from 1 to 2. This result is also evident from the areas $-\int v\, dP$ in Fig. 9.2.

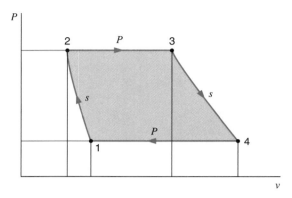

FIGURE 9.2 Four-process power cycle.

We conclude that it would be advantageous to have this difference in specific volume be as large as possible, as, for example, the difference between a vapor and a liquid.

If the four-process cycle shown in Fig. 9.2 were accomplished in a cylinder/piston system involving boundary movement work, then the net work output for this power system is given by

$$w_{net} = \int_1^2 P\,dv + \int_2^3 P\,dv + \int_3^4 P\,dv + \int_4^1 P\,dv$$

and from these four areas on Fig. 9.2, we note that the pressure is higher during any given change in volume in the two expansion processes than in the two compression processes, resulting in a net positive area and a net work output.

For either of the two cases just analyzed, it is noted from Fig. 9.2 that the net work output of the cycle is equal to the area enclosed by the process lines 1–2–3–4–1, and is equal for both, even though the work terms for the four individual processes are different for the two cases.

In the next several sections, we consider the Rankine cycle, which is the ideal, four-SSSF process cycle shown in Fig. 9.2, utilizing a phase change between vapor and liquid in order to maximize the difference in specific volume during expansion and compression. This is the idealized model for a steam powerplant system.

9.2 THE RANKINE CYCLE

We now consider the idealized four-SSSF-process cycle shown in Fig. 9.2, in which state 1 is saturated liquid and state 3 either saturated vapor or superheated vapor. This system is termed the Rankine cycle and is the model for the simple steam powerplant. It is convenient to show the states and processes on a *T–s* diagram, as given in Fig. 9.3. The four processes are

1–2:   Reversible adiabatic pumping process in the pump

2–3:   Constant-pressure transfer of heat in the boiler

3–4:   Reversible adiabatic expansion in the turbine (or other prime mover such as a steam engine)

4–1:   Constant-pressure transfer of heat in the condenser

As mentioned above, the Rankine cycle also includes the possibility of superheating the vapor, as cycle 1–2–3′–4′–1.

If changes of kinetic and potential energy are neglected, heat transfer and work may be represented by various areas on the *T–s* diagram. The heat transferred to the working fluid is represented by area *a*–2–2′–3–*b*–*a*, and the heat transferred from the working fluid by area *a*–1–4–*b*–*a*. From the first law we conclude that the area representing the work is the difference between these two areas—area 1–2–2′–3–4–1. The thermal efficiency is defined by the relation

$$\eta_{\text{th}} = \frac{w_{\text{net}}}{q_H} = \frac{\text{area } 1\text{-}2\text{-}2'\text{-}3\text{-}4\text{-}1}{\text{area } a\text{-}2\text{-}2'\text{-}3\text{-}b\text{-}a} \tag{9.1}$$

For analyzing the Rankine cycle, it is helpful to think of efficiency as depending on the average temperature at which heat is supplied and the average temperature at which heat is rejected. Any changes that increase the average temperature at which heat is supplied or decrease the average temperature at which heat is rejected will increase the Rankine-cycle efficiency.

It should be stated that in analyzing the ideal cycles in this chapter, the changes in kinetic and potential energies from one point in the cycle to another are neglected. In general, this is a reasonable assumption for the actual cycles.

It is readily evident that the Rankine cycle has a lower efficiency than a Carnot cycle with the same maximum and minimum temperatures as a Rankine cycle, because the average temperature between 2 and 2′ is less than the temperature during evaporation. We might well ask, why choose the Rankine cycle as the ideal cycle? Why not select the Carnot cycle 1′–2′–3–4–1′? At least two reasons can be given. The first reason concerns the pumping process. State 1′ is a mixture of liquid and vapor. Great difficulties are encountered in building a pump that will handle the mixture of liquid and vapor at 1′ and deliver saturated liquid at 2′. It is much easier to condense the vapor completely and handle only liquid in the pump: the Rankine cycle is based on this fact. The second reason concerns superheating the vapor. In the Rankine cycle the vapor is superheated at constant pressure, process 3–3′. In the Carnot cycle all the heat transfer is at constant temperature, and therefore the vapor is superheated in process 3–3″. Note, however, that during this process the pressure is dropping, which means that the heat must be transferred to the vapor as it undergoes an expansion process in which work is done. This heat transfer is also very difficult to achieve in practice. Thus, the Rankine cycle is the ideal cycle that can be approximated in practice. In the following sections we will consider some

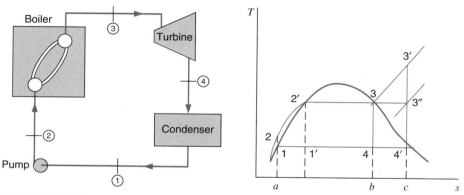

FIGURE 9.3 Simple steam power plant which operates on the Rankine cycle.

tions on the Rankine cycle that enable it to approach more closely the efficiency of the Carnot cycle.

Before we discuss the influence of certain variables on the performance of the Rankine cycle, an example is given.

EXAMPLE 9.1

Determine the efficiency of a Rankine cycle using steam as the working fluid in which the condenser pressure is 10 kPa. The boiler pressure is 2 MPa. The steam leaves the boiler as saturated vapor.

In solving Rankine-cycle problems, we let w_p denote the work into the pump per kilogram of fluid flowing, and q_L the heat rejected from the working fluid per kilogram of fluid flowing.

To solve this problem we consider, in succession, a control surface around the pump, the boiler, the turbine, and the condenser. For each the thermodynamic model is the steam tables, and the process is SSSF with negligible changes in kinetic and potential energies. Now, in turn,

Control volume: Pump.

Inlet state: P_1 known, saturated liquid; state fixed.

Exit state: P_2 known.

Analysis:

First law:

$$w_p = h_2 - h_1$$

Second law:

$$s_2 = s_1$$

Because

$$s_2 = s_1, \qquad h_2 - h_1 = \int_1^2 v\, dP$$

Solution

Assuming the liquid to be incompressible, we have

$$w_p = v(P_2 - P_1) = (0.001\,01)(2000 - 10) = 2.0 \text{ kJ / kg}$$

$$h_2 = h_1 + w_p = 191.8 + 2.0 = 193.8$$

Control volume: Boiler.

Inlet state: P_2, h_2 known; state fixed.

Exit state: P_3 known, saturated vapor; state fixed.

Analysis:

First law:

$$q_H = h_3 - h_2$$

Solution

$$q_H = h_3 - h_2 = 2799.5 - 193.8 = 2605.7 \text{ kJ / kg}$$

Control volume: Turbine.

Inlet state: State 3 known (above).

Exit state: P_4 known.

Analysis:

First law:

$$w_t = h_3 - h_4$$

Second law:

$$s_3 = s_4$$

Solution

We can determine the quality at state 4 as follows:

$$s_3 = s_4 = 6.3409 = 0.6493 + x_4 7.5009 \qquad x_4 = 0.7588$$

$$h_4 = 191.8 + 0.7588(2392.8) = 2007.5$$

$$w_t = 2799.5 - 2007.5 = 792.0 \text{ kJ / kg}$$

Control volume: Condenser.

Inlet state: State 4 known (as given).

Exit state: State 1 known (as given).

Analysis:

First law:

$$q_L = h_4 - h_1$$

Solution

$$q_L = h_4 - h_1 = 2007.5 - 191.8 = 1815.7 \text{ kJ / kg}$$

We can now calculate the thermal efficiency:

$$\eta_{th} = \frac{w_{net}}{q_H} = \frac{q_H - q_L}{q_H} = \frac{w_t - w_p}{q_H} = \frac{792.0 - 2.0}{2605.7} = 30.3\%$$

We could also write an expression for thermal efficiency in terms of properties at various points in the cycle.

$$\eta_{th} = \frac{(h_3 - h_2) - (h_4 - h_1)}{h_3 - h_2} = \frac{(h_3 - h_4) - (h_2 - h_1)}{h_3 - h_2}$$

$$= \frac{2605.7 - 1815.7}{2605.7} = \frac{792.0 - 2.0}{2605.7} = 30.3\%$$

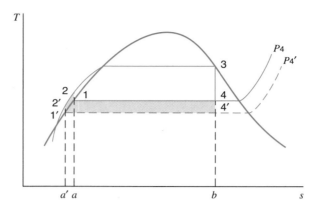

FIGURE 9.4 Effect of exhaust pressure on Rankine-cycle efficiency.

9.3 EFFECT OF PRESSURE AND TEMPERATURE ON THE RANKINE CYCLE

Let us first consider the effect of exhaust pressure and temperature on the Rankine cycle. This effect is shown on the *T-s* diagram of Fig. 9.4. Let the exhaust pressure drop from P_4 to P_4', with the corresponding decrease in temperature at which heat is rejected. The net work is increased by area 1–4–4′–1′–2′–2–1 (shown by the cross-hatching). The heat transferred to the steam is increased by area $a'–2'–2–a–a'$. Since these two areas are approximately equal, the net result is an increase in cycle efficiency. This is also evident from the fact that the average temperature at which heat is rejected is decreased. Note, however, that lowering the back pressure causes the moisture content of the steam leaving the turbine to increase. This is a significant factor because if the moisture in the low-pressure stages of the turbine exceeds about 10%, not only is there a decrease in turbine efficiency, but erosion of the turbine blades may also be a very serious problem.

Next, consider the effect of superheating the steam in the boiler, as shown in Fig. 9.5. It is readily evident that the work is increased by area 3–3′–4′–4–3, and the heat transferred in the boiler is increased by area 3–3′–b′–b–3. Since the ratio of these two areas is greater than the ratio of net work to heat supplied for the rest of the cycle, it is evident that for given pressures, superheating the steam increases the Rankine-cycle efficiency. This increase in efficiency would also follow from the fact that the average

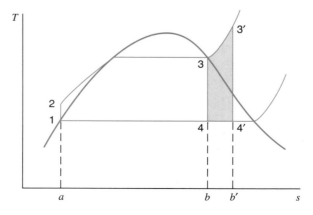

FIGURE 9.5 Effect of superheating on Rankine-cycle efficiency.

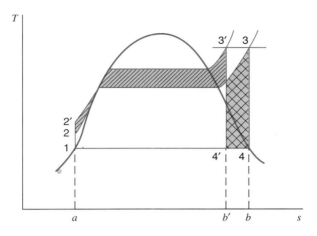

FIGURE 9.6 Effect of boiler pressure on Rankine-cycle efficiency.

temperature at which heat is transferred to the steam is increased. Note also that when the steam is superheated, the quality of the steam leaving the turbine increases.

Finally, the influence of the maximum pressure of the steam must be considered, and this is shown in Fig. 9.6. In this analysis the maximum temperature of the steam, as well as the exhaust pressure, is held constant. The heat rejected decreases by area $b'-4'-4-b-b'$. The net work increases by the amount of the single cross-hatching and decreases by the amount of the double cross-hatching. Therefore, the net work tends to remain the same, but the heat rejected decreases, and hence the Rankine-cycle efficiency increases with an increase in maximum pressure. Note that in this instance too the average temperature at which heat is supplied increases with an increase in pressure. The quality of the steam leaving the turbine decreases as the maximum pressure increases.

To summarize this section, we can say that the efficiency of the Rankine cycle can be increased by lowering the exhaust pressure, by increasing the pressure during heat addition, and by superheating the steam. The quality of the steam leaving the turbine is increased by superheating the steam and decreased by lowering the exhaust pressure and by increasing the pressure during heat addition.

EXAMPLE 9.2   In a Rankine cycle steam leaves the boiler and enters the turbine at 4 MPa, 400°C. The condenser pressure is 10 kPa. Determine the cycle efficiency.

To determine the cycle efficiency, we must calculate the turbine work, the pump work, and the heat transfer to the steam in the boiler. We do this by considering a control surface around each of these components in turn. In each case the thermodynamic model is the steam tables, and the process is SSSF with negligible changes in kinetic and potential energies.

Control volume: Pump.
Inlet state: P_1 known, saturated liquid; state fixed.
Exit state: P_2 known.

Analysis:
First law:

$$w_p = h_2 - h_1$$

Second law:

$$s_2 = s_1$$

Since $s_2 = s_1$,

$$h_2 - h_1 = \int_1^2 v \, dP = v(P_2 - P_1)$$

Solution

$$w_p = v(P_2 - P_1) = (0.001\,01)(4000 - 10) = 4.0 \text{ kJ / kg}$$

$$h_1 = 191.8$$

$$h_2 = 191.8 + 4.0 = 195.8$$

Control volume: Turbine.

Inlet state: P_3, T_3 known; state fixed.

Exit state: P_4 known.

Analysis:

First law:

$$w_t = h_3 - h_4$$

Second law:

$$s_4 = s_3$$

Solution

$$h_3 = 3213.6, \qquad s_3 = 6.7690$$

$$s_3 = s_4 = 6.7690 = 0.6493 + x_4 7.5009, \qquad x_4 = 0.8159$$

$$h_4 = 191.8 + 0.8159(2392.8) = 2144.1$$

$$w_t = h_3 - h_4 = 3213.6 - 2144.1 = 1069.5 \text{ kJ / kg}$$

$$\mathrm{w}_{\mathrm{net}} = w_t - w_p = 1069.5 - 4.0 = 1065.5 \text{ kJ / kg}$$

Control volume: Boiler.

Inlet state: P_2, h_2 known; state fixed.

Exit state: State 3 fixed (as given).

Analysis:

First law:

$$q_H = h_3 - h_2$$

Solution

$$q_H = h_3 - h_2 = 3213.6 - 195.8 = 3017.8 \text{ kJ / kg}$$

$$\eta_{\mathrm{th}} = \frac{w_{\mathrm{net}}}{q_H} = \frac{1065.5}{3017.8} = 35.3\%$$

The net work could also be determined by calculating the heat rejected in the condenser, q_L, and noting, from the first law, that the net work for the cycle is equal to the net heat transfer. Considering a control surface around the condenser, we have

$$q_L = h_4 - h_1 = 2144.1 - 191.8 = 1952.3 \text{ kJ} / \text{kg}$$

Therefore,

$$w_{net} = q_H - q_L = 3017.8 - 1952.3 = 1065.5 \text{ kJ} / \text{kg}$$

EXAMPLE 9.2E In a Rankine cycle, steam leaves the boiler and enters the turbine at 600 lbf/in.^{2}, 800 F. The condenser pressure is 1 lbf/in.^{2} Determine the cycle efficiency.

To determine the cycle efficiency, we must calculate the turbine work, the pump work, and the heat transfer to the steam in the boiler. We do this by considering a control surface around each of these components in turn. In each case the thermodynamic model is the steam tables, and the process is SSSF with negligible changes in kinetic and potential energies.

  Control volume: Pump.
  Inlet state: P_1 known, saturated liquid; state fixed.
  Exit state: P_2 known.

Analysis:

First law:

$$w_P = h_2 - h_1$$

Second law:

$$s_2 = s_1$$

Since $s_2 = s_1$,

$$h_2 - h_1 = \int_1^2 v \, dP = v \left(P_2 - P_1 \right)$$

Solution

$$w_p = v \left(P_2 - P_1 \right) = 0.01614 \left(600 - 1 \right) \times \tfrac{144}{778} = 1.8 \text{ Btu} / \text{1bm}$$

$$h_1 = 69.70$$

$$h_2 = 69.7 + 1.8 = 71.5 \text{ Btu} / \text{1bm}$$

  Control volume: Turbine.
  Inlet state: P_3, T_3 known; state fixed.
  Exit state: P_4 known.

Analysis:

First law:

$$w_t = h_3 - h_4$$

Second law:

$$s_4 = s_3$$

Solution

$$h_3 = 1407.6 \qquad s_3 = 1.6343$$

$$s_3 = s_4 = 1.6343 = 1.9779 - (1-x)_4 \, 1.8453$$

$$(1-x)_4 = 0.1861$$

$$h_4 = 1105.8 - 0.1861(1036.0) = 913.0$$

$$w_t = h_3 - h_4 = 1407.6 - 913.0 = 494.6 \text{ Btu / lbm}$$

$$w_{net} = w_t - w_p = 494.6 - 1.8 = 492.8 \text{ Btu / lbm}$$

Control volume: Boiler.

Inlet state: P_2, h_2 known; state fixed.

Exit state: State 3 fixed (above).

Analysis:

First law:

$$q_H = h_3 - h_2$$

Solution

$$q_H = h_3 - h_2 = 1407.6 - 71.5 = 1336.1 \text{ Btu / lbm}$$

$$\eta_{th} = \frac{w_{net}}{q_H} = \frac{492.8}{1336.1} = 36.9\%$$

The net work could also be determined by calculating the heat rejected in the condenser, q_L, and noting, from the first law, that the net work for the cycle is equal to the net heat transfer. Considering a control surface around the condenser, we have

$$q_L = h_4 - h_1 = 913.0 - 69.7 = 843.3 \text{ Btu / lbm}$$

Therefore,

$$w_{net} = q_H - q_L = 1336.1 - 843.3 = 492.8 \text{ Btu / lbm}$$

9.4 THE REHEAT CYCLE

In the last section we noted that the efficiency of the Rankine cycle could be increased by increasing the pressure during the addition of heat. However, the increase in pressure also increases the moisture content of the steam in the low-pressure end of the turbine. The reheat cycle has been developed to take advantage of the increased efficiency with higher pressures, and yet avoid excessive moisture in the low-pressure stages of the turbine. This cycle is shown schematically and on a *T–s* diagram in Fig. 9.7. The unique feature of this cycle is that the steam is expanded to some intermediate pressure in the turbine and is

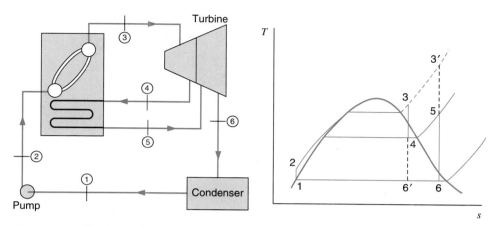

FIGURE 9.7 The ideal reheat cycle.

then reheated in the boiler, after which it expands in the turbine to the exhaust pressure. It is evident from the T–s diagram that there is very little gain in efficiency from reheating the steam, because the average temperature at which heat is supplied is not greatly changed. The chief advantage is in decreasing to a safe value the moisture content in the low-pressure stages of the turbine. If metals could be found that would enable us to superheat the steam to $3'$, the simple Rankine cycle would be more efficient than the reheat cycle, and there would be no need for the reheat cycle.

EXAMPLE 9.3   Consider a reheat cycle utilizing steam. Steam leaves the boiler and enters the turbine at 4 MPa, 400°C. After expansion in the turbine to 400 kPa, the steam is reheated to 400°C and then expanded in the low-pressure turbine to 10 kPa. Determine the cycle efficiency.

For each control volume analyzed, the thermodynamic model is the steam tables, the process is SSSF, and changes in kinetic and potential energies are negligible.

Control volume: High-pressure turbine.

Inlet state: P_3, T_3 known; state fixed.

Exit state: P_4 known.

Analysis:

First law:

$$w_{\text{h-p}} = h_3 - h_4$$

Second law:

$$s_3 = s_4$$

Solution

$$h_3 = 3213.6, \qquad s_3 = 6.7690$$
$$s_4 = s_3 = 6.7690 = 1.7766 + x_4\,5.1193, \qquad x_4 = 0.9752$$
$$h_4 = 604.7 + 0.9752(2133.8) = 2685.6$$

Control volume: Low-pressure turbine.

Inlet state: P_5, T_5 known; state fixed.

Exit state: P_6 known.

Analysis:

First law:

$$w_{1\text{-}p} = h_5 - h_6$$

Second law:

$$s_5 = s_6$$

Solution

$$h_5 = 3273.4 \qquad s_5 = 7.8985$$
$$s_6 = s_5 = 7.8985 = 0.6493 + x_6 7.5009, \qquad x_6 = 0.9664$$
$$h_6 = 191.8 + 0.9664(2392.8) = 2504.3$$

For the overall turbine, the total work output w_t is the sum of $w_{\text{h-p}}$ and $w_{1\text{-p}}$, so that

$$w_t = (h_3 - h_4) + (h_5 - h_6)$$
$$= (3213.6 - 2685.6) + (3273.4 - 2504.3)$$
$$= 1297.1 \text{ kJ / kg}$$

Control volume: Pump.

Inlet state: P_1 known, saturated liquid; state fixed.

Exit state: P_2 known.

Analysis:

First law:

$$w_p = h_2 - h_1$$

Second law:

$$s_2 = s_1$$

Since $s_2 = s_1$,

$$h_2 - h_1 = \int_1^2 v \, dP = v(P_2 - P_1)$$

Solution

$$w_p = v(P_2 - P_1) = (0.001\,01)(4000 - 10) = 4.0 \text{ kJ / kg}$$
$$h_2 = 191.8 + 4.0 = 195.8$$

Control volume: Boiler.

Inlet states: States 2 and 4 both known (above).

Exit states: States 3 and 5 both known (as given).

Analysis:

First law:

$$q_H = (h_3 - h_2) + (h_5 - h_4)$$

Solution

$$q_H = (h_3 - h_2) + (h_5 - h_4)$$
$$= (3213.6 - 195.8) + (3273.4 - 2685.6) = 3605.6 \text{ kJ / kg}$$

Therefore,

$$w_{\text{net}} = w_t - w_p = 1297.1 - 4.0 = 1293.1 \text{ kJ / kg}$$

$$\eta_{\text{th}} = \frac{w_{\text{net}}}{q_H} = \frac{1293.1}{3605.6} = 35.9\%$$

By comparing this example with Example 9.2, we find that through reheating the gain in efficiency is relatively small, but the moisture content of the vapor leaving the turbine is decreased from 18.4 to 3.4%.

9.5 THE REGENERATIVE CYCLE

Another important variation from the Rankine cycle is the regenerative cycle, which uses feedwater heaters. The basic concepts of this cycle can be demonstrated by considering the Rankine cycle without superheat as shown in Fig. 9.8. During the process between states 2 and 2′, the working fluid is heated while in the liquid phase, and the average temperature of the working fluid is much lower than during the vaporization process 2′–3.

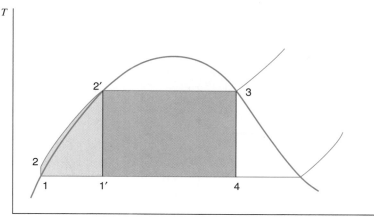

FIGURE 9.8
Temperature-entropy diagram showing the relationship between Carnot-cycle efficiency and Rankine-cycle efficiency.

The process between states 2 and 2′ causes the average temperature at which heat is supplied in the Rankine cycle to be lower than in the Carnot cycle 1′–2′–3–4–1′. Consequently, the efficiency of the Rankine cycle is lower than that of the corresponding Carnot cycle. In the regenerative cycle the working fluid enters the boiler at some state between 2 and 2′, and consequently the average temperature at which heat is supplied is higher.

Consider first an idealized regenerative cycle, as shown in Fig. 9.9. The unique feature of this cycle compared to the Rankine cycle is that after leaving the pump, the liquid circulates around the turbine casing, counterflow to the direction of vapor flow in the turbine. Thus, it is possible to transfer to the liquid flowing around the turbine the heat from the vapor as it flows through the turbine. Let us assume for the moment that this is a reversible heat transfer, that is, at each point the temperature of the vapor is only infinitesimally higher than the temperature of the liquid. In this instance line 4–5 on the *T–s* diagram of Fig. 9.9, which represents the states of the vapor flowing through the turbine, is exactly parallel to line 1–2–3, which represents the pumping process (1–2) and the states of the liquid flowing around the turbine. Consequently, areas 2–3–*b*–*a*–2 and 5–4–*d*–*c*–5 are not only equal but congruous, and these areas, respectively, represent the heat transferred to the liquid and from the vapor. Heat is also transferred to the working fluid at constant temperature in process 3–4, and area 3–4–*d*–*b*–3 represents this heat transfer. Heat is transferred from the working fluid in process 5–1, and area 1–5–*c*–*a*–1 represents this heat transfer. This area is exactly equal to area 1′–5′–*d*–*b*–1′, which is the heat rejected in the related Carnot cycle 1′–3–4–5′–1′. Thus, the efficiency of this idealized regenerative cycle is exactly equal to the efficiency of the Carnot cycle with the same heat supply and heat rejection temperatures.

Quite obviously this idealized regenerative cycle is impractical. First, it would be impossible to effect the necessary heat transfer from the vapor in the turbine to the liquid feedwater. Furthermore, the moisture content of the vapor leaving the turbine increases considerably as a result of the heat transfer. The disadvantage of this has been noted previously. The practical regenerative cycle extracts some of the vapor after it has partially expanded in the turbine and uses feedwater heaters, as shown in Fig. 9.10.

Steam enters the turbine at state 5. After expansion to state 6, some of the steam is extracted and enters the feedwater heater. The steam that is not extracted is expanded in the turbine to state 7 and is then condensed in the condenser. This condensate is pumped into the feedwater heater where it mixes with the steam extracted from the turbine. The

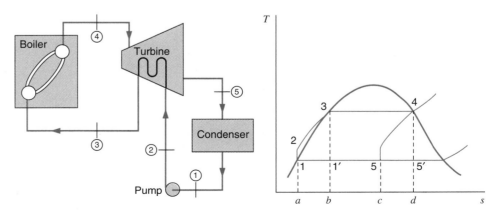

FIGURE 9.9 The ideal regenerative cycle.

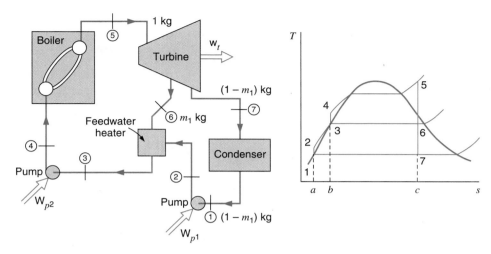

FIGURE 9.10 Regenerative cycle with open feedwater heater.

proportion of steam extracted is just sufficient to cause the liquid leaving the feedwater heater to be saturated at state 3. Note that the liquid has not been pumped to the boiler pressure, but only to the intermediate pressure corresponding to state 6. Another pump is required to pump the liquid leaving the feedwater heater to boiler pressure. The significant point is that the average temperature at which heat is supplied has been increased.

This cycle is somewhat difficult to show on a T–s diagram because the masses of steam flowing through the various components vary. The T–s diagram of Fig. 9.10 simply shows the state of the fluid at the various points.

Area 4–5–c–b–4 in Fig. 9.10 represents the heat transferred per kilogram of working fluid. Process 7–1 is the heat rejection process, but since not all the steam passes through the condenser, area 1–7–c–a–1 represents the heat transfer per kilogram flowing through the condenser, which does not represent the heat transfer per kilogram of working fluid entering the turbine. Between states 6 and 7 only part of the steam is flowing through the turbine. The example that follows illustrates the calculations for the regenerative cycle.

EXAMPLE 9.4   Consider a regenerative cycle using steam as the working fluid. Steam leaves the boiler and enters the turbine at 4 MPa, 400°C. After expansion to 400 kPa, some of the steam is extracted from the turbine for the purpose of heating the feedwater in an open feedwater heater. The pressure in the feedwater heater is 400 kPa and the water leaving it is saturated liquid at 400 kPa. The steam not extracted expands to 10 kPa. Determine the cycle efficiency.

The line diagram and T–s diagram for this cycle are shown in Fig. 9.10.

As in previous examples, the model for each control volume is the steam tables, the process is SSSF, and kinetic and potential energy changes are negligible.

From Examples 9.2 and 9.3 we have the following properties:

$$h_5 = 3213.6 \qquad h_6 = 2685.6$$
$$h_7 = 2144.1 \qquad h_1 = 191.8$$

Control volume: Low-pressure pump.

Inlet state: P_1 known, saturated liquid; state fixed.

Exit state: P_2 known.

Analysis:

First law:

$$w_{p1} = h_2 - h_1$$

Second law:

$$s_2 = s_1$$

Therefore,

$$h_2 - h_1 = \int_1^2 v \, dP = v \left(P_2 - P_1 \right)$$

Solution

$$w_{p1} = v \left(P_2 - P_1 \right) = \left(0.001\,01 \right) \left(400 - 10 \right) = 0.4 \text{ kJ / kg}$$

$$h_2 = h_1 + w_p = 191.8 + 0.4 = 192.2$$

Control volume: Turbine.

Inlet state: P_5, T_5 known; state fixed.

Exit state: P_6 known; P_7 known.

Analysis:

First law:

$$w_t = \left(h_5 - h_6 \right) + \left(1 - m_1 \right) \left(h_6 - h_7 \right)$$

Second law:

$$s_5 = s_6 = s_7$$

Solution

From the second law, the values for h_6 and h_7 given previously were calculated in Examples 9.2 and 9.3.

Control volume: Feedwater heater.

Inlet states: States 2 and 6 both known (as given).

Exit state: P_3 known, saturated liquid; state fixed.

Analysis:

First law:

$$m_1 \left(h_6 \right) + \left(1 - m_1 \right) h_2 = h_3$$

Solution

$$m_1(2685.6) + (1 - m_1)(192.2) = 604.7$$

$$m_1 = 0.1654$$

We can now calculate the turbine work.

$$\begin{aligned} w_t &= (h_5 - h_6) + (1 - m_1)(h_6 - h_7) \\ &= (3213.6 - 2685.6) + (1 - 0.1654)(2685.6 - 2144.1) \\ &= 979.9 \text{ kJ} / \text{kg} \end{aligned}$$

Control volume: High-pressure pump.

Inlet state: State 3 known (as given).

Exit state: P_4 known.

Analysis:

First law:

$$w_{p2} = h_4 - h_3$$

Second law:

$$s_4 = s_3$$

Solution

$$w_{p2} = v(P_4 - P_3) = (0.001\,084)(4000 - 400) = 3.9 \text{ kJ} / \text{kg}$$

$$h_4 = h_3 + w_{p2} = 604.7 + 3.9 = 608.6$$

Therefore,

$$\begin{aligned} w_{\text{net}} &= w_t - (1 - m_1)w_{p1} - w_{p2} \\ &= 979.9 - (1 - 0.1654)(0.4) - 3.9 = 975.7 \text{ kJ} / \text{kg} \end{aligned}$$

Control volume: Boiler.

Inlet state: P_4, h_4 known (as given); state fixed.

Exit state: State 5 known (as given).

Analysis:

First law:

$$q_H = h_5 - h_4$$

Solution

$$q_H = h_5 - h_4 = 3213.6 - 608.6 = 2605.0 \text{ kJ} / \text{kg}$$

$$\eta_{\text{th}} = \frac{w_{\text{net}}}{q_H} = \frac{975.7}{2605.0} = 37.5\%$$

Note the increase in efficiency over the efficiency of the Rankine cycle of Example 9.2.

Up to this point the discussion and examples have tacitly assumed that the extraction steam and feedwater are mixed in the feedwater heater. Another much-used type of feedwater heater, known as a closed heater, is one in which the steam and feedwater do not mix; rather heat is transferred from the extracted steam as it condenses on the outside of tubes while the feedwater flows through the tubes. In a closed heater, a schematic sketch of which is shown in Fig. 9.11, the steam and feedwater may be at considerably different pressures. The condensate may be pumped into the feedwater line, or it may be removed through a trap to a lower-pressure heater or to the condenser. (A trap is a device that permits liquid but not vapor to flow to a region of lower pressure.)

Open feedwater heaters have the advantage of being less expensive and having better heat-transfer characteristics compared to closed feedwater heaters. They have the disadvantage of requiring a pump to handle the feedwater between each heater.

In many power plants a number of stages of extraction are used, though only rarely more than five. The number is, of course, determined by economics. It is evident that using a very large number of extraction stages and feedwater heaters allows the cycle efficiency to approach that of the idealized regenerative cycle of Fig. 9.9, where the feedwater enters the boiler as saturated liquid at the maximum pressure. In practice, however, this could not be economically justified because the savings effected by the increase in efficiency would be more than offset by the cost of additional equipment (feedwater heaters, piping, and so forth).

A typical arrangement of the main components in an actual power plant is shown in Fig. 9.12. Note that one open feedwater heater is a deaerating feedwater heater; this heater has the dual purpose of heating and removing the air from the feedwater. Unless the air is removed, excessive corrosion occurs in the boiler. Note also that the condensate from the high-pressure heater drains (through a trap) to the intermediate heater, and the intermediate heater drains to the deaerating feedwater heater. The low-pressure heater drains to the condenser.

Many actual power plants combine one reheat stage with a number of extraction stages. The principles already considered are readily applied to such a cycle.

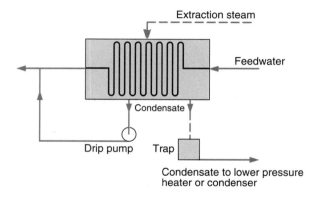

FIGURE 9.11 Schematic arrangement for a closed feedwater heater.

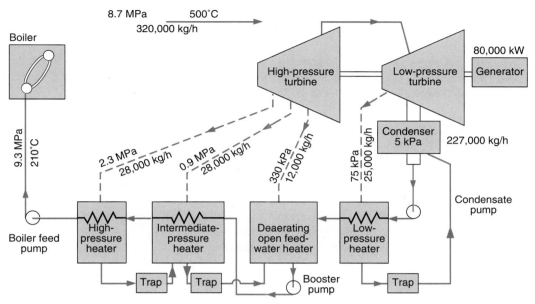

FIGURE 9.12 Arrangement of heaters in an actual power plant utilizing regenerative feedwater heaters.

9.6 DEVIATION OF ACTUAL CYCLES FROM IDEAL CYCLES

Before we leave the matter of vapor power cycles, a few comments are in order regarding the ways in which an actual cycle deviates from an ideal cycle. The losses associated with the combustion process are considered in a later chapter. The most important of these losses are the following.

Piping Losses

Pressure drops caused by frictional effects and heat transfer to the surroundings are the most important piping losses. Consider, for example, the pipe connecting the turbine to the boiler. If only frictional effects occur, states *a* and *b* in Fig. 9.13 would represent the states of the steam leaving the boiler and entering the turbine, respectively. Note that the frictional effects cause an increase in entropy. Heat transferred to the surroundings at constant pressure can be represented by process *bc*. This effect decreases entropy. Both

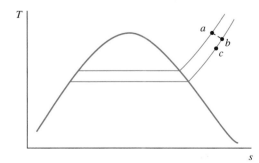

FIGURE 9.13 Temperature–entropy diagram showing effect of losses between boiler and turbine.

the pressure drop and heat transfer decrease the availability of the steam entering the turbine. The irreversibility of this process can be calculated by the methods outlined in Chapter 8.

A similar loss is the pressure drop in the boiler. Because of this pressure drop, the water entering the boiler must be pumped to a much higher pressure than the desired steam pressure leaving the boiler, and this requires additional pump work.

Turbine Losses

The losses in the turbine are primarily those associated with the flow of the working fluid through the turbine. Heat transfer to the surroundings also represents a loss, but this is usually of secondary importance. The effects of these two losses are the same as those outlined for piping losses. The process might be represented as shown in Fig. 9.14, where 4_s represents the state after an isentropic expansion and state 4 represents the actual state leaving the turbine. The governing procedures may also cause a loss in the turbine, particularly if a throttling process is used to govern the turbine.

The efficiency of the turbine was defined in Chapter 7 as

$$\eta_t = \frac{w_t}{h_3 - h_{4s}}$$

where the states are as designated in Fig. 9.14.

Pump Losses

The losses in the pump are similar to those of the turbine and are primarily due to the irreversibilities associated with the fluid flow. Heat transfer is usually a minor loss.

The pump efficiency is defined as

$$\eta_p = \frac{h_{2s} - h_1}{w_p}$$

where the states are as shown in Fig. 9.14, and w_p is the actual work input per kilogram of fluid.

Condenser Losses

The losses in the condenser are relatively small. One of these minor losses is the cooling below the saturation temperature of the liquid leaving the condenser. This represents a

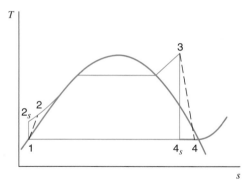

FIGURE 9.14 Temperature–entropy diagram showing effect of turbine and pump inefficiencies on cycle performance.

loss because additional heat transfer is necessary to bring the water to its saturation temperature.

The influence of these losses on the cycle is illustrated in the following example, which should be compared to Example 9.2.

EXAMPLE 9.5   A steam power plant operates on a cycle with pressures and temperatures as designated in Fig. 9.15. The efficiency of the turbine is 86% and the efficiency of the pump is 80%. Determine the thermal efficiency of this cycle.

As in previous examples, for each control volume the model used is the steam tables, and each process is SSSF with no changes in kinetic or potential energy. This cycle is shown on the T–s diagram of Fig. 9.16.

   Control volume: Turbine.
   Inlet state: P_5, T_5 known; state fixed.
   Exit state: P_6 known.

Analysis:

First law:

$$w_t = h_5 - h_6$$

Second law:

$$s_{6s} = s_5$$

$$\eta_t = \frac{w_t}{h_5 - h_{6s}} = \frac{h_5 - h_6}{h_5 - h_{6s}}$$

Solution

From the steam tables.

$$h_5 = 3169.1, \qquad s_5 = 6.7235$$
$$s_{6s} = s_5 = 6.7235 = 0.6493 + x_{6s}7.5009, \qquad x_{6s} = 0.8098$$
$$h_{6s} = 191.8 + 0.8098(2392.8) = 2129.5$$
$$w_t = \eta_t(h_5 - h_{6s}) = 0.86(3169.1 - 2129.5) = 894.1 \text{ kJ / kg}$$

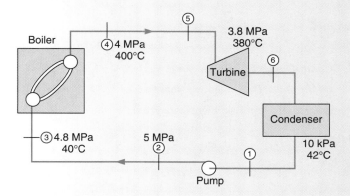

FIGURE 9.15 Schematic diagram for Example 9.5.

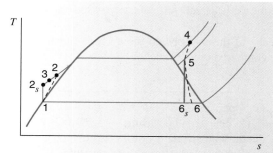

FIGURE 9.16 Temperature–entropy diagram for Example 9.5.

Control volume: Pump.

Inlet state: P_1, T_1 known; state fixed.

Exit state: P_2 known.

Analysis:

First law:

$$w_p = h_2 - h_1$$

Second law:

$$s_{2s} = s_1$$

$$\eta_p = \frac{h_{2s} - h_1}{w_p} = \frac{h_{2s} - h_1}{h_2 - h_1}$$

Since $s_{2s} = s_1$,

$$h_{2s} - h_1 = v(P_2 - P_1)$$

Therefore,

$$w_p = \frac{h_{2s} - h_1}{\eta_p} = \frac{v(P_2 - P_1)}{\eta_p}$$

Solution

$$w_p = \frac{v(P_2 - P_1)}{\eta_p} = \frac{(0.001\ 009)(5000 - 10)}{0.80} = 6.3 \text{ kJ / kg}$$

Therefore,

$$w_{\text{net}} = w_t - w_p = 894.1 - 6.3 = 887.8 \text{ kJ / kg}$$

Control volume: Boiler.

Inlet state: P_3, T_3 known; state fixed.

Exit state: P_4, T_4 known; state fixed.

Analysis:

First law:

$$q_H = h_4 - h_3$$

Solution

$$q_H = h_4 - h_3 = 3213.6 - 171.8 = 3041.8 \text{ kJ / kg}$$

$$\eta_{th} = \frac{887.8}{3041.8} = 29.2\%$$

This result compares to the Rankine efficiency of 35.3% for the similar cycle of Example 9.2.

EXAMPLE 9.5E A steam power plant operates on a cycle with pressure and temperatures as designated in Fig. 9.15E. The efficiency of the turbine is 86% and the efficiency of the pump is 80%. Determine the thermal efficiency of this cycle.

As in previous examples, for each control volume the model used is the steam tables, and each process is SSSF with no changes in kinetic or potential energy. This cycle is shown on the *T–s* diagram of Fig. 9.16.

Control volume: Turbine.

Inlet state: P_5, T_5 known; state fixed.

Exit state: P_6 known.

Analysis:

First law:

$$w_t = h_5 - h_6$$

Second law:

$$s_{6s} = s_5$$

$$\eta_t = \frac{w_t}{h_5 - h_{6s}} = \frac{h_5 - h_6}{h_5 - h_{6s}}$$

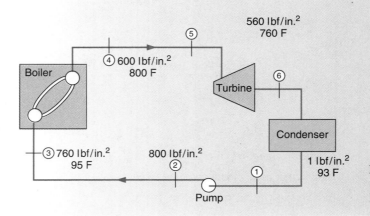

FIGURE 9.15E
Schematic diagram for Example 9.5E.

Solution

From the steam tables,

$$h_5 = 1386.8 \quad s_5 = 1.6248$$

$$s_{6s} = s_5 = 1.6248 = 1.9779 - (1-x)_{6s} \, 1.8453$$

$$(1-x)_{6s} = \frac{0.3531}{1.8453} = 0.1912$$

$$h_{6s} = 1105.8 - 0.1912(1036.0) = 907.6$$

$$w_t = \eta_t (h_5 - h_{6s}) = 0.86(1386.8 - 907.6)$$

$$= 0.86(479.2) = 412.1 \text{ Btu / lbm}$$

Control volume: Pump.

Inlet state: P_1, T_1 known; state fixed.

Exit state: P_2 known.

Analysis:

First law:

$$w_p = h_2 - h_1$$

Second law:

$$s_{2s} = s_1$$

$$\eta_p = \frac{h_{2s} - h_1}{w_p} = \frac{h_{2s} - h_1}{h_2 - h_1}$$

Since $s_{2s} = s_1$,

$$h_{2s} - h_1 = v(P_2 - P_1)$$

Therefore,

$$w_p = \frac{h_{2s} - h_1}{\eta_p} = \frac{v(P_2 - P_1)}{\eta_p}$$

Solution

$$w_p = \frac{v(P_2 - P_1)}{\eta_p} = \frac{0.016\ 15(800-1)144}{0.8 \times 778} = 3.0 \text{ Btu / lbm}$$

Therefore,

$$w_{net} = w_t - w_p = 412.1 - 3.0 = 409.1 \text{ Btu / lbm}$$

Control volume: Boiler.

Inlet state: P_3, T_3 known; state fixed.

Exit state: P_4, T_4 known; state fixed.

Analysis:

First law:

$$q_H = h_4 - h_3$$

Solution

$$q_H = h_4 - h_3 = 1407.6 - 65.1 = 1342.5 \text{ Btu / lbm}$$

$$\eta_{th} = \frac{409.1}{1342.5} = 30.4\%$$

This compares to an efficiency of 36.9% for the Rankine efficiency of the similar cycle of Example 9.2E.

9.7 COGENERATION

There are many occasions in industrial settings where the need arises for a specific source or supply of energy within the environment in which a steam powerplant is being used to generate electricity. In such cases, it is appropriate to consider supplying this source of energy in the form of steam that has already been expanded through the high-pressure section of the turbine in the powerplant cycle, thereby eliminating the construction and use of a second boiler or other energy source. Such an arrangement is shown in Fig. 9.17, in which the turbine is tapped at some intermediate pressure to furnish the necessary amount of process steam required for the particular energy need—perhaps to operate a special process in the plant, or in many cases simply for the purpose of space heating the facilities. This type of application is termed cogeneration, and if the system is designed as a package with both the electrical and the process steam requirements in mind, it is possible to achieve a substantial savings in capital cost of equipment and also

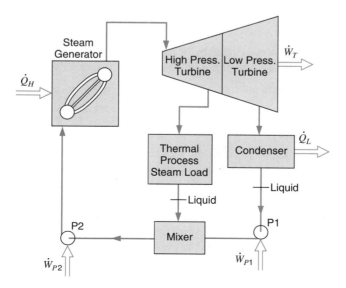

FIGURE 9.17 Example of a cogeneration system.

in the operating cost, through careful consideration of all the requirements and optimization of the various parameters involved. Specific examples of cogeneration systems are considered in the problems at the end of the chapter.

9.8 AIR-STANDARD POWER CYCLES

In Section 9.1 we considered idealized four-process cycles, including both SSSF-process and cylinder/piston boundary-movement cycles. The question of phase-change cycles and single-phase cycles was also mentioned. We then proceeded to examine the Rankine powerplant cycle in detail, the idealized model of a phase-change power cycle. However, many work-producing devices (engines) utilize a working fluid that is always a gas. The spark-ignition automotive engine is a familiar example, and so are the Diesel engine and the conventional gas turbine. In all these engines there is a change in the composition of the working fluid, because during combustion it changes from air and fuel to combustion products. For this reason these engines are called internal-combustion engines. In contrast, the steam power plant may be called an external combustion engine, because heat is transferred from the products of combustion to the working fluid. External-combustion engines using a gaseous working fluid (usually air) have been built. To date they have had only limited application, but the use of the gas-turbine cycle in conjunction with a nuclear reactor has been investigated extensively. Other external-combustion engines are currently receiving serious attention in an effort to combat air pollution.

Because the working fluid does not go through a complete thermodynamic cycle in the engine (even though the engine operates in a mechanical cycle) the internal-combustion engine operates on the so-called open cycle. However, for analyzing internal-combustion engines, it is advantageous to devise closed cycles that closely approximate the open cycles. One such approach is the air-standard cycle, which is based on the following assumptions.

1. A fixed mass of air is the working fluid throughout the entire cycle, and the air is always an ideal gas. Thus, there is no inlet process or exhaust process.

2. The combustion process is replaced by a process transferring heat from an external source.

3. The cycle is completed by heat transfer to the surroundings (in contrast to the exhaust and intake process of an actual engine).

4. All processes are internally reversible.

5. An additional assumption is often made, that air has a constant specific heat.

The principal value of the air-standard cycle is to enable us to examine qualitatively the influence of a number of variables on performance. The quantitative results obtained from the air-standard cycle, such as efficiency and mean effective pressure, will differ from those of the actual engine. Our emphasis, therefore, in our consideration of the air-standard cycle, will be primarily on the qualitative aspects.

The term "mean effective pressure" (mep), which is used in conjunction with reciprocating engines, is defined as the pressure that, if it acted on the piston during the entire power stroke, would do an amount of work equal to that actually done on the piston. The work for one cycle is found by multiplying this mean effective pressure by the area of the piston (minus the area of the rod on the crank end of a double-acting engine) and by the stroke.

9.9 THE BRAYTON CYCLE

In discussing idealized four-SSSF-process power cycles in Section 9.1, a cycle involving two constant-pressure and two isentropic processes was examined, and the results shown in Fig. 9.2. This cycle used with a condensing working fluid is the Rankine cycle, but when used with a single-phase, gaseous working fluid it is termed the Brayton cycle. The air-standard Brayton cycle is the ideal cycle for the simple gas turbine. The simple open-cycle gas turbine utilizing an internal-combustion process and the simple closed-cycle gas turbine, which utilizes heat-transfer processes, are both shown schematically in Fig. 9.18. The air-standard Brayton cycle is shown on the P–v and T–s diagrams of Fig. 9.19.

The efficiency of the air-standard Brayton cycle is found as follows:

$$\eta_{th} = 1 - \frac{Q_L}{Q_H} = 1 - \frac{C_p(T_4 - T_1)}{C_p(T_3 - T_2)} = 1 - \frac{T_1(T_4 / T_1 - 1)}{T_2(T_3 / T_2 - 1)}$$

We note, however, that

$$\frac{P_3}{P_4} = \frac{P_2}{P_1}$$

$$\frac{P_2}{P_1} = \left(\frac{T_2}{T_1}\right)^{k/(k-1)} = \frac{P_3}{P_4} = \left(\frac{T_3}{T_4}\right)^{k/(k-1)}$$

$$\frac{T_3}{T_4} = \frac{T_2}{T_1} \quad \therefore \frac{T_3}{T_2} = \frac{T_4}{T_1} \quad \text{and} \quad \frac{T_3}{T_2} - 1 = \frac{T_4}{T_1} - 1$$

$$\eta_{th} = 1 - \frac{T_1}{T_2} = 1 - \frac{1}{\left(P_2 / P_1\right)^{(k-1)/k}} \tag{9.2}$$

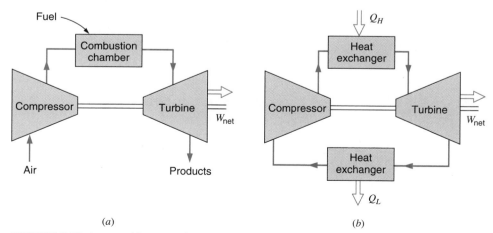

$$(a) \qquad\qquad\qquad\qquad\qquad (b)$$

FIGURE 9.18 A gas turbine operating on the Brayton cycle. (*a*) Open cycle. (*b*) Closed cycle.

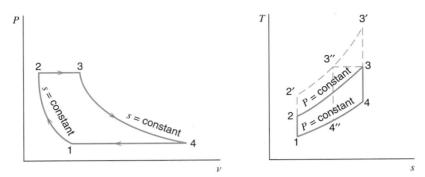

FIGURE 9.19 The air-standard Brayton cycle.

The efficiency of the air-standard Brayton cycle is therefore a function of the isentropic pressure ratio; Fig. 9.20 shows a plot of efficiency versus pressure ratio. The fact that efficiency increases with pressure ratio is evident from the T–s diagram of Fig. 9.19 because increasing the pressure ratio changes the cycle from 1–2–3–4–1 to 1–2′–3′–4–1. The latter cycle has a greater heat supply and the same heat rejected as the original cycle; therefore, it has a greater efficiency. Note that the latter cycle has a higher maximum temperature, T_3', than the original cycle, T_3. In the actual gas turbine the maximum temperature of the gas entering the turbine is fixed by material considerations. Therefore, if we fix the temperature T_3 and increase the pressure ratio, the resulting cycle is 1–2′–3″–4″–1. This cycle would have a higher efficiency than the original cycle, but the work per kilogram of working fluid is thereby changed.

With the advent of nuclear reactors, the closed-cycle gas turbine has become more important. Heat is transferred, either directly or via a second fluid, from the fuel in the nuclear reactor to the working fluid in the gas turbine. Heat is rejected from the working fluid to the surroundings.

The actual gas-turbine engine differs from the ideal cycle primarily because of irreversibilities in the compressor and turbine, and because of pressure drop in the flow passages and combustion chamber (or in the heat exchanger of a closed-cycle turbine). Thus, the state points in a simple open-cycle gas turbine might be as shown in Fig. 9.21.

The efficiencies of the compressor and turbine are defined in relation to isentropic processes. With the states designated as in Fig. 9.21, the definitions of compressor and

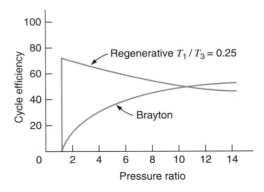

FIGURE 9.20 Cycle efficiency as a function of pressure ratio for the Brayton and regenerative cycles.

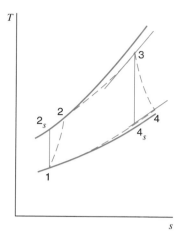

FIGURE 9.21 Effect of inefficiencies on the gas-turbine cycle.

turbine efficiencies are

$$\eta_{\text{comp}} = \frac{h_{2s} - h_1}{h_2 - h_1} \tag{9.3}$$

$$\eta_{\text{turb}} = \frac{h_3 - h_4}{h_3 - h_{4s}} \tag{9.4}$$

Another important feature of the Brayton cycle is the large amount of compressor work (also called back work) compared to turbine work. Thus, the compressor might require from 40 to 80% of the output of the turbine. This is particularly important when the actual cycle is considered, because the effect of the losses is to require a larger amount of compression work from a smaller amount of turbine work, and thus, the overall efficiency drops very rapidly with a decrease in the efficiencies of the compressor and turbine. In fact, if these efficiencies drop below about 60%, all the work of the turbine will be required to drive the compressor, and the overall efficiency will be zero. This is in sharp contrast to the Rankine cycle, where only 1 or 2% of the turbine work is required to drive the pump. This demonstrates the inherent advantage of the cycle utilizing a condensing working fluid, such that a much larger difference in specific volume between the expansion and compression processes is utilized effectively.

EXAMPLE 9.6   In an air-standard Brayton cycle the air enters the compressor at 0.1 MPa, 15°C. The pressure leaving the compressor is 1.0 MPa, and the maximum temperature in the cycle is 1100°C. Determine

1. The pressure and temperature at each point in the cycle

2. The compressor work, turbine work, and cycle efficiency

For each of the control volumes analyzed, the model is ideal gas with constant specific heat, value at 300 K, and each process is SSSF with no kinetic or potential energy changes. The diagram for this example is Fig. 9.19.

Control volume: Compressor.

Inlet state: P_1, T_1 known; state fixed.

Exit state: P_2 known.

Analysis:

First law:

$$w_c = h_2 - h_1$$

(Note that the compressor work w_c is here defined as work input to the compressor.)

Second law:

$$s_2 = s_1$$

so that

$$\frac{T_2}{T_1} = \left(\frac{P_2}{P_1}\right)^{(k-1)/k}$$

Solution

$$\left(\frac{P_2}{P_1}\right)^{(k-1)/k} = 10^{0.286} = 1.932 \qquad T_2 = 556.8 \text{ K}$$

$$w_c = h_2 - h_1 = C_p\left(T_2 - T_1\right)$$

$$= 1.0035\left(556.8 - 288.2\right) = 269.5 \text{ kJ / kg}$$

Control volume: Turbine.

Inlet state: P_3 ($= P_2$) known, T_3 known; state fixed.

Exit state: P_4 ($= P_1$) known.

Analysis:

First law:

$$w_t = h_3 - h_4$$

Second law:

$$s_3 = s_4$$

so that

$$\frac{T_3}{T_4} = \left(\frac{P_3}{P_4}\right)^{(k-1)/k}$$

Solution

$$\left(\frac{P_3}{P_4}\right)^{(k-1)/k} = 10^{0.286} = 1.932 \qquad T_4 = 710.8 \text{ K}$$

$$w_t = h_3 - h_4 = C_p\left(T_3 - T_4\right)$$

$$= 1.0035\left(1373.2 - 710.8\right) = 664.7 \text{ kJ} / \text{kg}$$

$$w_{\text{net}} = w_t - w_c = 664.7 - 269.5 = 395.2 \text{ kJ} / \text{kg}$$

Control volume: High-temperature heat exchanger.

Inlet state: State 2 fixed (as given).

Exit state: State 3 fixed (as given).

Analysis:

First law:

$$q_H = h_3 - h_2 = C_p\left(T_3 - T_2\right)$$

Solution

$$q_H = h_3 - h_2 = C_p\left(T_3 - T_2\right) = 1.0035\left(1373.2 - 556.8\right) = 819.3 \text{ kJ} / \text{kg}$$

Control volume: Low-temperature heat exchanger.

Inlet state: State 4 fixed (above).

Exit state: State 1 fixed (above).

Analysis:

First law:

$$q_L = h_4 - h_1 = C_p\left(T_4 - T_1\right)$$

Solution

$$q_L = h_4 - h_1 = C_p\left(T_4 - T_1\right) = 1.0035\left(710.8 - 288.2\right) = 424.1 \text{ kJ} / \text{kg}$$

Therefore,

$$\eta_{\text{th}} = \frac{w_{\text{net}}}{q_H} = \frac{395.2}{819.3} = 48.2\%$$

This may be checked by using Eq. 9.2.

$$\eta_{\text{th}} = 1 - \frac{1}{\left(P_2 / P_1\right)^{(k-1)/k}} = 1 - \frac{1}{10^{0.286}} = 48.2\%$$

EXAMPLE 9.7 Consider a gas turbine with air entering the compressor under the same conditions as in Example 9.5 and leaving at a pressure of 1.0 MPa. The maximum temperature is

1100°C. Assume a compressor efficiency of 80%, a turbine efficiency of 85%, and a pressure drop between the compressor and turbine of 15 kPa. Determine the compressor work, turbine work, and cycle efficiency.

As in the previous example, for each control volume the model is ideal gas, constant specific heat, value at 300 K, and each process is SSSF with no kinetic or potential energy changes. In this example the diagram is Fig. 9.21.

Control volume: Compressor.

Inlet state: P_1, T_1 known; state fixed.

Exit state: P_2 known.

Analysis:

First law, real process:

$$w_c = h_2 - h_1$$

Second law, ideal process:

$$s_{2s} = s_1$$

so that

$$\frac{T_{2s}}{T_1} = \left(\frac{P_2}{P_1}\right)^{(k-1)/k}$$

In addition,

$$\eta_c = \frac{h_{2s} - h_1}{h_2 - h_1} = \frac{T_{2s} - T_1}{T_2 - T_1}$$

Solution

$$\left(\frac{P_2}{P_1}\right)^{(k-1)/k} = \frac{T_{2s}}{T_1} = 10^{0.286} = 1.932 \qquad T_{2s} = 556.8 \text{ K}$$

$$\eta_c = \frac{h_{2s} - h_1}{h_2 - h_1} = \frac{T_{2s} - T_1}{T_2 - T_1} = \frac{556.8 - 288.2}{T_2 - T_1} = 0.80$$

$$T_2 - T_1 = \frac{556.8 - 288.2}{0.80} = 335.8 \qquad T_2 = 624.0 \text{ K}$$

$$w_c = h_2 - h_1 = C_p(T_2 - T_1)$$

$$= 1.0035(624.0 - 288.2) = 337.0 \text{ kJ/kg}$$

Control volume: Turbine.

Inlet state: P_3 (P_2 − drop) known, T_3 known; state fixed.

Exit state: P_4 known.

Analysis:

First law, real process:

$$w_t = h_3 - h_4$$

Second law, ideal process:

$$s_{4s} = s_3$$

So that

$$\frac{T_3}{T_{4s}} = \left(\frac{P_3}{P_4}\right)^{(k-1)/k}$$

In addition,

$$\eta_t = \frac{h_3 - h_4}{h_3 - h_{4s}} = \frac{T_3 - T_4}{T_3 - T_{4s}}$$

Solution

$$P_3 = P_2 - \text{pressure drop} = 1.0 - 0.015 = 0.985 \text{ MPa}$$

$$\left(\frac{P_3}{P_4}\right)^{(k-1)/k} = \frac{T_3}{T_{4s}} = 9.85^{0.286} = 1.9236 \qquad T_{4s} = 713.9 \text{ K}$$

$$\eta_t = \frac{h_3 - h_4}{h_3 - h_{4s}} = \frac{T_3 - T_4}{T_3 - T_{4s}} = 0.85$$

$$T_3 - T_4 = 0.85(1373.2 - 713.9) = 560.4$$

$$T_4 = 812.8 \text{ K}$$

$$w_t = h_3 - h_4 = C_p(T_3 - T_4)$$

$$= 1.0035(1373.2 - 812.8) = 562.4 \text{ kJ / kg}$$

$$w_{\text{net}} = w_t - w_c = 562.4 - 337.0 = 225.4 \text{ kJ / kg}$$

Control volume: High-temperature heat exchanger.
Inlet state: State 2 fixed (as given).
Exit state: State 3 fixed (as given).

Analysis:

First law:

$$q_H = h_3 - h_2$$

Solution

$$q_H = h_3 - h_2 = C_p(T_3 - T_2)$$

$$= 1.0035(1373.2 - 624.0) = 751.8 \text{ kJ / kg}$$

so that

$$\eta_{th} = \frac{w_{net}}{q_H} = \frac{225.4}{751.8} = 30.0\%$$

The following comparisons can be made between Examples 9.6 and 9.7.

	w_c	w_t	w_{net}	q_H	η_{th}
Example 9.6 (Ideal)	269.5	664.7	395.2	819.3	48.2
Example 9.7 (Actual)	337.0	562.4	225.4	751.8	30.0

As stated previously, the irreversibilities decrease the turbine work and increase the compressor work. Since the net work is the difference between these two, it decreases very rapidly as compressor and turbine efficiencies decrease. The development of compressors and turbines of high efficiency is therefore an important aspect of the development of gas turbines.

Note that in the ideal cycle (Example 9.6) about 41% of the turbine work is required to drive the compressor and 59% is delivered as net work. In the actual turbine (Example 9.7) 60% of the turbine work is required to drive the compressor and 40% is delivered as net work. Thus, if the net power of this unit is to be 10 000 kW, a 25 000-kW turbine and a 15 000-kW compressor are required. This result demonstrates that a gas turbine has a high back-work ratio.

9.10 THE SIMPLE GAS-TURBINE CYCLE WITH A REGENERATOR

The efficiency of the gas-turbine cycle may be improved by introducing a regenerator. The simple open-cycle gas-turbine cycle with a regenerator is shown in Fig. 9.22, and the corresponding ideal air-standard cycle with a regenerator is shown on the P–v and T–s diagrams. In cycle 1–2–x–3–4–y–1, the temperature of the exhaust gas leaving the turbine in state 4 is higher than the temperature of the gas leaving the compressor. Therefore, heat can be transferred from the exhaust gases to the high-pressure gases leaving the compressor. If this is done in a counterflow heat exchanger, which is known as a regenerator, the temperature of the high-pressure gas leaving the regenerator, T_x, may, in the ideal case, have a temperature equal to T_4, the temperature of the gas leaving the turbine. Heat transfer from the external source is necessary only to increase the temperature from T_x to T_3. Area x–3–d–b–x represents the heat transferred, and area y–1–a–c–y represents the heat rejected.

The influence of pressure ratio on the simple gas-turbine cycle with a regenerator is shown by considering cycle 1–2′–3′–4–1. In this cycle the temperature of the exhaust gas leaving the turbine is just equal to the temperature of the gas leaving the compressor; therefore, there is no possibility of utilizing a regenerator. This can be shown more exactly by determining the efficiency of the ideal gas-turbine cycle with a regenerator.

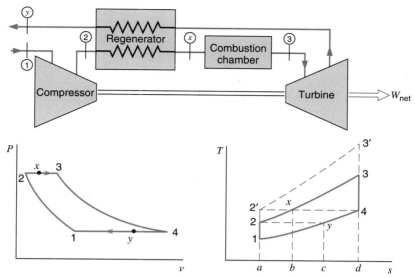

FIGURE 9.22 The ideal regenerative cycle.

The efficiency of this cycle with regeneration is found as follows, where the states are as given in Fig. 9.22.

$$\eta_{\text{th}} = \frac{w_{\text{net}}}{q_H} = \frac{w_t - w_c}{q_H}$$

$$q_H = C_p(T_3 - T_x)$$

$$w_t = C_p(T_3 - T_4)$$

But for an ideal regenerator, $T_4 = T_x$, and therefore $q_H = w_t$. Consequently,

$$\eta_{\text{th}} = 1 - \frac{w_c}{w_t} = 1 - \frac{C_p(T_2 - T_1)}{C_p(T_3 - T_4)}$$

$$= 1 - \frac{T_1(T_2/T_1 - 1)}{T_3(1 - T_4/T_3)} = 1 - \frac{T_1}{T_3} \frac{\left[(P_2/P_1)^{(k-1)/k} - 1\right]}{\left[1 - (P_1/P_2)^{(k-1)/k}\right]}$$

$$\eta_{\text{th}} = 1 - \frac{T_1}{T_3}\left(\frac{P_2}{P_1}\right)^{(k-1)/k}$$

Thus, for the ideal cycle with regeneration the thermal efficiency depends not only on the pressure ratio but also on the ratio of the minimum to maximum temperature. We note that, in contrast to the Brayton cycle, the efficiency decreases with an increase in pressure ratio. The thermal efficiency versus pressure ratio for this cycle is plotted in Fig. 9.20 for a value of

$$\frac{T_1}{T_3} = 0.25$$

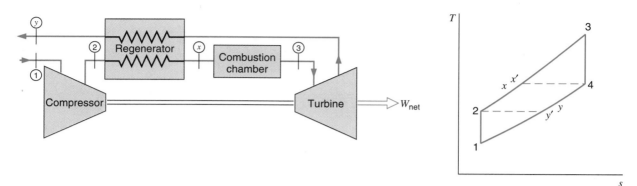

FIGURE 9.23 Temperature–entropy diagram to illustrate the definition of regenerator efficiency.

The effectiveness or efficiency of a regenerator is given by the regenerator efficiency, which can best be defined by reference to Fig. 9.23. State x represents the high-pressure gas leaving the regenerator. In the ideal regenerator there would be only an infinitesimal temperature difference between the two streams, and the high-pressure gas would leave the regenerator at temperature T_x', and $T_x' = T_4$. In an actual regenerator, which must operate with a finite temperature difference T_x, the actual temperature leaving the regenerator is therefore less than T_x'. The regenerator efficiency is defined by

$$\eta_{\text{reg}} = \frac{h_x - h_2}{h_x' - h_2}$$

If the specific heat is assumed to be constant, the regenerator efficiency is also given by the relation

$$\eta_{\text{reg}} = \frac{T_x - T_2}{T_x' - T_2}$$

It should be pointed out that a higher efficiency can be achieved by using a regenerator with a greater heat-transfer area. However, this also increases the pressure drop, which represents a loss, and both the pressure drop and the regenerator efficiency must be considered in determining which regenerator gives maximum thermal efficiency for the cycle. From an economic point of view, the cost of the regenerator must be weighed against the saving that can be effected by its use.

EXAMPLE 9.8   If an ideal regenerator is incorporated into the cycle of Example 9.6, determine the thermal efficiency of the cycle.

The diagram for this example is Fig. 9.23. Values are from Example 9.6. Therefore, for the analysis of the high-temperature heat exchanger (combustion chamber), from the first law,

$$q_H = h_3 - h_x$$

so that the solution is

$$T_x = T_4 = 710.8 \text{ K}$$

$$q_H = h_3 - h_x = C_p(T_3 - T_x) = 1.0035(1373.2 - 710.8) = 664.7 \text{ kJ / kg}$$

$$w_{\text{net}} = 395.2 \text{ kJ / kg} \quad \text{(from Example 9.6)}$$

$$\eta_{\text{th}} = \frac{395.2}{664.7} = 59.5\%$$

9.11 THE IDEAL GAS-TURBINE POWER CYCLE USING MULTISTAGE COMPRESSION WITH INTERCOOLING, MULTISTAGE EXPANSION WITH REHEATING, AND A REGENERATOR

The Brayton cycle, as the idealized model for the gas turbine powerplant, has a reversible, adiabatic compressor and a reversible, adiabatic turbine. In the following example, we consider the effect of replacing these components with reversible, isothermal processes.

EXAMPLE 9.9
An air-standard power cycle has the same states as given in Example 9.6. In this cycle, however, the compressor and turbine are both reversible, isothermal processes. Calculate the compressor work and the turbine work, and compare with the results of Example 9.6.

Control volumes: Compressor, turbine.

Analysis:

For each reversible, isothermal process, from Eq. 7.65,

$$w = -\int_i^e v\, dP = -P_i v_i \ln \frac{P_e}{P_i} = -RT_i \ln \frac{P_e}{P_i}$$

Solution

For the compressor,

$$w = -0.287 \times 288.2 \times \ln 10 = -190.5 \text{ kJ / kg}$$

compared with −269.5 kJ/kg in the adiabatic compressor.

For the turbine,

$$w = -0.287 \times 1373.2 \times \ln 0.1 = +907.5 \text{ kJ / kg}$$

compared with + 664.7 kJ/kg in the adiabatic turbine.

It is found that the isothermal process would be preferable to the adiabatic process in both the compressor and turbine. The resulting cycle, called the Ericsson cycle, consists of two reversible, constant-pressure processes and two reversible, constant-tempera-

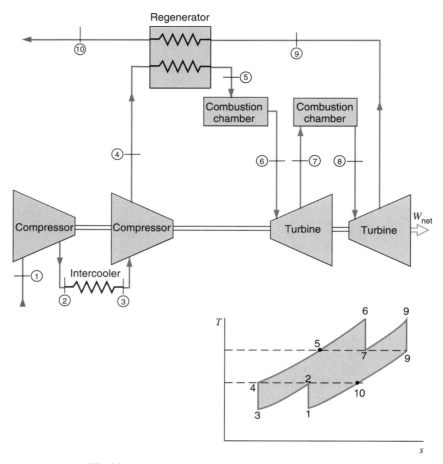

FIGURE 9.24 The ideal gas-turbine cycle utilizing intercooling, reheat, and a regenerator.

ture processes. The reason that the actual gas turbine does not attempt to emulate this cycle rather than the Brayton cycle, is that the compressor and turbine processes are both high-flow rate processes involving work-related devices in which it is not practical to attempt to transfer large quantities of heat. As a consequence, the processes tend to be essentially adiabatic, so that this becomes the process in the model cycle.

There is a modification of the Brayton/gas turbine cycle that tends to change its performance in the direction of the Ericsson cycle. This modification is to use multiple stages of compression with intercooling, and also multiple stages of expansion with reheat. Such a cycle with two stages of compression and expansion, and also incorporating a regenerator, is shown in Fig. 9.24. The air-standard cycle is given on the corresponding *T–s* diagram. It may be shown that for this cycle the maximum efficiency is obtained if equal pressure ratios are maintained across the two compressors and the two turbines. In this ideal cycle it is assumed that the temperature of the air leaving the intercooler, T_3, is equal to the temperature of the air entering the first stage of compression, T_1, and that the temperature after reheating, T_8, is equal to the temperature entering the first turbine, T_6. Furthermore, in the ideal cycle it is assumed that the temperature of the high-pressure air

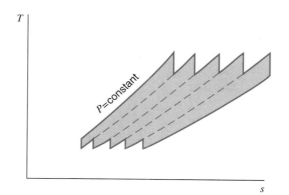

FIGURE 9.25 Temperature–entropy diagram that shows how the gas-turbine cycle with many stages approaches the Ericsson cycle.

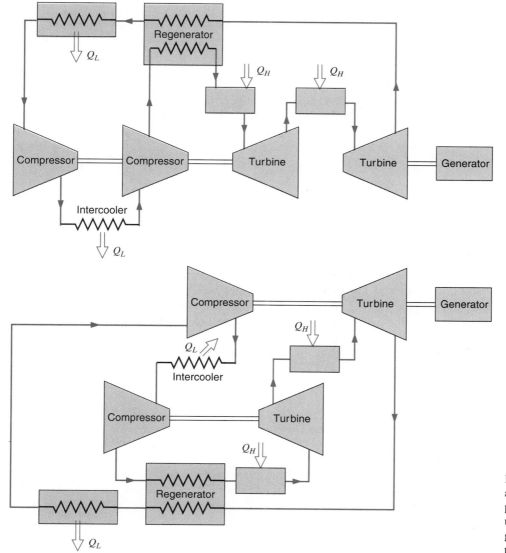

FIGURE 9.26 Some arrangements of components that may be utilized in stationary gas-turbine power plants.

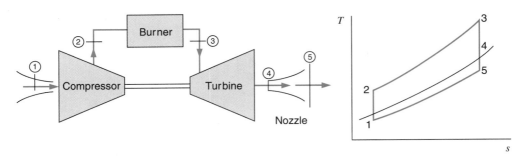

FIGURE 9.27 The ideal gas-turbine cycle for a jet engine.

leaving the regenerator, T_5, is equal to the temperature of the low-pressure air leaving the turbine, T_9.

If a large number of stages of compression and expansion are used, it is evident that the Ericsson cycle is approached. This is shown in Fig. 9.25. In practice, the economical limit to the number of stages is usually two or three. The turbine and compressor losses and pressure drops that have already been discussed would be involved in any actual unit employing this cycle.

There are a variety of ways in which the turbines and the compressors using this cycle can be utilized. Two possible arrangements for closed cycles are shown in Fig. 9.26. One advantage frequently sought in a given arrangement is ease of control of the unit under various loads. Detailed discussion of this point, however, is beyond the scope of this text.

9.12 THE AIR-STANDARD CYCLE FOR JET PROPULSION

The next air-standard power cycle we consider is utilized in jet propulsion. In this cycle the work done by the turbine is just sufficient to drive the compressor. The gases are expanded in the turbine to a pressure for which the turbine work is just equal to the compressor work. The exhaust pressure of the turbine will then be greater than that of the surroundings, and the gas can be expanded in a nozzle to the pressure of the surroundings. Since the gases leave at a high velocity, the change in momentum that the gases undergo gives a thrust to the aircraft in which the engine is installed. The air-standard cycle for this situation is shown in Fig. 9.27. The principles governing this cycle follow from the analysis of the Brayton cycle plus that for a reversible, adiabatic nozzle.

EXAMPLE 9.10 Consider an ideal jet propulsion cycle in which air enters the compressor at 0.1 MPa, 15°C. The pressure leaving the compressor is 1.0 MPa, and the maximum temperature is 1100°C. The air expands in the turbine to a pressure at which the turbine work is just equal to the compressor work. On leaving the turbine, the air expands in a nozzle to 0.1 MPa. The process is reversible and adiabatic. Determine the velocity of the air leaving the nozzle.

The model used is ideal gas, constant specific heat, value at 300 K, and each process is SSSF with no potential energy change. The only kinetic energy change occurs in the nozzle. The diagram is shown in Fig. 9.27.

The compressor analysis is the same as in Example 9.6. From the results of that solution,

$$P_1 = 0.1 \text{ MPa} \qquad T_1 = 288.2 \text{ K}$$
$$P_2 = 1.0 \text{ MPa} \qquad T_2 = 556.8 \text{ K}$$
$$w_c = 269.5 \text{ kJ / kg}$$

The turbine analysis is also the same as in Example 9.6. Here, however,

$$P_3 = 1.0 \text{ MPa} \qquad T_3 = 1373.2 \text{ K}$$
$$w_c = w_t = C_p(T_3 - T_4) = 269.5 \text{ kJ / kg}$$
$$T_3 - T_4 = \frac{269.5}{1.0035} = 268.6 \qquad T_4 = 1104.6 \text{ K}$$

so that

$$\frac{T_3}{T_4} = \left(\frac{P_3}{P_4} \right)^{(k-1)/k} = \frac{1373.2}{1104.6} = 1.2432$$

$$\frac{P_3}{P_4} = 2.142 \qquad P_4 = 0.4668 \text{ MPa}$$

Control volume: Nozzle.

Inlet state: State 4 fixed (above).

Exit state: P_5 known.

Analysis:

First law:

$$h_4 = h_5 + \frac{\mathbf{V}_5^2}{2}$$

Second law:

$$s_4 = s_5$$

Solution

Since P_5 is 0.1 MPa, from the second law we find that $T_5 = 710.8$ K.

$$\mathbf{V}_5^2 = 2C_{p0}(T_4 - T_5)$$
$$\mathbf{V}_5^2 = 2 \times 1000 \times 1.0035(1104.6 - 710.8)$$
$$\mathbf{V}_5 = 889 \text{ m / s}$$

9.13 THE OTTO CYCLE

In Section 9.1, we discussed power cycles incorporating either SSSF processes or cylinder/piston boundary work processes. In that section, it was noted that for the SSSF-process, there is no work in a constant-pressure process. Each of the SSSF power cycles presented in subsequent sections of this chapter incorporated two constant-pressure heat transfer processes. It was further noted in Section 9.1 that in a boundary movement work process, there is no work in a constant-volume process. In the next three sections, we will present ideal air-standard power cycles for cylinder/piston boundary movement work processes, each example of which includes either one or two constant-volume heat transfer processes.

The air-standard Otto cycle is an ideal cycle that approximates a spark-ignition internal-combustion engine. This cycle is shown on the P–v and T–s diagrams of Fig. 9.28. Process 1–2 is an isentropic compression of the air as the piston moves from crank-end dead center to head-end dead center. Heat is then added at constant volume while the piston is momentarily at rest at head-end dead center. (This process corresponds to the ignition of the fuel–air mixture by the spark and the subsequent burning in the actual engine.) Process 3–4 is an isentropic expansion, and process 4–1 is the rejection of heat from the air while the piston is at crank-end dead center.

The thermal efficiency of this cycle is found as follows, assuming constant specific heat of air.

$$\eta_{th} = \frac{Q_H - Q_L}{Q_H} = 1 - \frac{Q_L}{Q_H} = 1 - \frac{mC_v\left(T_4 - T_1\right)}{mC_v\left(T_3 - T_2\right)}$$

$$= 1 - \frac{T_1\left(T_4 / T_1 - 1\right)}{T_2\left(T_3 / T_2 - 1\right)}$$

We note further that

$$\frac{T_2}{T_1} = \left(\frac{V_1}{V_2}\right)^{k-1} = \left(\frac{V_4}{V_3}\right)^{k-1} = \frac{T_3}{T_4}$$

Therefore,

$$\frac{T_3}{T_2} = \frac{T_4}{T_1}$$

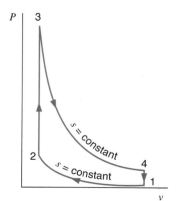

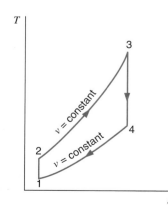

FIGURE 9.28 The air-standard Otto cycle.

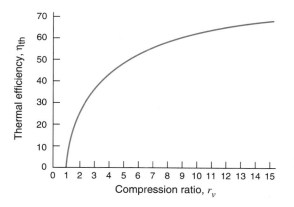

FIGURE 9.29 Thermal efficiency of the Otto cycle as a function of compression ratio.

and

$$\eta_{th} = 1 - \frac{T_1}{T_2} = 1 - (r_v)^{1-k} = 1 - \frac{1}{r_v^{k-1}} \tag{9.5}$$

where

$$r_v = \text{compression ratio} = \frac{V_1}{V_2} = \frac{V_4}{V_3}$$

The important thing to note is that the efficiency of the air-standard Otto cycle is a function only of the compression ratio, and that the efficiency is increased by increasing the compression ratio. Figure 9.29 shows a plot of the air-standard cycle thermal efficiency versus compression ratio. It is also true of an actual spark-ignition engine that the efficiency can be increased by increasing the compression ratio. The trend toward higher compression ratios is prompted by the effort to obtain higher thermal efficiency. In the actual engine there is an increased tendency for the fuel to detonate as the compression ratio is increased. After detonation the fuel burns rapidly, and strong pressure waves present in the engine cylinder give rise to the so-called spark knock. Therefore, the maximum compression ratio that can be used is fixed by the fact that detonation must be avoided. The advance in compression ratios over the years in the actual engine was originally made possible by developing fuels with better antiknock characteristics, primarily through the addition of tetraethyl lead. More recently, however, nonleaded gasolines with good antiknock characteristics have been developed in an effort to reduce atmospheric contamination.

Some of the most important ways in which the actual open-cycle spark-ignition engine deviates from the air-standard cycle are as follows.

1. The specific heats of the actual gases increase with an increase in temperature.

2. The combustion process replaces the heat-transfer process at high temperature, and combustion may be incomplete.

3. Each mechanical cycle of the engine involves an inlet and an exhaust process and, because of the pressure drop through the valves, a certain amount of work is required to charge the cylinder with air and exhaust the products of combustion.

4. There will be considerable heat transfer between the gases in the cylinder and the cylinder walls.

5. There will be irreversibilities associated with pressure and temperature gradients.

EXAMPLE 9.11   The compression ratio in an air-standard Otto cycle is 8. At the beginning of the compression stroke the pressure is 0.1 MPa and the temperature is 15°C. The heat transfer to the air per cycle is 1800 kJ/kg air. Determine

1. The pressure and temperature at the end of each process of the cycle
2. The thermal efficiency
3. The mean effective pressure

   Control mass: Air inside cylinder.

   Diagram: Fig. 9.28.

   State information: $P_1 = 0.1$ MPa, $T_1 = 288.2$ K.

   Process information: Four processes known (Fig. 9.28). Also $r_v = 8$ and $q_H = 1800$ kJ/kg.

   Model: Ideal gas, constant specific heat, value at 300 K.

Analysis:

Second law for compression process 1–2:

$$s_2 = s_1$$

so that

$$\frac{T_2}{T_1} = \left(\frac{V_1}{V_2}\right)^{k-1}$$

$$\frac{P_2}{P_1} = \left(\frac{V_1}{V_2}\right)^{k}$$

First law for heat addition process 2–3:

$$q_H = {}_2q_3 = u_3 - u_2 = C_v(T_3 - T_2)$$

Second law for expansion process 3–4:

$$s_4 = s_3$$

so that

$$\frac{T_3}{T_4} = \left(\frac{V_4}{V_3}\right)^{k-1}$$

$$\frac{P_3}{P_4} = \left(\frac{V_4}{V_3}\right)^{k}$$

In addition,

$$\eta_{th} = 1 - \frac{1}{r_v^{k-1}} \qquad \text{mep} = \frac{w_{net}}{v_1 - v_2}$$

Solution

$$v_1 = \frac{0.287 \times 288.2}{100} = 0.827 \text{ m}^3 / \text{kg}$$

$$\frac{T_2}{T_1} = \left(\frac{V_1}{V_2}\right)^{k-1} = 8^{0.4} = 2.3 \qquad T_2 = 662 \text{ K}$$

$$\frac{P_2}{P_1} = \left(\frac{V_1}{V_2}\right)^{k} = 8^{1.4} = 18.38 \qquad P_2 = 1.838 \text{ MPa}$$

$$v_2 = \frac{0.827}{8} = 0.1034 \text{ m}^3 / \text{kg}$$

$$_2q_3 = C_v(T_3 - T_2) = 1800 \text{ kJ / kg}$$

$$T_3 - T_2 = \frac{1800}{0.7165} = 2512 \qquad T_3 = 3174 \text{ K}$$

$$\frac{T_3}{T_2} = \frac{P_3}{P_2} = \frac{3174}{662} = 4.795 \qquad P_3 = 8.813 \text{ MPa}$$

$$\frac{T_3}{T_4} = \left(\frac{V_4}{V_3}\right)^{k-1} = 8^{0.4} = 2.3 \qquad T_4 = 1380 \text{ K}$$

$$\frac{P_3}{P_4} = \left(\frac{V_4}{V_3}\right)^{k} = 8^{1.4} = 18.38 \qquad P_4 = 0.4795 \text{ MPa}$$

$$\eta_{th} = 1 - \frac{1}{r_v^{k-1}} = 1 - \frac{1}{8^{0.4}} = 1 - \frac{1}{2.3} = 1 - 0.435 = 0.565 = 56.5\%$$

This can be checked by finding the heat rejected.

$$_4q_1 = C_v(T_1 - T_4) = 0.7165(288.2 - 1380) = -782.3 \text{ kJ/kg}$$

$$\eta_{th} = 1 - \frac{782.3}{1800} = 1 - 0.435 = 0.565 = 56.5\%$$

$$w_{net} = 1800 - 782.3 = 1017.7 \text{ kJ/kg} = (v_1 - v_2)\text{mep}$$

$$\text{mep} = \frac{1017.7}{(0.827 - 0.1034)} = 1406 \text{ kPa}$$

This is a high value for mean effective pressure, largely because the two constant-volume heat transfer processes keep the total volume change to a minimum (compared with a Brayton cycle, for example). Thus, the Otto cycle is a good model to emulate in the cylinder/piston internal combustion engine. At the other extreme, a low mean effective pressure means a large piston displacement for a given power output, which in turn means high frictional losses in an actual engine.

EXAMPLE 9.11E   The compression ratio in an air-standard Otto cycle is 8. At the beginning of the compression stroke the pressure is 14.7 lbf/in.² and the temperature is 60 F. The heat transfer to the air per cycle is 800 Btu/lbm air. Determine:

1. The pressure and temperature at the end of each process of the cycle
2. The thermal efficiency
3. The mean effective pressure

Control mass: Air inside cylinder.

Diagram: Fig. 9.28.

State information: $P_1 = 14.7$ lbf/in.² $T_1 = 520$ R.

Process of information: Four processes known (Fig. 9.28). Also, $r_v = 8$ and $q_H = 800$ Btu/lbm.

Model: Ideal gas, constant specific heat, value at 80 F.

Analysis:

Second law for compression process 1–2:

$$s_2 = s_1$$

so that

$$\frac{T_2}{T_1} = \left(\frac{V_1}{V_2}\right)^{k-1}$$

$$\frac{P_2}{P_1} = \left(\frac{V_1}{V_2}\right)^{k}$$

First law for heat addition process 2–3:

$$q_H = {_2}q_3 = u_3 - u_2 = C_v\left(T_3 - T_2\right)$$

Second law for expansion process 3–4:

$$s_4 = s_3$$

so that

$$\frac{T_3}{T_4} = \left(\frac{V_4}{V_3}\right)^{k-1}$$

$$\frac{P_3}{P_4} = \left(\frac{V_4}{V_3}\right)^{k}$$

In addition,

$$\eta_{th} = 1 - \frac{1}{r_v^{k-1}} \qquad mep = \frac{w_{net}}{(v_1 - v_2)}$$

Solution

$$v_1 = \frac{53.34 \times 520}{14.7 \times 144} = 13.08 \text{ ft}^3/\text{lbm}$$

$$\frac{T_2}{T_1} = \left(\frac{V_1}{V_2}\right)^{k-1} = 8^{0.4} = 2.3 \qquad T_2 = 2.3(520) = 1197 \text{ R}$$

$$\frac{P_2}{P_1} = \left(\frac{V_1}{V_2}\right)^{k} = 8^{1.4} = 18.4 \qquad P_2 = 18.4(14.7) = 270.3 \text{ lbf/in.}^2$$

$$v_2 = \frac{13.08}{8} = 1.637 \text{ ft}^3/\text{lbm}$$

$$_2q_3 = C_v(T_3 - T_2) = 800 \text{ Btu/lbm}$$

$$T_3 - T_2 = \frac{800}{0.171} = 4690 \text{ R} \qquad T_3 = 1197 + 4690 = 5887 \text{ R}$$

$$\frac{T_3}{T_2} = \frac{P_3}{P_2} = \frac{5887}{1197} = 4.92 \qquad P_3 = 4.92(270.3) = 1331 \text{ lbf/in.}^2$$

$$\frac{T_3}{T_4} = \left(\frac{V_4}{V_3}\right)^{k-1} = 8^{0.4} = 2.3 \qquad T_4 = \frac{5887}{2.3} = 2558 \text{ R}$$

$$\frac{P_3}{P_4} = \left(\frac{V_4}{V_3}\right)^{k} = 8^{1.4} = 18.4 \qquad P_4 = \frac{1331}{18.4} = 73.0 \text{ lbf/in.}^2$$

$$\eta_{th} = 1 - \frac{1}{r_v^{k-1}} = 1 - \frac{1}{8^{0.4}} = 1 - \frac{1}{2.3} = 1 - 0.435 = 0.565$$

This can be checked by finding the heat rejected.

$$_4q_1 = C_v(T_1 - T_4) = 0.171(520 - 2558) = -348 \text{ Btu/lbm}$$

$$\eta_{th} = 1 - \frac{348}{800} = 1 - 0.435 = 0.565$$

$$w_{net} = 800 - 348 = 452 \text{ Btu/lbm} = (v_1 - v_2)mep$$

$$mep = \frac{452 \times 778}{(13.08 - 1.637)144} = 213.5 \text{ lbf/in.}^2$$

This is a high value for mean effective pressure, largely because the two constant-volume heat transfer processes keeps the total volume change to a minimum (compared with a Brayton cycle, for example). Thus, the Otto cycle is a good model to emulate in the cylinder/piston internal combustion engine. At the other extreme, a low mean effective pressure means a large piston displacement for a given power output, which in turn means high frictional losses in an actual engine.

9.14 THE DIESEL CYCLE

The air-standard Diesel cycle is shown in Fig. 9.30. This is the ideal cycle for the Diesel engine, which is also called the compression-ignition engine.

In this cycle the heat is transferred to the working fluid at constant pressure. This process corresponds to the injection and burning of the fuel in the actual engine. Since the gas is expanding during the heat addition in the air-standard cycle, the heat transfer must be just sufficient to maintain constant pressure. When state 3 is reached the heat addition ceases and the gas undergoes an isentropic expansion, process 3–4, until the piston reaches crank-end dead center. As in the air-standard Otto cycle, a constant-volume rejection of heat at crank-end dead center replaces the exhaust and intake processes of the actual engine.

The efficiency of the Diesel cycle is given by the relation

$$\eta_{th} = 1 - \frac{Q_L}{Q_H} = 1 - \frac{C_v(T_4 - T_1)}{C_p(T_3 - T_2)} = 1 - \frac{T_1(T_4/T_1 - 1)}{kT_2(T_3/T_2 - 1)} \tag{9.6}$$

It is important to note that the isentropic compression ratio is greater than the isentropic expansion ratio in the Diesel cycle. In addition, for a given state before compression and a given compression ratio (that is, given states 1 and 2), the cycle efficiency decreases as the maximum temperature increases. This is evident from the T–s diagram, because the constant-pressure and constant-volume lines converge, and increasing the temperature from 3 to 3′ requires a large addition of heat (area 3–3′–c–b–3) and results in a relatively small increase in work (area 3–3′–4′–4–3).

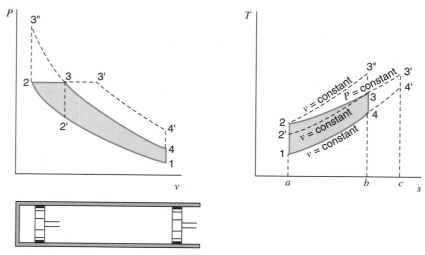

FIGURE 9.30 The air-standard diesel cycle.

There are a number of comparisons between the Otto cycle and the Diesel cycle, but here we will note only two. Consider Otto cycle 1–2–3″–4–1 and Diesel cycle 1–2–3–4–1, which have the same state at the beginning of the compression stroke and the same piston displacement and compression ratio. It is evident from the T–s diagram that the Otto cycle has the higher efficiency. In practice, however, the Diesel engine can operate on a higher compression ratio than the spark-ignition engine. The reason is that in the spark-ignition engine an air–fuel mixture is compressed, and detonation (spark knock) becomes a serious problem if too high a compression ratio is used. This problem does not exist in the Diesel engine because only air is compressed during the compression stroke.

Therefore, we might compare an Otto cycle with a Diesel cycle and in each case select a compression ratio that might be achieved in practice. Such a comparison can be made by considering Otto cycle 1–2′–3–4–1 and Diesel cycle 1–2–3–4–1. The maximum pressure and temperature are the same for both cycles, which means that the Otto cycle has a lower compression ratio than the Diesel cycle. It is evident from the T–s diagram that in this case the Diesel cycle has the higher efficiency. Thus, the conclusions drawn from a comparison of these two cycles must always be related to the basis on which the comparison has been made.

The actual compression-ignition open cycle differs from the air-standard Diesel cycle in much the same way that the spark-ignition open cycle differs from the air-standard Otto cycle.

EXAMPLE 9.12 An air-standard Diesel cycle has a compression ratio of 18, and the heat transferred to the working fluid per cycle is 1800 kJ/kg. At the beginning of the compression process the pressure is 0.1 MPa and the temperature is 15°C. Determine

1. The pressure and temperature at each point in the cycle

2. The thermal efficiency

3. The mean effective pressure

Control mass: Air inside cylinder.

Diagram: Figure 9.30.

State information: $P_1 = 0.1$ MPa, $T_1 = 288.2$ K.

Process information: Four processes known (Fig. 9.30). Also $r_v = 18$ and $q_H = 1800$ kJ/kg.

Model: Ideal gas, constant specific heat, value at 300 K.

Analysis:

Second law for compression process 1–2:

$$s_2 = s_1$$

so that

$$\frac{T_2}{T_1} = \left(\frac{V_1}{V_2}\right)^{k-1}$$

$$\frac{P_2}{P_1} = \left(\frac{V_1}{V_2}\right)^{k}$$

First law for heat addition process 2–3:

$$q_H = {}_2q_3 = C_p\left(T_3 - T_2\right)$$

Second law for expansion process 3–4:

$$s_4 = s_3$$

so that

$$\frac{T_3}{T_4} = \left(\frac{V_4}{V_3}\right)^{k-1}$$

In addition,

$$\eta_{th} = \frac{w_{net}}{q_H} \qquad mep = \frac{w_{net}}{v_1 - v_2}$$

Solution

$$v_1 = \frac{0.287 \times 288.2}{100} = 0.827 \text{ m}^3 / \text{kg}$$

$$v_2 = \frac{v_1}{18} = \frac{0.827}{18} = 0.045\,95 \text{ m}^3 / \text{kg}$$

$$\frac{T_2}{T_1} = \left(\frac{V_1}{V_2}\right)^{k-1} = 18^{0.4} = 3.1777 \qquad T_2 = 915.8 \text{ K}$$

$$\frac{P_2}{P_1} = \left(\frac{V_1}{V_2}\right)^{k} = 18^{1.4} = 57.2 \qquad P_2 = 5.72 \text{ MPa}$$

$$q_H = {}_2q_3 = C_p\left(T_3 - T_2\right) = 1800 \text{ kJ} / \text{kg}$$

$$T_3 - T_2 = \frac{1800}{1.0035} = 1794 \qquad T_3 = 2710 \text{ K}$$

$$\frac{V_3}{V_2} = \frac{T_3}{T_2} = \frac{2710}{915.8} = 2.959 \qquad v_3 = 0.135\,98 \text{ m}^3 / \text{kg}$$

$$\frac{T_3}{T_4} = \left(\frac{V_4}{V_3}\right)^{k-1} = \left(\frac{0.827}{0.135\,98}\right)^{0.4} = 2.0588 \qquad T_4 = 1316 \text{ K}$$

$$q_L = {}_4q_1 = C_v\left(T_1 - T_4\right) = 0.7165\left(288.2 - 1316\right) = -736.6 \text{ kJ} / \text{kg}$$

$$w_{net} = 1800 - 736.6 = 1063.4 \text{ kJ/kg}$$

$$\eta_{th} = \frac{w_{net}}{q_H} = \frac{1063.4}{1800} = 59.1\%$$

$$\text{mep} = \frac{w_{net}}{v_1 - v_2} = \frac{1063.4}{0.827 - 0.045\,95} = 1362 \text{ kPa}$$

9.15 THE STIRLING CYCLE

The final air-standard power cycle to be discussed is the Stirling cycle, which is shown on the P–v and T–s diagrams of Fig. 9.31. Heat is transferred to the working fluid during the constant-volume process 2–3 and also during the isothermal expansion process 3–4. Heat is rejected during the constant-volume process 4–1 and also during the isothermal compression process 1–2. Thus, this cycle is the same as the Otto cycle with the adiabatic processes of that cycle replaced with isothermal processes. Since the Stirling cycle includes two constant-volume heat transfer processes, keeping the total volume change during the cycle to a minimum, it is a good candidate for a cylinder/piston boundary work application; it should have a high mean effective pressure.

Stirling-cycle engines have been developed in recent years as external-combustion engines with regeneration. The significance of regeneration is noted from the ideal case shown in Fig. 9.31. Note that the heat transfer to the gas between states 2 and 3, area 2–3–b–a–2, is exactly equal to the heat transfer from the gas between states 4 and 1, area 1–4–d–c–1. Thus, in the ideal cycle, all external heat supplied Q_H takes place in the isothermal expansion process 3–4, and all external heat rejection Q_L takes place in the isothermal compression process 1–2. Since all heat is supplied and rejected isothermally, the efficiency of this cycle equals the efficiency of a Carnot cycle operating between the same temperatures. The same conclusions would be drawn in the case of an Ericsson cycle, which was discussed briefly in Section 9.11, if that cycle were to include a regenerator as well.

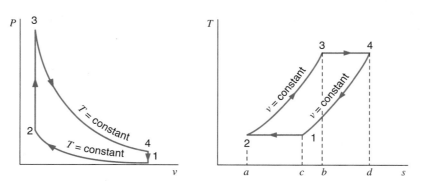

FIGURE 9.31 The air-standard Stirling cycle.

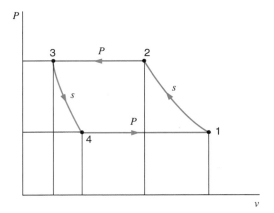

FIGURE 9.32 Four-process refrigeration cycle.

9.16 INTRODUCTION TO REFRIGERATION SYSTEMS

In Section 9.1, we discussed cyclic heat engines consisting of four separate processes, either SSSF or cylinder/piston boundary movement work devices. We further allowed for a working fluid that changes phase or for one that is single-phase throughout the cycle. We then considered a power system comprised of four reversible SSSF processes, two of which were constant-pressure heat transfer processes, for simplicity of equipment requirements, since these two processes involve no work. It was further assumed that the other two work-involved processes were adiabatic and therefore isentropic. The resulting power cycle appeared as Fig. 9.2.

We now consider the basic ideal refrigeration system cycle in exactly the same terms as those described above, except that each process is the reverse of that in the power cycle. The result is the ideal cycle shown in Fig. 9.32. Note that if the entire cycle takes place inside the two-phase liquid-vapor dome, the resulting cycle is, as with the power cycle, the Carnot cycle, since the two constant-pressure processes are also isothermal. Otherwise, this cycle is not a Carnot cycle. It is also noted, as before, that the net work input to the cycle is equal to the area enclosed by the process lines 1–2–3–4–1, independently of whether the individual processes are SSSF or cylinder/piston boundary movement.

In the next section, we make one modification to this idealized basic refrigeration system cycle in presenting and applying the model of refrigeration and heat pump systems.

9.17 THE VAPOR-COMPRESSION REFRIGERATION CYCLE

In this section, we consider the ideal refrigeration cycle for a working substance that changes phase during the cycle, in a manner equivalent to that done with the Rankine power cycle in Section 9.2. In doing so, we note that state 3 in Fig. 9.32 is saturated liquid at the condenser temperature and state 1 is saturated vapor at the evaporator temperature. This means that the isentropic expansion process from 3–4 will be in the two-phase

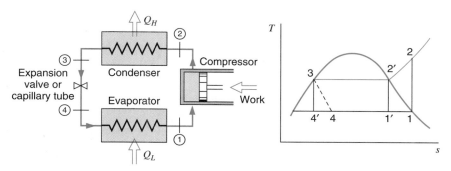

FIGURE 9.33 The ideal vapor-compression refrigeration cycle.

region, and mostly liquid. As a consequence, there will be very little work output from this process, such that it is not worth the cost of including this piece of equipment in the system. We therefore replace the turbine with a throttling device, usually a valve or a length of small-diameter tubing, by which the working fluid is throttled from the high-pressure to the low-pressure side. The resulting cycle becomes the ideal model for a vapor-compression refrigeration system, which is shown in Fig. 9.33. Saturated vapor at low pressure enters the compressor and undergoes a reversible adiabatic compression, process 1–2. Heat is then rejected at constant pressure in process 2–3, and the working fluid exits the condenser as saturated liquid. An adiabatic throttling process, 3–4, follows, and the working fluid is then evaporated at constant pressure, process 4–1, to complete the cycle.

The similarity of this cycle to the reverse of the Rankine cycle has already been noted. We also note the difference between this cycle and the ideal Carnot cycle, in which the working fluid always remains inside the two-phase region, 1′–2′–3–4′–1′. It is much more expedient to have a compressor handle only vapor than a mixture of liquid and vapor, as would be required in process 1′–2′ of the Carnot cycle. It is virtually impossible to compress, at a reasonable rate, a mixture such as that represented by state 1′ and still maintain equilibrium between liquid and vapor. The other difference, that of replacing the turbine by the throttling process, has already been discussed.

It should be pointed out that the system described in Fig. 9.33 can be used for either of two purposes. The first use is as a refrigeration system, in which case it is desired to maintain a space at a low temperature T_1 relative to the ambient temperature T_3. (In a real system, it would be necessary to allow a finite temperature difference in both the evaporator and condenser to provide a finite rate of heat transfer in each.) Thus, the reason for building the system in this case is the quantity q_L. The measure of performance of a refrigeration system is given in terms of the coefficient of performance, β, which was defined in Chapter 6 as

$$\beta = \frac{q_L}{w_c} \tag{9.7}$$

The second use of the system described in Fig. 9.33 is as a heat pump system, in which case it is desired to maintain a space at a temperature T_3 above that of the ambient (or other source) T_1. In this case, the reason for building the system is the quantity q_H, and the coefficient of performance for the heat pump, β', is now

$$\beta' = \frac{q_H}{w_c} \tag{9.8}$$

Refrigeration systems and heat pump systems are, of course, different in terms of design variables, but the analysis of the two is the same. When we discuss refrigerators in this and the following two sections, it should be kept in mind that the same comments generally apply to heat pump systems, as well.

EXAMPLE 9.13 Consider an ideal refrigeration cycle which uses R-12 as the working fluid. The temperature of the refrigerant in the evaporator is −20°C and in the condenser it is 40°C. The refrigerant is circulated at the rate of 0.03 kg/s. Determine the coefficient of performance and the capacity of the plant in rate of refrigeration.

The diagram for this example is as shown in Fig. 9.33. For each control volume analyzed, the thermodynamic model is the R-12 tables. Each process is SSSF with no changes in kinetic or potential energy.

    Control volume: Compressor.

    Inlet state: T_1 known, saturated vapor; state fixed.

    Exit state: P_2 known (saturation presssure at T_3).

Analysis:

First law:

$$w_c = h_2 - h_1$$

Second law:

$$s_2 = s_1$$

Solution

At $T_3 = 40°C$,

$$P_g = P_2 = 0.9607 \text{ MPa}$$

From the R-12 tables,

$$h_1 = 178.61 \qquad s_1 = 0.7082$$

Therefore,

$$s_2 = s_1 = 0.7082$$

so that

$$T_2 = 50.8°C \quad \text{and} \quad h_2 = 211.38$$

$$w_c = h_2 - h_1 = 211.38 - 178.61 = 32.77 \text{ kJ / kg}$$

    Control volume: Expansion valve.

    Inlet state: T_3 known, saturated liquid; state fixed.

    Exit state: T_4 known.

Analysis:

First law:

$$h_3 = h_4$$

Solution

$$h_4 = h_3 = 74.53$$

Control volume: Evaporator.

Inlet state: State 4 known (as given).

Exit state: State 1 known (as given).

Analysis:

First law:

$$q_L = h_1 - h_4$$

Solution

$$q_L = h_1 - h_4 = 178.61 - 74.53 = 104.08 \text{ kJ / kg}$$

Therefore,

$$\beta = \frac{q_L}{w_c} = \frac{104.08}{32.77} = 3.18$$

$$\text{Capacity} = 104.08 \times 0.03 = 3.12 \text{ kW}$$

9.18 WORKING FLUIDS FOR VAPOR-COMPRESSION REFRIGERATION SYSTEMS

A much larger number of different working fluids (refrigerants) are utilized in vapor-compression refrigeration systems than in vapor power cycles. Ammonia and sulfur dioxide were important in the early days of vapor-compression refrigeration, but both are highly toxic and therefore dangerous substances. For many years now, the principal refrigerants have been the halogenated hydrocarbons, which are marketed under the trade names of Freon and Genatron. For example, dichlorodifluoromethane (CCl_2F_2) is known as Freon-12 and Genatron-12, and therefore as refrigerant-12 or R-12. This group of substances, known commonly as chlorofluorocarbons or CFCs, are chemically very stable at ambient temperature, especially those lacking any hydrogen atoms. This characteristic is necessary for a refrigerant working fluid. This same characteristic, however, has devastating consequences if the gas, having leaked from an appliance into the atmosphere, spends many years slowly diffusing upward into the stratosphere. There it is broken down, releasing chlorine, which destroys the protective ozone layer of the stratosphere. It is therefore of overwhelming importance to us all to eliminate completely the widely used but life-threatening CFCs, particularly R-11 and R-12, and to develop suitable and acceptable replacements. The CFCs containing hydrogen (often termed HCFCs), such as R-22, have shorter atmospheric lifetimes, and therefore are not as likely to reach the stratosphere before being broken up and rendered harmless. The most desirable fluids, called HFCs, contain no chlorine atoms at all.

There are two important considerations when selecting refrigerant working fluids: the temperature at which refrigeration is needed and the type of equipment to be used.

As the refrigerant undergoes a change of phase during the heat transfer process, the pressure of the refrigerant will be the saturation pressure during the heat supply and heat rejection processes. Low pressures mean large specific volumes and correspondingly large equipment. High pressures mean smaller equipment, but it must be designed to withstand higher pressure. In particular, the pressures should be well below the critical pressure. For extremely low temperature applications a binary fluid system may be used by cascading two separate systems.

The type of compressor used has a particular bearing on the refrigerant. Reciprocating compressors are best adapted to low specific volumes, which means higher pressures, whereas centrifugal compressors are most suitable for low pressures and high specific volumes.

It is also important that the refrigerants used in domestic appliances be nontoxic. Other beneficial characteristics, in addition to being environmentally acceptable, are miscibility with compressor oil, dielectric strength, stability, and low cost. Refrigerants, however, have an unfortunate tendency to cause corrosion. For given temperatures during evaporation and condensation, not all refrigerants have the same coefficient of performance for the ideal cycle. It is, of course, desirable to use the refrigerant with the highest coefficient of performance, other factors permitting.

9.19 DEVIATION OF THE ACTUAL VAPOR-COMPRESSION REFRIGERATION CYCLE FROM THE IDEAL CYCLE

The actual refrigeration cycle deviates from the ideal cycle primarily because of pressure drops associated with fluid flow and heat transfer to or from the surroundings. The actual cycle might approach the one shown in Fig. 9.34.

The vapor entering the compressor will probably be superheated. During the compression process there are irreversibilities and heat transfer either to or from the surroundings, depending on the temperature of the refrigerant and the surroundings. Therefore, the entropy might increase or decrease during this process, for the irreversibility and the heat transferred to the refrigerant cause an increase in entropy, and the heat transferred from the refrigerant causes a decrease in entropy. These possibilities are represented by the two dashed lines 1–2 and 1–2′. The pressure of the liquid leaving the condenser will be less than the pressure of the vapor entering, and the temperature of the re-

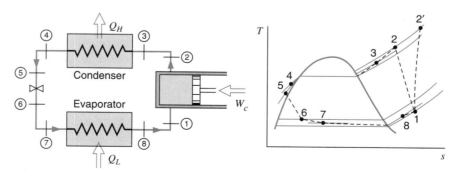

FIGURE 9.34 The actual vapor-compression refrigeration cycle.

frigerant in the condenser will be somewhat higher than that of the surroundings to which heat is being transferred. Usually the temperature of the liquid leaving the condenser is lower than the saturation temperature. It might drop somewhat more in the piping between the condenser and expansion valve. This represents a gain, however, because as a result of this heat transfer the refrigerant enters the evaporator with a lower enthalpy, which permits more heat to be transferred to the refrigerant in the evaporator.

There is some drop in pressure as the refrigerant flows through the evaporator. It may be slightly superheated as it leaves the evaporator, and through heat transferred from the surroundings its temperature will increase in the piping between the evaporator and the compressor. This heat transfer represents a loss, because it increases the work of the compressor, since the fluid entering it has an increased specific volume.

EXAMPLE 9.14 A refrigeration cycle utilizes R-12 as the working fluid. Following are the properties at various points of the cycle designated in Fig. 9.34.

$$P_1 = 125 \text{ kPa} \qquad T_1 = -10°C$$

$$P_2 = 1.2 \text{ MPa} \qquad T_2 = 100°C$$

$$P_3 = 1.19 \text{ MPa} \qquad T_3 = 80°C$$

$$P_4 = 1.16 \text{ MPa} \qquad T_4 = 45°C$$

$$P_5 = 1.15 \text{ MPa} \qquad T_5 = 40°C$$

$$P_6 = P_7 = 140 \text{ kPa} \qquad x_6 = x_7$$

$$P_8 = 130 \text{ kPa} \qquad T_8 = -20°C$$

The heat transfer from R-12 during the compression process is 4 kJ/kg. Determine the coefficient of performance of this cycle.

For each control volume, the model is the R-12 tables. Each process is SSSF with no changes in kinetic or potential energy.

Control volume: Compressor.

Inlet state: P_1, T_1 known; state fixed.

Exit state: P_2, T_2 known; state fixed.

Analysis:

First law:

$$q + h_1 = h_2 + w$$

$$w_c = -w = h_2 - h_1 - q$$

Solution

From the R-12 tables,

$$h_1 = 185.16 \qquad h_2 = 245.52$$

Therefore,

$$w_c = 245.52 - 185.16 - (-4) = 64.36 \text{ kJ}/\text{kg}$$

Control volume: Throttling valve plus line.

Inlet state: P_5, T_5 known; state fixed.

Exit state: $P_7 = P_6$ known, $x_7 = x_6$.

Analysis:

First law:

$$h_5 = h_6$$

Since $x_7 = x_6$, it follows that $h_7 = h_6$.

Solution

$$h_5 = h_6 = h_7 = 74.53$$

Control volume: Evaporator.

Inlet state: P_7, h_7 known (above).

Exit state: P_8, T_8 known; state fixed.

Analysis:

First law:

$$q_L = h_8 - h_7$$

Solution

$$q_L = h_8 - h_7 = 179.12 - 74.53 = 104.59 \text{kJ} / \text{kg}$$

Therefore,

$$\beta = \frac{q_L}{w_c} = \frac{104.59}{64.36} = 1.625$$

9.20 THE AMMONIA ABSORPTION REFRIGERATION CYCLE

The ammonia absorption refrigeration cycle differs from the vapor-compression cycle in the manner in which compression is achieved. In the absorption cycle the low-pressure ammonia vapor is absorbed in water and the liquid solution is pumped to a high pressure by a liquid pump. Figure 9.35 shows a schematic arrangement of the essential elements of such a system.

The low-pressure ammonia vapor leaving the evaporator enters the absorber where it is absorbed in the weak ammonia solution. This process takes place at a temperature slightly higher than that of the surroundings. Heat must be transferred to the surroundings during this process. The strong ammonia solution is then pumped through a heat exchanger to the generator where a higher pressure and temperature are maintained. Under these conditions ammonia vapor is driven from the solution as heat is transferred from a

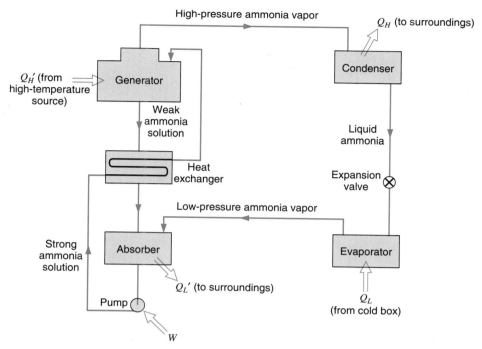

FIGURE 9.35 The ammonia-absorption refrigeration cycle.

high-temperature source. The ammonia vapor goes to the condenser where it is condensed, as in a vapor-compression system, and then to the expansion valve and evaporator. The weak ammonia solution is returned to the absorber through the heat exchanger.

The distinctive feature of the absorption system is that very little work input is required because the pumping process involves a liquid. This follows from the fact that for a reversible steady-flow process with negligible changes in kinetic and potential energy, the work is equal to $-\int v \, dP$ and the specific volume of the liquid is much less than the specific volume of the vapor. On the other hand, a relatively high-temperature source of heat must be available (100° to 200°C). There is more equipment in an absorption system than in a vapor-compression system, and it can usually be economically justified only when a suitable source of heat is available that would otherwise be wasted. In recent years, the absorption cycle has been given increased attention in connection with alternate energy sources, for example, solar energy or supplies of geothermal energy.

This cycle brings out the important principle that since the work in a reversible steady-flow process with negligible changes in kinetic and potential energy is $-\int v \, dP$, a compression process should take place with the smallest possible specific volume.

9.21 THE AIR-STANDARD REFRIGERATION CYCLE

If we consider the original ideal four-process refrigeration cycle of Fig. 9.32 with a non-condensing (gaseous) working fluid, then the work output during the isentropic expansion process is not negligibly small, as was the case with a condensing working fluid.

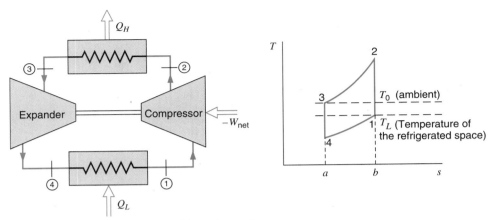

FIGURE 9.36 The air-standard refrigeration cycle.

Therefore, we retain the turbine in the four-SSSF process ideal air-standard refrigeration cycle shown in Fig. 9.36. This cycle is seen to be the reverse Brayton cycle, and is used in practice in the liquefaction of air and other gases and also in certain special situations that require refrigeration, such as aircraft cooling systems. After compression from states 1 to 2, the air is cooled as heat is transferred to the surroundings at temperature T_0. The air is then expanded in process 3–4 to the pressure entering the compressor, and the temperature drops to T_4 in the expander. Heat may then be transferred to the air until temperature T_L is reached. The work for this cycle is represented by area 1–2–3–4–1, and the refrigeration effect is represented by area 4–1–b–a–4. The coefficient of performance is the ratio of these two areas.

In practice, this cycle has been used to cool aircraft in an open cycle. A simplified form is shown in Fig. 9.37. Upon leaving the expander, the cool air is blown directly into the cabin, thus providing the cooling effect where needed.

When counterflow heat exchangers are incorporated, very low temperatures can be obtained. This is essentially the cycle used in low-pressure air liquefaction plants and in other liquefaction devices such as the Collins helium liquefier. The ideal cycle is as shown in Fig. 9.38. It is evident that the expander operates at very low temperature,

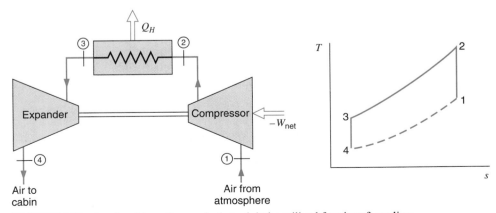

FIGURE 9.37 An air refrigeration cycle that might be utilized for aircraft cooling.

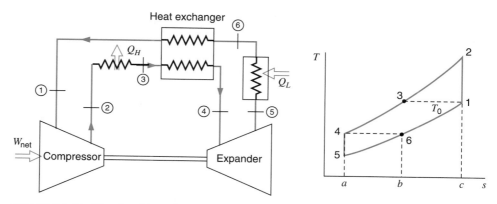

FIGURE 9.38 The air refrigeration cycle utilizing a heat exchanger.

which presents unique problems to the designer in providing lubrication and choosing materials.

EXAMPLE 9.15   Consider the simple air-standard refrigeration cycle of Fig. 9.36. Air enters the compressor at 0.1 MPa, −20°C, and leaves at 0.5 MPa. Air enters the expander at 15°C. Determine

1. The coefficient of performance for this cycle
2. The rate at which air must enter the compressor to provide 1 kW of refrigeration

For each control volume in this example, the model is ideal gas with constant specific heat, value at 300 K, and each process is SSSF with no kinetic or potential energy changes. The diagram for this example is Fig. 9.36.

> *Control volume:*   Compressor.
>
> *Inlet state:* P_1, T_1 known; state fixed.
>
> *Exit state:* P_2 known.

Analysis:

First law:

$$w_c = h_2 - h_1$$

(Here w_C designates work into the compressor.)

Second law:

$$s_1 = s_2$$

so that

$$\frac{T_2}{T_1} = \left(\frac{P_2}{P_1}\right)^{(k-1)/k}$$

Solution

$$\frac{T_2}{T_1} = \left(\frac{P_2}{P_1}\right)^{(k-1)/k} = 5^{0.286} = 1.5845 \qquad T_2 = 401.2 \text{ K}$$

$$w_c = h_2 - h_1 = C_p\left(T_2 - T_1\right)$$

$$= 1.0035\left(401.2 - 253.2\right) = 148.5 \text{ kJ/kg}$$

Control volume: Expander.

Inlet state: $P_3 (= P_2)$ known, T_3 known; state fixed.

Exit state: $P_4 (= P_1)$ known.

Analysis:

First law:

$$w_t = h_3 - h_4$$

Second law:

$$s_3 = s_4$$

so that

$$\frac{T_3}{T_4} = \left(\frac{P_3}{P_4}\right)^{(k-1)/k}$$

Solution

$$\frac{T_3}{T_4} = \left(\frac{P_3}{P_4}\right)^{(k-1)/k} = 5^{0.286} = 1.5845 \qquad T_4 = 181.9 \text{ K}$$

$$w_t = h_3 - h_4 = 1.0035\left(288.2 - 181.9\right) = 106.7 \text{ kJ/kg}$$

Control volume: High-temperature heat exchanger.

Inlet state: State 2 known (as given).

Exit state: State 3 known (as given).

Analysis:

First law:

$$q_H = h_2 - h_3 \quad \text{(heat rejected)}$$

Solution

$$q_H = h_2 - h_3 = C_p\left(T_2 - T_3\right) = 1.0035\left(401.2 - 288.2\right) = 113.4 \text{ kJ/kg}$$

Control volume: Low-temperature heat exchanger.

Inlet state: State 4 known (as given).

Exit state: State 1 known (as given).

Analysis:

First law:

$$q_L = h_1 - h_4$$

Solution

$$q_L = h_1 - h_4 = C_p(T_1 - T_4) = 1.0035(253.2 - 181.9) = 71.6 \text{ kJ/kg}$$

Therefore,

$$w_{net} = w_c - w_t = 148.5 - 106.7 = 41.8 \text{ kJ/kg}$$

$$\beta = \frac{q_L}{w_{net}} = \frac{71.6}{41.8} = 1.713$$

To provide 1 kW of refrigeration capacity, we have

$$\dot{m} = \frac{\dot{Q}_L}{q_L} = \frac{1}{71.6} = 0.014 \text{ kg/s}$$

EXAMPLE 9.15E Consider the simple air-standard refrigeration cycle of Fig. 9.36. Air enters the compressor at 14.7 lbf/in.^{2}, 0 F, and leaves at 80 lbf/in.^{2} Air enters the expander at 60 F. Determine

1. The coefficient of performance for this cycle
2. The rate at which air must enter the compressor in order to provide 1 kW of refrigeration

For each control volume in this example, the model is ideal gas with constant specific heat, value at 80 F, and each process is SSSF with no kinetic or potential energy changes. The diagram for this example is Fig. 9.36.

Control volume: Compressor.

Inlet state: P_1, T_1 known; state fixed.

Exit state: P_2 known.

Analysis

First law:

$$w_c = h_2 - h_1 \qquad \text{(W_c designates work into the compressor)}$$

Second law:

$$s_1 = s_2$$

so that

$$\frac{T_2}{T_1} = \left(\frac{P_2}{P_1}\right)^{(k-1)/k}$$

Solution

$$\frac{T_2}{T_1} = \left(\frac{P_2}{P_1}\right)^{(k-1)/k} = \left(\frac{80}{14.7}\right)^{0.286} = 1.624 \qquad T_2 = 747 \text{ R}$$

$$w_c = h_2 - h_1 = C_p(T_2 - T_1) = 0.24(747 - 460) = 68.9 \text{ Btu / lbm}$$

Control volume: Expander.

Inlet state: $P_3 (= P_2)$ known, T_3 known; state fixed.

Exit state: $P_4 (= P_1)$ known.

Analysis:

First law:

$$w_t = h_3 - h_4$$

Second law:

$$s_3 = s_4$$

so that

$$\frac{T_3}{T_4} = \left(\frac{P_3}{P_4}\right)^{(k-1)/k}$$

Solution

$$\frac{T_3}{T_4} = \left(\frac{P_3}{P_4}\right)^{(k-1)/k} = \left(\frac{80}{14.7}\right)^{0.286} = 1.624 \qquad T_4 = 320 \text{ R}$$

$$w_t = h_3 - h_4 = 0.24(520 - 320) = 48.0 \text{ Btu / lbm}$$

Control volume: High-temperature heat exchanger.

Inlet state: State 2 known (as given).

Exit state: State 3 known (as given).

Analysis:

First law:

$$q_H = h_2 - h_3 \qquad \text{(heat rejected)}$$

Solution

$$q_H = h_2 - h_3 = C_p(T_2 - T_3) = 0.24(747 - 520) = 54.5 \text{ Btu / lbm}$$

Control volume: Low-temperature heat exchanger.

Inlet state: State 4 known (as given).

Exit state: State 1 known (as given).

Analysis:

First law:

$$q_L = h_1 - h_4$$

Solution

$$q_L = h_1 - h_4 = C_p\left(T_1 - T_4\right) = 0.24\left(460 - 320\right) = 33.6 \text{ Btu} / \text{lbm}$$

Therefore,

$$w_{\text{net}} = w_c - w_t = 68.9 - 48.0 = 20.9 \text{ Btu} / \text{lbm}$$

$$\beta = \frac{q_L}{w_{\text{net}}} = \frac{33.6}{20.9} = 1.61$$

In order to provide 1 kW of refrigeration capacity (3412 Btu/h)

$$\dot{m} = \frac{\dot{Q}_L}{q_L} = \frac{3412}{33.6} = 101.5 \text{ lbm} / \text{h}$$

9.22 COMBINED-CYCLE POWER AND REFRIGERATION SYSTEMS

There are many situations in which it is desirable to combine two cycles in series, either power systems or refrigeration systems, in order to take advantage of a very wide temperature range or to utilize what would otherwise be waste heat to improve efficiency. One combined power cycle, shown in Fig. 9.39 as a simple steam cycle with a mercury-topping cycle, is often referred to as a binary cycle. The advantage of this combined system

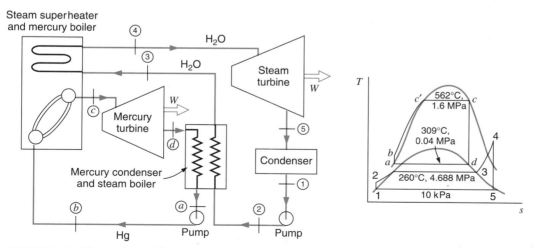

FIGURE 9.39 Mercury/water binary power system.

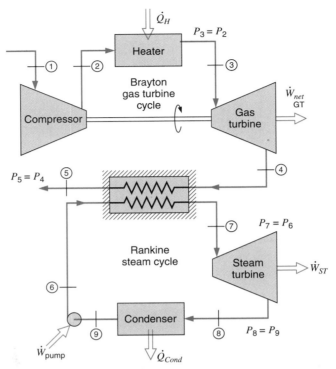

FIGURE 9.40 Combined Brayton/Rankine cycle power system.

is that mercury has a very low vapor pressure relative to that for water; therefore it is possible for an isothermal boiling process in the mercury to take place at a high temperature, much higher than the critical temperature of water, but still at a moderate pressure. The mercury condenser then provides an isothermal heat source as input to the steam boiler, such that the two cycles can be closely matched by proper selection of the cycle variables, with the resulting combined cycle then having a high thermal efficiency. Saturation pressures and temperatures for a typical mercury/water binary cycle are shown in the T–s diagram of Fig. 9.39.

A different type of combined cycle that has seen considerable attention is to use the "waste heat" exhaust from a Brayton cycle gas turbine engine (or another combustion engine such as a Diesel engine) as the heat source for a steam or other vapor power cycle, in which case the vapor cycle acts as a bottoming cycle for the gas engine, in order to improve the overall thermal efficiency of the combined power system. Such a system, utilizing a gas turbine and a steam Rankine cycle, is shown in Fig. 9.40. In such a combination, there is a natural mismatch using the cooling of a noncondensing gas as the energy source to effect an isothermal boiling process plus superheating the vapor, and careful design is required in order to avoid a pinch point, a condition at which the gas has cooled to the vapor boiling temperature without having provided sufficient energy to complete the boiling process.

One way to take advantage of the cooling exhaust gas in the Brayton cycle portion of the combined system is to utilize a mixture as the working fluid in the Rankine cycle.

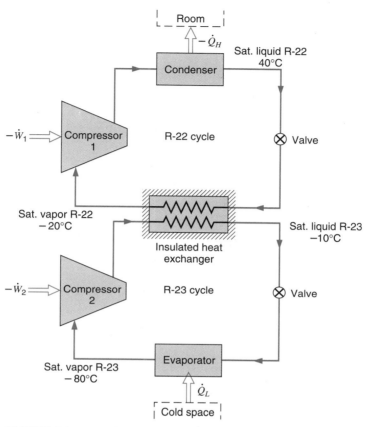

FIGURE 9.41 Combined-cycle cascade refrigeration system.

An example of this type of application is the Kalina cycle, which uses ammonia–water mixtures as the working fluid in the Rankine-type cycle. Such a cycle can be made very efficient, inasmuch as the temperature differences between the two fluid streams can be controlled through careful design of the combined system.

Combined cycles are used in refrigeration systems in cases in which there is a very large temperature difference between the ambient and the refrigerated space. Such a refrigeration system is often called a cascade system, an example of which is shown in Fig. 9.41. In this case, the refrigerant R-22 is used in the refrigeration system rejecting heat to the ambient, while its evaporator picks up the heat rejected in the low-temperature system condenser, the low temperature working fluid in this case being R-23, whose thermodynamic properties are suited to work as a refrigerant in this low-temperature range. As with the other combined-cycle systems, the working fluids and design variables must be considered very carefully, in order to optimize the performance of each unit.

We have described only a few combined-cycle systems here, as examples of the types of applications that can be dealt with, and the resulting improvement in overall performance that can occur. Obviously, there are many other combinations of power and refrigeration systems, and some of these are discussed in the problems at the end of the chapter.

PROBLEMS

9.1 A steam power plant has a boiler exit at 4 MPa, 500°C and a condenser exit temperature of 45°C. Assume all components are ideal and find the cycle efficiency and the specific work and heat transfer in the components.

9.2 Consider a solar-energy-powered ideal Rankine cycle that uses water as the working fluid. Saturated vapor leaves the solar collector at 175°C, and the condenser pressure is 10 kPa. Determine the thermal efficiency of this cycle.

9.3 A steam power plant operating in an ideal Rankine cycle has a high pressure of 5 MPa and a low pressure of 15 kPa. The turbine exhaust state should have a quality of at least 95% and the turbine power generated should be 7.5 MW. Find the necessary boiler exit temperature and the total mass flow rate.

9.4 A supply of geothermal hot water is to be used as the energy source in an ideal Rankine cycle, with R-134a as the cycle working fluid. Saturated vapor R-134a leaves the boiler at a temperature of 85°C, and the condenser temperature is 40°C.

   a. Calculate the thermal efficiency of this cycle.

   b. Repeat the calculation for R-22 as the working fluid.

9.5 Consider the ammonia Rankine-cycle power plant shown in Fig. P9.5, a plant that was designed to operate in a location where the ocean water temperature is 25°C near the surface and 5°C at some greater depth.

   a. Determine the turbine power output and the pump power input for the ammonia cycle.

   b. Determine the mass flow rate of water through each heat exchanger.

   c. What is the thermal efficiency of this power plant?

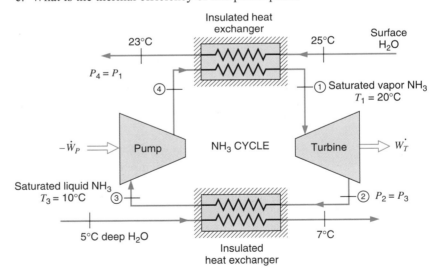

FIGURE P9.5

9.6 Consider the boiler in Problem 9.4 where the geothermal hot water should bring the R-134a (R-22) to saturated vapor. Assume a counter flowing heat exchanger arrangement. The geothermal water temperature should be equal to or greater than the R-134a (R-22) temperature at any location throughout the heat exchanger.

The point with the smallest temperature difference between the source and the working fluid is called the pinch point. If 2 kg/s of geothermal water is available at 95°C, what is the maximum power output of this cycle for R-134a as the working fluid?

9.7 Do the previous problem with R-22 as the working fluid.

9.8 A steam power plant as shown in Fig. 9.3 operating on the Rankine cycle has steam at 3.5 MPa, 400°C leaving the boiler. The turbine exhausts to the condenser operating at 10 kPa. Find the specific work and heat transfer in each of the ideal components and the cycle efficiency.

9.9 (Adv.) Find the availability of the water at all four states in the Rankine cycle described in Problem 9.8. Assume that the high-temperature source is 500°C and the low-temperature reservoir is at 25°C. Determine the flow of availability in or out of the reservoirs per kilogram of steam flowing in the cycle. What is the overall cycle second law efficiency?

9.10 The effect of turbine exhaust pressure on the performance of the ideal steam Rankine cycle given in Problem 9.8 is to be studied. Calculate the thermal efficiency of the cycle and the moisture content of the steam leaving the turbine for turbine exhaust pressures of 5, 10, 50, and 100 kPa. Plot the thermal efficiency versus turbine exhaust pressure for the specified turbine inlet pressure and temperature.

9.11 The effect of turbine inlet pressure on the performance of the ideal steam Rankine cycle given in Problem 9.8 is to be studied. Calculate the thermal efficiency of the cycle and the moisture content of the steam leaving the turbine for turbine inlet pressures of 1, 3.5, 6, and 10 MPa. Plot the thermal efficiency versus turbine inlet pressure for the specified turbine inlet temperature and exhaust pressure.

9.12 The effect of turbine inlet temperature on the performance of the ideal steam Rankine cycle given in Problem 9.8 is to be studied. Calculate the thermal efficiency of the cycle and the moisture content of the steam leaving the turbine for turbine inlet temperatures of 400°, 500°, 800°C, and saturated vapor (at 3.5 MPa). Plot the thermal efficiency versus turbine inlet temperature for the specified turbine inlet pressure and exhaust pressure.

9.13 Consider a simple ideal Rankine cycle that uses steam as the working fluid. The high-pressure side of the cycle is at a supercritical pressure. Such a cycle has a potential advantage of minimizing local temperature differences between the fluids in the steam generator, such as the instance in which the high-temperature energy source is the hot exhaust gas from a gas-turbine engine. Calculate the thermal efficiency of the cycle if the state entering the turbine is 25 MPa, 500°C, and the condenser pressure is 5 kPa. What is the steam quality at the turbine exit?

9.14 Consider an ideal steam reheat cycle in which the steam enters the high-pressure turbine at 3.5 MPa, 400°C, and then expands to 0.8 MPa. It is then reheated to 400°C and expands to 10 kPa in the low-pressure turbine. Calculate the thermal efficiency of the cycle and the moisture content of the steam leaving the low-pressure turbine.

9.15 The effect of the reheat pressure on the ideal steam reheat cycle is to be studied. Repeat Problem 9.14, using various values other than 0.8 MPa for the reheat pressure.

9.16 The effect of a number of reheat stages on the ideal steam reheat cycle is to be studied. Repeat Problem 9.14 using two reheat stages, one stage at 1.2 MPa and the second at 0.2 MPa, instead of the single reheat stage at 0.8 MPa.

9.17 A closed feedwater heater in a regenerative steam power cycle heats 20 kg/s of water from 100°C, 20 MPa to 250°C, 20 MPa. The extraction steam from the turbine enters the heater at 4 MPa, 275°C, and leaves as saturated liquid. What is the required mass flow rate of the extraction steam?

9.18 A power plant with one closed feedwater heater has a condenser temperature of 45°C, a maximum pressure of 5 MPa, and boiler exit temperature of 900°C. Extraction steam at 1 MPa to the feedwater heater condenses and is pumped up to the 5 MPa feedwater line where all the water goes to the boiler at 200°C. Find the fraction of extraction steam flow and the two specific pump work inputs.

9.19 Consider an ideal steam regenerative cycle in which steam enters the turbine at 3.5 MPa, 400°C, and exhausts to the condenser at 10 kPa. Steam is extracted from the turbine at 0.8 MPa and also at 0.2 MPa for heating the boiler feedwater in two open feedwater heaters. The feedwater leaves each heater at the temperature of the condensing steam. The appropriate pumps are used for the water leaving the condenser and the two feedwater heaters. Calculate the thermal efficiency of the cycle and the net work per kilogram of steam.

9.20 Repeat Problem 9.19, but assume closed instead of open feedwater heaters. A single pump is used to pump the water leaving the condenser up to the boiler pressure of 3.5 MPa. Condensate from the high-pressure heater is drained through a trap to the low-pressure heater, and that from the low-pressure heater is drained through a trap to the condenser.

9.21 The effect of a number of open feedwater heaters on the thermal efficiency of an ideal cycle is to be studied. Steam leaves the steam generator at 20 MPa, 600°C, and the cycle has a condenser pressure of 10 kPa. Determine the thermal efficiency for each of the following cases.

   a. No feedwater heater.

   b. One feedwater heater operating at 1 MPa.

   c. Two feedwater heaters, one operating at 3 MPa and the other at 0.2 MPa.

9.22 A steam power plant has high and low pressures of 25 MPa and 10 kPa, and one open feedwater heater operating at 1 MPa with the exit as saturated liquid. The maximum temperature is 800°C and the turbine has a total power output of 5 MW. Find the fraction of the flow for extraction to the feedwater and the total condenser heat transfer rate.

9.23 Consider an ideal steam combined reheat and regenerative cycle in which steam enters the high-pressure turbine at 3.5 MPa, 400°C, and is extracted for feedwater heating at 0.8 MPa. The remainder of the steam is reheated to 400°C at this pressure, 0.8 MPa, and is fed to the low-pressure turbine. Steam is extracted from the low-pressure turbine at 0.2 MPa for feedwater heating. The condenser pressure is 10 kPa. Both feedwater heaters are open heaters. Calculate the thermal efficiency of the cycle and the net work per kilogram of steam.

9.24 An ideal steam power plant is designed to operate on the combined reheat and regenerative cycle and to produce a net power output of 10 MW. Steam enters the high-pressure turbine at 8 MPa, 550°C, and is expanded to 0.6 MPa, at which pressure some of the steam is fed to an open feedwater heater, and the remainder is reheated to 550°C. The reheated steam is then expanded in the low-pressure turbine to 10 kPa.

  a. Determine the steam flow rate to the high-pressure turbine.

  b. Determine the size of motor required to drive each of the pumps.

  c. If the increase in the condenser cooling water temperature is restricted to a maximum of 10°C, what is the flow rate of the cooling water?

  d. If the steam velocity in the turbine-condenser connecting pipe is restricted to a maximum of 100 m/s, what is the diameter of the connecting pipe?

9.25 A steam power cycle has a high pressure of 3.5 MPa and a condenser exit temperature of 45°C. The turbine efficiency is 85%, and other cycle components are ideal. If the boiler superheats to 800°C, find the cycle thermal efficiency.

9.26 Repeat Problem 9.22 assuming the turbine has an isentropic efficiency of 85%.

9.27 Steam leaves a power plant steam generator at 3.5 MPa, 400°C, and enters the turbine at 3.4 MPa, 375°C. The isentropic turbine efficiency is 88%, and the turbine exhaust pressure is 10 kPa. Condensate leaves the condenser and enters the pump at 35°C, 10 kPa. The isentropic pump efficiency is 80%, and the discharge pressure is 3.7 MPa. The feedwater enters the steam generator at 3.6 MPa, 30°C. Calculate the following.

  a. The thermal efficiency of the cycle.

  b. The irreversibility of the process in the line between the steam generator exit and the turbine inlet, assuming an ambient temperature of 25°C.

9.28 For the steam power plant described in Problem 9.1, assume the isentropic efficiencies of the turbine and pump are 85% and 80%, respectively. Find the component specific work and heat transfers and the cycle efficiency.

9.29 Find the availability of the water at all the states in the steam power plant described in the previous problem. Assume the heat source in the boiler is at 600°C and the low-temperature reservoir is at 25°C. Give the second law efficiency of all the components.

9.30 In a particular reheat-cycle power plant, steam enters the high-pressure turbine at 5 MPa, 450°C and expands to 0.5 MPa, after which it is reheated to 450°C. The steam is then expanded through the low-pressure turbine to 7.5 kPa. Liquid water leaves the condenser at 30°C, is pumped to 5 MPa, and then returned to the steam generator. Each turbine is adiabatic with an isentropic efficiency of 87% and the pump efficiency is 82%. If the total power output of the turbines is 10 MW, determine

  a. The mass flow rate of steam

  b. The pump power input

  c. The thermal efficiency of the power plant

9.31 A supercritical steam power plant has a high pressure of 30 MPa and an exit condenser temperature of 50°C. The maximum temperature in the boiler is 1000°C. The turbine receives a flow rate of $\dot{m}_{tot}$, the exhaust is found to be saturated vapor and it produces 25 MW. There is one open feedwater heater receiving extraction from the turbine, $\dot{m}_1$ at 1MPa, and its exit is saturated liquid flowing to pump 2.

a. Find the overall turbine isentropic efficiency.

b. Find the ratio of the extraction mass flow to total flow, $\dot{m}_1/\dot{m}_{tot}$, assuming the same isentropic efficiency for the first section as the overall.

c. Find the mass flow rate out of the boiler, $\dot{m}_{tot}$.

d. What is the boiler inlet temperature with and without the feedwater heater?

9.32 In one type of nuclear power plant, heat is transferred in the nuclear reactor to liquid sodium. The liquid sodium is then pumped through a heat exchanger where heat is transferred to boiling water. Saturated vapor steam at 5 MPa exits this heat exchanger and is then superheated to 600°C in an external gas-fired superheater. The steam enters the turbine, which has one (open-type) feedwater extraction at 0.4 MPa. The isentropic turbine efficiency is 87%, and the condenser pressure is 7.5 kPa. Determine the heat transfer in the reactor and in the superheater to produce a net power output of 1 MW.

9.33 A certain industrial application has the following steam requirement: one 10-kg/s stream at a pressure of 0.5 MPa and one 5-kg/s stream at 1.4 MPa (both saturated or slightly superheated vapor). Instead of installing a low-pressure boiler to generate this steam, the designers plan is to obtain it by cogeneration, whereby the steam required will be the exhaust from a steam turbine. A high-pressure boiler generating steam at 10 MPa, 500°C, is therefore installed to supply steam to the turbine. The required amount is withdrawn at 1.4 MPa, and the remainder is expanded in the low-pressure end of the turbine to 0.5 MPa. Assuming that the high-pressure and the low-pressure turbines each have an isentropic efficiency of 85%, determine the following.

a. The power output of the turbine.

b. The heat-transfer rate required in the boiler compared with the rate needed were the steam generated in a low-pressure boiler without cogeneration. Assume that for each, 20°C liquid water is pumped to the required pressure and fed to the boiler.

9.34 In a cogenerating steam power plant the turbine receives steam from a high-pressure steam drum and a low-pressure steam drum as shown in Fig. P9.34. The condenser is made as two closed heat exchangers used to heat water running in a separate loop for district heating. The high-temperature heater adds 30 MW and the low-temperature heaters adds 31 MW to the district heating water flow. Find the power cogenerated by the turbine and the temperature in the return line to the deaerator.

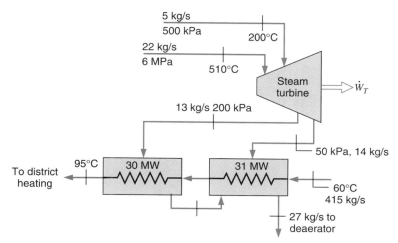

FIGURE P9.34

9.35 A boiler delivers steam at 10 MPa, 550°C to a two-stage turbine. After the first stage, 25% of the steam is extracted at 1.4 MPa for a process application and returned as compressed liquid at 1 MPa, 90°C to the feedwater line. The remainder of the steam continues through the low-pressure turbine stage, which exhausts to the condenser at 10 kPa. One pump after the condenser brings the feedwater to 1 MPa, where it is mixed with the return line from the process application and a second pump brings it up to 10 MPa. Assume the first and second stages in the steam turbine have isentropic efficiencies of 85% and 80%. If the process application requires 5 MW of power, how much power can then be cogenerated by the turbine? Assume that both pumps are ideal.

9.36 Consider an ideal air-standard Brayton cycle in which the pressure and temperature of the air entering the compressor are 100 kPa, 20°C, and the pressure ratio across the compressor is 12 to 1. The maximum temperature in the cycle is 1100°C, and the air flow rate is 10 kg/s. Assume constant specific heat for the air, with the value from Table A.10.

    a. Determine the compressor work, the turbine work, and the thermal efficiency of the cycle.

    b. If this cycle were to be used for a reciprocating machine, what would be the mean effective pressure? Would you recommend the Brayton cycle for a reciprocating machine?

9.37 Repeat Problem 9.36, but assume variable specific heat for the air.

9.38 An ideal regenerator is incorporated into the ideal air-standard Brayton cycle of Problem 9.36. Calculate the thermal efficiency of the cycle with this modification.

9.39 Consider a large stationary gas-turbine power plant that operates on the ideal Brayton cycle and delivers a power output of 100 MW to an electric generator. The minimum temperature in the cycle is 300 K, and the maximum temperature is 1600 K. The minimum pressure in the cycle is 100 kPa, and the compressor pressure ratio is 14 to 1.

a. Calculate the power output of the turbine. What fraction of the turbine output is required to drive the compressor?

b. What is the thermal efficiency of the cycle?

9.40 Repeat Problem 9.39, but assume that the compressor has an isentropic efficiency of 85% and the turbine an isentropic efficiency of 88%.

9.41 Repeat Problem 9.40, but include a regenerator with 75% efficiency in the cycle.

9.42 A gas turbine with air as the working fluid has two ideal turbine sections, as shown in Fig. P9.42, the first of which drives the ideal compressor, with the second producing the power output. The compressor input is at 290 K, 100 kPa, and the exit is at 450 kPa. A fraction of flow, x, bypasses the burner and the rest $(1 - x)$ goes through the burner where 1200 kJ/kg is added by combustion. The two flows then mix before entering the first turbine and continue through the second turbine, with exhaust at 100 kPa. If the mixing should result in a temperature of 1000 K into the first turbine find the fraction x. Find the required pressure and temperature into the second turbine and its specific power output.

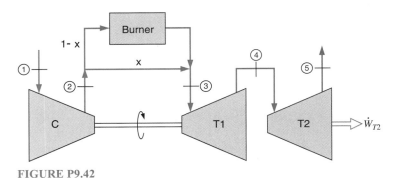

FIGURE P9.42

9.43 The gas-turbine cycle shown in Fig. P9.43 is to be used as an automotive engine. In the first turbine, the gas expands to pressure P_5, just low enough for this turbine to drive the compressor. The gas is then expanded through the second turbine connected to the drive wheels. The data for this engine are shown in the figure. Consider the working fluid to be air throughout the entire cycle, and assume that all processes are ideal. Determine

a. The intermediate pressure P_5.

b. The net specific work output of the engine, and the mass flow rate through the engine.

c. The air temperature entering the burner T_3, and the thermal efficiency of the engine.

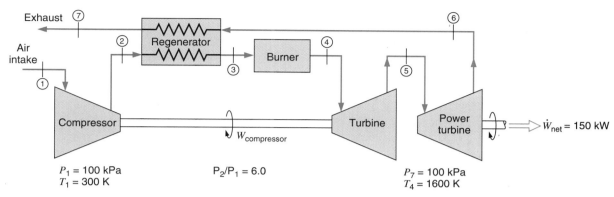

Exhaust ⑦

Air intake

② Regenerator

Burner

④

⑤

⑥

①

Compressor

$W_{compressor}$

Turbine

Power turbine

$\dot{W}_{net}$ = 150 kW

③

P_1 = 100 kPa
T_1 = 300 K

P_2/P_1 = 6.0

P_7 = 100 kPa
T_4 = 1600 K

FIGURE P9.43

9.44 Repeat Problem 9.43, but assume that the compressor has an efficiency of 82%, that both turbines have efficiencies of 87%, and that the regenerator efficiency is 70%. Also assume that friction causes pressure drops in the burner and on both sides of the regenerator. In each case, the pressure drop is estimated to be 2% of the inlet pressure to that component of the system.

9.45 Consider an ideal gas-turbine cycle with two stages of compression and two stages of expansion. The pressure ratio across each compressor stage and each turbine stage is 8 to 1. The pressure at the entrance to the first compressor is 100 kPa, the temperature entering each compressor is 20°C, and the temperature entering each turbine is 1100°C. An ideal regenerator is also incorporated into the cycle. Determine the compressor work, the turbine work, and the thermal efficiency of the cycle.

9.46 Repeat Problem 9.45, but assume that each compressor stage and each turbine stage has an isentropic efficiency of 85%. Also assume that the regenerator has an efficiency of 70%.

9.47 A two-stage air compressor has an intercooler between the two stages as shown in Fig. P9.47. The inlet state is 100 kPa, 290 K, and the final exit pressure is 1.6 MPa. Assume that the constant pressure intercooler cools the air to the inlet temperature, $T_3 = T_1$. It can be shown, see Problem 7.199, that the optimal pressure, $P_2 = \sqrt{P_1 P_4}$, for minimum total compressor work. Find the specific compressor works and the intercooler heat transfer for the optimal P_2.

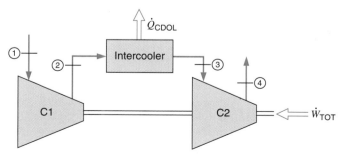

$\dot{Q}_{CDOL}$

① Intercooler ② ③ ④

C1

C2

$\dot{W}_{TOT}$

FIGURE P9.47

9.48 Repeat Problem 9.47 when the intercooler brings the air temperature to a known $T_3 = 320$ K. The corrected formula for the optimal pressure, see Problem 7.200, is

$$P_2 = \sqrt{P_1 P_4 (T_3 / T_1)^{n/(n-1)}}$$

where n is the exponent in the assumed polytropic process.

9.49 Consider an ideal air-standard Ericsson cycle that has an ideal regenerator as shown in Fig. P9.49. The high pressure is 1 MPa and the cycle efficiency is 70%. Heat is rejected in the cycle at a temperature of 300 K, and the cycle pressure at the beginning of the isothermal compression process is 100 kPa. Determine the high temperature, the compressor work, and the turbine work per kilogram of air.

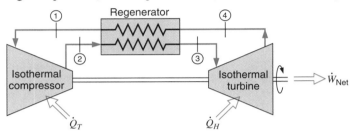

FIGURE P9.49

9.50 An air-standard Ericsson cycle has an ideal regenerator. Heat is supplied at 1000°C and heat is rejected at 20°C. Pressure at the beginning of the isothermal compression process is 70 kPa. The heat added is 600 kJ/kg. Find the compressor work, the turbine work, and the cycle efficiency.

9.51 Consider an ideal air-standard cycle for a gas-turbine, jet propulsion unit, such as that shown in Fig. 9.27. The pressure and temperature entering the compressor are 90 kPa, 290 K. The pressure ratio across the compressor is 14 to 1, and the turbine inlet temperature is 1500 K. When the air leaves the turbine, it enters the nozzle and expands to 90 kPa. Determine the pressure at the nozzle inlet and the velocity of the air leaving the nozzle.

9.52 The turbine in a jet engine receives air at 1250 K, 1.5 MPa. It exhausts to a nozzle at 250 kPa, which in turn exhausts to the atmosphere at 100 kPa. The isentropic efficiency of the turbine is 85% and the nozzle efficiency is 95%. Find the nozzle inlet temperature and the nozzle exit velocity. Assume negligible kinetic energy out of the turbine.

9.53 Repeat Problem 9.51, but assume that the isentropic compressor efficiency is 87%, the isentropic turbine efficiency is 89%, and the isentropic nozzle efficiency is 96%.

9.54 A jet aircraft is flying at an altitude of 4900 m, where the ambient pressure is approximately 55 kPa and the ambient temperature is −18°C. The velocity of the aircraft is 280 m/s, and the pressure ratio across the compressor is 14. Devise an air-standard cycle that approximates this engine, and determine the velocity (relative to the aircraft) of the air leaving the engine. Assume that the air has been expanded to the ambient pressure. In addition, assume a maximum gas temperature of 1450 K in the engine.

9.55 Air flows into a gasoline engine at 95 kPa, 300 K. The air is then compressed with a volumetric compression ratio of 8:1. In the combustion process 1300 kJ/kg of energy is released as the fuel burns. Find the temperature and pressure after combustion.

9.56 A gasoline engine has a volumetric compression ratio of 9. The state before compression is 290 K, 90 kPa, and the peak cycle temperature is 1800 K. Find the pressure after expansion, the cycle net work and the cycle efficiency using properties from Table A.12.

9.57 A stoichiometric mixture of gasoline and air has an energy release upon combustion of approximately 2800 kJ/kg of the mixture. To approximate an actual spark-ignition engine using such a mixture, consider an air-standard Otto cycle that has a heat addition of 2800 kJ/kg of air, a compression ratio of 7, and a pressure and temperature at the beginning of the compression process of 90 kPa, 10°C. Assuming constant specific heat, with the value from Table A.10, determine

a. The maximum pressure and temperature of the cycle.

b. The thermal efficiency of the cycle.

c. The mean effective pressure.

9.58 Repeat Problem 9.57, but assume variable specific heat. The ideal gas air tables, Table A.12, are recommended for this calculation (and the specific heat from Fig. 5.10 at high temperature).

9.59 When methanol produced from coal is considered as an alternative fuel to gasoline for automotive engines, it is recognized that the engine can be designed with a higher compression ratio, say 10 instead of 7, but that the energy release with combustion for a stoichiometric mixture with air is slightly smaller, about 2700 kJ/kg. Repeat Problem 9.57 using these values and compare the results.

9.60 It has been determined experimentally that the power stroke expansion in an internal combustion engine can be approximated with a polytropic process with a value of the polytropic exponent n somewhat larger than the specific heat ratio k. Repeat Problem 9.57 but assume that the expansion process is reversible and polytropic (instead of the isentropic expansion in the Otto cycle) with a value of n equal to 1.50.

9.61 In the air-standard Otto cycle all the heat transfer q_H occurs at constant volume. It would be more realistic to assume that part of q_H occurs after the piston has started its downward motion in the expansion stroke. Therefore, consider a cycle identical to the Otto cycle, except that the first two-thirds of the total q_H occurs at constant volume and the last one-third occurs at constant pressure. Assume that the total q_H is 2400 kJ/kg, that the pressure and temperature at the beginning of the compression process are 90 kPa, 20°C, and that the comparison ratio is 7. Calculate the maximum pressure and temperature and the thermal efficiency of this cycle. Compare the results with those of a conventional Otto cycle having the same given variables.

9.62 Consider an ideal air-standard diesel cycle in which the state before the compression process is 95 kPa, 290 K, and the compression ratio is 20. What maximum temperature must the cycle have to have a thermal efficiency of 60%?

9.63 Consider an ideal Stirling-cycle engine in which the pressure and temperature at the beginning of the isothermal compression process are 100 kPa, 25°C, the compression ratio is 6, and the maximum temperature in the cycle is 1100°C. Calculate

a. The maximum pressure in the cycle.

b. The thermal efficiency of the cycle with and without regenerators.

9.64 Consider an ideal air-standard Stirling cycle with an ideal regenerator. The minimum pressure and temperature in the cycle are 100 kPa, 25°C, the compression ratio is 10, and the maximum temperature in the cycle is 1000°C. Analyze each of the four processes in this cycle for work and heat transfer, and determine the overall performance of the engine.

9.65 The air-standard Carnot cycle was not shown in the text; show the T–s diagram for this cycle. In an air-standard Carnot cycle the low temperature is 280 K and the efficiency is 60%. If the pressure before compression and after heat rejection is 100 kPa, find the high temperature and the pressure just before heat addition.

9.66 Air in a piston/cylinder goes through a Carnot cycle in which $T_L = 26.8°C$ and the total cycle efficiency is $\eta = 2/3$. Find T_H, the specific work and volume ratio in the adiabatic expansion for constant C_p, C_v. Repeat the calculation for variable heat capacities.

9.67 Consider an ideal refrigeration cycle that has a condenser temperature of 45°C and an evaporator temperature of −15°C. Determine the coefficient of performance of this refrigerator for the working fluids R-12 and R-22.

9.68 The environmentally safe refrigerant R-134a is one of the replacements for R-12 in refrigeration systems. Repeat Problem 9.67 using R-134a and compare the result with that for R-12.

9.69 A refrigerator with R-12 as the working fluid has a minimum temperature of −10°C and a maximum pressure of 1 MPa. Assume an ideal refrigeration cycle as in Fig. 9.32. Find the specific heat transfer from the cold space and that to the hot space, and the coefficient of performance.

9.70 A refrigerator with R-12 as the working fluid has a minimum temperature of −10°C and a maximum pressure of 1 MPa. The actual compressor exit temperature is 60°C. Assume no pressure loss in the heat exchangers and no heat loss in the compressor. Find the specific heat transfer from the cold space and that to the hot space, the coefficient of performance and the isentropic efficiency of the compressor.

9.71 Consider an ideal heat pump that has a condenser temperature of 50°C and an evaporator temperature of 0°C. Determine the coefficient of performance of this heat pump for the working fluids R-12, R-22, and ammonia.

9.72 The effect of the evaporator temperature on the coefficient of performance of a heat pump is to be studied. Consider an ideal cycle with R-22 as the working fluid and a condenser temperature of 40°C. Plot a curve for the coefficient of performance versus the evaporator temperature for temperature values from 15 to −25°C.

9.73 How the difference between the temperature of the refrigerant in the condenser and the temperature of the surroundings affects the amount of cycle power required is to be studied. For this study, consider an ideal refrigeration cycle with R-22 as the working fluid, surroundings having a temperature of 30°C, and an evaporator whose temperature is −15°C. Plot a curve of power input per kilowatt of refrigeration for differences of 0 − 40°C between the temperatures of the surroundings and of the condenser.

9.74 How the difference between the temperature of the refrigerant in the condenser and the temperature of the surroundings affects the amount of cycle power required is to be studied. For this study, consider an ideal refrigeration cycle with R-22 as the working fluid, surroundings having a temperature of 50°C, and a cold space temperature of −15°C. Plot a curve of power input per kilowatt of refrigeration for differences of 0 to 30°C between the temperatures of the cold space and of the refrigerant in the evaporator.

9.75 A small heat pump unit is used to heat water for a hot-water supply. Assume that the unit uses R-22 and operates on the ideal refrigeration cycle. The evaporator temperature is 15°C and the condenser temperature is 60°C. If the amount of hot water needed is 0.1 kg/s, determine the amount of energy saved by using the heat pump instead of directly heating the water from 15 to 60°C.

9.76 The refrigerant R-22 is used as the working fluid in a conventional heat pump cycle. Saturated vapor enters the compressor of this unit at 10°C; its exit temperature from the compressor is measured and found to be 85°C. If the isentropic efficiency of the compressor is estimated to be 70%, what is the coefficient of performance of the heat pump?

9.77 In an actual refrigeration cycle using R-12 as the working fluid, the refrigerant flow rate is 0.05 kg/s. Vapor enters the compressor at 150 kPa, −10°C, and leaves at 1.2 MPa, 75°C. The power input to the compressor is measured and found be 2.4 kW. The refrigerant enters the expansion valve at 1.15 MPa, 40°C, and leaves the evaporator at 175 kPa, −15°C. Determine

a. The irreversibility during the compression process.

b. The refrigeration capacity.

c. The coefficient of performance for this cycle.

9.78 Consider a small ammonia absorption refrigeration cycle that is powered by solar energy and is to be used as an air conditioner. Saturated vapor ammonia leaves the generator at 50°C, and saturated vapor leaves the evaporator at 10°C. If 7000 kJ of heat is required in the generator (solar collector) per kilogram of ammonia vapor generated, determine the overall performance of this system.

9.79 The performance of an ammonia absorption cycle refrigerator is to be compared with that of a similar vapor-compression system. Consider an absorption system having an evaporator temperature of −10°C and a condenser temperature of 48°C. The generator temperature in this system is 150°C. In this cycle 0.42 kJ is transferred to the ammonia in the evaporator for each kilojoule transferred from the high-temperature source to the ammonia solution in the generator. To make the comparison, assume that a reservoir is available at 150°C, and that heat is transferred from this reservoir to a reversible engine that rejects heat to the surroundings at 25°C. This work is then used to drive an ideal vapor-compression system with ammonia as the refrigerant. Compare the amount of refrigeration that can be achieved per kilojoule from the high-temperature source with the 0.42 kJ that can be achieved in the absorption system.

9.80 A heat exchanger is incorporated into an ideal air-standard refrigeration cycle, as shown in Fig. P9.80. It may be assumed that both the compression and the expansion are reversible adiabatic processes in this ideal case. Determine the coefficient of performance for the cycle.

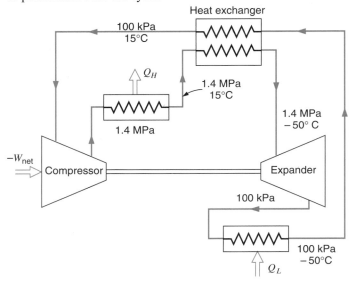

FIGURE P9.80

9.81 Repeat Problem 9.80, but assume an isentropic efficiency of 75% for both the compressor and the expander.

9.82 Repeat Problems 9.80 and 9.81, but assume that helium is the cycle working fluid instead of air. Discuss the significance of the results.

9.83 A binary system power plant uses mercury for the high-temperature cycle and water for the low-temperature cycle, as shown in Fig. 9.39. The temperatures and pressures are shown in the corresponding T–s diagram. The maximum temperature in the steam cycle is where the steam leaves the superheater at point 4 where it is 500°C. Determine

a. The ratio of the mass flow rate of mercury to the mass flow rate of water in the heat exchanger that condenses mercury and boils the water

b. The thermal efficiency of this ideal cycle

From tables of thermodynamic properties for mercury, the following saturation data are known.

P, MPa	T_g,°C	h_f, kJ / kg	h_g,kJ / kg	s_f, kJ / kg	s_g, kJ / kg
0.04	309	42.21	335.64	0.1034	0.6073
1.60	562	75.37	364.04	0.1498	0.4954

9.84 The power plant shown in Fig. 9.40 combines a gas-turbine cycle and a steam-turbine cycle. The following data are known for the gas-turbine cycle. Air enters the compressor at 100 kPa, 25°C, the compressor pressure ratio is 14, and the isentropic compressor efficiency is 87%; the heater input rate is 60 MW; the turbine inlet temperature is 1250°C, the exhaust pressure is 100 kPa, and the isentropic turbine efficiency is 87%; the cycle exhaust temperature from the heat exchanger is 200°C. The following data are known for the steam-turbine cycle. The pump inlet state is saturated liquid at 10 kPa, the pump exit pressure is 12.5 MPa, and the isentropic pump efficiency is 85%; turbine inlet temperature is 500°C and the isentropic turbine efficiency is 87%. Determine

a. The mass flow rate of air in the gas-turbine cycle.

b. The mass flow rate of water in the steam cycle.

c. The overall thermal efficiency of the combined cycle.

9.85 A Rankine steam power plant should operate with a high pressure of 3 MPa, a low pressure of 10 kPa, and the boiler exit temperature should be 500°C. The available high-temperature source is the exhaust of 175 kg/s air at 600°C from a gas turbine. If the boiler operates as a counterflowing heat exchanger where the temperature difference at the pinch point is 20°C, find the maximum water mass flow rate possible and the air exit temperature.

9.86 (Adv.) For Problem 9.85, determine the change of availability of the water flow and that of the air flow. Use these to determine a second law efficiency for the boiler heat exchanger.

9.87 A simple Rankine cycle with R-22 as the working fluid is to be used as a bottoming cycle for an electrical generating facility driven by a Diesel engine. The exhaust gas from the Diesel engine is used as the high temperature energy source in the R-22 boiler. Diesel inlet conditions are 100 kPa, 20°C, the compression ratio is 20, and the maximum temperature in the cycle is 2800°C. Saturated vapor R-22 leaves the bottoming cycle boiler at 110°C, and the condenser temperature is 30°C. The power output of the Diesel engine is 1 MW. Assuming ideal cycles throughout, determine

a. The flow rate required in the diesel engine.

b. The power output of the bottoming cycle, assuming that the diesel exhaust is cooled to 200°C in the R-22 boiler.

9.88 For a cryogenic experiment heat should be removed from a space at 75 K to a reservoir at 180 K. A heat pump is designed to use nitrogen and methane in a cascade arrangement, see Fig. 9.41, where the high temperature of the nitrogen condensation is at 10 K higher than the low-temperature evaporation of the methane. The two other phase changes take place at the listed reservoir temperatures. Find the saturation temperatures in the heat exchanger between the two cycles that gives the best coefficient of performance for the overall system.

9.89 In both power and refrigeration cycles that operate over a wide temperature range, it is frequently advantageous to use more than one working fluid. In refrigeration systems this is commonly referred to as a cascade system. Consider such a system composed of two ideal refrigeration cycles, as shown in Fig. 9.41. This system is designed to use environmentally safe refrigerants as the two working fluids. The high-temperature cycle uses R-22. Saturated liquid leaves the condenser at 40°C, and saturated vapor leaves the heat exchanger at −20°C. The low-temperature cycle uses a different refrigerant, R-23 (Fig. A.3 or the software). Saturated vapor leaves the evaporator at −80°C, and saturated liquid leaves the heat exchanger at −10°C. Calculate

a. The ratio of the mass flow rates through the two cycles.

b. The coefficient of performance of the system.

9.90 One means of improving the performance of a refrigeration system that operates over a wide temperature range is to use a two-stage compressor. Consider an ideal refrigeration system of this type that uses R-12 as the working fluid, as shown in Fig. P9.90. Saturated liquid leaves the condenser at 40°C and is throttled to −20°C. The liquid and vapor at this temperature are separated, and the liquid is throttled to the evaporator temperature, −70°C. Vapor leaving the evaporator is compressed to the saturation pressure corresponding to −20°C, after which it is mixed with the vapor leaving the flash chamber. It may be assumed that both the flash chamber and the mixing chamber are well insulated to prevent heat transfer from the ambient. Vapor leaving the mixing chamber is compressed in the second stage of the compressor to the saturation pressure corresponding to the condenser temperature, 40°C. Determine

a. The coefficient of performance of the system.

b. The coefficient of performance of a simple ideal refrigeration cycle operating over the same condenser and evaporator ranges as those of the two-stage compressor unit studied in this problem.

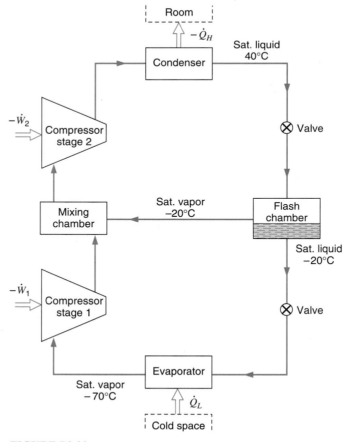

FIGURE P9.90

9.91 Consider an ideal dual-loop heat-powered refrigeration cycle using R-12 as the working fluid, as shown in Fig. P9.91. Saturated vapor at 105°C leaves the boiler and expands in the turbine to the condenser pressure. Saturated vapor at −15°C leaves the evaporator and is compressed to the condenser pressure. The ratio of the flows through the two loops is such that the turbine produces just enough power to drive the compressor. The two exiting streams mix together and enter the condenser. Saturated liquid leaving the condenser at 45°C is then separated into two streams in the necessary proportions. Determine

a. The ratio of mass flow rate through the power loop to that through the refrigeration loop.

b. The performance of the cycle, in terms of the ratio Q_L/Q_H.

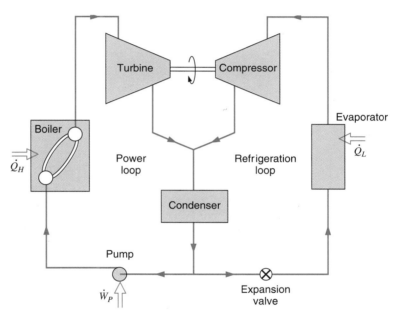

FIGURE P9.91

9.92 A jet ejector, a device with no moving parts, functions as the equivalent of a coupled turbine-compressor unit (see Problems 7.98 and 7.107). Thus, the turbine-compressor in the dual-loop cycle of Fig. P9.91 could be replaced by a jet ejector. The primary stream of the jet ejector enters from the boiler, the secondary stream enters from the evaporator, and the discharge flows to the condenser. Alternatively, a jet ejector may be used with water as the working fluid. The purpose of the device is to chill water, usually for an air-conditioning system. In this application the physical setup is as shown in Fig. P9.92. Using the data given on the diagram, evaluate the performance of this cycle in terms of the ratio Q_L/Q_H.

a. Assume an ideal cycle.

b. Assume an ejector efficiency of 20% (see Problem 7.107).

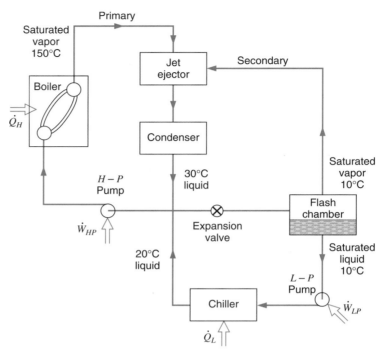

FIGURE P9.92

ENGLISH UNIT
PROBLEMS

9.93E A steam power plant, as shown in Fig. 9.3, has a boiler exit at 600 lbf/in.², 700 F and a condenser operating at 2 lbf/in.². Assume that all components are ideal and find the cycle efficiency and the specific work and heat transfer in the components.

9.94E Consider a solar-energy-powered ideal Rankine cycle that uses water as the working fluid. Saturated vapor leaves the solar collector at 350 F, and the condenser pressure is 1 lbf/in.². Determine the thermal efficiency of this cycle.

9.95E A supply of geothermal hot water is to be used as the energy source in an ideal Rankine cycle, with R-134a as the cycle working fluid. Saturated vapor R-134a leaves the boiler at a temperature of 180 F, and the condenser temperature is 100 F.

a. Calculate the thermal efficiency of this cycle.

b. Repeat the calculation for R-22 as the working fluid.

9.96E (Adv.) Find the availability of the water at all four states in the Rankine cycle described in Problem 9.93. Assume the high-temperature source is 900 F and the low-temperature reservoir is at 65 F. Determine the flow of availability in or out of the reservoirs per pound-mass of steam flowing in the cycle. What is the overall cycle second law efficiency?

9.97E The effect of turbine inlet pressure on the performance of the ideal steam Rankine cycle given in Problem 9.93 is to be studied. Calculate the thermal efficiency of the cycle and the moisture content of the steam leaving the turbine for turbine inlet pressures of 200, 600, 1000, and 1500 lbf/in.^{2}. Plot the thermal efficiency versus turbine inlet pressure for the specified turbine inlet temperature and exhaust pressure.

9.98E A closed feedwater heater in a regenerative steam power cycle heats 40 lbm/s of water from 200 F, 2000 lbf/in.^{2} to 450 F, 2000 lbf/in.^{2}. The extraction steam from the turbine enters the heater at 500 lbf/in.^{2}, 550 F and leaves as saturated liquid. What is the required mass flow rate of the extraction steam?

9.99E Consider a simple ideal Rankine cycle that uses steam as the working fluid. The high-pressure side of the cycle is at a supercritical pressure. Such a cycle has a potential advantage of minimizing local temperature differences between the fluids in the steam generator, such as the instance in which the high-temperature energy source is the hot exhaust gas from a gas-turbine engine. Calculate the thermal efficiency of the cycle if the state entering the turbine is 3500 lbf/in.^{2}, 1100 F, and the condenser pressure is 1 lbf/in.^{2}. What is the steam quality at the turbine exit?

9.100E Consider an ideal steam reheat cycle in which the steam enters the high-pressure turbine at 600 lbf/in.^{2}, 700 F, and then expands to 120 lbf/in.^{2}. It is then reheated to 700 F and expands to 2 lbf/in.^{2} in the low-pressure turbine. Calculate the thermal efficiency of the cycle and the moisture content of the steam leaving the low-pressure turbine.

9.101E Consider an ideal steam regenerative cycle in which steam enters the turbine at 600 lbf/in.^{2}, 700 F, and exhausts to the condenser at 2 lbf/in.^{2}. Steam is extracted from the turbine at 120 lbf/in.^{2} and also at 30 lbf/in.^{2} for heating the boiler feedwater in two open feedwater heaters. The feedwater leaves each heater at the temperature of the condensing steam. The appropriate pumps are used for the water leaving the condenser and the two feedwater heaters. Calculate the thermal efficiency of the cycle and the net work per pound-mass of steam.

9.102E Consider an ideal steam combined reheat and regenerative cycle in which steam enters the high-pressure turbine at 500 lbf/in.^{2}, 700 F, and is extracted for feedwater heating at 120 lbf/in.^{2}. The remainder of the steam is reheated to 700 F at this pressure, 120 lbf/in.^{2}, and is fed to the low-pressure turbine. Steam is extracted from the low-pressure turbine at 20 lbf/in.^{2} for feedwater heating. The condenser pressure is 2 lbf/in.^{2}. Both feedwater heaters are open heaters. Calculate the thermal efficiency of the cycle and the net work per pound-mass of steam.

9.103E A steam power cycle has a high pressure of 500 lbf/in.^{2} and a condenser exit temperature of 110 F. The turbine efficiency is 85%, and other cycle components are ideal. If the boiler superheats to 1400 F, find the cycle thermal efficiency.

9.104E A steam power plant has a boiler exit at 600 lbf/in.^{2}, 900 F and a condenser exit temperature of 110 F. Assume isentropic efficiency of the turbine is 85% and that for the pump it is 80%. Find the cycle efficiency and the specific work and heat transfer in the components.

9.105E Steam leaves a power plant steam generator at 500 lbf/in.^{2}, 650 F, and enters the turbine at 490 lbf/in.^{2}, 625 F. The isentropic turbine efficiency is 88%, and the turbine exhaust pressure is 1.7 lbf/in.^{2}. Condensate leaves the condenser and enters the pump at 110 F, 1.7 lbf/in.^{2}. The isentropic pump efficiency is 80%, and the discharge pressure is 520 lbf/in.^{2}. The feedwater enters the steam generator at 510 lbf/in.^{2}, 100 F. Calculate the following.

    a. The thermal efficiency of the cycle.

    b. The irreversibility of the process in the line between the steam generator exit and the turbine inlet, assuming an ambient temperature of 77 F.

9.106E In one type of nuclear power plant, heat is transferred in the nuclear reactor to liquid sodium. The liquid sodium is then pumped through a heat exchanger where heat is transferred to boiling water. Saturated vapor steam at 700 lbf/in.^{2} exits this heat exchanger and is then superheated to 1100 F in an external gas-fired superheater. The steam enters the turbine, which has one (open-type) feedwater extraction at 60 lbf/in.^{2}. The isentropic turbine efficiency is 87%, and the condenser pressure is 1 lbf/in.^{2}. Determine the heat transfer in the reactor and in the superheater to produce a net power output of 1000 Btu/s.

9.107E A boiler delivers steam at 1500 lbf/in.^{2}, 1000 F to a two-stage turbine. After the first stage, 25% of the steam is extracted at 200 lbf/in.^{2} for a process application and returned as compressed liquid at 150 lbf/in.^{2}, 190 F to the feedwater line. The remainder of the steam continues through the low-pressure turbine stage, which exhausts to the condenser at 2 lbf/in.^{2}. One pump after the condenser brings the feedwater to 150 lbf/in.^{2}, where it is mixed with the return line from the process application and a second pump brings it up to 1500 lbf/in.^{2}. Assume the first and second stages in the steam turbine have isentropic efficiencies of 85% and 80%. If the process application requires 5000 Btu/s of power how much power can then be cogenerated by the turbine? Assume that both pumps are ideal.

9.108E Consider a large stationary gas-turbine power plant that operates on the ideal Brayton cycle and delivers a power output of 100 000 hp to an electric generator. The minimum temperature in the cycle is 540 R, and the maximum temperature is 2900 R. The minimum pressure in the cycle is 1 atm, and the compressor pressure ratio is 14 to 1.

    a. Calculate the power output of the turbine. What fraction of the turbine output is required to drive the compressor?

    b. What is the thermal efficiency of the cycle?

9.109E An ideal regenerator is incorporated into the ideal air-standard Brayton cycle of Problem 9.108. Calculate the thermal efficiency of the cycle with this modification.

9.110E Consider an ideal gas-turbine cycle with two stages of compression and two stages of expansion. The pressure ratio across each compressor stage and each turbine stage is 8 to 1. The pressure at the entrance to the first compressor is 14 lbf/in.^{2}, the temperature entering each compressor is 70 F, and the temperature entering each turbine is 2000 F. An ideal regenerator is also incorporated into the cycle. Determine the compressor work, the turbine work, and the thermal efficiency of the cycle.

9.111E Repeat Problem 9.110, but assume that each compressor stage and each turbine stage has an isentropic efficiency of 85%. Also assume that the regenerator has an efficiency of 70%.

9.112E An air-standard Ericsson cycle has an ideal regenerator, as shown in Fig. P9.49. Heat is supplied at 1800 F and heat is rejected at 68 F. Pressure at the beginning of the isothermal compression process is 10 lbf/in.^{2}. The heat added is 275 Btu/lbm. Find the compressor work, the turbine work, and the cycle efficiency.

9.113E The turbine in a jet engine receives air at 2200 R, 220 lbf/in.^{2}. It exhausts to a nozzle at 35 lbf/in.^{2}, which in turn exhausts to the atmosphere at 14.7 lbf/in.^{2}. The isentropic efficiency of the turbine is 85% and the nozzle efficiency is 95%. Find the nozzle inlet temperature and the nozzle exit velocity. Assume negligible kinetic energy out of the turbine.

9.114E Air flows into a gasoline engine at 14 lbf/in.^{2}, 540 R. The air is then compressed with a volumetric compression ratio of 8 : 1. In the combustion process 560 Btu/lbm of energy is released as the fuel burns. Find the temperature and pressure after combustion.

9.115E A stoichiometric mixture of gasoline and air has an energy release upon combustion of approximately 1200 Btu/lbm of the mixture. To approximate an actual spark-ignition engine using such a mixture, consider an air-standard Otto cycle that has a heat addition of 1200 Btu/lbm of air, a compression ratio of 7, and a pressure and temperature at the beginning of the compression process of 13 lbf/in.^{2}, 50 F. Assuming constant specific heat, with the value from Table A.10E, determine

a. The maximum pressure and temperature of the cycle.

b. The thermal efficiency of the cycle.

c. The mean effective pressure.

9.116E In the air-standard Otto cycle all the heat transfer q_H occurs at constant volume. It would be more realistic to assume that part of q_H occurs after the piston has started its downwards motion in the expansion stroke. Therefore consider a cycle identical to the Otto cycle, except that the first two-thirds of the total q_H occurs at constant volume and the last one-third occurs at constant pressure. Assume the total q_H is 1100 Btu/lbm, that the pressure and temperature at the beginning of the compression process are 13 lbf/in.^{2}, 68 F, and that the compression ratio is 10. Calculate the maximum pressure and temperature and the thermal efficiency of this cycle. Compare the results with those of a conventional Otto cycle having the same given variables.

9.117E It has been determined experimentally that the power stroke expansion in an internal combustion engine can be approximated with a polytropic process with a value of the polytropic exponent n somewhat larger than the specific heat ratio k. Repeat Problem 9.115 but assume that the expansion process is reversible and polytropic (instead of the isentropic expansion in the Otto cycle) with a value of n equal to 1.50.

9.118E Consider an ideal air-standard Diesel cycle in which the state before the compression process is 14 lbf/in.^{2}, 63 F, and the compression ratio is 20. What maximum temperature must the cycle have to have a thermal efficiency of 60%?

9.119E Consider an ideal Stirling-cycle engine in which the pressure and temperature at the beginning of the isothermal compression process are 14.7 lbf/in.^{2}, 80 F, the compression ratio is 6, and the maximum temperature in the cycle is 2000 F. Calculate

a. The maximum pressure in the cycle.

b. The thermal efficiency of the cycle with and without regenerators.

9.120E Consider an ideal air-standard Stirling cycle with an ideal regenerator. The minimum pressure and temperature in the cycle are 14.7 lbf/in.^{2}, 70 F, the compression ratio is 10, and the maximum temperature in the cycle is 1800 F. Analyze each of the four processes in this cycle for work and heat transfer, and determine the overall performance of the engine.

9.121E The air-standard Carnot cycle was not shown in the text; show the T–s diagram for this cycle. In an air-standard Carnot cycle the low temperature is 500 R and the efficiency is 60%. If the pressure before compression and after heat rejection is 14.7 lbf/in.^{2}, find the high temperature and the pressure just before heat addition.

9.122E Air in a piston/cylinder goes through a Carnot cycle in which $T_L = 80.3$ F and the total cycle efficiency is $\eta = 2/3$. Find T_H, the specific work and volume ratio in the adiabatic expansion for constant C_p, C_v. Repeat the calculation for variable heat capacities.

9.123E Consider an ideal refrigeration cycle that has a condenser temperature of 110 F and an evaporator temperature of 5 F. Determine the coefficient of performance of this refrigerator for the working fluids R-12 and R-22.

9.124E The environmentally safe refrigerant R-134a is one of the replacements for R-12 in refrigeration systems. Repeat Problem 9.123 using R-134a and compare the result with that for R-12.

9.125E Consider an ideal heat pump that has a condenser temperature of 120 F and an evaporator temperature of 30 F. Determine the coefficient of performance of this heat pump for the working fluids R-12, R-22, and ammonia.

9.126E How the difference between the temperature of the refrigerant in the condenser and the temperature of the surroundings affects the amount of cycle power required is to be studied. For this study consider an ideal refrigeration cycle with R-22 as the working fluid, surroundings having a temperature of 85 F, and an evaporator whose temperature is 5 F. Plot a curve of power input per Btu/s of refrigeration for differences of 0 to 70 F between the temperatures of the surroundings and of the condenser.

9.127E The refrigerant R-22 is used as the working fluid in a conventional heat pump cycle. Saturated vapor enters the compressor of this unit at 50 F; its exit temperature from the compressor is measured and found to be 185 F. If the isentropic efficiency of the compressor is estimated to be 70%, what is the coefficient of performance of the heat pump?

9.128E Consider a small ammonia absorption refrigeration cycle that is powered by solar energy and is to be used as an air conditioner. Saturated vapor ammonia leaves the generator at 120 F, and saturated vapor leaves the evaporator at 50 F. If 3000 Btu of heat is required in the generator (solar collector) per pound-mass of ammonia vapor generated, determine the overall performance of this system.

9.129E The power plant shown in Fig. 9.40 combines a gas-turbine cycle and a steam-turbine cycle. The following data are known for the gas-turbine cycle. Air enters the compressor at 14.7 $lbf/in.^2$, 70 F, the compressor pressure ratio is 14, and the isentropic compressor efficiency is 87%; the heater input rate is 50 000 Btu/s; the turbine inlet temperature is 2250 F, the exhaust pressure is 14.7 $lbf/in.^2$, and the isentropic turbine efficiency is 87%; the cycle exhaust temperature from the heat exchanger is 390 F. The following data are known for the steam-turbine cycle. The pump inlet state is saturated liquid at 1 $lbf/in.^2$, the pump exit pressure is 1800 $lbf/in.^2$, and the isentropic pump efficiency is 85%; turbine inlet temperature is 900 F and the isentropic turbine efficiency is 87%. Determine

a. The mass flow rate of air in the gas-turbine cycle.

b. The mass flow rate of water in the steam cycle.

c. The overall thermal efficiency of the combined cycle.

9.130E Consider an ideal dual-loop heat-powered refrigeration cycle using R-12 as the working fluid, as shown in Fig. P9.91. Saturated vapor at 220 F leaves the boiler and expands in the turbine to the condenser pressure. Saturated vapor at 0 F leaves the evaporator and is compressed to the condenser pressure. The ratio of the flows through the two loops is such that the turbine produces just enough power to drive the compressor. The two exiting streams mix together and enter the condenser. Saturated liquid leaving the condenser at 110 F is then separated into two streams in the necessary proportions. Determine

a. The ratio of mass flow rate through the power loop to that through the refrigeration loop.

b. The performance of the cycle, in terms of the ratio Q_L/Q_H.

COMPUTER, DESIGN, AND OPEN-ENDED PROBLEMS

9.131 Write a program to solve the general case of Problems 9.10, 9.11, and 9.12. The effects of turbine exhaust pressure, inlet pressure, and inlet temperature should be studied for the ideal Rankine cycle.

9.132 Extend the program in Problem 9.131 to include the effects of turbine and pump isentropic efficiencies.

9.133 Write a program to solve the general case of Problems 9.14, 9.15, and 9.16. The effects of reheat pressure and the number of reheat stages should be studied for the ideal reheat Rankine cycle.

9.134 Extend the program in the previous problem to include the effects of turbine and pump isentropic efficiencies.

9.135 Write a program to solve the general case of Problems 9.19, 9.20, and 9.21. The effects of the feedwater heater extraction pressure and the number of heaters (open versus closed) should be studied for the ideal regenerative Rankine cycle.

9.136 Extend the program in the previous problem to include the effects of turbine and pump isentropic efficiencies.

9.137 Write a program to solve the general case of Problem 9.23. The effect of the feedwater heater extraction pressure on a combined reheat and regenerative cycle should be studied for the ideal reheat and regenerative Rankine cycle. Include also the effects of turbine and pump isentropic efficiencies.

9.138 Write a program to solve the general case of Problem 9.33, where the high pressure and temperature as well as the state of the two process streams are varied. Study the performance (w_T/q_H and q_{process}/q_H) for a variation of the parameters.

9.139 Write a program to solve the following problem. The effects of varying parameters on the performance of an air-standard Brayton cycle are to be determined. Consider a compressor inlet condition of 100 kPa, 20°C, and assume constant specific heat. The thermal efficiency of the cycle and the net specific work output should be determined for the combinations of the following variables.

a. Compressor pressure ratio of 6, 9, 12, and 15.

b. Maximum cycle temperature of 900, 1100, 1300, and 1500°C.

c. Compressor and turbine isentropic efficiencies each 100, 90, 80, and 70%.

9.140 The effect of variable specific heat on the result of the previous problem is to be studied. Write a modified program that uses the supplied software or use, as necessary, functions that are curvefits to some of the properties (enthalpy and relative pressure) listed in Table A.12.

9.141 Write a program to study the effect of the number of stages of an ideal gas-turbine cycle having intercooling, reheat, and a regenerator, as shown in Fig. 9.24 for two stages. Assume that the inlet condition, state 1, is 100 kPa, 20°C, and that the total pressure ratio, the maximum temperature in the cycle, and the number of stages (equal number of compressor and turbine stages) are all variable. Each stage has an equal pressure ratio. Determine the net work output per kilogram and the thermal efficiency of the cycle for various combinations of these variables.

9.142 Extend the program for the previous problem to use variable specific heat using curvefits for the properties of air based on Table A.12 or use the supplied software subroutine for air; see also Problem 9.140.

9.143 The effect of irreversibilities in the components of the gas-turbine cycle in the previous two problems is to be studied. Repeat one of these problems by including values for the isentropic efficiencies of the compressor and the turbine and for various values of the regenerator efficiency.

9.144 Write a program to simulate the Otto cycle using nitrogen as the working fluid. Use the variable specific heat as given in Table A.11. The beginning of compression has a state of 100 kPa, 20°C. Determine the net specific work output and the cycle thermal efficiency for various combinations of compression ratio and maximum cycle temperature. Compare the result with those found when constant specific heat is assumed.

9.145 Write a program to compare the performance of the Otto and Diesel cycles. Repeat the previous problem and add the Diesel cycle also using nitrogen as the working fluid. Use the specific heat equation given in Table A.11.

9.146 Write a program to study the ideal refrigeration heat pump cycle as shown in Problem 9.67 using ammonia as the working fluid. Vary the condenser and evaporator temperatures within a parameter space suitable for refrigerators, air-conditioning systems, and heat pumps.

9.147 Write a program to study the performance of the air-standard refrigeration cycle, incorporating a heat exchanger as shown in Fig. 9.38. The compressor inlet condition is 100 kPa, 15°C, and the pressure ratio across the compressor and expander is a variable. Include isentropic efficiencies for the compressor and expander.

9.148 Write a program to investigate the general case of the combined-cycle power plant in Problem 9.84, as shown in Fig. 9.40. The gas-turbine cycle compressor pressure ratio and the turbine inlet temperature are input variables together with the maximum pressure and temperature in the steam cycle.

9.149 A power plant is built to provide district heating of buildings that requires 90°C liquid water at 150 kPa. The district heating water is returned at 50°C, 100 kPa, in a closed loop in an amount such that 20 MW of power is delivered. This hot water is produced from a steam power cycle with a boiler making steam at 5 MPa, 600°C delivered to the steam turbine. The steam cycle could have its condenser operate at 90°C providing the power to the district heating. It could also be done in connection with extraction steam from the turbine. Suggest a system and evaluate its performance in terms of the cogenerated amount of turbine work.

9.150 In a particular nuclear power plant, the nuclear reactor is so designed that the maximum temperature in the steam cycle is 450°C. The steam-condensing temperature is fixed at 40°C. The plan is to build a steam power plant that has one open feedwater heater. Select what you consider to be a reasonable ideal steam cycle within these specifications, determine its thermal efficiency, and explain why the particular pressures involved were selected.

9.151 In Section 6.10, the maximum rate work out of a heat engine was examined as a function of finite heat transfer rates. Consider a basic steam Rankine cycle where the boiler delivers steam at 3.5 MPa, 400°C. The condenser has a heat transfer rate of $\dot{Q}_L = 15 + 0.1 \times \Delta T$ MW as a function of ΔT between the condensing steam and the cooling water at 20°C. Take the turbine inlet properties as constant assuming an infinite heat transfer rate is possible in the boiler and find the condenser operating temperature for maximum cycle work.

9.152 A hospital requires 2 kg/s steam at 200°C, 125 kPa for sterilization purposes and space heating requires 15 kg/s hot water at 90°C, 100 kPa. Both of these requirements are provided by the hospital's steam power plant. Discuss some arrangement that will accomplish this.

9.153 Consider the following preliminary design specifications for a supercritical steam power plant cycle. The maximum pressure will be 30 MPa, and the maximum temperature will be 600°C. The temperature of the cooling water is such that 10-kPa pressure can be maintained in the condensers. Turbine isentropic efficiencies of at least 87% can be expected.

a. Do you recommend any reheat for this cycle? If so, how many stages of reheat and at what pressures? Give reasons for your decisions.

b. Do you recommend feedwater heaters? If so, how many and at what pressures would they operate? Should they be open or closed feedwater heaters?

c. Estimate the thermal efficiency of the cycle that you recommend.

9.154 A gas-turbine engine is to be used for pumping natural gas through a cross-country pipeline. The required power input from this engine to the natural gas compressor is 1 MW. A simple open cycle with a regenerator will be used for this application. Since a supply of fuel is readily available, and since some of these units may be located in relatively isolated places, low maintenance costs are more important than efficiency. Making assumptions for the isentropic efficiencies of the compressor and turbine and also for the regenerator efficiency, recommend a cycle. For the gas-turbine engine unit, determine the power output of the turbine, the power input to the compressor, and the thermal efficiency of the engine.

9.155 The gasoline engine and diesel engine examples 9.11 and 9.12, show peak temperatures and efficiencies significantly larger than what is observed in real engines. Discuss the assumptions made for the ideal cycles and relate those to the real engines. Mention the intake and the exhaust, the finite combustion time, and heat transfer among other processes, and comment on how these make the real processes different.

9.156 Reconsider the Otto cycle described in Problem 9.57. Assume the expansion process is a polytropic process with an exponent n that is larger than $k \simeq 1.4$ so there is a heat loss. Investigate different exponents and list the predicted heat loss as a fraction of the energy from combustion together with the predicted cycle efficiency. (In a real engine the in-cylinder heat loss is about 10% of the fuel energy added).

9.157 (Adv.) To make a more realistic engine simulation the rate form of the energy equation must be integrated, see Problem 7.196. The volume is taken as

$$V = V_1 \left[1 + \frac{r_v - 1}{2} \left(1 - \cos\left(\frac{\theta \pi}{180} \right) \right) \right] / r_v$$

where r_v is the compression ratio and θ is the crank angle measured from top dead center (at V_{min}). The fraction of energy release is

$$x = \frac{1}{2} \left[1 - \cos\left(\frac{\pi(\theta - \theta_s)}{\theta_b} \right) \right]$$

where θ_s is the spark timing and θ_b is the burn duration in crank angle degrees. Integrate in crank angles, $d\theta = 6 \times \text{RPM} \times dt$, from −180 (state 1) to 0 (state

2–3) to +180 (state 4). RPM is the revolutions per minute. The energy equation

$$mC_v \frac{dT}{d\theta} = -\frac{C_H}{6\text{RPM}}\left(T - T_o\right) - P\frac{dV}{d\theta} + mq_H \frac{dx}{d\theta}$$

from Problem 7.196 becomes

and includes heat loss, work, and heat release. As temperature is integrated pressure is updated each step as $P = mRT/V$ and heat loss and work should also be integrated.

9.158 Read the previous problem and make a list of processes and properties that should be added for a more complete description. Make short comments for each item and list the subject(s) that must be studied to understand and model those.

9.159 To investigate the design of a refrigerator, first look at the effect of the substance used as the working fluid. Assume an ideal vapor compression cycle with temperatures so a 10°C difference to the surrounding space is used. For different refrigerants, R-12, R-22, R-134a, and ammonia, list the possible coefficients of performance that can be expected. Discuss the findings.

9.160 The compressor in a refrigerator receives R-22 as slightly superheated vapor at −10°C, 150 kPa. It is compressed to a pressure so the condensation can take place at 45°C. The exit temperature is found to be near 45°C so a heat loss must have occured. Assume that the process can be analyzed as a polytropic process and find the specific work and heat transfer for different exit temperatures. Compare this to the isentropic specific work. Is the heat loss beneficial or should it be prevented? Discuss the results.

9.161 Use the menu-driven software to get properties for the different refrigerants to solve the following problem. A heat pump should be designed to operate in the ideal refrigeration cycle where the high-temperature condensation is at T_H + 10°C and the low-temperature evaporation is at T_L − 10°C. Suggest a suitable refrigerant (from among the ones included in the software) to use for the cases where the temperatures are

a. High temperature is 30°C, low temperature is −10°C.

b. High temperature is −20°C, low temperature is −50°C.

9.162 Investigate the maximum power out of a steam power plant with operating conditions as in Problem 9.8. The energy source is 100 kg/s combustion products (air) at 125 kPa, 1200 K. Make sure the air temperature is higher than the water temperature throughout the boiler.

9.163 In Problem 9.85 a steam cycle was powered by the exhaust from a gas turbine. With a single water flow and air flow heat exchanger, the air is leaving with a relatively high temperature. Make an analysis of how some more of the energy in the air can be used before the air is flowing out to the chimney. Can it be used in a feedwater heater?

10 THERMODYNAMIC RELATIONS

We have already defined and used several thermodynamic properties. Among these are pressure, specific volume, density, temperature, mass, internal energy, enthalpy, entropy, constant-pressure and constant-volume specific heats, and the Joule–Thomson coefficient. Two other properties, the Helmholtz function and the Gibbs function, have been introduced and will be used more extensively in the following chapters. We have also had occasion to use tables of thermodynamic properties for a number of different substances.

One important question is now raised: Which of the thermodynamic properties can be experimentally measured? We can answer this question by considering the measurements we can make in the laboratory. Some of the properties such as internal energy and entropy cannot be measured directly and must be calculated from other experimental data. If we carefully consider all these thermodynamic properties, we conclude that there are only four that can be directly measured: pressure, temperature, volume, and mass.

This leads to a second question: How can values of the thermodynamic properties that cannot be measured be determined from experimental data on those properties that can be measured? In answering this question we will develop certain general thermodynamic relations. In view of the fact that there are millions of such equations that can be written, our study will be limited to certain basic considerations, with particular reference to the determination of thermodynamic properties from experimental data. We will also consider such related matters as generalized charts and equations of state.

10.1 TWO IMPORTANT RELATIONS

This chapter involves partial derivatives, and two important relations are reviewed here. Consider a variable z that is a continuous function of x and y.

$$z = f(x, y)$$

$$dz = \left(\frac{\partial z}{\partial x}\right)_y dx + \left(\frac{\partial z}{\partial y}\right)_x dy$$

It is convenient to write this function in the form

$$dz = M\, dx + N\, dy \tag{10.1}$$

$$M = \left(\frac{\partial z}{\partial x}\right)_y$$

= partial derivative of z with respect to x (the variable y being held constant)

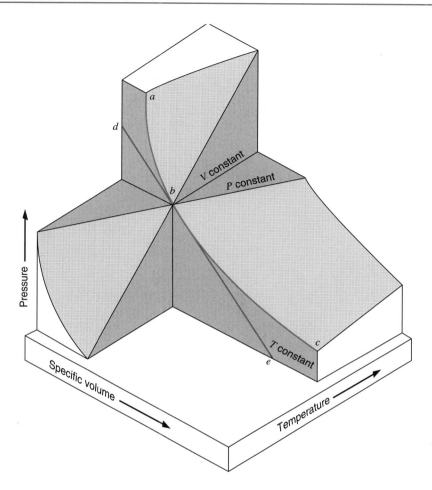

FIGURE 10.1
Schematic representation of partial derivatives.

$$N = \left(\frac{\partial z}{\partial y} \right)_x$$

= partial derivative of z with respect to y (the variable x being held constant)

The physical significance of partial derivatives as they relate to the properties of a pure substance can be explained by referring to Fig. 10.1, which shows a P–v–T surface of the superheated vapor region of a pure substance. It shows constant-temperature, constant-pressure, and constant-specific volume planes that intersect at point b on the surface. Thus, the partial derivative $(\partial P/\partial v)_T$ is the slope of curve abc at point b. Line de represents the tangent to curve abc at point b. A similar interpretation can be made of the partial derivatives $(\partial P/\partial T)_v$ and $(\partial v/\partial T)_p$.

If we wish to evaluate the partial derivative along a constant-temperature line the rules for ordinary derivatives can be applied. Thus, we can write for a constant-temperature process:

$$\left(\frac{\partial P}{\partial v} \right)_T = \frac{dP_T}{dv_T}$$

and the integration can be performed as usual. This point will be demonstrated later in a number of examples.

Let us return to the consideration of the relation

$$dz = M\,dx + N\,dy$$

If x, y, and z are all point functions (that is, quantities that depend only on the state and are independent of the path), the differentials are exact differentials. If this is the case, the following important relation holds:

$$\left(\frac{\partial M}{\partial y}\right)_x = \left(\frac{\partial N}{\partial x}\right)_y$$

The proof of this is

$$\left(\frac{\partial M}{\partial y}\right)_x = \frac{\partial^2 z}{\partial x\,\partial y}$$

$$\left(\frac{\partial N}{\partial x}\right)_y = \frac{\partial^2 z}{\partial y\,\partial x}$$

Since the order of differentiation makes no difference when point functions are involved, it follows that

$$\frac{\partial^2 z}{\partial x\,\partial y} = \frac{\partial^2 z}{\partial y\,\partial x}$$

$$\left(\frac{\partial M}{\partial y}\right)_x = \left(\frac{\partial N}{\partial x}\right)_y$$

The second important mathematical relation is

$$\left(\frac{\partial x}{\partial y}\right)_z \left(\frac{\partial y}{\partial z}\right)_x \left(\frac{\partial z}{\partial x}\right)_y = -1 \qquad (10.2)$$

The proof of this relation is as follows. Consider three variables x, y, and z. Suppose there exists a relation between the variables of the form

$$x = f(y, z)$$

Then

$$dx = \left(\frac{\partial x}{\partial y}\right)_z dy + \left(\frac{\partial x}{\partial z}\right)_y dz \qquad (10.3)$$

If this relationship between the three variables is written in the form

$$y = f(x, z)$$

it follows that

$$dy = \left(\frac{\partial y}{\partial x}\right)_z dx + \left(\frac{\partial y}{\partial z}\right)_x dz \qquad (10.4)$$

Substituting Eq. 10.4 into Eq. 10.3, we have

$$dx = \left(\frac{\partial x}{\partial y}\right)_z \left[\left(\frac{\partial y}{\partial x}\right)_z dx + \left(\frac{\partial y}{\partial z}\right)_x dz\right] + \left(\frac{\partial x}{\partial z}\right)_y dz$$

$$= \left(\frac{\partial x}{\partial y}\right)_z \left(\frac{\partial y}{\partial x}\right)_z dx + \left[\left(\frac{\partial x}{\partial y}\right)_z \left(\frac{\partial y}{\partial z}\right)_x + \left(\frac{\partial x}{\partial z}\right)_y\right] dz$$

There are two independent variables, and we select x and z as these variables. Suppose that $dz = 0$ and $dx \neq 0$. It then follows that

$$\left(\frac{\partial x}{\partial y}\right)_z \left(\frac{\partial y}{\partial x}\right)_z = 1 \tag{10.5}$$

Similarly, suppose that $dx = 0$ and $dz \neq 0$. It then follows that

$$\left(\frac{\partial x}{\partial y}\right)_z \left(\frac{\partial y}{\partial z}\right)_x + \left(\frac{\partial x}{\partial z}\right)_y = 0$$

$$\left(\frac{\partial x}{\partial y}\right)_z \left(\frac{\partial y}{\partial z}\right)_x = -\left(\frac{\partial x}{\partial z}\right)_y$$

$$\left(\frac{\partial x}{\partial y}\right)_z \left(\frac{\partial y}{\partial z}\right)_x \left(\frac{\partial z}{\partial x}\right)_y = -1$$

This is Eq. 10.2, which we set out to derive.

10.2 THE MAXWELL RELATIONS

Consider a simple compressible control mass of fixed chemical composition. The Maxwell relations, which can be written for such a system, are four equations relating the properties P, v, T, and s.

The Maxwell relations are most easily derived by considering four relations involving thermodynamic properties. Two of these relations have already been derived and are

$$du = T\,ds - P\,dv \tag{10.6}$$

$$dh = T\,ds + v\,dP \tag{10.7}$$

The other two are derived from the definition of the Helmholtz function, a, and the Gibbs function, g.

$$a = u - Ts$$

$$da = du - T\,ds - s\,dT$$

Substituting Eq. 10.6 into this relation gives the third relation.

$$da = -P\,dv - s\,dT \tag{10.8}$$

Similarly,

$$g = h - Ts$$

$$dg = dh - T\,ds - s\,dT$$

Substituting Eq. 10.7 yields the fourth relation.

$$dg = v\,dP - s\,dT \tag{10.9}$$

Since Eqs. 10.6, 10.7, 10.8, and 10.9 are relations involving properties, we conclude that these are exact differentials and, therefore, are of the general form

$$dz = M\,dx + N\,dy$$

Since

$$\left(\frac{\partial M}{\partial y}\right)_x = \left(\frac{\partial N}{\partial x}\right)_y \tag{10.10}$$

it follows from Eq. 10.6 that

$$\left(\frac{\partial T}{\partial v}\right)_s = -\left(\frac{\partial P}{\partial s}\right)_v \tag{10.11}$$

Similarly, from Eqs. 10.7, 10.8, and 10.9 we can write

$$\left(\frac{\partial T}{\partial P}\right)_s = \left(\frac{\partial v}{\partial s}\right)_P \tag{10.12}$$

$$\left(\frac{\partial P}{\partial T}\right)_v = \left(\frac{\partial s}{\partial v}\right)_T \tag{10.13}$$

$$\left(\frac{\partial v}{\partial T}\right)_P = -\left(\frac{\partial s}{\partial P}\right)_T \tag{10.14}$$

These four equations are known as the Maxwell relations for a simple compressible mass, and the great utility of these equations will be demonstrated in later sections of this chapter. In particular, it should be noted that pressure, temperature, and specific volume can be measured by experimental methods, whereas entropy cannot be determined experimentally. By using the Maxwell relations, changes in entropy can be determined from quantities that can be measured, that is, pressure, temperature, and specific volume.

There are a number of other very useful relations that can be derived from Eqs. 10.6 through 10.9. For example, from Eq. 10.6, we can write the relations

$$\left(\frac{\partial u}{\partial s}\right)_v = T \qquad \left(\frac{\partial u}{\partial v}\right)_s = -P \tag{10.15}$$

Similarly, from the other equations, we have the following:

$$\left(\frac{\partial h}{\partial s}\right)_P = T \qquad \left(\frac{\partial h}{\partial P}\right)_s = v$$

$$\left(\frac{\partial a}{\partial v}\right)_T = -P \qquad \left(\frac{\partial a}{\partial T}\right)_v = -s$$

$$\left(\frac{\partial g}{\partial P}\right)_T = v \qquad \left(\frac{\partial g}{\partial T}\right)_P = -s \tag{10.16}$$

As already noted, the Maxwell relations just presented are written for a simple compressible substance. It is readily evident, however, that similar Maxwell relations can be written for substances involving other effects, such as electrical and magnetic effects. For example, Eq. 7.9 can be written in the form

$$dU = T\, dS - P\, dV + \mathcal{T}\, dL + \mathcal{S}\, d\mathcal{A} + \mu_0\, \mathcal{H}\, d(V\,\mathcal{M}) + \mathcal{E}\, dZ + \ldots \qquad (10.17)$$

Thus, at constant volume for a substance involving only magnetic effects, we can write

$$dU = T\, dS + \mu_0\, V\, \mathcal{H}\, d\mathcal{M}$$

and it follows that for such a substance

$$\left(\frac{\partial T}{\partial \mathcal{M}}\right)_{S,V} = \mu_0 V \left(\frac{\partial \mathcal{H}}{\partial S}\right)_{\mathcal{M},V}$$

Other Maxwell relations similar to Eqs. 10.12–10.14 could be written for such a substance. The extension of this approach to other systems, as well as to the interrelation between the various effects that may occur in a given system, is readily evident. For example, suppose a system involved both magnetic and surface effects. For such a system we could consider a constant entropy process and write

$$\left(\frac{\partial \mathcal{S}}{\partial \mathcal{M}}\right)_{S,\mathcal{A},V} = \mu_0 V \left(\frac{\partial \mathcal{H}}{\partial \mathcal{A}}\right)_{S,\mathcal{M},V}$$

This matter becomes much more complex when we consider applying the property relation to a system of variable composition. This subject will be taken up in Chapter 11.

EXAMPLE 10.1 From an examination of the properties of compressed liquid water, as given in Table A.1.4 of the Appendix, we find that the entropy of compressed liquid is greater than the entropy of saturated liquid for a temperature of 0°C, and is less than that of saturated liquid for all the other temperatures listed. Explain why this follows from other thermodynamic data.

Control mass: Water.

Solution

Suppose we increase the pressure of liquid water that is initially saturated, while keeping the temperature constant. The change of entropy for the water during this process can be found by integrating the following Maxwell relation, Eq. 10.14:

$$\left(\frac{\partial s}{\partial P}\right)_T = -\left(\frac{\partial v}{\partial T}\right)_P$$

Therefore, the sign of the entropy change depends on the sign of the term $(\partial v/\partial T)_P$. The physical significance of this term is that it involves the change in specific volume of water as the temperature changes while the pressure remains constant. As water at moderate pressures and 0°C is heated in a constant-pressure process, the specific volume decreases until the point of maximum density is reached at approximately 4°C, after

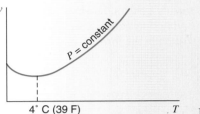

FIGURE 10.2 Sketch for Example 10.1.

which it increases. This is shown on a *v–T* diagram in Fig. 10.2. Thus, the quantity $(\partial v/\partial T)_P$ is the slope of the curve in Fig. 10.2. Since this slope is negative at 0°C, the quantity $(\partial s/\partial P)_T$ is positive at 0°C. At the point of maximum density the slope is zero and, therefore, the constant-pressure line shown in Fig. 7.7 crosses the saturated-liquid line at the point of maximum density.

10.3 CLAPEYRON EQUATION

The Clapeyron equation is an important relation concerning the saturation pressure and temperature, the change of enthalpy associated with a change of phase, and the specific volumes of the two phases. In particular it is an example of how a change in a property that cannot be measured directly, for example, the enthalpy, can be determined from measurements of pressure, temperature, and specific volume. It can be derived in a number of ways. Here we proceed by considering one of the Maxwell relations, Eq. 10.13.

$$\left(\frac{\partial P}{\partial T}\right)_v = \left(\frac{\partial s}{\partial v}\right)_T$$

Consider, for example, the change of state from saturated liquid to saturated vapor of a pure substance. This is a constant-temperature process, and therefore Eq. 10.13 can be integrated between the saturated-liquid and saturated-vapor state. We also note that when saturated states are involved, pressure and temperature are independent of volume. Therefore,

$$\left(\frac{dP}{dT}\right)_{\text{sat}} = \frac{s_g - s_f}{v_g - v_f} = \frac{s_{fg}}{v_{fg}} = \frac{h_{fg}}{Tv_{fg}} \tag{10.18}$$

The significance of this equation is that $(dP/dT)_{\text{sat}}$ is the slope of the vapor-pressure curve. Thus, h_{fg} at a given temperature can be determined from the slope of the vapor-pressure curve and the specific volume of saturated liquid and saturated vapor at the given temperature.

There are several different changes of phase that can occur at constant temperature and constant pressure. If we designate the two phases with superscripts ″ and ′, we can write the Clapeyron equation for the general case.

$$\left(\frac{dP}{dT}\right)_{\text{sat}} = \frac{s'' - s'}{v'' - v'}$$

We also note that $T(s'' - s') = h'' - h'$. Therefore,

$$\left(\frac{dP}{dT}\right)_{sat} = \frac{h'' - h'}{T(v'' - v')} \tag{10.19}$$

If the phase designated $''$ is vapor, then at low pressure the equation is usually simplified by assuming that $v'' \gg v'$ and that $v'' = RT/P$. The relation then becomes

$$\left(\frac{dP}{dT}\right)_{sat} = \frac{h_g - h'}{T(RT/P)}$$

$$\left(\frac{dP}{P}\right)_{sat} = \frac{(h_g - h')}{R}\left(\frac{dT}{T^2}\right)_{sat} \tag{10.20}$$

EXAMPLE 10.2 Determine the saturation pressure of water vapor at $-60°C$ using data available in the steam tables.

Control mass: Water.

Solution

Table 6 of the steam tables (Appendix Table A.1.5) does not give saturation pressures for temperatures less than $-40°C$. However, we do notice that h_{ig} is relatively constant in this range and, therefore, we proceed to Eq. 10.20 and integrate between the limits $-40°C$ and $-60°C$.

$$\int_1^2 \frac{dP}{P} = \int_1^2 \frac{h_{ig}}{R}\frac{dT}{T^2} = \frac{h_{ig}}{R}\int_1^2 \frac{dT}{T^2}$$

$$\ln\frac{P_2}{P_1} = \frac{h_{ig}}{R}\left(\frac{T_2 - T_1}{T_1 T_2}\right)$$

Let

$$P_2 = 0.0129 \text{ kPa} \qquad T_2 = 233.2 \text{ K} \qquad T_1 = 213.2 \text{ K}$$

Then

$$\ln\frac{P_2}{P_1} = \frac{2838.9}{0.461\,52}\left(\frac{233.2 - 213.2}{233.2 \times 213.2}\right) = 2.4744$$

$$P_1 = 0.001\,09 \text{ kPa}$$

EXAMPLE 10.2E Determine the saturation pressure of water vapor at -70 F using data available in the steam tables.

Control mass: Water.

Solution

Table 6 of the steam tables (Appendix Table A.1.5E) does not give saturation pressures for temperatures less than -40 F. However, we do notice that h_{ig} is relatively constant in this range and, therefore, we proceed to use Eq. 10.20 and integrate between the limits −40 F and −70 F.

$$\int_1^2 \frac{dP}{P} = \int_1^2 \frac{h_{ig}}{R} \frac{dT}{T^2} = \frac{h_{ig}}{R} \int_1^2 \frac{dT}{T^2}$$

$$\ln \frac{P_2}{P_1} = \frac{h_{ig}}{R} \left(\frac{T_2 - T_1}{T_1 T_2} \right)$$

Let

$$P_2 = 0.0019 \ \text{lbf / in.}^2 \qquad T_2 = 419.7 \ \text{R} \qquad T_1 = 389.7 \ \text{R}$$

Then,

$$\ln \frac{P_2}{P_1} = \frac{1218.7 \times 778}{85.76} \left(\frac{419.7 - 389.7}{419.7 \times 389.7} \right) = 2.0279$$

$$P_1 = 0.000 \ 25 \ \text{lbf / in.}^2$$

10.4 SOME THERMODYNAMIC RELATIONS INVOLVING ENTHALPY, INTERNAL ENERGY, AND ENTROPY

Let us first derive two equations, one involving C_p and the other involving C_v.

We have defined C_p as

$$C_p \equiv \left(\frac{\partial h}{\partial T} \right)_P$$

We have also noted that for a pure substance

$$T \, ds = dh - v \, dP$$

Therefore,

$$C_p = \left(\frac{\partial h}{\partial T} \right)_P = T \left(\frac{\partial s}{\partial T} \right)_P \tag{10.21}$$

Similarly, from the definition of C_v,

$$C_v \equiv \left(\frac{\partial u}{\partial T} \right)_v$$

and the relation

$$T \, ds = du + P \, dv$$

it follows that

$$C_v = \left(\frac{\partial u}{\partial T} \right)_v = T \left(\frac{\partial s}{\partial T} \right)_v \tag{10.22}$$

We will now derive a general relation for the change of enthalpy of a pure substance. We first note that for a pure substance

$$h = h(T, P)$$

Therefore,

$$dh = \left(\frac{\partial h}{\partial T} \right)_P dT + \left(\frac{\partial h}{\partial P} \right)_T dP$$

From the relation

$$T \, ds = dh - v \, dP$$

it follows that

$$\left(\frac{\partial h}{\partial P} \right)_T = v + T \left(\frac{\partial s}{\partial P} \right)_T$$

Substituting the Maxwell relation, Eq. 10.14, we have

$$\left(\frac{\partial h}{\partial P} \right)_T = v - T \left(\frac{\partial v}{\partial T} \right)_P \tag{10.23}$$

On substituting this equation and Eq. 10.21, we have

$$dh = C_p dT + \left[v - T \left(\frac{\partial v}{\partial T} \right)_P \right] dP \tag{10.24}$$

Along an isobar we have

$$dh_p = C_p dT_p$$

and along an isotherm,

$$dh_T = \left[v - T \left(\frac{\partial v}{\partial T} \right)_P \right] dP_T \tag{10.25}$$

The significance of Eq. 10.24 is that this equation can be integrated to give the change in enthalpy associated with a change of state

$$h_2 - h_1 = \int_1^2 C_p dT + \int_1^2 \left[v - T \left(\frac{\partial v}{\partial T} \right)_P \right] dP \tag{10.26}$$

The information needed to integrate the first term is a constant-pressure specific heat along one (and only one) isobar. The integration of the second integral requires that an equation of state giving the relation between P, v, and T be known. Furthermore, it is advantageous to have this equation of state explicit in v, for then the derivative $(\partial v / \partial T)_P$ is readily evaluated.

This matter can be further illustrated by reference to Fig. 10.3. Suppose we wish to know the change of enthalpy between states 1 and 2. We might determine this change along path 1–x–2, which consists of one isotherm, 1–x, and one isobar, x–2. Thus, we could integrate Eq. 10.26:

$$h_2 - h_1 = \int_{T_1}^{T_2} C_p dT + \int_{P_1}^{P_2} \left[v - T \left(\frac{\partial v}{\partial T} \right)_P \right] dP$$

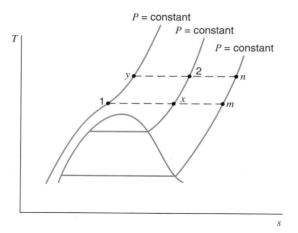

FIGURE 10.3 Sketch showing various paths by which a given change of state can take place.

Since $T_1 = T_x$ and $P_2 = P_x$, this can be written

$$h_2 - h_1 = \int_{T_x}^{T_2} C_p \, dT + \int_{P_1}^{P_x} \left[v - T \left(\frac{\partial v}{\partial T} \right)_P \right] dP$$

The second term in this equation gives the change in enthalpy along the isotherm 1–x and the first term the change in enthalpy along the isobar x–2. When these are added together, the result is the net change in enthalpy between 1 and 2. Therefore, the constant-pressure specific heat must be known along the isobar passing through 2 and x. The change in enthalpy could also be found by following path 1–y–2, in which case the constant-pressure specific heat must be known along the 1–y isobar. If the constant-pressure specific heat is known at another pressure, say, the isobar passing through m–n, the change in enthalpy can be found by following path 1–m–n–2. This involves calculating the change of enthalpy along two isotherms—1–m and n–2.

Let us now derive a similar relation for the change of internal energy. All the steps in this derivation are given, but without detailed comment. Note that the starting point is to write $u = u(T, v)$, whereas in the case of enthalpy the starting point was $h = h(T, P)$.

$$u = f(T, v)$$

$$du = \left(\frac{\partial u}{\partial T} \right)_v dT + \left(\frac{\partial u}{\partial v} \right)_T dv$$

$$T \, ds = du + P \, dv$$

Therefore,

$$\left(\frac{\partial u}{\partial v} \right)_T = T \left(\frac{\partial s}{\partial v} \right)_T - P \tag{10.27}$$

Substituting the Maxwell relation, Eq. 10.13, we have

$$\left(\frac{\partial u}{\partial v} \right)_T = T \left(\frac{\partial P}{\partial T} \right)_v - P$$

Therefore,

$$du = C_v dT + \left[T\left(\frac{\partial P}{\partial T}\right)_v - P \right] dv \qquad (10.28)$$

Along an isometric this reduces to

$$du_v = C_v dT_v$$

and along an isotherm we have

$$du_T = \left[T\left(\frac{\partial P}{\partial T}\right)_v - P \right] dv_T \qquad (10.29)$$

In a manner similar to that outlined above for changes in enthalpy, the change of internal energy for a given change of state for a pure substance can be determined from Eq. 10.28 if the constant-volume specific heat is known along one isometric and an equation of state explicit in P [to obtain the derivative $(\partial P/\partial T)_v$] is available in the region involved. A diagram similar to Fig. 10.3 could be drawn, with the isobars replaced with isometrics, and the same general conclusions would be reached.

To summarize, we have derived Eqs. 10.24 and 10.28:

$$dh = C_p dT + \left[v - T\left(\frac{\partial v}{\partial T}\right)_P \right] dP$$

$$du = C_v dT + \left[T\left(\frac{\partial P}{\partial T}\right)_v - P \right] dv$$

The first of these equations concerns the change of enthalpy, the constant-pressure specific heat, and is particularly suited to an equation of state explicit in v. The second equation concerns the change of internal energy and the constant-volume specific heat, and is particularly suited to an equation of state explicit in P. If the first of these equations is used to determine the change of enthalpy, the internal energy is readily found by noting that

$$u_2 - u_1 = h_2 - h_1 - (P_2 v_2 - P_1 v_1)$$

If the second equation is used to find changes of internal energy, the change of enthalpy is readily found from this same relation. Which of these two equations is used to determine changes in internal energy and enthalpy will depend on the information available for specific heat and an equation of state (or other P–v–T data).

Two parallel expressions can be found for the change of entropy.

$$s = s(T, P)$$

$$ds = \left(\frac{\partial s}{\partial T}\right)_P dT + \left(\frac{\partial s}{\partial P}\right)_T dP$$

Substituting Eqs. 10.21 and 10.14, we have

$$ds = C_p \frac{dT}{T} - \left(\frac{\partial v}{\partial T}\right)_P dP \qquad (10.30)$$

$$s_2 - s_1 = \int_1^2 C_p \frac{dT}{T} - \int_1^2 \left(\frac{\partial v}{\partial T}\right)_P dP \qquad (10.31)$$

Along an isobar we have

$$(s_2 - s_1)_P = \int_1^2 C_P \frac{dT_P}{T}$$

and along an isotherm

$$(s_2 - s_1)_T = -\int_1^2 \left(\frac{\partial v}{\partial T}\right)_P dP$$

Note from Eq. 10.31 that if a constant-pressure specific heat is known along one isobar and an equation of state explicit in v is available, the change of entropy can be evaluated. This is analogous to the expression for the change of enthalpy given in Eq. 10.24.

$$s = s(T, v)$$

$$ds = \left(\frac{\partial s}{\partial T}\right)_v dT + \left(\frac{\partial s}{\partial v}\right)_T dv$$

Substituting Eqs. 10.22 and 10.13 gives

$$ds = C_v \frac{dT}{T} + \left(\frac{\partial P}{\partial T}\right)_v dv \qquad (10.32)$$

$$s_2 - s_1 = \int_1^2 C_v \frac{dT}{T} + \int_1^2 \left(\frac{\partial P}{\partial T}\right)_v dv \qquad (10.33)$$

This expression for change of entropy concerns the change of entropy along an isometric where the constant-volume specific heat is known and along an isotherm where an equation of state explicit in P is known. Thus, it is analogous to the expression for change of internal energy given in Eq. 10.28.

EXAMPLE 10.3 Over a certain range of pressures and temperatures, the equation of state of a certain substance is given with considerable accuracy by the relation

$$\frac{Pv}{RT} = 1 - C' \frac{P}{T^4}$$

or

$$v = \frac{RT}{P} - \frac{C}{T^3}$$

where C and C' are constants.

Derive an expression for the change of enthalpy and entropy of this substance in an isothermal process.

Control mass: Gas.

Solution

Since the equation of state is explicit in v, Eq. 10.25 is particularly relevant to the change in enthalpy. On integrating this equation, we have

$$(h_2 - h_1)_T = \int_1^2 \left[v - T \left(\frac{\partial v}{\partial T} \right)_P \right] dP_T$$

From the equation of state,

$$\left(\frac{\partial v}{\partial T} \right)_P = \frac{R}{P} + \frac{3C}{T^4}$$

Therefore,

$$(h_2 - h_1)_T = \int_1^2 \left[v - T \left(\frac{R}{P} + \frac{3C}{T^4} \right) \right] dP_T$$

$$= \int_1^2 \left[\frac{RT}{P} - \frac{C}{T^3} - \frac{RT}{P} - \frac{3C}{T^3} \right] dP_T$$

$$(h_2 - h_1)_T = \int_1^2 -\frac{4C}{T^3} dP_T = -\frac{4C}{T^3} (P_2 - P_1)_T$$

For the change in entropy we use Eq. 10.31, which is particularly relevant for an equation of state explicit in v.

$$(s_2 - s_1)_T = -\int_1^2 \left(\frac{\partial v}{\partial T} \right)_P dP_T = -\int_1^2 \left(\frac{R}{P} + \frac{3C}{T^4} \right) dP_T$$

$$(s_2 - s_1)_T = -R \ln \left(\frac{P_2}{P_1} \right)_T - \frac{3C}{T^4} (P_2 - P_1)_T$$

10.5 SOME THERMODYNAMIC RELATIONS INVOLVING SPECIFIC HEAT

Some important relations concerning specific heats can also be developed. We have noted that the specific heat of an ideal gas is a function of the temperature only. For real gases the specific heat varies with pressure as well as temperature, and frequently we are interested in the variation of specific heat with pressure or volume. These relations can be derived as follows. Consider Eq. 10.30:

$$ds = \left(\frac{C_p}{T} \right) dT - \left(\frac{\partial v}{\partial T} \right)_p dP$$

Since this equation is of the general form $dz = M\, dx + N\, dy$, we can proceed as follows to find a relation that gives the variation of the constant-pressure specific heat with pressure at constant temperature.

$$\left(\frac{\partial (C_p/T)}{\partial P}\right)_T = -\left[\frac{\partial}{\partial T}\left(\frac{\partial v}{\partial T}\right)_P\right]_P$$

$$\left(\frac{\partial C_p}{\partial P}\right)_T = -T\left(\frac{\partial^2 v}{\partial T^2}\right)_P \tag{10.34}$$

The variation of the constant-volume specific heat with volume as the temperature remains constant can be found in a similar manner. Consider Eq. 10.32:

$$ds = \left(\frac{C_v}{T}\right)dT + \left(\frac{\partial P}{\partial T}\right)_v dv$$

Since this is of the form $dz = M\,dx + N\,dy$,

$$\left(\frac{\partial (C_v/T)}{\partial v}\right)_T = \left[\frac{\partial}{\partial T}\left(\frac{\partial P}{\partial T}\right)_v\right]_v$$

$$\left(\frac{\partial C_v}{\partial v}\right)_T = -T\left(\frac{\partial^2 P}{\partial T^2}\right)_v \tag{10.35}$$

The important thing to note about Eqs. 10.34 and 10.35 is that the variation of the constant-volume and constant-pressure specific heats at constant temperature can be found from the equation of state.

EXAMPLE 10.4 Determine the variation of C_p with pressure at constant temperature for a substance such as the one in Example 10.3 over the range where the equation of state is given by the relation

$$v = \frac{RT}{P} - \frac{C}{T^3}$$

Control mass: Gas.

Solution

Using Eq. 10.34, we have

$$\left(\frac{\partial C_p}{\partial P}\right)_T = -T\left(\frac{\partial^2 v}{\partial T^2}\right)_P$$

$$\left(\frac{\partial v}{\partial T}\right)_P = \frac{R}{P} + \frac{3C}{T^4}$$

$$\left(\frac{\partial^2 v}{\partial T^2}\right)_P = -\frac{12C}{T^5}$$

$$\left(\frac{\partial C_p}{\partial P}\right)_T = -T\left(-\frac{12C}{T^5}\right) = \frac{12C}{T^4}$$

A final interesting and useful relation involving the difference between C_p and C_v can be derived by equating Eqs. 10.30 and 10.32.

$$C_p \frac{dT}{T} - \left(\frac{\partial v}{\partial T}\right)_P dP = C_v \frac{dT}{T} + \left(\frac{\partial P}{\partial T}\right)_v dv$$

$$dT = \frac{T(\partial P / \partial T)_v}{C_p - C_v} dv + \frac{T(\partial v / \partial T)_P}{C_p - C_v} dP$$

But

$$T = f(v, P)$$

$$dT = \left(\frac{\partial T}{\partial v}\right)_P dv + \left(\frac{\partial T}{\partial P}\right)_v dP$$

Therefore,

$$\left(\frac{\partial T}{\partial v}\right)_P = \frac{T(\partial P / \partial T)_v}{C_p - C_v}$$

and

$$\left(\frac{\partial T}{\partial P}\right)_v = \frac{T(\partial v / \partial T)_P}{C_p - C_v}$$

When these equations are solved for $C_p - C_v$, they yield the same result.

$$C_p - C_v = T \left(\frac{\partial v}{\partial T}\right)_P \left(\frac{\partial P}{\partial T}\right)_v \tag{10.36}$$

But, from Eq. 10.2,

$$\left(\frac{\partial P}{\partial T}\right)_v = -\left(\frac{\partial v}{\partial T}\right)_P \left(\frac{\partial P}{\partial v}\right)_T$$

Therefore,

$$C_p - C_v = -T \left(\frac{\partial v}{\partial T}\right)_P^2 \left(\frac{\partial P}{\partial v}\right)_T \tag{10.37}$$

We draw several conclusions from this equation.

1. For liquids and solids, $(\partial v/\partial T)_P$ is usually relatively small and, therefore, for these phases the difference between the constant-pressure and constant-volume specific heats is small. For this reason many tables simply give the specific heat of a solid or a liquid without designating that it is at constant volume or pressure. Further, $C_p = C_v$ exactly when $(\partial v/\partial T)_P = 0$, as is true at the point of maximum density of water.
2. Because $C_p \rightarrow C_v$ as $T \rightarrow 0$ we therefore conclude that the constant-pressure and constant-volume specific heats are equal at absolute zero.
3. The difference between C_p and C_v is always positive because $(\partial v/\partial T)_P^2$ is always positive and $(\partial P/\partial v)_T$ is negative for all known substances.

10.6 VOLUME EXPANSIVITY AND ISOTHERMAL AND ADIABATIC COMPRESSIBILITY

The student has most likely encountered the coefficient of linear expansion in his or her studies of strength of materials. This coefficient indicates how the length of a solid body is influenced by a change in temperature while the pressure remains constant. In terms of the notation of partial derivatives, the *coefficient of linear expansion*, δ_T, is defined as

$$\delta_T = \frac{1}{L}\left(\frac{\partial L}{\partial T}\right)_P \tag{10.38}$$

A similar coefficient can be defined for changes in volume. Such a coefficient is applicable to liquids and gases as well as to solids. This coefficient of volume expansion, α_P, also called the volume expansivity, is an indication of the change in volume as temperature changes while the pressure remains constant. The definition of *volume expansivity* is

$$\alpha_P \equiv \frac{1}{V}\left(\frac{\partial V}{\partial T}\right)_P = \frac{1}{v}\left(\frac{\partial v}{\partial T}\right)_P \tag{10.39}$$

The isothermal compressibility, β_T, is an indication of the change in volume as pressure changes while the temperature remains constant. The definition of the *isothermal compressibility* is

$$\beta_T \equiv -\frac{1}{V}\left(\frac{\partial V}{\partial P}\right)_T = -\frac{1}{v}\left(\frac{\partial v}{\partial P}\right)_T \tag{10.40}$$

The reciprocal of the isothermal compressibility is called the *isothermal bulk modulus, B_T.*

$$B_T \equiv -v\left(\frac{\partial P}{\partial v}\right)_T \tag{10.41}$$

The *adiabatic compressibility*, β_s, is an indication of the change in volume as pressure changes while the entropy remains constant; it is defined as

$$\beta_s \equiv -\frac{1}{v}\left(\frac{\partial v}{\partial P}\right)_s \tag{10.42}$$

The *adiabatic bulk modulus, B_s,* is the reciprocal of the adiabatic compressibility.

$$B_s \equiv -v\left(\frac{\partial P}{\partial v}\right)_s \tag{10.43}$$

Both the volume expansivity and isothermal compressibility are thermodynamic properties of a substance, and for a simple compressible substance are functions of two independent properties. Values of these properties are found in the standard handbooks of physical properties. The following examples give an indication of the use and significance of the volume expansivity and isothermal compressibility.

EXAMPLE 10.5 Show that $C_p - C_v$ can be expressed in terms of the volume expansivity α_p, the specific volume v, the temperature T, and the isothermal compressibility β_T, by the relation

$$C_p - C_v = \frac{\alpha_p^2 v T}{\beta_T}$$

Control mass: Pure substance.

Solution

From Eq. 10.37

$$C_p - C_v = -T \left(\frac{\partial v}{\partial T} \right)_P^2 \left(\frac{\partial P}{\partial v} \right)_T$$

$$\alpha_p \equiv \frac{1}{v} \left(\frac{\partial v}{\partial T} \right)_P \qquad \left(\frac{\partial v}{\partial T} \right)^2 = \alpha_p^2 v^2$$

$$\beta_T \equiv -\frac{1}{v} \left(\frac{\partial v}{\partial P} \right)_T \qquad \left(\frac{\partial P}{\partial v} \right)_T = -\frac{1}{\beta_T v}$$

Therefore

$$C_p - C_v = -T(\alpha_p^2 v^2) \left(-\frac{1}{\beta_T v} \right) = \frac{\alpha_p^2 v T}{\beta_T}$$

EXAMPLE 10.6 The pressure on a block of copper having a mass of 1 kg is increased in a reversible process from 0.1 to 100 MPa while the temperature is held constant at 15°C. Determine the work done on the copper during this process, the change in entropy per kilogram of copper, the heat transfer, the change of internal energy per kilogram, and $C_p - C_v$ for this change of state.

Over the range of pressure and temperature in this problem, the following data can be used:

Volume expansivity = $\alpha_p = 5.0 \times 10^{-5} \, \text{K}^{-1}$

Isothermal compressibility = $\beta_T = 8.6 \times 10^{-12} \, \text{m}^2 / \text{N}$

Specific volume = 0.000 114 m^3 / kg

Control mass: Copper block.

States: Initial and final states known.

Process: Constant temperature, reversible.

Analysis:

The work done during the isothermal compression is

$$w = \int P \, dv_T$$

The isothermal compressibility has been defined as

$$\beta_T = -\frac{1}{v}\left(\frac{\partial v}{\partial P}\right)_T$$

$$v\beta_T dP_T = -dv_T$$

Therefore, for this isothermal process,

$$w = -\int_1^2 v\beta_T P\, dP_T$$

Since v and β_T remain essentially constant, this is readily integrated:

$$w = -\frac{v\beta_T}{2}(P_2^2 - P_1^2)$$

The change of entropy can be found by considering the Maxwell relation, Eq. 10.14, and the definition of volume expansivity.

$$\left(\frac{\partial s}{\partial P}\right)_T = -\left(\frac{\partial v}{\partial T}\right)_P = -\frac{v}{v}\left(\frac{\partial v}{\partial T}\right)_P = -v\alpha_P$$

$$ds_T = v\alpha_P dP_T$$

This equation can be readily integrated, if we assume that v and α_P remain constant:

$$(s_2 - s_1)_T = -v\alpha_p(P_2 - P_1)_T$$

The heat transfer for this reversible isothermal process is

$$q = T(s_2 - s_1)$$

The change in internal energy follows directly from the first law.

$$(u_2 - u_1) = q - w$$

From Example 10.5,

$$C_p - C_v = \frac{\alpha_p^2 vT}{\beta_T}$$

Solution

$$w = -\frac{v\beta_T}{2}(P_2^2 - P_1^2)$$

$$= -\frac{0.000\,114 \times 8.6\times10^{-12}}{2}(100^2 - 0.1^2)\times10^{12}$$

$$= -4.9\ \text{J/kg}$$

$$(s_2 - s_1)_T = -v\alpha_p(P_2 - P_1)_T$$

$$= -0.000\,114 \times 5.0\times10^{-5}(100-0.1)\times10^6$$

$$= -0.5694\ \text{J/kg K}$$

$$q = T(s_2 - s_2) = -288.2 \times 0.5694 = -164.1 \text{ J/kg}$$
$$(u_2 - u_1) = q - w = -164.1 - (-4.9) = -159.2 \text{ J/kg}$$
$$C_p - C_v = (5.0 \times 10^{-5})^2 \times \frac{0.000\ 114 \times 288.2}{8.6 \times 10^{-12}}$$
$$= 9.55 \text{ J/kg K}$$

This is consistent with the earlier observation that $C_p - C_v$ is relatively small for solids.

10.7 DEVELOPING TABLES OF THERMODYNAMIC PROPERTIES FROM EXPERIMENTAL DATA

There are many ways in which tables of thermodynamic properties can be developed from experimental data. The purpose of this section is to convey some general principles and concepts by considering only the liquid and vapor phases.

Let us assume that the following data for a pure substance have been obtained in the laboratory.

1. Vapor-pressure data. That is, saturation pressures and temperatures have been measured over a wide range.
2. Pressure, specific volume, and temperature data in the vapor region. These data are usually obtained by determining the mass of the substance in a closed vessel (which means a fixed specific volume) and then measuring the pressure as the temperature is varied. This is done for a large number of specific volumes.
3. Density of the saturated liquid and the critical pressure and temperature.
4. Zero-pressure specific heat for the vapor. This might be obtained either calorimetrically or from spectroscopic data.

From these data a complete set of thermodynamic tables for the saturated liquid, saturated vapor, and superheated vapor can be calculated. The first step is to determine an equation for the vapor-pressure curve that accurately fits the data. It may be necessary to use one equation for one portion of the vapor-pressure curve and a different equation for another portion of the curve.

One form of equation that has been used is

$$\ln P_{\text{sat}} = A + \frac{B}{T} + C \ln T + DT$$

Once an equation has been found that accurately represents the data, the saturation pressure for any given temperature can be found by solving this equation. Thus, the saturation pressures in Table 1 of the Steam Tables would be determined for the given temperatures. The second step is to determine an equation of state for the vapor region that accurately represents the P–v–T data. There are many possible forms of the equation of state that may be selected. The important considerations are that the equation of state accurately represents the data, and that it be of such a form that the differentiations required can be performed (that is, in some cases it may be desirable to have an equation of state

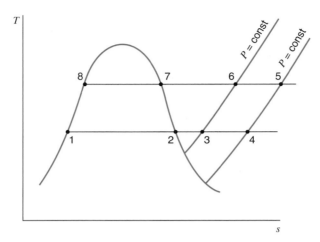

FIGURE 10.4 Sketch showing procedure for developing a table of thermodynamic properties from experimental data.

explicit in v, whereas on other occasions an equation of state that is explicit in P may be more desirable).

Once an equation of state has been determined, the specific volume of superheated vapor at given pressures and temperatures can be determined by solving the equation and tabulating the results as in the superheat tables for steam, ammonia, and the other substances listed in the appendix. The specific volume of saturated vapor at a given temperature may be found by finding the saturation pressure from the vapor-pressure curve and substituting this saturation pressure and temperature into the equation of state.

The procedure followed in determining enthalpy and entropy is best explained with the aid of Fig. 10.4. Let us assume that the enthalpy and entropy of saturated liquid in state 1 are zero. The enthalpy of saturated vapor in state 2 can be found from the Clapeyron equation.

$$\left(\frac{dP}{dT}\right)_{sat} = \frac{h_{fg}}{T(v_g - v_f)}$$

The left side of this equation is found by differentiating the vapor-pressure curve. The specific volume of the saturated vapor is found by the procedure outlined in the last paragraph, and it is assumed the specific volume of the saturated liquid has been measured. Thus the enthalpy of evaporation, h_{fg}, can be found for this particular temperature, and the enthalpy at state 2 is equal to the enthalpy of evaporation (since the enthalpy in state 1 is assumed to be zero). The entropy at state 2 is readily found, since

$$s_{fg} = \frac{h_{fg}}{T}$$

From state 2 we proceed along this isotherm into the superheated vapor region. The specific volume at 3 is found from the equation of state at this pressure, while the enthalpy and entropy are determined by integrating Eqs. 10.25 and 10.31:

$$h_3 - h_2 = \int_2^3 \left[v - T\left(\frac{\partial v}{\partial T}\right)_P\right] dP_T$$

$$s_3 - s_2 = \int_2^3 -\left(\frac{\partial v}{\partial T}\right)_P dP_T$$

The properties at point 4 are found in exactly the same manner. Pressure P_4 is sufficiently low that the real superheated vapor behaves essentially as an ideal gas (perhaps 10 kPa). Thus we use this constant-pressure line to make all temperature changes for our calculations, as for example to point 5. Since the specific heat C_{po} is known as a function of temperature, the enthalpy and entropy at 5 are found by integrating the ideal-gas relations

$$(h_5 - h_4)_P = \int_4^5 C_{po} dT_P$$

$$(s_5 - s_4)_P = \int_4^5 C_{po} \frac{dT_P}{T}$$

The properties at points 6 and 7 are found from those at 5 in the same manner as those at points 3 and 4 were found from 2 (the saturation pressure P_7 is calculated from the vapor-pressure equation). Finally, the enthalpy and entropy for saturated liquid at point 8 are found from the properties at point 7 by applying the Clapeyron equation.

Thus values for the pressure, temperature, specific volume, enthalpy, entropy, and internal energy of saturated liquid, saturated vapor, and superheated vapor can be tabulated for the entire region for which experimental data were obtained. The accuracy of such a table depends both on the accuracy of the experimental data and the degree to which the equation for the vapor pressure and the equation of state represent the experimental data.

10.8 THE IDEAL GAS

The ideal gas has been discussed at various relevant points in prior chapters, particularly in Sections 3.4, 5.7, and 7.10. Since an understanding of the properties of an ideal gas is essential to a study of the behavior of real substances, equations of state, and tables of thermodynamic properties, a brief summary of the behavior of ideal gases is presented here.

The ideal gas was defined in Section 3.4 as a gas that follows the equation of state, Eq. 3.1,

$$P\bar{v} = \bar{R}T$$

From a microscopic viewpoint, such an equation results when there are no intermolecular forces, in other words when the molecules are widely separated from one another. Thus, the ideal gas can be a reasonable model of real gas behavior only at very low density.

In Section 5.7 it was stated that the internal energy of an ideal gas is a function only of temperature, as stated by Eq. 5.19. At this point it is appropriate to demonstrate that this is in fact the case. For an ideal gas, from Eq. 3.2,

$$P = \frac{RT}{v}$$

Therefore,

$$\left(\frac{\partial P}{\partial T} \right)_v = \frac{R}{v}$$

Substituting this relation into Eq. 10.29, we have

$$du_T = \left[T\left(\frac{\partial P}{\partial T}\right)_v - P \right] dv_T = \left[T\left(\frac{R}{v}\right) - P \right] dv_T = 0$$

and we find no variation of internal energy along an isotherm. In other words, for the ideal gas Eq. 10.28 reduces to Eq. 5.20,

$$du = C_{vo} dT$$

We can also examine various other general thermodynamic relations for the special case of the ideal gas. The procedure is in each case similar to that followed above for the internal energy, and the details are left as an exercise. To summarize the results:

Eqs. 10.34 and 10.35 demonstrate the validity of Eq. 5.26,

$$C_{vo} = f(T) \qquad C_{po} = f(T)$$

Eq. 10.24 reduces to Eq. 5.24,

$$dh = C_{po} dT$$

Eq. 10.37 reduces to Eq. 5.27,

$$C_{po} - C_{vo} = R$$

Eq. 10.32 reduces to Eq. 7.19,

$$ds = C_{vo} \frac{dT}{T} + R \frac{dv}{v}$$

and Eq. 10.30 reduces to Eq. 7.21,

$$ds = C_{po} \frac{dT}{T} - R \frac{dP}{P}$$

In discussing the accuracy of the ideal gas equation of state in Section 3.4, we introduced the compressibility factor

$$Z = \frac{P\bar{v}}{\bar{R}T} \tag{10.44}$$

as a useful and important parameter in expressing the nonideality of gases (or other phases). The compressibility factor is always unity (for all pressures and temperatures) for an ideal gas.

Another useful parameter in describing the behavior of a real gas relative to the ideal gas is the residual volume α, which is defined as

$$\alpha = \frac{\bar{R}T}{P} - \bar{v} \tag{10.45}$$

We note that α is always zero for an ideal gas.

The Joule–Thomson coefficient μ_J was defined in Section 5.13 as

$$\mu_J = \left(\frac{\partial T}{\partial P}\right)_h \tag{10.46}$$

Since h is a function only of T for an ideal gas, it follows that the Joule–Thomson coefficient is always zero for an ideal gas.

10.9 THE BEHAVIOR OF REAL GASES

In Section 3.4 we discussed briefly the nonideality in *P–v–T* behavior of a gas, using nitrogen as an example, and in connection with that discussion constructed a skeleton compressibility diagram, Fig. 3.7. A more detailed compressibility diagram for N_2 is shown in Fig. 10.5. In examining this diagram, we note that *Z* approaches unity for all isotherms as *P* approaches zero, that the nonideality of the gas phase is especially severe in the vicinity of the critical point (*Z* at the critical point is about 0.29), and that at high pressures *Z* increases to values greater than unity for all isotherms.

If we examine compressibility diagrams for other pure substances, we find that the diagrams are all similar in the characteristics described above for nitrogen, at least in a qualitative sense. Quantitatively the diagrams are all different, since the critical temperatures and pressures of different substances vary over wide ranges, as evidenced from the values listed in Table A.8. Is there a way in which we can put all of these substances on a common basis? To do so, we "reduce" the properties with respect to the values at the critical point. The reduced properties are defined as

$$\text{Reduced pressure} = P_r = \frac{P}{P_c} \qquad P_c = \text{Critical pressure}$$

$$\text{Reduced temperature} = T_r = \frac{T}{T_c} \qquad T_c = \text{Critical temperature}$$

$$\text{Reduced specific volume} = v_r = \frac{v}{v_c} \qquad v_c = \text{Critical specific volume}$$

These equations state that the reduced property for a given state is the value of this property in this state divided by the value of this same property at the critical point.

If lines of constant T_r are plotted on a *Z* versus P_r diagram, a plot such as that in Fig. A.7 is obtained. The striking fact is that when such *Z* versus P_r diagrams are prepared for a number of different substances, all of them very nearly coincide, especially when the substances have simple, essentially spherical molecules. Correlations for substances with more complicated molecules are reasonably close, except near or at saturation or at high density. Thus, Figure A.7 is actually a generalized diagram for simple molecules, which means that it represents the average behavior for a number of different simple substances. When such a diagram is used for a particular substance, the results will generally be somewhat in error. On the other hand, if *P–v–T* information is required for a substance in a region where no experimental measurements have been made, this generalized compressibility diagram will give reasonably accurate results. We need know only the critical pressure and critical temperature to use this basic generalized chart.

In this text, the simple-fluid compressibilities represented in Fig. A.7 are calculated from a generalized equation of state called the Lee–Kesler equation, which is an extension of the Benedict–Webb–Rubin equation (Eq. 3.7). Both equations are discussed in greater detail in the following section. These calculated compressibilities are also presented in tabular form in Table A.15.

A third parameter is often introduced to improve the accuracy of correlation of the generalized compressibility factor. One such parameter is termed the acentric factor ω, which is intended to account for molecular shape, geometric complexity, and polarity. Values for the acentric factor are listed with the critical constants in Table A.8, and its use in adding a correction to *Z*, as well as to other thermodynamic properties, is discussed in Appendix C. The examples in this chapter will include only the simple-fluid

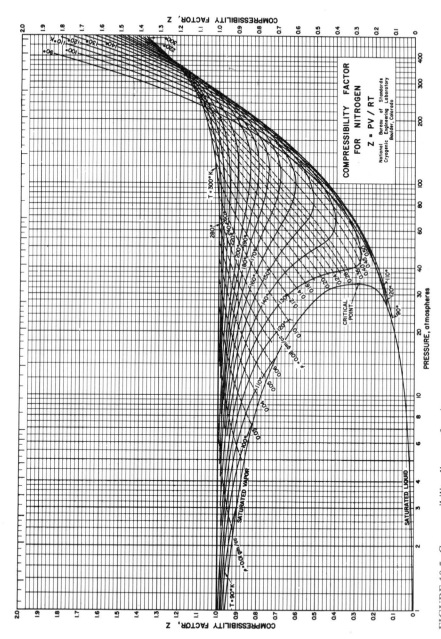

FIGURE 10.5 Compressibility diagram for nitrogen.

terms, although several problems at the end of the chapter do include the additional corrections from Appendix C. Use of the generalized table A.15 (or chart, Fig. A.7) is illustrated by the following example.

EXAMPLE 10.7

1. Volume unknown. Calculate the specific volume of propane at a pressure of 7 MPa and a temperature of 150°C, and compare this with the specific volume given by the ideal-gas equation of state.

   Control mass: Propane.

   State: P, T known.

   Model: Generalized compressibility tables, A.15.

Solution

For propane

$$T_c = 369.8 \text{ K} \qquad P_c = 4.25 \text{ MPa}$$
$$R = 0.188\ 55 \text{ kJ / kg K}$$
$$T_r = \frac{423.2}{369.8} = 1.144 \qquad P_r = \frac{7}{4.25} = 1.647$$

From the compressibility tables,

$$Z = 0.5233$$

(This is the value from the software at the given reduced temperature and pressure, and with zero acentric factor. A linear interpolation among the printed values in A.15 will be slightly different.)

$$v = \frac{ZRT}{P} = \frac{0.5233 \times 0.188\ 55 \times 423.2}{7000} = 0.005\ 965 \text{ m}^3 / \text{kg}$$

The ideal-gas equation would give the value

$$v = \frac{0.188\ 55 \times 423.2}{7000} = 0.0114 \text{ m}^3 / \text{kg}$$

2. Pressure unknown. What pressure is required in order that propane have a specific volume of 0.005 965 m³/kg at a temperature of 150°C?

   State: T, v known.

Solution

$$T_r = \frac{423.2}{369.8} = 1.144$$
$$P_r = \frac{P}{P_c} = \frac{ZRT}{vP_c} = Z \times \frac{0.188\ 55 \times 423.2}{0.005\ 965 \times 4250} = 3.1476\ Z$$

By a trial-and-error procedure, or by plotting a few points and drawing the curve representing this equation, P_r is found to be 1.647 at the point where $T_r = 1.144$.

Therefore,

$$P = P_r P_c = 1.647 \times 4250 = 7000 \text{ kPa} = 7 \text{ MPa}$$

3. Temperature unknown. What will be the temperature of propane when it has a specific volume of 0.005 965 m³/kg and a pressure of 7 MPa?

State: P, v known.

Solution

$$P_r = \frac{7}{4.25} = 1.647$$

$$T_r = \frac{T}{T_c} = \frac{Pv}{ZRT_c} = \frac{7000 \times 0.005\ 965}{Z \times 0.188\ 55 \times 369.8} = \frac{0.598\ 85}{Z}$$

By trial and error the reduced temperature at which this expression is satisfied for the reduced pressure of 1.647 is found to be 1.144. Therefore,

$$T_r = 1.144$$

$$T = T_r \times T_c = 1.144 \times 369.8 = 423.2 \text{ K}$$

To gain additional insight into the behavior of gases at low density, let us examine the low-pressure portion of the generalized compressibility chart in greater detail. This behavior is as shown in Fig. 10.6. The isotherms are essentially straight lines in this region, and their slope is of particular importance. Note that the slope increases as T_r increases until a maximum value is reached at a T_r of about 5, and then the slope decreases toward the $Z = 1$ line for higher temperatures. That single temperature, about 2.5 times the critical temperature, for which

$$\lim_{P \to 0} \left(\frac{\partial Z}{\partial P} \right)_T = 0 \tag{10.47}$$

is defined as the Boyle temperature of the substance. This is the only temperature at which a gas really behaves exactly as an ideal gas at low, but finite pressures, since all

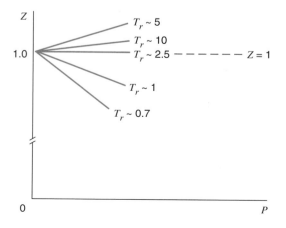

FIGURE 10.6 Low-pressure region of compressibility chart.

other isotherms go to zero pressure on Fig. 10.6 with a nonzero slope. To amplify this point, let us consider the residual volume α, which was defined according to Eq. 10.45,

$$\alpha = \frac{\bar{R}T}{P} - \bar{v}$$

Multiplying this equation by P we have

$$\alpha P = \bar{R}T - P\bar{v} \tag{10.48}$$

Thus the quantity αP is the difference between $\bar{R}T$ and $P\bar{v}$. Now as $P \to 0$, $P\bar{v} \to \bar{R}T$. However, it does not necessarily follow that $\alpha \to 0$ as $P \to 0$. Instead, it is only required that α remain finite. The derivative in Eq. 10.47 can be written as

$$\lim_{P \to 0} \left(\frac{\partial Z}{\partial P} \right)_T = \lim_{P \to 0} \left(\frac{Z-1}{P-0} \right)$$

$$= \lim_{P \to 0} \frac{1}{\bar{R}T} \left(\bar{v} - \frac{\bar{R}T}{P} \right)$$

$$= -\frac{1}{\bar{R}T} \lim_{P \to 0} (\alpha) \tag{10.49}$$

from which we find that α tends to zero as $P \to 0$ only at the Boyle temperature, since that is the only temperature for which the isothermal slope is zero on Fig. 10.6. It is perhaps a somewhat surprising result that in the limit as $P \to 0$, $P\bar{v} \to \bar{R}T$ but in general the quantity $(\bar{R}T/P - \bar{v})$ does not go to zero, but is instead a small difference between two large values. This does have an effect on certain other properties of the gas.

Another aspect of generalized behavior of gases is the behavior of isotherms in the vicinity of the critical point. If we plot experimental data on P–v coordinates, it is found that the critical isotherm is unique in that it goes through a horizontal inflection point at the critical point as shown in Fig. 10.7. Mathematically, this means that the first two derivatives are zero at the critical point,

$$\left(\frac{\partial P}{\partial v} \right)_{T_c} = 0 \qquad \text{at C.P.} \tag{10.50}$$

$$\left(\frac{\partial^2 P}{\partial v^2} \right)_{T_c} = 0 \qquad \text{at C.P.} \tag{10.51}$$

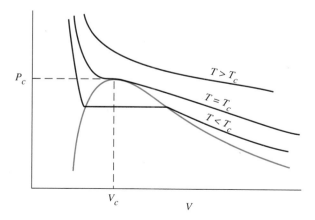

FIGURE 10.7 Plot of isotherms in the region of the critical point on pressure–volume coordinates for a typical pure substance.

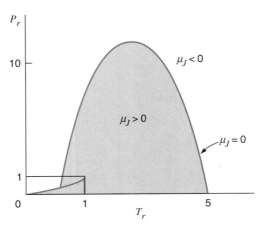

FIGURE 10.8 Joule–Thomson inversion curve.

a feature that is used to constrain many equations of state, as will be discussed in Section 10.10.

The Joule–Thomson coefficient, defined in connection with the throttling process, can be expressed in terms of $P–v–T$ behavior as

$$\mu_J = \left(\frac{\partial T}{\partial P}\right)_h = \frac{T\left(\dfrac{\partial v}{\partial T}\right)_P - v}{C_p} = \frac{RT^2}{PC_p}\left(\frac{\partial Z}{\partial T}\right)_P \tag{10.52}$$

From Eq. 10.52 and Fig. 10.6 it is evident that at low pressure μ_J is zero only at one temperature, T_r about 5, and is positive at lower temperatures and negative at higher temperatures. By examining higher density $P–v–T$ data, we can determine the locus of points at which μ_J is zero, which is called the Joule–Thomson inversion curve. The result is as shown in Fig. 10.8. Inside the dome μ_J is positive, which means that a substance cools upon being throttled to a lower pressure. Outside the dome μ_J is negative, and the temperature increases during a throttling process.

There are also many other aspects of generalized behavior of gases, some of which are discussed in connection with equations of state in the next section.

10.10 EQUATIONS OF STATE

An accurate equation of state, which is an analytical representation of $P–v–T$ behavior, is often desirable from a computational standpoint. Many different equations of state have been developed. Most of these are accurate only to some density less than the critical density, though a few are reasonably accurate to approximately 2.5 times the critical density. All equations of state fail badly when the density exceeds the maximum density for which the equation was developed.

Three broad classifications of equations of state can be identified, namely, generalized, empirical, and theoretical.

The best known of the generalized equations of state is also the oldest, namely, the van der Waals equation, which was presented in 1873 as a semitheoretical improvement over the ideal gas equation. The van der Waals equation of state is:

$$P = \frac{RT}{v-b} - \frac{a}{v^2}$$ (10.53)

The constant b is intended to correct for the volume occupied by the molecules, and the term a/v^2 is a correction that accounts for the intermolecular forces of attraction. As might be expected in the case of a generalized equation, the constants a and b are evaluated from the general behavior of gases. In particular these constants are evaluated by noting that the critical isotherm passes through a point of inflection at the critical point, and that the slope is zero at this point. Thus, for the van der Waals equation of state we have

$$\left(\frac{\partial P}{\partial v}\right)_T = -\frac{RT}{(v-b)^2} + \frac{2a}{v^3}$$ (10.54)

$$\left(\frac{\partial^2 P}{\partial v^2}\right)_T = \frac{2RT}{(v-b)^3} - \frac{6a}{v^4}$$ (10.55)

Since both of these derivatives are equal to zero at the critical point we can write

$$-\frac{RT_c}{(v_c-b)^2} + \frac{2a}{v_c^3} = 0$$

$$\frac{2RT_c}{(v_c-b)^3} - \frac{6a}{v_c^4} = 0$$ (10.56)

$$P_c = \frac{RT_c}{(v_c-b)} - \frac{a}{v_c^2}$$

Solving these three equations we find

$$v_c = 3b$$

$$a = \frac{27}{64} \frac{R^2 T_c^2}{P_c}$$ (10.57)

$$b = \frac{RT_c}{8P_c}$$

The compressibility factor at the critical point for the van der Waals equation is

$$Z_c = \frac{P_c v_c}{RT_c} = \frac{3}{8}$$

which is considerably higher than the actual value for any substance.

The van der Waals equation can be written in terms of the compressibility factor and the reduced pressure and reduced temperature as follows:

$$Z^3 - \left(\frac{P_r}{8T_r} + 1\right)Z^2 + \left(\frac{27P_r}{64T_r^2}\right)Z - \frac{27P_r^2}{512T_r^3} = 0$$ (10.58)

It is significant that this is of the same form as the generalized compressibility chart, namely, $Z = f(P_r, T_r)$, although the functional relation is quite different from that of the generalized chart. This concept that different substances will have the same compressibility factor at the same reduced pressure and reduced temperature is another way of expressing the rule of corresponding states.

A simple equation of state that is considerably more accurate than the van der Waals equation is that proposed by Redlich and Kwong in 1949.

$$P = \frac{\overline{R}T}{\overline{v}-b} - \frac{a}{\overline{v}(\overline{v}+b)T^{1/2}} \tag{10.59}$$

with

$$a = 0.427\ 48 \frac{\overline{R}^2 T_c^{5/2}}{P_c} \tag{10.60}$$

$$b = 0.086\ 64 \frac{\overline{R}T_c}{P_c} \tag{10.61}$$

The numerical values in the constants have been determined by a procedure similar to that followed in the van der Waals equation. Because of its simplicity this equation could not be expected to be sufficiently accurate to find use in the calculation of precision tables of thermodynamic properties. It has, however, been used frequently for mixture calculations and phase equilibrium correlations with reasonably good success. A number of modified versions of this equation have also been utilized in recent years.

One of the best-known empirical equations of state is the Benedict–Webb–Rubin equation, often termed the BWR equation. First proposed in 1940, it has since then been very widely used in its original version and in modified versions. The original form of this equation, given in Chapter 3, is

$$P = \frac{RT}{v} + \frac{RTB_0 - A_0 - C_0/T^2}{v^2} + \frac{RTb - a}{v^3} + \frac{a\alpha}{v^6} + \frac{c}{v^3 T^2}\left(1 + \frac{\gamma}{v^2}\right)e^{-\gamma/v^2} \tag{10.62}$$

with eight empirical constants. The values of the constants are given for a number of substances in Table 3.3.

One particularly interesting modification of the BWR equation of state is the Lee–Kesler equation, which was proposed in 1975. This equation has 12 constants and is written in terms of generalized properties as

$$Z = \frac{P_r v_r'}{T_r} = 1 + \frac{B}{v_r'} + \frac{C}{v_r'^2} + \frac{D}{v_r'^5} + \frac{c_4}{T_r^3 v_r'^2}\left(\beta + \frac{\gamma}{v_r'^2}\right)\exp\left(-\frac{\gamma}{v_r'^2}\right)$$

$$B = b_1 - \frac{b_2}{T_r} - \frac{b_3}{T_r^2} - \frac{b_4}{T_r^3}$$

$$C = c_1 - \frac{c_2}{T_r} + \frac{c_3}{T_r^3} \tag{10.63}$$

$$D = d_1 + \frac{d_2}{T_r}$$

in which the variable v_r' is not the true reduced specific volume, but is instead defined as

$$v_r' = \frac{v}{RT_c/P_c} \tag{10.64}$$

Two sets of empirical constants are given for this equation in Table A.15. The first set of constants is based on observed generalized behavior of simple fluids and therefore yields, from Eq. 10.63, the corresponding simple-fluid compressibility factors that are listed in Table A.15. The second set of constants is for a relatively complicated reference fluid, which allows for a correction factor to the simple-fluid result. The details of this procedure and values for the correction factor are listed in Appendix C, and are not discussed further in this chapter.

A different approach to this problem is from the theoretical point of view. The theoretical equation of state, which is derived from kinetic theory or statistical thermodynamics, is written here in the form of a power series in reciprocal volume:

$$Z = \frac{P\bar{v}}{\bar{R}T} = 1 + \frac{B(T)}{\bar{v}} + \frac{C(T)}{\bar{v}^2} + \frac{D(T)}{\bar{v}^3} + \cdots \tag{10.65}$$

where $B(T)$, $C(T)$, $D(T)$ are temperature dependent and are called virial coefficients. $B(T)$ is termed the second virial coefficient and is due to binary interactions on the molecular level. The general temperature dependence of the second virial coefficient is as shown for nitrogen in Fig. 10.9. If we multiply Eq. 10.65 by $\bar{R}T/P$, the result can be rearranged to the form

$$\frac{\bar{R}T}{P} - \bar{v} = \alpha = -B(T)\frac{\bar{R}T}{P\bar{v}} - C(T)\frac{\bar{R}T}{P\bar{v}^2} \cdots \tag{10.66}$$

In the limit, as $P \to 0$,

$$\lim_{P \to 0} \alpha = -B(T) \tag{10.67}$$

and we conclude from Eqs. 10.47 and 10.49 that the single temperature at which $B(T) = 0$, Fig. 10.9, is the Boyle temperature. The second virial coefficient can be viewed as the first-order correction for nonideality of the gas, and consequently becomes of considerable importance and interest. In fact, the low-density behavior of the isotherms shown in Fig. 10.6 is directly attributable to the second virial coefficient. This is of further interest, since the virial coefficients can be expressed in terms of intermolecular forces from statistical mechanics, and evaluated upon selection of an empirical potential function model. The best known of these is the Lennard–Jones (6–12) potential. This function has two force constants, ε/k and b_0, values of which are listed for various substances in Table

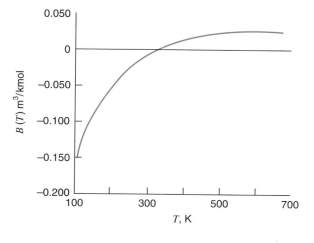

FIGURE 10.9 The second virial coefficient for nitrogen.

A.14. The expressions for $B(T)$ and $C(T)$ can be put into dimensionless form in terms of the parameter

$$T^* = \frac{T}{\varepsilon/k}$$ (10.68)

and evaluated for the dimensionless second virial coefficient $B^*(T^*)$, defined from the relation

$$B^*(T^*) = \frac{B(T)}{b_0}$$ (10.69)

and also for the dimensionless third virial coefficient $C^*(T^*)$, defined in a similar manner by

$$C^*(T^*) = \frac{C(T)}{b_0^2}$$ (10.70)

Values of the dimensionless parameters $B^*(T^*)$ and $C^*(T^*)$, as calculated using the Lennard–Jones (6–12) potential, are also given in Table A.14.

EXAMPLE 10.8

Use the virial equation of state to determine the error in specific volume if carbon dioxide is assumed to be an ideal gas at 300 K, 1 MPa pressure.

Control mass: Carbon dioxide.

State: P, T known.

Model: Virial equation with second and third virial coefficient; compare with ideal-gas model.

Solution

From Eq. 10.65 at low density,

$$Z = \frac{P\bar{v}}{\bar{R}T} = 1 + \frac{B(T)}{\bar{v}} + \frac{C(T)}{\bar{v}^2}$$

For carbon dioxide, from Table A.14,

$$\varepsilon/k = 186 \text{ K} \qquad b_0 = 0.118 \text{ m}^3/\text{kmol}$$

Therefore,

$$T^* = \frac{T}{\varepsilon/k} = \frac{300}{186} = 1.612$$

From Table A.14,

$$B^*(T^*) = -1.0365 \qquad C^*(T^*) = +0.5145$$

Using Eqs. 10.68 and 10.70,

$$B(T) = b_0 B^*(T^*) = 0.118(-1.0365) = -0.122 \text{ m}^3/\text{kmol}$$

$$C(T) = b_0^2 C^*(T^*) = (0.118)^2 \times 0.5145 = 0.007\,16 \text{ (m}^3/\text{kmol})^2$$

Substituting and solving for $\bar{v}$,

$$\frac{1000\bar{v}}{8.3145 \times 300} = 1 - \frac{0.122}{\bar{v}} + \frac{0.007\,16}{\bar{v}^2}$$

$$\bar{v} = 2.369 \text{ m}^3 / \text{kmol}$$

For an ideal gas,

$$\bar{v} = \frac{\bar{R}\,T}{P} = \frac{8.3145 \times 300}{1000} = 2.494 \text{ m}^3 / \text{kmol}$$

which is in error by 5.3%.

10.11 THE GENERALIZED TABLE/CHART FOR CHANGES OF ENTHALPY AT CONSTANT TEMPERATURE

In Section 10.5, Eq. 10.25 was derived for the change of enthalpy at constant temperature.

$$(h_2 - h_1)_T = \int_1^2 \left[v - T\left(\frac{\partial v}{\partial T}\right)_P \right] dP_T$$

This equation is appropriately used when a volume-explicit equation of state is known. Otherwise, it is more convenient to calculate the isothermal change in internal energy from Eq. 10.29,

$$(u_2 - u_1)_T = \int_1^2 \left[T\left(\frac{\partial P}{\partial T}\right)_v - P \right] dv_T$$

and then calculate the change in enthalpy from its definition as

$$(h_2 - h_1) = (u_2 - u_1) + (P_2 v_2 - P_1 v_1)$$

$$= (u_2 - u_1) + RT(Z_2 - Z_1)$$

To determine the change in enthalpy behavior consistent with the generalized tables of Table A.15 (and as represented in chart form in Fig. A.7), we follow the second of these approaches, since the Lee–Kesler generalized equation of state, Eq. 10.63, is a pressure-explicit form in terms of specific volume and temperature. Equation 10.63 is expressed in terms of the compressibility factor Z, so we write

$$P = \frac{ZRT}{v}, \qquad \left(\frac{\partial P}{\partial T}\right)_v = \frac{ZR}{v} + \frac{RT}{v}\left(\frac{\partial Z}{\partial T}\right)_v$$

Therefore, substituting into Eq. 10.29, we have

$$du = \frac{RT^2}{v}\left(\frac{\partial Z}{\partial T}\right)_v dv$$

But

$$\frac{dv}{v} = \frac{dv'_r}{v'_r} \qquad \frac{dT}{T} = \frac{dT_r}{T_r}$$

so that, in terms of reduced variables,

$$\frac{1}{RT_c} du = \frac{T_r^2}{v'_r} \left(\frac{\partial Z}{\partial T_r} \right)_{v'_r} dv'_r$$

This expression is now integrated at constant temperature from any given state (P_r, v'_r) to the ideal-gas limit $(P_r^* \to 0, v'_r{}^* \to \infty)$ (the superscript * will always denote an ideal-gas state or property), causing an internal energy change or departure from the ideal-gas value at the given state,

$$\frac{u^* - u}{RT_c} = \int_{v'_r}^{\infty} \frac{T_r^2}{v'_r} \left(\frac{\partial Z}{\partial T_r} \right)_{v'_r} dv'_r \tag{10.71}$$

The integral on the right-hand side of Eq. 10.71 can be evaluated from the Lee–Kesler equation, Eq. 10.63. The corresponding enthalpy departure at the given state (P_r, v'_r) is then found from integrating Eq. 10.71 to be

$$\frac{h^* - h}{RT_c} = \frac{u^* - u}{RT_c} + T_r(1 - Z) \tag{10.72}$$

Following the same procedure as for the compressibility factor, we can evaluate Eq. 10.72 with the set of Lee–Kesler simple-fluid constants to give a simple-fluid enthalpy departure. The values for the enthalpy departure are listed in Table A.15 and are shown graphically in Fig. A.8. Use of the enthalpy departure function is illustrated in the following example.

EXAMPLE 10.9 Nitrogen is throttled from 20 MPa, −70°C, to 2 MPa in an adiabatic, steady-state, steady-flow process. Determine the final temperature of the nitrogen.

Control volume: Throttling valve.

Inlet state: P_1, T_1 known; state fixed.

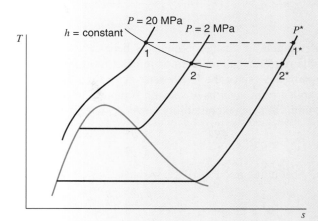

FIGURE 10.10 Sketch for Example 10.9.

Exit state: P_2 known.

Process: SSSF, throttling process.

Diagram: Figure 10.10.

Model: Generalized tables, Table A.15.

Analysis:

First law:

$$h_1 = h_2$$

Solution

Using values from Table A.8, we have

$$P_1 = 20 \text{ MPa} \qquad P_{r1} = \frac{20}{3.39} = 5.9$$

$$T_1 = 203.2 \text{ K} \qquad T_{r1} = \frac{203.2}{126.2} = 1.61$$

$$P_2 = 2 \text{ MPa} \qquad P_{r2} = \frac{2}{3.39} = 0.59$$

From the generalized tables, Table A.15, for the change in enthalpy at constant temperature, we have

$$\frac{h_1^* - h_1}{RT_c} = 2.0314$$

$$h_1^* - h_1 = 2.0314 \times 0.2968 \times 126.2 = 76.1 \text{ kJ / kg}$$

It is now necessary to assume a final temperature and to check whether the net change in enthalpy for the process is zero. Let us assume that $T_2 = 148$ K. Then the change in enthalpy between 1* and 2* can be found from the zero-pressure, specific-heat data.

$$h_1^* - h_2^* = C_{p0}(T_1^* - T_2^*) = 1.0416(203.2 - 148) = +57.5 \text{ kJ / kg}$$

(The variation in C_{p0} with temperature can be taken into account when necessary.)
We now find the enthalpy change between 2* and 2.

$$T_{r2} = \frac{148}{126.2} = 1.173 \qquad P_{r2} = 0.59$$

Therefore, from the enthalpy departure table, Table A.15, at this state

$$\frac{h_2^* - h_2}{RT_c} = 0.4901$$

$$h_2^* - h_2 = 0.4901 \times 0.2968 \times 126.2 = 18.4 \text{ kJ / kg}$$

We now check to see whether the net change in enthalpy for the process is zero.

$$h_1 - h_2 = 0 = -(h_1^* - h_1) + (h_1^* - h_2^*) + (h_2^* - h_2)$$

$$= -76.1 + 57.5 + 18.4 = -0.2 \approx 0$$

It essentially checks. We conclude that the final temperature is approximately 148 K. It is interesting that the thermodynamic tables for nitrogen, Table A.6, give essentially this same value for the final temperature.

10.12 The Generalized Table/Chart for Changes of Entropy at Constant Temperature

In this section we wish to develop a generalized table or chart giving entropy departures from ideal gas values at a given temperature and pressure, in a manner similar to that followed for enthalpy in the previous section. Once again, we have two alternatives. From Eq. 10.30, at constant temperature,

$$ds_T = -\left(\frac{\partial v}{\partial T}\right)_P dP_T$$

which is convenient for use with a volume-explicit equation of state. The Lee–Kesler expression, Eq. 10.63, is, however, a pressure-explicit equation. It is therefore more appropriate to use Eq. 10.32, which is, along an isotherm,

$$ds_T = \left(\frac{\partial P}{\partial T}\right)_v dv_T$$

In the Lee–Kesler form, in terms of reduced properties, this equation becomes

$$\frac{ds}{R} = \left(\frac{\partial P_r}{\partial T_r}\right)_{v'_r} dv'_r$$

When this expression is integrated from a given state (P_r, v'_r) to the ideal-gas limit $(P_r^* \to 0, v_r'^* \to \infty)$, there is a problem because ideal-gas entropy is a function of pressure and approaches infinity as the pressure approaches zero. We can eliminate this problem with a two-step procedure. First, the integral is taken only to a certain finite $P_r^*, v_r'^*$, which gives the entropy change

$$\frac{s_{p*}^* - s_p}{R} = \int_{v'_r}^{v'_r*} \left(\frac{\partial P_r}{\partial T_r}\right)_{v'_r} dv'_r \tag{10.73}$$

This integration by itself is not entirely acceptable, because it contains the entropy at some arbitrary, low reference pressure. A value for the reference pressure would have to be specified. Let us now repeat the integration over the same change of state, except this time for a hypothetical ideal gas. The entropy change for this integration is

$$\frac{s_{p*}^* - s_p^*}{R} = +\ln\frac{P}{P^*} \tag{10.74}$$

If we now subtract Eq. 10.74 from Eq. 10.73, the result is the difference in entropy of a hypothetical ideal gas at a given state (T_r, P_r) and that of the real substance at the same

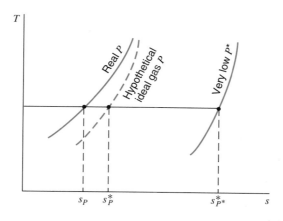

FIGURE 10.11 Real and ideal gas states and entropies.

state, or

$$\frac{s_p^* - s_p}{R} = -\ln\frac{P}{P^*} + \int_{v_r'}^{v_r'^* \to \infty} \left(\frac{\partial P_r}{\partial T_r}\right)_{v'} dv_r' \tag{10.75}$$

Here the values associated with the arbitrary reference state P_r^*, $v_r'^*$, cancel out of the right-hand side of the equation. (The first term of the integral includes the term $+\ln(P/P^*)$, which cancels the other term. The three different states associated with the development of Eq. 10.75 are shown in Fig. 10.11.

The same procedure that was given in Section 10.11 for enthalpy departure values is followed for generalized entropy departure values. The Lee–Kesler simple-fluid constants are used in evaluating the integral of Eq. 10.75 and yield a simple-fluid entropy departure. The values for the entropy departure are listed in Table A.15 and shown graphically in Fig. A.9.

EXAMPLE 10.10   Nitrogen at 8 MPa, 150 K, is throttled to 0.5 MPa. After the gas passes through a short length of pipe, its temperature is measured and found to be 125 K. Determine the heat transfer and the change of entropy using the generalized charts. Compare these results with those obtained by using the nitrogen tables.

   Control volume: Throttle and pipe.

   Inlet state: P_1, T_1 known; state fixed.

   Exit state: P_2, T_2 known; state fixed.

   Process: SSSF.

   Diagram: Figure 10.12.

   Model: Generalized tables, results to be compared with those obtained with nitrogen tables.

Analysis:

There is no work done, and we neglect changes in kinetic and potential energies.

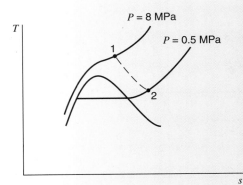

FIGURE 10.12 Sketch for Example 10.10.

Therefore, per kilogram,

First law:

$$q + h_1 = h_2$$

$$q = h_2 - h_1 = -(h_2^* - h_2) + (h_2^* - h_1^*) + (h_1^* - h_1)$$

Solution

Using values from Table A.8, we have

$$P_{r1} = \frac{8}{3.39} = 2.36 \qquad T_{r1} = \frac{150}{126.2} = 1.189$$

$$P_{r2} = \frac{0.5}{3.39} = 0.147 \qquad T_{r2} = \frac{125}{126.2} = 0.99$$

From Table A.15,

$$\frac{h_1^* - h_1}{RT_c} = 2.5085$$

$$h_1^* - h_1 = 2.5085 \times 0.2968 \times 126.2 = 94.0 \text{ kJ / kg}$$

$$\frac{h_2^* - h_2}{RT_c} = 0.1562$$

$$h_2^* - h_2 = 0.1562 \times 0.2968 \times 126.2 = 5.9 \text{ kJ / kg}$$

Assuming a constant specific heat for the ideal gas, we have

$$h_2^* - h_1^* = C_{p0}(T_2 - T_1) = 1.0416(125 - 150) = -26.0 \text{ kJ / kg}$$

$$q = -5.9 - 26.0 + 94.0 = 62.1 \text{ kJ / kg}$$

From the nitrogen tables, Table A.6, we can find the change of enthalpy directly.

$$q = h_2 - h_1 = 123.77 - 61.92 = 61.85 \text{ kJ / kg}$$

To calculate the change of entropy using the generalized charts, we proceed as follows:

$$s_2 - s_1 = -(s^*_{P_2,T_2} - s_2) + (s^*_{P_2,T_2} - s^*_{P_1,T_1}) + (s^*_{P_1,T_1} - s_1)$$

From Table A.15

$$\frac{s^*_{P_1,T_1} - s_{P_1,T_1}}{R} = 1.5905$$

$$s^*_{P_1,T_1} - s_{P_1,T_1} = 1.5905 \times 0.2968 = 0.4721 \text{ kJ / kg K}$$

$$\frac{s^*_{P_2,T_2} - s_{P_2,T_2}}{R} = 0.1064$$

$$s^*_{P_2,T_2} - s_{P_2,T_2} = 0.1064 \times 0.2968 = 0.0316 \text{ kJ / kg K}$$

Assuming a constant specific heat for the ideal gas, we have

$$s^*_{P_2,T_2} - s^*_{P_1,T_1} = C_{p0} \ln \frac{T_2}{T_1} - R \ln \frac{P_2}{P_1}$$

$$= 1.0416 \ln \frac{125}{150} - 0.2968 \ln \frac{0.5}{8}$$

$$= 0.6330 \text{ kJ / kg K}$$

$$s_2 - s_1 = -0.0316 + 0.6330 + 0.4721$$

$$= 1.0735 \text{ kJ / kg K}$$

From the nitrogen tables, Table A.6,

$$s_2 - s_1 = -5.4282 - 4.3522 = 1.0760 \text{ kJ / kg K}$$

10.13 FUGACITY AND THE GENERALIZED FUGACITY TABLE/CHART

At this point a new thermodynamic property, fugacity, f, is introduced. Fugacity is particularly important when considering mixtures and equilibrium, which is discussed in Chapter 13. A generalized fugacity table and chart can be developed, however. For this reason fugacity is introduced here.

Fugacity is essentially a pseudopressure. When fugacity is substituted for pressure, we can, in effect use the same equations for real gases that we normally use for ideal gases. The concept of fugacity is introduced as follows. Consider the relation

$$dg = -s \, dT + v \, dP$$

At constant temperature

$$dg_T = v \, dP_T \qquad (10.76)$$

For an ideal gas, this last equation can be written

$$dg_T = \frac{RT}{P} dP_T = RT \, d(\ln P)_T \qquad (10.77)$$

and for a real gas, with the equation of state $Pv = ZRT$, this equation becomes

$$dg_T = ZRT \frac{dP_T}{P} = ZRT\, d(\ln P)_T \tag{10.78}$$

Fugacity, f, is defined as

$$dg_T = RT\, d(\ln f)_T \tag{10.79}$$

with the requirement that

$$\lim_{P \to 0}\left(\frac{f}{P}\right) = 1 \tag{10.80}$$

Therefore, as $P \to 0$, $f \to 0$.

Let us consider the change in the Gibbs function of a real gas during an isothermal process at temperature T that changes pressure from very low P^* (where ideal-gas behavior can be assumed), to a higher pressure P. Let the Gibbs function at this low pressure be designated g^*. The value of the Gibbs function at this temperature T and at the pressure P can be found in terms of the fugacity at P and T and g^*. This is evident if we integrate Eq. 10.79 from the very low pressure P^* to the pressure P,

$$\int_{g^*}^{g} dg_T = \int_{f^*}^{f} RT(d\ln f)_T$$

$$g_P = g_{P*}^* + RT\ln\frac{f}{P^*} \tag{10.81}$$

Let us repeat this integration of Eq. 10.79 at constant temperature from pressure P^* to P, except now for a hypothetical ideal gas. The expression is

$$g_P^* = g_{P*}^* + RT\ln\frac{P}{P^*} \tag{10.82}$$

Subtracting Eq. 10.82 from Eq. 10.81, gives

$$\tag{10.83}$$

$$g_P - g_P^* = RT\ln\frac{f}{P}$$

Using the definition of the Gibbs function, we have

$$g_P - g_P^* = (h_P - h_P^*) - T(s_P - s_P^*) \tag{10.84}$$

Substituting Eq. 10.83 and rearranging terms, we obtain the following expression for the fugacity coefficient:

$$\ln\frac{f}{P} = -\frac{1}{T_r}\frac{h_P^* - h_P}{RT_c} + \frac{s_P^* - s_P}{R} \tag{10.85}$$

The two terms on the right-hand side of this equation are the enthalpy departure and the entropy departure, which were considered in Sections 10.11 and 10.12. Therefore, we can make up a generalized table and chart for the fugacity coefficient using those values already calculated for the simple-fluid model. The results are listed in Table A.15, and shown graphically in Fig. A.10.

Another development related to fugacity is often very useful, particularly when pressure is an independent variable. Comparing Eqs. 10.76 and 10.79 at a given temperature, we note that

$$d \ln f_T = \frac{v}{RT} dP_T = \frac{Z}{P} dP_T$$

If we subtract the identity

$$d \ln P_T = \frac{dP_T}{P}$$

from this expression, we obtain

$$d \ln \left(\frac{f}{P} \right)_T = (Z-1) \frac{dP_T}{P} = (Z-1) d \ln P_T$$

This expression can now be integrated along the isotherm from very low pressure $P^* \rightarrow 0$ to pressure P, which gives for the fugacity coefficient at P

$$\ln \frac{f}{P} = \int_0^P (Z-1) d \ln P_T \qquad (10.86)$$

EXAMPLE 10.11

Calculate the work of compression and the heat transfer per kilogram when ethane is compressed reversibly and isothermally from 0.1 to 7 MPa at a temperature of 45°C in a steady-state steady-flow process.

Control volume: Compressor.

Inlet state: P_1, T known; state fixed.

Exit state: P_2, T known; state fixed.

Process: SSSF, isothermal and reversible.

Diagram: Figure 10.13.

Model: Generalized tables.

Analysis:

First law:

$$q + h_1 + \text{KE}_1 + \text{PE}_1 = h_2 + \text{KE}_2 + \text{PE}_2 + w$$

Since the process is reversible and isothermal,

$$q = T(s_2 - s_1) = T_2 s_2 - T_1 s_1$$

Therefore,

$$-w = h_2 - h_1 - (T_2 s_2 - T_1 s_1) + \Delta\text{KE} + \Delta\text{PE}$$

$$-w = g_2 - g_1 + \Delta\text{KE} + \Delta\text{PE}$$

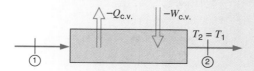

FIGURE 10.13 Sketch for Example 10.11.

We assume that the changes in kinetic and potential energy are negligible.

$$-w = g_2 - g_1 = RT \ln(f_2 / f_1)$$

Solution

The fugacity in states 1 and 2 can be found from the generalized fugacity table, Table A.15.

For ethane,

$$P_c = 4.88 \text{ MPa} \qquad T_c = 305.4 \text{ K}$$

Therefore,

$$P_{r1} = \frac{0.1}{4.88} = 0.0205 \qquad P_{r2} = \frac{7.0}{4.88} = 1.434$$

$$T_{r1} = T_{r2} = \frac{318.2}{305.4} = 1.042$$

From the generalized enthalpy departure table, Table A.15,

$$\frac{h_1^* - h_1}{RT_c} = 0.0191$$

$$h_1^* - h_1 = 0.0191 \times 0.2765 \times 305.4 = 1.6 \text{ kJ / kg}$$

$$\frac{h_2^* - h_2}{RT_c} = 3.0614$$

$$h_2^* - h_2 = 3.0614 \times 0.2765 \times 305.4 = 258.5 \text{ kJ / kg}$$

Since $h_2^* - h_1^* = 0$,

$$h_2 - h_1 = -258.5 + 0 + 1.6 = -256.9 \text{ kJ / kg}$$

From the generalized fugacity table,

$$\ln\left(\frac{f}{P}\right)_1 = -0.0060$$

$$f_1 = 0.994 \times 0.1 = 0.099 \text{ MPa}$$

$$\ln\left(\frac{f}{P}\right)_2 = -0.5494$$

$$f_2 = 0.5773 \times 7 = 4.041 \text{ MPa}$$

Therefore,

$$-w = g_2 - g_1 = RT \ln \frac{f_2}{f_1}$$

$$= 0.2765 \times 318.2 \ln \frac{4.041}{0.099} = 326.3 \text{ kJ / kg}$$

$$g_2 - g_1 = (h_2 - h_1) - T(s_2 - s_1) = (h_2 - h_1) - q$$

$$q = -256.9 - 326.3 = -583.2 \text{ kJ / kg}$$

EXAMPLE 10.11E Calculate the work of compression and the heat transfer per pound when ethane is compressed reversibly and isothermally from 14.7 lbf/in.^{2} to 1000 lbf/in.^{2} at a temperature of 110 F in a steady-state steady-flow process.

  Control volume: Compressor.

  Inlet state: P_1, T known; state fixed.

  Exit state: P_2, T known; state fixed.

  Process: SSSF, isothermal and reversible.

  Diagram: Figure 10.13.

  Model: Generalized tables.

Analysis:

First law:

$$q + h_1 + \text{KE}_1 + \text{PE}_1 = h_2 + \text{KE}_2 + \text{PE}_2 + w$$

Since the process is reversible and isothermal,

$$q = T(s_2 - s_1) = T_2 s_2 - T_1 s_1$$

Therefore,

$$-w = h_2 - h_1 - (T_2 s_2 - T_1 s_1) + \Delta\text{KE} + \Delta\text{PE}$$

$$-w = g_2 - g_1 + \Delta\text{KE} + \Delta\text{PE}$$

We assume that the changes in kinetic and potential energy are negligible.

$$-w = g_2 - g_1 = RT \ln(f_2 / f_1)$$

Solution

The fugacity in states 1 and 2 can be found from the generalized fugacity table, Table A.15.

  For ethane,

$$P_c = 708 \text{ lbf / in}^2 \qquad R = \frac{51.38}{778} = 0.066 \text{ Btu / lbm R} \qquad T_c = 549.7 \text{ R}$$

Therefore,

$$P_{r1} = \frac{14.7}{708} = 0.021 \qquad P_{r2} = \frac{1000}{708} = 1.412$$

$$T_{r1} = T_{r2} = \frac{570}{549.7} = 1.037$$

From the generalized enthalpy departure table, Table A.15,

$$\frac{h_1^* - h_1}{RT_c} = 0.0197$$

$$h_1^* - h_1 = 0.0197 \times 0.066 \times 549.7 = 0.7 \text{ Btu / lbm}$$

$$\frac{h_2^* - h_2}{RT_c} = 3.1143$$

$$h_2^* - h_2 = 3.1143 \times 0.066 \times 549.7 = 113.0 \text{ Btu / lbm}$$

Since $h_2^* - h_1^* = 0$,

$$h_2 - h_1 = -113.0 + 0 + 0.7 = -112.3 \text{ Btu / lbm}$$

From the generalized fugacity table,

$$\ln\left(\frac{f}{P}\right)_1 = -0.0063$$

$$f_1 = 0.994 \times 14.7 = 14.6 \text{ lbf / in.}^2$$

$$\ln\left(\frac{f}{P}\right)_2 = -0.5527$$

$$f_2 = 0.575 \times 1000 = 575 \text{ lbf / in.}^2$$

Therefore,

$$-w = g_2 - g_1 = RT \ln \frac{f_2}{f_1}$$

$$= 0.066 \times 570 \ln \frac{575}{14.6} = 138.2 \text{ Btu / lbm}$$

$$g_2 - g_1 = (h_2 - h_1) - T(s_2 - s_1) = (h_2 - h_1) - q$$

$$q = -112.3 - 138.2 = -250.5 \text{ Btu / lbm}$$

PROBLEMS

10.1 A special application requires R-12 at $-140°C$. It is known that the triple-point temperature is $-157°C$. Find the pressure and specific volume of the saturated vapor at the required condition.

10.2 Ice (solid water) at $-3°C$, 100 kPa is compressed isothermally until it becomes liquid. Find the required pressure.

10.3 Calculate the values h_{fg} and s_{fg} for nitrogen at 70 K and at 110 K from the Clapeyron equation, using the necessary pressure and specific volume values from Table A.6.1SI.

10.4 Using thermodynamic data for water from Tables A.1.1SI and A.1.5SI, estimate the freezing temperature of liquid water at a pressure of 30 MPa.

10.5 Helium boils at 4.22 K at atmospheric pressure, 101.3 kPa, with $\bar{h}_{fg} = 83.3$ kJ/kmol. By pumping a vacuum over liquid helium, the pressure can be lowered and it may then boil at a lower temperature. Estimate the necessary pressure to produce a boiling temperature of 1 K and one of 0.5 K.

10.6 A certain refrigerant vapor enters an SSSF constant pressure condenser at 150 kPa, 70°C, at a rate of 1.5 kg/s, and it exits as saturated liquid. Calculate the rate of heat transfer from the condenser. It may be assumed that the vapor is an ideal gas, and also that at saturation, $v_f \ll v_g$. The following quantities are known for this refrigerant:

$$\ln P_g = 8.15 - 1000 / T \qquad C_{po} = 0.7 \text{ kJ} / \text{kg K}$$

with pressure in kPa and temperature in K. The molecular weight is 100.

10.7 A container has a double wall where the wall cavity is filled with carbon dioxide at room temperature and pressure. When the container is filled with a cryogenic liquid at 100 K the carbon dioxide will freeze so the wall cavity has a mixture of solid and vapor carbon dioxide at the sublimation pressure. Assume that we do not have data for CO_2 at 100 K, but it is known that at $-90°C$: $P_{sat} = 38.1$ kPa, $h_{ig} = 574.5$ kJ/kg. Estimate the pressure in the wall cavity at 100 K.

10.8 An experiment is conducted at $-100°C$ inside a rigid sealed tank containing liquid R-22 with a small amount of vapor at the top. When the experiment is done the container and the R-22 warms up to room temperature of 20°C. What is the pressure inside the tank during the experiment? If the pressure at room temperature should not exceed 1 MPa, what is the maximum percent of liquid by volume that can be used during the experiment?

10.9 Small solid particles formed in combustion should be investigated. We would like to know the sublimation pressure as a function of temperature. The only information available is T, h_{fg} for boiling at 101.3 kPa and T, h_{if} for melting at 101.3 kPa. Develop a procedure that will allow a determination of the sublimation pressure, $P_{sat}(T)$.

10.10 In a Carnot heat engine, the heat addition changes the working fluid from saturated liquid to saturated vapor at T, P. The heat rejection process occurs at lower temperature and pressure $(T - \Delta T)$, $(P - \Delta P)$. The cycle takes place in a piston cylinder arrangement where the work is boundary work. Apply both the first and second law with simple approximations for the integral equal to work. Then show that the relation between ΔP and ΔT results in the Clapeyron equation in the limit $\Delta T \to dT$.

10.11 Repeat the previous problem for a heat engine where the cycle takes place in steady-state steady-flow devices, where the work is shaft work.

10.12 Derive expressions for $(\partial T / \partial v)_u$ and for $(\partial h / \partial s)_v$ that do not contain the properties h, u, or s.

10.13 Derive expressions for $(\partial h / \partial v)_T$ and for $(\partial h / \partial T)_v$ that do not contain the properties h, u, or s.

10.14 Develop an expression for the variation in temperature with pressure in a constant entropy process, $(\partial T / \partial P)_s$, that only includes the properties P–v–T and the specific heat, C_p.

10.15 Determine the volume expansivity, α_P, and the isothermal compressibility, β_T, for water at 20°C, 5 MPa and at 300°C, and 15 MPa using the steam tables.

10.16 Sound waves propagate through a media as pressure waves that cause the media to go through isentropic compression and expansion processes. The speed of sound c is defined by $c^2 = (\partial P / \partial \rho)_s$ and it can be related to the adiabatic compressibility, which for liquid ethanol at 20°C is 940 $\mu m^2/N$. Find the speed of sound at this temperature.

10.17 Consider the speed of sound as defined in Problem 10.16. Calculate the speed of sound for liquid water at 20°C, 2.5 MPa and for water vapor at 200°C, 300 kPa using the steam tables.

10.18 Find the speed of sound for air at 20°C, 100 kPa using the definition in Problem 10.16 and relations for polytropic processes in ideal gases.

10.19 A cylinder fitted with a piston contains liquid methanol at 20°C, 100 kPa and volume 10 L. The piston is moved, compressing the methanol to 20 MPa at constant temperature. Calculate the work required for this process. The isothermal compressibility of liquid methanol at 20°C is 1220 μm²/N.

10.20 (Adv.) Suppose the following information is available for a given pure substance

   a. The liquid-vapor saturation pressure, $P_{sat}(T)$

   b. An equation of state for the vapor, $P = Fct(T, v)$

   c. The saturated liquid specific volume, $v_f(T)$

   d. Critical pressure and temperature, P_c, T_c

   e. The constant volume specific heat for vapor, C_v, at v_x

Outline the procedure that should be followed to develop a table of thermodynamic properties comparable to Tables 1, 2, and 3 of the steam tables.

10.21 A piston/cylinder contains 5 kg of butane gas at 500 K, 5 MPa. The butane expands in a reversible polytropic process with polytropic exponent, $n = 1.05$, until the final pressure is 3 MPa. Determine the final temperature and the work done during the process.

10.22 Show that the two expressions for the Joule–Thomson coefficient μ_J given by Eq. 10.54 are valid.

10.23 A 200-L rigid tank contains propane at 9 MPa, 280°C. The propane is then allowed to cool to 50°C as heat is transferred with the surroundings. Determine the quality at the final state and the mass of liquid in the tank, using the generalized compressibility tables or the chart, Fig. A.7. Assume that the propane behaves as a "simple" fluid.

10.24 A rigid tank contains 5 kg of ethylene at 3 MPa, 30°C. It is cooled until the ethylene reaches the saturated vapor curve. What is the final temperature?

10.25 Two uninsulated tanks of equal volume are connected by a valve. One tank contains a gas at a moderate pressure P_1, and the other tank is evacuated. The valve is opened and remains open for a long time. Is the final pressure P_2 greater than, equal to, or less than $P_1/2$?

10.26 (Adv.) Develop expressions for isothermal changes in enthalpy and in entropy for each of the following equations of state:

   a. The van der Waals equation

   b. The Redlich–Kwong equation

   c. The Bennedict–Webb–Rubin equation

10.27 Determine the Boyle temperature as predicted by an equation of state, using

   a. The van der Waals equation

   b. The Redlich–Kwong equation

10.28 Consider a straight line connecting the point $P = 0$, $Z = 1$ to the critical point $P = P_c$, $Z = Z_c$ on a Z versus P compressibility diagram. This straight line will be tangent to one particular isotherm at low pressure. Determine what value of reduced temperature is predicted by an equation of state, using

a. The van der Waals equation

b. The Redlich–Kwong equation

10.29 Determine the low-pressure Joule–Thomson inversion temperature

$$\lim_{P \to 0} \mu_J = 0$$

as predicted by an equation of state, using

a. The van der Waals equation

b. The Redlich–Kwong equation

10.30 (Adv.) One early attempt to improve on the van der Waals equation of state was an expression of the form

$$P = \frac{RT}{v - b} - \frac{a}{v^2 T}$$

Solve for the constants a, b, and v_c using the same procedure as for the van der Waals equation.

10.31 Use the equation of state from the previous problem and determine the Boyle temperature and the low pressure Joule–Thomson inversion temperature.

10.32 A liquid argon storage facility boils off some vapor that is collected in a 100-L rigid tank at 120 K, 500 kPa. Use the virial equation of state to determine the mass of argon inside the tank. Explain why one of two possible answers cannot be correct.

10.33 Calculate the difference in internal energy of the ideal-gas value and the real-gas value for carbon dioxide at the state 20°C, 1 MPa, as determined using the virial equation of state.

10.34 Calculate the difference in entropy of the ideal-gas value and the real-gas value for carbon dioxide at the state 20°C, 1 MPa, as determined using the virial equation of state.

10.35 Refrigerant-123, dichlorotrifluoroethane, which is currently under development as a potential replacement for environmentally hazardous refrigerants, undergoes an isothermal SSSF process in which the R-123 enters a heat exchanger as saturated liquid at 40°C and exits at 100 kPa. Calculate the heat transfer per kilogram of R-123, using the generalized tables, Table A.15.

10.36 Calculate the heat transfer during the process described in Problem 10.21.

10.37 Saturated vapor R-22 at 30°C is throttled to 200 kPa in an SSSF process. Calculate the exit temperature assuming no changes in the kinetic energy, using

a. The generalized tables, Table A.15

b. The R-22 tables, Table A.4

10.38 An uninsulated cylinder with a frictionless piston contains propene, C_3H_6, at ambient temperature, 19°C, with a quality of 50% and a volume of 10 L. The propene now expands very slowly until the pressure in the cylinder drops to 460 kPa. Calculate the mass of propene, the work and heat transfer for this process.

10.39 A 250-L tank contains propane at 30°C, 90% quality. The tank is heated to 300°C. Calculate the heat transfer during the process.

10.40 A cylinder contains ethylene, C_2H_4, at 1.536 MPa, −13°C. It is now compressed in a reversible isobaric (constant P) process to saturated liquid. Find the specific work and heat transfer.

10.41 An ordinary lighter is nearly full of liquid propane with a small amount of vapor, the volume is 5 cm³ and temperature is 23°C. The propane is now discharged slowly such that heat transfer keeps the propane and valve flow at 23°C. Find the initial pressure and mass of propane and the total heat transfer to empty the lighter.

10.42 Nitrogen gas is to be injected into an oil field at very high pressure as part of an enhanced oil recovery process. The nitrogen enters a compressor from storage at 17 MPa, 15°C, and is compressed to 34 MPa in a compressor having an isentropic efficiency of 85%. Determine the required work per kilogram of nitrogen and the compressor exit temperature.

10.43 How much less work would be required if the compression process in Problem 10.42 is achieved in two stages, of equal pressure ratios, with intercooling back to 15°C between the stages?

10.44 A newly developed compound is being considered for use as the working fluid in a small Rankine-cycle power plant driven by a supply of waste heat. Assume the cycle is ideal, with saturated vapor at 200°C entering the turbine and saturated liquid at 20°C exiting the condenser. The only properties known for this compound are molecular weight of 80 kg/kmol, ideal gas heat capacity $C_{po} = 0.80$ kJ/kg K and $T_c = 500$ K, $P_c = 5$ MPa. Calculate the work input, per kilogram, to the pump and the cycle thermal efficiency.

10.45 A 200-L rigid tank contains propane at 400 K, 3.5 MPa. A valve is opened, and propane flows out until half the initial mass has escaped, at which point the valve is closed. During this process the mass remaining inside the tank expands according to the relation $Pv^{1.4} = $ constant. Calculate the heat transfer to the tank during the process.

10.46 The refrigerant R-152a, difluoroethane, is under study as a possible replacement for other refrigerants. Several tanks are to be filled with R-152a for testing in another location. The procedure is as follows. A 10-L evacuated tank is connected to a line flowing saturated-vapor R-152a at 40°C. The valve is then opened, and the fluid flows in rapidly, so that the process is essentially adiabatic. The valve is to be closed when the pressure reaches a certain value P_2, and the tank will then be disconnected from the line. After a period of time, the temperature inside the tank will return to ambient temperature, 25°C, through heat transfer with the surroundings. At this time, the pressure inside the tank must be 500 kPa. What is the pressure P_2 at which the valve should be closed during the filling process? The ideal gas constant-pressure specific heat of R-152a is 0.996 kJ/kg K.

10.47 A geothermal power plant on the Raft river uses isobutane as the working fluid. The fluid enters the reversible adiabatic turbine, as shown in Fig. P10.47, at 160°C, 5.475 MPa and the condenser exit condition is saturated liquid at 33°C. Isobutane has the properties $T_c = 408.14$ K, $P_c = 3.65$ MPa, $C_{po} = 1.664$ kJ/kg K and ratio of specific heats $k = 1.094$ with a molecular weight as 58.124. Find the specific turbine work and the specific pump work.

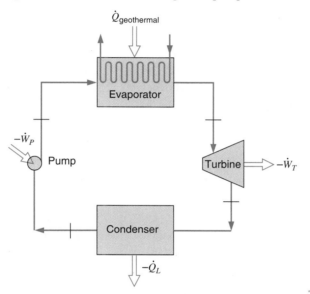

FIGURE P10.47

10.48 Carbon dioxide collected from a fermentation process at 5°C, 100 kPa should be brought to 243 K, 4 MPa in an SSSF process. Find the minimum amount of work required and the heat transfer. What devices are needed to accomplish this change of state?

10.49 An evacuated 100-L rigid tank is connected to a line flowing R-142b gas, chlorodifluoroethane, at 2 MPa, 100°C. The valve is opened, allowing the gas to flow into the tank for a period of time and then it is closed. Eventually, the tank cools to ambient temperature, 20°C, at which point it contains 50% liquid, 50% vapor, by volume. Calculate the heat transfer for this process. The ideal-gas constant-pressure specific heat of R-142b is 0.787 kJ/kg K.

10.50 An insulated cylinder has a piston loaded with a linear spring (spring constant of 600 kN/m) and held by a pin, as shown in Fig. P10.50. The cylinder cross-sectional area is 0.2 m², the initial volume is 0.1 m³, and it contains carbon dioxide at 2.5 MPa, 0°C. The piston mass and outside atmosphere adds a force per unit area of 250 kPa, and the spring force would be zero at a cylinder volume of 0.05 m³. Now the pin is pulled out; what is the final pressure inside the cylinder?

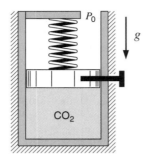

FIGURE P10.50

10.51 Saturated liquid ethane at 2.44 MPa enters (SSSF) a heat exchanger and is brought to 611 K at constant pressure, after which it enters a reversible adiabatic turbine where it expands to 100 kPa. Find the heat transfer in the heat exchanger, the turbine exit temperature and turbine work.

10.52 A 10-m^3 storage tank contains methane at low temperature. The pressure inside is 700 kPa, and the tank contains 25% liquid and 75% vapor, on a volume basis. The tank warms very slowly because heat is transferred from the ambient.

a. What is the temperature of the methane when the pressure reaches 10 MPa?

b. Calculate the heat transferred in the process, using the generalized tables or charts.

c. Repeat parts (a) and (b), using the methane tables, Table A.7. Discuss the differences in the results.

10.53 A flow of oxygen at 230 K, 5 MPa is throttled to 100 kPa in an SSSF process. Find the exit temperature and the entropy generation.

10.54 Carbon dioxide enters a nozzle at 15 MPa, 400 K, at a velocity of 10 m/s. The cross-sectional area at the nozzle entrance is 2000 mm^2. At the nozzle exit, the area is 200 mm^2 and the pressure is 5 MPa. There is heat rejection at the rate of 200 kW from the nozzle to the surroundings. What is the exit velocity of the carbon dioxide?

10.55 A cylinder contains ethylene, C_2H_4, at 1.536 MPa, −13°C. It is now compressed isothermally in a reversible process to 5.12 MPa. Find the specific work and heat transfer.

10.56 A control mass of 10 kg butane gas initially at 80°C, 500 kPa, is compressed in a reversible isothermal process to one-fifth of its initial volume. What is the heat transfer in the process?

10.57 Carbon dioxide gas enters a turbine at 5 MPa, 100°C, and exits at 1 MPa. If the isentropic efficiency of the turbine is 75%, determine the exit temperature and the second-law efficiency.

10.58 A 4-m^3 uninsulated storage tank, initially evacuated, is connected to a line flowing ethane gas at 10 MPa, 100°C. The valve is opened, and ethane flows into the tank for a period of time, after which the valve is closed. Eventually, the whole system cools to ambient temperature, 0°C, at which time the it contains one-fourth liquid and three-fourths vapor, by volume. For the overall process, calculate the heat transfer from the tank and the net change of entropy.

10.59 The working fluid in a power cycle is *n*-butane which exits the boiler as saturated vapor at 80°C and the temperature in the condenser is 30°C. Assume the pump and turbine both have an isentropic efficiency of 80%. Find the pump work, the turbine work and the cycle efficiency.

10.60 An insulated hollow steel sphere with inside diameter of 1 m and a wall thickness of 4 mm. It contains saturated-vapor ethane at ambient temperature, 300 K. A valve on the top of the sphere is opened, and ethane flows out rapidly until the pressure reaches 500 kPa, at which point the valve is closed. During this process the ethane remaining inside the sphere may be assumed to undergo a reversible adiabatic expansion. Now, after a period of time, the sphere and its contents come to a uniform temperature. Assuming no heat transfer with the surroundings, determine the final state of the ethane inside the sphere.

10.61 An uninsulated compressor delivers ethylene, C_2H_4, to a pipe, $D = 10$ cm, at 10.24 MPa, 94°C and velocity 30 m/s. The ethylene enters the compressor at 6.4 MPa, 20.5°C and the work input required is 300 kJ/kg. Find the mass flow rate, the total heat transfer and entropy generation, assuming the surroundings are at 25°C.

10.62 The environmentally safe refrigerant R-142b (see Problem 10.49) is to be evaluated as the working fluid in a portable, closed-cycle power plant, as shown in Fig. P10.62. The air-cooled condenser temperature is fixed at 50°C, and the maximum cycle temperature is fixed at 180°C, because of concerns about thermal stability. The isentropic efficiency of the expansion engine is estimated to be 80%, and the minimum allowable quality of the fluid exiting the expansion engine is 90%. Calculate the heat transfer from the condenser, assuming saturated liquid at the exit. Determine the maximum cycle pressure, based on the specifications listed.

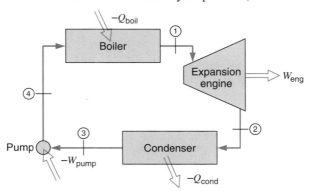

FIGURE P10.62

10.63 A distributor of bottled propane, C_3H_8, needs to bring propane from 350 K, 100 kPa to saturated liquid at 290 K in an SSSF process. If this should be accomplished in a reversible setup given the surroundings at 300 K, find the ratio of the volume flow rates $\dot{V}_{in}/\dot{V}_{out}$, the heat transfer and the work involved in the process.

10.64 A spring- and atmosphere-loaded piston/cylinder arrangement contains 0.1 m^3 of ethane, C_2H_6, at 229 K, $x = 0.25$. When the piston is at the bottom of the cylinder the balancing pressure below it is 0 kPa. Heat from a 600°C reservoir now brings the ethane to 1.464 MPa. Find the final volume and the work. Verify that final temperature is about 414 K and find the total entropy generated in this process.

10.65 One kilogram per second water enters a solar collector at 40°C and exits at 190°C, as shown in Fig. P10.65. The hot water is sprayed into a direct-contact heat exchanger (no mixing of the two fluids) used to boil the liquid butane. Pure saturated-vapor butane exits at the top at 80°C and is fed to the turbine. If the butane condenser temperature is 30°C and the turbine and pump isentropic efficiencies are each 80%, determine the net power output of the cycle.

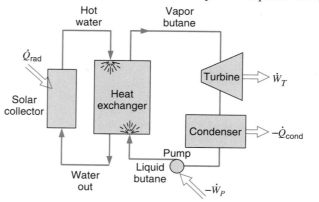

FIGURE P10.65

10.66 A line with a steady supply of octane, C_8H_{18}, is at 400°C, 3 MPa. What is your best estimate for the availability in an SSSF setup where changes in potential and kinetic energies may be neglected?

10.67 The environmentally safe refrigerant R-152a (see Problem 10.46) is to be evaluated as the working fluid for a heat pump system that will heat winter households in two different climates. In the colder climate the cycle evaporator temperature is −20°C, and the more moderate climate the evaporator temperature is 0°C. In both climates the cycle condenser temperature is 30°C. For this study assume all processes are ideal. Determine the cycle coefficient of performance for the two climates.

10.68 Repeat the calculation for the coefficient of performance of the heat pump in the two climates as described in Problem 10.67 using R-12 as the working fluid. Compare the two results.

10.69 In Problem 10.23, propane was treated as a simple fluid. Redo the problem using the acentric factor with the corrections shown in Appendix C.

10.70 Calculate the difference in internal energy of the ideal-gas value and the real-gas value for carbon dioxide at the state 20°C, 1 MPa, as determined using the acentric factor and the table in Appendix C.

10.71 Saturated liquid nitrogen dioxide, NO_2, at 300 K is brought to a state of 300 K, 100 kPa in an expansion process. Find the change in enthalpy of the nitrogen dioxide for this process using

a. The "simple" fluid approximation.

b. The acentric factor correction and the tables in Appendix C.

10.72 A line with a steady supply of octane, C_8H_{18}, is at 400°C, 3 MPa. Find the availability in a SSSF setup using the acentric factor corrections and the tables in Appendix C. Assume changes in potential and kinetic energies are small.

10.73 Ethanol, C_2H_5OH, is supplied to an engine as liquid at 308 K, 600 kPa. It is evaporated in a SSSF heater at constant pressure and brought to 460 K before entering the cylinder of the engine. Calculate the heat transfer per kilogram using the acentric factor correction and the table in Appendix C.

10.74E A special application requires R-12 at −200 F. It is known that the triple-point temperature is −250 F. Find the pressure and specific volume of the saturated vapor at the required condition.

10.75E Ice (solid water) at 27 F, 1 atm is compressed isothermally until it becomes liquid. Find the required pressure.

10.76E Using thermodynamic data for water from Tables A.1.1E and A.1.5E, estimate the freezing temperature of liquid water at a pressure of 5000 lbf/in.^{2}.

10.77E An experiment is conducted at −150 F inside a rigid sealed tank containing liquid R-22 with a small amount of vapor at the top. When the experiment is done the container and the R-22 warms up to room temperature of 70 F. What is the pressure inside the tank during the experiment? If the pressure at room temperature should not exceed 150 lbf/in.^{2}, what is the maximum percent of liquid by volume that can be used during the experiment?

10.78E Determine the volume expansivity, α_P, and the isothermal compressibility, β_T, for water at 50 F, 500 lbf/in.^{2} and at 500 F, 1500 lbf/in.^{2} using the steam tables.

10.79E Sound waves propagate through a media as pressure waves that causes the media to go through isentropic compression and expansion processes. The speed of sound c is defined by $c^2 = (\partial P/\partial \rho)_s$ and it can be related to the adiabatic compressibility, which for liquid ethanol at 70 F is 6.4 in.^{2}/lbf. Find the speed of sound at this temperature.

10.80E Consider the speed of sound as defined in Problem 10.79. Calculate the speed of sound for liquid water at 50 F, 250 lbf/in.^{2} and for water vapor at 400 F, 80 lbf/in.^{2} using the steam tables.

10.81E A rigid tank contains 5 lbm of ethylene at 450 lbf/in.^{2}, 90 F. It is cooled until the ethylene reaches the saturated vapor curve. What is the final temperature?

10.82E A cylinder fitted with a piston contains liquid methanol at 70 F, 15 lbf/in.^{2} and volume 1 ft^3. The piston is moved, compressing the methanol to 3000 lbf/in.^{2} at constant temperature. Calculate the work required for this process. The isothermal compressibility of liquid methanol at 70 F is 8.3×10^{-3} in.^{2}/lbf.

10.83E A piston/cylinder contains 10 lbm of butane gas at 900 R, 750 lbf/in.^{2}. The butane expands in a reversible polytropic process with polytropic exponent, $n = 1.05$, until the final pressure is 450 lbf/in.^{2}. Determine the final temperature and the work done during the process.

10.84E A 7-ft^3 rigid tank contains propane at 1300 lbf/in.^{2}, 540 F. The propane is then allowed to cool to 120 F as heat is transferred with the surroundings. Determine the quality at the final state and the mass of liquid in the tank, using the generalized compressibility tables or the chart, Fig. A.7. Assume that the propane behaves as a "simple" fluid.

10.85E A liquid argon storage facility boils off some vapor that is collected in a 4-ft³ rigid tank at 217 R, 75 lbf/in.². Use the virial equation of state to determine the mass of argon inside the tank. Explain why one of two possible answers cannot be correct.

10.86E Calculate the difference in internal energy of the ideal-gas value and the real-gas value for carbon dioxide at the state 70 F, 150 lbf/in.², as determined using the virial equation of state.

10.87E Calculate the heat transfer during the process described in Problem 10.83.

10.88E Saturated vapor R-22 at 90 F is throttled to 30 lbf/in.² in a SSSF process. Calculate the exit temperature assuming no changes in the kinetic energy, using

a. The generalized tables, Table A.15.

b. The R-22 tables, Table A.4E.

10.89E A 10-ft³ tank contains propane at 90 F, 90% quality. The tank is heated to 600 F. Calculate the heat transfer during the process.

10.90E A newly developed compound is being considered for use as the working fluid in a small Rankine-cycle power plant driven by a supply of waste heat. Assume the cycle is ideal, with saturated vapor at 400 F entering the turbine and saturated liquid at 70 F exiting the condenser. The only properties known for this compound are molecular weight of 80 lbm/lbmol, ideal gas heat capacity $C_{po} = 0.20$ Btu/lbm R and $T_c = 900$ R, $P_c = 750$ lbf/in.². Calculate the work input, per lbm, to the pump and the cycle thermal efficiency.

10.91E A 7-ft³ rigid tank contains propane at 730 R, 500 lbf/in.². A valve is opened, and propane flows out until half the initial mass has escaped, at which point the valve is closed. During this process the mass remaining inside the tank expands according to the relation $Pv^{1.4} = $ constant. Calculate the heat transfer to the tank during the process.

10.92E A geothermal power plant on the Raft river uses isobutane as the working fluid as shown in Fig. P10.47. The fluid enters the reversible adiabatic turbine at 320 F, 805 lbf/in.² and the condenser exit condition is saturated liquid at 91 F. Isobutane has the properties $T_c = 734.65$ R, $P_c = 537$ lbf/in.², $C_{po} = 0.3974$ Btu/lbm R and ratio of specific heats $k = 1.094$ with a molecular weight as 58.124. Find the specific turbine work and the specific pump work.

10.93E Carbon dioxide collected from a fermentation process at 40 F, 15 lbf/in.² should be brought to 438 R, 590 lbf/in.² in a SSSF process. Find the minimum amount of work required and the heat transfer. What devices are needed to accomplish this change of state?

10.94E An insulated cylinder has a piston loaded with a linear spring (spring constant of 3500 lbf/in.²) and held by a pin, as shown in Fig. P10.50. The cylinder cross-sectional area is 2 ft², the initial volume is 3.5 ft³, and it contains carbon dioxide at 350 lbf/in.², 35 F. The piston mass and outside atmosphere adds a force per unit area of 35 lbf/in.², and the spring force would be zero at a cylinder volume of 1.5 ft³. Now the pin is pulled out; what is the final pressure inside the cylinder?

10.95 E Carbon dioxide enters a nozzle at 2200 lbf/in.^{2}, 720 R, at a velocity of 30 ft/s. The cross-sectional area at the nozzle entrance is 3 in.^{2}. At the nozzle exit, the area is 0.3 in.^{2} and the pressure is 735 lbf/in.^{2}. There is heat rejection at the rate of 190 Btu/s from the nozzle to the surroundings. What is the exit velocity of the carbon dioxide?

10.96 E A control mass of 10 lbm butane gas initially at 180 F, 75 lbf/in.^{2}, is compressed in a reversible isothermal process to one-fifth of its initial volume. What is the heat transfer in the process?

10.97 E A cylinder contains ethylene, C_2H_4, at 222.6 lbf/in.^{2}, 8 F. It is now compressed isothermally in a reversible process to 742 lbf/in.^{2}. Find the specific work and heat transfer.

10.98 E A cylinder contains ethylene, C_2H_4, at 222.6 lbf/in.^{2}, 8 F. It is now compressed in a reversible isobaric (constant P) process to saturated liquid. Find the specific work and heat transfer.

10.99 E A distributor of bottled propane, C_3H_8, needs to bring propane from 630 R, 14.7 lbf/in.^{2} to saturated liquid at 520 R in a SSSF process. If this should be accomplished in a reversible setup given the surroundings at 540 R, find the ratio of the volume flow rates $\dot{V}_{in}/\dot{V}_{out}$, the heat transfer and the work involved in the process.

10.100 E Carbon dioxide gas enters a turbine at 50 atm, 200 F, and exits at 10 atm. If the isentropic efficiency of the turbine is 75%, determine the exit temperature and the second-law efficiency.

10.101 E A line with a steady supply of octane, C_8H_{18}, is at 750 F, 440 lbf/in.^{2}. What is your best estimate for the availability in an SSSF setup where changes in potential and kinetic energies may be neglected?

10.102 E In Problem 10.84 propane was treated as a simple fluid. Redo the problem using the acentric factor with the corrections shown in Appendix C.

10.103 E Saturated liquid nitrogen dioxide, NO_2, at 540 R is brought to a state of 540 R, 15 lbf/in.^{2} in an expansion process. Find the change in enthalpy of the nitrogen dioxide for this process using

a. The "simple" fluid approximation.

b. The acentric factor correction and the tables in Appendix C.

10.104 E A line with a steady supply of octane, C_8H_{18}, is at 750 F, 440 lbf/in.^{2}. Find the availability in a SSSF setup using the acentric factor corrections and the tables in Appendix C. Assume that potential and kinetic energies are small and can be neglected.

COMPUTER, DESIGN, AND OPEN-ENDED PROBLEMS

10.105 Write a program to obtain a plot of pressure versus specific volume at various temperatures (all on a generalized reduced basis) as predicted by the van der Waals equation of state. Temperatures less than the critical temperature should be included in the results.

10.106 Write a program to solve the following problem. For one of the substances listed in Table 3.3, calculate enthalpy changes along several isotherms at various integral pressures, using the Benedict–Webb–Rubin equation of state.

10.107 (Adv.) Write a program to curve-fit a polynomial in reduced temperature $T*$ to values of the reduced second virial coefficient $B*(T*)$ given in Table A.14.1 over a specific range of reduced temperatures. For one of the substances listed in Table A.14.2. plot Z versus P at low pressure for various isotherms.

10.108 (Adv.) Extend the program of Problem 10.107 to include the calculation of isothermal enthalpy and entropy changes. For one of the substances listed in Table A.14.2. plot these changes versus P at low pressure for various isotherms.

10.109 Write a program to plot the entire Joule–Thomson inversion locus as shown in Fig. 10.8, using both the van der Waals equation of state and the Redlich–Kwong equation of state.

10.110 We wish to determine the isothermal compressibility, β_T, for a range of states of liquid water. Use the menu-driven software or write a program to determine this at a pressure of 1 MPa and at 25 MPa for temperatures of 0°C, 100°C, and 300°C.

10.111 Do Problem 10.23 using the generalized compressibility tables in the menu-driven software and include the acentric factor correction term.

10.112 (Adv.) Consider the small Rankine-cycle power plant in Problem 10.44. What single change would you suggest to make the power plant more realistic?

10.113 (Adv.) Write a program to calculate the enthalpy and entropy changes between two arbitrary gas-phase states, using the Redlich–Kwong equation of state to represent the real-gas behavior and an ideal-gas specific-heat equation from either Table A.11 (Problem 5.205) or from the result of Problem 5.214. Input parameters should include the specific-heat information, the critical pressure and temperature, and the two independent properties specifying each of the two states.

10.114 (Adv.) Supercritical fluid chromatography is an experimental technique for analyzing compositions of mixtures. It utilizes a carrier fluid, often carbon dioxide, in the dense fluid region just above the critical temperature. Write a program to express the fluid density as a function of reduced temperature and pressure and acentric factor, in the region of $1.0 \leq T_r \leq 1.2$ in reduced temperature and $2 \leq P_r \leq 8$ in reduced pressure. The relation should be an expression curve-fitted to values consistent with the generalized compressibility tables, Table A.15.

10.115 List a number of requirements for a substance that should be used as the working fluid in a refrigerator. Discuss the choices and explain the requirements.

10.116 The speed of sound is used in many applications. Make a list of the speed of sound at P_o, T_o for gases, liquids, and solids. Find at least 3 different substances for each phase. List a number of applications where the knowledge about the speed of sound can be used to estimate other quantities of interest.

10.117 Propane is used as a fuel distributed to the end consumer in a steel bottle. Make a list of design specifications for these bottles and give characteristic sizes and the amount of propane they can hold.

10.118 Carbon dioxide is used in soft drinks and comes in a separate bottle for large volume users such as restaurants. Find typical sizes of these, the pressure they should withstand, and the amount of carbon dioxide they can hold.

MIXTURES AND SOLUTIONS 11

Up to this point in our development of thermodynamics we have considered primarily pure substances. A large number of thermodynamic problems involve mixtures of different pure substances. Sometimes these mixtures are referred to as solutions, particularly in the liquid and solid phases.

In this chapter we shall turn our attention to various thermodynamic considerations of mixtures and solutions. We begin with a consideration of a rather simple problem, mixtures of ideal gases. This leads to a consideration of a simplified but very useful model of certain mixtures, such as air and water vapor, which may involve a condensed (solid or liquid) phase of one of the components. This is followed by certain considerations of mixtures and solutions in general.

An understanding of the materials in this chapter is a necessary foundation for the consideration of chemical reactions and chemical and phase equilibrium. These topics are covered in subsequent chapters.

11.1 GENERAL CONSIDERATIONS AND MIXTURES OF IDEAL GASES

Let us consider a general mixture of N components, each a pure substance, so the total mass and the total number of moles are

$$m_{\text{tot}} = m_1 + m_2 + \cdots + m_N = \sum m_i$$

$$n_{\text{tot}} = n_1 + n_2 + \cdots + n_N = \sum n_i$$

The mixture is usually described by a mass fraction (concentration)

$$c_i = \frac{m_i}{m_{\text{tot}}} \tag{11.1}$$

or a mole fraction for each component as

$$y_i = \frac{n_i}{n_{\text{tot}}} \tag{11.2}$$

which are related through the molecular weight, M_i, as $m_i = n_i M_i$. We may then convert from a mole basis to a mass basis as

$$c_i = \frac{m_i}{m_{\text{tot}}} = \frac{n_i M_i}{\sum n_j M_j} = \frac{n_i M_i / n_{\text{tot}}}{\sum n_j M_j / n_{\text{tot}}} = \frac{y_i M_i}{\sum y_j M_j} \tag{11.3}$$

491

and from a mass basis to a mole basis as

$$y_i = \frac{n_i}{n_{\text{tot}}} = \frac{m_i / M_i}{\sum m_j / M_j} = \frac{m_i / (M_i m_{\text{tot}})}{\sum m_j / (M_j m_{\text{tot}})} = \frac{c_i / M_i}{\sum c_j / M_j} \qquad (11.4)$$

The molecular weight for the mixture becomes

$$M_{\text{mix}} = \frac{m_{\text{tot}}}{n_{\text{tot}}} = \frac{\sum n_i M_i}{n_{\text{tot}}} = \sum y_i M_i \qquad (11.5)$$

Consider a mixture of two gases (not necessarily ideal gases) such as shown in Fig. 11.1. What properties can we experimentally measure for such a mixture? Certainly we can measure the pressure, temperature, volume, and mass of the mixture. We can also experimentally measure the composition of the mixture, and thus determine the mole and mass fractions.

Suppose that this mixture undergoes a process or a chemical reaction and we wish to perform a thermodynamic analysis of this process or reaction. What type of thermodynamic data would we use in performing such an analysis? One possibility would be to have tables of thermodynamic properties of mixtures. However, the number of different mixtures that is possible, both as regards the substances involved and the relative amounts of each, is such that we would need a library full of tables of thermodynamic properties to handle all possible situations. It would be much simpler if we could determine the thermodynamic properties of a mixture from the properties of the pure components. This is in essence the approach that is used in dealing with ideal gases and certain other simplified models of mixtures.

One exception to this procedure is the case where a particular mixture is encountered very frequently, the most familiar being air. Tables and charts of the thermodynamics properties of air are available. However, even in this case it is necessary to define the composition of the "air" for which the tables are given, because the composition of the atmosphere varies with altitude, with the number of pollutants, and with other variables at a given location. The composition of air on which air tables are usually based is as follows:

Component	%on Mole Basis
Nitrogen	78.10
Oxygen	20.95
Argon	0.92
CO_2 & trace elements	0.03

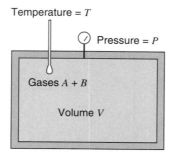

FIGURE 11.1 A mixture of two gases.

Consider again the mixture of Fig. 11.1. For the general case the properties of the mixture are defined in terms of the partial molal properties of the individual components. A partial molal property is defined as the value of a property, such as internal energy, for a given component as it exists in the mixture. With this definition, the internal energy of the mixture of Fig. 11.1 would be

$$U_{\text{mix}} = n_A \overline{U}_A + n_B \overline{U}_B \qquad (11.6)$$

where $\overline{U}$ designates the partial molal internal energy. Similar equations can be written for other properties, and this matter will be further developed in Section 11.9.

In this section we focus on mixtures of ideal gases. We assume that each component is uninfluenced by the presence of the other components, and that each component can be treated as an ideal gas. In an actual case of a gaseous mixture at high pressure this assumption would probably not be true because of the nature of the interaction between the molecules of the different components.

Two models are used in analyzing the mixtures of gases, namely, the Dalton model and the Amagat model.

Dalton Model

For the Dalton model, the properties of each component are considered as though each component existed separately at the volume and temperature of the mixture, as shown in Fig. 11.2.

Consider this model for the special case in which both the mixture and the separated components can be considered an ideal gas.

For the mixture:

$$PV = n\overline{R}T$$

$$n = n_A + n_B \qquad (11.7)$$

For the components:

$$P_A V = n_A \overline{R}T$$
$$P_B V = n_B \overline{R}T \qquad (11.8)$$

On substituting, we have

$$n = n_A + n_B$$

$$\frac{PV}{\overline{R}T} = \frac{P_A V}{\overline{R}T} + \frac{P_B V}{\overline{R}T} \qquad (11.9)$$

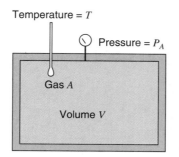
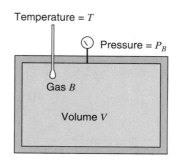

FIGURE 11.2 The Dalton model.

or

$$P = P_A + P_B$$

where P_A and P_B are referred to as partial pressures.

Thus for a mixture of ideal gases, the pressure is the sum of the partial pressures of the individual components.

It should be stressed that the term partial pressure is relevant only for ideal gases. This concept assumes that the molecules of each component are uninfluenced by the other components, and that the total pressure is the sum of partial pressures of the individual components. It should also be noted that partial pressure is not a partial molal property in the sense as defined by Eq. 11.3, since partial molal properties relate only to extensive properties.

Amagat Model

In the Amagat model the properties of each component are considered as though each component existed separately at the pressure and temperature of the mixture, as shown in Fig. 11.3. The volumes of A and B under these conditions are V_A and V_B, respectively.

In the general case, the sum of the volumes when separated, namely, $V_A + V_B$, need not be equal to the volume of the mixture. However, let us consider the special case in which both the separated components and the mixture are considered to be ideal gases. In this case we can write:

For the mixture:
$$PV = n\overline{R}T$$
$$n = n_A + n_B \tag{11.10}$$

For the components:
$$PV_A = n_A\overline{R}T$$
$$PV_B = n_B\overline{R}T \tag{11.11}$$

On substituting we have
$$n = n_A + n_B$$
$$\frac{PV}{\overline{R}T} = \frac{PV_A}{\overline{R}T} + \frac{PV_B}{\overline{R}T}$$

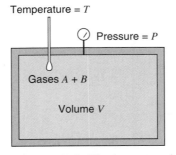
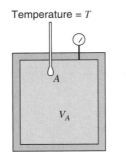
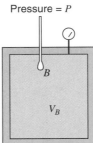

FIGURE 11.3 The Amagat model.

Therefore

$$V_A + V_B = V$$

or

$$\frac{V_A}{V} + \frac{V_B}{V} = 1 \tag{11.12}$$

where V_A/V and V_B/V are referred to as the volume fractions.

Thus for ideal gases, Amagat's model leads to the conclusion that the sum of the volume fractions is unity, and that there would be no volume change if the components were mixed while holding the temperature and pressure constant.

From Eqs. 11.7, 11.8, 11.10, and 11.11 it is evident that

$$\frac{V_A}{V} = \frac{n_A}{n} = \frac{P_A}{P}$$

$$\frac{V_A}{V} = y_A = \frac{P_A}{P} \tag{11.13}$$

That is, for each component of a mixture of ideal gases, the volume fraction, the mole fraction, and the ratio of the partial pressure to the total pressure are equal.

In determining the internal energy, enthalpy, and entropy of a mixture of ideal gases, the Dalton model proves useful because the assumption is made that each constituent behaves as though it occupies the entire volume by itself. Thus, the internal energy, enthalpy, and entropy can be evaluated as the sum of the respective properties of the constituent gases at the condition at which the component exists in the mixture. Since for ideal gases the internal energy and enthalpy are functions only of temperature, it follows that

$$U = n\bar{u} = n_A \bar{u}_A + n_B \bar{u}_B \tag{11.14}$$

$$H = n\bar{h} = n_A \bar{h}_A + n_B \bar{h}_B \tag{11.15}$$

where $\bar{u}_A$ and $\bar{h}_A$ are the internal energy and enthalpy per mole for pure A and $\bar{u}_B$ and $\bar{h}_B$ are the same quantities for pure B, all at the temperature of the mixture.

The entropy of an ideal gas is a function of pressure as well as temperature. Since each component exists in the mixture at its partial pressure,

$$S = n\bar{s} = n_A \bar{s}_A + n_B \bar{s}_B \tag{11.16}$$

where $\bar{s}_A$ is the entropy per mole for pure A at T and P_A (the partial pressure of A), and $\bar{s}_B$ is the entropy per mole for pure B at T and P_B.

EXAMPLE 11.1   A volumetric analysis of a gaseous mixture yields the following results:

CO_2	12.0%
O_2	4.0
N_2	82.0
CO	2.0

TABLE 11.1

Constituent	Percent by Volume	Mole Fraction		Molecular Weight		Mass kg per kmol of Mixture	Analysis on Mass Basis, Percent
CO_2	12	0.12	×	44.0	=	5.28	$\dfrac{5.28}{30.08} = 17.55$
O_2	4	0.04	×	32.0	=	1.28	$\dfrac{1.28}{30.08} = 4.26$
N_2	82	0.82	×	28.0	=	22.96	$\dfrac{22.96}{30.08} = 76.33$
CO	2	0.02	×	28.0	=	$\dfrac{0.56}{30.08}$	$\dfrac{0.56}{30.08} = 1.86$
							100.00

Determine the analysis on a mass basis, and the molecular weight and the gas constant on a mass basis for the mixture. Assume ideal gas behavior.

Control mass: Gas mixture.

State: Composition known.

Solution

It is convenient to set up and solve the problem as shown below in Table 11.1
From this table, we note that

$$\text{Molecular weight of mixture} = 30.08$$

$$R \text{ for mixture} = \frac{\overline{R}}{M} = \frac{8.3145}{30.08} = 0.2764 \text{ kJ / kg K}$$

If the analysis has been given on a mass basis, and the mole fraction or volumetric analysis is desired, the procedure shown in Table 11.2 can be used.

$$M = \frac{1}{\text{kmol / kg mixture}} = \frac{1}{0.033\,24} = 30.08$$

$$R = \frac{\overline{R}}{M} = \frac{8.3145}{30.08} = 0.2764 \text{ kJ / kg K}$$

TABLE 11.2

Constituent	Mass Fraction		Molecular Weight		kmol per kg of Mixture	Mole Fraction	Volumetric Analysis, Percent
CO_2	0.1755	÷	44.0	=	0.003 99	0.120	12.0
O_2	0.0426	÷	32.0	=	0.001 33	0.040	4.0
N_2	0.7633	÷	28.0	=	0.027 26	0.820	82.0
CO	0.0186	÷	28.0	=	0.000 66	0.020	2.0
					0.033 24	1.000	100.0

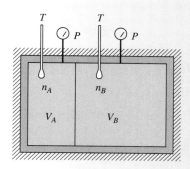
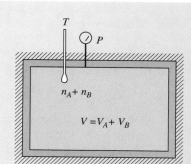

FIGURE 11.4
Sketch for Example 11.2.

EXAMPLE 11.2

Let n_A moles of gas A at a given pressure and temperature be mixed with n_B moles of gas B at the same pressure and temperature in an adiabatic constant-volume process, as shown in Fig. 11.4. Determine the increase in entropy for this process.

Control mass: All gas (A and B).

Initial states: P, T known for A and B.

Final state: P, T of mixture known.

Sketch: Figure 11.4.

Analysis and Solution

The final partial pressure of gas A is P_A and for gas B it is P_B. Since there is no change in temperature, Eq. 7.23 reduces to

$$\left(S_2 - S_1\right)_A = -n_A \overline{R} \ln \frac{P_A}{P} = -n_A \overline{R} \ln y_A$$

$$\left(S_2 - S_1\right)_B = -n_B \overline{R} \ln \frac{P_B}{P} = -n_B \overline{R} \ln y_B$$

The total change in entropy is the sum of the entropy changes for gases A and B.

$$S_2 - S_1 = -\overline{R}\left(n_A \ln y_A + n_B \ln y_B\right)$$

The result of Example 11.2 can readily be generalized to account for the mixing of any number of components at the same temperature and pressure. The result is

$$S_2 - S_1 = -\overline{R}\sum_k n_k \ln y_k \qquad (11.17)$$

The interesting thing about this equation is that the increase in entropy depends only on the number of moles of component gases, and is independent of the composition of the gas. For example, when 1 mol of oxygen and 1 mol of nitrogen are mixed, the increase in entropy is the same as when 1 mol of hydrogen and 1 mol of nitrogen are mixed. But we also know that if 1 mol of nitrogen is "mixed" with another mole of nitrogen there is no increase in entropy. The question that arises is how dissimilar must the gases be in order to have an increase in entropy? The answer lies in our ability to distinguish between the two gases. The entropy increases whenever we can distinguish between the gases being mixed. When we cannot distinguish between the gases, there is no increase in entropy.

11.2 A SIMPLIFIED MODEL OF A MIXTURE INVOLVING GASES AND A VAPOR

Let us now consider a simplification, which is often a reasonable one, of the problem involving a mixture of ideal gases that is in contact with a solid or liquid phase of one of the components. The most familiar example is a mixture of air and water vapor in contact with liquid water or ice, such as encountered in air conditioning or in drying. We are all familiar with the condensation of water from the atmosphere when it cools on a summer day.

This problem and a number of similar problems can be analyzed quite simply and with considerable accuracy if the following assumptions are made:

1. The solid or liquid phase contains no dissolved gases.
2. The gaseous phase can be treated as a mixture of ideal gases.
3. When the mixture and the condensed phase are at a given pressure and temperature, the equilibrium between the condensed phase and its vapor is not influenced by the presence of the other component. This means that when equilibrium is achieved, the partial pressure of the vapor will be equal to the saturation pressure corresponding to the temperature of the mixture.

Since this approach is used extensively and with considerable accuracy, let us give some attention to the terms that have been defined and the type of problems for which this approach is valid and relevant. In our discussion we will refer to this as a gas–vapor mixture.

The dew point of a gas–vapor mixture is the temperature at which the vapor condenses or solidifies when it is cooled at constant pressure. This is shown on the *T–s* diagram for the vapor shown in Fig. 11.5. Suppose that the temperature of the gaseous mixture and the partial pressure of the vapor in the mixture are such that the vapor is initially superheated at state 1. If the mixture is cooled at constant pressure, the partial pressure of the vapor remains constant until point 2 is reached, and then condensation begins. The temperature at state 2 is the dew-point temperature. Lines 1–3 on the diagram indicates

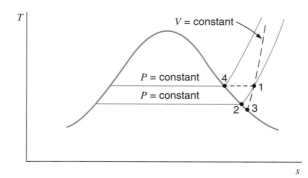

FIGURE 11.5 Temperature–entropy diagram to show definition of the dew point.

that if the mixture is cooled at constant volume the condensation begins at point 3, which is slightly lower than the dew-point temperature.

If the vapor is at the saturation pressure and temperature, the mixture is referred to as a saturated mixture, and for an air–water vapor mixture, the term "saturated air" is used.

The relative humidity ϕ is defined as the ratio of the mole fraction of the vapor in the mixture to the mole fraction of vapor in a saturated mixture at the same temperature and total pressure. Since the vapor is considered an ideal gas, the definition reduces to the ratio of the partial pressure of the vapor as it exists in the mixture, P_v, to the saturation pressure of the vapor at the same temperature, P_g.

$$\phi = \frac{P_v}{P_g}$$

In terms of the numbers on the T–s diagram of Fig. 11.5, the relative humidity ϕ would be

$$\phi = \frac{P_1}{P_4}$$

Since we are considering the vapor to be an ideal gas, the relative humidity can also be defined in terms of specific volume or density.

$$\phi = \frac{P_v}{P_g} = \frac{\rho_v}{\rho_g} = \frac{v_g}{v_v} \qquad (11.18)$$

The humidity ratio ω of an air-water vapor mixture is defined as the ratio of the mass of water vapor m_v to the mass of dry air m_a. The term "dry air" is used to emphasize that this refers only to air and not to the water vapor. The term "specific humidity" is used synonymously with humidity ratio.

$$\omega = \frac{m_v}{m_a} \qquad (11.19)$$

This definition is identical for any other gas–vapor mixture, and the subscript a refers to the gas, exclusive of the vapor. Since we consider both the vapor and the mixture to be ideal gases, a very useful expression for the humidity ratio in terms of partial pressures and molecular weights can be developed.

$$m_v = \frac{P_v V}{R_v T} = \frac{P_v V M_v}{\overline{R} T} \qquad m_a = \frac{P_a V}{R_a T} = \frac{P_a V M_a}{\overline{R} T}$$

Then

$$\omega = \frac{P_v V / R_v T}{P_a V / R_a T} = \frac{R_a P_v}{R_v P_a} = \frac{M_v P_v}{M_a P_a} \qquad (11.20)$$

For an air–water vapor mixture, this reduces to

$$\omega = 0.622 \frac{P_v}{P_a} \qquad (11.21)$$

The degree of saturation is defined as the ratio of the actual humidity ratio to the humidity ratio of a saturated mixture at the same temperature and total pressure.

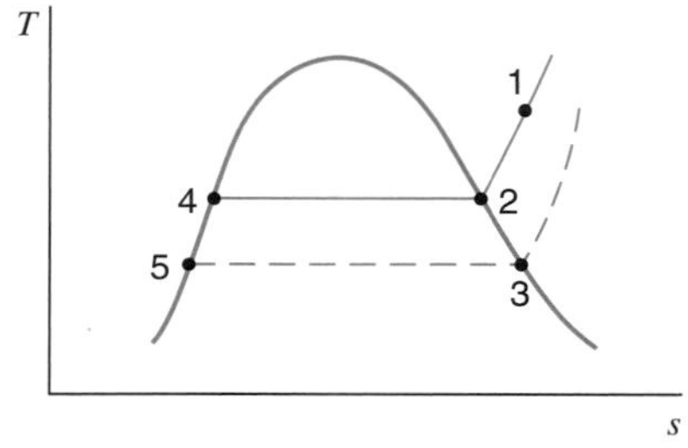

FIGURE 11.6 Temperature–entropy diagram to show the cooling of a gas–vapor mixture at a constant pressure.

An expression for the relation between the relative humdity ϕ and the humidity ratio ω can be found by solving Eqs. 11.18 and 11.21 for P_v and equating them. The resulting relation for an air–water vapor mixture is

$$\phi = \frac{\omega P_a}{0.622 P_g} \qquad (11.22)$$

A few words should also be said about the nature of the process that occurs when a gas–vapor mixture is cooled at constant pressure. Suppose that the vapor is initially superheated at state 1 in Fig. 11.6. As the mixture is cooled at constant pressure, the partial pressure of the vapor remains constant until the dew point is reached at point 2, where the vapor in the mixture is saturated. The initial condensate is at state 4, and is in equilibrium with the vapor at state 2. As the temperature is lowered further, more of the vapor condenses, which lowers the partial pressure of the vapor in the mixture. The vapor that remains in the mixture is always saturated, and the liquid or solid is in equilibrium with it. For example, when the temperature is reduced to T_3, the vapor in the mixture is at state 3, and its partial pressure is the saturation pressure corresponding to T_3. The liquid in equilibrium with it is at state 5.

EXAMPLE 11.3 Consider 100 m³ of an air–water vapor mixture at 0.1 MPa, 35°C, 70% relative humidity. Calculate the humidity ratio, dew point, mass of air, and mass of vapor.

Control mass: Mixture.

State: P, T, ϕ known; state fixed.

Analysis and Solution

From Eq. 11.18 and the steam tables,

$$\phi = 0.70 = \frac{P_v}{P_g}$$

$$P_v = 0.70(5.628) = 3.94 \text{ kPa}$$

The dew point is the saturation temperature corresponding to this pressure, which is 28.6°C.

The partial pressure of the air is

$$P_a = P - P_v = 100 - 3.94 = 96.06 \text{ kPa}$$

The humidity ratio can be calculated from Eq. 11.21.

$$\omega = 0.622 \times \frac{P_v}{P_a} = 0.622 \times \frac{3.94}{96.06} = 0.0255$$

The mass of air is

$$m_a = \frac{P_a V}{R_a T} = \frac{96.06 \times 100}{0.287 \times 308.2} = 108.6 \text{ kg}$$

The mass of the vapor can be calculated by using the humidity ratio or by using the ideal gas equation of state.

$$m_v = \omega m_a = 0.0255(108.6) = 2.77 \text{ kg}$$

$$m_v = \frac{3.94 \times 100}{0.461\,52 \times 308.2} = 2.77 \text{ kg}$$

EXAMPLE 11.3E Consider 2000 ft³ of an air–water vapor mixture at 14.7 lbf/in.², 90 F, 70% relative humidity. Calculate the humidity ratio, dew point, mass of air, and mass of vapor.

Control mass: Mixture.

State: P, T, ϕ known; state fixed.

Analysis and Solution

From Eq. 11.18 and the steam tables,

$$\phi = 0.70 = \frac{P_v}{P_g}$$

$$P_v = 0.70(0.6988) = 0.4892 \text{ lbf / in}^2$$

The dew point is the saturation temperature corresponding to this pressure, which is 78.9 F.

The partial pressure of the air is

$$P_a = P - P_v = 14.70 - 0.49 = 14.21 \text{ lbf / in}^2$$

The humidity ratio can be calculated from Eq. 11.21.

$$\omega = 0.622 \times \frac{P_v}{P_a} = 0.622 \times \frac{0.4892}{14.21} = 0.02135$$

The mass of air is

$$m_a = \frac{P_a V}{R_a T} = \frac{14.21 \times 144 \times 2000}{53.34 \times 550} = 139.6 \text{ lbm}$$

The mass of the vapor can be calculated by using the humidity ratio or by using the ideal gas equation of state.

$$m_v = \omega m_a = 0.02135(139.6) = 2.98 \text{ lbm}$$

$$m_v = \frac{0.4892 \times 144 \times 2000}{85.7 \times 550} = 2.98 \text{ lbm}$$

EXAMPLE 11.4   Calculate the amount of water vapor condensed if the mixture of Example 11.3 is cooled to 5°C in a constant-pressure process.

     Control mass: Mixture.

     Initial state: Known (Example 11.3).

     Final state: T known.

     Process: Constant pressure.

Analysis:

At the final temperature, 5°C, the mixture is saturated, since this is below the dew-point temperature. Therefore,

$$P_{v2} = P_{g2}, \qquad P_{a2} = P - P_{v2}$$

and

$$\omega_2 = 0.622 \frac{P_{v2}}{P_{a2}}$$

From the conservation of mass, it follows that the amount of water condensed is equal to the difference between the initial and final mass of water vapor, or

$$\text{Mass of vapor condensed} = m_a(\omega_1 - \omega_2)$$

Solution

$$P_{v2} = P_{g2} = 0.8721 \text{ kPa}$$

$$P_{a2} = 100 - 0.8721 = 99.128 \text{ kPa}$$

$$\omega_2 = 0.622 \times \frac{0.8721}{99.128} = 0.0055$$

$$\text{Mass of vapor condensed} = m_a(\omega_1 - \omega_2) = 108.6(0.0255 - 0.0055)$$
$$= 2.172 \text{ kg}$$

EXAMPLE 11.4E   Calculate the amount of water vapor condensed if the mixture of Example 11.3E is cooled to 40 F in a constant-pressure process.

     Control mass: Mixture.

     Initial state: Known (Example 11.3E).

     Final state: T known.

     Process: Constant pressure.

Analysis:

At the final temperature, 40 F, the mixture is saturated, since this is below the dew-point temperature. Therefore,

$$P_{v2} = P_{g2}, \qquad P_{a2} = P - P_{v2}$$

and

$$\omega_2 = 0.622 \frac{P_{v2}}{P_{a2}}$$

From the conservation of mass, it follows that the amount of water condensed is equal to the difference between the initial and final mass of water vapor, or

$$\text{Mass of vapor condensed} = m_a(\omega_1 - \omega_2)$$

Solution

$$P_{v2} = P_{g2} = 0.1217 \text{ lbf / in.}^2$$

$$P_{a2} = 14.7 - 0.12 = 14.58 \text{ lbf / in.}^2$$

$$\omega_2 = 0.622 \times \frac{0.1217}{14.58} = 0.00520$$

$$\text{Mass of vapor condensed} = m_a(\omega_1 - \omega_2) = 139.6(0.02135 - 0.0052)$$
$$= 2.25 \text{ lbm}$$

11.3 THE FIRST LAW APPLIED TO GAS–VAPOR MIXTURES

In applying the first law of thermodynamics to gas–vapor mixtures, it is helpful to realize that because of our assumption that ideal gases are involved, the various components can be treated separately when calculating changes of internal energy and enthalpy. Therefore, in dealing with air–water vapor mixtures, the changes in enthalpy of the water vapor can be found from the steam tables and the ideal-gas relations can be applied to the air. This is illustrated by the examples that follow.

EXAMPLE 11.5   An air-conditioning unit is shown in Fig. 11.7, with pressure, temperature, and relative humidity data. Calculate the heat transfer per kilogram of dry air, assuming that changes in kinetic energy are negligible.

Control volume: Duct, excluding cooling coils.

Inlet state: Known (Fig. 11.7).

Exit state: Known (Fig. 11.7).

Process: SSSF, no kinetic or potential energy changes.

Model: Air—ideal gas, constant specific heat, value at 300 K. Water—steam tables. (Since the water vapor at these low pressures is being considered an ideal gas, the enthalpy of the water vapor is a function of the temperature only. Therefore the enthalpy of slightly superheated water vapor is equal to the enthalpy of saturated vapor at the same temperature.)

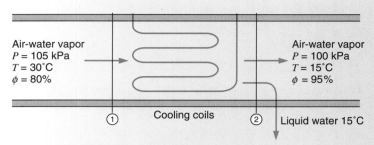

FIGURE 11.7 Sketch for Example 11.5.

Analysis:

Continuity, air and water:

$$\dot{m}_{a1} = \dot{m}_{a2}$$

$$\dot{m}_{v1} = \dot{m}_{v2} + \dot{m}_{l2}$$

First law:

$$\dot{Q}_{c.v.} + \sum \dot{m}_i h_i = \sum \dot{m}_e h_e$$

$$\dot{Q}_{c.v.} + \dot{m}_a h_{a1} + \dot{m}_{v1} h_{v1} = \dot{m}_a h_{a2} + \dot{m}_{v2} h_{v2} + \dot{m}_{l2} h_{l2}$$

If we divide this equation by $\dot{m}_a$, introduce the continuity equation for the water, and note that $\dot{m}_v = \omega \dot{m}_a$, we can write the first law in the form

$$\frac{\dot{Q}_{c.v.}}{\dot{m}_a} + h_{a1} + \omega_1 h_{v1} = h_{a2} + \omega_2 h_{v2} + (\omega_1 - \omega_2) h_{l2}$$

Solution

$$P_{v1} = \phi_1 P_{g1} = 0.80(4.246) = 3.397 \text{ kPa}$$

$$\omega_1 = \frac{R_a}{R_v} \frac{P_{v1}}{P_{a1}} = 0.622 \times \left(\frac{3.397}{105 - 3.4} \right) = 0.0208$$

$$P_{v2} = \phi_2 P_{g2} = 0.95(1.7051) = 1.620 \text{ kPa}$$

$$\omega_2 = \frac{R_a}{R_v} \times \frac{P_{v2}}{P_{a2}} = 0.622 \times \left(\frac{1.62}{100 - 1.62} \right) = 0.0102$$

Substituting:

$$\dot{Q}_{c.v.} / \dot{m}_a + h_{a1} + \omega_1 h_{v1} = h_{a2} + \omega_2 h_{v2} + (\omega_1 - \omega_2) h_{l2}$$

$$\dot{Q}_{c.v.} / \dot{m}_a = 1.0035(15 - 30) + 0.0102(2528.9)$$

$$- 0.0208(2556.3) + (0.0208 - 0.0102)(62.99)$$

$$= -41.76 \text{ kJ} / \text{kg dry air}$$

EXAMPLE 11.6   A tank has a volume of 0.5 m³ and contains nitrogen and water vapor. The temperature of the mixture is 50°C and the total pressure is 2 MPa. The partial pressure of the water vapor is 5 kPa. Calculate the heat transfer when the contents of the tank are cooled to 10°C.

Control mass: Nitrogen and water.
Initial state: P_1, T_1 known; state fixed.
Final state: T_2 known.
Process: Constant volume.
Model: Ideal gas mixture; constant specific heat for nitrogen; steam tables for water.

Analysis:

This is a constant-volume process. Since the work is zero, the first law reduces to

$$Q = U_2 - U_1 = m_{N_2} C_{v(N_2)}(T_2 - T_1) + (m_2 u_2)_v + (m_2 u_2)_l - (m_1 u_1)_v$$

This equation assumes that some of the vapor condensed. This assumption must be checked, however, as shown in the solution.

Solution

The mass of nitrogen and water vapor can be calculated using the ideal-gas equation of state.

$$m_{N_2} = \frac{P_{N_2}V}{R_{N_2}T} = \frac{1995 \times 0.5}{0.2968 \times 323.2} = 10.39 \text{ kg}$$

$$m_{v_1} = \frac{P_{v1}V}{R_v T} = \frac{5 \times 0.5}{0.461\,52 \times 323.2} = 0.016\,76 \text{ kg}$$

If condensation takes place, the final state of the vapor will be saturated vapor at 10°C. Therefore,

$$m_{v2} = \frac{P_{v2}V}{R_v T} = \frac{1.2276 \times 0.5}{0.461\,52 \times 283.2} = 0.004\,70 \text{ kg}$$

Since this amount is less than the original mass of vapor, there must have been condensation.

The mass of liquid that is condensed, m_{l2}, is

$$m_{l2} = m_{v_1} - m_{v_2} = 0.016\,76 - 0.004\,70 = 0.012\,06 \text{ kg}$$

The internal energy of the water vapor is equal to the internal energy of saturated water vapor at the same temperature. Therefore,

$$u_{v_1} = 2443.5 \text{ kJ / kg}$$
$$u_{v_2} = 2389.2 \text{ kJ / kg}$$
$$u_{l2} = 42.0 \text{ kJ / kg}$$
$$Q_{c.v.} = 10.39 \times 0.7448(10 - 50) + 0.0047(2389.2)$$
$$+ 0.012\,06(42.0) - 0.016\,76(2443.5)$$
$$= -338.8 \text{ kJ}$$

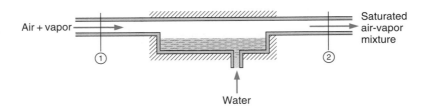

FIGURE 11.8 The adiabatic saturation process.

11.4 THE ADIABATIC SATURATION PROCESS

An important process for an air–water vapor mixture is the adiabatic saturation process. In this process, an air–vapor mixture comes in contact with a body of water in a well-insulated duct (Fig. 11.8). If the initial relative humidity is less than 100%, some of the water will evaporate and the temperature of the air–vapor mixture will decrease. If the mixture leaving the duct is saturated and if the process is adiabatic, the temperature of the mixture on leaving is known as the adiabatic saturation temperature. For this to take place as a steady-flow process, make-up water at the adiabatic saturation temperature is added at the same rate at which water is evaporated. The pressure is assumed to be constant.

Considering the adiabatic saturation process to be a steady-state, steady-flow process, and neglecting changes in kinetic and potential energy, the first law reduces to

$$h_{a1} + \omega_1 h_{v1} + (\omega_2 - \omega_1)h_{l2} = h_{a2} + \omega_2 h_{v2}$$

$$\omega_1(h_{v1} - h_{l2}) = C_{pa}(T_2 - T_1) + \omega_2(h_{v2} - h_{l2})$$

$$\omega_1(h_{v1} - h_{l2}) = C_{pa}(T_2 - T_1) + \omega_2 h_{fg2} \tag{11.23}$$

The most significant point to be made about the adiabatic saturation process is that the adiabatic saturation temperature, the temperature of the mixture when it leaves the duct, is a function of the pressure, temperature, and relative humidity of the entering air–vapor mixture and of the exit pressure. Thus, the relative humidity and the humidity ratio of the entering air–vapor mixture can be determined from the measurements of the pressure and temperature of the air–vapor mixture entering and leaving the adiabatic saturator. Since these measurements are relatively easy to make, this is one means of determining the humidity of an air–vapor mixture.

EXAMPLE 11.7
The pressure of the mixture entering and leaving the adiabatic saturator is 0.1 MPa, the entering temperature is 30°C, and the temperature leaving is 20°C, which is the adiabatic saturation temperature. Calculate the humidity ratio and relative humidity of the air–water vapor mixture entering.

Control volume: Adiabatic saturator.

Inlet state: P_1, T_1 known.

Exit state: P_2, T_2 known; $\phi_2 = 100\%$; state fixed.

Process: SSSF, adiabatic saturation (Fig. 11.8).

Model: Ideal-gas mixture; constant specific heat for air; steam tables for water.

Analysis:

Continuity and first law, Eq. 11.23.

Solution

Since the water vapor leaving is saturated, $P_{v2} = P_{g2}$ and ω_2 can be calculated.

$$\omega_2 = 0.622 \times \left(\frac{2.339}{100 - 2.34} \right) = 0.0149$$

ω_1 can be calculated using Eq. 11.23.

$$\omega_1 = \frac{C_{pa}(T_2 - T_1) + \omega_2 h_{fg2}}{(h_{v1} - h_{f2})}$$

$$\omega_1 = \frac{1.0035(20 - 30) + 0.0149 \times 2454.1}{2556.3 - 83.96} = 0.0107$$

$$\omega_1 = 0.0107 = 0.622 \times \left(\frac{P_{v1}}{100 - P_{v1}} \right)$$

$$P_{v1} = 1.691 \text{ kPa}$$

$$\phi_1 = \frac{P_{v1}}{P_{g1}} = \frac{1.691}{4.246} = 0.398$$

EXAMPLE 11.7E

The pressure of the mixture entering and leaving the adiabatic saturator is 14.7 lbf/in.^{2}, the entering temperature is 84 F, and the temperature leaving is 70 F, which is the adiabatic saturation temperature. Calculate the humidity ratio and relative humidity of the air–water vapor mixture entering.

Control volume: Adiabatic saturator.

Inlet state: P_1, T_1 known.

Exit state: P_2, T_2 known; $\phi_2 = 100\%$; state fixed.

Process: SSSF, adiabatic saturation (Fig. 11.8).

Model: Ideal-gas mixture; constant specific heat for air; steam tables for water.

Analysis:

Continuity and first law, Eq. 11.23.

Solution

Since the water vapor leaving is saturated, $P_{v2} = P_{g2}$ and ω_2 can be calculated.

$$\omega_2 = 0.622 \times \frac{0.3632}{14.7 - 0.36} = 0.01573$$

ω_1 can be calculated using Eq. 11.23.

$$\omega_1 = \frac{C_{pa}(T_2 - T_1) + \omega_2 h_{fg2}}{(h_{v1} - h_{l2})}$$

$$\omega_1 = \frac{0.24(70 - 84) + 0.01573 \times 1054.0}{1098.1 - 38.1} = \frac{-3.36 + 16.60}{1060.0} = 0.0125$$

$$\omega_1 = 0.622 \times \left(\frac{P_{v1}}{14.7 - P_{v1}}\right) = 0.0125$$

$$P_{v1} = 0.289$$

$$\phi_1 = \frac{P_{v1}}{P_{g1}} = \frac{0.289}{0.584} = 0.495$$

11.5 WET-BULB AND DRY-BULB TEMPERATURES

The humidity of air–water vapor mixtures has traditionally been measured with a device called a psychrometer, which uses the flow of air past wet-bulb and dry-bulb thermometers. The bulb of the wet-bulb thermometer is covered with a cotton wick that is saturated with water. The dry-bulb thermometer is used simply to measure the temperature of the air. The airflow can be maintained by a fan, as shown in the continuous-flow psychrometer depicted in Fig. 11.9.

The processes that take place at the wet-bulb thermometer are somewhat complicated. First, if the air–water vapor mixture is not saturated, some of the water in the wick evaporates and diffuses into the surrounding air, which cools the water in the wick. As soon as the temperature of the water drops, however, heat is transferred to the water from both the air and the thermometer, with corresponding cooling. A steady state, determined by heat and mass transfer rates, will be reached, in which the wet-bulb thermometer temperature is lower than the dry-bulb temperature.

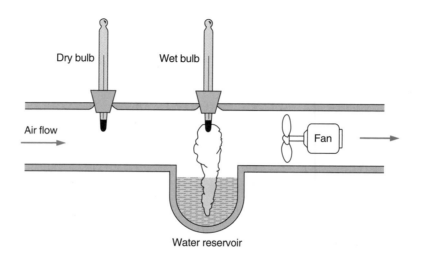

FIGURE 11.9
Steady-flow apparatus for measuring wet- and dry-bulb temperatures.

It can be argued that this evaporative cooling process is very similar, but not identical, to the adiabatic saturation process described and analyzed in Section 11.4. In fact, the adiabatic saturation temperature is often termed the thermodynamic wet-bulb temperature. It is clear, however, that the wet-bulb temperature as measured by a psychrometer is influenced by heat and mass transfer rates, which depend, for example, on the airflow velocity and not simply on thermodynamic equilibrium properties. It does happen that the two temperatures are very close for air–water vapor mixtures at atmospheric temperature and pressure, and they will be assumed to be equivalent in this text.

In recent years, humidity measurements have been made using other phenomena and other devices, primarily electronic devices for convenience and simplicity. For example, some substances tend to change in length, in shape, or in electrical capacitance, or in a number of other ways, when they absorb moisture. They are therefore sensitive to the amount of moisture in the atmosphere. An instrument making use of such a substance can be calibrated to measure the humidity of air–water vapor mixtures. The instrument output can be programmed to furnish any of the desired parameters, such as relative humidity, humidity ratio, or wet-bulb temperature.

11.6 THE PSYCHROMETRIC CHART

Properties of air–water vapor mixtures are given in graphical form on psychrometric charts. These are available in a number of different forms, and only the main features are considered here. It should be recalled that three independent properties will describe the state of this binary mixture such as pressure, temperature and mixture composition.

A simplified version of the chart included in Appendix, Fig. A.4, is shown in Fig. 11.10. This basic psychrometric chart is a plot of humidity ratio (ordinate) as a function of dry-bulb temperature (abscissa) with relative humidity, wet-bulb temperature and mix-

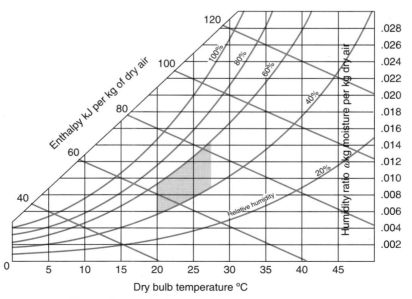

FIGURE 11.10 Psychrometric chart.

ture enthalpy per mass of dry air as parameters. If we fix the total pressure for which the chart is to be constructed (which in our chart is 1 bar, or 100 kPa), lines of constant relative humidity and wet-bulb temperature can be drawn on the chart, because for a given dry-bulb temperature, total pressure, and humidity ratio, the relative humidity and wet-bulb temperature are fixed. The partial pressure of the water vapor is fixed by the humidity ratio and the total pressure, and therefore a second ordinate scale that indicates the partial pressure of the water vapor could be constructed. Likewise it would also be possible to include the mixture specific volume and entropy on the chart.

Most psychrometric charts give the enthalpy of an air–vapor mixture per kilogram of dry air. The values given assume that the enthalpy of the dry air is zero at −20°C, and the enthalpy of the vapor is taken from the steam tables (which are based on the assumption that the enthalpy of saturated liquid is zero at 0°C). The value used in the psychrometric chart is then

$$\tilde{h} \equiv h_a - h_a(-20°C) + \omega h_v$$

This procedure is satisfactory because we are usually concerned only with differences in enthalpy. The fact that the lines of constant enthalpy are essentially parallel to lines of constant wet-bulb temperature is evident from the fact that the wet-bulb temperature is essentially equal to the adiabatic saturation temperature. Thus in Fig. 11.8, if we neglect the enthalpy of the liquid entering the adiabatic saturator, the enthalpy of the air–vapor mixture leaving, and a given adiabatic saturation temperature fixes the enthalpy of the mixture entering.

The chart plotted in Fig. 11.10 also indicates the human comfort zone, as the range of conditions most agreeable for human well being. An air-conditioner should then be able to maintain an environment within the comfort zone regardless of the outside atmospheric conditions to be considered adequate. Some charts are available that give corrections for variation from standard atmospheric pressures. Before using a given chart one should fully understand the assumptions made in constructing it, and that it is applicable to the particular problem at hand.

11.7 INTRODUCTION TO REAL MIXTURES AND SOLUTIONS

In the remainder of this chapter, we will develop certain concepts and relations concerning the behavior of real mixtures and solutions. These considerations are not necessarily restricted to gaseous mixtures alone; they are also valid for solid, liquid, and gaseous solutions. In later chapters we will also include the effects of chemical reactions as well as the equilibrium between different phases, but for the present will concern ourselves with nonreacting mixtures or solutions in a single, homogeneous phase.

In a thermodynamic analysis involving any real mixture or solution, there are two distinct alternatives in the approach to the problem, as far as the expression of extensive thermodynamic properties at a given state (that is, at a given pressure and temperature). The first approach is to treat the mixture as a pseudo-pure substance having its own set of properties, as, for example, has often been used previously for air in cases in which it could be assumed an ideal gas and there was no question about saturation or change of phase. The second approach is to treat the mixture properties as a summation of the contributions of the mixture components, at the conditions at which the components exist in the mixture, that is, using the concept of partial properties. From what we have considered up to this point in the present chapter, it is apparent that ideal gas mixtures can be analyzed using either of these two alternative approaches. Real mixture models, however, follow one of these alternatives or the other. The remainder of this chapter develops these two alternatives and also several related topics concerning the general thermodynamics of real mixtures. In the next section, the matter of pseudo-pure substance models is developed, and it is also necessary to consider the matter of reference states in connection with this approach to mixture analysis. Subsequent sections then deal with partial properties, including the development of various thermodynamic relations and a number of expressions relating various thermodynamic properties, as well as models for expressing mixture behavior.

11.8 PSEUDOPURE SUBSTANCE MODELS FOR REAL GAS MIXTURES

A basic prerequisite to the treatment of real gas mixtures in terms of pseudopure substance models is the concept and use of appropriate reference states. As an introduction to this topic, let us consider several preliminary reference state questions for a pure substance undergoing a change of state, for which it is desired to calculate the entropy change. We can express the entropy at the initial state 1 and also at the final state 2 in terms of a reference state 0, in a manner similar to that followed when dealing with the generalized-charts corrections in Chapter 10. It follows that

$$s_1 = s_0 + (s^*_{P_0 T_0} - s_0) + (s^*_{P_1 T_1} - s^*_{P_0 T_0}) + (s_1 - s^*_{P_1 T_1}) \tag{11.24}$$

$$s_2 = s_0 + (s^*_{P_0 T_0} - s_0) + (s^*_{P_2 T_2} - s^*_{P_0 T_0}) + (s_2 - s^*_{P_2 T_2}) \tag{11.25}$$

These are entirely general expressions for the entropy at each state in terms of an arbitrary reference state value and a set of consistent calculations from that state to the actual desired state. One simplification of these equations would result from choosing the reference state to be a hypothetical ideal gas state at P_0 and T_0, thereby making the term

$$(s^*_{P_0 T_0} - s_0) = 0 \tag{11.26}$$

in each equation, which results in

$$s_0 = s^*_0 \tag{11.27}$$

It should be apparent that this choice is a reasonable one, since whatever value is chosen for the correction term, Eq. 11.26, it will cancel out of the two equations when the change $s_2 - s_1$ is calculated, and the simplest value to choose is zero. In a similar manner, the simplest value to choose for the ideal gas reference value, Eq. 11.27, is zero, and we would commonly do that if there are no restrictions on choice, such as occur in the case of a chemical reaction.

Another point to be noted concerning reference states is related to the choice of P_0 and T_0, and for this purpose, let us substitute Eqs. 11.26 and 11.27 into Eqs. 11.24 and 11.25, and also assume constant specific heat, such that those equations can be written in the form

$$s_1 = s^*_0 + C_{po} \ln\left(\frac{T_1}{T_0}\right) - R \ln\left(\frac{P_1}{P_0}\right) + (s_1 - s^*_{P_1 T_1}) \tag{11.28}$$

$$s_2 = s^*_0 + C_{po} \ln\left(\frac{T_2}{T_0}\right) - R \ln\left(\frac{P_2}{P_0}\right) + (s_2 - s^*_{P_2 T_2}) \tag{11.29}$$

FIGURE 11.11 Example of mixing process.

Since the choice for P_0 and T_0 is arbitrary if there are no restrictions, such as would be the case with chemical reactions, it should be apparent from examining Eqs. 11.28 and 11.29 that the simplest choice would be for

$$P_0 = P_1 \text{ or } P_2 \qquad T_0 = T_1 \text{ or } T_2$$

It should be emphasized that inasmuch as the reference state was chosen as a hypothetical ideal gas at P_0, T_0, Eq. 11.26, it is immaterial how the real substance behaves at that pressure and temperature. As a result, there is no need to select a low value for the reference state pressure P_0.

Let us now extend these reference state developments to include real gas mixtures. Consider the mixing process shown in Fig. 11.11, with the states and amounts of each substance as given on the diagram. Proceeding with entropy expressions as was done above, we have

$$\bar{s}_1 = \bar{s}_{A_0}^* + \overline{C}_{po_A} \ln\left(\frac{T_1}{T_0}\right) - \bar{R} \ln\left(\frac{P_1}{P_0}\right) + (\bar{s}_1 - \bar{s}_{P_1 T_1}^*)_A \tag{11.30}$$

$$\bar{s}_2 = \bar{s}_{B_0}^* + \overline{C}_{po_B} \ln\left(\frac{T_2}{T_0}\right) - \bar{R} \ln\left(\frac{P_2}{P_0}\right) + (\bar{s}_2 - \bar{s}_{P_2 T_2}^*)_B \tag{11.31}$$

$$\bar{s}_3 = \bar{s}_{\text{mix}_0}^* + \overline{C}_{po_{\text{mix}}} \ln\left(\frac{T_3}{T_0}\right) - \bar{R} \ln\left(\frac{P_3}{P_0}\right) + (\bar{s}_3 - \bar{s}_{P_3 T_3}^*)_{\text{mix}} \tag{11.32}$$

in which

$$\bar{s}_{\text{mix}_0}^* = y_A \bar{s}_{A_0}^* + y_B \bar{s}_{B_0}^* - \bar{R}(y_A \ln y_A + y_B \ln y_B) \tag{11.33}$$

$$\overline{C}_{po_{\text{mix}}} = y_A \overline{C}_{po_A} + y_B \overline{C}_{po_B} \tag{11.34}$$

We note that when Eqs. 11.30–11.32 are substituted into the equation for the entropy change,

$$n_3 \bar{s}_3 - n_1 \bar{s}_1 - n_2 \bar{s}_2$$

the arbitrary reference values $\bar{s}_{A0}^*$, $\bar{s}_{B0}^*$, P_0, and T_0 all cancel out of the result, which is, of course, necessary in view of their arbitrary nature. An ideal gas entropy of mixing expression, the final term in Eq. 11.33, remains in the result, establishing, in effect, the mixture reference value relative to its components. The remarks made earlier concerning the choices for reference state and the reference state entropies apply in this situation as well.

To summarize the development to this point, we find that a calculation of real mixture properties, as, for example, using Eq. 11.32, requires the establishment of a hypothetical ideal gas reference state, a consistent ideal gas calculation to the conditions of the real mixture, and finally a correction that accounts for the real behavior of the mixture at that state. We note that this last term is the only place that the real behavior is introduced, and this is therefore the term that must be calculated by the pseudopure substance model to be used.

In treating a real gas mixture as a pseudopure substance, there are two approaches that we will follow to represent the P–v–T behavior: use of the generalized tables/charts, and use of an analytical equation of state. With the generalized tables/charts, we need to have a model that provides a set of pseudocritical pressure and temperature in terms of the mixture component values. Many such models have been proposed and utilized over the years, but the simplest is that suggested by W. B. Kay in 1936, in which

$$(P_c)_{mix} = \sum_i y_i P_{ci}, \qquad (T_c)_{mix} = \sum_i y_i T_{ci} \tag{11.35}$$

This is the only pseudocritical model that we will consider in this chapter. Another model, discussed in Appendix C in connection with corrections to the compressibility tables, A.15, is somewhat more complicated to evaluate and use, but is considerably more accurate.

The other approach to be considered is that of using an analytical equation of state, in which the equation for the mixture must be developed from that for the components. In other words, for an equation in which the constants are known for each of the components, we must develop a set of empirical combining rules that will then give a set of constants for the mixture as though it were a pseudopure substance. This problem has been studied for many equations of state, using experimental data for the real gas mixtures, and various empirical rules have been proposed. For example, for both the van der Waals equation, Eq. 10.53, and the Redlich–Kwong equation, Eq. 10.59, the two pure substance constants a and b are commonly combined according to the relations

$$a_m = \left(\sum_i y_i a_i^{1/2} \right)^2 \qquad b_m = \sum_i y_i b_i \tag{11.36}$$

The following example illustrates the use of these two approaches to treating real gas mixtures as pseudopure substances.

EXAMPLE 11.8   A mixture of 59.39% CO_2 and 40.61% CH_4 (mole basis) is maintained at 310.94 K, 86.19 bar, at which condition the specific volume has been measured as 0.2205 m^3/kmol. Calculate the percent deviation if the specific volume had been calculated by (a) Kay's rule and (b) van der Waals' equation of state.

Control mass: Gas mixture.

State: P, v, T known.

Model: (a) Kay's rule. (b) van der Waals' equation.

Solution

(a) For convenience, let

$$CO_2 = A \qquad CH_4 = B$$

Then

$$T_{c_A} = 304.1 \text{ K} \qquad P_{c_A} = 7.38 \text{ MPa}$$
$$T_{c_B} = 190.4 \text{ K} \qquad P_{c_B} = 4.60 \text{ MPa}$$

For Kay's rule, Eq. 11.35,

$$T_{c_m} = \sum_i y_i T_{c_i} = y_A T_{c_A} + y_B T_{c_B}$$
$$= 0.5939(304.1) + 0.4061(190.4)$$
$$= 257.9 \text{ K}$$
$$P_{c_m} = \sum_i y_i P_{c_i} = y_A P_{c_A} + y_B P_{c_B}$$
$$= 0.5939(7.38) + 0.4061(4.60)$$
$$= 6.257 \text{ MPa}$$

Therefore, the pseudoreduced properties of the mixture are

$$T_{r_m} = \frac{T}{T_{c_m}} = \frac{310.94}{257.9} = 1.206$$

$$P_{r_m} = \frac{P}{P_{c_m}} = \frac{8.619}{6.251} = 1.379$$

From the generalized tables, A.15

$$Z_m = 0.698$$

and

$$\bar{v} = \frac{Z_m \bar{R} T}{P} = \frac{0.698 \times 8.3145 \times 310.94}{8619}$$
$$= 0.2094 \text{ m}^3 / \text{kmol}$$

The percent deviation from the experimental value is

$$\text{Percent deviation} = \left(\frac{0.2205 - 0.2094}{0.2205} \right) \times 100 = 5.03\%$$

The major factor contributing to this 5% error is the use of the linear Kay's rule pseudocritical model, Eq. 11.35. Use of an accurate pseudocritical model and the generalized Table A.15 would reduce the error to approximately 1%.

(b) For van der Waals' equation, the pure substance constants are

$$a_A = \frac{27 \bar{R}^2 T_{cA}^2}{64 P_{cA}} = 365.454 \frac{\text{kPa m}^6}{\text{kmol}^2}$$

$$b_A = \frac{\bar{R} T_{cA}}{8 P_{cA}} = 0.042\,83 \text{ m}^3 / \text{kmol}$$

and

$$a_B = \frac{27 \bar{R}^2 T_{cB}^2}{64 P_{cB}} = 229.843 \frac{\text{kPa m}^6}{\text{kmol}^2}$$

$$b_B = \frac{\bar{R} T_{cB}}{8 P_{cB}} = 0.043\,02 \text{ m}^3 / \text{kmol}$$

Therefore, for the mixture, from Eq. 11.36,

$$a_m = \left(y_A\sqrt{a_A} + y_B\sqrt{a_B}\right)^2$$

$$= \left(0.5939\sqrt{365.454} + 0.4061\sqrt{229.843}\right)^2$$

$$= 306.607\frac{\text{kPa m}^6}{\text{kmol}^2}$$

$$b_m = y_A b_A + y_B b_B$$

$$= (0.5939\times0.042\,83 + 0.4061\times0.04302)$$

$$= 0.042\,91 \text{ m}^3/\text{kmol}$$

The equation of state for the mixture of this composition is

$$P = \frac{\overline{R}T}{\overline{v} - b_m} - \frac{a_m}{\overline{v}^2}$$

$$8619 = \frac{8.3145\times310.94}{\overline{v} - 0.042\,91} - \frac{306.607}{\overline{v}^2}$$

Solving for $\overline{v}$ by trial and error,

$$\overline{v} = 0.2063 \text{ m}^3/\text{kmol}$$

$$\text{Percent deviation} = \left(\frac{0.2205 - 0.2063}{0.2205}\right)\times100 = 6.4\%$$

As a point of interest from the ideal-gas law, $\overline{v} = 0.300$ m^3/kmol, which is a deviation of 36% from the measured value. Also, if we use the Redlich–Kwong equation of state, and follow the same procedure as for the van der Waals equation, the calculated specific volume of the mixture is 0.2127 m^3/kmol, which is in error by 3.5%.

We must be careful not to draw too general a conclusion from the results of this example. We have calculated percent deviation in v at only a single point for only one mixture. We do note, however, that the various methods used give quite different results. From a more general study of these models for a number of mixtures, we find that the results found here are fairly typical, at least qualitatively. Kay's rule is very useful because it is fairly accurate and yet relatively simple. The van der Waals equation is too simplified an expression to accurately represent P–v–T behavior except at moderate densities, but it is useful to demonstrate the procedures followed in utilizing more complex analytical equations of state. The Redlich–Kwong equation is considerably better, and is still relatively simple to use.

As noted in the example, the more sophisticated generalized behavior models and empirical equations of state will represent mixture P–v–T behavior to within about 1 percent over a wide range of density, but they are of course more difficult to use than the methods considered in Example 11.8. The generalized models have the advantage of being easier to use, and they are suitable for hand computations. Calculations with the complex empirical equations of state become very involved, but have the advantage of expressing the P–v–T composition relations in analytical form, which is of great value when using a digital computer for such calculations.

The theoretical virial equation of state, Eq. 10.65, can be applied to mixtures, in which case the virial coefficients are dependent on composition as well as on temperature. For a mixture of A and B, the second virial coefficient becomes

$$B(T, y) = y_A^2 B_A(T) + 2 y_A y_B B_{AB}(T) + y_B^2 B_B(T) \tag{11.37}$$

in which the interaction coefficient $B_{AB}(T)$ results from forces between unlike molecules. For the Lennard–Jones (6-12) potential, this can be found from the parameters

$$\varepsilon_{AB} = (\varepsilon_A \varepsilon_B)^{1/2}$$
$$b_{0AB} = \tfrac{1}{8}(b_{0A}^{1/3} + b_{0B}^{1/3})^3 \tag{11.38}$$

and the values in Table A.14. Representation of the third virial coefficient for a mixture involves two interaction coefficients, and becomes more complicated than that for the second virial coefficient. This topic will not be treated in detail in this text.

11.9 Partial Molal Properties

In this section we begin to examine the second approach to the analysis of real mixtures that was mentioned in Section 11.7, that in which mixture extensive properties are treated as a summation of the contributions of the components, at the conditions at which the components exist in the mixture. This requires a detailed examination of the concept of partial molal properties, which is the subject of the following paragraphs. The term partial property was first used in connection with the expression for the internal energy of a mixture, Eq. 11.6. In the general case, any extensive property X is a function of the temperature and pressure of the mixture and the number of moles of each component. Thus, for a mixture of two components,

$$X = f(T, P, n_A, n_B)$$

Therefore,

$$dX_{T,P} = \left(\frac{\partial X}{\partial n_A}\right)_{T,P,n_B} dn_A + \left(\frac{\partial X}{\partial n_B}\right)_{T,P,n_A} dn_B \tag{11.39}$$

Since at constant temperature and pressure an extensive property is directly proportional to the mass, Eq. 11.39 can be integrated to give

$$X_{T,P} = \overline{X}_A n_A + \overline{X}_B n_B \tag{11.40}$$

where

$$\overline{X}_A = \left(\frac{\partial X}{\partial n_A}\right)_{T,P,n_B} \qquad \overline{X}_B = \left(\frac{\partial X}{\partial n_B}\right)_{T,P,n_A} \tag{11.41}$$

$\overline{X}$ is defined as the partial molal property for a component in a mixture. It is particularly important to note that the partial molal property is defined under conditions of constant temperature and pressure. We also note that the general expression Eq. 11.40 is of the same form as Eq. 11.6.

The general extensive property X discussed above can be any of the properties V, H, U, S, A or G. As an example of the characteristics of partial properties, consider it to be the volume V. Then,

$$V = n_A \overline{V}_A + n_B \overline{V}_B \tag{11.42}$$

where, by definition,

$$\overline{V}_A \equiv \left(\frac{\partial V}{\partial n_A} \right)_{T,P,n_B} \qquad \overline{V}_B \equiv \left(\frac{\partial V}{\partial n_B} \right)_{T,P,n_A} \tag{11.43}$$

For the special case of pure substance A,

$$\overline{V}_A = \left(\frac{\partial V}{\partial n_A} \right)_{T,P} = \overline{v}_A$$

and

$$V_A = \overline{v}_A n_A \tag{11.44}$$

Thus the partial volume of A when no B is present reduces to the specific volume of pure A as would naturally be expected.

Let us now consider the general case of a real gas mixture, for which we can express the specific volume by dividing Eq. 11.42 by n, resulting in

$$\overline{v} = \frac{V}{n} = y_A \overline{V}_A + y_B \overline{V}_B$$

$$= \overline{V}_B + y_A (\overline{V}_A - \overline{V}_B) \tag{11.45}$$

Suppose that the specific volume of this mixture at a given temperature and pressure varies with mixture composition in the manner shown in Fig. 11.12. From the relationships in Eqs. 11.43–11.45, we note that the specific volume varies from the value $\overline{v}_B$ for pure B to the value $\overline{v}_A$ for pure A, and that at any mixture composition, the partial molal volumes of A and B are equal to the respective intercepts of the tangent to the volume curve at that mixture composition point.

For the special case of an ideal gas mixture of A and B,

$$\overline{V}_A = \left(\frac{\partial V}{\partial n_A} \right)_{T,P,n_B} = \frac{\overline{R}T}{P} = \overline{v}$$

and

$$\overline{V}_B = \left(\frac{\partial V}{\partial n_B} \right)_{T,P,n_A} = \frac{\overline{R}T}{P} = \overline{v}$$

Since $\overline{v}$, $\overline{v}_A$, $\overline{v}_B$ are the same per mole,

$$\overline{V}_A = \overline{V}_B = \overline{v} = \overline{v}_A = \overline{v}_B = \frac{\overline{R}T}{P} \tag{11.46}$$

Therefore, for a mixture of ideal gases we conclude from Eq. 11.42 that

$$V = n_A \overline{v}_A + n_B \overline{v}_B = V_A + V_B$$

which is Amagat's rule of additive volumes. We also note from Eq. 11.46 that a plot corresponding to that of Fig. 11.12 for an ideal gas mixture would be a horizontal straight line, where all the volumetric values in Eq. 11.46 are equal and are the same, regardless of the mixture composition.

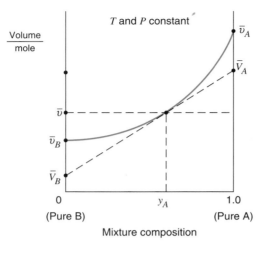

FIGURE 11.12 Specific volume–composition diagram for a mixture of two components.

The physical significance of the partial molal volume may be visualized by considering a reasonably large volume of a mixture of components A and B. Let a very small amount of A be added while the temperature, pressure and moles of B remain constant. The composition of the mixture will be essentially unchanged. The volume of the mixture will be increased by an amount equal to the specific volume of component A in the mixture (the partial molal volume of A) multiplied by the number of moles of component A added. Thus, in this case,

$$dV = \left(\frac{\partial V}{\partial n_A}\right)_{T,P,n_B} dn_A = \overline{V}_A dn_A$$

Similar analyses can of course be performed for the other extensive thermodynamic properties mentioned above. It can also be shown that the thermodynamic relations that apply to a mixture as a whole also hold for relations between partial molal properties for a given component. For example, consider the relation

$$dG = dH - T\,dS - S\,dT$$

Let us again consider a mixture of components A and B, and let T, P, and n_B be held constant, and let n_A change by the amount dn_A. In this case, the preceding equation becomes

$$\left(\frac{\partial G}{\partial n_A}\right)_{T,P,n_B} = \left(\frac{\partial H}{\partial n_A}\right)_{T,P,n_B} - T\left(\frac{\partial S}{\partial n_A}\right)_{T,P,n_B}$$

or

$$\overline{G}_A = \overline{H}_A - T\overline{S}_A \tag{11.47}$$

Similarly

$$\overline{H}_A = \overline{U}_A + P\overline{V}_A \tag{11.48}$$

$$\overline{A}_A = \overline{U}_A - T\overline{S}_A \tag{11.49}$$

11.10 CHANGE IN PROPERTIES UPON MIXING

In this section, we will develop expressions for the changes in volume, enthalpy, and entropy upon mixing of two pure substances at constant temperature and pressure. We have previously seen that there is no volume change upon mixing of ideal gases at constant temperature and pressure. If two real gases each at T, P were mixed, however, the volume of the mixture would not necessarily equal the sum of the volumes of the constituents. The same is true for solutions involving liquids and solids. The change of volume on mixing, ΔV_{mix}, can be expressed as

$$\Delta V_{mix} = V_{mixture} - V_{components}$$

For a mixture of components A and B

$$\Delta V_{mix} = \left(\overline{V}_A n_A + \overline{V}_B n_B\right) - \left(\overline{v}_A n_A + \overline{v}_B n_B\right)$$

$$= \left(\overline{V}_A - \overline{v}_A\right)n_A + \left(\overline{V}_B - \overline{v}_B\right)n_B \tag{11.50}$$

Similarly the enthalpy change of mixing is given by

$$\Delta H_{mix} = \left(\overline{H}_A - \overline{h}_A\right)n_A + \left(\overline{H}_B - \overline{h}_B\right)n_B \tag{11.51}$$

and the entropy change of mixing by

$$\Delta S_{mix} = \left(\overline{S}_A - \overline{s}_A\right)n_A + \left(\overline{S}_B - \overline{s}_B\right)n_B \tag{11.52}$$

For a mixture of ideal gases, the special case previously discussed, we concluded (Eq. 11.46) that

$$\left(\overline{V}_A - \overline{v}_A\right) = 0$$

$$\left(\overline{V}_B - \overline{v}_B\right) = 0$$

Therefore,

$$\Delta V_{mix} = 0 \tag{11.53}$$

Also, for a mixture of ideal gas

$$\left(\overline{H}_A - \overline{h}_A\right) = 0$$

$$\left(\overline{H}_B - \overline{h}_B\right) = 0$$

and therefore

$$\Delta H_{mix} = 0 \tag{11.54}$$

But, the entropy of mixing for real gases is given by the relation

$$\left(\overline{S}_A - \overline{s}_A\right) = -\overline{R}\ln y_A$$

$$\left(\overline{S}_B - \overline{s}_B\right) = -\overline{R}\ln y_B$$

$$\Delta S_{mix} = -\overline{R}\left(n_A \ln y_A + n_B \ln y_B\right) \tag{11.55}$$

in which $\overline{S}_A$ is the partial entropy of A in the mixture at T, P, while $\overline{s}_A$ is that for pure A at the same temperature and pressure, and similarly for component B. This is the expression that was derived in Example 11.2.

11.11 THE THERMODYNAMIC PROPERTY RELATION FOR VARIABLE COMPOSITION

In Section 10.2 we considered several forms of the property relation for systems of fixed mass, such as

$$dU = T \, dS - P \, dV$$

We note that in this equation temperature is the intensive property or potential function associated with entropy and pressure is the intensive property associated with volume. Now suppose we have a chemical reaction in which the amounts of components A and B change. How would we modify the property relation for this situation? Intuitively we might write the equation

$$dU = T \, dS - P \, dV + \mu_A \, dn_A + \mu_B \, dn_B \qquad (11.56)$$

where μ_A is the intensive property or potential function associated with n_A, and similarly μ_B for n_B. This potential function is called the chemical potential.

To derive an expression for the chemical potential, we examine Eq. 11.56 and conclude that it is reasonable to write an expression for U in the form

$$U = f(S, V, n_A, n_B)$$

Therefore,

$$dU = \left(\frac{\partial U}{\partial S}\right)_{V, n_A, n_B} dS + \left(\frac{\partial U}{\partial V}\right)_{S, n_A, n_B} dV + \left(\frac{\partial U}{\partial n_A}\right)_{S, V, n_B} dn_A + \left(\frac{\partial U}{\partial n_B}\right)_{S, V, n_A} dn_B$$

Since the expressions

$$\left(\frac{\partial U}{\partial S}\right)_{V, n_A, n_B} \qquad \text{and} \qquad \left(\frac{\partial U}{\partial V}\right)_{S, n_A, n_B}$$

imply constant composition, it follows from Eq. 10.15 that

$$\left(\frac{\partial U}{\partial S}\right)_{V, n_A, n_B} = T \qquad \text{and} \qquad \left(\frac{\partial U}{\partial V}\right)_{S, n_A, n_B} = -P$$

Thus

$$dU = T \, dS - P \, dV + \left(\frac{\partial U}{\partial n_A}\right)_{S, V, n_B} dn_A + \left(\frac{\partial U}{\partial n_B}\right)_{S, V, n_A} dn_B \qquad (11.57)$$

On comparing this equation with Eq. 11.56 it follows that the chemical potential can be defined by the relation

$$\mu_A = \left(\frac{\partial U}{\partial n_A}\right)_{S, V, n_B}, \qquad \mu_B = \left(\frac{\partial U}{\partial n_B}\right)_{S, V, n_A} \qquad (11.58)$$

We can also relate the chemical potential to the partial molal Gibbs function. To do so we proceed as follows.

$$G = U + PV - TS$$

$$dG = dU + P \, dV + V \, dP - T \, dS - S \, dT$$

Substituting Eq. 11.56 into this relation we have

$$dG = -S \, dT + V \, dP + \mu_A dn_A + \mu_B dn_B \tag{11.59}$$

This equation suggests that we write an expression for G in the following form.

$$G = f(T, P, n_A, n_B)$$

Proceeding as we did in the case of a similar expression for internal energy, Eq. 11.57, we have

$$dG = \left(\frac{\partial G}{\partial T}\right)_{P,n_A,n_B} dT + \left(\frac{\partial G}{\partial P}\right)_{T,n_A,n_B} dP + \left(\frac{\partial G}{\partial n_A}\right)_{T,P,n_B} dn_A + \left(\frac{\partial G}{\partial n_B}\right)_{T,P,n_A} dn_B$$

$$= -S \, dT + V \, dP + \left(\frac{\partial G}{\partial n_A}\right)_{T,P,n_B} dn_A + \left(\frac{\partial G}{\partial n_B}\right)_{T,P,n_A} dn_B$$

Comparing this with Eq. 11.59 it follows that

$$\mu_A = \left(\frac{\partial G}{\partial n_A}\right)_{T,P,n_B}, \qquad \mu_B = \left(\frac{\partial G}{\partial n_B}\right)_{T,P,n_A}$$

Note that, since partial molal properties are defined at constant temperature and pressure, the quantities $(\partial G/\partial n_A)_{T,P,n_B}$ and $(\partial G/\partial n_B)_{T,P,n_A}$ are the partial molal Gibbs functions for the two components. That is, the chemical potential is equal to the partial molal Gibbs function.

$$\mu_A = \overline{G}_A = \left(\frac{\partial G}{\partial n_A}\right)_{T,P,n_B}, \qquad \mu_B = \overline{G}_B = \left(\frac{\partial G}{\partial n_B}\right)_{T,P,n_A} \tag{11.60}$$

In terms of the enthalpy

$$H = U + PV$$

Equation 11.56 becomes

$$dH = T \, dS + V \, dP + \mu_A dn_A + \mu_B dn_B \tag{11.61}$$

Following the same procedure as before, we find that

$$\mu_A = \left(\frac{\partial H}{\partial n_A}\right)_{S,P,n_B} \tag{11.62}$$

In terms of the Helmholtz function

$$A = U - TS$$

the property relation is

$$dA = -S \, dT - P \, dV + \mu_A dn_A + \mu_B dn_B \tag{11.63}$$

As before, we find an expression for the chemical potential

$$\mu_A = \left(\frac{\partial A}{\partial n_A}\right)_{T,V,n_B} \tag{11.64}$$

Thus, we have found four different expressions for the chemical potential in terms of other properties, Eqs. 11.58, 11.60, 11.62, and 11.64. Of these, only the one, Eq. 11.60

satisfies the definition of a partial property. The partial molal Gibbs function is an extremely important property in the thermodynamic analysis of chemical reactions because at constant temperature and pressure (the conditions under which many chemical reactions occur) it is a measure of the chemical potential or the driving force tending to cause a chemical reaction to take place.

Finally, we also realize that the surface, magnetic, and other terms considered in Chapter 10 may be added to the property relation, Eq. 11.56, thereby giving a completely general form of the relation. This introduces a problem with identification of the chemical potential, and this matter is discussed in the following section.

11.12 A GENERAL DEFINITION OF GIBBS FUNCTION AND ENTHALPY

In the previous section we extended the thermodynamic property relation to systems of variable composition, and found that the chemical potential can be identified with the partial molal Gibbs function according to Eq. 11.60. Let us now consider the general situation, in which we have a substance that is not a simple compressible substance, as in Eq. 10.17, in addition to allowing variable composition, as in Eq. 11.56. We can then write the general thermodynamic property relation

$$dU = T\ dS - P\ dV + \mathcal{T}\ dL + \mathcal{S}\ d\mathcal{A} + \mu_0 \mathcal{H}\ d(V\mathcal{M}) + \mathcal{E}\ dZ$$
$$+ \sum_i \mu_i\ dn_i + \cdots \qquad (11.65)$$

The problem that arises in connection with this general relationship is that Eq. 11.60 is not valid if the Gibbs function is defined in the usual manner, according to

$$G = U + PV - TS \qquad (11.66)$$

To demonstrate that this is in fact the case, it is sufficient to consider only one of the additional effects. Let us analyze a system of two components A and B in which surface effects are of importance. For this case Eq. 11.65 reduces to

$$dU = T\ dS - P\ dV + \mathcal{S}\ d\mathcal{A} + \mu_A\ dn_A + \mu_B\ dn_B \qquad (11.67)$$

From Eqs. 11.66 and 11.67

$$dG = -S\ dT - V\ dP + \mathcal{S}\ d\mathcal{A} + \mu_A\ dn_A + \mu_B\ dn_B \qquad (11.68)$$

which suggests a functional relationship of the form

$$G = f(T, P, \mathcal{A}, n_A, n_B)$$

Proceeding as in Section 11.11, we conclude that

$$\mu_A = \left(\frac{\partial G}{\partial n_A}\right)_{T,P,\mathcal{A},n_B} \qquad (11.69)$$

There is nothing incorrect about this result, but we do realize that it is not consistent with the concept of a partial molal property since the extensive parameter $\mathcal{A}$ is held constant. Instead, the corresponding intensive property $\mathcal{S}$ should be held constant as are the intensive properties T and P if this quantity is to be a partial molal property.

We can explain why this difficulty occurs by analogy to the situation of boundary movement work when the mass is not constant. In that case we recall a boundary movement work (flow work) associated with pushing mass across a control surface, this being in addition to the ordinary boundary movement work. In the property relation, that effect $P\overline{V}_A$ appears as part of the chemical potential. In the present case, in addition to the ordinary surface extension work $\mathscr{S}\,d\mathscr{A}$, there is then an additional work associated with the creation of new surface, $\mathscr{S}\,\mathscr{A}_A dn_A$, as the mass of component A changes. Thus, it is appropriate to treat the $\mathscr{S}\,\mathscr{A}$ term as analogous to the PV term. Let us therefore define a general Gibbs function for this situation as

$$G = U + PV - TS - \mathscr{S}\,\mathscr{A} \tag{11.70}$$

Then

$$dG = dU + P\,dV + V\,dP - T\,dS - S\,dT - \mathscr{S}\,d\mathscr{A} - \mathscr{A}\,d\mathscr{S}$$

and from Eq. 11.68,

$$dG = -S\,dT + V\,dP - \mathscr{A}\,d\mathscr{S} + \mu_A\,dn_A + \mu_B\,dn_B \tag{11.71}$$

We note that Eq. 11.71 suggests the relation

$$G = f(T, P, \mathscr{S}, n_A, n_B)$$

which, proceeding as before, leads to

$$\mu_A = \left(\frac{\partial G}{\partial n_A}\right)_{T,P,\mathscr{S},n_B} = \overline{G}_A \tag{11.72}$$

and we conclude that if G is defined according to Eq. 11.70, then the chemical potential in Eqs. 11.67 and 11.71 is equal to the partial molal Gibbs function. It also follows for this case that

$$\mu_A = \overline{G}_A = \overline{U}_A + P\overline{V}_A - T\overline{S}_A - \mathscr{S}\,\overline{\mathscr{A}}_A \tag{11.73}$$

where the partial properties are all defined holding the intensive property $\mathscr{S}$ constant.

Having defined a general Gibbs function by Eq. 11.70 for the case of surface tension and variable mass, it is necessary to define enthalpy in a similar manner by

$$H = G + TS = U + PV - \mathscr{S}\,\mathscr{A} \tag{11.74}$$

and the property relation Eq. 11.67 can then be written in terms of H in a consistent manner. It should also be noted that there is no effect on the Helmholtz function, as that property does not include the energy terms associated with mass transfer modes of work.

It is not difficult to extend our analysis to the general case associated with Eq. 11.65. That expression contains the various quasiequilibrium work modes represented in Eq. 4.13, which can be written in general variables as

$$\delta W = -\sum_k F_k\,dX_k \tag{11.75}$$

In this equation, F_k is the intensive property or driving force, and X_k is the corresponding extensive property affected by F_k. A general definition of the Gibbs function is then

$$G = U - TS - \sum_k F_k X_k \tag{11.76}$$

and the enthalpy is

$$H = G + TS = U - \sum_k F_k X_k \tag{11.77}$$

The general thermodynamic property relation, Eq. 11.65, is

$$dU = T \, dS + \sum_k F_k \, dX_k + \sum_i \mu_i \, dn_i \qquad (11.78)$$

or, in terms of G, from Eq. 11.76

$$dG = -S \, dT - \sum_k X_k \, dF_k + \sum_i \mu_i \, dn_i \qquad (11.79)$$

The chemical potential of component i in Eqs. 11.78 and 11.79 is

$$\mu_i = \left(\frac{\partial G}{\partial n_i} \right)_{T, F_k, n_{j \neq i}} \qquad (11.80)$$

The property relation can similarly be expressed in terms of enthalpy, using its definition Eq. 11.77, or in terms of the Helmholtz function.

11.13 Fugacity in a Mixture and Its Relation to Other Properties

Let us define the fugacity of component A in a mixture by a similar procedure to that for a pure substance in Section 10.12. At constant temperature we define $\bar{f}_A$ by

$$(d\bar{G}_A)_T = \bar{R} T \, d(\ln \bar{f}_A)_T \qquad (11.81)$$

along with the requirement that

$$\lim_{p \to 0} \left(\frac{\bar{f}_A}{y_A P} \right) = 1 \qquad (11.82)$$

so that as pressure approaches zero, the fugacity of component A approaches the ideal gas mixture partial pressure of component A. The fugacity of a component in a mixture as defined by these equations is not a true partial property according to the definition of Eq. 11.41. Nevertheless, we include the bar above the symbol as a reminder that the substance involved is a component of a mixture. Note that in the sense that f is essentially a pseudopressure, $\bar{f}$ can be considered a pseudopartial pressure. The fugacity of component B in the mixture is defined by a pair of equations analogous to Eqs. 11.81 and 11.82.

To determine the relation of fugacity of a component to other properties, consider first the property relation written in terms of the Gibbs function. For a mixture of components A and B, this relation is given by Eq. 11.59.

At constant T and n_B this expression reduces to

$$dG_{T,n_B} = V \, dP_{T,n_B} + \bar{G}_A \, dn_{A \, T, n_B}$$

Taking a Maxwell cross-partial derivative we have

$$\left(\frac{\partial \bar{G}_A}{\partial P} \right)_{T, n_A, n_B} = \left(\frac{\partial V}{\partial n_A} \right)_{T, P, n_B} = \bar{V}_A \qquad (11.83)$$

Therefore, from Eqs. 11.81 and 11.83 at constant temperature and composition,

$$d \left(\bar{G}_A \right)_{T, n_A, n_B} = \bar{R} T \, d \left(\ln \bar{f}_A \right)_{T, n_A, n_B} = \bar{V}_A \, dP_{T, n_A, n_B} \qquad (11.84)$$

For pure substance A at the same constant temperature,

$$(d \bar{g}_A) = \bar{R} T d(\ln f_A)_T = \bar{v}_A dP_T \qquad (11.85)$$

In order to derive an expression that permits us to evaluate the fugacity of component A in the mixture from measurable quantities, let us subtract Eq. 11.85 from Eq. 11.84, and integrate the resultant expression from P^* to P (where P^* is a very low pressure) at constant temperature and composition.

$$\int_{\ln \bar{f}_A^*/f_A^*}^{\ln \bar{f}_A/f_A} \bar{R}Td\left(\ln \frac{\bar{f}_A}{f_A}\right) = \int_{P^* \to 0}^{P} \left(\bar{V}_A - \bar{v}_A\right)dP \tag{11.86}$$

At this low pressure P^*, $\bar{f}_A^* = y_A P^*$ (Eq. 11.82) and $f_A^* = P^*$. Therefore, as $P^* \to 0$,

$$\left(\ln \frac{\bar{f}_A^*}{f_A^*}\right) \to \ln\left(\frac{y_A P^*}{P^*}\right) = \ln y_A$$

Therefore we can write

$$\int_{\ln y_A}^{\ln \bar{f}_A/f_A} \bar{R}Td\left(\ln \frac{\bar{f}_A}{f_A}\right) = \int_{P^* \to 0}^{P} \left(\bar{V}_A - \bar{v}_A\right)dP$$

This integration results in the expression

$$\bar{R}T\ln\left(\frac{\bar{f}_A}{y_A f_A}\right) = \int_{0}^{P} \left(\bar{V}_A - \bar{v}_A\right)dP \tag{11.87}$$

in which $\bar{f}_A$ is the fugacity of component A in the mixture of given composition at the given temperature and pressure P, whereas f_A is the fugacity of pure A at the same temperature and pressure. Equation 11.87 expresses the relation between $\bar{f}_A$ and f_A in terms of the difference between $\bar{V}_A$ and $\bar{v}_A$, a measurable quantity. It is also convenient to find a relation between $\bar{f}_A$ and f_A in terms of the other quantities discussed in Sections 11.19 and 11.10, namely, $\bar{H}_A - \bar{h}_A$ and $\bar{S}_A - \bar{s}_A$, for such relations will permit evaluation of one or more of these quantities in terms of $\bar{V}_A - \bar{v}_A$.

First we shall determine the partial molal Gibbs function of component A in a mixture at T, P with respect to a state at which the Gibbs function is known. Let us integrate Eq. 11.81 at constant T from P^* to P, where again P^* is sufficiently low to assume ideal gas behavior,

$$\int_{\bar{G}_A^*}^{\bar{G}_A} d\bar{G}_A = \int_{\bar{f}_A^* \to y_A P^*}^{\bar{f}_A} \bar{R}Td\left(\ln \bar{f}_A\right)_T \tag{11.88}$$

This integration yields the result

$$\bar{G}_A = \bar{G}_A^* + \bar{R}T\ln\frac{\bar{f}_A}{y_A P^*} = \bar{G}_A^* - \bar{R}T\ln y_A + \bar{R}T\ln\frac{\bar{f}_A}{P^*}$$

We have noted, Eq. 11.47, that

$$\bar{G}_A = \bar{H}_A - T\bar{S}_A$$

Therefore,

$$\bar{G}_A = \bar{H}_A^* - T\bar{S}_A^* - \bar{R}T\ln y_A + \bar{R}T\ln\frac{\bar{f}_A}{P^*} \tag{11.89}$$

But, at low pressures,

$$\bar{H}_A^* = \bar{h}_A^*$$

Also, from Eq. 11.55,

$$\bar{S}_A^* + \bar{R}\ln y_A = \bar{s}_A^*$$

Therefore,

$$\overline{G}_A = \bar{h}_A^* - T\bar{s}_A^* + \bar{R}T\ln\frac{\bar{f}_A}{P*}$$

$$\overline{G}_A = \bar{g}_A^* + \bar{R}T\ln\frac{\bar{f}_A}{P*} \tag{11.90}$$

To determine the change in $\overline{G}_A$ with respect to temperature, we take another Maxwell cross-partial derivative of Eq. 11.59,

$$\left(\frac{\partial \overline{G}_A}{\partial T}\right)_{P,n_A,n_B} = -\left(\frac{\partial S}{\partial n_A}\right)_{T,P,n_B} = -\overline{S}_A \tag{11.91}$$

or, at constant pressure and composition,

$$d\overline{G}_A = -\overline{S}_A \, dT \tag{11.92}$$

But, from Eq. 11.47,

$$\overline{S}_A = \frac{\overline{H}_A - \overline{G}_A}{T}$$

Substituting,

$$(d\overline{G}_A)_{P,n_A,n_B} = -\frac{\overline{H}_A - \overline{G}_A}{T} dT_{P,n_A,n_B}$$

which may be rearranged to the form

$$\frac{T \, d\overline{G}_A - \overline{G}_A \, dT}{T^2} = -\frac{\overline{H}_A}{T^2} dT_{P,n_A,n_B} \tag{11.93}$$

We note that Eq. 11.93 is, in effect,

$$d\left(\frac{\overline{G}_A}{T}\right)_{P,n_A,n_B} = -\frac{\overline{H}_A}{T^2} dT_{P,n_A,n_B} \tag{11.94}$$

By the same procedure for pure A at constant pressure, we can write

$$d\left(\frac{\bar{g}_A}{T}\right)_P = -\frac{\bar{h}_A}{T^2} dT_P \tag{11.95}$$

Now, let us divide Eq. 11.90 by T and differentiate at constant pressure and composition,

$$d\left(\frac{\overline{G}_A}{T}\right)_{P,n_A,n_B} = d\left(\frac{\bar{g}_A^*}{T}\right)_{P,n_A,n_B} + \overline{R}\,d(\ln\bar{f}_A)_{P,n_A,n_B} \tag{11.96}$$

Substituting Eqs. 11.94 and 11.95, we see that at constant pressure and composition,

$$d(\ln\bar{f}_A)_{P,n_A,n_B} = \frac{\bar{h}_A^* - \overline{H}_A}{\overline{R}T^2} dT_{P,n_A,n_B} \tag{11.97}$$

Again, by the same procedure for pure A at constant pressure,

$$d(\ln f_A)_P = \frac{\overline{h}_A^* - \overline{h}_A}{\overline{R} T^2} dT_P \tag{11.98}$$

Combining Eqs. 11.97 and 11.98,

$$d\left(\ln \frac{\overline{f}_A}{f_A}\right)_{P, n_A, n_B} = \frac{\overline{h}_A - \overline{H}_A}{\overline{R} T^2} dT_{P, n_A, n_B} \tag{11.99}$$

which gives another relation between $\overline{f}_A$ and f_A at a given T, P, in this case in terms of the difference between $(\overline{h}_A - \overline{H}_A)$.

Finally, let us determine the relation between $\overline{f}_A$ and f_A in terms of the difference $(\overline{S}_A - \overline{s}_A)$. For pure component A,

$$\overline{g}_A = \overline{h}_A - T\overline{s}_A$$

Also,

$$\overline{G}_A = \overline{H}_A - T\overline{S}_A$$

Combining these equations,

$$(\overline{S}_A - \overline{s}_A) = \left(\frac{\overline{H}_A - \overline{h}_A}{T}\right) - \left(\frac{\overline{G}_A - \overline{g}_A}{T}\right) \tag{11.100}$$

Substituting Eqs. 10.94 and 11.90, we have

$$(\overline{S}_A - \overline{s}_A) = \left(\frac{\overline{H}_A - \overline{h}_A}{T}\right) - \overline{R} \ln\left(\frac{\overline{f}_A}{f_A}\right) \tag{11.101}$$

which can also be written in the form

$$(\overline{S}_A - \overline{s}_A) = \left(\frac{\overline{H}_A - \overline{h}_A}{T}\right) - \overline{R} \ln\left(\frac{\overline{f}_A}{y_a f_A}\right) - \overline{R} \ln y_A \tag{11.102}$$

so that the second term on the right side of the equation is similar to the first term of Eq. 11.87.

11.14 THE IDEAL SOLUTION

There are many mixtures and solutions for which the change of volume on mixing is negligibly small. These are referred to as ideal solutions. By this definition a mixture of ideal gases is an ideal solution, and this is quite acceptable terminology. However, ideal solutions also include certain solid solutions and liquid solutions as well as mixtures of nonideal gases. A number of significant simplications result from this assumption.

We have noted (Eq. 11.46) that for a mixture of ideal gases consisting of components A and B,

$$(\overline{V}_A - \overline{v}_A) = (\overline{V}_B - \overline{v}_B) = 0$$

Let us define an ideal solution as any solution or mixture for which

$$(\overline{V}_A - \overline{v}_A) = 0, \qquad (\overline{V}_B - \overline{v}_B) = 0 \tag{11.103}$$

at the pressure of the mixture and for all lower pressures as well. This definition therefore requires that

$$\Delta V_{\text{mix}} = 0$$

as in the case of mixtures of ideal gases.

Since, in accordance with Eq. 11.42, the volume of a mixture of components A and B is

$$V = \overline{V}_A n_A + \overline{V}_B n_B$$

the assumption of an ideal solution, Eq. 11.107, is equivalent to assuming Amagat's rule of additive volumes,

$$V = \overline{v}_A n_A + \overline{v}_B n_B$$

where $\overline{v}_A$ and $\overline{v}_B$ are the molal specific volumes of pure A and pure B each in the same phase as the mixture, and each at P and T, the pressure and temperature of the mixture. It should be noted that in the case of an ideal solution, $\overline{v}_A$ and $\overline{v}_B$ are the actual specific volumes of these components, and no assumption of ideal gas behavior is made.

The fugacity of a component in an ideal solution is readily determined by reference to Eq. 11.87.

$$\overline{R} T \ln\left(\frac{\overline{f}_A}{y_A f_A} \right) = \int_0^P \left(\overline{V}_A - \overline{v}_A \right) dP$$

From the definition of an ideal solution, $(\overline{V}_A - \overline{v}_A) = 0$, it follows that

$$\overline{f}_A = y_A f_A \tag{11.104}$$

Similarly, for component B,

$$\overline{f}_B = y_B f_B \tag{11.105}$$

For the general case of an ideal solution we can write

$$\overline{f}_i = y_i f_i \tag{11.106}$$

Eq. 11.106 constitutes the Lewis–Randall rule, which holds for an ideal solution. Note the similar appearance of these equations and those for the partial pressures of ideal gas mixture components as discussed in Section 11.1. This is consistent with the concept that fugacity is a pseudopressure.

The enthalpy of mixing for a mixture of components A and B is given by Eq. 11.51.

$$\Delta H_{\text{mix}} = (\overline{H}_A - \overline{h}_A) n_A + (\overline{H}_B - \overline{h}_B) n_B$$

The fact that the enthalpy of mixing is zero for an ideal solution can be demonstrated by consideration of Eq. 11.99, which was written for constant composition.

$$d\left(\ln \frac{\overline{f}_A}{f_A} \right)_{P, n_A, n_B} = \left(\frac{\overline{h}_A - \overline{H}_A}{\overline{R} T^2} \right) dT_{P, n_A, n_B}$$

Substituting Eq. 11.87 it follows that for an ideal solution,

$$(\overline{h}_A - \overline{H}_A) = 0 \tag{11.107}$$

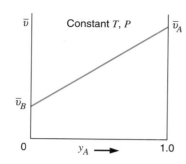 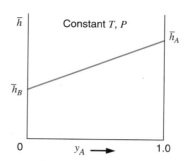

FIGURE 11.13
Typical specific volume–composition and enthalpy–composition diagrams for an ideal solution.

A similar expression can be written for component B, and therefore one concludes that the enthalpy of mixing for an ideal solution is equal to zero.

$$\Delta H_{mix} = 0$$

We had reached the same conclusion for a mixture of ideal gases, but for an ideal solution $\bar{h}_A$ and $\bar{h}_B$ are actual enthalpies, and no assumption of ideal gas behavior is made.

The entropy of mixing for an ideal solution can be found by consideration of Eq. 11.102.

$$\bar{S}_A - \bar{s}_A = \left(\frac{\overline{H}_A - \overline{h}_A}{T} \right) - \overline{R} \ln \left(\frac{\bar{f}_A}{y_A f_A} \right) - \overline{R} \ln y_A$$

Since $\overline{H}_A - \overline{h}_A = 0$ and $\bar{f}_A = y_A f_A$, it follows that for an ideal solution

$$(\bar{S}_A - \bar{s}_A) = -\overline{R} \ln y_A$$
$$(\bar{S}_B - \bar{s}_B) = -\overline{R} \ln y_B \qquad (11.108)$$

in which $\bar{s}_A$ and $\bar{s}_B$ are for pure A and B at T, P, the pressure and temperature of the mixture.

The entropy of mixing, Eq. 11.52 is

$$\Delta S_{mix} = (\bar{S}_A - \bar{s}_A)n_A + (\bar{S}_B - \bar{s}_B)n_B$$

Therefore, for an ideal solution the entropy of mixing is given by the relation

$$\Delta S_{mix} = -\overline{R}(n_A \ln y_A + n_B \ln y_B)$$

Note that the same expression for the entropy of mixing was obtained for a mixture of ideal gases, Eq. 11.17, although for an ideal solution $\bar{s}_A$ and $\bar{s}_B$ are actual entropies and not ideal gas entropies.

It also follows that since $\overline{V}_A - \overline{v}_A = 0$ and $\overline{V}_B - \overline{v}_B = 0$ for an ideal solution, the specific volume–composition diagram is a straight line for an ideal solution. The same would be true for an enthalpy–composition diagram. These are shown in Fig. 11.13.

EXAMPLE 11.9   A gas mixture consisting of 75% Ar and 25% C_2H_4 on a mole basis is stored at 25°C, 8.25 MPa in a cylinder having a volume of 0.5 m³. Determine the mass of gas in the tank assuming an ideal solution, and compare these results with those obtained from assuming a mixture of ideal gases.

Solution

For the ideal solution model, from Eqs. 11.40 and 11.101,

$$V = (n\bar{v})_{Ar} + (n\bar{v})_{C_2H_4}$$

or

$$\bar{v} = (y\bar{v})_{Ar} + (y\bar{v})_{C_2H_4}$$

If we had tables of thermodynamic properties that gave the specific volume of Ar and C_2H_4 at 25°C and 8.25 MPa, we would simply find the gas-phase molal specific volume $\bar{v}$ of each component at this temperature and pressure, and proceed with the solution. In the absence of these data we can use the generalized tables, which will yield fairly accurate results. For the Ar at 25°C and 8.25 MPa,

$$T_r = \frac{298.2}{150.8} = 1.977 \qquad P_r = \frac{8.25}{4.87} = 1.694$$

For the C_2H_4 at 25°C and 8.25 MPa,

$$T_r = \frac{298.2}{282.4} = 1.06 \qquad P_r = \frac{8.25}{5.04} = 1.637$$

From the generalized compressibility table, A.15,

$$Z_{Ar} = 0.961 \qquad Z_{C_2H_4} = 0.330$$

Therefore,

$$\bar{v}_{Ar} = \frac{Z\bar{R}T}{P} = \frac{0.961 \times 8.3145 \times 298.2}{8250} = 0.2888 \text{ m}^3/\text{kmol}$$

$$\bar{v}_{C_2H_4} = \frac{Z\bar{R}T}{P} = \frac{0.33 \times 8.3145 \times 298.2}{8250} = 0.0992 \text{ m}^3/\text{kmol}$$

$$\bar{v} = 0.75(0.2888) + 0.25(0.0992)$$

$$= 0.2414 \text{ m}^3/\text{kmol}$$

$$M = (yM)_{Ar} + (yM)_{C_2H_4} = 0.75(39.95) + 0.25(28.0) = 36.96$$

$$v = \frac{\bar{v}}{M} = \frac{0.2414}{36.96} = 0.006\,531 \text{ m}^3/\text{kg}$$

$$m = \frac{V}{v} = \frac{0.5}{0.006\,531} = 76.6 \text{ kg}$$

Assuming this to be a mixture of ideal gases we can use either the partial pressure or additive volume approach. Let us use the former.

$$P_{Ar} = y_{Ar}P = 0.75(8.25) = 6.188 \text{ MPa}$$

$$m_{Ar} = \frac{P_{Ar}V}{RT} = \frac{6188 \times 0.5}{0.208\,13 \times 298.2} = 49.85 \text{ kg}$$

$$P_{C_2H_4} = y_{C_2H_4}P = 0.25(8.25) = 2.063 \text{ MPa}$$

$$m_{C_2H_4} = \frac{P_{C_2H_4}V}{RT} = \frac{2063 \times 0.5}{0.296\,37 \times 298.2} = 11.67 \text{ kg}$$

$$m = 49.85 + 11.67 = 61.52 \text{ kg}$$

This compares with 76.6 kg for the assumption of an ideal solution.

We note two difficulties in particular from the results of this example. First, it is quite tedious to utilize the ideal solution model with the generalized tables, as it requires evaluating the entire problem separately for each of the components. Second, it required that we use the specific volume $\bar{v}_A$ of pure gaseous A at the temperature and pressure of the gas mixture, and similarly for component B. However, there are certain pressures and temperatures at which the mixture would exist as a gas, but one of the components would, as a pure substance, be a solid or liquid. A familiar example is an air–water vapor mixture at ambient conditions. This poses the question of how to evaluate the properties of this component as a gas at the pressure and temperature of the mixture. There is no simple solution for this problem. In view of these difficulties, we may very well question the use of this model, even though the results are certainly more accurate than assuming ideal gas mixture behavior. The answer to this question is that there are other mixture models, described earlier in Section 11.8, that are much simpler to use with the generalized charts, and which are at least as accurate as the ideal solution for P–v–T, enthalpy, and entropy calculations. The ideal solution does find important application, however, in connection with phase and chemical equilibria correlations, as will be noted in Chapter 13. It should also be pointed out that examination of the ideal solution is beneficial from an educational standpoint, as it is the general model of mixtures of which the ideal gas mixture is a special case. Thus, we are able to develop an intuitive feeling for the validity of assuming ideal gas when analyzing gas mixture behavior.

11.15 ACTIVITY

In our consideration of equilibrium in a following chapter we will find it convenient to make use of an additional property, the activity. It is appropriate, however, to introduce this property at this point.

We have noted that the partial molal Gibbs function for a component in a mixture can be expressed in terms of the low-pressure ideal gas state, as was done in Eq. 11.90.

$$\overline{G}_A = \overline{g}_A^* + \overline{R}T \ln \left(\bar{f}_A / P^* \right)$$

A more general approach would be to express the partial Gibbs functions in terms of a state at which we consider the mixture to behave as an ideal solution instead of an ideal gas mixture. The temperature of this state is the same as that of the mixture, and the pressure is the standard-state pressure P°, a reference pressure at the system temperature. Standard-state values were mentioned briefly in Chapter 7 in connection with entropy, and will be discussed more generally at a later point. It should be pointed out that this state may very well be a hypothetical one. That is, the mixture at T and P° may not actually behave as an ideal solution, but as long as we consistently calculate the mixture properties and develop the equations as though it did, the fact that it is a hypothetical reference state will introduce no difficulties. The significance of this statement will become evident as we proceed further.

The procedure for expressing $\overline{G}_A$ is similar to that followed in the development of Eq. 11.90. In this case Eq. 11.81 is integrated at constant temperature and composition from the pressure P° to the mixture pressure P,

$$\int_{\overline{G}_A^\circ}^{\overline{G}_A} (d\overline{G}_A)_T = \int_{\bar{f}_A^\circ = y_A f_A^\circ}^{\bar{f}_A} \overline{R}T \, d(\ln \bar{f}_A)_T$$

in which f_A° is the fugacity of pure A at T, P°. Therefore,

$$\overline{G}_A = \overline{G}_A^\circ + \overline{R}T\ln\left(\frac{\bar{f}_A}{y_A f_A^\circ}\right)$$

$$= \overline{G}_A^\circ - \overline{R}T\ln y_A + \overline{R}T\ln\left(\frac{\bar{f}_A}{f_A^\circ}\right)$$

$$= \overline{H}_A^\circ - T\overline{S}_A^\circ - \overline{R}T\ln y_A + \overline{R}T\ln\left(\frac{\bar{f}_A}{f_A^\circ}\right) \tag{11.109}$$

Since we assume an ideal solution at P°, from Eq. 11.107,

$$\overline{H}_A^\circ = \bar{h}_A^\circ \tag{11.110}$$

and from Eq. 11.108,

$$\overline{S}_A^\circ + \overline{R}\ln y_A = \bar{s}_A^\circ \tag{11.111}$$

$$\overline{G}_A = \bar{h}_A^\circ - T\bar{s}_A^\circ + \overline{R}T\ln\left(\frac{\bar{f}_A}{f_A^\circ}\right)$$

Therefore,

$$\overline{G}_A = \bar{g}_A^\circ + \overline{R}T\ln\left(\frac{\bar{f}_A}{f_A^\circ}\right) \tag{11.112}$$

Thus we have an expression for the partial molal Gibbs function in terms of known values. It is from this important equation that the equilibrium constant is defined in Chapter 13. Consequently Eq. 11.112 is a very powerful relation. The values f_A° and $\bar{g}_A^\circ$ are for pure substance A at the temperature T and pressure P°, which is referred to as the standard state pressure. For gaseous mixtures, P° is commonly taken as 0.1 MPa. For liquid and vapor phases in two-phase systems (discussed in Chapter 13), the standard state for each component is usually taken as the pure substance in that phase at the pressure of the mixture.

Examining Eq. 11.112, we find it convenient to define a quantity called activity. The activity a_A of component A in a mixture at T, P is defined as

$$a_A = \frac{\bar{f}_A}{f_A^\circ} \tag{11.113}$$

The activity of component B is, of course, similarly defined. Equation 11.112 may now be written in the convenient form

$$\overline{G}_A = \bar{g}_A^\circ + \overline{R}T\ln a_A \tag{11.114}$$

Several special cases for evaluating $\overline{G}_A$ from this equation have already been considered. For example, in the case for which the mixture can be assumed an ideal gas mixture at the standard-state pressure, the equation reduces to Eq. 11.90. If the mixture can be assumed an ideal solution at pressure P as well as at P°, then from Eq. 11.104,

$$a_A = \frac{y_A f_A}{f_A^\circ} \tag{11.115}$$

If conditions are such that an ideal gas mixture can be assumed at both $P°$ and P, then from Eq. 11.82,

$$a_A = \frac{y_A P}{P°} \qquad (11.116)$$

PROBLEMS

11.1 A carbureted internal combustion engine is converted to run on methane gas (natural gas). The air–fuel ratio in the cylinder is to be 20 to 1 on a mass basis. How many moles of oxygen per mole of methane are there in the cylinder?

11.2 The mixture specified in Problem 11.1 is at 90 kPa, 30°C, at the beginning of the compression stroke in the engine cycle. The engine compression ratio (volume ratio) is 9 to 1. Assuming that the compression is reversible and adiabatic, determine the pressure and temperature at the end of the compression stroke. Determine also the compression work input per kilogram of mixture.

11.3 At a certain point in a coal gasification process, a sample of the gas is taken and stored in a 1-L cylinder. An analysis of the mixture yields the following results:

Component	H_2	CO	CO_2	N_2
Percent by volume	25	40	15	20

Determine

a. The mixture composition on a mass basis.

b. The mass of the sample in the cylinder at 100 kPa, 20°C.

c. The amount of heat that must be transferred to heat the sample at constant volume from the initial state to 100°C.

11.4 Carbon dioxide gas at 320 K is mixed with nitrogen at 280 K in a SSSF insulated mixing chamber. Both flows are at 100 kPa and the mole ratio of carbon dioxide to nitrogen is 2:1. Find the exit temperature and the total entropy generation per mole of the exit mixture.

11.5 The gas mixture from Problem 11.3 is compressed in a reversible adiabatic process from the initial state in the sample cylinder to a volume of 0.2 L. Determine the final temperature of the mixture and the work done during the process.

11.6 A mixture of 60% helium and 40% nitrogen by volume enters a turbine at 1 MPa, 800 K at a rate of 2 kg/s. The adiabatic turbine has an exit pressure of 100 kPa and an isentropic efficiency of 85%. Find the turbine work.

11.7 A mixture of 50% carbon dioxide and 50% water by mass is brought from 1500 K, 1 MPa to 500 K, 200 kPa in a polytropic process through a SSSF device. Find the necessary heat transfer and work involved.

11.8 Two insulated tanks A and B are connected by a valve. Tank A has a volume of 1 m^3 and initially contains argon at 300 kPa, 10°C. Tank B has a volume of 2 m^3 and initially contains ethane at 200 kPa, 50°C. The valve is opened and remains open until the resulting gas mixture comes to a uniform state. Determine

a. The final pressure and temperature.

b. The entropy change for the process.

11.9 (Adv.) Recent developments in semipermeable membrane technology have been successfully applied to a number of industrial gas separation processes. One particular application is the partial removal of oxygen from air that is to be fed through a grain elevator storage facility. Ambient air (79% nitrogen, 21% oxygen on a mole basis) is to be compressed to an appropriate pressure, cooled to ambient temperature, 25°C, and then fed through a bundle of hollow polymer fibers that selectively absorb oxygen, so the mixture leaving at 120 kPa, 25°C, contains only 5% oxygen. The absorbed oxygen is bled off through the fiber walls at 40 kPa, 25°C, to a vacuum pump. What is the minimum inlet air pressure to the fiber bundle?

11.10 A cylinder with a frictionless piston contains helium at 110 kPa at ambient temperature 20°C, and initial volume of 20 L as shown in Fig. P11.10. The stops are mounted to give a maximum volume of 25 L and the nitrogen line conditions are 300 kPa, 30°C. The valve is now opened which allows nitrogen to flow in and mix with the helium. The valve is closed when the pressure inside reaches 200 kPa, at which point the temperature inside is 40°C. Is this process consistent with the second law of thermodynamics?

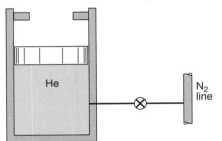

FIGURE P11.10

11.11 A spherical balloon has an initial diameter of 1 m and contains argon gas at 200 kPa, 40°C. The balloon is connected by a valve to a 500-L rigid tank containing carbon dioxide at 100 kPa, 100°C. The valve is opened, and eventually the balloon and tank reach a uniform state in which the pressure is 185 kPa. The balloon pressure is directly proportional to its diameter. Take the balloon and tank as a control volume, and calculate the final temperature and the heat transfer for the process.

11.12 A mixture of 2 kg oxygen and 2 kg of argon is in an insulated piston cylinder arrangement at 100 kPa, 300 K. The piston now compresses the mixture to half its initial volume. Find the final pressure, temperature and the piston work.

11.13 A large SSSF air separation plant takes in ambient air (79% N_2, 21% O_2 by volume) at 100 kPa, 20°C, at a rate of 1 kmol/s. It discharges a stream of pure O_2 gas at 200 kPa, 100°C, and a stream of pure N_2 gas at 100 kPa, 20°C. The plant operates on an electrical power input of 2000 kW. Calculate the net rate of entropy change for the process.

11.14 A 100-L insulated tank contains nitrogen gas at 200 kPa and ambient temperature 25°C. The tank is connected by a valve to a supply line flowing carbon dioxide at 1.2 MPa, 90°C. A mixture of 50 mole percent nitrogen and 50 mole percent carbon dioxide is to be obtained by opening the valve and allowing flow into the tank until an appropriate pressure is reached and the valve is closed. What is the pressure? The tank eventually cools to ambient temperature. Calculate the net entropy change for the overall process.

11.15 The only known sources of helium are the atmosphere (mole fraction approximately 5×10^{-6}) and natural gas. A large unit is being constructed to separate 100 m^3/s of natural gas, assumed to be 0.001 He mole fraction and 0.999 CH_4. The gas enters the unit at 150 kPa, 10°C. Pure helium exits at 100 kPa, 20°C, and pure methane exits at 140 kPa, 30°C. Any heat transfer is with the surroundings at 20°C. Is an electrical power input of 3000 kW sufficient to drive this unit?

11.16 A tank has two sides initially separated by a diaphragm. Side A contains 1 kg of water and side B contains 1.2 kg of air, both at 20°C, 100 kPa. The diaphragm is now broken and the whole tank is heated to 600°C by a 700°C reservoir. Find the final total pressure, heat transfer and total entropy generation.

11.17 A 0.2 m^3 insulated, rigid vessel is divided into two equal parts A and B by an insulated partition, as shown in Fig. P11.17. The partition will support a pressure difference of 400 kPa before breaking. Side A contains methane and side B contains carbon dioxide. Both sides are initially at 1 MPa, 30°C. A valve on side B is opened, and carbon dioxide flows out. The carbon dioxide that remains in B is assumed to undergo a reversible adiabatic expansion while there is flow out. Eventually the partition breaks, and the valve is closed. Calculate the net entropy change for the process that begins when the valve is closed.

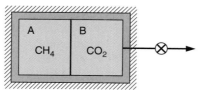

FIGURE P11.17

11.18 Consider a room of dimensions $4 \times 6 \times 2.5$ m that contains an air–water vapor mixture at 100 kPa, 15°C, and 20% relative humidity. How much water would have to be added to this space for the room to be at 100 kPa, 25°C, and have a relative humidity of 50%?

11.19 The products of combustion of a hydrocarbon fuel with air is analyzed with the following results:

Component	CO_2	CO	H_2O	O_2	N_2
Percent by volume	10.4	0.1	11.8	3.4	74.3

The gas mixture passes through a SSSF heat exchanger at ambient pressure at the rate 0.1 kg/s. What is the dew-point temperature? If the mixture is cooled 10°C below the dew-point temperature, how long will it take to collect 10 kg of liquid water?

11.20 A new high-efficiency home heating system includes an air-to-air heat exchanger which uses energy from outgoing stale air to heat the fresh incoming air. If the outside ambient temperature is −10°C and the relative humidity is 30%, how much water will have to be added to the incoming air, if it flows in at the rate of 1 m^3/s and must eventually be conditioned to 20°C and 40% relative humidity?

11.21 Ambient air at 100 kPa, 30°C, 40% relative humidity goes through a constant pressure heat exchanger in a SSSF process. In one case it is heated to 45°C and in another case it is cooled until it reaches saturation. For both cases find the exit relative humidity and the amount of heat transfer per kilogram dry air.

11.22 A flow of air at 5°C, $\Phi = 90\%$, is brought into a house, where it is conditioned to 25°C, 60% relative humidity. This is done in a SSSF process with a combined heater-evaporator where any liquid water is at 10°C. Find any flow of liquid, and the necessary heat transfer, both per kilogram dry air flowing. Find the dew point for the final mixture.

11.23 Consider a 500-L rigid tank containing an air–water vapor mixture at 100 kPa, 35°C, with a 70% relative humidity. The system is cooled until the water just begins to condense. Determine the final temperature in the tank and the heat transfer for the process.

11.24 One hundred kilograms of saturated moist air at 100 kPa, 5°C is in a piston/cylinder arrangement. If it is heated to 45°C in an isobaric process, find $_1q_2$ and the final relative humidity. If it is compressed from the initial state to 200 kPa in an isothermal process, find the mass of water condensing.

11.25 Atmospheric air at 35°C, relative humidity of 10%, is too warm and also too dry. An air conditioner should deliver air at 21°C and 50% relative humidity in the amount of 3600 m³ per hour. Sketch a setup to accomplish this, find any amount of liquid (at 20°C) that is needed or discarded and any heat transfer.

11.26 Ambient moist air enters a steady-flow air-conditioning unit at 102 kPa, 30°C, with a 60% relative humidity. The volume flow rate entering the unit is 100 L/s. The moist air leaves the unit at 95 kPa, 15°C, with a relative humidity of 100%. Liquid condensate also leaves the unit at 15°C. Determine the rate of heat transfer for this process.

11.27 Compare the weather two places where it is cloudy and breezy. At beach A it is 20°C, 103.5 kPa, relative humidity 90% and beach B has 25°C, 99 kPa, relative humidity 20%. Suppose you just took a swim and came out of the water. Where would you feel more comfortable and why?

11.28 A saturated air–water vapor mixture at 20°C, 100 kPa, is contained in a 5-m³ closed tank in equilibrium with 1 kg of liquid water. The tank is then heated to 80°C. Is there any liquid water at the final state? Calculate the heat transfer for the process.

11.29 Outdoor ambient moist air enters a steady-flow heater-humidifier unit at 0°C, 100 kPa, 50% relative humidity, at the volumetric flow rate of 0.1 m³/s. Liquid water at 10°C is sprayed into the mixture, and enough heat is transferred that the moist air leaves the unit at 30°C, 100 kPa, 35% relative humidity. Determine the mass flow rate of liquid water sprayed into the mixture and the rate of heat transfer to the unit.

11.30 Moist air at 24°C, 100 kPa and relative humidity of 70%, flows such that a fraction $^1/_3$ of it goes through a cooler shown in Fig. P11.30. There water condenses out as it cools to 6°C after which the flow is mixed with the remaining fraction that bypassed the cooler in an adiabatic process. Find the liquid flow out of the cooler and the heat transfer, both per kilogram dry air. Find the relative and specific humidity at the cooler exit and after the final mixing process.

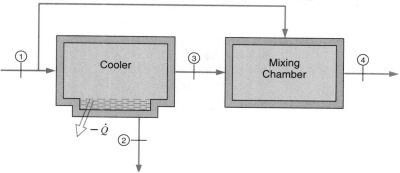

FIGURE P11.30

11.31 A high-efficiency home heating system uses an air-to-air heat exchanger, in which energy from outgoing stale air heats the fresh incoming air in a counterflow SSSF heat exchanger. The unit is advertised as "recovering 80% of the energy" of the outgoing moist air. Using a reasonable assumption for the meaning of this statement, determine the temperature and the relative humidity of the incoming air as it exits the heat exchanger. Assume the room is at 100 kPa, 20°C, 40% relative humidity, and the outdoor ambient conditions are 100 kPa, −10°C, 40% relative humidity.

11.32 Air in a piston cylinder is at 35°C, 100 kPa and a relative humidity of 80%. It is now compressed to a pressure of 500 kPa in a constant temperature process. Find the final relative and specific humidity and the volume ratio V_2/V_1.

11.33 A cylinder/piston loaded with a linear spring contains saturated moist air at 120 kPa, 0.1 m^3 volume and ambient temperature, 20°C. The piston area is 0.2 m^2, and the spring constant is 20 kN/m. This cylinder is attached by a valve to a line flowing dry air at 800 kPa, 40°C. The valve is opened, and air flows into the cylinder until the pressure reaches 200 kPa, at which point the temperature is 40°C. Determine

   a. The relative humidity at the final state.

   b. The mass of air entering the cylinder.

   c. The work done during the process. Consider the cylinder a control volume.

11.34 Consider the previous problem and additionally determine the heat transfer. Show that the process does not violate the second law.

11.35 One means of air-conditioning hot summer air is by evaporative cooling, which is a process similar to the SSSF adiabatic saturation process. Consider outdoor ambient air at 35°C, 100 kPa, 30% relative humidity. What is the maximum amount of cooling that can be achieved by such a technique? What disadvantage is there to this approach? Solve the problem using

   a. A first law analysis.

   b. The psychrometric chart, Fig. A.6.

11.36 The air-conditioning by evaporative cooling in the previous problem is modified by adding a dehumidification process before the water spray cooling process. This dehumidification is achieved as shown in Fig. P11.36 by using a desiccant material, which absorbs water on one side of a rotating drum heat exchanger. The desiccant is regenerated by heating on the other side of the drum to drive the water out. The pressure is 100 kPa everywhere and other properties are on the diagram. Calculate the relative humidity of the cool air supplied to the room at state 4, and the heat transfer per unit mass of air that needs to be supplied to the heater unit.

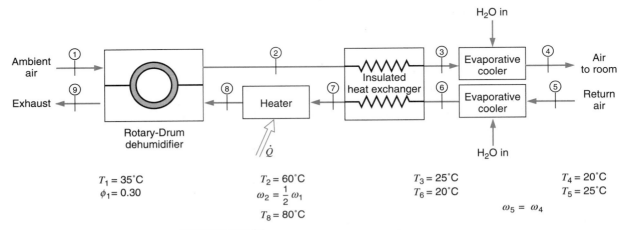

FIGURE P11.36

11.37 A combination air cooler and dehumidification unit receives outside ambient air at 35°C, 100 kPa, 90% relative humidity. The moist air is first cooled to a low temperature to condense the proper amount of water. The moist air is then heated and leaves the unit at 20°C, 100 kPa, 30% relative humidity. The volume flow rate of the moist air at the exit is 0.01 m³/s.

a. Find the temperature to which the mixture must be cooled and the mass of water condensed out per kilogram of dry air. Show the process for the water in a T–s diagram.

b. Assuming all the liquid condensed out leaves at the minimum temperature, calculate the overall heat transfer rate.

11.38 In a car's defrost/defog system atmospheric air, 21°C, relative humidity 80%, is taken in and cooled such that liquid water drips out. The now dryer air is heated to 41°C and then blown onto the windshield, where it should have a maximum of 10% relative humidity to remove water from the windshield. Find the dew point of the atmospheric air, specific humidity of air onto the windshield, the lowest temperature and the specific heat transfer in the cooler.

11.39 A water-filled reactor of 1 m³ is at 20 MPa, 360°C and located inside an insulated containment room of 100 m³ that contains air at 100 kPa and 25°C. Due to a failure the reactor ruptures and the water fills the containment room. Find the final pressure.

11.40 A 300-L rigid vessel initially contains moist air at 150 kPa, 40°C, with a relative humidity of 10%. A supply line connected to this vessel by a valve carries steam at 600 kPa, 200°C. The valve is opened, and steam flows into the vessel until the relative humidity of the resultant moist air mixture is 90%. Then the valve is closed. Sufficient heat is transferred from the vessel so the temperature remains at 40°C during the process. Determine the heat transfer for the process, the mass of steam entering the vessel, and the final pressure inside the vessel.

11.41 Steam power plants often utilize large cooling towers to cool the condenser cooling water so it can be recirculated; see Fig. P11.41. The process is essentially evaporative adiabatic cooling, in which part of the water is lost and must therefore be replenished. Consider the setup shown in Fig. P11.41, in which 1000 kg/s of warm water at 32°C from the condenser enters the top of the cooling tower and the cooled water leaves the bottom at 20°C. The moist ambient air enters the bottom at 100 kPa, dry bulb temperature of 18°C and a wet bulb temperature of 10°C. The moist air leaves the tower at 95 kPa, 30°C, and relative humidity of 85%. Determine the required mass flow rate of dry air, and the fraction of the incoming water that evaporates and is lost.

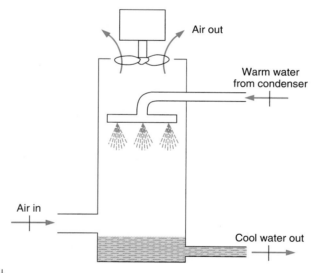

Air out

Warm water from condenser

Air in

Cool water out

FIGURE P11.41

11.42 A vertical cylinder is fitted with a piston held in place by a pin, as shown in Fig. P11.42. The initial volume is 200 L and the cylinder contains moist air at 100 kPa, 25°C, with wet-bulb temperature of 15°C. The pin is removed, and at the same time a valve on the bottom of the cylinder is opened, allowing the mixture to flow out. A cylinder pressure of 150 kPa is required to balance the piston. The valve is closed when the cylinder volume reaches 100 L, at which point the temperature is that of the surroundings, 15°C.

   a. Is there any liquid water in the cylinder at the final state?

   b. Calculate the heat transfer to the cylinder during the process.

   c. Take a control volume around the cylinder, calculate the entropy change of the control volume and that of the surroundings.

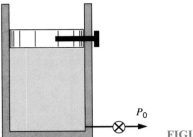

FIGURE P11.42

11.43 Consider two states of atmospheric air. (1) 35°C, $T_{wet} = 18°C$ and (2) 26.5°C, $\Phi = 60\%$. Suggest a system of devices that will allow air in a SSSF process to change from (1) to (2) and from (2) to (1). Heaters, coolers (de)humidifiers, liquid traps etc. are available and any liquid/solid flowing is assumed to be at the lowest temperature seen in the process. Find the specific and relative humidity for state 1, dew point for state 2 and the heat transfer per kilogram dry air in each component in the systems.

11.44 An insulated tank has an air inlet, $\omega_1 = 0.0084$, and an outlet, $T_2 = 22°C$, $\Phi_2 = 90\%$ both at 100 kPa. A third line sprays 0.25 kg/s of water at 80°C, 100 kPa. For a SSSF operation find the outlet specific humidity, the mass flow rate of air needed and the required air inlet temperature, T_1.

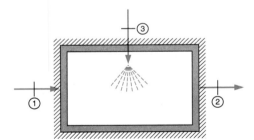

FIGURE P11.44

11.45 A steady supply of air is available from a line at 200 kPa, 30°C. This line is supplied by compressing ambient air (20°C, 100 kPa, 15°C wet-bulb temperature). No provision has been made to remove water vapor. The line is connected by a valve to a 500-L tank that initially contains water at 80°C, 99% quality. The valve is opened until the pressure inside reaches 200 kPa, at which point the temperature inside is 40°C.

  a. How much liquid water is in the tank in the final state?

  b. Calculate the heat transfer from the tank.

  c. Calculate the net entropy change for the tank filling process (do not include the air-line compressor).

11.46 A rigid container, 10 m³ in volume, contains moist air at 45°C, 100 kPa, $\Phi = 40\%$. The container is now cooled to 5°C. Neglect the volume of any liquid that might be present and find the final mass of water vapor, final total pressure and the heat transfer.

11.47 You have just washed your hair and now blow dry it in a room with 23°C, Φ = 60%, (1). The dryer, 500 W, heats the air to 49°C, (2), blows it through your hair where the air becomes saturated (3), and then flows on to hit a window where it cools to 15°C (4). Find the relative humidity at state 2, the heat transfer per kilogram of dry air in the dryer, the air flow rate, and the amount of water condensed on the window, if any.

11.48 A water-cooling tower for a power plant cools 45°C liquid water by evaporation. The tower receives air at 19.5°C, Φ = 30%, 100 kPa that is blown through/over the water such that it leaves the tower at 25°C, Φ = 70%. The remaining liquid water flows back to the condenser at 30°C having given off 1 MW. Find the mass flow rate of air, and the amount of water that evaporates.

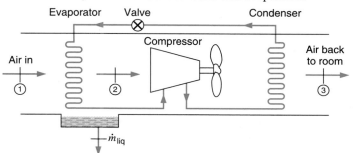

FIGURE P11.49

11.49 An indoor pool evaporates 1.512 kg/h of water, which is removed by a dehumidifier to maintain 21°C, Φ = 70% in the room. The dehumidifier, shown in Fig. P11.49, is a refrigeration cycle in which air flowing over the evaporator cools such that liquid water drops out, and the air continues flowing over the condenser. For an air flow rate of 0.1 kg/s the unit requires 1.4 kW input to a motor driving a fan and the compressor and it has a coefficient of performance, $\beta = \dot{Q}_L/\dot{W}_c = 2.0$. Find the state of the air as it returns to the room and the compressor work input.

11.50 Two moist air streams with 85% relative humidity, both flowing at a rate of 0.1 kg/s of dry air are mixed in a SSSF setup. One inlet flowstream is at 32.5°C and the other at 16°C. Find the exit relative humidity.

11.51 Ambient air is at a condition of 100 kPa, 35°C, 50% relative humidity. A steady stream of air at 100 kPa, 23°C, 70% relative humidity, is to be produced by first cooling one stream to an appropriate temperature to condense out the proper amount of water and then mix this stream adiabatically with the second one at ambient conditions. What is the ratio of the two flow rates? To what temperature must the first stream be cooled?

11.52 To refresh air in a room, a counterflow heat exchanger, see Fig. P11.52, is mounted in the wall, drawing in outside air at 0.5°C, 80% relative humidity and pushing out room air, 40°C, 50% relative humidity. Assume an exchange of 3 kg/min dry air in a SSSF device, and also that the room air exits the heat exchanger to the atmosphere at 23°C. Find the net amount of water removed from the room, any liquid flow in the heat exchanger and (T, Φ) for the fresh air entering the room.

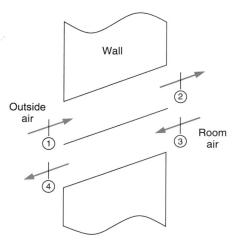

FIGURE P11.52

11.53 A liquid mixture of 75% octane and 25% methanol by mass is at 300 K, 500 kPa. The mixture is heated in a constant pressure SSSF process to 600 K. Determine the heat transfer using Kay's rule and the simple fluid generalized chart corrections.

11.54 A gas mixture of 75% ethylene and 25% ethane on a mole basis enters a SSSF compressor at 50°C, 500 kPa, at a rate of 0.01 m^3/s. The power input to the compressor is 8 kW and the gas mixture exits at 75°C, 2 MPa. Assuming Kay's rule for the gas mixture, determine the mass flow rate and the heat transfer rate from the compressor.

11.55 One kmol/s of saturated liquid methane, CH_4, at 1 MPa and 2 kmol/s of ethane, C_2H_6, at 250°C, 1 MPa are fed to a mixing chamber with the resultant mixture exiting at 50°C, 1 MPa. Assume that Kay's rule applies to the mixture and determine the heat transfer in the process.

11.56 Consider the following reference state conditions: the entropy of real saturated liquid methane at −100°C is to be taken as 100 kJ/kmol K, and the entropy of hypothetical ideal gas ethane at −100°C is to be taken as 200 kJ/kmol K. Calculate the entropy per kmol of a real gas mixture of 50% methane, 50% ethane (mole basis) at 20°C, 4 MPa, in terms of the specified reference state values, and assuming Kay's rule for the real mixture behavior.

11.57 A cylinder/piston contains a gas mixture, 50% CO_2 and 50% C_2H_6 (mole basis) at 700 kPa, 35°C, at which point the cylinder volume is 5 L. The mixture is now compressed to 5.5 MPa in a reversible isothermal process. Calculate the heat transfer and work for the process, using the following model for the gas mixture:

a. Ideal gas mixture.

b. Kay's rule and the generalized charts.

c. The van der Waals equation of state.

11.58 Consider a gas mixture of 50% nitrogen and 50% argon, on a mole basis, contained in a 1-m^3 volume at 180 K, 2 MPa. Calculate the mass of this mixture, using the virial equation of state, and determine the percent error if ideal gas behavior is assumed.

11.59 Take the gas mixture described in the previous problem. Calculate the mass of this mixture assuming Kay's rule and the generalized charts, and also using the Redlich–Kwong equation of state to represent the mixture behavior.

11.60 A gas mixture of ethane (component A) and propane (component B) is contained in a cylinder as shown in Fig. P11.60. The initial state is 30°C (room temperature), 70 kPa, 600 L, and a mole fraction of ethane of 0.20. The line contains pure ethane at 30°C, 7 MPa. The mass of the piston and weights is such that a cylinder pressure of 3.5 MPa would be required to float the piston. The valve is now cracked open and the pin is pulled from the piston. When the valve is closed, the temperature inside the cylinder is 65°C and the mole fraction of ethane is 60%. Calculate the final volume in the cylinder and the net entropy change for the process.

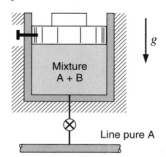

FIGURE P11.60

11.61 Consider a gas mixture, 60% nitrogen and 40% oxygen on a mole basis, flowing through a long pipe. At the pipe inlet, the mixture is at 4 MPa, 175 K, with a velocity of 30 m/s. If the pressure drop in the pipe is estimated at 680 kPa, determine the exit temperature and velocity. Assume that the mixture behavior is represented by Kay's rule and the generalized charts and the pipe is insulated.

11.62 A gas mixture of a known composition is frequently required for different purposes, e.g., in the calibration of gas analyzers. It is desired to prepare a gas mixture of 80% ethylene and 20% carbon dioxide (mole basis) at 10 MPa, 25°C in an uninsulated, rigid 50-L tank. The tank is initially to contain CO_2 at 25°C and some pressure P_1. The valve to a line flowing C_2H_4 at 25°C, 10 MPa, is now opened slightly, and remains open until the tank reaches 10 MPa, at which point the temperature can be assumed to be 25°C. Assume that the gas mixture so prepared can be represented by Kay's rule and the generalized charts. Given the desired final state, what is the initial pressure of the carbon dioxide, P_1? Determine the heat transfer and the net entropy change for the process of charging ethylene into the tank.

11.63 The following data have been measured for carbon dioxide, methane, and their mixtures at 8.618 MPa, 37.8°C:

y_{CO_2}	1.0000	0.7961	0.5939	0.3944	0.1528	0.0000
y_{CH_2}	0.0000	0.2039	0.4061	0.6056	0.8472	1.0000
Z	0.3117	0.6262	0.7350	0.8084	0.8634	0.8892

a. Determine the partial molal volumes of the components for a mixture of 50% CO_2, 50% CH_4 at this pressure and temperature.

b. Evaluate the assumption of ideal solution model (and the generalized charts) instead of by Kay's rule.

11.64 Repeat Problem 11.56, assuming that the real gas mixture behavior is represented by the ideal solution model (and the generalized charts) instead of by Kay's rule.

11.65 An insulated rigid 500-L tank contains a gas mixture, 85% methane and 15% ethane on a mole basis, at 3.5 MPa, 40°C. A valve on the tank is now opened accidentally, and the pressure inside quickly drops to 2 MPa before the valve is closed. Assume that the gas mixture is homogeneous and that its behavior can be represented by Kay's rule and the generalized charts.

a. Calculate the mass that has escaped from the tank.

b. Eventually, the temperature inside the tank returns to that of the surroundings, 40°C. What is the pressure at that time?

11.66 Repeat the previous problem using the Redlich–Kwong equation of state to represent the real gas mixture behavior, instead of Kay's rule and the generalized charts.

11.67 (Adv.) Two heavily insulated, rigid tanks, A and B, each of 20-L volume are connected by a valve. Tank A initially contains methane at 25°C, 25 MPa, and tank B initially contains carbon dioxide at 25°C, 1.4 MPa. The valve is now opened and methane flows quickly from A to B until the pressure inside A drops to 17 MPa, at which time the valve is closed. Assuming that tank B now contains a homogeneous mixture and that tank A contains only methane, determine the final temperature in A, and the final temperature, pressure and composition in B. Specify any assumptions and thermodynamic models used in the analysis and solution.

11.68 Repeat Problem 11.61, assuming ideal solution for the gas mixture, instead of Kay's rule.

11.69 Capsule A containing 1 L of saturated liquid butane at 110°C is located inside a rigid tank B, which has a total volume (including A) of 51 L. Tank B initially contains ethane at 1.4 MPa, and is at room temperature, 25°C. The capsule now breaks and the two substances eventually form a homogeneous gas mixture at 25°C. Determine the final pressure in the tank and the irreversibility of the process.

11.70 (Adv.) A 100-L insulated vessel contains a mixture of 70 mole percent methane and 30 mole percent butane at 310 K, 6 MPa. A valve on the vessel is opened, giving access to a semipermeable membrane that allows methane but not butane to flow out. The valve is closed when the pressure inside the vessel reaches 3 MPa. Determine the final state inside the vessel (T_2, y_{CH_4}).

11.71 A gas mixture, 80% ethylene and 20% propane on a mole basis, enters a SSSF device at 30°C, 3.5 MPa, at the rate of 4 L/s. Two separate streams leave the device; one is pure ethylene gas at 30°C, 2.8 MPa, and the other is pure liquid propane at 65°C, 3.5 MPa. Determine the minimum power input, in kilowatts, required to operate this device.

11.72 (Adv.) Boil-off vapor from a liquid propane storage tank at 0°C is throttled into an adiabatic mixing chamber and mixed with nitrogen in the ratio 4 kmol C_3H_8/1 kmol N_2, as shown in Fig. P11.72. The resulting mixture then enters a compressor, with an exit state of 4.2 MPa, 60°C.

State	1	2	3	4	5	
$T\,[°C]$	0	–	0	–	60	
$P\,[MPa]$		–	0.35	0.35	0.35	4.2

It is claimed that the compression process is adiabatic. Using the second law, show whether or not this is possible.

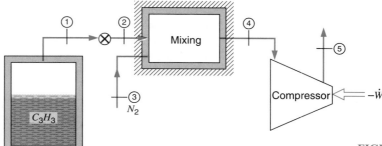

FIGURE P11.72

11.73 A liquid mixture of 75% octane and 25% methanol by mass is at 300 K, 500 kPa. The mixture is heated in a constant pressure SSSF process to 600 K. This problem is identical to Problem 11.53, where it was solved using Kay's rule. Determine the heat transfer using Lee–Kesler pseudocritical constants for the mixture and the acentric factor correction as shown in Appendix C.

11.74 Solve Problem 11.57 using Lee–Kesler pseudocritical constants for the mixture and the acentric factor correction as shown in Appendix C.

11.75 A mixture of 40% nitrogen dioxide, NO_2, and 60% nitrogen by mass is charged into a 100-L cylinder so it is at 5 MPa, 25°C. Find the mass of nitrogen dioxide using the Lee–Kesler pseudocritical constants and the acentric factor correction as shown in Appendix C.

11.76 A mixture of 1 kmol of ethanol and 1 kmol of methanol is at 20°C, 1 MPa. It is heated up in a SSSF process at constant pressure to a temperature of 400 K. Find the heat transfer using the Lee–Kesler pseudocritical constants and the acentric factor correction as shown in Appendix C.

ENGLISH UNIT PROBLEMS

11.77E Carbon dioxide gas at 580 R is mixed with nitrogen at 500 R in a SSSF insulated mixing chamber. Both flows are at 14.7 lbf/in.^{2} and the mole ratio of carbon dioxide to nitrogen is 2:1. Find the exit temperature and the total entropy generation per mole of the exit mixture.

11.78E A mixture of 60% helium and 40% nitrogen by volume enters a turbine at 150 lbf/in.^{2}, 1500 R at a rate of 4 lbm/s. The adiabatic turbine has an exit pressure of 15 lbf/in.^{2} and an isentropic efficiency of 85%. Find the turbine work.

11.79E A mixture of 50% carbon dioxide and 50% water by mass is brought from 2800 R, 150 lbf/in.^{2} to 900 R, 30 lbf/in.^{2} in a polytropic process through a SSSF device. Find the necessary heat transfer and work involved.

11.80E Two insulated tanks A and B are connected by a valve. Tank A has a volume of 30 ft^3 and initially contains argon at 50 lbf/in.^{2}, 50 F. Tank B has a volume of 60 ft^3 and initially contains ethane at 30 lbf/in.^{2}, 120 F. The valve is opened and remains open until the resulting gas mixture comes to a uniform state. Determine

a. The final pressure and temperature.

b. The entropy change for the process.

11.81E A spherical balloon has an initial diameter of 3 ft and contains argon gas at 30 lbf/in.^{2}, 100 F. The balloon is connected by a valve to a 20-ft^3 rigid tank containing carbon dioxide at 15 lbf/in.^{2}, 200 F. The valve is opened, and eventually the balloon and tank reach a uniform state in which the pressure is 25 lbf/in.^{2}. The balloon pressure is directly proportional to its diameter. Take the balloon and tank as a control volume, calculate the final temperature and the heat transfer for the process.

11.82E A mixture of 4 lbm oxygen and 4 lbm of argon is in an insulated piston cylinder arrangement at 14.7 lbf/in.^{2}, 540 R. The piston now compresses the mixture to half its initial volume. Find the final pressure, temperature and the piston work.

11.83E A 4-ft^3 insulated tank contains nitrogen gas at 30 lbf/in.^{2} and ambient temperature 77 F. The tank is connected by a valve to a supply line flowing carbon dioxide at 180 lbf/in.^{2}, 190 F. A mixture of 50 mole percent nitrogen and 50 mole percent carbon dioxide is to be obtained by opening the valve and allowing flow into the tank until an appropriate pressure is reached and the valve is closed. What is the pressure? The tank eventually cools to ambient temperature. Calculate the net entropy change for the overall process.

11.84E A large SSSF air separation plant takes in ambient air (79% N$_2$, 21% O$_2$ by volume) at 14.7 lbf/in.^{2}, 70 F, at a rate of 2 lb mol/s. It discharges a stream of pure O$_2$ gas at 30 lbf/in.^{2}, 200 F, and a stream of pure N$_2$ gas at 14.7 lbf/in.^{2}, 70 F. The plant operates on an electrical power input of 2000 kW. Calculate the net rate of entropy change for the process.

11.85E Consider a room of dimensions 12 × 18 × 8 ft that contains an air-water vapor mixture at 14.7 lbf/in.^{2}, 60 F, and 20% relative humidity. How much water would have to be added to this space for the room to be at 14.7 lbf/in.^{2}, 75 F, and have a relative humidity of 50%?

11.86E A tank has two sides initially separated by a diaphragm. Side A contains 2 lbm of water and side B contains 2.4 lbm of air, both at 68 F, 14.7 lbf/in.^{2}. The diaphragm is now broken and the whole tank is heated to 1100 F by a 1300 F reservoir. Find the final total pressure, heat transfer, and total entropy generation.

11.87E Consider a 10-ft^3 rigid tank containing an air-water vapor mixture at 14.7 lbf/in.^{2}, 90 F, with a 70% relative humidity. The system is cooled until the water just begins to condense. Determine the final temperature in the tank and the heat transfer for the process.

11.88E Atmospheric air at 95 F, relative humidity of 10%, is too warm and also too dry. An air conditioner should deliver air at 70 F and 50% relative humidity in the amount of 3600 ft^3 per hour. Sketch a setup to accomplish this, find any amount of liquid (at 68 F) that is needed or discarded and any heat transfer.

11.89E Outdoor ambient moist air enters a steady-flow heater-humidifier unit at 32 F, 14.7 lbf/in.^{2}, 50% relative humidity, at the volumetric flow rate of 5 ft^3/s. Liquid water at 50 F is sprayed into the mixture, and enough heat is transferred that the moist air leaves the unit at 85 F, 14.7 lbf/in.^{2}, 35% relative humidity. Determine

a. The mass flow rate of liquid water sprayed into the mixture.

b. The rate of heat transfer to the unit.

11.90E Moist air at 75 F, 14.7 lbf/in.^{2} and relative humidity of 70%, flows such that a fraction $^1/_3$ of it goes through a cooler. There water condenses out as it cools to 43 F after which the flow is mixed with the remaining fraction that bypassed the cooler in an adiabatic process, shown in Fig. P11.30. Find the liquid flow out of the cooler and the heat transfer, both per poundmass of dry air. Find the relative and specific humidity at the cooler exit and after the final mixing process.

11.91E Air in a piston cylinder is at 95 F, 15 lbf/in.^{2} and a relative humidity of 80%. It is now compressed to a pressure of 75 lbf/in.^{2} in a constant temperature process. Find the final relative and specific humidity and the volume ratio V_2/V_1.

11.92E A combination air cooler and dehumidification unit receives outside ambient air at 100 F, 14.7 lbf/in.^{2}, 90% relative humidity. The moist air is first cooled to a low temperature to condense the proper amount of water. The moist air is then heated and leaves the unit at 70 F, 14.7 lbf/in.^{2}, 30% relative humidity. The volume flow rate of the moist air at the exit is 0.2 ft^3/s.

a. Find the temperature to which the mixture must be cooled and the mass of water condensed out per pound mass of dry air. Show the process for the water in a T–s diagram.

b. Assuming all the liquid condensed out leaves at the minimum temperature, calculate the overall heat transfer rate.

11.93E A 10-ft^3 rigid vessel initially contains moist air at 20 lbf/in.^{2}, 100 F, with a relative humidity of 10%. A supply line connected to this vessel by a valve carries steam at 100 lbf/in.^{2}, 400 F. The valve is opened, and steam flows into the vessel until the relative humidity of the resultant moist air mixture is 90%. Then the valve is closed. Sufficient heat is transferred from the vessel so the temperature remains at 100 F during the process. Determine the heat transfer for the process, the mass of steam entering the vessel, and the final pressure inside the vessel.

11.94E A water-filled reactor of 50 ft^3 is at 2000 lbf/in.^{2}, 550 F and located inside an insulated containment room of 5000 ft^3 that has air at 1 atm. and 77 F. Due to a failure the reactor ruptures and the water fills the containment room. Find the final pressure.

11.95E A vertical cylinder is fitted with a piston held in place by a pin, as shown in Fig. P11.42. The initial volume is 4 ft^3 and the cylinder contains moist air at 14.7 lbf/in.^{2}, 75 F, with wet-bulb temperature of 60 F. The pin is removed, and at the same time a valve on the bottom of the cylinder is opened, allowing the mixture to flow out. A cylinder pressure of 20 lbf/in.^{2} is required to balance the piston. The valve is closed when the cylinder volume reaches 2 ft^3, at which point the temperature is 60 F.

a. Is there any liquid water in the cylinder at the final state?

b. Calculate the heat transfer to the cylinder during the process.

c. Take a control volume around the cylinder, and calculate the entropy change of the control volume and that of the surroundings.

11.96E An indoor pool evaporates 3 lbm/h of water, which is removed by a dehumidifier to maintain 70 F, $\Phi = 70\%$ in the room. The dehumidifier is a refrigeration cycle in which air flowing over the evaporator cools such that liquid water drops out, and the air continues flowing over the condenser, as shown in Fig. P11.49. For an air flow rate of 0.2 lbm/s the unit requires 1.2 Btu/s input to a motor driving a fan and the compressor and it has a coefficient of performance, $\beta = Q_L/W_c = 2.0$. Find the state of the air after the evaporator, T_2, ω_2, Φ_2 and the heat rejected. Find the state of the air as it returns to the room and the compressor work input.

11.97E Two moist air streams with 85% relative humidity, both flowing at a rate of 0.2 lbm/s of dry air are mixed in a SSSF setup. One inlet flowstream is at 90 F and the other at 61 F. Find the exit relative humidity.

11.98E Ambient air is at a condition of 14.7 lbf/in.^{2}, 95 F, 50% relative humidity. A steady stream of air at 14.7 lbf/in.^{2}, 73 F, 70% relative humidity, is to be produced by first cooling one stream to an appropriate temperature to condense out the proper amount of water and then mix this stream adiabatically with the second one at ambient conditions. What is the ratio of the two flow rates? To what temperature must the first stream be cooled?

11.99E To refresh air in a room, a counterflow heat exchanger is mounted in the wall, as shown in Fig. P11.52. It draws in outside air at 33 F, 80% relative humidity and draws room air, 104 F, 50% relative humidity, out. Assume an exchange of 6 lbm/min dry air in a SSSF device, and also that the room air exits the heat exchanger to the atmosphere at 72 F. Find the net amount of water removed from room, any liquid flow in the heat exchanger and (T, Φ) for the fresh air entering the room.

11.100E A liquid mixture of 75% octane and 25% methanol by mass is at 560 R, 75 lbf/in.^{2}. The mixture is heated in a constant pressure SSSF process to 1100 R. Determine the heat transfer using Kay's rule and the simple fluid generalized chart corrections.

11.101E One pound mole per second of saturated liquid methane, CH_4, at 150 lbf/in.^{2} and 2 lb mol/s of ethane, C_2H_6, at 480 F, 150 lbf/in.^{2} are fed to a mixing chamber with the resultant mixture exiting at 120 F, 150 lbf/in.^{2}. Assume that Kay's rule applies to the mixture and determine the heat transfer in the process.

11.102E An insulated rigid 5-ft³ tank contains a gas mixture, 85% methane and 15% ethane on a mole basis, at 500 lbf/in.², 100 F. A valve on the tank is now opened accidentally, and the pressure inside quickly drops to 300 lbf/in.² before the valve is closed. Assume that the gas mixture is homogeneous and that its behavior can be represented by Kay's rule and the generalized charts.

    a. Calculate the mass that has escaped from the tank.

    b. Eventually, the temperature inside the tank returns to that of the surroundings, 100 F. What is the pressure at that time?

11.103E Repeat the previous problem using the Redlich–Kwong equation of state to represent the real gas mixture behavior, instead of Kay's rule and the generalized charts.

11.104E (Adv.) Two heavily insulated, rigid tanks, A and B, each of 1-ft³ volume are connected by a valve. Tank A initially contains methane at 80 F, 3500 lbf/in.², and tank B initially contains carbon dioxide at 80 F, 200 lbf/in.². The valve is now opened and methane flows quickly from A to B until the pressure inside A drops to 2500 lbf/in.², at which time the valve is closed. Assuming that tank B now contains a homogeneous mixture and that tank A contains only methane, determine the final temperature in A, and the final temperature, pressure and composition in B. Specify any assumptions and thermodynamic models used in the analysis and solution.

11.105E (Adv.) Boil-off vapor from a liquid propane storage tank at 32 F is throttled into an adiabatic mixing chamber and mixed with nitrogen in the ratio 4 lb mol C_3H_8/1 lb mol N_2, as shown in Fig. P11.72 with the properties as shown in the following table. The resulting mixture then enters a compressor.

State	1	2	3	4	5
T [F]	32	–	32	–	140
P [lbf / in.²]	–	50	50	50	600

It is claimed that the compression process is adiabatic. Using the second law, show whether or not this is possible.

11.106E A liquid mixture of 75% octane and 25% methanol by mass is at 540 R, 75 lbf/in.². The mixture is heated in a constant pressure SSSF process to 1100 R. This problem is identical to Problem 11.100, where it was solved using Kay's rule. Determine the heat transfer using Lee–Kesler pseudocritical constants for the mixture and the acentric factor correction as shown in Appendix C.

11.107E A mixture of 40% nitrogen dioxide, NO_2, and 60% nitrogen by mass is charged into a 4-ft³ cylinder so it is at 750 lbf/in.², 77 F. Find the mass of nitrogen dioxide using the Lee–Kesler pseudocritical constants and the acentric factor correction as shown in Appendix C.

11.108E A mixture of 1 lb mol of ethanol and 1 lb mol of methanol is at 68 F, 150 lbf/in.². It is heated up in a SSSF process at constant pressure to a temperature of 720 R. Find the heat transfer using the Lee–Kesler pseudocritical constants and the acentric factor correction as shown in Appendix C.

COMPUTER,
DESIGN, AND
OPEN-ENDED
PROBLEMS

11.109 Write a program to solve the general case of Problem 11.8 in which the two volumes and the initial state properties of the argon and the ethane are input variables.

11.110 Write a program to solve the general case of Problem 11.14, in which the nitrogen initial state, the carbon dioxide supply line state, and the desired mixture percentages are input variables.

11.111 Write a program to solve the general case of Problem 11.20 (only for ambient temperatures above 0°C), in which the outdoor ambient state and the desired indoor state are program input variables. For this and later program convenience and independence, curve fit an equation to represent the water liquid–vapor saturation pressure as a function of temperature. Use steam table data for the range 0.01°C to 40°C. The form of the equation should be log P_g as a polynomial in T, powers −1 to +4.

11.112 Extend the solution range of Problem 11.111 to allow for outdoor ambient temperatures as low as −30°C.

11.113 Use the equation developed in Problem 11.111 as the starting point in writing a program that will perform all the functions of the psychrometric chart, Fig. A.6. For program convenience, curve-fit equations for the water saturated-liquid enthalpy and saturated-vapor enthalpy as polynomial functions of temperature (powers 0 to +3), as was done earlier for saturation pressure. The final "psychrometric chart" program input parameters are to be pressure, temperature, and one of the following: wet-bulb temperature, relative humidity, or specific humidity.

11.114 Write a program to solve the general case of Problem 11.29, using the program developed in the previous problem. The outdoor ambient state and the desired indoor state are to be input variables.

11.115 Extend the previous problem to include a second law analysis. Curve fit water saturated-liquid entropy and saturated-vapor entropy as polynomial functions of temperature, as was done for the enthalpies.

11.116 (Adv.) A heat pump clothes dryer, the principle of which is shown in Fig. P11.116, is to be designed. Warm (and increasing with time) saturated air comes from the clothes dryer and is cooled by flowing over the evaporator coils, thereby condensing out water. The dry air is then heated by passing it over the condenser coils. The hot, dry air returns to the clothes dryer. An electrically controlled bypass loop allows control of the air temperature to a maximum of 70°C, and it is also possible to flow air over the evaporator and condenser coils at different rates. Consider the heat pump cycle to be an ideal cycle using ammonia as the working fluid. Evaluate the performance of this clothes dryer as a function of time for a variety of operating conditions.

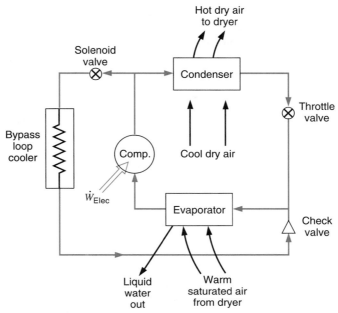

FIGURE P11.116

11.117 Write a program that will generate a set of Lee–Kesler pseudocritical constants for a gas mixture, given the critical constants of the (variable number of) mixture components and the mixture composition.

11.118 (Adv.) Extend the programs developed in Problems 10.107 and 10.108 to include binary gas mixture calculations. Input parameters should include the Lennard–Jones force constants of the components and the composition, in addition to the thermodynamic state variables.

11.119 (Adv.) Extend the program developed in Problem 10.113 to include binary gas mixture calculations using the Redlich–Kwong equation of state. Input parameters should include the critical pressure and temperature of the components and the composition, in addition to the thermodynamic state variables.

11.120 The mixing of carbon dioxide and nitrogen in a SSSF setup was given in Problem 11.4. If the temperatures were very different an assumption about constant specific heat is inappropriate. Study the problem assuming the carbon dioxide enters at 300 K, 100 kPa as a function of the nitrogen inlet temperature using the ideal gas tables in Table A.13. Give the nitrogen inlet temperature for which the constant specific heat assumption starts to be more than 1%, 5%, and 10% wrong for the exit mixture temperature.

11.121 The temperature in the standard atmosphere drops by 6.5°C/1000 m altitude from sea level. When warm moist air near the ground is forced up it cools down so clouds form due to condensation at some altitude. For a range of atmospheric ground conditions predict the cloud cover altitude. Should you also include the variation in the total pressure? See Problem 3.80 for this.

11.122 Reconsider the car air conditioner in Problem 11.38. Assume the cooling is done by a refrigeration cycle using R-134a as the working substance. Select a reasonable cycle and investigate the coefficient of performance for the cycle. Study the system under worst case conditions and estimate total power required by the compressor.

11.123 The setup in Problem 11.44 is similar to a process which can be used to produce dry powder from a slurry of water and dry material as coffee or milk. The water flow at state 3 is a mixture of 80% liquid water and 20% dry material on a mass basis with $C_{dry} = 0.4$ kJ/kg K. After the water is evaporated the dry material falls to the bottom and is removed in an additional line, m_{dry} exit at state 4. Assume a reasonable T_4 and that state 1 is heated atmospheric air. Investigate the inlet flow temperature as a function of the humidity ratio of the atmosphere.

11.124 A dehumidifier for household applications is similar to the system shown in Fig. P11.49. Study the requirements to the refrigeration cycle as a function of the atmospheric conditions and include a worst case estimation.

11.125 Consider the process described in Problem 11.51 where the inlet flow is split into two flow streams and then mixed after the cooling and water removal process. Is it more energy efficient (power input to a refrigeration cycle) to do this or should the whole flow be cooled down? Examine the problem when the coefficient of performance of the refrigeration cycle is included. This cycle provides the cooling and has heating available from the condenser that may be partly bypassed.

11.126 A clothes dryer heats air up and tumbles the wet clothes in this hot relatively dry air and the air then leaves with the water as vapor. Weigh a typical load of wet laundry and then weigh it again when dry to determine the amount of water removed. Assume the parameters in the problem as mass flow rate of air, an inlet and exit temperatures and humidities. Study the time required to dry the clothes and the power needed for a range of these parameters.

11.127 A clothes dryer has a 60°C, $\Phi = 90\%$ air flow out at a rate of 3 kg/min. The atmospheric conditions are 20°C, relative humidity of 50%. How much water is carried away and how much power is needed? To increase the efficiency a counterflow heat exchanger is installed to preheat the incoming atmospheric air up with the hot exit flow. Estimate suitable exit temperatures from the heat exchanger and investigate the design changes to the clothes dryer (what happens to the condensed water?). How much energy can be saved this way?

11.128 Addition of steam to combustors in gas turbines and to internal combustion engines reduces the peak temperatures and lowers emission of NO_x. Consider a modification to a gas turbine as shown in Fig. P11.128, where the modified cycle is called the Cheng cycle. In this example it is used for a cogenerating power plant. Assume 12 kg/s air with state 2 at 1.25 Mpa, unknown temperature, is mixed with 2.5 kg/s water at 450°C at constant pressure before the inlet to the turbine. The turbine exit temperature is $T_4 = 500°C$ and the pressure is 125 kPa. For a reasonable turbine efficiency estimate the required air temperature at state 2. Compare the result to the case where no steam is added to the mixing chamber and only air runs through the turbine.

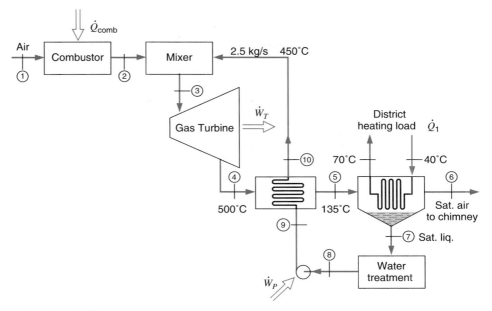

FIGURE P11.128

11.129 Consider the district water heater acting as the condenser for part of the water between states 5 and 6 in Fig. P11.128. If the temperature of the mixture (12 kg/s air, 2.5 kg/s steam) at state 5 is 135°C make a study of the district heating load, $\dot{Q}_1$, as a function of the exit temperature T_6. Study also the sensitivity of the results with respect to the assumption that state 6 is saturated moist air.

11.130 The cogeneration gas turbine cycle can be augmented with a heat pump to extract more energy from the turbine exhaust gas as shown in Fig. P11.130. The heat pump upgrades the energy to be delivered at the 70°C line for district heating. In the modified application the first heat exchanger has exit temperature, $T_{6a} = T_{7a} = 45°C$ and the second one has, $T_{6b} = T_{7b} = 36°C$. Assume the district heating line has the same exit temperature as before so this arrangement allows for a higher flow rate. Estimate the increase in the district heating load that can be obtained and the necessary work input to the heat pump.

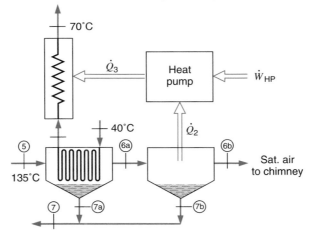

FIGURE P11.130

11.131 (Adv.) Several applications of dehumidification do not rely on water condensation by cooling. A desiccant has a greater affinity to water and can absorb water directly from the air accompanied by a heat release. The desiccant is then regenerated by heating driving the water out. Make a list of several such materials as liquids, gels, and solids and show examples where they are used.

CHEMICAL REACTIONS 12

Many thermodynamic problems involve chemical reactions. Among the most familiar of these is the combustion of hydrocarbon fuels, for this process is utilized in most of our power-generating devices. However, we can all think of a host of other processes involving chemical reactions, including those that occur in the human body.

It is our purpose in this chapter to consider a first and second law analysis of systems undergoing a chemical reaction. In many respects, this chapter is simply an extension of our previous consideration of the first and second laws. However, a number of new terms are introduced, and it will also be necessary to introduce the third law of thermodynamics.

In this chapter the combustion process is considered in detail. There are two reasons for this emphasis. The first reason is that the combustion process is of great significance in many problems and devices with which the engineer is concerned. The second reason is that the combustion process provides an excellent vehicle for teaching the basic principles of the thermodynamics of chemical reactions. The student should keep both of these objectives in mind as the study of this chapter progresses.

Chemical equilibrium will be considered in Chapter 13 and, therefore, the matter of dissociation will be deferred until then.

12.1 FUELS

A thermodynamics textbook is not the place for a detailed treatment of fuels. However, some knowledge of them is a prerequisite to a consideration of combustion, and this section is therefore devoted to a brief discussion of some of the hydrocarbon fuels. Most fuels fall into one of three categories—coal, liquid hydrocarbons, or gaseous hydrocarbons.

Coal consists of the remains of vegetation deposits of past geologic ages, after subjection of biochemical actions, high pressure, temperature, and submersion. The characteristics of coal vary considerably with location, and even within a given mine there is some variation in composition.

The analysis of a sample of coal is given on one of two bases: the proximate analysis specifies, on a mass basis, the relative amounts of moisture, volatile matter, fixed carbon, and ash; the ultimate analysis specifies, on a mass basis, the relative amounts of carbon, sulfur, hydrogen, nitrogen, oxygen, and ash. The ultimate analysis may be given on an "as received" basis or on a dry basis. In the latter case the ultimate analysis does not include the moisture as determined by the proximate analysis.

There are also a number of other properties of coal that are important in evaluating a coal for a given use. Some of these are the fusibility of the ash, the grindability or ease of pulverization, the weathering characteristics, and size.

Table 12.1

Characteristics of Some of the Hydrocarbon Families

Family	Formula	Structure	Saturated
Paraffin	C_nH_{2n+2}	Chain	Yes
Olefin	C_nH_{2n}	Chain	No
Diolefin	C_nH_{2n-2}	Chain	No
Naphthene	C_nH_{2n}	Ring	Yes
Aromatic			
Benzene	C_nH_{2n-6}	Ring	No
Naphthene	C_nH_{2n-12}	Ring	No

Most liquid and gaseous hydrocarbon fuels are a mixture of many different hydrocarbons. For example, gasoline consists primarily of a mixture of about 40 hydrocarbons, with many others present in very small quantities. In discussing hydrocarbon fuels, therefore, brief consideration should be given to the most important families of hydrocarbons, which are summarized in Table 12.1.

Three concepts should be defined. The first pertains to the structure of the molecule. The important types are the ring and chain structures; the difference between the two is illustrated in Fig. 12.1. The same figure illustrates the definition of saturated and unsaturated hydrocarbons. An unsaturated hydrocarbon has two or more adjacent carbon atoms joined by a double or triple bond, whereas in a saturated hydrocarbon all the carbon atoms are joined by a single bond. The third term to be defined is an isomer. Two hydrocarbons with the same number of carbon and hydrogen atoms and different structures are called isomers. Thus there are several different octanes (C_8H_{18}), each having 8 carbon atoms and 18 hydrogen atoms, but each with a different structure.

The various hydrocarbon families are identified by a common suffix. The compounds comprising the paraffin family all end in "-ane" (as propane and octane). Similarly, the compounds comprising the olefin family end in "-ylene" or "-ene" (as propene and octene), and the diolefin family ends in "-diene" (as butadiene). The naphthene family has the same chemical formula as the olefin family, but has a ring rather than chain structure. The hydrocarbons in the naphthene family are named by adding the prefix "cyclo-" (as cyclopentane).

The aromatic family includes the benzene series (C_nH_{2n-6}) and the naphthalene series (C_nH_{2n-12}). The benzene series has a ring structure and is unsaturated.

Alcohols are sometimes used as a fuel in internal combustion engines. The characteristic feature of the alcohol family is that one of the hydrogen atoms is replaced by an OH radical. Thus methyl alcohol, also called methanol, is CH_3OH.

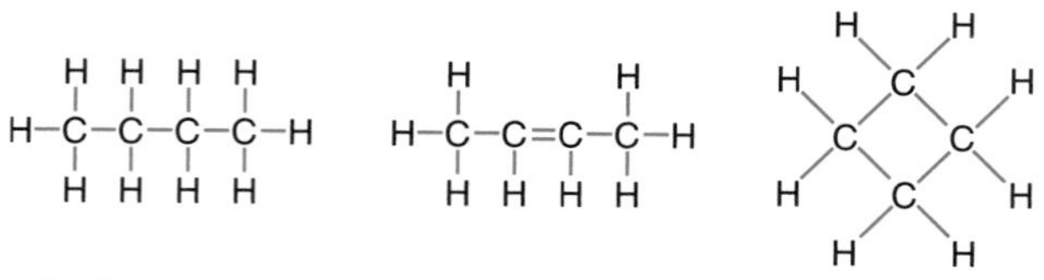

Chain structure saturated   Chain structure unsaturated   Ring structure saturated

FIGURE 12.1 Molecular structure of some hydrocarbon fuels.

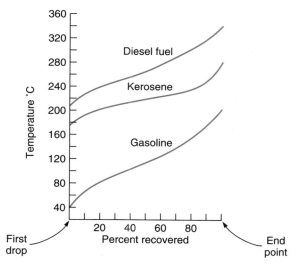

FIGURE 12.2 Typical distillation curves of some hydrocarbon fuels.

Most liquid hydrocarbon fuels are mixtures of hydrocarbons that are derived from crude oil through distillation and cracking processes. Thus, from a given crude oil, a variety of different fuels can be produced, some of the common ones being gasoline, kerosene, diesel fuel, and fuel oil. Within each of these classifications there is a wide variety of grades, and each is made up of a large number of different hydrocarbons. The important distinction between these fuels is the distillation curve, Fig. 12.2. The distillation curve is obtained by slowly heating a sample of fuel so that it vaporizes. The vapor is

TABLE 12.2
Volumetric Analyses of Some Typical Gaseous Fuels

| Constituent | Various Natural Gases | | | | Producer Gas from Bituminous Coal | Carbureted Water Gas | Coke-Oven Gas |
	A	B	C	D			
Methane	93.9	60.1	67.4	54.3	3.0	10.2	32.1
Ethane	3.6	14.8	16.8	16.3			
Propane	1.2	13.4	15.8	16.2			
Butanes plus[a]	1.3	4.2		7.4			
Ethene						6.1	3.5
Benzene						2.8	0.5
Hydrogen					14.0	40.5	46.5
Nitrogen		7.5		5.8	50.9	2.9	8.1
Oxygen					0.6	0.5	0.8
Carbon monoxide					27.0	34.0	6.3
Carbon dioxide					4.5	3.0	2.2

[a]This includes butane and all heavier hydrocarbons

then condensed and the amount measured. The more volatile hydrocarbons are vaporized first, and thus the temperature of the nonvaporized fraction increases during the process. The distillation curve, which is a plot of the temperature of the nonvaporized fraction versus the amount of vapor condensed, is an indication of the volatility of the fuel.

For the combustion of liquid fuels, it is convenient to express the composition in terms of a single hydrocarbon, even though it is a mixture of many hydrocarbons. Thus gasoline is usually considered to be octane, C_8H_{18}, and diesel fuel is considered to be dodecane, $C_{12}H_{26}$. The composition of a hydrocarbon fuel may also be given in terms of percentage of carbon and hydrogen.

The two primary sources of gaseous hydrocarbon fuels are natural gas wells and certain chemical manufacturing processes. Table 12.2 gives the composition of a number of gaseous fuels. The major constituent of natural gas is methane, which distinguishes it from manufactured gas.

At the present time, there is a considerable effort being devoted to develop more economical processes for producing gaseous and also liquid hydrocarbon fuels from coal, and also from oil shale and tar sands deposits. Several alternative techniques have been demonstrated to be feasible, and these resources promise to provide an increasing proportion of our fuel supply in future years.

12.2 THE COMBUSTION PROCESS

The combustion process consists of the oxidation of constituents in the fuel that are capable of being oxidized, and can therefore be represented by a chemical equation. During a combustion process the mass of each element remains the same. Thus, writing chemical equations and solving problems concerning quantities of the various constituents basically involves the conservation of mass of each element. A brief review of this subject, particularly as it applies to the combustion process, is presented in this chapter.

Consider first the reaction of carbon with oxygen.

$$\text{Reactants} \quad \text{Products}$$

$$C + O_2 \rightarrow CO_2$$

This equation states that 1 kmol of carbon reacts with 1 kmol of oxygen to form 1 kmol of carbon dioxide. This also means that 12 kg of carbon react with 32 kg of oxygen to form 44 kg of carbon dioxide. All the initial substances that undergo the combustion process are called the reactants, and the substances that result from the combustion process are called the products.

When a hydrocarbon fuel is burned, both the carbon and the hydrogen are oxidized. Consider the combustion of methane as an example.

$$CH_4 + 2O_2 \rightarrow CO_2 + 2H_2O \tag{12.1}$$

Here the products of combustion include both carbon dioxide and water. The water may be in the vapor, liquid, or solid phases, depending on the temperature and pressure of the products of combustion.

It should be pointed out that in the combustion process many intermediate products are formed during the chemical reaction. In this book we are concerned with the initial and final products and not with the intermediate products, but this aspect is very important in a detailed consideration of combustion.

In most combustion processes the oxygen is supplied as air rather than as pure oxygen. The composition of air on a molal basis is approximately 21% oxygen, 78% nitrogen, and 1% argon. We assume that the nitrogen and the argon do not undergo chemical reaction (except for dissociation, which will be considered in Chapter 13). They do leave at the same temperature as the other products, however, and therefore undergo a change of state if the products are at a temperature other than the original air temperature. It should be pointed out that at the high temperatures achieved in internal-combustion engines, there is actually some reaction between the nitrogen and oxygen, and this gives rise to the air pollution problem associated with the oxides of nitrogen in the engine exhaust.

In combustion calculations concerning air, the argon is usually neglected, and the air is considered to be composed of 21% oxygen and 79% nitrogen by volume. When this assumption is made, the nitrogen is sometimes referred to as atmospheric nitrogen. Atmospheric nitrogen has a molecular weight of 28.16 (which takes the argon into account) as compared to 28.013 for pure nitrogen. This distinction will not be made in this text, and we will consider the 79% nitrogen to be pure nitrogen.

The assumption that air is 21.0% oxygen and 79.0% nitrogen by volume leads to the conclusion that for each mole of oxygen, $79.0/21.0 = 3.76$ moles of nitrogen are involved. Therefore, when the oxygen for the combustion of methane is supplied as air, the reaction can be written

$$CH_4 + 2O_2 + 2(3.76)N_2 \rightarrow CO_2 + 2H_2O + 7.52N_2 \tag{12.2}$$

The minimum amount of air that supplies sufficient oxygen for the complete combustion of all the carbon, hydrogen, and any other elements in the fuel that may oxidize is called the theoretical air. When complete combustion is achieved with theoretical air, the products contain no oxygen. A general combustion reaction with a hydrocarbon fuel and air is thus written

$$C_xH_y + v_{O_2}(O_2 + 3.76N_2) \rightarrow v_{CO_2}CO_2 + v_{H_2O}H_2O + v_{N_2}N_2 \tag{12.3}$$

with the coefficients to the substances called stoichiometric coefficients. The balance of atoms yields the theoretical amount of air as

$$\text{C:} \quad v_{CO_2} = x$$
$$\text{H:} \quad 2v_{H_2O} = y$$
$$\text{N}_2\text{:} \quad v_{N_2} = 3.76 \times v_{O_2}$$
$$\text{O}_2\text{:} \quad v_{O_2} = v_{CO_2} + v_{H_2O}/2 = x + y/4$$

and the total number of moles of air for 1 mole of fuel becomes

$$n_{air} = v_{O_2} \times 4.76 = 4.76(x + y/4)$$

This amount of air is equal to 100% theoretical air. In practice, complete combustion is not likely to be achieved unless the amount of air supplied is somewhat greater than the theoretical amount. Two important parameters often used to express the ratio of fuel and air are the air–fuel ratio (designated AF) and its reciprocal, the fuel–air ratio (designated FA). These ratios are usually expressed on a mass basis, but a mole basis is used at times.

$$AF_{mass} = \frac{m_{air}}{m_{fuel}} \tag{12.4}$$

$$AF_{mole} = \frac{n_{air}}{n_{fuel}} \tag{12.5}$$

They are related through the molecular weights as

$$AF_{mass} = \frac{m_{air}}{m_{fuel}} = \frac{n_{air}M_{air}}{n_{fuel}M_{fuel}} = AF_{mole}\frac{M_{air}}{M_{fuel}}$$

and a subscript s is used to indicate the ratio for 100% theoretical air, also called a stoichiometric mixture. In an actual combustion process, an amount of air is expressed as a fraction of the theoretical amount, called percent theoretical air. A similar ratio named the equivalence ratio equals the actual fuel–air ratio divided by the theoretical fuel–air ratio as

$$\Phi = FA / FA_s = AF_s / AF \tag{12.6}$$

the reciprocal of percent theoretical air. Since the percent theoretical air and the equivalence ratio both are ratios of the stoichiometric air–fuel ratio and the actual air–fuel ratio the molecular weights cancel out and they are the same whether a mass basis or a mole basis is used.

Thus, 150% theoretical air means that the air actually supplied is 1.5 times the theoretical air and the equivalence ratio is $2/3$. The complete combustion of methane with 150% theoretical air is written

$$CH_4 + 1.5 \times 2(O_2 + 3.76N_2) \rightarrow CO_2 + 2H_2O + O_2 + 11.28N_2 \tag{12.7}$$

having balanced all the stoichiometric coefficients from conservation of all the atoms.

The amount of air actually supplied may also be expressed in terms of percent excess air. The excess air is the amount of air supplied over and above the theoretical air. Thus, 150% theoretical air is equivalent to 50% excess air. The terms "theoretical air," "excess air" and "equivalence ratio" are all in current usage and give an equivalent information about the reactant mixture of fuel and air.

When the amount of air supplied is less than the theoretical air required, the combustion is incomplete. If there is only a slight deficiency of air, the usual result is that some of the carbon unites with the oxygen to form carbon monoxide (CO) instead of carbon dioxide (CO_2). If the air supplied is considerably less than the theoretical air, there may also be some hydrocarbons in the products of combustion.

Even when some excess air is supplied there may be small amounts of carbon monoxide present, the exact amount depending on a number of factors including the mixing and turbulence during combustion. Thus the combustion of methane with 110% theoretical air might be as follows:

$$CH_4 + 2(1.1)O_2 + 2(1.1)3.76N_2 \rightarrow$$
$$+ 0.95CO_2 + 0.05CO + 2H_2O + 0.225O_2 + 8.27N_2 \tag{12.8}$$

The material covered so far in this section is illustrated by the following examples.

EXAMPLE 12.1   Calculate the theoretical air–fuel ratio for the combustion of octane, C_8H_{18}.

Solution

The combustion equation is

$$C_8H_{18} + 12.5O_2 + 12.5(3.76)N_2 \rightarrow 8CO_2 + 9H_2O + 47.0N_2$$

The air–fuel ratio on a mole basis is

$$AF = \frac{12.5 + 47.0}{1} = 59.5 \text{ kmol air / kmol fuel}$$

The theoretical air–fuel ratio on a mass basis is found by introducing the molecular weight of the air and fuel.

$$AF = \frac{59.5(28.97)}{114.2} = 15.0 \text{ kg air / kg fuel}$$

EXAMPLE 12.2 Determine the molal analysis of the products of combustion when octane, C_8H_{18}, is burned with 200% theoretical air, and determine the dew point of the products if the pressure is 0.1 MPa.

Solution

The equation for the combustion of octane with 200% theoretical air is

$$C_8H_{18} + 12.5(2)O_2 + 12.5(2)(3.76)N_2 \rightarrow 8CO_2 + 9H_2O + 12.5O_2 + 94.0N_2$$

Total kmols of product = 8 + 9 + 12.5 + 94.0 = 123.5
Molal analysis of products:

$$
\begin{aligned}
CO_2 &= 8/123.5 &=& \quad 6.47\% \\
H_2O &= 9/123.5 &=& \quad 7.29 \\
O_2 &= 12.5/123.5 &=& \quad 10.12 \\
N_2 &= 94/123.5 &=& \quad 76.12 \\
&& & \overline{\quad 100.00\%}
\end{aligned}
$$

The partial pressure of the water is 100(0.0729) = 7.29 kPa.

The saturation temperature corresponding to this pressure is 39.7°C, which is also the dew-point temperature.

The water condensed from the products of combustion usually contains some dissolved gases and therefore may be quite corrosive. For this reason the products of combustion are often kept above the dew point until discharged to the atmosphere.

EXAMPLE 12.2E Determine the molal analysis of the products of combustion when octane, C_8H_{18}, is burned with 200% theoretical air, and determine the dew point of the products if the pressure is 14.7 lbf/in.^{2}.

Solution

The equation for the combustion of octane with 200% theoretical air is

$$C_8H_{18} + 12.5(2)O_2 + 12.5(2)(3.76)N_2 \rightarrow 8CO_2 + 9H_2O + 12.5O_2 + 94.0N_2$$

Total moles of product = 8 + 9 + 12.5 + 94.0 = 123.5

Molal analysis of products:

$$CO_2 = 8/123.5 \quad = \quad 6.47\%$$
$$H_2O = 9/123.5 \quad = \quad 7.29$$
$$O_2 = 12.5/123.5 = \quad 10.12$$
$$N_2 = 94/123.5 \quad = \quad 76.12$$
$$\overline{100.00\%}$$

The partial pressure of the H_2O is $14.7(0.0729) = 1.072$ lbf/in.^{2}

The saturation temperature corresponding to this pressure is 104 F, which is also the dew-point temperature.

The water condensed from the products of combustion usually contains some dissolved gases and therefore may be quite corrosive. For this reason the products of combustion are often kept above the dew point until discharged to the atmosphere.

EXAMPLE 12.3   Producer gas from bituminous coal (see Table 12.2) is burned with 20% excess air. Calculate the air–fuel ratio on a volumetric basis and on a mass basis.

Solution

To calculate the theoretical air requirement, let us write the combustion equation for the combustible substances in 1 kmol of fuel.

$$0.14H_2 + 0.070O_2 \rightarrow 0.14H_2O$$
$$0.27CO + 0.135O_2 \rightarrow 0.27CO_2$$
$$0.03CH_4 + 0.06O_2 \rightarrow 0.03CO_2 + 0.06H_2O$$
$$\overline{0.265} = \text{kmol oxygen required / kmol fuel}$$
$$-0.006 = \text{oxygen in fuel / kmol fuel}$$
$$\overline{0.259} = \text{kmol oxygen required from air / kmol fuel}$$

Therefore, the complete combustion equation for 1 kmol of fuel is

$$\overbrace{0.14H_2 + 0.27CO + 0.03CH_4 + 0.006O_2 + 0.509N_2 + 0.045CO_2}^{\text{fuel}}$$

$$\overbrace{+0.259O_2 + 0.259(3.76)N_2}^{\text{air}} \rightarrow 0.20H_2O + 0.345CO_2 + 1.482N_2$$

$$\left(\frac{\text{kmol air}}{\text{kmol fuel}}\right)_{\text{theo}} = 0.259 \times \frac{1}{0.21} = 1.233$$

If the air and fuel are at the same pressure and temperature, this also represents the ratio of the volume of air to the volume of fuel.

$$\text{For 20\% excess air,} \quad \frac{\text{kmol air}}{\text{kmol fuel}} = 1.233 \times 1.200 = 1.48$$

The air–fuel ratio on a mass basis is

$$AF = \frac{1.48(28.97)}{0.14(2)+0.27(28)+0.03(16)+0.006(32)+0.509(28)+0.045(44)}$$
$$= \frac{1.48(28.97)}{24.74} = 1.73 \text{ kg air / kg fuel}$$

An analysis of the products of combustion affords a very simple method for calculating the actual amount of air supplied in a combustion process. There are various experimental methods by which such an analysis can be made. Some yield results on a "dry" basis; that is, the fractional analysis of all the components, except for water vapor. Other experimental procedures give results that include the water vapor. In this presentation we are not concerned with the experimental devices and procedures, but rather with the use of such information in a thermodynamic analysis of the chemical reaction. The following examples illustrate how an analysis of the products can be used to determine the chemical reaction and the composition of the fuel.

The basic principle in using the analysis of the products of combustion to obtain the actual fuel–air ratio is conservation of the mass of each of the elements. Thus, in changing from reactants to products, we can make a carbon balance, hydrogen balance, oxygen balance, and nitrogen balance (plus any other elements that may be involved). Furthermore, we recognize that there is a definite ratio between the amounts of some of these elements. Thus, the ratio between the nitrogen and oxygen supplied in the air is fixed, as well as the ratio between carbon and hydrogen if the composition of a hydrocarbon fuel is known.

EXAMPLE 12.4   Methane (CH_4) is burned with atmospheric air. The analysis of the products on a dry basis is as follows:

CO_2	10.00%
O_2	2.37
CO	0.53
N_2	87.10
	100.00%

Calculate the air–fuel ratio and the percent theoretical air, and determine the combustion equation.

Solution

The solution consists of writing the combustion equation for 100 kmol of dry products, introducing letter coefficients for the unknown quantities, and then solving for them.

From the analysis of the products, the following equation can be written, keeping in mind that this analysis is on a dry basis.

$$a\text{CH}_4 + b\text{O}_2 + c\text{N}_2 \rightarrow 10.0\text{CO}_2 + 0.53\text{CO} + 2.37\text{O}_2 + d\text{H}_2\text{O} + 87.1\text{N}_2$$

A balance for each of the elements will enable us to solve for all the unknown coefficients:

$$\text{Nitrogen balance: } c = 87.1$$

Since all the nitrogen comes from the air,

$$\frac{c}{b} = 3.76 \qquad b = \frac{87.1}{3.76} = 23.16$$

Carbon balance: $a = 10.00 + 0.53 = 10.53$
Hydrogen balance: $d = 2a = 21.06$
Oxygen balance: All the unknown coefficients have been solved for, and therefore the oxygen balance provides a check on the accuracy. Thus, b can also be determined by an oxygen balance

$$b = 10.00 + \frac{0.53}{2} + 2.37 + \frac{21.06}{2} = 23.16$$

Substituting these values for a, b, c, and d we have

$$10.53\text{CH}_4 + 23.16\text{O}_2 + 87.1\text{N}_2$$
$$\rightarrow 10.0\text{CO}_2 + 0.53\text{CO} + 2.37\text{O}_2 + 21.06\text{H}_2\text{O} + 87.1\text{N}_2$$

Dividing through by 10.53 yields the combustion equation per kmol of fuel.

$$\text{CH}_4 + 2.2\text{O}_2 + 8.27\text{N}_2 \rightarrow 0.95\text{CO}_2 + 0.05\text{CO} + 2\text{H}_2\text{O} + 0.225\text{O}_2 + 8.27\text{N}_2$$

The air–fuel ratio on a mole basis is

$$2.2 + 8.27 = 10.47 \text{ kmol air / kmol fuel}$$

The air–fuel ratio on a mass basis is found by introducing the molecular weights.

$$AF = \frac{10.47 \times 28.97}{16.0} = 18.97 \text{ kg air / kg fuel}$$

The theoretical air–fuel ratio is found by writing the combustion equation for theoretical air.

$$\text{CH}_4 + 2\text{O}_2 + 2(3.76)\text{N}_2 \rightarrow \text{CO}_2 + 2\text{H}_2\text{O} + 7.52\text{N}_2$$
$$AF_{\text{theo}} = \frac{(2 + 7.52)28.97}{16.0} = 17.23 \text{ kg air / kg fuel}$$

The percent theoretical air is $\dfrac{18.97}{17.23} = 110\%$

EXAMPLE 12.5 Coal from Jenkin, Kentucky, has the following ultimate analysis on a dry basis, percent by mass:

Component	Percent by Mass
Sulfur	0.6
Hydrogen	5.7
Carbon	79.2
Oxygen	10.0
Nitrogen	1.5
Ash	3.0

This coal is to be burned with 30% excess air. Calculate the air–fuel ratio on a mass basis.

Solution

One approach to this problem is to write the combustion equation for each of the combustible elements per 100 kg of fuel. The molal composition per 100 kg of fuel is found first.

$$\text{kmol S}/100 \text{ kg fuel} = \frac{0.6}{32} = 0.02$$

$$\text{kmol H}_2/100 \text{ kg fuel} = \frac{5.7}{2} = 2.85$$

$$\text{kmol C}/100 \text{ kg fuel} = \frac{79.2}{12} = 6.60$$

$$\text{kmol O}_2/100 \text{ kg fuel} = \frac{10}{32} = 0.31$$

$$\text{kmol N}_2/100 \text{ kg fuel} = \frac{1.5}{28} = 0.05$$

The combustion equations for the combustible elements are now written, which enables us to find the theoretical oxygen required.

$$0.02S + 0.02O_2 \rightarrow 0.02SO_2$$
$$2.85H_2 + 1.42O_2 \rightarrow 2.85H_2O$$
$$6.60C + 6.60O_2 \rightarrow 6.60CO_2$$

$$8.04 \text{ kmol O}_2 \text{ required}/100 \text{ kg fuel}$$
$$-0.31 \text{ kmol O}_2 \text{ in fuel}/100 \text{ kg fuel}$$
$$7.73 \text{ kmol O}_2 \text{ from air}/100 \text{ kg fuel}$$

$$AF_{\text{theo}} = \frac{[7.73 + 7.73(3.76)]28.97}{100} = 10.63 \text{ kg air}/\text{kg fuel}$$

For 30% excess air the air–fuel ratio is

$$AF = 1.3 \times 10.63 = 13.82 \text{ kg air / kg fuel}$$

12.3 ENTHALPY OF FORMATION

In the first eleven chapters of this book the problems always concerned a fixed chemical composition and never a change of composition through a chemical reaction. Therefore, in dealing with a thermodynamic property, we used tables of thermodynamic properties for the given substance, and in each of these tables the thermodynamic properties were given relative to some arbitrary base. In the steam tables, for example, the internal energy of saturated liquid at 0.01°C is assumed to be zero. This procedure is quite adequate when there is no change in composition, because we are concerned with the changes in the properties of a given substance. The properties at the condition of the reference state cancel out in the calculation. When dealing with the matter of reference states in Section 11.8, we noted that for a given substance (perhaps a component of a mixture), we are free to choose a reference state condition, for example, a hypothetical ideal gas, as long as we then carry out a consistent calculation from that state and condition to the real desired state. We also noted that we are free to choose a reference state value, as long as there is no subsequent inconsistency in the calculation of the change in a property because of a chemical reaction with a resulting change in the amount of a given substance. Now that we are to include the possibility of a chemical reaction, it will become necessary to choose these reference state values on a common and consistent basis. We will use as our reference state a temperature of 25°C, a pressure of 0.1 MPa, and a hypothetical ideal gas condition for those substances that are gases.

Consider the simple SSSF combustion process shown in Fig. 12.3. This idealized reaction involves the combustion of solid carbon with gaseous (ideal gas) oxygen, each of which enters the control volume at the reference state, 25°C and 0.1 MPa. The carbon dioxide (ideal gas) formed by the reaction leaves the chamber at the reference state, 25°C and 0.1 MPa. If the heat transfer could be accurately measured, it would be found to be −393 522 kJ/kmol of carbon dioxide formed. The chemical reaction can be written

$$C + O_2 \rightarrow CO_2$$

Applying the first law to this process we have

$$Q_{c.v.} + H_R = H_P \qquad (12.9)$$

where the subscripts R and P refer to the reactants and products, respectively. We will

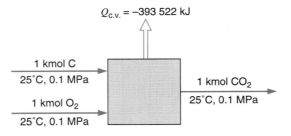

FIGURE 12.3 Example of combustion process.

find it convenient to also write the first law for such a process in the form

$$Q_{c.v.} + \sum_R n_i \bar{h}_i = \sum_P n_e \bar{h}_e \qquad (12.10)$$

where the summations refer, respectively, to all the reactants or all the products.

Thus, a measurement of the heat transfer would give us the difference between the enthalpy of the products and the reactants, where each is at the reference state condition. Suppose, however, that we assign the value of zero to the enthalpy of all the elements at the reference state. In this case, the enthalpy of the reactants is zero, and

$$Q_{c.v.} = H_P = -393\ 522\ \text{kJ} / \text{kmol}$$

The enthalpy of (hypothetical) ideal gas carbon dioxide at 25°C, 0.1 MPa pressure (with reference to this arbitrary base in which the enthalpy of the elements is chosen to be zero), is called the enthalpy of formation. We designate this with the symbol $\bar{h}_f^\circ$. Thus, for carbon dioxide

$$\bar{h}_f^\circ = -393\ 522\ \text{kJ} / \text{kmol}$$

The enthalpy of carbon dioxide in any other state, relative to this base in which the enthalpy of the elements is zero, would be found by adding the change of enthalpy between ideal gas at 25°C, 0.1 MPa, and the given state to the enthalpy of formation. That is, the enthalpy at any temperature and pressure, $\bar{h}_{T,P}$, is

$$\bar{h}_{T,P} = (\bar{h}_f^\circ)_{298,\,0.1\text{MPa}} + (\Delta\bar{h})_{298,\,0.1\text{MPa} \to T,P} \qquad (12.11)$$

where the term $(\Delta\bar{h})_{298,\,0.1\text{MPa} \to T,P}$ represents the difference in enthalpy between any given state and the enthalpy of ideal gas at 298.15 K, 0.1 MPa. For convenience we usually drop the subscripts in the examples that follow.

The procedure that we have demonstrated for carbon dioxide can be applied to any compound.

Table A.16 gives values of the enthalpy of formation for a number of substances in the units kJ/kmol (or Btu/lb mol in A.16E).

Three further observations should be made in regard to enthalpy of formation.

1. We have demonstrated the concept of enthalpy of formation in terms of the measurement of the heat transfer in an idealized chemical reaction in which a compound is formed from the elements. Actually, the enthalpy of formation is usually found by the application of statistical thermodynamics, using observed spectroscopic data.

2. The justification of this procedure of arbitrarily assigning the value of zero to the enthalpy of the elements at 25°C, 0.1 MPa, rests on the fact that in the absence of nuclear reactions the mass of each element is conserved in a chemical reaction. No conflicts or ambiguities arise with this choice of reference state, and it proves to be very convenient in studying chemical reactions from a thermodynamic point of view.

3. In certain cases an element or compound can exist in more than one state at 25°C, 0.1 MPa. Carbon, for example, can be in the form of graphite or diamond. It is essential that the state to which a given value is related be clearly identified. Thus, in Table A.16, the enthalpy of formation of graphite is given the value of zero, and the enthalpy of each substance that contains carbon is given relative to this base. Another example is that oxygen may exist in the monatomic or diatomic form, and

also as ozone, O_3. The value chosen as zero is for the form that is chemically stable at the reference state, which in the case of oxygen is the diatomic form. Then each of the other forms must have an enthalpy of formation consistent with the chemical reaction and heat transfer for the reaction that produces that form of oxygen.

It will be noted from Table A.16 that two values are given for the enthalpy of formation for water; one is for liquid water and the other for gaseous (hypothetical ideal gas) water, both at the reference state of 25°C, 0.1 MPa. It is convenient to use the hypothetical ideal-gas reference in connection with the ideal-gas table property changes given in Table A.13, and to use the real liquid reference in connection with real water property changes as given in the steam tables, Table A.1. The real-liquid reference state properties are obtained from those at the hypothetical ideal-gas reference by following the procedure of calculation described in Section 11.8. The same procedure can be followed for other substances that have a saturation pressure less than 0.1 MPa at the reference temperature of 25°C.

Frequently students are bothered by the minus sign when the enthalpy of formation is negative. For example, the enthalpy of formation of carbon dioxide is negative. This is quite evident because the heat transfer is negative during the steady-flow chemical reaction, and the enthalpy of the carbon dioxide must be less than the sum of enthalpy of the carbon and oxygen initially, both of which are assigned the value of zero. This is quite analogous to the situation we would have in the steam tables if we let the enthalpy of saturated vapor be zero at 0.1 MPa pressure, for in this case the enthalpy of the liquid would be negative, and we would simply use the negative value for the enthalpy of the liquid when solving problems.

12.4 FIRST LAW ANALYSIS OF REACTING SYSTEMS

The significance of the enthalpy of formation is that it is most convenient in performing a first law analysis of a reacting system, for the enthalpies of different substances can be added or subtracted, since they are all given relative to the same base.

In such problems we will write the first law for a steady-state, steady-flow process in the form

$$Q_{c.v.} + H_R = W_{c.v.} + H_P$$

or

$$Q_{c.v.} + \sum_R n_i \bar{h}_i = W_{c.v.} + \sum_P n_e \bar{h}_e$$

where R and P refer to the reactants and products, respectively. In each problem it is necessary to choose one parameter as the basis of the solution. Usually this is taken as 1 kmol of fuel.

EXAMPLE 12.6   Consider the following reaction, which occurs in a steady-state, steady-flow process.

$$CH_4 + 2O_2 \rightarrow CO_2 + 2H_2O(l)$$

The reactants and products are each at a total pressure of 0.1 MPa and 25°C. Determine the heat transfer per kilomole of fuel entering the combustion chamber.

Control volume: Combustion chamber.

Inlet state: P and T known; state fixed.

Exit state: P and T known; state fixed.

Process: SSSF.

Model: Three gases ideal gases; real liquid water.

Analysis:

First law:

$$Q_{c.v.} + \sum_R n_i \bar{h}_i = \sum_P n_e \bar{h}_e$$

Solution

Using values from Table A.16, we have

$$\sum_R n_i \bar{h}_i = (\bar{h}_f^\circ)_{CH_4} = -74\,873 \text{ kJ}$$

$$\sum_P n_e \bar{h}_e = (\bar{h}_f^\circ)_{CO_2} + 2(\bar{h}_f^\circ)_{H_2O(l)}$$

$$= -393\,522 + 2(-285\,830) = -965\,182 \text{ kJ}$$

$$Q_{c.v.} = -965\,182 - (-74\,873) = -890\,309 \text{ kJ}$$

In most instances, however, the substances that comprise the reactants and products in a chemical reaction are not at a temperature of 25°C and a pressure of 0.1 MPa (the state at which the enthalpy of formation is given). Therefore, the change of enthalpy between 25°C and 0.1 MPa and the given state must be known. For a solid or liquid, this change of enthalpy can usually be found from a table of thermodynamic properties or from specific heat data. For gases, this change of enthalpy can usually be found by one of the following procedures.

1. Assume ideal-gas behavior between 25°C, 0.1 MPa, and the given state. In this case, the enthalpy is a function of the temperature only, and can be found by an equation of C_{po} or from tabulated values of enthalpy as a function of temperature (which assumes ideal gas behavior). Table A.11 gives an equation for $\bar{C}_{po}$ for a number of substances and Table A.13 gives values of $(\bar{h}^\circ - \bar{h}^\circ_{298})$ (that is, the $\Delta\bar{h}$ of Eq. 12.11) in kJ/kmol, ($\bar{h}^\circ_{298}$ refers to 25°C or 298.15 K. For simplicity this is designated $\bar{h}^\circ_{298}$.) The superscript $^\circ$ is used to designate that this is the enthalpy at 0.1 MPa pressure, based on ideal-gas behavior, that is, the standard-state enthalpy.

2. If a table of thermodynamic properties is available, $\Delta\bar{h}$ can be found directly from these tables if a real substance behavior reference state is being used, such as that described above for liquid water. If a hypothetical ideal gas reference state is being

used, then it is necessary to account for the real substance correction to properties at that state to gain entry to the tables.

3. If the deviation from ideal-gas behavior is significant, but no tables of thermodynamic properties are available, the value for $\Delta \bar{h}$ can be found from the generalized tables or charts and the values for $\bar{C}_{po}$ or $\Delta \bar{h}$ at 0.1 MPa pressure as indicated above.

Thus, in general, for applying the first law to a steady-state, steady-flow process involving a chemical reaction and negligible changes in kinetic and potential energy, we can write

$$Q_{c.v.} + \sum_R n_i (\bar{h}_f^\circ + \Delta \bar{h})_i = W_{c.v.} + \sum_P n_e (\bar{h}_f^\circ + \Delta \bar{h})_e \qquad (12.12)$$

EXAMPLE 12.7

Calculate the enthalpy of water (on a kilomole basis) at 3.5 MPa, 300°C, relative to the 25°C and 0.1 MPa base, using the following procedures.

1. Assume the steam to be an ideal gas with the value of $\bar{C}_{po}$ given in the Appendix, Table A.11.

2. Assume the steam to be an ideal gas with the value for $\Delta \bar{h}$ as given in the Appendix, Table A.13.

3. The steam tables.

4. The specific heat behavior given in 2 above and the generalized tables or charts.

Solution

For each of these procedures, we can write

$$\bar{h}_{T,P} = (\bar{h}_f^\circ + \Delta \bar{h})$$

The only difference is in the procedure by which we calculate $\Delta \bar{h}$. From Table A.16 we note that

$$(\bar{h}_f^\circ)_{H_2O(g)} = -241\,826 \text{ kJ / kmol}$$

1. Using the specific heat equation for $H_2O(g)$ from Table A.11,

$$\bar{C}_{po} = 143.05 - 183.54 \, \theta^{0.25} + 82.751 \, \theta^{0.5} - 3.6989 \, \theta \text{ kJ / kmol K}$$

where

$$\theta = T / 100$$

Therefore,

$$\Delta \bar{h} = \int_{298.15}^{573.15} \bar{C}_{po}(T)dT$$

$$= \int_{2.9815}^{5.7315} \bar{C}_{po}(\theta)100 d\theta$$

$$= 9517 \text{ kJ / kmol}$$

$$\bar{h}_{T,P} = -241\,826 + 9517 = -232\,309 \text{ kJ / kmol}$$

2. Using Table A.13 for $H_2O(g)$,

$$\Delta \bar{h} = 9494 \text{ kJ / kmol}$$

$$\bar{h}_{T,P} = -241\,826 + 9494 = -232\,332 \text{ kJ / kmol}$$

3. Using the steam tables, either the liquid reference or the gaseous reference state may be used.

For the liquid,

$$\Delta \bar{h} = 18.015(2877.5 - 104.9) = 51\,750 \text{ kJ / kmol}$$

$$\bar{h}_{T,P} = -285\,830 + 51\,750 = -234\,080 \text{ kJ / kmol}$$

For the gas,

$$\Delta \bar{h} = 18.015(2877.5 - 2547.2) = 7752 \text{ kJ / kmol}$$

$$\bar{h}_{T,P} = -241\,826 + 7752 = -234\,074 \text{ kJ / kmol}$$

The very small difference results from using the enthalpy of saturated vapor at 25°C (which is almost but not exactly an ideal gas) in calculating the Δh.

4. When using the generalized tables or charts we use the notation introduced in Chapter 10.

$$\bar{h}_{T,P} = \bar{h}_f^\circ - (\bar{h}_f^* - \bar{h}_2) + (\bar{h}_2^* - \bar{h}_1^*) + (\bar{h}_1^* - \bar{h}_1)$$

where subscript 2 refers to the state at 3.5 MPa, 300°C, and state 1 refers to the state at 0.1 MPa, 25°C.

From part 2, $\bar{h}_2^* - \bar{h}_1^* = 9494$ kJ/kmol.

$$\bar{h}_1^* - \bar{h}_1 = 0 \quad \text{(ideal gas reference)}$$

$$P_{r2} = \frac{3.5}{22.09} = 0.158 \qquad T_{r2} = \frac{573.2}{647.3} = 0.886$$

From the generalized enthalpy table, A.15,

$$\frac{\bar{h}_2^* - \bar{h}_2}{\bar{R}T_c} = 0.2113, \quad \bar{h}_2^* - \bar{h}_2 = 0.2113 \times 8.3145 \times 647.3 = 1137 \text{ kJ / kmol}$$

$$\bar{h}_{T,P} = -241\,826 - 1137 + 9494 = -233\,469 \text{ kJ / mol}$$

The particular approach that is used in a given problem will depend on the data available for the given substance.

EXAMPLE 12.8 A small gas turbine uses $C_8H_{18}(l)$ for fuel, and 400% theoretical air. The air and fuel enter at 25°C and the products of combustion leave at 900 K. The output of the engine and the fuel consumption are measured and it is found that the specific fuel consumption is 0.25 kg/s of fuel per megawatt output. Determine the heat transfer from the engine per kilomole of fuel. Assume complete combustion.

Control volume: Gas turbine engine.

Inlet states: *T* known for fuel and air.

Exit state: *T* known for combustion products.

Process: SSSF.

Model: All gases ideal gases, Table A.13; liquid octane, Table A.16.

Analysis:

The combustion equation is

$$C_8H_{18}(l) + 4(12.5)O_2 + 4(12.5)(3.76)N_2 \rightarrow 8CO_2 + 9H_2O + 37.5O_2 + 188.0N_2$$

First law:

$$Q_{c.v.} + \sum_R n_i (\bar{h}_f^\circ + \Delta\bar{h})_i = W_{c.v.} + \sum_P n_e (\bar{h}_f^\circ + \Delta\bar{h})_e$$

Solution

Since the air is composed of elements and enters at 25°C, the enthalpy of the reactants is equal to that of the fuel,

$$\sum_R n_i (\bar{h}_f^\circ + \Delta\bar{h})_i = (\bar{h}_f^\circ)_{C_8H_{18}(l)} = -250\,105 \text{ kJ / kmol fuel}$$

Considering the products, we have

$$\sum_P n_e (\bar{h}_f^\circ + \Delta\bar{h})_e = n_{CO_2}(\bar{h}_f^\circ + \Delta\bar{h})_{CO_2} + n_{H_2O}(\bar{h}_f^\circ + \Delta\bar{h})_{H_2O}$$

$$+ n_{O_2}(\Delta\bar{h})_{O_2} + n_{N_2}(\Delta\bar{h})_{N_2}$$

$$= 8(-393\,522 + 28\,030) + 9(-241\,826 + 21\,892)$$

$$+ 37.5(19\,249) + 188(18\,222)$$

$$= -755\,769 \text{ kJ / kmol fuel}$$

$$W_{c.v.} = \frac{1000 \text{ kJ / s}}{0.25 \text{ kg / s}} \times \frac{114.23 \text{ kg}}{\text{kmol}} = 456\,920 \text{ kJ / kmol fuel}$$

Therefore, from the first law,

$$Q_{c.v.} = -755\,769 + 456\,920 - (-250\,105)$$

$$= -48\,744 \text{ kJ / kmol fuel}$$

EXAMPLE 12.8E A small gas turbine uses $C_8H_{18}(l)$ for fuel, and 400% theoretical air. The air and fuel enter at 77 F, and the products of combustion leave at 1100 F. The output of the engine and the fuel consumption are measured and it is found that the specific fuel consumption is one pound of fuel per horsepower-hour. Determine the heat transfer from the engine per pound mole of fuel. Assume complete combustion.

Control volume: Gas turbine engine.

Inlet states: *T* known for fuel and air.

Exit state: T known for combustion products.

Process: SSSF.

Model: All gases ideal gases, Table A.13E; liquid octane, Table A.16E.

Analysis:

The combustion equation is

$$C_8H_{18}(l) + 4(12.5)O_2 + 4(12.5)(3.76)N_2 \rightarrow 8CO_2 + 9H_2O + 37.5O_2 + 188.0N_2$$

First law:

$$Q_{c.v.} + \sum_R n_i(\bar{h}_f^\circ + \Delta\bar{h})_i = W_{c.v.} + \sum_P n_e(\bar{h}_f^\circ + \Delta\bar{h})_e$$

Solution

Since the air is composed of elements and enters at 77 F, the enthalpy of the reactants is equal to that of the fuel.

$$\sum_R n_i\left[\bar{h}_f^\circ + \Delta\bar{h}\right]_i = (\bar{h}_f^\circ)_{C_8H_{18}(l)} = -107\,526 \text{ Btu / lb mol.}$$

Considering the products

$$\sum_P n_e(\bar{h}_f^\circ + \Delta\bar{h})_e = n_{CO_2}(\bar{h}_f^\circ + \Delta\bar{h})_{CO_2} + n_{H_2O}(\bar{h}_f^\circ + \Delta\bar{h})_{H_2O} + n_{O_2}(\Delta\bar{h})_{O_2} + n_{N_2}(\Delta\bar{h})_{N_2}$$

$$= 8(-169\,184 + 11\,291) + 9(-103\,966 + 8858)$$

$$+ 37.5(7778) + 188(7374)$$

$$= -441\,129 \text{ Btu / lb mol fuel.}$$

$$W_{c.v.} = 2544 \times 114.23 = 290\,601 \text{ Btu / lb mol fuel.}$$

Therefore, from the first law,

$$Q_{c.v.} = -441\,129 + 290\,601 - (-107\,526)$$

$$= -43\,002 \text{ Btu / lb mol fuel.}$$

EXAMPLE 12.9   A mixture of 1 kmol of gaseous ethene and 3 kmol of oxygen at 25°C reacts in a constant-volume bomb. Heat is transferred until the products are cooled to 600 K. Determine the amount of heat transfer from the system.

Control mass: Constant-volume bomb.

Initial state: T known.

Final state: T known.

Process: Constant volume.

Model: Ideal gas mixtures, Tables A.13, A.16,

Analysis:

The chemical reaction is

$$C_2H_4 + 3O_2 \rightarrow 2CO_2 + 2H_2O(g)$$

First law:

$$Q + U_R = U_P$$

$$Q + \sum_R n(\bar{h}_f^\circ + \Delta\bar{h} - \bar{R}T) = \sum_P n(\bar{h}_f^\circ + \Delta\bar{h} - \bar{R}T)$$

Solution

Using values from Tables A.13 and A.16, gives

$$\sum_R n(\bar{h}_f^\circ + \Delta\bar{h} - \bar{R}T) = (\bar{h}_f^\circ - \bar{R}T)_{C_2H_4} - n_{O_2}(\bar{R}T)_{O_2} = (\bar{h}_f^\circ)_{C_2H_4} - 4\bar{R}T$$

$$= 52\,467 - 4 \times 8.3145 \times 298.2 = 42\,550 \text{ kJ}$$

$$\sum_P n(\bar{h}_f^\circ + \Delta\bar{h} - \bar{R}T) = 2\left[(\bar{h}_f^\circ)_{CO_2} + \Delta\bar{h}_{CO_2}\right] + 2\left[(\bar{h}_f^\circ)_{H_2O(g)} + \Delta\bar{h}_{H_2O(g)}\right] - 4\bar{R}T$$

$$= 2(-393\,522 + 12\,899) + 2(-241\,826 + 10\,463)$$

$$-4 \times 8.3145 \times 600$$

$$= -1\,243\,927 \text{ kJ}$$

Therefore,

$$Q = -1\,243\,927 - 42\,550 = -1\,286\,477 \text{ kJ}$$

For a real gas mixture, a pseudocritical method such as Kay's rule, Eq. 11.35, could be used to evaluate the nonideal gas contribution to enthalpy at the temperature and pressure of the mixture and this value added to the ideal gas mixture enthalpy at that temperature, as in the procedure developed in Section 11.8.

12.5 ADIABATIC FLAME TEMPERATURE

Consider a given combustion process that takes place adiabatically and with no work or changes in kinetic or potential energy involved. For such a process the temperature of the products is referred to as the adiabatic flame temperature. With the assumptions of no work and no changes in kinetic or potential energy, this is the maximum temperature that can be achieved for the given reactants because any heat transfer from the reacting substances and any incomplete combustion would tend to lower the temperature of the products.

For a given fuel and given pressure and temperature of the reactants, the maximum adiabatic flame temperature that can be achieved is with a stoichiometric mixture. The adiabatic flame temperature can be controlled by the amount of excess air that is used. This is important, for example, in gas turbines, where the maximum permissible temperature is determined by metallurgical considerations in the turbine, and close control of the temperature of the products is essential.

Example 12.10 shows how the adiabatic flame temperature may be found. The dissociation that takes place in the combustion products, which has a significant effect on the adiabatic flame temperature, will be considered in the next chapter.

EXAMPLE 12.10   Liquid octane at 25°C is burned with 400% theoretical air at 25°C in a steady-flow process. Determine the adiabatic flame temperature.

Control volume: Combustion chamber.

Inlet states: T known for fuel and air.

Process: SSSF.

Model: Gases ideal gases, Table A.13; liquid octane, Table A.16.

Analysis:

The reaction is

$$C_8H_{18}(l) + 4(12.5)O_2 + 4(12.5)(3.76)N_2 \rightarrow$$

$$8CO_2 + 9H_2O(g) + 37.5O_2 + 188.0N_2$$

First law: Since the process is adiabatic,

$$H_R = H_P$$

$$\sum_R n_i(\bar{h}_f^\circ + \Delta\bar{h})_i = \sum_P n_e(\bar{h}_f^\circ + \Delta\bar{h})_e$$

where $\Delta\bar{h}_e$ refers to each constituent in the products at the adiabatic flame temperature.

Solution

From Tables A.13 and A.16,

$$H_R = \sum_R n_i(\bar{h}_f^\circ + \Delta\bar{h})_i = (\bar{h}_f^\circ)_{C_8H_{18}(l)} = -250\,105 \text{ kJ / kmol fuel}$$

$$H_P = \sum_P n_e(\bar{h}_f^\circ + \Delta\bar{h})_e$$

$$= 8(-393\,522 + \Delta\bar{h}_{CO_2}) + 9(-241\,826 + \Delta\bar{h}_{H_2O}) + 37.5\Delta\bar{h}_{O_2} + 188.0\Delta\bar{h}_{N_2}$$

By trial-and-error solution, a temperature of the products is found that satisfies this equation. Assume that

$$T_P = 900 \text{ K}$$

$$H_P = \sum_P n_e(\bar{h}_f^\circ + \Delta\bar{h})_e$$

$$= 8(-393\,522 + 28\,030) + 9(-241\,826 + 21\,892)$$

$$+ 37.5(19\,249) + 188(18\,222)$$

$$= -755\,769 \text{ kJ / kmol fuel}$$

Assume that

$$T_P = 1000 \text{ K}$$

$$
\begin{aligned}
H_P &= \sum_P n_e (\bar{h}_f^\circ + \Delta \bar{h})_e \\
&= 8(-393\,522 + 33\,400) + 9(-241\,826 + 25\,956) \\
&\quad + 37.5(22\,710) + 188(21\,461) \\
&= 62\,487 \text{ kJ / kmol fuel}
\end{aligned}
$$

Since $H_P = H_R = -250\,105$ kJ, we find by linear interpolation that the adiabatic flame temperature is 961.8 K. Because the ideal-gas enthalpy is not really a linear function of temperature, the true answer will be slightly different from this value.

12.6 Enthalpy and Internal Energy of Combustion; Heat of Reaction

The enthalpy of combustion, h_{RP}, is defined as the difference between the enthalpy of the products and the enthalpy of the reactants when complete combustion occurs at a given temperature and pressure. That is,

$$\bar{h}_{RP} = H_P - H_R$$

$$\bar{h}_{RP} = \sum_P n_e (\bar{h}_f^\circ + \Delta \bar{h})_e - \sum_R n_i (\bar{h}_f^\circ + \Delta \bar{h})_i \tag{12.13}$$

The usual parameter for expressing the enthalpy of combustion is a unit mass of fuel, such as a kilogram (h_{RP}) or a kilomole ($\bar{h}_{RP}$) of fuel.

The tabulated values of the enthalpy of combustion of fuels are usually given for a temperature of 25°C and a pressure of 0.1 MPa. The enthalpy of combustion for a number of hydrocarbon fuels at this temperature and pressure, which we designate h_{RP0}, is given in Table 12.3.

The internal energy of combustion is defined in a similar manner.

$$\bar{u}_{RP} = U_P - U_R$$

$$= \sum_P n_e (\bar{h}_f^\circ + \Delta \bar{h} - P\bar{v})_e - \sum_R n_i (\bar{h}_f^\circ + \Delta \bar{h} - P\bar{v})_i \tag{12.14}$$

When all the gaseous constituents can be considered as ideal gases, and the volume of the liquid and solid constituents is negligible compared to the value of the gaseous constituents, this relation for $\bar{u}_{RP}$ reduces to

$$\bar{u}_{RP} = \bar{h}_{RP} - \bar{R} T (n_{\text{gaseous products}} - n_{\text{gaseous reactants}}) \tag{12.15}$$

Frequently the term "heating value" or "heat of reaction" is used. This represents the heat transferred from the chamber during combustion or reaction at constant temperature. In the case of a constant pressure or steady-flow process, we conclude from the first law of thermodynamics that it is equal to the negative of the enthalpy of combustion. For this reason this heat transfer is sometimes designated the constant-pressure heating value for combustion processes.

In the case of a constant-volume process, the heat transfer is equal to the negative of the internal energy of combustion. This is sometimes designated the constant-volume heating value in the case of combustion.

When the term heating value is used, the terms "higher" and "lower" heating value are used. The higher heating value is the heat transfer with liquid water in the products, and the lower heating value is the heat transfer with vapor water in the products.

EXAMPLE 12.11   Calculate the enthalpy of combustion of propane at 25°C on both a kilomole and kilogram basis under the following conditions.

1. Liquid propane with liquid water in the products.
2. Liquid propane with gaseous water in the products.
3. Gaseous propane with liquid water in the products.
4. Gaseous propane with gaseous water in the products.

This example is designed to show how the enthalpy of combustion can be determined from enthalpies of formation. The enthalpy of evaporation of propane is 370 kJ/kg.

Analysis and Solution

The basic combustion equation is

$$C_3H_8 + 5O_2 \rightarrow 3CO_2 + 4H_2O$$

From Table A.16, $(\bar{h}_f^\circ)_{C_3H_8(g)} = -103\,900$ kJ/kmol. Therefore,

$$(\bar{h}_f^\circ)_{C_3H_8(l)} = -103\,900 - 44.097(370) = -120\,216 \text{ kJ / kmol}$$

1.   Liquid propane–liquid water:

$$\bar{h}_{RP_0} = 3(\bar{h}_f^\circ)_{CO_2} + 4(\bar{h}_f^\circ)_{H_2O(l)} - (\bar{h}_f^\circ)_{C_3H_8(l)}$$

$$= 3(-393\,522) + 4(-285\,830) - (-120\,216)$$

$$= -2\,203\,670 \text{ kJ / kmol} = -\frac{2\,203\,670}{44.097} = -49\,973 \text{ kJ / kg}$$

The higher heating value of liquid propane is 49 973 kJ/kg.

2.   Liquid propane–gaseous water:

$$\bar{h}_{RP_0} = 3(\bar{h}_f^\circ)_{CO_2} + 4(\bar{h}_f^\circ)_{H_2O(g)} - (\bar{h}_f^\circ)_{C_3H_8(l)}$$

$$= 3(-393\,522) + 4(-241\,826) - (-120\,216)$$

$$= -2\,027\,654 \text{ kJ / kmol} = -\frac{2\,027\,654}{44.097} = -45\,982 \text{ kJ / kg}$$

The lower heating value of liquid propane is 45 982 kJ/kg.

Table 12.3

Enthalpy of Combustion of Some Hydrocarbons at 25°C

Hydrocarbon	Formula	Liquid H_2O in Products		Gas H_2O in Products	
		Liq. HC	Gas HC	Liq. HC	Gas HC
Paraffins	C_nH_{2n+2}				
Methane	CH_4		−55 496		−50 010
Ethane	C_2H_6		−51 875		−47 484
Propane	C_3H_8	−49 973	−50 343	−45 982	−46 352
n-Butane	C_4H_{10}	−49 130	−49 500	−45 344	−45 714
n-Pentane	C_5H_{12}	−48 643	−49 011	−44 983	−45 351
n-Hexane	C_6H_{14}	−48 308	−48 676	−44 733	−45 101
n-Heptane	C_7H_{16}	−48 071	−48 436	−44 557	−44 922
n-Octane	C_8H_{18}	−47 893	−48 256	−44 425	−44 788
n-Decane	$C_{10}H_{22}$	−47 641	−48 000	−44 239	−44 598
n-Dodecane	$C_{12}H_{26}$	−47 470	−47 828	−44 109	−44 467
n-Cetane	$C_{16}H_{34}$	−47 300	−47 658	−44 000	−44 358
Olefins	C_nH_{2n}				
Ethene	C_2H_4		−50 296		−47 158
Propene	C_3H_6		−48 917		−45 780
Butene	C_4H_8		−48 453		−45 316
Pentene	C_5H_{10}		−48 134		−44 996
Hexene	C_6H_{12}		−47 937		−44 800
Heptene	C_7H_{14}		−47 800		−44 662
Octene	C_8H_{16}		−47 693		−44 556
Nonene	C_9H_{18}		−47 612		−44 475
Decene	$C_{10}H_{20}$		−47 547		−44 410
Alkylbenzenes	$C_{6+n}H_{6+2n}$				
Benzene	C_6H_6	−41 831	−42 266	−40 141	−40 576
Methylbenzene	C_7H_8	−42 437	−42 847	−40 527	−40 937
Ethylbenzene	C_8H_{10}	−42 997	−43 395	−40 924	−41 322
Propylbenzene	C_9H_{12}	−43 416	−43 800	−41 219	−41 603
Butylbenzene	$C_{10}H_{14}$	−43 748	−44 123	−41 453	−41 828
Other fuels					
Gasoline	C_7H_{17}	−48 201	−48 582	−44 506	−44 886
Diesel	$C_{14.4}H_{24.9}$	−45 700	−46 074	−42 934	−43 308
Methanol	CH_3OH	−22 657	−23 840	−19 910	−21 093
Ethanol	C_2H_5OH	−29 676	−30 596	−26 811	−27 731
Nitromethane	CH_3NO_2	−11 618	−12 247	−10 537	−11 165
Phenol	C_6H_5OH	−32 520	−33 176	−31 117	−31 774

3. Gaseous propane–liquid water:

$$\bar{h}_{RP_0} = 3(\bar{h}_f^\circ)_{CO_2} + 4(\bar{h}_f^\circ)_{H_2O(l)} - (\bar{h}_f^\circ)_{C_3H_8(g)}$$

$$= 3(-393\,522) + 4(-285\,830) - (-103\,900)$$

$$= -2\,219\,986 \text{ kJ / kmol} = -\frac{2\,219\,986}{44.097} = -50\,343 \text{ kJ / kg}$$

The higher heating value of gaseous propane is 50 343 kJ/kg.

4. Gaseous propane–gaseous water:

$$\bar{h}_{RP_0} = 3(\bar{h}_f^\circ)_{CO_2} + 4(\bar{h}_f^\circ)_{H_2O(g)} - (\bar{h}_f^\circ)_{C_3H_8(g)}$$

$$= 3(-393\,522) + 4(-241\,826) - (-103\,900)$$

$$= -2\,043\,970 \text{ kJ / kmol} = -\frac{2\,043\,970}{44.097} = -46\,352 \text{ kJ / kg}$$

The lower heating value of gaseous propane is 46 352 kJ/kg.

Each of the four values calculated in this example corresponds to the appropriate value given in Table 12.3.

EXAMPLE 12.12   Calculate the enthalpy of combustion of gaseous propane at 500 K. (At this temperature all the water formed during combustion will be vapor.) This example will demonstrate how the enthalpy of combustion of propane varies with temperature. The average constant-pressure specific heat of propane between 25°C and 500 K is 2.1 kJ/kg K.

Analysis:

The combustion equation is

$$C_3H_8(g) + 5O_2 \rightarrow 3CO_2 + 4H_2O(g)$$

The enthalpy of combustion is, from Eq. 12.13,

$$(\bar{h}_{RP})_T = \sum_P n_e(\bar{h}_f^\circ + \Delta\bar{h})_e - \sum_R n_i(\bar{h}_f^\circ + \Delta\bar{h})_i$$

Solution

$$\bar{h}_{R_{500}} = \left[\bar{h}_f^\circ + \bar{C}_{p\,av}(\Delta T)\right]_{C_3H_8(g)} + n_{O_2}(\Delta\bar{h})_{O_2}$$

$$= -103\,900 + 2.1 \times 44.097(500 - 298.2) + 5(6095)$$

$$= -54\,738 \text{ kJ}$$

$$\bar{h}_{P_{500}} = n_{CO_2}(\bar{h}_f^\circ + \Delta\bar{h})_{CO_2} + n_{H_2O}(\bar{h}_f^\circ + \Delta\bar{h})_{H_2O}$$

$$= 3(-393\,522 + 8297) + 4(-241\,826 + 6896)$$

$$= -2\,095\,395 \text{ kJ}$$

$$\bar{h}_{RP_{500}} = -2\,095\,395 - (-54\,738) = -2\,040\,657 \text{ kJ / kmol}$$

$$\bar{h}_{RP_{500}} = \frac{-2\,040\,657}{44.097} = -46\,277 \text{ kJ / kg}$$

This compares with a value of $-46\,352$ at 25°C.

This problem could also have been solved using the given value of the enthalpy of combustion at 25°C by noting that

$$\bar{h}_{RP_{500}} = (H_P)_{500} - (H_R)_{500}$$

$$= n_{CO_2}(\bar{h}_f^\circ + \Delta\bar{h})_{CO_2} + n_{H_2O}(\bar{h}_f^\circ + \Delta\bar{h})_{H_2O}$$

$$-\left[\bar{h}_f^\circ + \overline{C}_{p\,av}(\Delta T)\right]_{C_3H_8(g)} - n_{O_2}(\Delta\bar{h})_{O_2}$$

$$= \bar{h}_{RP_0} + n_{CO_2}(\Delta\bar{h})_{CO_2} + n_{H_2O}(\Delta\bar{h})_{H_2O}$$

$$-\overline{C}_{p\,av}(\Delta T)_{C_3H_8(g)} - n_{O_2}(\Delta\bar{h})_{O_2}$$

$$\bar{h}_{RP_{500}} = -46\,352 \times 44.097 + 3(8297) + 4(6896)$$

$$-2.1 \times 44.097(500 - 298.2) - 5(6095)$$

$$= -2\,040\,657 \text{ kJ / kmol}$$

$$h_{RP_{500}} = \frac{-2\,040\,657}{44.097} = -46\,277 \text{ kJ / kg}$$

12.7 THE THIRD LAW OF THERMODYNAMICS AND ABSOLUTE ENTROPY

As we consider a second-law analysis of chemical reactions, we face the same problem we had with the first law: What base should be used for the entropy of the various substances? This problem leads directly to a consideration of the third law of thermodynamics.

The third law of thermodynamics was formulated during the early part of the twentieth century. The initial work was done primarily by W. H. Nernst (1864–1941) and Max Planck (1858–1947). The third law deals with the entropy of substances at the absolute zero of temperature, and in essence states that the entropy of a perfect crystal is zero at absolute zero. From a statistical point of view, this means that the crystal structure has the maximum degree of order. Further, because the temperature is absolute zero, the thermal energy is minimum. It also follows that a substance that does not have a perfect crystalline structure at absolute zero, but instead has a degree of randomness, such as a solid solution or a glassy solid, has a finite value of entropy at absolute zero. The experimental evidence on which the third law rests is primarily data on chemical reactions at low temperatures and measurements of heat capacity at temperatures approaching absolute zero. In contrast to the first and second laws, which lead, respectively, to the properties of internal energy and entropy, the third law deals only with the question of entropy at ab solute zero. However, the implications of the third law are quite profound, particularly in respect to chemical equilibrium.

The particular relevance of the third law is that it provides an absolute base from which to measure the entropy of each substance. The entropy relative to this base is termed the absolute entropy. The increase in entropy between absolute zero and any given state can be found either from calorimetric data or by procedures based on statistical thermodynamics. The calorimetric method gives precise measurements of specific-heat data over the temperature range, as well as of the energy associated with phase transformations. These measurements are in agreement with the calculations based on statistical thermodynamics and observed molecular data.

Table A.16 gives the absolute entropy at 25°C and 0.1-MPa pressure for a number of substances. Table A.13 gives the absolute entropy for a number of gases at 0.1-MPa pressure and various temperatures. For gases the numbers in all these tables are the hypothetical ideal-gas values. The pressure $P°$ of 0.1 MPa is termed the standard-state pressure, and the absolute entropy as given in these tables is designated $\bar{s}°$. The temperature is designated in kelvins with a subscript such as $\bar{s}°_{1000}$.

If the value of the absolute entropy is known at the standard-state pressure of 0.1 MPa and a given temperature, it is a straightforward procedure to calculate the entropy change from this state (whether hypothetical ideal gas or real substance) to another desired state following the procedure described in Section 11.8. If the substance is listed in Table A.13, then

$$\bar{s}_{T,P} = \bar{s}°_T - \bar{R} \ln \frac{P}{P°} + (\bar{s}_{T,P} - \bar{s}*_{T,P}) \tag{12.16}$$

In this expression, the first term on the right side is the value from Table A.13, the second is the ideal-gas term to account for a change in pressure from $P°$ to P, and the third is the term that corrects for real-substance behavior, as given in the generalized entropy chart or table in the Appendix. If the real-substance behavior is to be evaluated from an equation of state or thermodynamic table of properties, the term for the change in pressure should be made to a low pressure $P*$, at which ideal-gas behavior is a reasonable assumption, but it is also listed in the tables. Then

$$\bar{s}_{T,P} = \bar{s}°_T - \bar{R} \ln \frac{P*}{P°} + (\bar{s}_{T,P} - \bar{s}*_{T,P*}) \tag{12.17}$$

If the substance is not one of those listed in Table A.13, and the absolute entropy is known only at one temperature T_0, as given in Table A.16 for example, then it will be necessary to calculate $\bar{s}°_T$ from

$$\bar{s}°_T = \bar{s}°_{T_0} + \int_{T_0}^{T} \frac{\overline{C}_{p0}}{T} dT \tag{12.18}$$

and then proceed with the calculation of Eq. 12.16 or 12.17.

It should be noted that if Eq. 12.16 is being used to calculate the absolute entropy of a substance in a region in which the ideal-gas model is a valid representation of the behavior of that substance, then the last term on the right-side of Eq. 12.16 simply drops out of the calculation.

For calculation of the absolute entropy of a mixture of ideal gases at T, P, the mixture entropy is given in terms of the component partial entropies as

$$\bar{s}*_{mix} = \sum_i y_i \bar{S}*_i \tag{12.19}$$

where

$$\overline{S}_i^* = \overline{s}_{Ti}^\circ - \overline{R}\ln\frac{P}{P^\circ} - \overline{R}\ln y_i = \overline{s}_{Ti}^\circ - \overline{R}\ln\frac{y_i P}{P^\circ} \qquad (12.20)$$

For a real-gas mixture, a correction can be added to the ideal-gas entropy calculated from Eqs. 12.19 and 12.20 by using a pseudocritical method such as was discussed in Section 11.8. The corrected expression is

$$\overline{s}_{\text{mix}} = \overline{s}_{\text{mix}}^* + (\overline{s} - \overline{s}^*)_{T,P} \qquad (12.21)$$

in which the second term on the right side is the correction term from the generalized entropy chart or table.

12.8 SECOND-LAW ANALYSIS OF REACTING SYSTEMS

The concepts of reversible work, irreversibility, and availability were introduced in Chapter 8. These concepts included both the first and second laws of thermodynamics. At the conclusion of Chapter 8, the Gibbs function was introduced, and we made the statement that this property is particularly relevant in dealing with chemical reactions. We proceed now to develop this matter further, and we will be particularly concerned with determining the maximum work (availability) that can be done through a combustion process and with examining the irreversibilities associated with such processes.

The reversible work for a steady-state, steady-flow process in which there is no heat transfer with reservoirs other than the surroundings, and also in the absence of changes in kinetic and potential energy is, from Eq. 8.20,

$$W^{\text{rev}} = \sum m_i(h_i - T_0 s_i) - \sum m_e(h_e - T_0 s_e)$$

Applying this equation to an SSSF process that involves a chemical reaction, and introducing the symbols from this chapter, we have

$$W^{\text{rev}} = \sum_R n_i(\overline{h}_f^\circ + \Delta\overline{h} - T_0\overline{s})_i - \sum_P n_e(\overline{h}_f^\circ + \Delta\overline{h} - T_0\overline{s})_e \qquad (12.22)$$

Similarly, the irreversibility for such a process can be written as

$$I = W^{\text{rev}} - W = \sum_P n_e T_0 \overline{s}_e - \sum_R n_i T_0 \overline{s}_i - Q_{\text{c.v.}} \qquad (12.23)$$

The availability, ψ, for an SSSF process, in the absence of kinetic and potential energy changes, was defined in Eq. 8.28 as

$$\psi = (h - T_0 s) - (h_0 - T_0 s_0)$$

It was also indicated in Chapter 8 that when an SSSF chemical reaction takes place in such a manner that both the reactants and products are in temperature equilibrium with the surroundings, the Gibbs function ($g = h - Ts$) becomes a significant variable, Eq. 8.38. For such a process, in the absence of changes in kinetic and potential energy, the reversible work is given by the relation

$$W^{\text{rev}} = \sum_R n_i \overline{g}_i - \sum_P n_e \overline{g}_e \qquad (12.24)$$

Since the Gibbs function is so relevant for processes that involve chemical reactions, the Gibbs function of formation, $\bar{g}_f^\circ$, has been defined by a procedure similar to that used in defining the enthalpy of formation. That is, the Gibbs function of each of the elements at 25°C and 0.1-MPa pressure is assumed to be zero, and the Gibbs function of each substance is found relative to this base. Table A.16 lists the Gibbs function of formation of a number of substances at 25°C and 0.1-MPa pressure. It is also evident that the Gibbs function can be found directly from data for $\bar{h}_f^\circ$ and $\bar{s}^\circ$ at the given temperature, as indicated in the following example.

EXAMPLE 12.13

Determine the Gibbs function of formation of carbon dioxide in SI units.

Analysis:

Consider the reaction

$$C + O_2 \rightarrow CO_2$$

Assume that the carbon and oxygen are each initially at 25°C and 0.1-MPa pressure, and that the carbon dioxide is finally at 25°C and 0.1-MPa pressure.

The change in Gibbs function for this reaction is found first.

$$G_P - G_R = (H_P - H_R) - T_0(S_P - S_R)$$

$$\sum_P n_e (\bar{g}_f^\circ)_e - \sum_R n_i (\bar{g}_f^\circ)_i = \sum_P n_e (\bar{h}_f^\circ)_e - \sum_R n_i (\bar{h}_f^\circ)_i - T_0 \left[\sum_P n_e (\bar{s}_{298}^\circ)_e - \sum_R n_i (\bar{s}_{298}^\circ)_i \right]$$

Solution

$$G_P - G_R = (\bar{h}_f^\circ)_{CO_2} - 298.15 \left[(\bar{s}_{298}^\circ)_{CO_2} - (\bar{s}_{298}^\circ)_C - (\bar{s}_{298}^\circ)_{O_2} \right]$$

$$= -393\,522 - 298.15(213.795 - 5.740 - 205.148)$$

$$= -394\,389 \text{ kJ} / \text{mol}$$

Since the Gibbs function of the reactants, G_R, is zero (in accordance with the assumption that the Gibbs function of the elements in zero at 25°C and 0.1-MPa pressure), it follows that

$$G_P = (\bar{g}_f^\circ)_{CO_2} = -394\,389 \text{ kJ} / \text{kmol}$$

This is the value given in Table A.16.

EXAMPLE 12.13E

Determine the Gibbs function of formation of carbon dioxide in English units.

Analysis:

Consider the reaction

$$C + O_2 \rightarrow CO_2$$

Assume that the carbon and oxygen are each initially at 77 F and 1 atm pressure, and that the CO_2 is finally at 77 F and 1 atm pressure.

The change in Gibbs function for this reaction is found first.

$$G_P - G_R = (H_P - H_R) - T_0(S_P - S_R)$$

$$\sum_P n_e(\bar{g}_f^\circ)_e - \sum_R n_i(\bar{g}_f^\circ)_i = \sum_P n_e(\bar{h}_f^\circ)_e - \sum_R n_i(\bar{h}_f^\circ)_i - T_0\left[\sum_P n_e(\bar{s}_{537}^\circ)_e - \sum_R n_i(\bar{s}_{537}^\circ)_i\right]$$

Solution

$$G_P - G_R = (\bar{h}_f^\circ)_{CO_2} - 536.67\left[(\bar{s}_{537}^\circ)_{CO_2} - (\bar{s}_{537}^\circ)_C - (\bar{s}_{537}^\circ)_{O_2}\right]$$

$$= -169\,184 - 536.67(51.038 - 1.371 - 48.973)$$

$$= -169\,184 - 372 = -169\,556 \text{ Btu / lb mol.}$$

Since the Gibbs function of the reactants, G_R, is zero (in accordance with the assumption that the Gibbs function of the elements is zero at 77 F and 1 atm pressure), it follows that

$$G_P = (\bar{g}_f^\circ)_{CO_2} = -169\,556 \text{ Btu / lb mol}$$

This is the value given in Table A.16E.

Let us now consider the question of the maximum work that can be done during a chemical reaction. For example, consider 1 kmol of hydrocarbon fuel and the necessary air for complete combustion, each at 0.1-MPa pressure and 25°C, the pressure and temperature of the surroundings. What is the maximum work that can be done as this fuel reacts with the air? From the considerations covered in Chapter 8, we conclude that the maximum work would be done if this chemical reaction took place reversibly and the products were finally in pressure and temperature equilibrium with the surroundings. We conclude that this reversible work could be calculated from the relation in Eq. 12.24,

$$W^{\text{rev}} = \sum_R n_i\bar{g}_i - \sum_P n_e\bar{g}_e$$

However, since the final state is in equilibrium with the surroundings, we could consider this amount of work to be the availability of the fuel and air.

EXAMPLE 12.14   Ethene (g) at 25°C and 0.1-MPa pressure is burned with 400% theoretical air at 25°C and 0.1-MPa pressure. Assume that this reaction takes place reversibly at 25°C and that the products leave at 25°C and 0.1-MPa pressure. To simplify this problem further, assume that the oxygen and nitrogen are separated before the reaction takes place (each at 0.1 MPa, 25°C), that the constituents in the products are separated, and that each is at 25°C and 0.1 MPa. Thus, the reaction takes place as shown in Fig. 12.4. For purposes of comparison between this and the two subsequent examples, we consider all the water in the products to be a gas (a hypothetical situation in this example and Example 12.16).

Determine the reversible work for this process (that is, the work that would be done if this chemical action took place reversibly and isothermally).

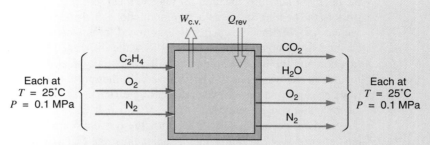

FIGURE 12.4 Sketch for Example 12.14.

Control volume: Combustion chamber.

Inlet states: P, T known for each gas.

Exit states: P, T known for each gas.

Model: All ideal gases, Tables A.13 and A.16.

Sketch: Figure 12.4.

Analysis:

The equation for this chemical reaction is

$$C_2H_4(g) + 3(4)O_2 + 3(4)(3.76)N_2 \rightarrow 2CO_2 + 2H_2O(g) + 9O_2 + 45.1N_2$$

The reversible work for this process is equal to the decrease in Gibbs function during this reaction, Eq. 12.24. The values for the Gibbs function can be taken directly from Table A.16, since each is at 25°C and 0.1-MPa pressure. From Eq. 12.24, we have

$$W^{\text{rev}} = \sum_R n_i \bar{g}_i - \sum_P n_e \bar{g}_e$$

Solution

Since each of the reactants and each of the products are at 0.1-MPa pressure and 25°C, this equation reduces to

$$W^{\text{rev}} = (\bar{g}_f^\circ)_{C_2H_4} - 2(\bar{g}_f^\circ)_{CO_2} - (\bar{g}_f^\circ)_{H_2O(g)}$$

$$= 68\,421 - 2(-394\,389) - 2(-228\,582)$$

$$= 1\,314\,363 \text{ kJ / mol } C_2H_4$$

$$= \frac{1\,314\,363}{28.054} = 46\,851 \text{ kJ / kg}$$

Therefore, we might say that when 1 kg of ethene is at 25°C, 0.1-MPa pressure, the temperature and pressure of the surroundings, it has an availability of 46 851 kJ.

Thus, it would seem logical to rate the efficiency of a device designed to do work by utilizing a combustion process, such as an internal-combustion engine or a steam power plant, as the ratio of the actual work to the reversible work, or in Example 12.14, the decrease in Gibbs function for the chemical reaction, instead of comparing the actual work to the heating value, as is commonly done. This is, in fact, the basic principle of the second-law efficiency, which was introduced in connection with availability analysis in Chapter 8. As noted from Example 12.14, the difference between the decrease in Gibbs function and the heating value is small, which is typical for hydrocarbon fuels. The difference in the two types of efficiencies will, therefore, not usually be large. We must always be careful, however, when discussing efficiencies, to note the definition of the efficiency under consideration.

It is of particular interest to study the irreversibility that takes place during a combustion process. The following examples illustrate this matter. We consider the same hydrocarbon fuel that was used in Example 12.14—ethene (*g*) at 25°C and 0.1 MPa. We determined its availability and found it to be 46 851 kJ/kg. Now let us burn this fuel with 400% theoretical air in a steady-state, steady-flow adiabatic process. We can determine the irreversibility of this process in two ways. The first way is to calculate the increase in entropy for the process. Since the process is adiabatic, the increase in entropy is due entirely to the irreversibilities for the process, and we can find the irreversibility from Eq. 12.23. We can also calculate the availabilities of the products of combustion at the adiabatic flame temperature, and note that they are less than the availability of the fuel and air before the combustion process. The difference is the irreversibility that occurs during the combustion process.

EXAMPLE 12.15   Consider the same combustion process as in Example 12.14, but let it take place adiabatically. Assume that each constituent in the products is at 0.1-MPa pressure and at the adiabatic flame temperature. This combustion process is shown schematically in Fig. 12.5. The temperature of the surroundings is 25°C.

For this combustion process, determine (1) the increase in entropy during combustion and (2) the availability of the products of combustion.

Control volume: Combustion chamber.

Inlet states: P, T known for each gas.

Exit states: P known for each gas.

Sketch: Figure 12.5.

Model: All ideal gases, Table A.13; Table A.16 for ethene.

Analysis:

The combustion equation is

$$C_2H_4(g) + 12O_2 + 12(3.76)N_2 \rightarrow 2CO_2 + 2H_2O(g) + 9O_2 + 45.1N_2$$

The adiabatic flame temperature is determined first.

First law:

$$H_R = H_P$$

$$\sum_R n_i(\bar{h}_f^\circ)_i = \sum_P n_e(\bar{h}_f^\circ + \Delta\bar{h})_e$$

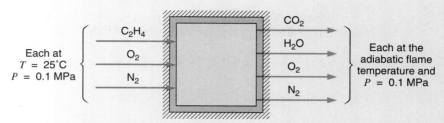

FIGURE 12.5 Sketch for Example 12.15

Solution

$$52\,467 = 2(-393\,522 + \Delta\bar{h}_{CO_2}) + 2(-241\,826 + \Delta\bar{h}_{H_2O(g)}) + 9\Delta\bar{h}_{O_2} + 45.1\Delta\bar{h}_{N_2}$$

By a trial-and-error solution we find the adiabatic flame temperature to be 1016 K.

We now proceed to find the change in entropy during this adiabatic combustion process.

$$S_R = \sum_R (n_i\bar{s}_i^\circ)_{298} = (\bar{s}_{C_2H_4}^\circ + 12\bar{s}_{O_2}^\circ + 45.1\bar{s}_{N_2}^\circ)_{298}$$

$$= 219.330 + 12(205.147) + 45.1(191.610)$$

$$= 11\,322.705 \text{ kJ}/\text{kmol (fuel) K}$$

$$S_P = \sum_P (n_e\bar{s}_e^\circ)_{1016} = (2\bar{s}_{CO_2}^\circ + 2\bar{s}_{H_2O(g)}^\circ + 9\bar{s}_{O_2}^\circ + 45.1\bar{s}_{N_2}^\circ)_{1016}$$

$$= 2(270.194) + 2(233.355) + 9(244.135) + 45.1(228.691)$$

$$= 13\,518.277 \text{ kJ}/\text{kmol (fuel) K}$$

$$S_P - S_R = 2195.572 \text{ kJ}/\text{kmol (fuel) K}$$

Since this is an adiabatic process, the increase in entropy indicates the irreversibility of the adiabatic combustion process. This irreversibility can be found from Eq. 12.23.

$$I = T_0\left(\sum_P n_e\bar{s}_e - \sum_R n_i\bar{s}_i\right)$$

$$= 298.15 \times 2195.572 = 654\,610 \text{ kJ}/\text{kmol}$$

$$= \frac{654\,610}{28.054} = 23\,334 \text{ kJ}/\text{kg fuel}$$

Therefore, the availability after the combustion process is

$$\psi_P = 46\,851 - 23\,334 = 23\,517 \text{ kJ}/\text{kg}$$

The availability of the products can also be found from the relation

$$\psi_P = \sum_P\left[(\bar{h}_e - T_0\bar{s}_e) - (\bar{h}_0 - T_0\bar{s}_0)\right]$$

Since in this problem the products are separated, and each is at 0.1-MPa pressure and the adiabatic flame temperature of 1016 K, this equation can be evaluated, yielding

$$\psi_P = \sum_P n_e \left[(\bar{h}_e^\circ - \bar{h}_0^\circ) - T_0 (\bar{s}_e^\circ - \bar{s}_0^\circ) \right]$$

$$= 2(34\ 271) + 2(26\ 618) + 9(23\ 268) + 45.1(21\ 985)$$

$$-298.15 \left[2(270.194 - 213.795) + 2(233.355 - 188.834) \right.$$

$$+9(244.135 - 205.147) + 45.1(228.691 - 191.610) \right]$$

$$= 659\ 746\ \text{kJ} / \text{kmol} = 23\ 517\ \text{kJ} / \text{kg}$$

In other words, if every process after the adiabatic combustion process is reversible, the maximum amount of work that could be done is 23 517 kJ/kg fuel. This compares to a value of 46 851 kJ/kg for the reversible isothermal reaction. This means that if we had an engine with the indicated adiabatic combustion process, and if all other processes were completely reversible, the efficiency would be about 50%.

In the two prior examples we assumed, for purposes of simplifying the calculation, that the constituents in the reactants and products were separated, and each was at 0.1-MPa pressure. This of course is not a realistic problem. In the following example, Example 12.14 is repeated with the assumption that the reactants and products each consist of a mixture at 0.1-MPa pressure.

EXAMPLE 12.16

Consider the same combustion process as in Example 12.14, but assume that the reactants consist of a mixture at 0.1-MPa pressure and 25°C and that the products also consist of a mixture at 0.1 MPa and 25°C. Thus, the combustion process is as shown in Fig. 12.6.

Determine the work that would be done if this combustion process took place reversibly and in pressure and temperature equilibrium with the surroundings.

Control volume: Combustion chamber.

Inlet state: P, T known.

Exit state: P, T known.

Sketch: Figure 12.6.

Model: Reactants—ideal-gas mixture, Table A.13. Products—ideal-gas mixture, Table A.13.

Analysis:

The combustion equation, as noted previously, is

$$C_2H_4(g) + 3(4)O_2 + 3(4)(3.76)N_2 \rightarrow 2CO_2 + 2H_2O(g) + 9O_2 + 45.1N_2$$

In this case we must find the entropy of each substance as it exists in the mixture; that is, at its partial pressure and the given temperature of 25°C. Because the absolute

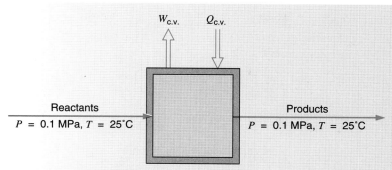

FIGURE 12.6 Sketch for Example 12.16.

entropies given in Tables A.13 and A.16 are at 0.1-MPa pressure and 25°C, the entropy of each constituent in the mixture can be found, using Eq. 12.30 from the relation

$$\bar{S}* = \bar{s}° - \bar{R}\ln y\,\frac{P}{P°}$$

where $\bar{S}*$ = partial entropy of the constituent in the mixture
$\quad\bar{s}°$ = absolute entropy at the same temperature and 0.1-MPa pressure
$\quad P$ = pressure of the mixture
$\quad P°$ = 0.1-MPa pressure
$\quad y$ = mole fraction of the constituent

Since $P°$ and the pressure of the mixture are both 0.1 MPa, the partial entropy of each constituent can be found by the relation

$$\bar{S}* = \bar{s}° - \bar{R}\ln y = \bar{s}° + \bar{R}\ln\frac{1}{y}$$

Solution

For the reactants:

	n	$1/y$	$\bar{R}\ln 1/y$	$\bar{s}°$	$\bar{S}*$
C_2H_4	1	58.1	33.774	219.330	253.104
O_2	12	4.842	13.114	205.147	218.261
N_2	45.1	1.288	2.104	191.610	193.714
	58.1				

For the products:

	n	$1/y$	$\bar{R}\ln 1/y$	$\bar{s}°$	$\bar{S}*$
CO_2	2	29.05	28.011	213.795	241.806
H_2O	2	29.05	28.011	188.834	216.845
O_2	9	6.456	15.506	205.147	220.653
N_2	45.1	1.288	2.104	191.610	193.714
	58.1				

With the assumption of ideal-gas behavior, the enthalpy of each constituent is equal to the enthalpy of formation at 25°C. The values of entropy are as just calculated. Therefore, from Eq. 12.22,

$$W^{\mathrm{rev}} = \sum_R n_i (\bar{h}_f^\circ)_i - \sum_P n_e (\bar{h}_f^\circ)_e - T_0 \left(\sum_R n_i \bar{s}_i - \sum_P n_e \bar{s}_e \right)$$

$$= (\bar{h}_f^\circ)_{\mathrm{C_2H_4}} - 2(\bar{h}_f^\circ)_{\mathrm{CO_2}} - 2(\bar{h}_f^\circ)_{\mathrm{H_2O}(g)}$$

$$- 298.15 \left(\bar{S}_{\mathrm{C_2H_4}}^* + 12\bar{S}_{\mathrm{O_2}}^* + 45.1\bar{S}_{\mathrm{N_2}}^* - 2\bar{S}_{\mathrm{CO_2}}^* - 2\bar{S}_{\mathrm{H_2O}(g)}^* - 9\bar{S}_{\mathrm{O_2}}^* - 45.1\bar{S}_{\mathrm{N_2}}^* \right)$$

$$= 52\,467 - 2(-393\,522) - 2(-241\,826)$$

$$- 298.15[253.104 + 12(218.264) + 45.1(193.714)$$

$$- 2(241.806) - 2(216.845) - 9(220.653) - 45.1(193.714)]$$

$$= 1\,332\,378 \text{ kJ / kmol}$$

$$= \frac{1\,332\,378}{28.054} = 47\,493 \text{ kJ / kg}$$

Note that this value is essentially the same as the value that was obtained in Example 12.14, when the reactants and products were each separated and at 0.1-MPa pressure.

These examples raise the question of the possibility of a reversible chemical reaction. Some reactions can be made to approach reversibility by having them take place in an electrolytic cell, as described in Chapter 1. When a potential exactly equal to the electromotive force of the cell is applied, no reaction takes place. When the applied potential is increased slightly, the reaction proceeds in one direction, and if the applied potential is decreased slightly, the reaction proceeds in the opposite direction. The work done is the electrical energy supplied or delivered.

Consider a reversible reaction occuring at constant temperature equal to that of its environment. The work output of the fuel cell is

$$W = -\left(\sum n_e \bar{g}_e - \sum n_i \bar{g}_i \right) = -\Delta G$$

where ΔG is the change in Gibbs function for the overall chemical reaction. We also realize that the work is given in terms of the charged electrons flowing through an electrical potential $\mathscr{E}$ as

$$W = \mathscr{E} n_e N_0 e$$

in which n_e is the number of kilomoles of electrons flowing through the external circuit, and

$$N_0 e = 6.022\,136 \times 10^{26} \text{ elec / kmol} \times 1.602\,177 \times 10^{-22} \text{ kJ / elec V}$$

$$= 96\,485 \text{ kJ / kmol V}$$

Thus, for a given reaction, the maximum (reversible reaction) electrical potential $\mathscr{E}^\circ$ of a fuel cell at a given temperature is

$$\mathscr{E}^\circ = \frac{-\Delta G}{96\ 485 n_e} \tag{12.25}$$

EXAMPLE 12.17 Calculate the reversible electromotive force (EMF) at 25°C for the hydrogen–oxygen fuel cell described in Section 1.2.

Solution

The anode side reaction was stated to be

$$2H_2 \rightarrow 4H^+ + 4e^-$$

and the cathode side reaction is

$$4H^+ + 4e^- + O_2 \rightarrow 2H_2O$$

Therefore, the overall reaction is, in kilomoles,

$$2H_2 + O_2 \rightarrow 2H_2O$$

for which 4 kmol of electrons flow through the external circuit. Let us assume that each component is at its standard-state pressure of 0.1 MPa and that the water formed is liquid. Then

$$\begin{aligned}
\Delta G_{25°C} &= 2(\overline{g}_f^\circ)_{H_2O} - 2(\overline{g}_f^\circ)_{H_2} - (\overline{g}_f^\circ)_{O_2} \\
&= 2(-237\ 146) - 2(0) - 1(0) \\
&= -474\ 292\ \text{kJ}
\end{aligned}$$

Therefore, from Eq. 12.25,

$$\mathscr{E}^\circ = \frac{-(-474\ 292)}{96\ 485 \times 4} = 1.229\ \text{V}$$

Much effort is being directed toward the development of fuel cells in which carbon, hydrogen, or hydrocarbons will react with oxygen and produce electricity directly. If a fuel cell can be developed that utilizes a hydrocarbon fuel and has a sufficiently high efficiency and capacity (for a given volume or weight), our techniques for generating electricity on a commercial scale will undergo drastic changes. At the present time, however, fuel cells cannot compete with conventional power plants for the large-scale production of electricity.

12.9 EVALUATION OF ACTUAL COMBUSTION PROCESSES

A number of different parameters can be defined for evaluating the performance of an actual combustion process, depending on the nature of the process and the system considered. In the combustion chamber of a gas turbine, for example, the objective is to raise the temperature of the products to a given temperature (usually the maximum temperature the metals in the turbine can withstand). If we had a combustion process that achieved complete combustion and that was adiabatic, the temperature of the products would be the adiabatic flame temperature. Let us designate the fuel–air ratio needed to reach a given temperature under these conditions as the ideal fuel–air ratio. In the actual combustion chamber the combustion will be incomplete to some extent, and there will be some heat transfer to the surroundings. Therefore, more fuel will be required to reach the given temperature, and this we designate as the actual fuel–air ratio. The combustion efficiency, η_{comb}, is defined here as

$$\eta_{comb} = \frac{FA_{ideal}}{FA_{actual}} \tag{12.26}$$

On the other hand, in the furnace of a steam generator (boiler), the purpose is to transfer the maximum possible amount of heat to the steam (water). In practice, the efficiency of a steam generator is defined as the ratio of the heat transferred to the steam to the higher heating value of the fuel. For a coal this is the heating value as measured in a bomb calorimeter, which is the constant-volume heating value, and it corresponds to the internal energy of combustion. We observe a minor inconsistency, since the boiler involves a flow process, and the change in enthalpy is the significant factor. In most cases, however, the error thus introduced is less than the experimental error involved in measuring the heating value, and the efficiency of a steam generator is defined by the relation

$$\eta_{\text{steam generator}} = \frac{\text{heat transferred to steam} / \text{kg fuel}}{\text{higher heating value of the fuel}} \tag{12.27}$$

In an internal-combustion engine the purpose is to do work. The logical way to evaluate the performance of an internal-combustion engine would be to compare the actual work done to the maximum work that would be done by a reversible change of state from the reactants to the products. This, as we noted previously, is called the second-law efficiency.

However, in practice the efficiency of an internal-combustion engine is defined as the ratio of the actual work to the negative of the enthalpy of combustion of the fuel (that is, the constant-pressure heating value). This ratio is usually called the thermal efficiency, η_{th}:

$$\eta_{th} = \frac{w}{-h_{RP_0}} = \frac{w}{\text{heating value}} \tag{12.28}$$

The overall efficiency of a gas turbine or steam power plant is defined in the same way. It should be pointed out that in an internal-combustion engine or fuel-burning steam power plant, the fact that the combustion is itself irreversible is a significant factor in the relatively low thermal efficiency of these devices.

One other factor should be pointed out regarding efficiency. We have noted that the enthalpy of combustion of a hydrocarbon fuel varies considerably with the phase of the water in the products, which leads to the concept of higher and lower heating values. Therefore, when we consider the thermal efficiency of an engine, the heating value used to determine this efficiency must be borne in mind. Two engines made by different manufacturers may have identical performance, but if one manufacturer bases his or her efficiency on the higher heating value and the other on the lower heating value, the latter will be able to claim a higher thermal efficiency. This claim is not significant, of course, as the performance is the same, and a consideration of how the efficiency was defined would reveal this.

The whole matter of the efficiencies of devices that undergo combustion processes is treated in detail in textbooks dealing with particular applications, and our discussion is intended only as an introduction to the subject. Two examples are given, however, to illustrate these remarks.

EXAMPLE 12.18

The combustion chamber of a gas turbine uses a liquid hydrocarbon fuel that has an approximate composition of C_8H_{18}. During testing the following data are obtained.

$$T_{air} = 400 \text{ K} \qquad T_{products} = 1100 \text{K}$$

$$\mathbf{V}_{air} = 100 \text{ m/s} \qquad \mathbf{V}_{products} = 150 \text{ m/s}$$

$$T_{fuel} = 50°C \qquad FA_{actual} = 0.0211 \text{ kg fuel/kg air}$$

Calculate the combustion efficiency for this process.

Control volume: Combustion chamber.

Inlet states: T known for air and fuel.

Exit state: T known.

Model: Air and products—ideal gas, Table A.13. Fuel—Table A.16.

Analysis:

For the ideal chemical reaction the heat transfer is zero. Therefore, writing the first law for a control volume that includes the combustion chamber, we have

$$H_R + KE_R = H_P + KE_P$$

$$H_R + KE_R = \sum_R n_i \left(\bar{h}_f^° + \Delta\bar{h} + \frac{M\mathbf{V}^2}{2} \right)_i$$

$$= \left[\bar{h}_f^° + \bar{C}_p(50-25) \right]_{C_8H_{18}(l)} + n_{O_2}\left(\Delta\bar{h} + \frac{M\mathbf{V}^2}{2} \right)_{O_2}$$

$$+ 3.76 n_{O_2}\left(\Delta\bar{h} + \frac{M\mathbf{V}^2}{2} \right)_{N_2}$$

$$H_P + KE_P = \sum_P n_e \left(\bar{h}_f^° + \Delta\bar{h} + \frac{M\mathbf{V}^2}{2} \right)_e$$

$$= 8\left(\bar{h}_f^° + \Delta\bar{h} + \frac{M\mathbf{V}^2}{2} \right)_{CO_2} + 9\left(\bar{h}_f^° + \Delta\bar{h} + \frac{M\mathbf{V}^2}{2} \right)_{H_2O}$$

$$+ (n_{O_2} - 12.5)\left(\Delta\bar{h} + \frac{M\mathbf{V}^2}{2} \right)_{O_2} + 3.76 n_{O_2}\left(\Delta\bar{h} + \frac{M\mathbf{V}^2}{2} \right)_{N_2}$$

Solution

$$H_R + KE_R = -250\,105 + 1.7113 \times 114.23(50-25)$$

$$+ n_{O_2}\left[3034 + \frac{32 \times (100)^2}{2 \times 1000}\right]$$

$$+ 3.76 n_{O_2}\left[2971 + \frac{28.02 \times (100)^2}{2 \times 1000}\right]$$

$$= -245\,218 + 14\,892 n_{O_2}$$

$$H_P + KE_P = 8\left[-393\,522 + 38\,891 + \frac{44.01 \times (150)^2}{2 \times 1000}\right]$$

$$+ 9\left[-241\,826 + 30\,147 + \frac{18.02 \times (150)^2}{2 \times 1000}\right]$$

$$+ (n_{O_2} - 12.5)\left[26\,218 + \frac{32 \times (150)^2}{2 \times 1000}\right]$$

$$+ 3.76 n_{O_2}\left[24\,758 + \frac{28.02 \times (150)^2}{2 \times 1000}\right]$$

$$= -5\,068\,599 + 120\,853 n_{O_2}$$

Therefore,

$$-245\,218 + 14\,892 n_{O_2} = -5\,068\,599 + 120\,853 n_{O_2}$$

$$n_{O_2} = 45.52 \text{ kmol O}_2 \text{ / kmol fuel}$$

$$\text{kmol air / kmol fuel} = 4.76(45.52) = 216.675$$

$$FA_{\text{ideal}} = \frac{114.23}{216.675 \times 28.97} = 0.0182 \text{ kg fuel / kg air}$$

$$\eta_{\text{comb}} = \frac{0.0182}{0.0211} \times 100 = 86.2 \text{ percent}$$

EXAMPLE 12.19 In a certain steam power plant 325 000 kg of water per hour enter the boiler at a pressure of 12.5 MPa and a temperature of 200°C. Steam leaves the boiler at 9 MPa, 500°C. The power output of the turbine is 81 000 kW. Coal is used at the rate of 26 700 kg/h and has a higher heating value of 33 250 kJ/kg. Determine the efficiency of the steam generator and the overall thermal efficiency of the plant.

In power plants the efficiency of both the boiler and the overall efficiency of the plant are based on the higher heating value of the fuel.

Solution

The efficiency of the boiler is defined by Eq. 12.27 as

$$\eta_{\text{steam generator}} = \frac{\text{heat transferred to } H_2O / \text{kg fuel}}{\text{higher heating value}}$$

Therefore

$$\eta_{\text{steam generator}} = \frac{325\,000(3386.1 - 857.1)}{26\,700 \times 33\,250} \times 100 = 92.6\%$$

The thermal efficiency is defined by Eq. 12.28,

$$\eta_{\text{th}} = \frac{w}{\text{heating value}} = \frac{81\,000 \times 3600}{26\,700 \times 33\,250} \times 100 = 32.8\%$$

PROBLEMS

12.1 A certain Pennsylvania coal contains 74.2% C, 5.1% H, 6.7% O, (dry basis, mass percent) plus ash and small percentages of N and S. This coal is fed into a gasifier along with oxygen and steam, as shown in Fig. P12.1. The exiting product gas composition is measured and is found to include, on a mole basis, 39.9% CO, 30.8% H_2, 11.4% CO_2, 16.4% H_2O plus small percentages of CH_4, N_2, and H_2S. How many kilograms of coal are required to produce 100 kmol of product gas? How much oxygen and steam are required?

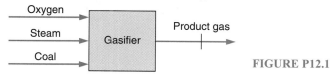

FIGURE P12.1

12.2 Repeat Problem 12.1 for a certain Utah coal that contains, according to the coal analysis, 68.2% C, 4.8% H, 15.7% O on a mass basis. The exiting product gas contains 30.9% CO, 26.7% H_2, 15.9% CO_2 and 25.7% H_2O on a mole basis.

12.3 In a combustion process with decane, $C_{10}H_{22}$, and air, the dry product mole fractions are 86.9% nitrogen, 1.163% oxygen, 10.975% carbon dioxide and 0.954% carbon monoxide. Find the equivalence ratio and the percent theoretical air of the reactants.

12.4 Liquid propane is burned with dry air. A volumetric analysis of the products of combustion yields the following percent composition on a dry basis:

Product	CO_2	CO	O_2	N_2
Percent by volume	8.6	0.6	7.2	83.6

Determine the percent of theoretical air used in this combustion process.

12.5 A fuel, C_xH_y, is burned with dry air and the product composition is measured on a dry basis to be: 9.6% CO_2, 7.3% O_2 and 83.1% N_2. Find the fuel composition (x/y) and the percent theoretical air used.

12.6 A sample of pine bark has the following ultimate analysis on a dry basis, percent by mass:

Component	H	C	S	N	O	ash
Percent by mass	5.6	53.4	0.1	0.1	37.9	2.9

This bark will be used as a fuel by burning it with 100% theoretical air in a furnace. Determine the air–fuel ratio on a mass basis.

12.7 Pentane is burned with 120% theoretical air in a constant pressure process at 100 kPa. The products are cooled to ambient temperature, 20°C.

   a. How many kilograms of water are condensed per kilogram of fuel?

   b. Repeat part (a), assuming that the air used in the combustion has a relative humidity of 90%.

12.8 The coal gasifier in an integrated gasification combined cycle (IGCC) power plant produces a gas mixture with the following volumetric percent composition:

Product	CH_2	H_2	CO	CO_2	N_2	H_2O	H_2S	NH_3
Percent by volume	0.3	29.6	41.0	10.0	0.8	17.0	1.1	0.2

This gas is cooled to 40°C, 3 MPa, and the H_2S and NH_3 are removed in water scrubbers. Assuming that the resulting mixture, which is sent to the combustors, is saturated with water, determine the mixture composition and the theoretical air–fuel ratio in the combustors.

12.9 The hot exhaust gas from an internal combustion engine is analyzed and found to have the following percent composition on a volumetric basis at the engine exhaust manifold.

Product	CO_2	CO	H_2O	O_2	N_2
Percent by volume	10	2	13	3	72

This gas is fed to an exhaust gas reactor and mixed with a certain amount of air to eliminate the carbon monoxide, as shown in Fig. P12.9. It has been determined that a mole fraction of 10% oxygen in the mixture at state 3 will ensure that no CO remains. What must the ratio of flows be entering the reactor?

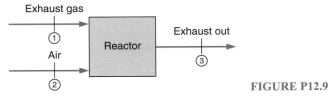

FIGURE P12.9

12.10 Many coals from the western United States have a high moisture content. Consider the following sample of Wyoming coal, for which the ultimate analysis on an as-received basis is, by mass:

Component	Moisture	H	C	S	N	O	ash
Percent by mass	28.9	3.5	48.6	0.5	0.7	12.0	5.8

This coal is burned in the steam generator of a large power plant with 150% theoretical air. Determine the air–fuel ratio on a mass basis.

12.11 Methanol, CH_3OH, is burned with 200% theoretical air in an engine and the products are brought to 100 kPa, 30°C. How much water is condensed per kilogram of fuel?

12.12 The output gas mixture of a certain air–blown coal gasifier has the composition of producer gas as listed in Table 12.2. Consider the combustion of this gas with 120% theoretical air at 100 kPa pressure.

a. Determine the dew point of the products.

b. How many kilograms of water will be condensed per kilogram of fuel if the products are cooled 10°C below the dew-point temperature?

12.13 Pentene, C_5H_{10} is burned with pure oxygen in an SSSF process. The products at one point are brought to 700 K and used in a heat exchanger, where they are cooled to 25°C. Find the specific heat transfer in the heat exchanger.

12.14 A gas mixture of 50% ethane and 50% propane by volume enters a combustion chamber at 350 K, 10 MPa. Determine the enthalpy per kilomole of this mixture relative to the thermochemical base of enthalpy.

12.15 Liquid pentane is burned with dry air and the products are measured on a dry basis as: 10.1% CO_2, 0.2% CO, 5.9% O_2 remainder N_2. Find the enthalpy of formation for the fuel and the actual equivalence ratio.

12.16 A rigid vessel initially contains 2 kmol of carbon and 2 kmol of oxygen at 25°C, 200 kPa. Combustion occurs, and the resulting products consist of 1 kmol of carbon dioxide, 1 kmol of carbon monoxide, and excess oxygen at a temperature of 1000 K. Determine the final pressure in the vessel and the heat transfer from the vessel during the process.

12.17 Propylbenzene, C_9H_{12}, is listed in Table 12.3, but not in Table A.16. No molecular weight is listed in the book. Find the molecular weight, the enthalpy of formation for the liquid fuel and the enthalpy of evaporation.

12.18 In a test of rocket propellant performance, liquid hydrazine (N_2H_4) at 100 kPa, 25°C, and oxygen gas at 100 kPa, 25°C, are fed to a combustion chamber in the ratio of 0.5 kg O_2/kg N_2H_4. The heat transfer from the chamber to the surroundings is estimated to be 100 kJ/kg N_2H_4. Determine the temperature of the products exiting the chamber. Assume that only H_2O, H_2, and N_2 are present. The enthalpy of formation of liquid hydrazine is +50417 kJ/kmol.

12.19 Repeat the previous problem, but assume that saturated-liquid oxygen at 90 K is used instead of 25°C oxygen gas in the combustion process. Use the generalized tables or charts to determine the properties of liquid oxygen.

12.20 Ethene, C_2H_4, and propane, C_3H_8, in a 1:1 mole ratio as gases are burned with 120% theoretical air in a gas turbine. Fuel is added at 25°C, 1 MPa and the air comes from the atmosphere, 25°C, 100 kPa through a compressor to 1 MPa and mixed with the fuel. The turbine work is such that the exit temperature is 800 K with an exit pressure of 100 kPa. Find the mixture temperature before combustion, and also the work, assuming an adiabatic turbine.

12.21 One alternative to using petroleum or natural gas as fuels is ethanol (C_2H_5OH), which is commonly produced from grain by fermentation. Consider a combustion process in which liquid ethanol is burned with 120% theoretical air in an SSSF process. The reactants enter the combustion chamber at 25°C, and the products exit at 60°C, 100 kPa. Calculate the heat transfer per kilomole of ethanol, using the enthalpy of formation of ethanol gas plus the generalized tables or charts.

12.22 Another alternative to using petroleum or natural gas as fuels is methanol (CH_3OH), which can be produced from coal. Both methanol and ethanol have been used in automotive engines. Repeat the previous problem using liquid methanol as the fuel instead of ethanol.

12.23 Another alternative fuel to be seriously considered is hydrogen. It can be produced from water by various techniques that are under extensive study. Its biggest problem at the present time is cost, storage, and safety. Repeat Problem 12.21 using hydrogen gas as the fuel instead of ethanol.

12.24 Hydrogen peroxide, H_2O_2, enters a gas generator at 25°C, 500 kPa at the rate of 0.1 kg/s and is decomposed to steam and oxygen exiting at 800 K, 500 kPa. The resulting mixture is expanded through a turbine to atmospheric pressure, 100 kPa, as shown in Fig. P12.24. Determine the power output of the turbine, and the heat transfer rate in the gas generator. The enthalpy of formation of liquid H_2O_2 is −187583 kJ/kmol.

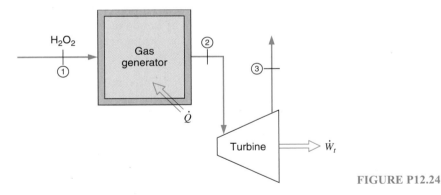

FIGURE P12.24

12.25 In a new high-efficiency furnace, natural gas, assumed to be 90% methane and 10% ethane (by volume) and 110% theoretical air each enter at 25°C, 100 kPa, and the products (assumed to be 100% gaseous) exit the furnace at 40°C, 100 kPa. What is the heat transfer for this process? Compare this to an older furnace where the products exit at 250°C, 100 kPa.

12.26 Repeat the previous problem, but take into account the actual phase behavior of the products exiting the furnace.

12.27 Methane, CH_4, is burned in an SSSF process with two different oxidizers:

a. Pure oxygen, O_2.

b. A mixture of $O_2 + x$ Ar.

   The reactants are supplied at T_0, P_0 and the products should for both cases be at 1800 K. Find the required equivalence ratio in case (a) and the amount of Argon, x, for a stoichiometric ratio in case (b).

12.28 Butane gas at 25°C is mixed with 150% theoretical air at 600 K and is burned in an adiabatic SSSF combustor. What is the temperature of the products exiting the combustor?

12.29 Liquid butane at 25°C is mixed with 150% theoretical air at 600 K and is burned in an adiabatic SSSF combustor. Use the generalized tables or charts for the liquid fuel and find the temperature of the products exiting the combustor.

12.30 A stoichiometric mixture of benzene, C_6H_6, and air is mixed from the reactants flowing at 25°C, 100 kPa. Find the adiabatic flame temperature. What is the error if constant specific heat at T_0 for the products from Table A.10 are used?

12.31 Liquid n-butane at T_0, is sprayed into a gas turbine with primary air flowing at 1.0 MPa, 400 K in a stoichiometric ratio. After complete combustion, the products are at the adiabatic flame temperature, which is too high so secondary air at 1.0 MPa, 400 K is added, with the resulting mixture being at 1400 K. Show that $T_{ad} >$ 1400 K and find the ratio of secondary to primary air flow.

12.32 Consider the gas mixture fed to the combustors in the integrated gasification combined cycle power plant, as described in Problem 12.8. If the adiabatic flame temperature should be limited to 1500 K, what percent theoretical air should be used in the combustors?

12.33 Acetylene gas at 25°C, 100 kPa is fed to the head of a cutting torch. Calculate the adiabatic flame temperature if the acetylene is burned with

a. 100% theoretical air at 25°C.

b. 100% theoretical oxygen at 25°C.

12.34 Ethene, C_2H_4, burns with 150% theoretical air in an SSSF constant-pressure process with reactants entering at P_0, T_0. Find the adiabatic flame temperature.

12.35 Gaseous propane at 25°C is burned with moist air at 400 K in a steady-state, steady-flow process. The combustion process is adiabatic, and the exiting temperature is measured to 1200 K. A sample of the products is tested and found to have a dew-point temperature of 70°C. Determine the percentage of theoretical air used and the relative humidity of this air. Assume the combustion is complete and that the pressure is 100 kPa throughout the process.

12.36 Solid carbon is burned with stoichiometric air in an SSSF process. The reactants at T_0, P_0 are heated in a preheater to 500 K before the combustion chamber and the products deliver the required energy before flowing to a second heat exchanger, which they leave at T_0. Find the temperature of the products out of the first into the second heat exchanger, and the heat transfer per kmol of fuel in the second heat exchanger.

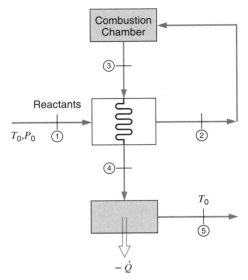

FIGURE P12.36

12.37 A study is to be made using liquid ammonia as the fuel in a gas-turbine engine. Consider the compression and combustion processes of this engine.

a. Air enters the compressor at 100 kPa, 25°C, and is compressed to 1600 kPa, where the isentropic compressor efficiency is 87%. Determine the exit temperature and the work input per kilomole.

b. Two kilomoles of liquid ammonia at 25°C and x times theoretical air from the compressor enter the combustion chamber. What is x if the adiabatic flame temperature is to be fixed at 1600 K?

12.38 A mixture of 80% ethane and 20% methane on a mole basis is throttled from 10 MPa, 65°C, to 100 kPa and is fed to a combustion chamber where it undergoes complete combustion with air, which enters at 100 kPa, 600 K. The amount of air is such that the products of combustion exit at 100 kPa, 1200 K. Assuming that the combustion process is adiabatic and that all components behave as ideal gases except the fuel mixture, which behaves according to the generalized tables or charts, with pseudocritical constants, determine

a. The percentage of theoretical air used in the process.

b. The dew-point temperature of the products.

12.39 A closed, insulated container is charged with a stoichiometric ratio of oxygen and hydrogen at 25°C and 150 kPa. After combustion, liquid water at 25°C is sprayed in such that the final temperature is 1200 K. What is the final pressure?

12.40 Wet biomass waste from a food-processing plant is fed to a catalytic reactor, where in an SSSF process it is converted into a low-energy fuel gas suitable for firing the processing plant boilers. The fuel gas has a composition of 50% methane, 45% carbon dioxide, and 5% hydrogen on a volumetric basis. Determine the lower heating value of this fuel gas mixture per unit volume.

12.41 Gaseous propane mixes with air, both supplied at 500 K, 0.1 MPa. The mixture goes into a combustion chamber and products of combustion exit at 1300 K, 0.1 MPa. The products analyzed on a dry basis are 11.42% CO_2, 0.79% CO, 2.68%

O_2, and 85.11% N_2 on a volume basis. Find the equivalence ratio and the heat transfer per kmol of fuel.

12.42 Determine the lower heating value of the gas generated from coal as described in Problem 12.8. Do not include the components removed by the water scrubbers.

12.43 Determine the higher heating value of the sample Wyoming coal as specified in Problem 12.10.

12.44 Consider natural gas A and natural gas D, both of which are listed in Table 12.2. Calculate the enthalpy of combustion of each gas at 25°C, assuming that the products include

a. Vapor water.

b. Liquid water.

12.45 A stoichiometric mixture of CO and O_2 at T_0 and total pressure P_0 burns to CO_2 in a constant volume bomb. How much heat transfer per kmol CO is necessary for the product to be at 2500 K? With no heat transfer, what would the final pressure and temperature $(T > 5600$ K) be?

12.46 Blast furnace gas in a steel mill is available at 250°C to be burned for the generation of steam. The composition of this gas is, on a volumetric basis,

Component	CH_4	H_2	CO	CO_2	N_2	H_2O
Percent by volume	0.1	2.4	23.3	14.4	56.4	3.4

Determine the higher heating value (kJ/m^3) of this gas at ambient temperature and pressure.

12.47 In an engine a mixture of liquid octane and ethanol, mole ratio 9:1, and stoichiometric air are taken in at T_0, P_0. In the engine the enthalpy of combustion is used so that 30% goes out as work, 30% goes out as heat loss and the rest goes out the exhaust. Find the work and heat transfer per kilogram of fuel mixture and also the exhaust temperature.

12.48 Consider the same situation as in the previous problem. Find the dew point temperature of the products. If the products in the exhaust are cooled to 10°C, find the mass of water condensed per kilogram of fuel mixture.

12.49 The enthalpy of formation of magnesium oxide, MgO(s), is −601 827 kJ/kmol at 25°C. The melting point of magnesium oxide is approximately 3000 K, and the increase in enthalpy between 298 and 3000 K is 128 449 kJ/kmol. The enthalpy of sublimation at 3000 K is estimated at 418 000 kJ/kmol, and the specific heat of magnesium oxide vapor above 3000 K is estimated at 37.24 kJ/kmol K.

a. Determine the enthalpy of combustion per kilogram of magnesium.

b. Estimate the adiabatic flame temperature when magnesium is burned with theoretical oxygen.

12.50 A rigid container is charged with butene, C_4H_8, and air in a stoichiometric ratio at P_0, T_0. The charge burns in a short time with no heat transfer to state 2. The products then cool with time to 1200 K, state 3. Find the final pressure, P_3, the total heat transfer, $_1Q_3$, and the temperature immediately after combustion, T_2.

12.51 Pyrolysis, or partial oxidation, of a low-grade petroleum distillate yields a syngas of carbon monoxide and hydrogen in the molar ratio $1:2$. The gas is fed directly to a steady-flow combustor at 800 K along with 120% theoretical air preheated to 600 K. Products exit the combustor at the adiabatic flame temperature. Calculate the irreversibility for the process.

12.52 Calculate the irreversibility for the process described in Problem 12.16.

12.53 A mixture of propane and 150% theoretical air enters a combustion chamber at 25°C, 200 kPa, and the products of combustion exit the chamber at 1400 K, 200 kPa. Assuming complete combustion, determine the heat transfer from the combustion chamber and the net entropy change for the process, both per kilomole of propane.

12.54 A closed rigid container is charged with propene, C_3H_6, and 150% theoretical air at 100 kPa, 298 K. The mixture is ignited and burns with complete combustion. Heat is transferred to a reservoir at 500 K so the final temperature of the products is 700 K. Find the final pressure, the heat transfer per kmole fuel and the total entropy generated per kmol fuel in the process.

12.55 A flame torch uses gaseous acetylene, C_2H_2, and the oxidizer is a mixture of oxygen and nitrogen in mole ratio $1:2$. Reactants are supplied at T_0, P_0 and combustion is at P_0. If the products exit at T_0, what is the ratio of the mass flows $(\dot{m}_{fu}/\dot{m}_{ox})$ that will give the highest flame temperature, and what is the change in availability?

12.56 Consider the combustion of methanol, CH_3OH, with 25% excess air. The combustion products are passed through a heat exchanger and exit at 200 kPa, 40°C. Calculate the absolute entropy of the products exiting the heat exchanger per kilomole of methanol burned, using appropriate reference states as needed.

12.57 The turbine in Problem 12.20 is adiabatic. Is it reversible, irreversible, or impossible?

12.58 An inventor claims to have built a device that will take 0.001 kg/s of water from the faucet at 10°C, 100 kPa, and produce separate streams of hydrogen and oxygen gas, each at 400 K, 175 kPa. It is stated that this device operates in a 25°C room on 10-kW electrical power input. How do you evaluate this claim?

12.59 Two kilomoles of ammonia are burned in an SSSF process with x kmol of oxygen. The products, consisting of H_2O, N_2, and the excess O_2, exit at 200°C, 7 MPa.

   a. Calculate x if half the water in the products is condensed.

   b. Calculate the absolute entropy of the products at the exit conditions.

12.60 Consider one cylinder of a spark-ignition, internal-combustion engine. Before the compression stroke, the cylinder is filled with a mixture of air and methane. Assume that 110% theoretical air has been used, that the state before compression is 100 kPa, 25°C. The compression ratio of the engine is 9 to 1.

   a. Determine the pressure and temperature after compression, assuming a reversible adiabatic process.

   b. Assume that complete combustion takes place while the piston is at top dead center (at minimum volume) in an adiabatic process. Determine the temperature and pressure after combustion, and the increase in entropy during the combustion process.

c. What is the irreversibility for this process?

12.61 Consider the combustion process described in Problem 12.38.

a. Calculate the absolute entropy of the fuel mixture before it is throttled into the combustion chamber.

b. Calculate the irreversibility for the overall process.

12.62 Liquid acetylene, C_2H_2, is stored in a high-pressure storage tank at ambient temperature, 25°C. The liquid is fed to an insulated combustor/steam boiler at the steady rate of 1 kg/s, along with 140% theoretical oxygen, O_2, which enters at 500 K, as shown in Fig. P12.62. The combustion products exit the unit at 500 kPa, 350 K. Liquid water enters the boiler at 10°C, at the rate of 15 kg/s, and superheated steam exits at 200 kPa.

a. Calculate the absolute entropy, per kmol, of liquid acetylene at the storage tank state.

b. Determine the phase(s) of the combustion products exiting the combustor/boiler unit, and the amount of each, if more than one.

c. Determine the temperature of the steam at the boiler exit.

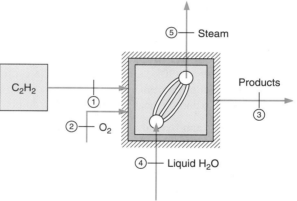

FIGURE P12.62

12.63 A compressor takes ambient air, P_0, T_0, up to 1 MPa where liquid benzene, C_6H_6, is added in a stoichiometric ratio. After combustion liquid water at 25°C, 1 MPa is added to give a temperature of 1500 K. This mixture enters a reversible adiabatic turbine that expands to P_0. What is the turbine exit temperature and work per kmol of fuel?

12.64 A piston cylinder, shown in Fig. P12.64, is divided into two rooms. Room A contains 1 kmol of saturated-liquid butane at 320 K, and room B contains 100% theoretical oxygen at 275 K. The loading of the piston maintains a pressure of 10 MPa on the oxygen. The partition separating the rooms is broken, and the two substances react to form carbon dioxide and water. During the reaction process, there is heat transfer of 2.5×10^6 kJ out of the cylinder to the surroundings.

a. What is the final temperature? Make an initial rough estimate of the temperature, and use this value to decide on detailed models, assumptions, and approximations to be used in the final solution.

b. Calculate the absolute entropy of the cylinder contents at the final state.

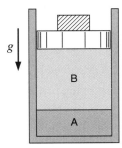

FIGURE P12.64

12.65 Graphite, C, at P_0, T_0 is burned with air coming in at P_0, 500 K in a ratio so the products exit at P_0, 1200 K. Find the equivalence ratio, the percent theoretical air, and the total irreversibility.

12.66 A small air-cooled gasoline engine is tested, and the output is found to be 1.0 kW. The temperature of the products is measured and found to be 600 K. The products are analyzed on a dry volumetric basis, with the following result:

Product	CO_2	CO	O_2	N_2
Percent by volume	11.4	2.9	1.6	84.1

The fuel may be considered to be liquid octane. The fuel and air enter the engine at 25°C, and the flow rate of fuel to the engine is 1.5×10^{-4} kg/s. Determine the rate of heat transfer from the engine and its thermal efficiency.

12.67 Consider the SSSF combustion of propane at 25°C with air at 400 K. The products exit the combustion chamber at 1200 K. It may be assumed that the combustion efficiency is 90%, and that 95% of the carbon in the propane burns to form carbon dioxide; the remaining 5% forms carbon monoxide. Determine the ideal fuel–air ratio and the heat transfer from the combustion chamber.

12.68 Natural gas (approximate it as methane) at a ratio of 0.3 kg/s is burned with 250% theoretical air in a combustor at 1 MPa where the reactants are supplied at T_0. Steam at 1 MPa, 450°C at a rate of 2.5 kg/s is added to the products before they enter an adiabatic turbine with an exhaust pressure of 150 kPa. Determine the turbine inlet temperature and the turbine work assuming the turbine is reversible.

12.69 A gasoline engine uses liquid octane and air, both supplied at P_0, T_0, in a stoichiometric ratio. The products (complete combustion) flow out of the exhaust valve at 1100 K. Assume that the heat loss carried away by the cooling water, at 100°C, is equal to the work output. Find the efficiency of the engine expressed as (work/lower heating value) and the second law efficiency.

12.70 In Example 12.17, a basic hydrogen–oxygen fuel cell reaction was analyzed at 25°C, 100 kPa. Repeat this calculation, assuming that the fuel cell operates on air at 25°C, 100 kPa, instead of on pure oxygen at this state.

12.71 Consider a methane-oxygen fuel cell in which the reaction at the anode is

$$CH_4 + 2H_2O \rightarrow CO_2 + 8e^- + 8H^+$$

The electrons produced by the reaction flow through the external load, and the positive ions migrate through the electrolyte to the cathode, at which the reaction is

$$8e^- + 8H^+ + 2O_2 \rightarrow 4H_2O$$

a. Calculate the reversible work and the reversible EMS for the fuel cell operating at 25°C, 100 kPa.

b. Repeat part (a), but assume that the fuel cell operates at 600 K instead of at room temperature.

ENGLISH UNIT
PROBLEMS

12.72E Pentane is burned with 120% theoretical air in a constant pressure process at 14.7 lbf/in.^{2}. The products are cooled to ambient temperature, 70 F.

a. How many pounds-mass of water are condensed per pound-mass of fuel?

b. Repeat part (a), assuming that the air used in the combustion has a relative humidity of 90%.

12.73E The output gas mixture of a certain air-blown coal gasifier has the composition of producer gas as listed in Table 12.2. Consider the combustion of this gas with 120% theoretical air at 14.7 lbf/in.^{2} pressure.

a. Determine the dew point of the products.

b. How many pounds-mass of water will be condensed per pound-mass of fuel if the products are cooled 20 F below the dew point temperature?

12.74E Pentene, C_5H_{10} is burned with pure oxygen in an SSSF process. The products at one point are brought to 1300 R and used in a heat exchanger, where they are cooled to 77 F. Find the specific heat transfer in the heat exchanger.

12.75E A gas mixture of 50% ethane and 50% propane by volume enters a combustion chamber at 650 R, 1500 lbf/in.^{2}. Determine the enthalpy per pound mole of this mixture relative to the thermochemical base of enthalpy.

12.76E A rigid vessel initially contains 2 pound mole of carbon and 2 pound mole of oxygen at 77 F, 30 lbf/in.^{2}. Combustion occurs, and the resulting products consist of 1 pound mole of carbon dioxide, 1 pound mole of carbon monoxide, and excess oxygen at a temperature of 1800 R. Determine the final pressure in the vessel and the heat transfer from the vessel during the process.

12.77E In a test of rocket propellant performance, liquid hydrazine (N_2H_4) at 14.7 lbf/in.^{2}, 77 F, and oxygen gas at 14.7 lbf/in.^{2}, 77 F, are fed to a combustion chamber in the ratio of 0.5 lbm O_2/lbm N_2H_4. The heat transfer from the chamber to the surroundings is estimated to be 45 Btu/lbm N_2H_4. Determine the temperature of the products exiting the chamber. Assume that only H_2O, H_2, and N_2 are present. The enthalpy of formation of liquid hydrazine is +21647 Btu/lb mole.

12.78E Repeat the previous problem, but assume that saturated-liquid oxygen at 170 R is used instead of 77 F oxygen gas in the combustion process. Use the generalized tables or charts to determine the properties of liquid oxygen.

12.79E Ethene, C_2H_4, and propane, C_3H_8, in a 1 : 1 mole ratio as gases are burned with 120% theoretical air in a gas turbine. Fuel is added at 77 F, 150 lbf/in.^{2} and the air comes from the atmosphere, 77 F, 15 lbf/in.^{2} through a compressor to 150 lbf/in.^{2} and mixed with the fuel. The turbine work is such that the exit temperature is 1500 R with an exit pressure of 14.7 lbf/in.^{2}. Find the mixture temperature before combustion, and also the work, assuming an adiabatic turbine.

12.80E One alternative to using petroleum or natural gas as fuels is ethanol (C_2H_5OH), which is commonly produced from grain by fermentation. Consider a combustion process in which liquid ethanol is burned with 120% theoretical air in an SSSF process. The reactants enter the combustion chamber at 77 F, and the products exit at 140 F, 14.7 lbf/in.^{2}. Calculate the heat transfer per pound mole of ethanol, using the enthalpy of formation of ethanol gas plus the generalized tables or charts.

12.81E Hydrogen peroxide, H_2O_2, enters a gas generator at 77 F, 75 lbf/in.^{2} at the rate of 0.2 lbm/s and is decomposed to steam and oxygen exiting at 1500 R, 75 lbf/in.^{2}. The resulting mixture is expanded through a turbine to atmospheric pressure, 14.7 lbf/in.^{2}, as shown in Fig. P12.24. Determine the power output of the turbine, and the heat transfer rate in the gas generator. The enthalpy of formation of liquid H_2O_2 is −80541 Btu/lb mol.

12.82E In a new high-efficiency furnace, natural gas, assumed to be 90% methane and 10% ethane (by volume) and 110% theoretical air each enter at 77 F, 14.7 lbf/in.^{2}, and the products (assumed to be 100% gaseous) exit the furnace at 100 F, 14.7 lbf/in.^{2}. What is the heat transfer for this process? Compare this to an older furnace where the products exit at 450 F, 14.7 lbf/in.^{2}.

12.83E Repeat the previous problem, but take into account the actual phase behavior of the products exiting the furnace.

12.84E Methane, CH_4, is burned in an SSSF process with two different oxidizers:

a. Pure oxygen, O_2.

b. A mixture of $O_2 + x$ Ar.

The reactants are supplied at T_0, P_0 and the products should for both cases be at 3200 R. Find the required equivalence ratio in case (a) and the amount of Argon, x, for a stoichiometric ratio in case (b).

12.85E Butane gas at 77 F is mixed with 150% theoretical air at 1000 R and is burned in an adiabatic SSSF combustor. What is the temperature of the products exiting the combustor?

12.86E Liquid n-butane at T_0, is sprayed into a gas turbine with primary air flowing at 150 lbf/in.^{2}, 700 R in a stoichiometric ratio. After complete combustion, the products are at the adiabatic flame temperature, which is too high so secondary air at 150 lbf/in.^{2}, 700 R is added, with the resulting mixture being at 2500 R. Show that $T_{ad} > 2500$ R and find the ratio of secondary to primary air flow.

12.87E Acetylene gas at 77 F, 14.7 lbf/in.^{2} is fed to the head of a cutting torch. Calculate the adiabatic flame temperature if the acetylene is burned with

a. 100% theoretical air at 77 F.

b. 100% theoretical oxygen at 77 F.

12.88E Ethene, C_2H_4, burns with 150% theoretical air in an SSSF constant-pressure process with reactants entering at P_0, T_0. Find the adiabatic flame temperature.

12.89E Solid carbon is burned with stoichiometric air in an SSSF process, as shown in Fig. P12.36. The reactants at T_0, P_0 are heated in a preheater to 900 R before the combustion chamber and the products deliver the required energy before flowing to a second heat exchanger, which they leave at T_0. Find the temperature of the products out of the first into the second heat exchanger, and the heat transfer per lb mol of fuel in the second heat exchanger.

12.90E A closed, insulated container is charged with a stoichiometric ratio of oxygen and hydrogen at 77 F and 20 lbf/in.^{2}. After combustion, liquid water at 77 F is sprayed in such that the final temperature is 2100 R. What is the final pressure?

12.91E A stoichiometric mixture of CO and O_2 at T_0 and total pressure P_0 burns to CO_2 in a constant volume bomb. How much heat transfer per lb mol CO is necessary for the product to be at 4500 R? With no heat transfer, what would the final pressure and temperature ($T > 10\,000$ R) be?

12.92E Blast furnace gas in a steel mill is available at 500 F to be burned for the generation of steam. The composition of this gas is, on a volumetric basis,

Component	CH_4	H_2	CO	CO_2	N_2	H_2O
Percent by volume	0.1	2.4	23.3	14.4	56.4	3.4

Determine the higher heating value (Btu/ft^3) of this gas at ambient temperature and pressure.

12.93E A mixture of propane and 150% theoretical air enters a combustion chamber at 77 F, 30 lbf/in.^{2}, and the products of combustion exit the chamber at 2060 F, 30 lbf/in.^{2}. Assuming complete combustion, determine the heat transfer from the combustion chamber and the net entropy change for the process, both per pound mole of propane.

12.94E A closed rigid container is charged with propene, C_3H_6, and 150% theoretical air at 14.7 lbf/in.^{2}, 77 F. The mixture is ignited and burns with complete combustion. Heat is transferred to a reservoir at 900 R so the final temperature of the products is 1300 R. Find the final pressure, the heat transfer per pound mole fuel, and the total entropy generated per pound mol of fuel in the process.

12.95E A flame torch uses gaseous acetylene, C_2H_2, and the oxidizer is a mixture of oxygen and nitrogen in mole ratio 1:2. Reactants are supplied at T_0, P_0 and combustion is at P_0. If the products exit at T_0 what is the ratio of the mass flows ($\dot{m}_{fu}/\dot{m}_{ox}$) that will give the highest flame temperature, and what is the change in availability?

12.96E Two pound moles of ammonia are burned in an SSSF process with x lb mol of oxygen. The products, consisting of H_2O, N_2, and the excess O_2, exit at 400 F, 1000 lbf/in.^{2}.

a. Calculate x if half the water in the products is condensed.

b. Calculate the absolute entropy of the products at the exit conditions.

12.97E Graphite, C, at P_0, T_0 is burned with air coming in at P_0, 900 R in a ratio so the products exit at P_0, 2200 R. Find the equivalence ratio, the percent theoretical air and the total irreversibility.

12.98E A small air-cooled gasoline engine is tested, and the output is found to be 2.0 hp. The temperature of the products is measured and found to be 730 F. The products are analyzed on a dry volumetric basis, with the following result:

Product	CO_2	CO	O_2	N_2
Percent by volume	11.4	2.9	1.6	84.1

The fuel may be considered to be liquid octane. The fuel and air enter the engine at 77 F, and the flow rate of fuel to the engine is 1.8 lbm/h. Determine the rate of heat transfer from the engine and its thermal efficiency.

12.99E Natural gas (approximate it as methane) at a rate of 0.65 lbm/s is burned with 250% theoretical air in a combustor at 150 lbf/in.^{2} where the reactants are supplied at T_0. Steam at 150 lbf/in.^{2}, 850 F at a rate of 5.5 lbm/s is added to the products before they enter an adiabatic turbine with an exhaust pressure of 20 lbf/in.^{2}. Determine the turbine inlet temperature and the turbine work assuming the turbine is reversible.

12.100E A gasoline engine uses liquid octane and air, both supplied at P_0, T_0, in a stoichiometric ratio. The products (complete combustion) flow out of the exhaust valve at 2000 R. Assume that the heat loss carried away by the cooling water, at 200 F, is equal to the work output. Find the efficiency of the engine expressed as (work/lower heating value) and the second law efficiency.

12.101 In Example 12.17, a basic hydrogen–oxygen fuel cell reaction was analyzed at 25°C, 100 kPa. Repeat this calculation, assuming that the fuel cell operates on air at 77 F, 14.7 lbf/in.^{2}, instead of on pure oxygen at this state.

COMPUTER, DESIGN, AND OPEN-ENDED PROBLEMS

12.102 Write a program to solve the general case of Problem 12.4 for any hydrocarbon fuel C_xH_y, where x and y are input parameters. We wish to calculate the percentage of theoretical air for any given percentages of combustion products.

12.103 Write a program to solve the general case of Problem 12.8 for different proportions of the components listed and for different temperatures and pressures of the product gas, including water saturation.

12.104 Write a program to solve the general case of Problem 12.10 for different percentages of the components given in the ultimate analysis of the coal.

12.105 Write a subroutine that can calculate the enthalpy and entropy of the products from the combustion of hydrocarbon fuels with air. Assume that all species are ideal gases and use the supplied software for the individual component properties. Let the stoichiometric coefficients, temperature and pressure be the input variables.

12.106 Write a program to solve the general case of Problem 12.16, in which the initial amounts of C and O_2, the final fractions of CO and CO_2, and the final temperature are all input parameters.

12.107 Write a program to solve the general case of the set of Problems 12.21, 12.22, and 12.23, in which the percentage of theoretical air and the product temperature and pressure are all input variables.

12.108 Write a program that will find the adiabatic flame temperature for a fuel, $C_xH_yO_z$, being burned with air from stoichiometric ratio and up so it is never fuel rich (excess fuel). Assume adiabatic SSSF, the reactants coming in at reference conditions and complete combustion. Let x, y, z, the fuel enthalpy of formation, and the percent theoretical air be program input variables. Use the software supplied to give the product properties (assuming ideal gases).

12.109 Write a program to study the following problem. We wish to study the effect of the percentage of theoretical air on the adiabatic flame temperature for a (variable) hydrocarbon fuel. Assume that the fuel and air enter the combustion chamber at 25°C, and assume complete combustion. Use equations from Table A.11 to represent the specific heat of the various products of combustion.

12.110 Write a program that will find the adiabatic flame temperature for the combustion of a fuel and air in a combustion bomb (constant volume). Use the same input and conditions as described in Problem 12.108.

12.111 Write a program to solve the general case of Problem 12.37, in which the compressor pressure ratio, the isentropic efficiency, and the adiabatic flame temperature are all input parameters.

12.112 Write a program to solve the general case of Problem 12.53, in which the percentage of theoretical air and the inlet and exit temperatures and pressures are all input variables.

12.113 Write a program to study the turbine work in Problem 12.68 as a function of the amount of steam added and include an isentropic efficiency for the turbine (use 85%). Use the software to provide the properties assuming they are ideal gases, which means an iteration for the turbine inlet/exit temperatures is necessary. Determine also the second law efficiency of the turbine.

12.114 Write a program to study the performance of the methane–oxygen fuel cell of Problem 12.71, as a function of temperature, for any value above ambient temperature.

12.115 Power plants may use off-peak power to compress air into a large storage facility, see Problem 7.90. The compressed air is then used as the air supply to a gas-turbine system where it is burned with some fuel, usually natural gas. The system is then used to produce power at peak load times. Investigate such a setup and estimate the power generated with conditions given in Problem 7.90 and combustion with 200–300% theoretical air and exhaust to the atmosphere.

12.116 A car that runs on natural gas has it stored in a heavy tank with a maximum pressure of 3600 psi (25 MPa). Size the tank for a range of 300 miles (500 km) assuming a car engine that has a 30% efficiency requiring about 25 hp (20 kW) to drive the car at 55 mi/h (90 km/h).

12.117 The Cheng cycle, shown in Fig. P11.128, is powered by the combustion of natural gas (essentially methane) being burned with 250–300% theoretical air. In the case with a single water-condensing heat exchanger, where $T_6 = 40°C$ and $\Phi_6 = 100\%$, is any make-up water needed at state 8 or is there a surplus? Does the humidity in the compressed atmospheric air at state 1 make any difference? Study the problem over a range of air–fuel ratios.

12.118 The cogenerating powerplant shown in Problem 9.34 burns 170 kg/s air with natural gas, CH_4. The setup is shown in Fig. P12.118 where a fraction of the air flow out of the compressor with compression ratio 15.8:1 is used to preheat the feedwater in the steam cycle. The fuel flow rate is 3.2 kg/s. Make an analysis of the system determining the total heat transfer to the steam cycle from the turbine exhaust gases, the heat transfer in the preheater, and the gas turbine inlet temperature.

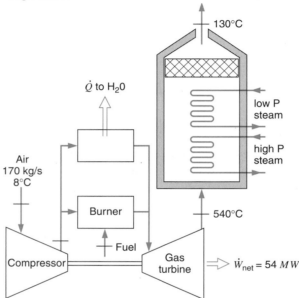

FIGURE P12.118

12.119 Consider the combustor in the Cheng cycle, see Problems 11.128 and 12.68. Atmospheric air is compressed to 1.25 MPa, state 1. It is burned with natural gas, CH_4, with the products leaving at state 2. The fuel should add a total of about 15 MW to the cycle, with an air flow of 12 kg/s. For a compressor with an intercooler estimate the temperatures T_1, T_2 and the fuel flow rate.

12.120 Study the coal gasification process that will produce methane, CH_4, or methanol CH_3OH. What is involved in such a process? Compare the heating values of the gas products with that of the original coal. Discuss the merits of this conversion.

12.121 Ethanol, C_2H_5OH, can be produced from corn or biomass. Investigate the process and the chemical reactions that occurs. For different raw materials estimate the amount of ethanol that can be obtained per mass of the raw material.

12.122 A Diesel engine is used as a stationary power plant in remote locations such as a ship, oil drilling rig, farm, etc. Assume diesel fuel is used with 300% theoretical air in a 1000-hp diesel engine. Estimate the amount of fuel used, the efficiency and the potential use of the exhaust gases for heating of rooms or water. Investigate if other fuels can be used.

12.123 When a power plant burns coal or some blends of oil, the combustion process can generate pollutants as SO_x and NO_x. Investigate the use of scrubbers to remove these. Explain the processes that take place and the effect on the power plant operation (energy, exhaust pressures, etc.)

12.124 For a number of fuels listed in Table 12.3 estimate their adiabatic flame temperature when they are burned with 200% theoretical air. Assume a power generating device like a gasoline or diesel engine or a gas-turbine with reasonable choices for their operating conditions. Find the power that can be generated as a fraction of the enthalpy of combustion. Does a ranking of the fuels follow the magnitude of the enthalpy of combustion?

13 INTRODUCTION TO PHASE AND CHEMICAL EQUILIBRIUM

Up to this point we have assumed that we are dealing either with systems that are in equilibrium or with those in which the deviation from equilibrium is infinitesimal, as in a quasiequilibrium or reversible process. For irreversible processes we made no attempt to describe the state of the system during the process but dealt only with the initial and final states of the system, in the case of a control mass, or the inlet and exit states as well in the case of a control volume. For any case, we either considered the system to be in equilibrium throughout, or at least made the assumption of local equilibrium.

In this chapter we examine the criteria for equilibrium and from them derive certain relations that will enable us, under certain conditions, to determine the properties of a system when it is in equilibrium. The specific cases we will consider are those involving equilibrium between phases and those involving chemical equilibrium in a single phase (homogeneous equilibrium) as well as certain related topics.

13.1 REQUIREMENTS FOR EQUILIBRIUM

As a general requirement for equilibrium we postulate that a system is in equilibrium when there is no possibility that it can do any work when it is isolated from its surroundings. In applying this criterion it is helpful to divide the system into two or more subsystems, and consider the possibility of doing work by any conceivable interaction between these two subsystems. For example, in Fig. 13.1 a system has been divided into two systems and an engine, of any conceivable variety, placed between these subsystems. A system may be so defined as to include the immediate surroundings. In this case we can let the immediate surroundings be a subsystem and thus consider the general case of the equilibrium between a system and its surroundings.

The first requirement for equilibrium is that the two subsystems have the same temperature, for otherwise we could operate a heat engine between the two systems and do work. Thus we conclude that one requirement for equilibrium is that a system must be at a uniform temperature to be in equilibrium. It is also evident that there must be no un-

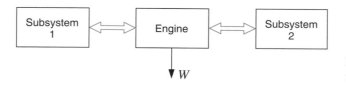

FIGURE 13.1   Two subsystems that communicate

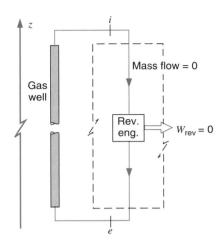

FIGURE 13.2 Illustration showing the relation between reversible work and the criteria for equilibrium.

balanced mechanical forces between the two systems, or else one could operate a turbine or piston engine between the two systems and do work.

However, we would like to establish general criteria for equilibrium that would apply to all simple compressible substances, including those that undergo chemical reactions. We will find that the Gibbs function is a particularly significant property in defining the criteria for equilibrium.

Let us first consider a qualitative example to illustrate this point. Consider a natural gas well that is 1 km deep, and let us assume that the temperature of the gas is constant throughout the gas well. Suppose we have analyzed the composition of the gas at the top of the well, and we would like to know the composition of the gas at the bottom of the well. Furthermore, let us assume that equilibrium conditions prevail in the well. If this is true we would expect that an engine such as that shown in Fig. 13.2 (which operates on the basis of the pressure and composition change with elevation and does not involve combustion) would not be capable of doing any work.

If we consider a steady-state, steady-flow process for a control volume around this engine, we could apply Eq. 8.40:

$$\dot{W}^{\text{rev}} = \dot{m}_i \left(g_i + \frac{\mathbf{V}_i^2}{2} + gZ_i \right) - \dot{m}_e \left(g_e + \frac{\mathbf{V}_e^2}{2} + gZ_e \right)$$

However,

$$\dot{W}^{\text{rev}} = 0, \qquad \dot{m}_i = \dot{m}_e \qquad and \qquad \frac{\mathbf{V}_i^2}{2} = \frac{\mathbf{V}_e^2}{2}$$

Then we can write

$$g_i + gZ_i = g_e + gZ_e$$

and the requirement for equilibrium in the well between two levels that are a distance dZ apart would be

$$dg_T + g \, dZ_T = 0$$

In contrast to a deep gas well, most of the systems that we consider are of such size that ΔZ is negligibly small, and therefore we consider the pressure to be uniform throughout.

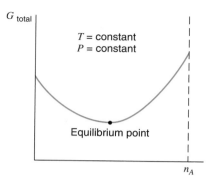

FIGURE 13.3 Illustration of the requirement for chemical equilibrium.

This leads to the general statement of equilibrium that applies to simple compressible substances that may undergo a change in chemical composition, namely, that at equilibrium

$$dG_{T,P} = 0 \tag{13.1}$$

In the case of a chemical reaction, it is helpful to think of the equilibrium state as the state in which the Gibbs function is a minimum. For example, consider a control mass consisting initially of n_A moles of substance A and n_B moles of substance B, which react in accordance with the relation

$$v_A A + v_B B \rightleftharpoons v_C C + v_D D$$

Let the reaction take place at constant pressure and temperature. If we plot G for this control mass as a function of n_A, the number of moles of A present, we would have a curve as shown in Fig. 13.3. At the minimum point on the curve, $dG_{T,P} = 0$, and this will be the equilibrium composition for this system at the given temperature and pressure. The subject of chemical equilibrium will be developed further in Section 13.6.

13.2 EQUILIBRIUM BETWEEN TWO PHASES OF A PURE SUBSTANCE

As another example of this requirement for equilibrium, let us consider the equilibrium between two phases of a pure substance. Consider a control mass consisting of two phases of a pure substance at equilibrium. We know that under these conditions the two phases are at the same pressure and temperature. Consider the change of state associated with a transfer of dn moles from phase 1 to phase 2 while the temperature and pressure remain constant. That is,

$$dn^1 = -dn^2$$

The Gibbs function of this control mass is given by

$$G = f(T.P, n^1, n^2)$$

where n^1 and n^2 designate the number of moles in each phase. Therefore,

$$dG = \left(\frac{\partial G}{\partial T}\right)_{P,n^1,n^2} dT + \left(\frac{\partial G}{\partial P}\right)_{T,n^1,n^2} dP + \left(\frac{\partial G}{\partial n^1}\right)_{T,P,n^2} dn^1 + \left(\frac{\partial G}{\partial n^2}\right)_{T,P,n^1} dn^2$$

By definition,

$$\left(\frac{\partial G}{\partial n^1}\right)_{T,P,n^2} = \bar{g}^1 \qquad \left(\frac{\partial G}{\partial n^2}\right)_{T,P,n^1} = \bar{g}^2$$

Therefore, at constant temperature and pressure,

$$dG = \bar{g}^1 \, dn^1 + \bar{g}^2 \, dn^2 = dn^1(\bar{g}^1 - \bar{g}^2)$$

Now at equilibrium (Eq. 13.1)

$$dG_{T,P} = 0$$

Therefore, at equilibrium, we have

$$\bar{g}^1 = \bar{g}^2 \tag{13.2}$$

That is, under equilibrium conditions, the Gibbs function of each phase of a pure substance is equal. Let us check this by determining the Gibbs function of saturated liquid (water) and saturated vapor (steam) at 300 kPa. From the steam tables:

For the liquid:

$$g_f = h_f - Ts_f = 561.47 - 406.7 \times 1.6718 = -118.4 \text{ kJ / kg}$$

For the vapor:

$$g_g = h_g - T_g = 2725.3 - 406.7 \times 6.9919 = -118.4 \text{ kJ / kg}$$

Equation 13.2 can also be derived by applying the relation

$$T \, ds = dh - v \, dP$$

to the change of phase that takes place at constant pressure and temperature. For this process this relation can be integrated as follows:

$$\int_f^g T \, ds = \int_f^g dh$$

$$T(s_g - s_f) = (h_g - h_f)$$

$$h_f - Ts_f = h_g - Ts_g$$

$$g_f = g_g$$

The Clapeyron equation, which was derived in Section 10.3, can be derived by an alternate method by considering the fact that the Gibbs functions of two phases in equilibrium are equal. In Chapter 10 we considered the relation (Eq. 10.9) for a simple compressible substance:

$$dg = v \, dP - s \, dT$$

Consider a control mass that consists of a saturated liquid and a saturated vapor in equilibrium, and let this system undergo a change of pressure dP. The corresponding change in temperature, as determined from the vapor-pressure curve, is dT. Both phases will undergo the change in Gibbs function, dg, but since the phases always have the same value of the Gibbs function when they are in equilibrium, it follows that

$$dg_f = dg_g$$

But, from Eq. 10.9,

$$dg = v\, dP - s\, dT$$

it follows that

$$dg_f = v_f dP - s_f dT$$

$$dg_g = v_g dP - s_g dT$$

Since

$$dg_f = dg_g$$

it follows that

$$v_f dP - s_f dT = v_g dP - s_g dT$$

$$dP(v_g - v_f) = dT(s_g - s_f) \tag{13.3}$$

$$\frac{dP}{dT} = \frac{s_{fg}}{v_{fg}} = \frac{h_{fg}}{Tv_{fg}}$$

In summary, when different phases of a pure substance are in equilibrium, each phase has the same value of the Gibbs function per unit mass. This fact is relevant to different solid phases of a pure substance and is important in metallurgical applications of thermodynamics. Example 13.1 illustrates this principle.

EXAMPLE 13.1 What pressure is required to make diamonds from graphite at a temperature of 25°C? The following data are given for a temperature of 25°C and a pressure of 0.1 MPa.

	Graphite	Diamond
g	0	2867.8 kJ/mol
v	0.000 444 m³/kg	0.000 284 m³/kg
β_T	0.304×10^{-6} 1/MPa	0.016×10^{-6} 1/MPa

Analysis and Solution

The basic principle in the solution is that graphite and diamond can exist in equilibrium when they have the same value of the Gibbs function. At 0.1 MPa pressure the Gibbs function of the diamond is greater than that of the graphite. However, the rate of increase in Gibbs function with pressure is greater for the graphite than for the diamond and, therefore, at some pressure they can exist in equilibrium, and our problem is to find this pressure.

We have already considered the relation

$$dg = v\, dP - s\, dT$$

Since we are considering a process that takes place at constant temperature, this reduces to

$$dg_T = v\, dP_T \tag{a}$$

Now at any pressure P and the given temperature, the specific volume can be found from the following relation, which utilizes the isothermal compressibility factor.

$$v = v° + \int_{P=0.1}^{P} \left(\frac{\partial v}{\partial P}\right)_T dP = v° + \int_{P=0.1}^{P} \frac{v}{v}\left(\frac{\partial v}{\partial P}\right)_T dP$$

$$= v° - \int_{P=0.1}^{P} v\beta_T dP \qquad \text{(b)}$$

The superscript ° will be used in this example to indicate the properties at a pressure of 0.1 MPa and a temperature of 25°C.

The specific volume changes only slightly with pressure, so that $v \approx v°$. Also, we assume that β_T is constant and that we are considering a very high pressure. With these assumptions this equation can be integrated to give

$$v = v° - v°\beta_T P = v°(1-\beta_T P) \qquad \text{(c)}$$

We can now substitute this into Eq. a to give the relation

$$dg_T = [v°(1-\beta_T P)]dP_T$$

$$g - g° = v°(P-P°) - v°\beta_T \frac{(P^2 - P°^2)}{2} \qquad \text{(d)}$$

If we assume that $P° \ll P$, this reduces to

$$g - g° = v°\left(P - \frac{\beta_T P^2}{2}\right) \qquad \text{(e)}$$

For the graphite, $g° = 0$ and we can write

$$g_G = v°_G\left[P - (\beta_T)_G \frac{P^2}{2}\right]$$

For the diamond, $g°$ has a definite value and we have

$$g_D = g°_D + v°_D\left[P - (\beta_T)_D \frac{P^2}{2}\right]$$

But, at equilibrium the Gibbs function of the graphite and diamond are equal

$$g_G = g_D$$

Therefore,

$$v°_G\left[P - (\beta_T)_G \frac{P^2}{2}\right] = g°_D + v°_D\left[P - (\beta_T)_D \frac{P^2}{2}\right]$$

$$(v°_G - v°_D)P - [v°_G(\beta_T)_G - v°_D(\beta_T)_D]\frac{P^2}{2} = g°_D$$

$$(0.000\,444 - 0.000\,284)P$$

$$- (0.000\,444 \times 0.304 \times 10^{-6} - 0.000\,284 \times 0.016 \times 10^{-6})P^2/2 = \frac{2867.8}{12.011 \times 1000}$$

Solving this for P we find

$$P = 1493 \text{ MPa}$$

That is, at 1493 MPa, 25°C, graphite and diamond can exist in equilibrium, and the possibility exists for conversion from graphite to diamonds.

The preceding example could also have been evaluated in terms of the fugacities instead of the Gibbs functions for the two forms of carbon. Similarly, the equilibrium requirement for liquid and vapor water discussed earlier could be expressed in terms of the fugacities of the two phases. That is,

$$f^L = f^V = f^{\text{sat}}$$

where f^{sat} is the fugacity of the substance determined at the given temperature and its saturation pressure P^{sat}.

The fugacity of a pure compressed liquid or solid phase at pressures moderately greater than the saturation pressure can be calculated with considerable accuracy as follows. For a pure substance at constant temperature,

$$dg_T = RT(d \ln f)_T = v \, dP_T$$

Integrating at constant temperature between the saturation state and the pressure P, we have

$$\int_{f^{\text{sat}}}^{f} RT(d \ln f)_T = \int_{P^{\text{sat}}}^{P} v \, dP_T$$

$$RT \ln \frac{f}{f^{\text{sat}}} = \int_{P^{\text{sat}}}^{P} v \, dP_T$$

If we assume that v is a constant, which often holds with considerable accuracy for liquids and solids, we have

$$RT \ln \frac{f}{f^{\text{sat}}} \approx v(P - P^{\text{sat}}) \tag{13.4}$$

Equation 13.4 is used to determine the fugacity coefficients of compressed liquid presented in the generalized chart given in Appendix Fig. A.7.

Since v is small for the liquid and solid phases, the quantity $v(P - P^{\text{sat}})$ is small for moderate changes of pressure. Therefore, for liquids and solids at moderate pressures, we conclude that

$$f^L \approx f^{\text{sat}} \qquad \text{and} \qquad f^S \approx f^{\text{sat}} \tag{13.5}$$

13.3 EQUILIBRIUM OF A MULTICOMPONENT, MULTIPHASE SYSTEM

As an introduction to a consideration of equilibrium in multicomponent, multiphase systems, let us consider a control mass consisting of two components and two phases. To show the general characteristics of such a system, consider a mixture of oxygen and nitrogen at a pressure of 0.1 MPa and in the range of temperature where both the liquid and vapor phases are present.

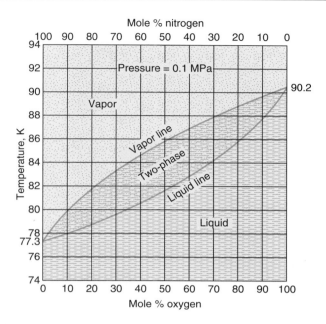

FIGURE 13.4 Equilibrium diagram for liquid–vapor phases of the nitrogen–oxygen system at a pressure of 0.1 MPa.

Figure 13.4 shows the composition of the liquid and vapor phases that are in equilibrium as a function of temperature at a pressure of 0.1 MPa. The upper line, marked "vapor line," gives the composition of the vapor phase, and the lower line gives the composition of the liquid phase. If we have pure nitrogen, the boiling point is 77.3 K. If we have pure oxygen, the boiling point is 90.2 K. If we have a mixture of nitrogen and oxygen with both phases present at a temperature of 84 K, the vapor will have a composition of 64% nitrogen and 36% oxygen. The liquid will have a composition of 30% nitrogen and 70% oxygen.

Consider a mixture consisting of 21% oxygen and 79% nitrogen (approximately the composition of air) and at a pressure of 0.1 MPa and an initial temperature of 74 K. At this state the mixture will be in the liquid phase. Let this liquid be slowly heated while the pressure remains constant. By referring to Fig. 13.4 we note that when the temperature reaches 78.8 K, the first bubble will form. This vapor will have a composition of approximately 6% oxygen and 94% nitrogen. As more heat is added, the temperature increases and the mole fraction of oxygen in the liquid increases. When the last drop of liquid remains, the vapor will have a composition of 21% oxygen and 79% nitrogen, the temperature will be 81.9 K, and the liquid composition will be 54% oxygen and 46% nitrogen.

To understand the phase behavior of a system such as that described here, it may be helpful to briefly discuss the more general phase diagram. We recall that the phase behavior for a pure substance is dictated by the saturation line on *P–T* coordinates, such as in Fig. 3.2. For a two-component system *A* and *B*, overall composition is now an additional independent property, and phase behavior is represented in terms of surfaces on a pressure–composition–temperature diagram, as shown in Fig. 13.5*a*. This diagram includes three constant-temperature planes to indicate that there is a lower vapor surface and an upper liquid surface that together form an equilibrium envelope between the two pure substance saturation lines. Only within this envelope is it possible to have two phases present in equilibrium. Beneath the envelope the two-component system is all superheated vapor and above the envelope the system is all compressed liquid.

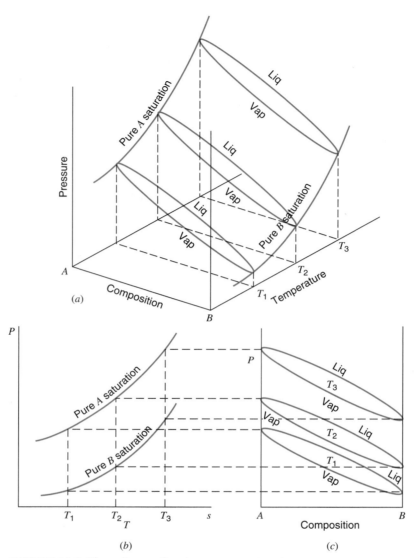

FIGURE 13.5 Phase diagram for a two-component system.

We realize now that a constant-pressure plane (0.1 MPa) of the three-dimensional phase diagram, as viewed from above, results in the temperature-composition diagram, Fig. 13.4, discussed in some detail previously. The other two projections of such a diagram are as shown in Figs. 13.5*b* and 13.5*c*.

Let us now consider the situation in a two-component system as shown in Fig. 13.5 when the temperature is increased to a value greater than the critical temperature of one of the components. The pressure–temperature and pressure–composition projections now appear as shown in Figs. 13.6*a* and 13.6*b* for the two temperatures T_4 and T_5. We note that as the temperature is increased toward the critical temperature of component B, there is an increasing range of composition for which the two-phase envelope is not intersected as the pressure is increased at a constant temperature.

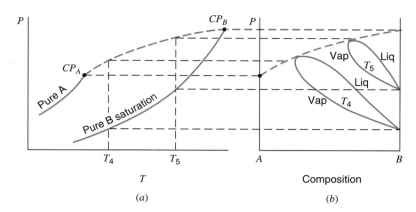

FIGURE 13.6 Two-component system in the critical region.

(a)

(b)

T

Composition

In studying the characteristics of a two-component system, we must first ask the question, "What is the requirement for phase equilibrium in this system?" We can proceed in a manner analogous to that followed in Section 13.2 for a pure substance, with superscripts 1 and 2 denoting the two phases, and in this case subscripts A and B referring to the two components. It follows from applying Eq. 11.57 to each phase that

$$dG^1 = -S^1 dT + V^1 dP + \mu_A^1 dn_A^1 + \mu_B^1 dn_B^1$$
$$dG^2 = -S^2 dT + V^2 dP + \mu_A^2 dn_A^2 + \mu_B^2 dn_B^2 \tag{13.6}$$

Let us consider this two-phase, two-component mixture that is in equilibrium as a system, and let each phase be considered as a subsystem. One possible change of state that might occur is for a very small amount of component A to be transferred from phase 1 to phase 2 while the moles of B in each phase and the temperature and pressure remain constant. This is shown schematically in Fig. 13.7.

For this change of state

$$dn_A^2 = -dn_A^1 \tag{13.7}$$

Since this system is in equilibrium, $dG = 0$. Therefore,

$$dG = dG^1 + dG^2 = 0 \tag{13.8}$$

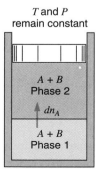

FIGURE 13.7 An equilibrium mixture involving two components and two phases.

Since T, P, n_B^1, and n_B^2 are all constant, it follows from Eqs. 13.6 and 13.8 that

$$dG = \mu_A^1 dn_A^1 + \mu_A^2 dn_A^2$$
$$= \mu_A^1 dn_A^1 + \mu_A^2 dn_A^1 = dn_A^1 (\mu_A^1 - \mu_A^2)$$
$$= 0$$

Therefore, at equilibrium

$$\mu_A^1 = \mu_A^2 \tag{13.9}$$

Thus the requirement for equilibrium is that the chemical potential of each component is the same in all phases. If the chemical potential is not the same in all phases, there will be a tendency for mass to pass from one phase to the other. When the chemical potential of a component is the same in both phases, there is no tendency for a net transfer of mass from one phase to the other.

This requirement for equilibrium is readily extended to multicomponent, multiphase systems, for the chemical potential of each component must be the same in all phases.

$$\mu_A^1 = \mu_A^2 = \mu_A^3 = \cdots \text{for all phases}$$

$$\mu_B^1 = \mu_B^2 = \mu_B^3 = \cdots \text{for all phases}$$

$$\begin{array}{ccc} - & - & - \\ & & \\ - & - & - \\ & & \\ - & - & - \\ & & \\ - & - & - \end{array} \tag{13.10}$$

for all components

Let us now consider alternate means of expressing the requirement for phase equilibrium at a given temperature and pressure. We found in Section 11.11, Eq. 11.60, that the chemical potential of component A is identical to the partial Gibbs function $\overline{G}_A$ of that component in the mixture. Thus, the requirement for equilibrium, Eq. 13.9, can also be expressed as

$$\overline{G}_A^1 = \overline{G}_A^2 \tag{13.11}$$

That is, at equilibrium the partial Gibbs function of a given component is the same in each phase. In this form the result is analogous to that for pure substance phase equilibrium, Eq. 13.2. It also follows from the definition of fugacity of a component in a mixture, Eqs. 11.81 and 11.82,

$$(d\overline{G}_A) = \overline{R}Td(\ln \bar{f}_A)_T$$

$$\lim_{P \to 0}(\bar{f}_A / y_A P) = 1$$

that the requirement for phase equilibrium at a given temperature and pressure, Eq. 13.9, can be expressed in terms of fugacity as

$$\bar{f}_A^1 = \bar{f}_A^2 \tag{13.12}$$

That is, at equilibrium the fugacity of a given component is the same in each phase. This is perhaps the most useful formulation of the requirement for equilibrium between phases at a given temperature and pressure, as we have previous experience in evaluating fugacities. A more general concept, however, is that of activity, which was defined and discussed in Section 11.15. Thus, the definition of activity, Eq. 11.113, could be substituted into Eq. 13.12 for the fugacity in each phase.

We find that the requirement for equilibrium can be expressed in numerous ways, in terms of different thermodynamic variables. The question of interest to us at this point is whether we can use such expressions to predict the equilibrium compositions of the different phases from the known properties of the pure substances that make up the system. To be able to do this, we need to utilize a model for each phase. In this section we will consider two models, namely, the Ideal Solution model and the Raoult's rule–Ideal Gas model. For each model, we will restrict our discussion to a two-phase (liquid and vapor), two-component system as shown in Fig. 13.7. The results of this analysis are then readily extended to multiphase or multicomponent systems. In the analysis, we will distinguish liquid phase mole fractions from those for the vapor phase by using the symbol x for the liquid phase.

Ideal Solution

A frequently used mixture model is the Ideal Solution, in which the fugacity of component A in a mixture is expressed as the product of the mole fraction of A and the fugacity of pure A in the same phase as the mixture and at the pressure and temperature of the mixture (and similarly for component B). These specified conditions for pure A may very well result in a hypothetical state. The Ideal Solution has been discussed in detail in Section 11.14, but for our purposes here we need only be familiar with the fugacity equation described in Eq. 11.104.

Let us assume that both the liquid and vapor phases of a two-component system behave according to the Ideal Solution model. The requirement for equilibrium for component A is, from Eqs. 13.12 and 11.104,

$$x_A f_A^L = y_A f_A^V \tag{13.13}$$

Similarly, for component B,

$$x_B f_B^L = y_B f_B^V \tag{13.14}$$

We also realize that for the liquid phase

$$x_A + x_B = 1 \tag{13.15}$$

and for the vapor phase

$$y_A + y_B = 1 \tag{13.16}$$

Now, at a given temperature and pressure, the four pure substance fugacities in Eqs. 13.13 and 13.14 are fixed values. Therefore, Eqs. 13.13 through 13.16 comprise a set of four equations in four unknowns, which can be solved for the equilibrium composition of the liquid and vapor phases.

The problem that arises in evaluating the Ideal Solution model is that connected with hypothetical states. This is most easily recognized by referring to the phase diagram, Fig. 13.5b. For an equilibrium two-phase mixture of A and B, the system pressure P is less than P_A^{sat} and greater than P_B^{sat} at the given temperature T. We realize from examining Fig. 13.5b that at the conditions of P and T, pure A is a superheated vapor and pure B is a compressed liquid. As a result, the fugacities f_A^V and f_B^L are readily found by conven-

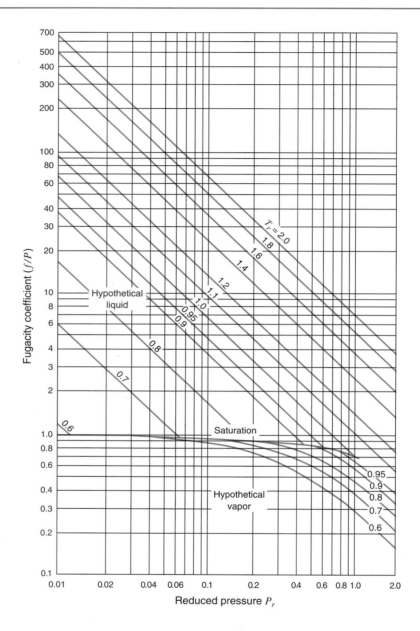

FIGURE 13.8
Hypothetical liquid
and vapor fugacities.

tional methods, as for example from Fig. A.10. However, the fugacities f_A^L and f_B^V cannot be found, as pure liquid A and pure vapor B do not exist at the conditions of P and T. These are the hypothetical states. To determine the fugacities for these hypothetical states, vapor-phase fugacities have been extrapolated into the liquid region and liquid-phase fugacities extrapolated into the vapor region in a manner so as to give reasonable generalized correlations from Eqs. 13.13 and 13.14 with experimental phase equilibrium data. These extrapolated hypothetical values are presented in Fig. 13.8.

The following example should help to clarify the determination of fugacities in the Ideal Solution model.

EXAMPLE 13.2   Calculate the phase compositions in the liquid-vapor system methane–carbon monoxide at −100°C, 4 MPa, assuming Ideal Solution in both phases.

Analysis and Solution

Let CO be component A and CH_4 be component B. Using critical point properties from Table A.8,

$$T_{rA} = \frac{173.2}{132.9} = 1.303 \qquad P_{rA} = \frac{4}{3.50} = 1.143$$

$$T_{rB} = \frac{173.2}{190.4} = 0.910 \qquad P_{rB} = \frac{4}{4.60} = 0.870$$

From Fig. A.10, it is seen that pure A is a gas and pure B is a liquid at the given state. Therefore, from Fig. A.10 or Table A.15,

$$f_A^V = 0.841 \times 4 = 3.36 \text{ MPa}$$

$$f_B^L = 0.518 \times 4 = 2.07 \text{ MPa}$$

Pure liquid A and pure gaseous B are the hypothetical states. From Fig. 13.8,

$$f_A^L = 1.8 \times 4 = 7.2 \text{ MPa}$$

$$f_B^V = 0.65 \times 4 = 2.6 \text{ MPa}$$

Substituting these values into the equilibrium equations, Eqs. 13.13 and 13.14, and solving,

$$x_A = \frac{3.36}{7.2} y_A = 0.4667 y_A$$

$$x_B = \frac{2.6}{2.07} y_B = 1.256 y_B$$

$$x_A + x_B = 0.4667 y_A + 1.256(1 - y_A) = 1$$

$$y_A = 0.324, \qquad x_A = 0.151$$

When the temperature is greater than the critical temperature of one of the components, the situation shown in Fig. 13.6, the hypothetical liquid-phase fugacity can become very large, as can be seen from the values for T_r greater than unity in Fig. 13.8. The result of a very large value for f_A^L is to drive the liquid-phase mole fraction x_A to a very small value. In an extreme case, the liquid phase becomes essentially pure B, and Eqs. 13.13 and 13.15 drop out of the equilibrium set, leaving Eqs. 13.14 and 13.16 to specify the two gas-phase mole fractions at a given temperature and pressure.

Raoult's Rule—Ideal Gas

This model is a special case of the Ideal Solution, and while not as accurate, it is nevertheless quite reasonable for the behavior of many systems at low pressure. For the liquid phase, we make two simplifying assumptions:

1. The fugacity of pure liquid A at T and P of the system is equal to the fugacity of saturated A (liquid or vapor) at the same T, and its corresponding saturation pressure P_A^{sat}, or

$$f_A^L = f_A^{sat} \qquad (13.17)$$

This is equivalent to assuming that the $\int v\, dP$ correction of Eq. 13.4 is negligibly small.

2. Pure saturated vapor A at T and $P^{sat}{}_A$ behaves as an ideal gas, or

$$f_A^{sat} = P_A^{sat} \qquad (13.18)$$

Combining Eqs. 13.17 and 13.18, we obtain, for the two components A and B,

$$f_A^L = P_A^{sat}, \qquad f_B^L = P_B^{sat} \qquad (13.19)$$

which, when combined with the Ideal Solution model, Eq. 11.104, is termed Raoult's rule.

For the vapor phase, we assume that pure gas A and pure gas B behave as ideal gases at T and P, so that

$$f_A^V = P, \qquad f_B^V = P \qquad (13.20)$$

Now, substituting Eqs. 13.19 and 13.20 for each component into the result for the Ideal Solution model, Eqs. 13.13 and 13.14, we obtain the result for the Raoult's rule–Ideal Gas model:

$$x_A P_A^{sat} = y_A P \qquad (13.21)$$

$$x_B P_B^{sat} = y_B P \qquad (13.22)$$

which together with

$$x_A + x_B = 1 \qquad (13.23)$$

$$y_A + y_B = 1 \qquad (13.24)$$

forms a set of four equations in four unknowns at a given T and P (P_A^{sat} and P_B^{sat} depend only on T). This set of equations can then be solved to determine the equilibrium compositions in each phase.

The three additional assumptions made in this model should clearly indicate its limitations as compared with the Ideal Solution, which is the more accurate, especially at higher pressures.

EXAMPLE 13.3   Air (assumed to be 21% O_2, 79% N_2) is cooled to 80 K, 0.1 MPa pressure. Calculate the composition of the liquid and vapor phases at this condition, assuming the Raoult's rule–Ideal Gas model, and compare the results with Fig. 13.4. At 80 K,

$$P_{N_2}^{sat} = 0.1370 \text{ MPa} \qquad P_{O_2}^{sat} = 0.030\,06 \text{ MPa}$$

Analysis and Solution

For convenience let $N_2 = A$ and $O_2 = B$ in the calculations. Using Eqs. 13.21 and 13.22,

$$x_A P_A^{sat} = y_A P$$

$$x_B P_B^{sat} = y_B P$$

Therefore, at 80 K, 0.1 MPa pressure,

$$x_A(0.137) = y_A(0.1)$$

$$x_B(0.030\,06) = y_B(0.1)$$

But,

$$x_A + x_B = 1$$

$$y_A + y_B = 1$$

Substituting for y_A, y_B in the last equation,

$$\frac{0.137}{0.1}x_A + \frac{0.030\,06}{0.1}x_B = 1.37x_A + 0.3006(1 - x_A) = 1$$

$$x_A = 0.654, \qquad y_A = 0.896$$

From Fig. 13.4, which is based on experimental data, we find that at 80 K, 0.1 MPa,

$$x_A = 0.66, \qquad y_A = 0.89$$

Thus we conclude that for this system at this temperature and pressure, Raoult's rule gives quite accurate results.

13.4 THE GIBBS PHASE RULE (WITHOUT CHEMICAL REACTION)

The Gibbs phase rule, which was derived by Professor J. Willard Gibbs of Yale University in 1875, ranks among the truly significant contributions to physical science. In this section we consider the Gibbs phase rule for a control mass that does not involve a chemical reaction. For such a system the Gibbs phase rule is

$$\mathscr{P} + \mathscr{V} = \mathscr{C} + 2 \tag{13.25}$$

where $\mathscr{P}$ is the number of phases present, $\mathscr{V}$ is the variance, and $\mathscr{C}$ the number of components present. The term variance designates the number of intensive properties that must be specified to completely fix the state of the control mass. For example, in Example 13.3 we considered a two-phase mixture of oxygen and nitrogen. For this system the number of phases $\mathscr{P}$ equals 2, the number of components $\mathscr{C}$ is 2, and therefore the variance $\mathscr{V}$ is 2. This is evident from Fig. 13.4, for this diagram is for a fixed pressure (0.1 MPa), and the fixing of one additional intensive property, such as temperature, mole fraction of a given component in the liquid phase, or mole fraction of a given component in the vapor phase, will completely determine all other intensive properties and thus fix the state of the control mass (although not the relative amounts of the two phases).

Let us consider further the application of the Gibbs phase rule to a pure substance. In this case $\mathscr{C} = 1$. When we have one phase present, such as superheated vapor, $\mathscr{P} = 1$, and we conclude that $\mathscr{V} = \mathscr{C} + 2 - 1 = 2$. That is, two intensive properties must be specified to fix the state. We are already familiar with the superheated vapor tables for a number of substances and recognize that these tables are presented with pressure and temperature as the two independent properties. Such a system is referred to as a bivariant system.

Suppose we have two phases of a pure substance in equilibrium, such as saturated liquid and saturated vapor. In this case $\mathscr{C} = 1$, $\mathscr{P} = 2$, and $\mathscr{V} = 1 + 2 - 2 = 1$. That is, one intensive property fixes the value of all other intensive properties and thus determines the state of each phase. Again, from our familiarity with tables of thermodynamic properties, we recall that either pressure or temperature is used as the independent intensive property for tabulating liquid–vapor equilibrium data for a pure substance. Such a system is known as monovariant.

Consider also the triple point of a pure substance. In this case $\mathscr{C} = 1$ and $\mathscr{P} = 3$. Therefore $\mathscr{V} = 1 + 2 - 3 = 0$. That is, at the triple point all intensive properties are fixed, and if any intensive property is varied, we are no longer at the triple point. This is known as an invariant system.

It might be well at this point to again emphasize that a pure substance can have several phases in the solid state. For example, Fig. 3.6 shows the various phases of water. Note that there are several states at which three phases exist in equilibrium, and each of these is a triple point.

The application of the phase rule to a two-component, two-phase system has already been made. As a final application let us consider a two-component, three-phase control mass. For such a system $\mathscr{V} = 2 + 2 - 3 = 1$. That is, it is a monovariant system, and specifying one property, such as pressure or temperature, fixes the state.

The validity of the Gibbs phase rule can be outlined by considering a control mass consisting of $\mathscr{C}$ components and $\mathscr{P}$ phases in equilibrium at a given temperature and pressure. Assuming that each component exists in each phase, the state of the system could be completely specified if the concentration of each component in each phase and the temperature and pressure were specified. This would be a total of $\mathscr{C}\mathscr{P} + 2$ intensive properties.

We know, however, that these are not all independent intensive properties. If we determine the number of equations we have between these intensive properties, we can subtract this from the $\mathscr{C}\mathscr{P} + 2$ intensive properties and find the number of independent intensive properties or, as it has been defined, the variance. The fact that at equilibrium the chemical potential of each component is the same in all phases gives rise to a total of $\mathscr{C}(\mathscr{P} - 1)$ equations between the intensive properties of the mixture.

Furthermore, the fact that the sum of the mole fractions equals unity in each of the $\mathscr{P}$ phases gives $\mathscr{P}$ additional equations. Therefore the variance is

$$\mathscr{V} = \mathscr{C}\mathscr{P} + 2 - \mathscr{P} - \mathscr{C}(\mathscr{P} - 1) = \mathscr{C} + 2 - \mathscr{P}$$

13.5 METASTABLE EQUILIBRIUM

Although the limited scope of this book precludes an extensive treatment of metastable equilibrium, a brief introduction to the subject is presented in this section. Let us first consider an example of metastable equilibrium.

Consider a slightly superheated vapor, such as steam, expanding in a convergent-

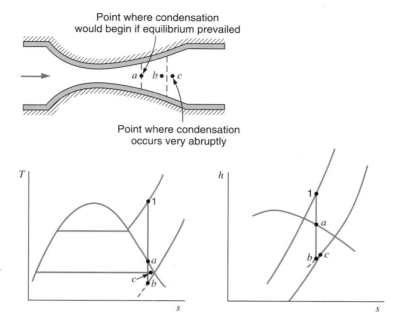

FIGURE 13.9
Illustration of the phenomenon of supersaturation in a nozzle.

divergent nozzle, as shown in Fig. 13.9. Assuming the process is reversible and adiabatic, the steam will follow path 1-*a* on the *T*–*s* diagram, and at point *a* we would expect condensation to occur. However, if point *a* is reached in the divergent section of the nozzle, it is observed that no condensation occurs until point *b* is reached, and at this point the condensation occurs very abruptly in what is referred to as a condensation shock. Between points *a* and *b* the steam exists as a vapor, but the temperature is below the saturation temperature for the given pressure. This is known as a metastable state. The possibility of a metastable state exists with any phase transformation. The dotted lines on the equilibrium diagram shown in Fig. 13.10 represent possible metastable states for solid–liquid–vapor equilibrium.

The nature of a metastable state is often pictured schematically by the kind of diagram shown in Fig. 13.11. The ball is in a stable position (the "metastable state") for small displacements, but with a large displacement it moves to a new equilibrium position. The steam expanding in the nozzle is in a metastable state between *a* and *b*. This means that droplets smaller than a certain critical size will reevaporate, and only when droplets of larger than this critical size have formed (this corresponds to moving the ball out of the depression) will the new equilibrium state appear.

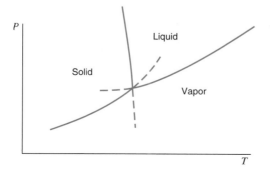

FIGURE 13.10 Metastable states for solid–liquid–vapor equilibrium.

FIGURE 13.11 Schematic diagram illustrating a metastable state.

13.6 CHEMICAL EQUILIBRIUM

We now turn our attention to chemical equilibrium and consider first a chemical reaction involving only one phase. This is referred to as a homogeneous chemical reaction. It may be helpful to visualize this as a gaseous phase, but the basic considerations apply to any phase.

Consider a vessel, Fig. 13.12, that contains four compounds, A, B, C, and D, which are in chemical equilibrium at a given pressure and temperature. For example, these might consist of CO_2, H_2, CO, and H_2O in equilibrium. Let the number of moles of each component be designated n_A, n_B, n_C, and n_D. Further, let the chemical reaction that takes place between these four constituents be

$$v_A A + v_B B \rightleftharpoons v_C C + v_D D \qquad (13.26)$$

where the v's are the stoichiometric coefficients. It should be emphasized that there is a very definite relation between the v's (the stoichiometric coefficients), whereas the n's (the number of moles present) for any constituent can be varied simply by varying the amount of that component in the reaction vessel.

Let us now consider how the requirement for equilibrium, namely, that $dG_{T,P} = 0$ at equilibrium, applies to a homogeneous chemical reaction. When we considered phase equilibrium (Section 13.3) we proceeded by assuming that the two phases were in equilibrium at a given temperature and pressure, and then let a small quantity of one component be transferred from one phase to the other. In a similar manner, let us assume that the four components are in chemical equilibrium and then assume that from this equilibrium state, while the temperature and pressure remain constant, the reaction proceeds an infinitesimal amount toward the right as Eq. 13.26 is written. This results in a decrease in the moles of A and B and an increase in the moles of C and D. Let us designate the degree of reaction by ε, and define the degree of reaction by the relations

$$dn_A = -v_A d\varepsilon$$
$$dn_B = -v_B d\varepsilon$$
$$dn_C = +v_C d\varepsilon$$
$$dn_D = +v_D d\varepsilon \qquad (13.27)$$

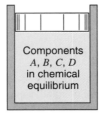

Components
A, B, C, D
in chemical
equilibrium

FIGURE 13.12 Schematic diagram for consideration of chemical equilibrium.

That is, the change in the number of moles of any component during a chemical reaction is given by the product of the stoichiometric coefficients (the ν's) and the degree of reaction.

Let us evaluate the change in the Gibbs function associated with this chemical reaction that proceeds to the right in the amount $d\varepsilon$. In doing so we use, as would be expected, the Gibbs function of each component in the mixture—the partial molal Gibbs function (or its equivalent, the chemical potential):

$$dG_{T,P} = \overline{G}_C \, dn_C + \overline{G}_D \, dn_D + \overline{G}_A \, dn_A + \overline{G}_B \, dn_B$$

Substituting Eq. 13.27, we have

$$dG_{T,P} = (\nu_C \overline{G}_C + \nu_D \overline{G}_D - \nu_A \overline{G}_A - \nu_B \overline{G}_B) \, d\varepsilon \tag{13.28}$$

We have already considered (Section 11.15) how to evaluate the partial molal Gibbs function of a given component in terms of the Gibbs function of the pure component at the standard state and the activity of that component. For the component i, we can write (Eq. 11.114)

$$\overline{G}_i = \overline{g}_i^\circ + \overline{R}T \ln a_i$$

Substituting this relation into Eq. 13.28 we have

$$dG_{T,P} = [\nu_C(\overline{g}_C^\circ + \overline{R}T \ln a_C) + \nu_D(\overline{g}_D^\circ + \overline{R}T \ln a_D)$$
$$- \nu_A(\overline{g}_A^\circ + \overline{R}T \ln a_A) - \nu_B(\overline{g}_B^\circ + \overline{R}T \ln a_B)] \, d\varepsilon \tag{13.29}$$

Let us define ΔG° as follows:

$$\Delta G^\circ = \nu_C \overline{g}_C^\circ + \nu_D \overline{g}_D^\circ - \nu_A \overline{g}_A^\circ - \nu_B \overline{g}_B^\circ \tag{13.30}$$

That is, ΔG° is the change in the Gibbs function that would occur if the chemical reaction given by Eq. 13.26 (which involves the stoichiometric amounts of each component) proceeded completely from left to right, with the reactants A and B initially separated and at temperature T and the standard state pressure and the products C and D finally separated and at temperature T and the standard state pressure. Note also that ΔG° for a given reaction is a function of only the temperature. This will be most important to bear in mind as we proceed with our developments of homogeneous chemical equilibrium. Let us now digress from our development to consider an example involving the calculation of ΔG°.

EXAMPLE 13.4 Determine the value of ΔG° for the reaction $2H_2O \rightleftharpoons 2H_2 + O_2$ at 298 K and at 2000 K, with the water in the gaseous phase.

Solution

We take 0.1 MPa as our standard-state pressure, and recall that $\overline{g}$ for all the elements is assumed to be zero at 0.1-MPa pressure and 298 K. Therefore, at 298 K,

$$(\overline{g}_f^\circ)_{H_2} = 0 \qquad (\overline{g}_f^\circ)_{O_2} = 0$$

From Table A.16, at 298 K,

$$(\overline{g}_f^\circ)_{H_2O} = -228\,582 \text{ kJ / kmol}$$

$$\Delta G^\circ = 2(\overline{g}_f^\circ)_{H_2} + (\overline{g}_f^\circ)_{O_2} - 2(\overline{g}_f^\circ)_{H_2O}$$

$$= 0 + 0 - 2(-228\,582) = 457\,164 \text{ kJ}$$

At 2000 K, from Table A.13,

$$\overline{g}_{H_2}^\circ = \overline{g}_{2000}^\circ - \overline{g}_{298}^\circ = (\overline{h}_{2000}^\circ - \overline{h}_{298}^\circ) - (2000\overline{s}_{2000}^\circ - 298.15\overline{s}_{298}^\circ)$$

$$= 52\,942 - (2000 \times 188.419 - 298.15 \times 130.678)$$

$$= -284\,934 \text{ kJ / kmol}$$

Similarly, for the oxygen at 2000 K,

$$\overline{g}_{O_2}^\circ = (\overline{h}_{2000}^\circ - \overline{h}_{298}^\circ) - (2000\overline{s}_{2000}^\circ - 298.15\overline{s}_{298}^\circ)$$

$$= 59\,176 - (2000 \times 268.748 - 298.15 \times 205.148)$$

$$= -417\,155 \text{ kJ / kmol}$$

For the water,

$$\overline{g}_{H_2O}^\circ = \overline{g}_f^\circ + \overline{g}_{2000}^\circ - \overline{g}_{298}^\circ$$

$$= -228\,582 + 72\,788 - (2000 \times 264.769 - 298.15 \times 188.835)$$

$$= -629\,031 \text{ kJ / kmol}$$

Therefore, at 2000 K,

$$\Delta G^\circ = 2(-284\,934) + (-417\,155) - 2(-629\,031) = 271\,039 \text{ kJ}$$

The value for ΔG° at 2000 K can alternately be evaluated by using the fact that at constant T,

$$\Delta G^\circ = \Delta H^\circ - T\Delta S^\circ$$

At 2000 K,

$$\Delta H^\circ = 2(\overline{h}_{2000}^\circ - \overline{h}_{298}^\circ)_{H_2} + (\overline{h}_{2000}^\circ - \overline{h}_{298}^\circ)_{O_2} - 2(\overline{h}_f^\circ + \overline{h}_{2000}^\circ - \overline{h}_{298}^\circ)_{H_2O}$$

$$= 2(52\,942) + (59\,176) - 2(-241\,826 + 72\,788)$$

$$= 503\,136 \text{ kJ}$$

$$\Delta S^\circ = 2(\overline{s}_{2000}^\circ)_{H_2} + (\overline{s}_{2000}^\circ)_{O_2} - 2(\overline{s}_{2000}^\circ)_{H_2O}$$

$$= 2(188.419) + 268.748) - 2(264.769)$$

$$= 116.048 \text{ kJ / K}$$

Therefore,

$$\Delta G^\circ = 503\,136 - 2000 \times 116.048 = 271\,040 \text{ kJ}$$

Note that although the two methods are identical in result, the second method requires fewer calculations and is therefore simpler to use.

Returning now to our development, substituting Eq. 13.30 into Eq. 13.29 and rearranging we can write

$$dG_{T,P} = \left\{ \Delta G^\circ + \overline{R}T \ln \left[\frac{a_C^{v_C} a_D^{v_D}}{a_A^{v_A} a_B^{v_B}} \right] \right\} d\varepsilon \qquad (13.31)$$

At equilibrium $dG_{T,P} = 0$. Therefore, since $d\varepsilon$ is arbitrary,

$$\ln \left[\frac{a_C^{v_C} a_D^{v_D}}{a_A^{v_A} a_B^{v_B}} \right] = -\frac{\Delta G^\circ}{\overline{R}T} \qquad (13.32)$$

For convenience, we define the equilibrium constant K as

$$\ln K = -\frac{\Delta G^\circ}{\overline{R}T} \qquad (13.33)$$

which we note must be a function of temperature only for a given reaction, since ΔG° is given by Eq. 13.30 in terms of the properties of the pure substances at a given temperature and the standard-state pressure.

Combining Eqs. 13.32 and 13.33, we have

$$K = \frac{a_C^{v_C} a_D^{v_D}}{a_A^{v_A} a_B^{v_B}} \qquad (13.34)$$

which is the chemical equilibrium equation corresponding to the reaction equation, Eq. 13.26.

EXAMPLE 13.5   Determine the equilibrium constant K, expressed as in K, for the reaction $2H_2O \rightleftharpoons 2H_2 + O_2$ at 298 K and at 2000 K.

Solution

We have already found, in Example 13.4, ΔG° for this reaction at these two temperatures. Therefore, at 298 K,

$$(\ln K)_{298} = -\frac{\Delta G_{298}^\circ}{\overline{R}T} = \frac{-457\,155}{8.3145 \times 298.15} = -184.42$$

At 2000 K, we have

$$(\ln K)_{2000} = -\frac{\Delta G_{2000}^\circ}{\overline{R}T} = \frac{-271\,040}{8.3145 \times 2000} = -16.299$$

Table A.17 gives the values of the equilibrium constant for a number of reactions. Note again that for each reaction the value of the equilibrium constant is determined from the properties of each of the pure constituents at the standard-state pressure and is a function of temperature only.

For other reaction equations, the chemical equilibrium constant can be calculated as in Example 13.5 or can be determined analytically in the following manner. Consider the general reaction Eq. 13.26. The standard-state Gibbs function for each constituent can be expressed by writing Eq. 11.93 at the pressure P°. For component A

$$d\left(\frac{\overline{g}_A^\circ}{T}\right)_{P^\circ} = -\frac{\overline{h}_A^\circ}{T^2}dT_{P^\circ} \tag{13.35}$$

Therefore, for the reaction given by Eq. 13.26,

$$d\left(\frac{\Delta G^\circ}{T}\right)_{P^\circ} = -\frac{\Delta H^\circ}{T^2}dT_{P^\circ} \tag{13.36}$$

where ΔG° is defined by Eq. 13.30 and ΔH° similarly by

$$\Delta H^\circ = v_C \overline{h}_C^\circ + v_D \overline{h}_D^\circ - v_A \overline{h}_A^\circ - v_B \overline{h}_B^\circ \tag{13.37}$$

Now, substituting the definition of the equilibrium constant Eq. 13.33 into Eq. 13.36,

$$d\ln K = \frac{\Delta H^\circ}{\overline{R}T^2}dT_{P^\circ} \tag{13.38}$$

which is termed the van't Hoff equation. In integrating this equation, we must be careful to note that ΔH° as defined by Eq. 13.37 is a function of temperature.

In applying the concept of the equilibrium constant to the determination of the equilibrium composition for a chemical reaction, we must be able to determine the activity of the various constituents in the mixture. The most general model that we will utilize in this text is the Ideal Solution. For this model, the activity of each component is, from Eq. 11.115,

$$a_i = \frac{y_i f_i}{f_i^\circ}$$

If we further assume that all of the standard-state values are for gases, then the standard-state fugacity f_i° is equal to the standard-state pressure P°. The chemical equilibrium equation, Eq. 13.34, then can be rewritten for an ideal solution as

$$K = \frac{a_C^{v_C} a_D^{v_D}}{a_A^{v_A} a_B^{v_B}} = \frac{y_C^{v_C} y_D^{v_D}}{y_A^{v_A} y_B^{v_B}}\left(\frac{P}{P^\circ}\right)^{v_C+v_D-v_A-v_B}\left[\frac{(f/P)_C^{v_C}(f/P)_D^{v_D}}{(f/P)_A^{v_A}(f/P)_B^{v_B}}\right] \tag{13.39}$$

We have written the equation in this form because it demonstrates quite clearly the influence of various factors on the equilibrium composition (the y's). That is, we know that temperature and pressure both influence the equilibrium composition. From Eq. 13.39 it can be seen that the influence of temperature enters through the value of K (which is a function of temperature only) and the influence of pressure through the term $(P/P^\circ)^{v_C+v_D-v_A-v_B}$ (both temperature and pressure influence the various fugacity coefficients f/P).

Let us now consider a special case of the ideal solution, the ideal gas model, which will be found to be appropriate for most of our examples and applications. If each component behaves as an ideal gas, then each fugacity coefficient f/P in Eq. 13.39 is unity, and the chemical equilibrium equation reduces to the form

$$K = \frac{y_C^{v_C} y_D^{v_D}}{y_A^{v_A} y_B^{v_B}}\left(\frac{P}{P^\circ}\right)^{v_C+v_D-v_A-v_B} \tag{13.40}$$

We now consider a number of examples that illustrate the procedure for determining the equilibrium composition for a homogeneous reaction and the influence of certain variables on the equilibrium composition.

EXAMPLE 13.6

One kilomole of carbon at 25°C and 0.1 MPa pressure reacts with 1 kmol of oxygen at 25°C and 0.1 MPa pressure to form an equilibrium mixture of CO_2, CO, and O_2 at 3000 K, 0.1 MPa pressure, in a steady-flow process. Determine the equilibrium composition and the heat transfer for this process.

    Control volume: Combustion chamber.

    Inlet states: *P, T* known for carbon and for oxygen.

    Exit state: *P, T* known.

    Process: SSSF.

    Sketch: Fig. 13.13.

    Model: Table A.16 for carbon; ideal gases, Tables A.13 and A.16.

Analysis and Solution

It is convenient to view the overall process as though it occurs in two separate steps, a combustion process followed by a heating and dissociation of the combustion product carbon dioxide, as indicated in Fig. 13.13. This two-step process is represented as

    Combustion: $C + O_2 \rightarrow CO_2$

    Dissociation reaction: $2CO_2 \rightleftharpoons 2CO + O_2$

That is, the energy released by the combustion of C and O_2 heats the CO_2 formed to high temperature, which causes dissociation of part of the CO_2 to CO and O_2. Thus, the overall reaction can be written

$$C + O_2 \rightarrow aCO_2 + bCO + dO_2$$

where the unknown coefficients *a*, *b*, and *d* must be found by solution of the equilibrium equation associated with the dissociation reaction. Once this is accomplished, we

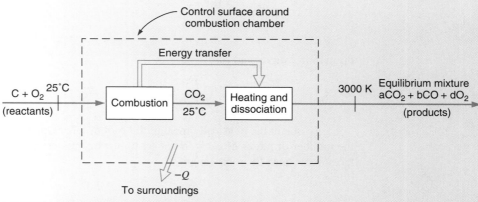

FIGURE 13.13 Sketch for Example 13.6.

can write the first law for a control volume around the combustion chamber to calculate the heat transfer.

From the combustion equation we find that the initial composition for the dissociation reaction is 1 kmol CO_2. Therefore, letting $2z$ be the number of kilomoles of CO_2 dissociated, we find

$$2CO_2 \rightleftharpoons 2CO + O_2$$

Initial:	1	0	0
Change:	$-2z$	$+2z$	$+z$
At equilibrium:	$(1-2z)$	$2z$	z

Therefore, the overall reaction is

$$C + O_2 \rightarrow (1-2z)CO_2 + 2z\,CO + z\,O_2$$

and the total number of kilomoles at equilibrium is

$$n = (1-2z) + 2z + z = 1 + z$$

The equilibrium mole fractions are

$$y_{CO_2} = \frac{1-2z}{1+z} \qquad y_{CO} = \frac{2z}{1+z} \qquad y_{O_2} = \frac{z}{1+z}$$

From Table A.17 we find that the value of the equilibrium constant at 3000 K for the dissociation reaction considered here is

$$\ln K = -2.217 \qquad K = 0.1089$$

Substituting these quantities along with $P = 0.1$ MPa into Eq. 13.40, we have the equilibrium equation,

$$K = 0.1089 = \frac{y_{CO}^2 y_{O_2}}{y_{CO_2}^2}\left(\frac{P}{P^\circ}\right)^{2+1-2} = \frac{\left(\dfrac{2z}{1+z}\right)^2\left(\dfrac{z}{1+z}\right)}{\left(\dfrac{1-2z}{1+z}\right)^2} \quad (1)$$

or, in more convenient form,

$$\frac{K}{P/P^\circ} = \frac{0.1089}{1} = \left(\frac{2z}{1-2z}\right)^2\left(\frac{z}{1+z}\right)$$

To obtain the physically meaningful root of this mathematical relation, we note that the number of moles of each component must be greater than zero. Thus, the root of interest to us must lie in the range

$$0 \le z \le 0.5$$

Solving the equilibrium equation by trial and error, we find

$$z = 0.2189$$

Therefore, the overall process is

$$C + O_2 \rightarrow 0.5622CO_2 + 0.4378CO + 0.2189O_2$$

where the equilibrium mole fractions are

$$y_{CO_2} = \frac{0.5622}{1.2189} = 0.4612$$

$$y_{CO} = \frac{0.4378}{1.2189} = 0.3592$$

$$y_{O_2} = \frac{0.2189}{1.2189} = 0.1796$$

The heat transfer from the combustion chamber to the surroundings can be calculated using the enthalpies of formation and Table A.13. For this process

$$H_R = (\overline{h}_f^\circ)_C + (\overline{h}_f^\circ)_{O_2} = 0 + 0 = 0$$

The equilibrium products leave the chamber at 3000 K. Therefore,

$$\begin{aligned}
H_P &= n_{CO_2}(\overline{h}_f^\circ + \overline{h}_{3000}^\circ - \overline{h}_{298}^\circ)_{CO_2} \\
&\quad + n_{CO}(\overline{h}_f^\circ + \overline{h}_{3000}^\circ - \overline{h}_{298}^\circ)_{CO} \\
&\quad + n_{O_2}(\overline{h}_f^\circ + \overline{h}_{3000}^\circ - \overline{h}_{298}^\circ)_{O_2} \\
&= 0.5622(-393\,522 + 152\,853) \\
&\quad + 0.4378(-110\,527 + 93\,504) \\
&\quad + 0.2189(98\,013) \\
&= -121\,302 \text{ kJ}
\end{aligned}$$

Substituting into the first law gives

$$\begin{aligned}
Q_{c.v.} &= H_P - H_g \\
&= -121\,302 \text{ kJ / kmol C burned}
\end{aligned}$$

EXAMPLE 13.7   One kilomole of carbon at 25°C reacts with 2 kmol of oxygen at 25°C to form an equilibrium mixture of CO_2, CO, and O_2 at 3000 K, 0.1 MPa pressure. Determine the equilibrium composition.

Control volume: Combustion chamber.

Inlet states: T known for carbon and for oxygen.

Exit state: P, T known.

Process: SSSF.

Model: Ideal gas mixture at equilibrium.

Analysis and Solution

The overall process can be imagined to occur in two steps as in the previous example. The combustion process is

$$C + 2O_2 \rightarrow CO_2 + O_2$$

and the subsequent dissociation reaction is

	$2CO_2 \rightleftharpoons$	$2CO +$	O_2
Initial:	1	0	1
Change:	$-2z$	$+2z$	$+z$
At equilibrium:	$(1-2z)$	$2z$	$(1+z)$

We find that in this case the overall process is

$$C + 2O_2 \rightarrow (1-2z)CO_2 + 2z\,CO + (1+z)O_2$$

and the total number of kilomoles at equilibrium is

$$n = (1-2z) + 2z + (1+z) = 2 + z$$

The mole fractions are

$$y_{CO_2} = \frac{1-2z}{2+z} \qquad y_{CO} = \frac{2z}{2+z} \qquad y_{O_2} = \frac{1+z}{2+z}$$

The equilibrium constant for the reaction $2CO_2 \rightleftharpoons 2CO + O_2$ at 3000 K was found in Example 13.6 to be 0.1089. Therefore, with these expressions, quantities, and $P = 0.1$ MPa substituted, the equilibrium equation is

$$k = 0.1089 = \frac{y_{CO}^2 y_{O_2}}{y_{CO_2}^2} \left(\frac{P}{P^\circ} \right)^{2+1-2}$$

$$= \frac{\left(\dfrac{2z}{2+z} \right)^2 \left(\dfrac{1+z}{2+z} \right)}{\left(\dfrac{1-2z}{2+z} \right)^2} \quad (1)$$

or

$$\frac{K}{P/P^\circ} = \frac{0.1089}{1} = \left(\frac{2z}{1-2z} \right)^2 \left(\frac{1+z}{2+z} \right)$$

We note that in order for the number of kilomoles of each component to be greater than zero,

$$0 \le z \le 0.5$$

Solving the equilibrium equation for z, we find

$$z = 0.1553$$

so that the overall process is

$$C + 2O_2 \rightarrow 0.6894CO_2 + 0.3106CO + 1.1553O_2$$

The mole fractions of the components in the equilibrium mixture are

$$y_{CO_2} = \frac{0.6894}{2.1553} = 0.320$$

$$y_{CO} = \frac{0.3106}{2.1553} = 0.144$$

$$y_{O_2} = \frac{1.1553}{2.1553} = 0.536$$

The heat transferred from the chamber in this process could be found by the same procedure followed in Example 13.6, considering the overall process.

13.7 SIMULTANEOUS REACTIONS

In developing the equilibrium equation and equilibrium constant expressions of Section 13.6, it was assumed that there was only a single chemical reaction equation relating the substances present in the system. To demonstrate the more general situation in which there is more than one chemical reaction, we will now analyze a case involving two simultaneous reactions by a procedure analogous to that followed in Section 13.6. These results are then readily extended to systems involving several simultaneous reactions.

Consider a mixture of substances A, B, C, D, L, M, and N as indicated in Fig. 13.14. These substances are assumed to exist at a condition of chemical equilibrium at temperature T and pressure P, and are related by the two independent reactions

$$(1) \quad v_{A_1}A + v_B B \rightleftharpoons v_C C + v_D D \tag{13.41}$$

$$(2) \quad v_{A_2}A + v_L L \rightleftharpoons v_M M + v_N N \tag{13.42}$$

We have considered the situation where one of the components (substance A) is involved in each of the reactions in order to demonstrate the effect of this condition on the resulting equations. As in the previous section, the changes in amounts of the components are related by the various stoichiometric coefficients (which are not the same as the number of moles of each substance present in the vessel). We also realize that the coefficients v_{A1} and v_{A2} are not necessarily the same. That is, substance A does not in general take part in each of the reactions to the same extent.

Development of the requirement for equilibrium is completely analogous to that of

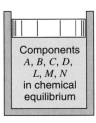

FIGURE 13.14 Sketch demonstrating simultaneous reactions.

Section 13.6. We consider that each reaction proceeds an infinitesimal amount toward the right side. This results in a decrease in the number of moles of A, B, and L, and an increase in the moles of C, D, M, and N. Letting the degrees of reaction be ε_1 and ε_2 for reactions 1 and 2, respectively, the changes in the number of moles are, for infinitesimal shifts from the equilibrium composition,

$$
\begin{aligned}
dn_A &= -v_{A_1} d\varepsilon_1 - v_{A_2} d\varepsilon_2 \\
dn_B &= -v_B d\varepsilon_1 \\
dn_L &= -v_L d\varepsilon_2 \\
dn_C &= +v_C d\varepsilon_1 \\
dn_D &= +v_D d\varepsilon_1 \\
dn_M &= +v_M d\varepsilon_2 \\
dn_N &= +v_N d\varepsilon_2
\end{aligned}
\tag{13.43}
$$

The change in Gibbs function for the mixture in the vessel at constant temperature and pressure is

$$
dG_{T,P} = \overline{G}_A dn_A + \overline{G}_B dn_B + \overline{G}_C dn_C + \overline{G}_D dn_D + \overline{G}_L dn_L + \overline{G}_M dn_M + \overline{G}_N dn_N
$$

Substituting the expressions of Eq. 13.43 and collecting terms,

$$
\begin{aligned}
dG_{T,P} &= (v_C \overline{G}_C + v_D \overline{G}_D - v_{A_1} \overline{G}_A - v_B \overline{G}_B) d\varepsilon_1 \\
&\quad + (v_M \overline{G}_M + v_N \overline{G}_N - v_{A_2} \overline{G}_A - v_L \overline{G}_L) d\varepsilon_2
\end{aligned}
\tag{13.44}
$$

It is convenient to again express each of the partial molal Gibbs functions in terms of the activity as

$$
G_i = \overline{g}_i^\circ + \overline{R} T \ln a_i
$$

Equation 13.44 written in this form becomes

$$
\begin{aligned}
dG_{T,P} &= \left[\Delta G_1^\circ + \overline{R} T \ln \frac{a_C^{v_C} a_D^{v_D}}{a_A^{v_{A_1}} a_B^{v_B}} \right] d\varepsilon_1 \\
&\quad + \left[\Delta G_2^\circ + \overline{R} T \ln \frac{a_M^{v_M} a_N^{v_N}}{a_A^{v_{A_2}} a_L^{v_L}} \right] d\varepsilon_2
\end{aligned}
\tag{13.45}
$$

In this equation the standard-state change in Gibbs function for each reaction is defined as

$$
\Delta G_1^\circ = v_C \overline{g}_C^\circ + v_D \overline{g}_D^\circ - v_{A_1} \overline{g}_A^\circ - v_B \overline{g}_B^\circ
\tag{13.46}
$$

$$
\Delta G_2^\circ = v_M \overline{g}_M^\circ + v_N \overline{g}_N^\circ - v_{A_2} \overline{g}_A^\circ - v_L \overline{g}_L^\circ
\tag{13.47}
$$

Equation 13.45 expresses the change in Gibbs function of the system at constant T, P, for infinitesimal degrees of reaction of both reactions 1 and 2, Eqs. 13.41 and 13.42. The requirement for equilibrium is that $dG_{T,P} = 0$. Therefore, since reactions 1 and 2 are independent, $d\varepsilon_1$ and $d\varepsilon_2$ can be independently varied. It follows that at equilibrium each of the bracketed terms of Eq. 13.45 must be zero. Defining equilibrium constants for the two reactions by

$$
\ln K_1 = -\frac{\Delta G_1^\circ}{\overline{R} T}
\tag{13.48}
$$

and

$$\ln K_2 = -\frac{\Delta G_2^\circ}{\bar{R}T} \tag{13.49}$$

We find that at equilibrium

$$K_1 = \frac{a_C^{v_C} a_D^{v_D}}{a_A^{v_A} a_B^{v_B}} \tag{13.50}$$

and

$$K_2 = \frac{a_M^{v_M} a_N^{v_N}}{a_A^{v_{A_2}} a_L^{v_L}} \tag{13.51}$$

can be expressed in terms of the mole fractions for the appropriate model of the mixture, after which these expressions must be solved simultaneously for the equilibrium composition of the mixture. The following example is presented to demonstrate and clarify this procedure.

EXAMPLE 13.8   One kilomole of water vapor is heated to 3000 K, 0.1 MPa pressure. Determine the equilibrium composition, assuming that H_2O, H_2, O_2, and OH are present.

   Control volume: Heat exchanger.

   Exit state: P, T known.

   Model: Ideal gas mixture at equilibrium.

Analysis and Solution

There are two independent reactions relating the four components of the mixture at equilibrium. These can be written as

$$(1)\ \ 2H_2O \rightleftharpoons 2H_2 + O_2$$
$$(2)\ \ 2H_2O \rightleftharpoons H_2 + 2OH$$

Let $2a$ be the number of kilomoles of water dissociating according to reaction (1) during the heating, and $2b$ the number of kilomoles of water dissociating according to reaction (2). Since the initial composition is 1 kmol water, the changes according to the two reactions are

$$(1)\ \ 2H_2O \rightleftharpoons 2H_2 + O_2$$

Change: $-2a\ \ \ +2a\ \ \ +a$

$$(2)\ \ 2H_2O \rightleftharpoons H_2 + 2OH$$

Change: $-2b\ \ \ +b\ \ \ +2b$

Therefore, the number of kilomoles of each component at equilibrium is its initial number plus the change, so that at equilibrium

$$n_{H_2O} = 1 - 2a - 2b$$

$$n_{H_2} = 2a + b$$

$$n_{O_2} = a$$

$$\underline{n_{OH} = 2b}$$

$$n = 1 + a + b$$

The overall chemical reaction that occurs during the heating process can be written

$$H_2O \rightarrow (1 - 2a - 2b)H_2O + (2a + b)H_2 + aO_2 + 2bOH$$

The right-hand side of this expression is the equilibrium composition of the system. Since the number of kilomoles of each substance must necessarily be greater than zero, we find that the possible values of a and b are restricted to

$$a \geq 0$$

$$b \geq 0$$

$$(a + b) \leq 0.5$$

The two equilibrium equations are, assuming that the mixture behaves as an ideal gas,

$$K_1 = \frac{y_{H_2}^2 y_{O_2}}{y_{H_2O}^2} \left(\frac{P}{P^\circ} \right)^{2+1-2}$$

$$K_2 = \frac{y_{H_2} y_{OH}^2}{y_{H_2O}^2} \left(\frac{P}{P^\circ} \right)^{1+2-2}$$

Since the mole fraction of each component is the ratio of the number of kilomoles of the component to the total number of kilomoles of the mixture, these equations can be written in the form

$$K_1 = \frac{\left(\dfrac{2a+b}{1+a+b} \right)^2 \left(\dfrac{a}{1+a+b} \right)}{\left(\dfrac{1-2a-2b}{1+a+b} \right)^2} \left(\frac{P}{P^\circ} \right)$$

$$= \left(\frac{2a+b}{1-2a-2b} \right)^2 \left(\frac{a}{1+a+b} \right) \left(\frac{P}{P^\circ} \right)$$

and

$$K_2 = \frac{\left(\dfrac{2a+b}{1+a+b} \right) \left(\dfrac{2b}{1+a+b} \right)^2}{\left(\dfrac{1-2a-2b}{1+a+b} \right)^2} \left(\frac{P}{P^\circ} \right)$$

$$= \left(\frac{2a+b}{1+a+b} \right) \left(\frac{2b}{1-2a-2b} \right)^2 \left(\frac{P}{P^\circ} \right)$$

giving two equations in the two unknowns a and b, since $P = 0.1$ MPa and the values of K_1, K_2 are known. From Table A.16 at 3000 K, we find

$$K_1 = 0.002\,062 \qquad K_2 = 0.002\,893$$

Therefore, the equations can be solved simultaneously for a and b. The values satisfying the equations are

$$a = 0.0534 \qquad b = 0.0551$$

Substituting these values into the expressions for the number of kilomoles of each component and of the mixture, we find the equilibrium mole fractions to be

$$y_{H_2O} = 0.7063$$

$$y_{H_2} = 0.1461$$

$$y_{O_2} = 0.0482$$

$$y_{OH} = 0.0994$$

The methods used in this section can readily be extended to equilibrium systems having more than two independent reactions. In each case, the number of simultaneous equilibrium equations is equal to the number of independent reactions. The solution of a large set of nonlinear simultaneous equations naturally becomes quite difficult, however, and is not easily accomplished by hand calculations. These problems are normally solved using iterative procedures on a digital computer.

13.8 IONIZATION

In the final section of this chapter, we consider the equilibrium of systems that are made up of ionized gases, or plasmas, a field that has been studied and applied increasingly in recent years. In previous sections we discussed chemical equilibrium, with a particular emphasis on molecular dissociation, as for example the reaction

$$N_2 \rightleftharpoons 2N$$

which occurs to an appreciable extent for most molecules only at high temperature, of the order of magnitude 3000 to 10 000 K. At still higher temperatures, such as those found in electric arcs, the gas becomes ionized. That is, some of the atoms lose an electron, according to the reaction

$$N \rightleftharpoons N^+ + e^-$$

where N^+ denotes a singly ionized nitrogen atom, one that has lost one electron and consequently has a positive charge, and e^- represents the free electron. As the temperature rises still higher, many of the ionized atoms lose another electron, according to the reaction

$$N^+ \rightleftharpoons N^{++} + e^-$$

and thus becomes doubly ionized. As the temperature continues to rise, the process continues until a temperature is reached at which all the electrons have been stripped from the nucleus.

Ionization generally is appreciable only at high temperature. However, dissociation and ionization both tend to occur to greater extents at low pressure, and consequently dissociation and ionization may be appreciable in such environments as the upper atmosphere, even at moderate temperature. Other effects such as radiation will also cause ionization, but these effects are not considered here.

The problems of analyzing the composition in a plasma become much more difficult than for an ordinary chemical reaction, for in an electric field the free electrons in the mixture do not exchange energy with the positive ions and neutral atoms at the same rate that they do with the field. Consequently, in a plasma in an electric field, the electron gas is not at exactly the same temperature as the heavy particles. However, for moderate fields, assuming a condition of thermal equilibrium in the plasma is a reasonable approximation, at least for preliminary calculations. Under this condition we can treat the ionization equilibrium in exactly the same manner as an ordinary chemical equilibrium analysis.

At these extremely high temperatures, we may assume that the plasma behaves as an ideal-gas mixture of neutral atoms, positive ions, and electron gas. Thus, for the ionization of some atomic species A,

$$A \rightleftharpoons A^+ + e^- \tag{13.52}$$

we may write the ionization equilibrium equation in the form

$$K = \frac{y_{A^+} y_{e^-}}{y_A} \left(\frac{P}{P^\circ} \right)^{1+1-1} \tag{13.53}$$

The ionization-equilibrium constant K is defined in the ordinary manner

$$\ln K = -\frac{\Delta G^\circ}{\overline{R} T} \tag{13.54}$$

and is a function of temperature only. The standard-state Gibbs function change for reaction 13.52 is found from

$$\Delta G^\circ = \overline{g}^\circ_{A^+} + \overline{g}^\circ_{e^-} - \overline{g}^\circ_A \tag{13.55}$$

The standard-state Gibbs function for each component at the given plasma temperature can be calculated using the procedures of statistical thermodynamics, so that ionization–equilibrium constants can be tabulated as functions of temperature.

Solution of the ionization–equilibrium equation, Eq. 13.53, is then accomplished in the same manner as for an ordinary chemical-reaction equilibrium.

EXAMPLE 13.9   Calculate the equilibrium composition if argon gas is heated in an arc to 10 000 K, 1 kPa, assuming the plasma to consist of Ar, Ar^+, e^-. The ionization–equilibrium constant for the reaction

$$Ar \rightleftharpoons Ar^+ + e^-$$

at this temperature is 0.000 42.

Control volume: Heating arc.

Exit state: P, T known.

Model: Ideal gas mixture at equilibrium.

Analysis and Solution

Consider an initial composition of 1 kmol neutral argon, and let z be the number of kilomoles ionized during the heating process. Therefore,

$$\text{Ar} \rightleftharpoons \text{Ar}^+ + e^-$$

Initial:	1	0	0
Change:	$-z$	$+z$	$+z$
Equilibrium:	$(1-z)$	z	z

and

$$n = (1-z) + z + z = 1 + z$$

Since the number of kilomoles of each component must be positive, the variable z is restricted to the range

$$0 \le z \le 1$$

The equilibrium mole fractions are

$$y_{\text{Ar}} = \frac{n_{\text{Ar}}}{n} = \frac{1-z}{1+z}$$

$$y_{\text{Ar}^+} = \frac{n_{\text{Ar}^+}}{n} = \frac{z}{1+z}$$

$$y_{e^-} = \frac{n_{e^-}}{n} = \frac{z}{1+z}$$

The equilibrium equation is

$$K = \frac{y_{\text{Ar}^+} y_{e^-}}{y_{\text{Ar}}} \left(\frac{P}{P^\circ}\right)^{1+1-1} = \frac{\left(\dfrac{z}{1+z}\right)\left(\dfrac{z}{1+z}\right)}{\left(\dfrac{1-z}{1+z}\right)} \left(\frac{P}{P^\circ}\right)$$

so that, at 10 000 K, 1 kPa,

$$0.000\,42 = \left(\frac{z^2}{1-z^2}\right)(0.01)$$

Solving,

$$z = 0.2008$$

and the composition is found to be

$$y_{\text{Ar}} = 0.6656$$

$$y_{\text{Ar}^+} = 0.1672$$

$$y_{e^-} = 0.1672$$

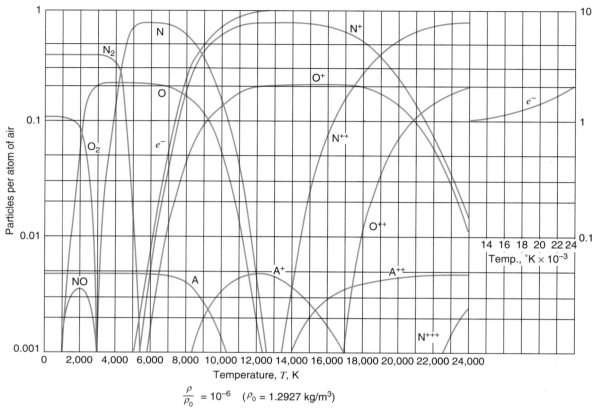

Particles per atom of air

$$\frac{\rho}{\rho_0} = 10^{-6} \quad (\rho_0 = 1.2927 \text{ kg/m}^3)$$

FIGURE 13.15 Equilibrium composition of air [W. E. Moeckel and K. C. Weston, NACA TN 4265 (1958)].

Simultaneous reactions, such as simultaneous molecular dissociation and ionization reactions or multiple ionization reactions, can be analyzed in the same manner as the ordinary simultaneous chemical reactions of Section 13.7. In doing so, we again make the assumption of thermal equilibrium in the plasma, which, as mentioned before, is, in many cases, a reasonable approximation. Figure 13.15 shows the equilibrium composition of air at high temperature and low density, and indicates the overlapping regions of the various dissociation and ionization processes.

PROBLEMS

13.1 Carbon dioxide at 15 MPa is injected into the top of a 5-km deep well in connection with an enhanced oil-recovery process. The fluid column standing in the well is at a uniform temperature of 40°C. What is the pressure at the bottom of the well?

13.2 Consider a 2-km-deep gas well containing a gas mixture of methane and ethane at a uniform temperature of 30°C. The pressure at the top of the well is 14 MPa, and the composition on a mole basis is 90% methane, 10% ethane. Determine the pressure and composition at the bottom of the well, assuming an ideal gas mixture, and also assuming an ideal solution.

13.3 On the basis of the most recent observations, Titan, the large moon of the planet Saturn, is believed to have an atmosphere made up largely of nitrogen and methane, with surface conditions of about 94 K, 0.13 MPa. Assuming equilibrium between a gaseous atmosphere and a liquid ocean of these two substances at this temperature and pressure, what is the composition of each?

13.4 Ethylene gas is burned with air, and the products exit at 120°C, 3.5 MPa. Assuming complete combustion, what is the minimum percent theoretical air if there is to be no condensation of the products?

13.5 Consider the liquid–vapor equilibrium system of nitrogen–methane at a temperature of 100 K. Plot to scale the liquid- and vapor–phase boundary lines on a pressure-versus-composition diagram. Repeat this process for a temperature of 120 K.

13.6 (Adv.) Ambient air at 20°C, 100 kPa, 60% relative humidity, enters a large multi-stage compressor at the rate of 5.0 m^3/s, and exits the unit at 100°C, 10 MPa. What fraction of the water is condensed at the exit? What is the change of thermodynamic availability of the air plus water for this change of state?

13.7 A mixture of 3 kmol hydrogen and 1 kmol carbon monoxide is contained in a rigid vessel at 25°C and an unknown pressure, P_1. A chemical reaction occurs, and the products consist solely of methane and water, at 200°C and 4 MPa. Determine the initial pressure P_1 and the change $S_2 - S_1$ for the content of the vessel.

13.8 Consider an SSSF process for the production of a synthetic fuel, methanol, from waste material that has initially been partially oxidized to produce a gas mixture of one-third hydrogen and two-thirds carbon monoxide, by mole. This mixture enters the reactor at 600 K, 4 MPa, at the rate of 3 kmol/s, and steam at 300°C, 4 MPa, enters the reactor separately at the rate of 1 kmol/s. Exiting the reactor is a mixture of methanol and carbon dioxide at 450 K, 4 MPa. What is the rate of entropy change for this process?

13.9 Hydrogen gas enters an SSSF combustion chamber at 25°C, 5 MPa. A separate line supplies 80% theoretical air, which is throttled from 25°C, 15 MPa, into the chamber. The products, consisting of H_2O, H_2, and N_2, exit from the chamber at 175°C, 5 MPa. What is the irreversibility of this process?

13.10 At a certain point in a synthetic fuel production process, the product, which consists of equal amounts of ethanol and carbon dioxide, is stored in a 1-m^3 tank at 400 K, 2 MPa, before being processed further. Determine the total amount of product in the tank and its absolute entropy per kmol of product at this state.

13.11 Calculate the equilibrium constant for the reaction $O_2 \rightleftharpoons 2O$ at temperatures of 298 K and 6000 K.

13.12 Plot to scale the values of ln K versus $1/T$ for the reaction $2CO_2 \rightleftharpoons 2CO + O_2$. Write an equation for ln K as a function of temperature.

13.13 Pure oxygen is heated from 25°C to 3200 K in an SSSF process at a constant pressure of 200 kPa. Find the exit composition and the heat transfer.

13.14 Pure oxygen is heated from 25°C, 100 kPa to 3200 K in a constant volume container. Find the final pressure, composition, and the heat transfer.

13.15 Air (assumed to be 79% nitrogen and 21% oxygen) is heated in an SSSF process at a constant pressure of 100 kPa, and some NO is formed. At what temperature will the mole fraction of $\dot{N}O$ be 0.001?

13.16 Hydrogen gas is heated from room temperature to 4000 K, 500 kPa, at which state the diatomic species has partially dissociated to the monatomic form. Determine the equilibrium composition at this state.

13.17 One kilomole Ar and one kilomole O_2 is heated up at a constant pressure of 100 kPa to 3200 K, where it comes to equilibrium. Find the final mole fractions for Ar, O_2, and O.

13.18 The combustion products from burning pentane, C_5H_{12}, with pure oxygen in a stoichiometric ratio exists at 2400 K. Consider the dissociation of only CO_2 and find the equilibrium mole fraction of CO.

13.19 Complete combustion of hydrogen and pure oxygen in a stoichiometric ratio at P_0, T_0 to form water would result in a computed adiabatic flame temperature of 4990 K for an SSSF setup.

    a. How should the adiabatic flame temperature be found if the equilibrium reaction involving water, hydrogen, and oxygen is considered? Disregard all other possible reactions (dissociations) and show the final equation(s) to be solved.

    b. Find the equilibrium composition at 3800 K, again disregarding all other reactions.

    c. Which other reactions should be considered and which components will be present in the final mixture?

13.20 Which of the nitrogen oxides, NO or NO_2, is the more stable at ambient conditions? What about at 2000 K?

13.21 A mixture of 1 kmol carbon dioxide, 2 kmol carbon monoxide, and 2 kmol oxygen, at 25°C, 150 kPa, is heated in a constant pressure SSSF process to 3000 K. Assuming that only these same substances are present in the exiting chemical equilibrium mixture, determine the composition of that mixture.

13.22 Repeat the previous problem for an initial mixture that also includes 2 kmol of nitrogen, which does not dissociate during the process.

13.23 Gasification of char (primarily carbon) with steam following coal pyrolysis yields a gas mixture of 1 kmol CO and 1 kmol H_2. We wish to upgrade the hydrogen content of this syngas fuel mixture, so it is fed to an appropriate catalytic reactor along with 1 kmol of H_2O. Exiting the reactor is a chemical equilibrium gas mixture of CO, H_2, H_2O, and CO_2 at 600 K, 500 kPa. Determine the equilibrium composition.

13.24 A gas mixture of 1 kmol carbon monoxide, 1 kmol nitrogen, and 1 kmol oxygen at 25°C, 150 kPa, is heated in a constant pressure SSSF process. The exit mixture can be assumed to be in chemical equilibrium with CO_2, CO, O_2, and N_2 present. The mole fraction of CO_2 at this point is 0.176. Calculate the heat transfer for the process.

13.25 A rigid container initially contains 2 kmol of carbon monoxide and 2 kmol of oxygen at 25°C, 100 kPa. The content is then heated to 3000 K at which point an equilibrium mixture of CO_2, CO, and O_2 exists. Disregard other possible species and determine the final pressure, the equilibrium composition and the heat transfer for the process.

13.26 (Adv.) The following system is proposed to produce hydrogen gas, for use as an alternate fuel. Sulfuric acid, H_2SO_4, is fed to a solar collector, and an ideal-gas mixture of H_2SO_4, H_2O, SO_2, and O_2 exits at 1000 K, 150 kPa.

a. Determine the equilibrium composition of the mixture exiting the solar collector.

b. Under different operating conditions, the SO_2 produced in the solar collector reaction undergoes at electrolytic reaction with the appropriate amount of water to re-form sulfuric acid (to return to the collector) and also produce hydrogen gas, H_2. Show why it may be thermodynamically desirable to utilize this process to produce hydrogen rather than by direct production from water electrolytically.

The properties of sulfuric acid and sulfur dioxide are

Ideal gas	$\overline{h}_f^\circ$	$\overline{s}_{25°C}^\circ$	$\overline{C}_{p0,650\text{ K}}$
	[kJ / kmol]	[kJ / kmol K]	[kJ / kmol K]
H_2SO_4	−735129	298.796	118.5
SO_2	−296842	248.212	50.0

13.27 One approach to using hydrocarbon fuels in a fuel cell is to "reform" the hydrocarbon to obtain hydrogen, which is then fed to the fuel cell. As a part of the analysis of such a procedure, consider the reforming section

$$CH_4 + H_2O \rightleftharpoons 3H_2 + CO$$

a. Determine the equilibrium constant for this reaction at a temperature of 800 K.

b. One kilomole each of methane and water are fed to a catalytic reformer. A mixture of CH_4, H_2O, H_2, and CO exits in chemical equilibrium at 800 K, 100 kPa; determine the equilibrium composition of this mixture.

13.28 In a test of a gas-turbine combustor, saturated-liquid methane at 115 K is to be burned with excess air to hold the adiabatic flame temperature to 1600 K. It is assumed that the products consist of a mixture of CO_2, H_2O, N_2, O_2, and NO in chemical equilibrium. Determine the percent excess air used in the combustion, and the percentage of NO in the products.

13.29 Catalytic gas generators are frequently used to decompose a liquid, providing a desired gas mixture (spacecraft control systems, fuel cell gas supply, and so forth). Consider feeding pure liquid hydrazine, N_2H_4, to a gas generator, from which exits a gas mixture of N_2, H_2, and NH_3 in chemical equilibrium at 100°C, 350 kPa. Calculate the mole fractions of the species in the equilibrium mixture.

13.30 Acetylene gas at 25°C is burned with 140% theoretical air, which enters the burner at 25°C, 100 kPa, 80% relative humidity. The combustion products form a mixture of CO_2, H_2O, N_2, O_2, and NO in chemical equilibrium at 2200 K, 100 kPa. This mixture is then cooled to 1000 K very rapidly, so that the composition does not change. Determine the mole fraction of NO in the products and the heat transfer for the overall process.

13.31 A step in the production of a synthetic liquid fuel from organic waste matter is the following conversion process: 1 kmol of ethylene gas (converted from the waste) at 25°C, 5 MPa, and 2 kmol of steam at 300°C, 5 MPa, enter a catalytic reactor. An ideal gas mixture of ethanol, ethylene, and water in chemical equilibrium leaves the reactor at 700 K, 5 MPa.

a. Determine the composition of the mixture.

b. Calculate the heat transfer for the reactor.

c. Would it be more desirable to operate this system with different inlet proportions?

13.32 Methane at 25°C, 100 kPa, is burned with 200% theoretical oxygen at 400 K, 100 kPa, in an adiabatic SSSF process, and the products of combustion exit at 100 kPa. Assume that the only significant dissociation reaction in the products is that of carbon dioxide going to carbon monoxide and oxygen. Determine the equilibrium composition of the products and also their temperature at the combustor exit.

13.33 Calculate the irreversibility for the adiabatic combustion process described in the previous problem.

13.34 An important step in the manufacture of chemical fertilizer is the production of ammonia, according to the reaction

$$N_2 + 3H_2 \rightleftharpoons 2NH_3$$

a. Calculate the equilibrium constant for this reaction at 150°C.

b. For an initial composition of 25% nitrogen, 75% hydrogen, on a mole basis, calculate the equilibrium composition at 150°C, 5 MPa.

13.35 In rich (too much fuel) combustion the excess fuel may be broken down to give H_2 and CO may form. In the products at 1200 K, 200 kPa the reaction called the water gas reaction may take place

$$CO_2 + H_2 \rightleftharpoons H_2O + CO$$

Find the equilibrium constant for this reaction from the elementary reactions.

13.36 One kilomole of carbon dioxide, CO_2, and 1 kmol of hydrogen, H_2 at room temperature, 200 kPa is heated to 1200 K at 200 kPa. Use the water gas reaction, see previous problem, to determine the mole fraction of CO. Neglect dissociations of H_2 and O_2.

13.37 Consider the following simplification of the production of a synthetic fuel (methanol) from coal. A gas mixture of 50% carbon monoxide and 50% hydrogen leaves a coal gasifier at 500 K, 1 MPa, and enters a catalytic converter. A gas mixture of methanol, carbon monoxide and hydrogen in chemical equilibrium leaves the converter at the same temperature and pressure.

a. Calculate the equilibrium constant for an appropriate reaction equation, and determine the equilibrium composition of the mixture leaving the converter.

b. Would it be more desirable to operate the converter at ambient pressure?

c. Would it be more desirable to operate the converter at ambient temperature?

d. Would it be desirable to change the composition of the inlet mixture before feeding it into the catalytic converter?

13.38 Consider the following coal gasifier proposed for supplying a syngas fuel to a gas turbine power plant. Fifty kilograms per second of dry coal (represented as 48 kg C plus 2 kg H) enter the gasifier, along with 4.76 kmol/s of air and 2 kmol/s of steam. The output stream from this unit is a gas mixture containing H_2, CO, N_2, CH_4, and CO_2 in chemical equilibrium at 900 K, 1 MPa.

a. Set up the reaction and equilibrium equation(s) for this system, and calculate the appropriate equilibrium constant(s).

b. Determine the composition of the gas mixture leaving the gasifier.

13.39 Ethane is burned with 150% theoretical air in a gas turbine combustor. The products exiting consist of a mixture of CO_2, H_2O, O_2, N_2, and NO in chemical equilibrium at 1800 K, 1 MPa. Determine the mole fraction of NO in the products. Is it reasonable to ignore CO in the products?

13.40 One kilomole of liquid oxygen, O_2, at 93 K, and x kmol of gaseous hydrogen, H_2, at 25°C, are fed to an SSSF combustion chamber. x is greater than 2, such that there is excess hydrogen for the combustion process. There is a heat loss from the chamber of 1000 kJ per kmol of reactants. Products exit the chamber at chemical equilibrium at 3800 K, 400 kPa, and are assumed to include only H_2O, H_2, and O.

a. Determine the equilibrium composition of the products and also x, the amount of H_2 entering the combustion chamber.

b. Should another substance(s) have been included in part (a) as being present in the products? Justify your answer.

13.41 Butane is burned with 200% theoretical air, and the products of combustion, an equilibrium mixture containing only CO_2, H_2O, O_2, N_2, NO, and NO_2, exit from the combustion chamber at 1400 K, 2 MPa. Determine the equilibrium composition at this state.

13.42 A mixture of 1 kmol water and 1 kmol oxygen at 400 K is heated to 3000 K, 200 kPa, in an SSSF process. Determine the equilibrium composition at the outlet of the heat exchanger, assuming that the mixture consists of H_2O, H_2, O_2, and OH.

13.43 One kilomole of air (assumed to be 78% nitrogen, 21% oxygen, and 1% argon) at room temperature is heated to 4000 K, 200 kPa. Find the equilibrium composition at this state, assuming that only N_2, O_2, NO, O, and Ar are present.

13.44 Dry air is heated from 25°C to 4000 K in a 100-kPa constant-pressure process. List the possible reactions that may take place and determine the equilibrium composition. Find the required heat transfer.

13.45 Acetylene gas and x times theoretical air ($x > 1$) at room temperature and 500 kPa are burned at constant pressure in an adiabatic SSSF process. The flame temperature is 2600 K, and the combustion products are assumed to consist of N_2, O_2, CO_2, H_2O, CO, and NO. Determine the value of x.

13.46 The char residue (assumed to be pure carbon) from a coal pyrolysis process producing methane is to be gasified to produce a syngas fuel. At the gasifier inlet, 0.5 kmol of oxygen and x kmol of water enter for each kilomole of char. Exiting the gasifier is a homogeneous ideal-gas mixture in chemical equilibrium at T and P. The mixture contains CH_4, H_2, H_2O, CO and CO_2.

a. Determine the number of independent reaction equations for this system.

b. Set up the equilibrium equations, and specify the procedure to solve for the composition.

c. Comment on any restrictions on x, and on whether it is desirable to have x as small or large as possible.

13.47 One kilomole of water vapor at 100 kPa, 400 K, is heated to 3000 K in a constant pressure SSSF process. Determine the final composition, assuming that H_2O, H_2, H, O_2, and OH are present at equilibrium.

13.48 Methane is burned with theoretical oxygen in an SSSF process, and the products exit the combustion chamber at 3200 K, 700 kPa. Calculate the equilibrium composition at this state, assuming that only CO_2, CO, H_2O, H_2, O_2, and OH are present.

13.49 Operation of an MHD converter requires an electrically conducting gas. It is proposed to use helium gas "seeded" with 1.0 mole percent cesium, as shown in Fig. P13.49. The cesium is partly ionized ($Cs \rightleftharpoons Cs^+ + e^-$) by heating the mixture to 2500 K, 1 MPa, in a nuclear reactor to provide free electrons. No helium is ionized in this process, so that the mixture entering the converter consists of He, Cs, Cs^+, and e^-. To analyze the converter process, we need to know the precise mole fraction of electrons in the mixture. Determine this fraction. At 2500 K, $\ln K = -13.4$ for the cesium ionization reaction described.

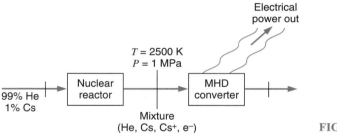

FIGURE P13.49

13.50 One kilomole of argon gas at room temperature is heated to 20 000 K, 100 kPa. Assume that the plasma in this condition consists of an equilibrium mixture of Ar, Ar^+, Ar^{++}, and e^- according to the simultaneous reactions

$$(1) \qquad Ar \rightleftharpoons Ar^+ + e^-$$

$$(2) \qquad Ar^+ \rightleftharpoons Ar^{++} + e^-$$

The ionization equilibrium constants for these reactions at 20 000 K have been calculated from spectroscopic data as

$$\ln K_1 = 3.11 \qquad \ln K_2 = -4.92$$

Determine the equilibrium composition of the plasma.

13.51 Plot to scale the equilibrium composition of nitrogen at 10 kPa over the temperature range 5000 K to 15 000 K, assuming that N_2, N, N^+, and e^- are present. For the ionization reaction $N \rightleftharpoons N^+ + e^-$, the ionization equilibrium constant K has been calculated from spectroscopic data as

T[K]	10 000	12 000	14 000	16 000
$100K$	6.26×10^{-2}	1.51	15.1	92

13.52 Hydrides are rare earth metals, M, that have the ability to react with hydrogen to form a different substance MH_x with a release of energy. The hydrogen can then be released, the reaction reversed, by heat addition to the MH_x. In this reaction only the hydrogen is a gas so the formula developed for the chemical equilibrium is inappropriate. Show that the proper expression to be used instead of Eq. 13.34 is

$$\ln(P_{H_2} / P_0) = \frac{\Delta G^\circ}{\overline{R}T}$$

when the reaction is scaled to 1 kmol of H_2.

13.53E Carbon dioxide at 2200 lbf/in.^{2} is injected into the top of a 3-mi deep well in connection with an enhanced oil recovery process. The fluid column standing in the well is at a uniform temperature of 100 F. What is the pressure at the bottom of the well?

13.54E Ethylene gas is burned with air, and the products exit at 250 F, 520 lbf/in.^{2}. Assuming complete combustion, what is the minimum percent theoretical air if there is to be no condensation of the products?

13.55E (Adv.) Ambient air at 68 F, 14.7 lbf/in.^{2}, 60% relative humidity, enters a large multistage compressor at the rate of 200 ft^3/s, and exits the unit at 210 F, 1500 lbf/in.^{2}. What fraction of the water is condensed at the exit? What is the change of thermodynamic availability of the air plus water for this change of state?

13.56E A mixture of 3 lb mol hydrogen and 1 lb mol carbon monoxide is contained in a rigid vessel at 77 F and an unknown pressure, P_1. A chemical reaction occurs, and the products consist solely of methane and water, at 400 F and 40 atm. Determine the initial pressure P_1 and the change $S_2 - S_1$ for the content of the vessel.

13.57E Hydrogen gas enters an SSSF combustion chamber at 77 F, 50 atm. A separate line supplies 80% theoretical air, which is throttled from 77 F, 150 atm, into the chamber. The products, consisting of H_2O, H_2, and N_2, exit from the chamber at 350 F, 50 atm. What is the irreversibility of this process?

13.58E Pure oxygen is heated from 77 F to 5300 F in an SSSF process at a constant pressure of 30 lbf/in.^{2}. Find the exit composition and the heat transfer.

13.59E Pure oxygen is heated from 77 F, 14.7 lbf/in.^{2} to 5300 F in a constant volume container. Find the final pressure, composition, and the heat transfer.

13.60E Air (assumed to be 79% nitrogen and 21% oxygen) is heated in an SSSF process at a constant pressure of 14.7 lbf/in.^{2}, and some NO is formed. At what temperature will the mole fraction of NO be 0.001?

13.61E The combustion products from burning pentane, C_5H_{12}, with pure oxygen in a stoichiometric ratio exists at 4400 R. Consider the dissociation of only CO_2 and find the equilibrium mole fraction of CO.

13.62E A gas mixture of 1 pound mol carbon monoxide, 1 pound mol nitrogen, and 1 pound mol oxygen at 77 F, 20 lbf/in.^{2}, is heated in a constant pressure SSSF process. The exit mixture can be assumed to be in chemical equilibrium with CO_2, CO, O_2, and N_2 present. The mole fraction of CO_2 at this point is 0.176. Calculate the heat transfer for the process.

13.63 In a test of a gas-turbine combustor, saturated-liquid methane at 210 R is to be burned with excess air to hold the adiabatic flame temperature to 2880 R. It is assumed that the products consist of a mixture of CO_2, H_2O, N_2, O_2, and NO in chemical equilibrium. Determine the percent excess air used in the combustion, and the percentage of NO in the products.

13.64 Acetylene gas at 77 F is burned with 140% theoretical air, which enters the burner at 77 F, 14.7 lbf/in.^{2}, 80% relative humidity. The combustion products form a mixture of CO_2, H_2O, N_2, O_2, and NO in chemical equilibrium at 3500 F, 14.7 lbf/in.^{2}. This mixture is then cooled to 1340 F very rapidly, so that the composition does not change. Determine the mole fraction of NO in the products and the heat transfer for the overall process.

13.65 An important step in the manufacture of chemical fertilizer is the production of ammonia, according to the reaction

$$N_2 + 3H_2 \rightleftharpoons 2NH_3$$

a. Calculate the equilibrium constant for this reaction at 300 F.

b. For an initial composition of 25% nitrogen, 75% hydrogen, on a mole basis, calculate the equilibrium composition at 300 F, 750 lbf/in.^{2}.

13.66 In rich (too much fuel) combustion the excess fuel may be broken down to give H_2 and CO may form. In the products at 2160 R, 30 lbf/in.^{2} the reaction called the water gas reaction may take place

$$CO_2 + H_2 \rightleftharpoons H_2O + CO$$

Find the equilibrium constant for this reaction from the elementary reactions.

13.67 Ethane is burned with 150% theoretical air in a gas turbine combustor. The products exiting consist of a mixture of CO_2, H_2O, O_2, N_2, and NO in chemical equilibrium at 2800 F, 150 lbf/in.^{2}. Determine the mole fraction of NO in the products. Is it reasonable to ignore CO in the products?

13.68 One pound mole of air (assumed to be 78% nitrogen, 21% oxygen, and 1% argon) at room temperature is heated to 7200 R, 30 lbf/in.^{2}. Find the equilibrium composition at this state, assuming that only N_2, O_2, NO, O, and Ar are present.

13.69 Dry air is heated from 77 F to 7200 R in a 14.7 lbf/in.^{2} constant-pressure process. List the possible reactions that may take place and determine the equilibrium composition. Find the required heat transfer.

13.70 Acetylene gas and x times theoretical air ($x > 1$) at room temperature and 75 lbf/in.^{2} are burned at constant pressure in an adiabatic SSSF process. The flame temperature is 4600 R, and the combustion products are assumed to consist of N_2, O_2, CO_2, H_2O, CO, and NO. Determine the value of x.

13.71 One pound mole of water vapor at 14.7 lbf/in.^{2}, 720 R, is heated to 5400 R in a constant pressure SSSF process. Determine the final composition, assuming that H_2O, H_2, H, O_2, and OH are present at equilibrium.

13.72E Methane is burned with theoretical oxygen in an SSSF process, and the products exit the combustion chamber at 5300 F, 100 lbf/in.^{2}. Calculate the equilibrium composition at this state, assuming that only CO_2, CO, H_2O, H_2, O_2, and OH are present.

COMPUTER, DESIGN, AND OPEN-ENDED PROBLEMS

13.73 Write a program to solve the general case of Problem 13.23. The effects of varying the relative amount of steam input and the effects of equilibrium temperature and pressure on the equilibrium composites are to be studied.

13.74 Write a program to solve the following problem. One kmol of carbon at 25°C is burned with b kmol of oxygen in a constant pressure adiabatic process. The products consist of an equilibrium mixture of CO_2, CO, and O_2. We wish to determine the flame temperature for various combinations of b and the pressure P, assuming variable specific heat for the components and equations from Table A.11.

13.75 Write a program to solve the general case of Problem 13.30, in which the percent theoretical air and the combustion products temperature are program input variables.

13.76 Write a program to solve the general case of Problem 13.31, in which the relative amount of steam input and the reactor temperature and pressure are program input variables.

13.77 Write a program to solve the general case of Problem 13.46. The effects of different operating temperatures and pressures of the gasifier are to be studied.

13.78 Write a program to solve the general case of Problem 13.48. The effects of the percent theoretical oxygen input and the exit temperature and pressure are to be studied.

13.79 Write a program to solve the following problem. One kilomole of water is heated to temperature T and pressure P, at which exists a mixture of H_2O, H_2, and O_2 in chemical equilibrium. We wish to find the equilibrium composition at this T and P by directly determining the minimum value of the system's Gibbs function in terms of the amount of water that has dissociated (see Fig. 13.3), and to compare the result with that found using the analytical procedure developed in Section 13.6.

13.80 Extend the program written for the solution of Problem 13.79 to include OH in the equilibrium mixture, in addition to the other components given in that problem.

13.81 Use the menu-driven equilibrium program to solve for the adiabatic flame temperature including chemical equilibrium in Problem 13.19.

13.82 Use the supplied menu-driven equilibrium program to do Problem 13.42 for a range of exit temperatures.

13.83 Use the supplied menu-driven equilibrium program to do Problem 13.48 for a range of exit temperatures.

13.84 Use the supplied menu-driven equilibrium program to calculate the adiabatic flame temperature in Problem 13.48 assuming all the reactions included in the program are activated.

13.85 Study the chemical reactions that take place when CFC-type refrigerants are released to the atmosphere. The chlorine may create compounds as HCl and $ClONO_2$ that react with the ozone O_3.

13.86 Examine the chemical equilibrium that takes place in an engine where CO and various nitrogen oxygen compounds summarized as NO_x may be formed. Study the processes for a range of air–fuel ratios and temperatures for typical fuels. Are there important reactions not listed in the book?

13.87 A number of products may be produced from the conversion of organic waste that can be used as fuel, see Problem 13.31. Study the subject and make a list of the major products that are formed and the conditions at which they are formed in desirable concentrations.

13.88 Using the supplied menu-driven software for chemical equilibrium calculations study the formation of CO as a function of temperature as done in Problem 13.27 for a single temperature.

13.89 The hydrides as explained in Problem 13.52 can store large amounts of hydrogen. The penalty for the storage is the energy must be supplied when the hydrogen should be released. Investigate the literature for quantitative information about the quantities and energy involved in such a hydrogen storage.

13.90 The hydrides explained in Problem 13.52 can be used in a chemical heat pump. The energy involved in the chemical reaction can be added and removed at different temperatures. For some hydrides these temperatures are low enough to make them feasible for heat pumps for heat upgrade, refrigerators, and air conditioners. Investigate the literature for such applications and give some typical values for these systems.

13.91 Power plants and engines have high peak temperatures in the combustion products where NO is produced. The equilibrium NO level at the high temperature is frozen at that level during the rapid drop in temperature with the expansion. The final exhaust therefore contains NO at a level much higher than the equilibrium value at the exhaust temperature. Study the NO level at equilibrium when natural gas, CH_4, is burned adiabatically with air (in at T_0) in various ratios.

13.92 Excess air or steam addition is often used to lower the peak temperature in combustion to limit formation of pollutants like NO. Study the steam addition to the combustion of natural gas as in the Cheng cycle, see Problem 11.128, assuming the steam is added before the combustion. How does this affect the peak temperature and the NO concentration?

FLOW THROUGH NOZZLES AND BLADE PASSAGES 14

This chapter deals with the thermodynamic aspects of one-dimensional flow through nozzles and blade passages. In addition, the momentum equation for the control volume is developed and applied to these same problems. The sonic velocity is defined in terms of thermodynamic properties, and the importance of the Mach number as a variable in compressible flow is noted.

14.1 STAGNATION PROPERTIES

In dealing with problems involving flow, many discussions and equations can be simplified by introducing the concept of the isentropic stagnation state and the properties associated with it. The isentropic stagnation state is the state a flowing fluid would attain if it underwent a reversible adiabatic deceleration to zero velocity. This state is designated in this chapter with the subscript 0. From the first law for a steady-state, steady-flow process we conclude that

$$h + \frac{\mathbf{V}^2}{2} = h_0 \tag{14.1}$$

The actual and the isentropic stagnation states for a typical gas or vapor are shown on the h–s diagram of Fig. 14.1. Sometimes it is advantageous to make a distinction between the actual and the isentropic stagnation states. The actual stagnation state is the state achieved after an actual deceleration to zero velocity (as at the nose of a body placed in a fluid stream), and there may be irreversibilities associated with the decelera-

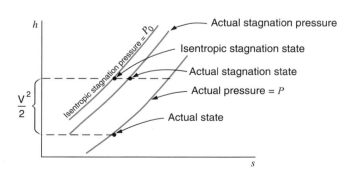

FIGURE 14.1
Enthalpy–entropy diagram illustrating the definition of stagnation state.

tion process. Therefore, the term stagnation property is sometimes reserved for the properties associated with the actual state, and the term total property is used for the isentropic stagnation state.

It is evident from Fig. 14.1 that the enthalpy is the same for both the actual and isentropic stagnation states (assuming that the actual process is adiabatic). Therefore, for an ideal gas, the actual stagnation temperature is the same as the isentropic stagnation temperature. However, the actual stagnation pressure may be less than the isentropic stagnation pressure and for this reason the term total pressure (meaning isentropic stagnation pressure) has particular meaning compared to the actual stagnation pressure.

EXAMPLE 14.1 Air flows in a duct at a pressure of 150 kPa with a velocity of 200 m/s. The temperature of the air is 300 K. Determine the isentropic stagnation pressure and temperature.

Analysis and Solution

If we assume that the air is an ideal gas with contant specific heat as given in Table A.10, the calculation is as follows. From Eq. 14.1

$$\frac{\mathbf{V}^2}{2} = h_0 - h = C_{p0}(T_0 - T)$$

$$\frac{(200)^2}{2 \times 1000} = 1.0035(T_0 - 300)$$

$$T_0 = 319.9 \text{ K}$$

The stagnation pressure can be found from the relation

$$\frac{T_0}{T} = \left(\frac{P_0}{P}\right)^{(k-1)/k}$$

$$\frac{319.9}{300} = \left(\frac{P_0}{150}\right)^{0.286}$$

$$P_0 = 187.8 \text{ kPa}$$

The Air Tables, Table A.12, which are calculated from Table A.13, could also have been used, and then the variation of specific heat with temperature would have been taken into account. Since the actual and stagnation states have the same entropy, we proceed as follows: Using Table A.12,

$$T = 300 \text{ K} \qquad h = 300.47 \qquad P_r = 1.1146$$

$$h_0 = h + \frac{\mathbf{V}^2}{2} = 300.47 + \frac{(200)^2}{2 \times 1000} = 320.47$$

$$T_0 = 320 \text{ K} \qquad P_{r0} = 1.3956$$

$$P_0 = 150 \times \frac{1.3956}{1.1146} = 187.8 \text{ kPa}$$

14.2 THE MOMENTUM EQUATION FOR THE CONTROL VOLUME

Before proceeding it will be advantageous to develop the momentum equation for the control volume. Newton's second law states that the sum of the external forces acting on a body in a given direction is proportional to the rate of change of momentum in the given direction. Writing this in equation form for the x-direction we have

$$\sum F_x \propto \frac{d(m\mathbf{V}_x)}{dt}$$

For the system of units used in this book, this proportionality can be written directly as an equality.

$$\sum F_x = \frac{d(m\mathbf{V}_x)}{dt} \tag{14.2}$$

Equation 14.2 has been written for a body of fixed mass, or in thermodynamic parlance, for a control mass. We now proceed to write the momentum equation for a control volume, and follow a procedure similar to that used in writing the continuity equation and the first and second laws of thermodynamics for a control volume.

Consider the control mass and control volume shown in Fig. 14.2. Let the control volume be fixed relative to its coordinate frame. During the time interval δt, the mass δm_i enters the control volume with a velocity $(\mathbf{V}_r)_i$ and velocity components $(\mathbf{V}_x)_i$, $(\mathbf{V}_y)_i$, and $(\mathbf{V}_z)_i$. During this same time interval the mass δm_e leaves the control volume, with velocity $(\mathbf{V}_r)_e$ and velocity components $(\mathbf{V}_x)_e$, $(\mathbf{V}_y)_e$, and $(\mathbf{V}_z)_e$.

If we write the x-momentum equation for the control mass during this time interval we have

$$\left(\sum F_x\right)_{av} = \frac{\Delta(m\mathbf{V}_x)}{\delta t} = \frac{(m\mathbf{V}_x)_2 - (m\mathbf{V}_x)_1}{\delta t} \tag{14.3}$$

Let $(m\mathbf{V}_x)_t = x$ - momentum in the control volume at time t

$(m\mathbf{V}_x)_{t+\delta t} = x$ - momentum in the control volume at time $t + \delta t$

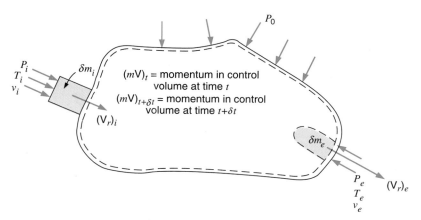

FIGURE 14.2 Schematic diagram for the development of the momentum equation for a control volume.

Then

$$(m\mathbf{V}_x)_1 = (m\mathbf{V}_x)_t + (\mathbf{V}_x)_i \,\delta m_i = x\text{-momentum of the control mass at time } t.$$

$$(m\mathbf{V}_x)_2 = (m\mathbf{V}_x)_{t+\delta t} + (\mathbf{V}_x)_e \,\delta m_e = x\text{-momentum of the control mass at time } t+\delta t.$$

It follows that

$$(m\mathbf{V}_x)_2 - (m\mathbf{V}_x)_1 = [(m\mathbf{V}_x)_{t+\delta t} - (\mathbf{V}_x)_t] + [(m\mathbf{V}_x)_e \,\delta m_e - (\mathbf{V}_x)_i \,\delta m_i] \qquad (14.4)$$

The first bracketed term on the right side of Eq. 14.4 represents the change of x-momentum within the control volume during the time interval δt, and the second bracketed term in that equation represents the x-directional momentum flow across the control surface during δt. Now, dividing Eq. 14.4 by δt and substituting into Eq. 14.3,

$$\left(\sum F_x\right)_{av} = \frac{(m\mathbf{V}_x)_{t+\delta t} - (m\mathbf{V}_x)_t}{\delta t} + \frac{(\mathbf{V}_x)_e \,\delta m_e - (\mathbf{V}_x)_i \,\delta m_i}{\delta t} \qquad (14.5)$$

We now establish the limit for each of the terms in this expression as $\delta t \to 0$.

$$\lim_{\delta t \to 0} \left(\sum F_x\right)_{av} = \sum F_x$$

$$\lim_{\delta t \to 0} \left[\frac{(m\mathbf{V}_x)_{t+\delta t} - (m\mathbf{V}_x)_t}{\delta t} \right] = \frac{d(m\mathbf{V}_x)_{c.v.}}{dt}$$

$$\lim_{\delta t \to 0} \left[\frac{(\mathbf{V}_x)_e \,\delta m_e - (\mathbf{V}_x)_i \,\delta m_i}{\delta t} \right] = \sum \dot{m}_e (\mathbf{V}_e)_x - \sum \dot{m}_i (\mathbf{V}_i)_x \qquad (14.6)$$

Thus, as $\delta t \to 0$, we have a rate form of the momentum equation for the control volume.

$$\sum F_x = \frac{d(m\mathbf{V}_x)_{c.v.}}{dt} + \sum \dot{m}_e (\mathbf{V}_e)_x - \sum \dot{m}_i (\mathbf{V}_i)_x \qquad (14.7)$$

Similar equations can also be written for the y- and z-directions.

$$\sum F_y = \frac{d(m\mathbf{V}_y)_{c.v.}}{dt} + \sum \dot{m}_e (\mathbf{V}_e)_y - \sum \dot{m}_i (\mathbf{V}_i)_y \qquad (14.8)$$

$$\sum F_z = \frac{d(m\mathbf{V}_x)_{c.v.}}{dt} + \sum \dot{m}_e (\mathbf{V}_e)_z - \sum \dot{m}_i (\mathbf{V}_i)_z \qquad (14.9)$$

In this chapter we will be concerned primarily with steady-state, steady-flow processes in which there is a single flow with uniform properties into the control volume, and a single flow with uniform properties out of the control volume. The SSSF assumption means that the rate of momentum change for the control volume terms in Eqs. 14.7, 14.8, and 14.9 are equal to zero. That is,

$$\frac{d(m\mathbf{V}_x)_{c.v.}}{dt} = 0 \qquad \frac{d(m\mathbf{V}_y)_{c.v.}}{dt} = 0 \qquad \frac{d(m\mathbf{V}_z)_{c.v.}}{dt} = 0 \qquad (14.10)$$

Therefore for the SSSF process the momentum equation for the control volume, assuming uniform properties at each state, reduces to the form

$$\sum F_x = \sum \dot{m}_e (\mathbf{V}_e)_x - \sum \dot{m}_i (\mathbf{V}_i)_x \qquad (14.11)$$

$$\sum F_y = \sum \dot{m}_e (\mathbf{V}_e)_y - \sum \dot{m}_i (\mathbf{V}_i)_y \qquad (14.12)$$

$$\sum F_z = \sum \dot{m}_e (\mathbf{V}_e)_z - \sum \dot{m}_i (\mathbf{V}_i)_z \qquad (14.13)$$

Further, for the special case in which there is a single flow into and out of the control volume, these equations reduce to

$$\sum F_x = \dot{m}[(\mathbf{V}_e)_x - (\mathbf{V}_i)_x] \tag{14.14}$$

$$\sum F_y = \dot{m}[(\mathbf{V}_e)_y - (\mathbf{V}_i)_y] \tag{14.15}$$

$$\sum F_z = \dot{m}[(\mathbf{V}_e)_z - (\mathbf{V}_i)_z] \tag{14.16}$$

EXAMPLE 14.2   On a level floor a man is pushing a wheelbarrow (Fig. 14.3) into which sand is falling at the rate of 1 kg/s. The man is walking at the rate of 1 m/s and the sand has a velocity of 10 m/s as it falls into the wheelbarrow. Determine the force the man must exert on the wheelbarrow and the force the floor exerts on the wheelbarrow due to the falling sand.

Analysis and Solution

Consider a control surface around the wheelbarrow. Consider first the x-direction. From Eq. 14.7

$$\sum F_x = \frac{d(m\mathbf{V}_x)_{\text{c.v.}}}{dt} + \sum \dot{m}_e(\mathbf{V}_e)_x - \sum \dot{m}_i(\mathbf{V}_i)_x$$

Let us analyze this problem from the point of view of an observer riding on the wheelbarrow. For this observer, $\mathbf{V}_x$ of the material in the wheelbarrow is zero and therefore,

$$\frac{d(m\mathbf{V}_x)_{\text{c.v.}}}{dt} = 0$$

However, for this observer the sand crossing the control surface has an x-component velocity of −1 m/s, and ṁ, the mass flow out of the control volume, is −1 kg/s. Therefore,

$$F_x = (1 \text{ kg}/\text{s}) \times (1 \text{ m}/\text{s}) = 1 \text{ N}$$

If one considers this from the point of view of an observer who is stationary on the earth's surface we conclude that $\mathbf{V}_x$ of the falling sand is zero and therefore

$$\sum \dot{m}_e(\mathbf{V}_e)_x - \sum \dot{m}_i(\mathbf{V}_i)_x = 0$$

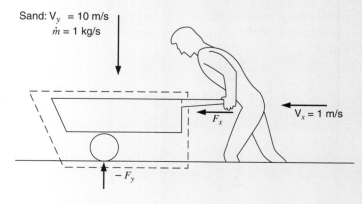

Sand: $V_y = 10$ m/s
$\dot{m} = 1$ kg/s

F_x

$V_x = 1$ m/s

$-F_y$

FIGURE 14.3
Sketch for Example 14.2.

However, for this observer there is a change of momentum within the control volume, namely,

$$\sum F_x = \frac{d(m\mathbf{V}_x)_{\text{c.v.}}}{dt} = (1 \text{ m/s}) \times (1 \text{ kg/s}) = 1 \text{ N}$$

Next consider the vertical (y) direction.

$$\sum F_y = \frac{d(m\mathbf{V}_y)_{\text{c.v.}}}{dt} + \sum \dot{m}_e (\mathbf{V}_e)_y - \sum \dot{m}_i (\mathbf{V}_i)_y$$

For both the stationary and moving observer the first term drops out because $\mathbf{V}_y$ of the mass within the control volume is zero. However, for the mass crossing the control surface, $\mathbf{V}_y = 10$ m/s and

$$\dot{m} = -1 \text{ kg/s}$$

Therefore,

$$F_y = (10 \text{ m/s}) \times (-1 \text{ kg/s}) = -10 \text{ N}$$

The minus sign indicates that the force is in the opposite direction to $\mathbf{V}_y$.

14.3 FORCES ACTING ON A CONTROL SURFACE

In the last section we considered the momentum equation for the control volume. We now wish to evaluate the net force on a control surface that causes this change in momentum. Let us do this by considering the control mass shown in Fig. 14.4, which involves a pipe bend. The control surface is designated by the dotted lines, and is so chosen that at the point where the fluid crosses the boundary of the control mass are assumed to be negligible. Figure 14.4a shows the velocities and Fig. 14.4b shows the forces involved. The force R is the result of all external forces on the control mass, except for the pressure of all surroundings. The pressure of the surroundings, P_0, acts on the entire boundary except at $\mathcal{A}_i$ and $\mathcal{A}_e$, where the fluid crosses the control surface, P_i and P_e represent the absolute pressures at these points.

The net forces acting on the system in the x- and y-directions, F_x and F_y, are the sum of the pressure forces and the external force R in their respective directions. The influence of the pressure of the surroundings, P_0, is most easily taken into account by noting that it acts over the entire control mass boundary except at $\mathcal{A}_i$ and $\mathcal{A}_e$. Therefore, we can write

$$\sum F_x = (P_i \mathcal{A}_i)_x - (P_0 \mathcal{A}_i)_x + (P_e \mathcal{A}_e)_x - (P_0 \mathcal{A}_e)_x + R_x$$

$$\sum F_y = (P_i \mathcal{A}_i)_y - (P_0 \mathcal{A}_i)_y + (P_e \mathcal{A}_e)_y - (P_0 \mathcal{A}_e)_y + R_y$$

This equation may be simplified by combining the pressure terms.

$$\sum F_x = [(P_i - P_0)\mathcal{A}_i]_x + [(P_e - P_0)\mathcal{A}_e]_x + R_x$$

$$\sum F_y = [(P_i - P_0)\mathcal{A}_i]_y + [(P_e - P_0)\mathcal{A}_e]_y + R_y \qquad (14.17)$$

The proper sign for each pressure and force must of course be used in all calculations.

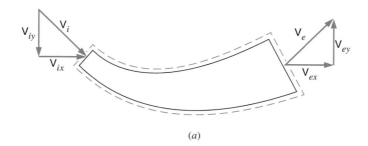

(a)

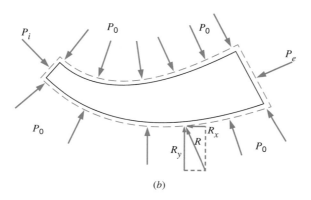

(b)

FIGURE 14.4
Forces acting on a
control surface.

Equations 14.11, 14.12, and 14.17 may be combined to give

$$\sum F_x = \sum \dot{m}_e (\mathbf{V}_e)_x - \sum \dot{m}_i (\mathbf{V}_i)_x$$

$$= \sum [(P_i - P_0)\mathcal{A}_i]_x + \sum [(P_e - P_0)\mathcal{A}_e]_x + R_x$$

$$\sum F_y = \sum \dot{m}_e (\mathbf{V}_e)_y - \sum \dot{m}_i (\mathbf{V}_i)_y$$

$$= \sum [(P_i - P_0)\mathcal{A}_i]_y + \sum [(P_e - P_0)\mathcal{A}_e]_y + R_y \qquad (14.18)$$

If there is a single flow across the control surface, Eqs. 14.14, 14.15, and 14.17 can be
combined to give

$$\sum F_x = \dot{m}(\mathbf{V}_e - \mathbf{V}_i)_x = [(P_i - P_0)\mathcal{A}_i]_x + [(P_e - P_0)\mathcal{A}_e]_x + R_x$$

$$\sum F_y = \dot{m}(\mathbf{V}_e - \mathbf{V}_i)_y = [(P_i - P_0)\mathcal{A}_i]_y + [(P_e - P_0)\mathcal{A}_e]_y + R_y \qquad (14.19)$$

A similar equation could be written for the z-direction. These equations are very useful in
analyzing the forces involved in a control volume analysis.

EXAMPLE 14.3 A jet engine is being tested on a test stand (Fig. 14.5). The inlet area to the compressor
is 0.2 m^2 and air enters the compressor at 95 kPa, 100 m/s. The pressure of the atmosphere is 100 kPa. The exit area of the engine is 0.1 m^2, and the products of combustion
leave the exit plane at a pressure of 125 kPa and a velocity of 450 m/s. The air–fuel
ratio is 50 kg air/kg fuel, and the fuel enters with a low velocity. The rate of air flow
entering the engine is 20 kg/s. Determine the thrust on the engine.

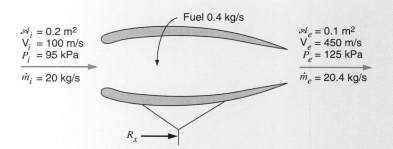

FIGURE 14.5
Sketch for
Example 14.3.

Analysis and Solution

In the solution that follows it is assumed that forces and velocities to the right are positive.

Using Eq. 14.19

$$R_x + [(P_i - P_0)\mathscr{A}_i]_x + [(P_e - P_0)\mathscr{A}_e]_x = (\dot{m}_e\mathbf{V}_e - \dot{m}_i\mathbf{V}_i)_x$$

$$R_x + [(95-100)\times 0.2] - [(125-100)\times 0.1] = \frac{20.4\times 450 - 20\times 100}{1000}$$

$$R_x = 10.68 \text{ kN}$$

(Note that the momentum of the fuel entering has been neglected.)

14.4 ADIABATIC, ONE-DIMENSIONAL, STEADY-STATE STEADY FLOW OF AN INCOMPRESSIBLE FLUID THROUGH A NOZZLE

A nozzle is a device in which the kinetic energy of a fluid is increased in an adiabatic process. This increase involves a decrease in pressure and is accomplished by the proper change in flow area. A diffuser is a device that has the opposite function, namely, to increase the pressure by decelerating the fluid. In this section we discuss both nozzles and diffusers, but to minimize words we shall use only the term nozzle.

Consider the nozzle shown in Fig. 14.6, and assume an adiabatic, one-dimensional, steady-state steady-flow process of an incompressible fluid. From the continuity equation

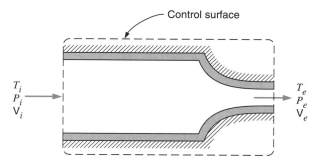

FIGURE 14.6 Schematic sketch of a nozzle.

we conclude that

$$\dot{m}_e = \dot{m}_i = \rho \mathcal{A}_i \mathbf{V}_i = \rho \mathcal{A}_e \mathbf{V}_e$$

or

$$\frac{\mathcal{A}_i}{\mathcal{A}_e} = \frac{\mathbf{V}_e}{\mathbf{V}_i} \qquad (14.20)$$

The first law for this process is

$$h_e - h_i + \frac{\mathbf{V}_e^2 - \mathbf{V}_i^2}{2} + (Z_e - Z_i)g = 0 \qquad (14.21)$$

From the second law we conclude that $s_e \geq s_i$, where the equality holds for a reversible process. Therefore, from the relation

$$T \, ds = dh - v \, dP$$

we conclude that for the reversible process

$$h_e - h_i = \int_i^e v \, dP \qquad (14.22)$$

If we assume that the fluid is incompressible, Eq. 14.22 can be integrated to give

$$h_e - h_i = v(P_e - P_i) \qquad (14.23)$$

Substituting this in Eq. 14.21 we have

$$v(P_e - P_i) + \frac{\mathbf{V}_e^2 - \mathbf{V}_i^2}{2} + (Z_e - Z_i)g = 0 \qquad (14.24)$$

This is of course the Bernoulli equation, which was derived in Section 7.14, Eq. 7.62, and for the reversible, adiabatic, one-dimensional, steady-state steady flow of an incompressible fluid through a nozzle the Bernoulli equation represents a combined statement of the first and second laws of thermodynamics.

EXAMPLE 14.4 Water enters the diffuser in a pump casing with a velocity of 30 m/s, a pressure of 350 kPa, and a temperature of 25°C. It leaves the diffuser with a velocity of 7 m/s and a pressure of 600 kPa. Determine the exit pressure for a reversible diffuser with these inlet conditions and exit velocity. Determine the increase in enthalpy, internal energy, and entropy for the actual diffuser.

Analysis and Solution

Consider first a control surface around a reversible diffuser with the given inlet conditions and exit velocity. Equation 14.24, the Bernoulli equation, is a statement of the first and second laws of thermodynamics for this process. Since there is no change in elevation this equation reduces to

$$v[(P_e)_s - P_i] + \frac{\mathbf{V}_e^2 - \mathbf{V}_i^2}{2} = 0$$

where $(P_e)_s$ represents the exit pressure for the reversible diffuser. From the steam tables, $v = 0.001\,003$ m³/kg.

$$P_{es} - P_i = \frac{(30)^2 - (7)^2}{0.001\,003 \times 2 \times 1000} = 424 \text{ kPa}$$

$$P_{es} = 774 \text{ kPa}$$

Next consider a control surface around the actual diffuser. The change in enthalpy can be found from the first law for this process, Eq. 14.21.

$$h_e - h_i = \frac{\mathbf{V}_i^2 - \mathbf{V}_e^2}{2} = \frac{(30)^2 - (7)^2}{2 \times 1000} = 0.4255 \text{ kJ / kg}$$

The change in internal energy can be found from the definition of enthalpy, $h_e - h_i = (u_e - u_i) + (P_e v_e - P_i v_i)$.

Thus, for an incompressible fluid

$$u_e - u_i = h_e - h_i - v(P_e - P_i)$$

$$= 0.4255 - 0.001\,003(600 - 350)$$

$$= 0.174\,75 \text{ kJ / kg}$$

The change of entropy can be approximated from the familiar relation

$$T\,ds = du + P\,dv$$

by assuming that the temperature is constant (which is approximately true in this case) and noting that for an incompressible fluid $dv = 0$. With these assumptions

$$s_e - s_i = \frac{u_e - u_i}{T} = \frac{0.174\,75}{298.2} = 0.000\,586 \text{ kJ / kg K}$$

Since this is an irreversible adiabatic process, the entropy will increase, as the above calculation indicates.

14.5 VELOCITY OF SOUND IN AN IDEAL GAS

When a pressure disturbance occurs in a compressible fluid, the disturbance travels with a velocity that depends on the state of the fluid. A sound wave is a very small pressure disturbance; the velocity of sound, also called the sonic velocity, is an importance parameter in compressible-fluid flow. We proceed now to determine an expression for the sonic velocity of an ideal gas in terms of the properties of the gas.

Let a disturbance be set up by the movement of the piston at the end of the tube, Fig. 14.7a. A wave travels down the tube with a velocity c, which is the sonic velocity. Assume that after the wave has passed the properties of the gas have changed an infinitesimal amount and that the gas is moving with the velocity $d\mathbf{V}$ toward the wave front.

In Fig. 14.7b this process is shown from the point of view of an observer who travels with the wave front. Consider the control surface shown in Fig. 14.7b. From the first

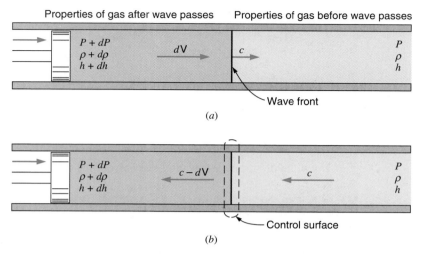

Properties of gas after wave passes   Properties of gas before wave passes

(a)

(b)

FIGURE 14.7   Diagram illustrating sonic velocity. (*a*) Stationary observer. (*b*) Observer traveling with wave front.

law for this steady-state steady-flow process we can write

$$h + \frac{c^2}{2} = (h + dh) + \frac{(c - d\mathbf{V})^2}{2}$$

$$dh - c \, d\mathbf{V} = 0 \tag{14.25}$$

From the continuity equation we can write

$$\rho \mathcal{A} c = (\rho + d\rho)\mathcal{A}(c - d\mathbf{V})$$

$$c \, d\rho - \rho \, d\mathbf{V} = 0 \tag{14.26}$$

Consider also the relation between properties

$$T \, ds = dh - \frac{dP}{\rho}$$

If the process is isentropic, $ds = 0$, and this equation can be combined with Eq. 14.25 to give the relation

$$\frac{dP}{\rho} - c \, d\mathbf{V} = 0 \tag{14.27}$$

This can be combined with Eq. 14.26 to give the relation

$$\frac{dP}{d\rho} = c^2$$

Since we have assumed the process to be isentropic this is better written as a partial derivative.

$$\left(\frac{\partial P}{\partial \rho} \right)_s = c^2 \tag{14.28}$$

An alternate derivation is to introduce the momentum equation. For the control volume of Fig. 14.7b the momentum equation is

$$P\mathcal{A} - (P + dP)\mathcal{A} = \dot{m}(c - d\mathbf{V} - c) = \rho \mathcal{A} c(c - d\mathbf{V} - c)$$

$$d\mathbf{V} = \rho c\, dP \tag{14.29}$$

On combining this with Eq. 14.26 we obtain Eq. 14.28.

$$\left(\frac{\partial P}{\partial \rho}\right)_s = c^2$$

It will be of particular advantage to solve Eq. 14.28 for the velocity of sound in an ideal gas.

When an ideal gas undergoes an isentropic change of state, we found in Chapter 7 that, for this process, assuming constant specific heat

$$\frac{dP}{P} - k\frac{d\rho}{\rho} = 0$$

or

$$\left(\frac{\partial P}{\partial \rho}\right)_s = \frac{kP}{\rho}$$

Substituting this equation in Eq. 14.28 we have an equation for the velocity of sound in an ideal gas,

$$c^2 = \frac{kP}{\rho} \tag{14.30}$$

Since for an ideal gas

$$\frac{P}{\rho} = RT$$

this equation may also be written

$$c^2 = kRT \tag{14.31}$$

EXAMPLE 14.5   Determine the velocity of sound in air at 300 K and at 1000 K.

Analysis and Solution

Using Eq. 14.31

$$c = \sqrt{kRT}$$

$$= \sqrt{1.4 \times 0.287 \times 300 \times 1000} = 347.2 \text{ m/s}$$

Similarly, at 1000 K, using $k = 1.4$,

$$c = \sqrt{1.4 \times 0.287 \times 1000 \times 1000} = 633.9 \text{ m/s}$$

Note the significant increase in sonic velocity as the temperature increases.

The Mach number, M, is defined as the ratio of the actual velocity $\mathbf{V}$ to the sonic velocity c.

$$M = \frac{\mathbf{V}}{c} \tag{14.32}$$

When $M > 1$ the flow is supersonic; when $M < 1$ the flow is subsonic; and when $M = 1$ the flow is sonic. The importance of the Mach number as a parameter in fluid-flow problems will be evident in the paragraphs that follow.

14.6 REVERSIBLE, ADIABATIC, ONE-DIMENSIONAL STEADY FLOW OF AN IDEAL GAS THROUGH A NOZZLE

A nozzle or diffuser with both a converging and diverging section is shown in Fig. 14.8. The minimum cross-sectional area is called the throat.

Our first consideration concerns the conditions that determine whether a nozzle or diffuser should be converging or diverging, and the conditions that prevail at the throat. For the control volume shown the following relations can be written.

First law:

$$dh + \mathbf{V}\, d\mathbf{V} = 0 \tag{14.33}$$

Property relation:

$$T\, ds = dh - \frac{dP}{\rho} = 0 \tag{14.34}$$

Continuity equation:

$$\rho \mathscr{A} \mathbf{V} = \dot{m} = \text{constant}$$

$$\frac{d\rho}{\rho} + \frac{d\mathscr{A}}{\mathscr{A}} + \frac{d\mathbf{V}}{\mathbf{V}} = 0 \tag{14.35}$$

Combining Eqs. 14.33 and 14.34 we have

$$dh = \frac{dP}{\rho} = -\mathbf{V}\, d\mathbf{V}$$

$$d\mathbf{V} = -\frac{1}{\rho \mathbf{V}}\, dP$$

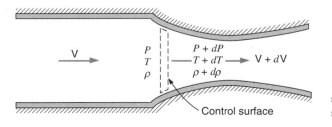

FIGURE 14.8 One-dimensional, reversible, adiabatic steady flow through a nozzle.

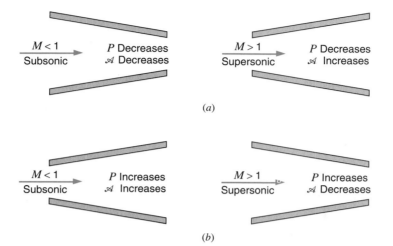

FIGURE 14.9
Required area
changes for (a) noz-
zles and (b) diffusers.

Substituting this in Eq. 14.35

$$\frac{d\mathcal{A}}{\mathcal{A}} = \left(-\frac{d\rho}{\rho} - \frac{d\mathbf{V}}{\mathbf{V}} \right) = -\frac{d\rho}{\rho} \left(\frac{dP}{dP} \right) + \frac{1}{\rho\mathbf{V}^2} dP$$

$$= \frac{-dP}{\rho} \left(\frac{d\rho}{dP} - \frac{1}{\mathbf{V}^2} \right) = \frac{dP}{\rho} \left(-\frac{1}{(dP/d\rho)} + \frac{1}{\mathbf{V}^2} \right)$$

Since the flow is isentropic

$$\frac{dP}{d\rho} = c^2 = \frac{\mathbf{V}^2}{M^2}$$

and therefore,

$$\frac{d\mathcal{A}}{\mathcal{A}} = \frac{dP}{\rho\mathbf{V}^2}(1 - M^2) \tag{14.36}$$

This is a very significant equation, for from it we can draw the following conclusions about the proper shape for nozzles and diffusers:

For a nozzle, $dP < 0$. Therefore,

for a subsonic nozzle, $M < 1$, $d\mathcal{A} < 0$, and the nozzle is converging;

for a supersonic nozzle, $M > 1$, $d\mathcal{A} > 0$, and the nozzle is diverging.

For a diffuser, $dP > 0$. Therefore,

for a subsonic diffuser, $M < 1$, $\delta\mathcal{A} > 0$, and the diffuser is diverging;

for a supersonic diffuser, $M > 1$, $d\mathcal{A} < 0$, and the diffuser is converging.

When $M = 1$, $d\mathcal{A} = 0$, which means that sonic velocity can be achieved only at the throat of a nozzle or diffuser. These conclusions are summarized in Fig. 14.9.

We will now develop a number of relations between the actual properties, stagnation properties, and Mach number. These relations are very useful in dealing with isentropic flow of an ideal gas in a nozzle.

Equation 14.1 gives the relation between enthalpy, stagnation enthalpy, and kinetic energy.

$$h + \frac{\mathbf{V}^2}{2} = h_0$$

For an ideal gas with constant specific heat Eq. 14.1 can be written

$$\mathbf{V}^2 = 2C_{po}(T_0 - T) = 2\frac{kRT}{k-1}\left(\frac{T_0}{T} - 1\right)$$

Since

$$c^2 = kRT$$

$$\mathbf{V}^2 = \frac{2c^2}{k-1}\left(\frac{T_0}{T} - 1\right)$$

$$\frac{\mathbf{V}^2}{c^2} = M^2 = \frac{2}{k-1}\left(\frac{T_0}{T} - 1\right)$$

$$\frac{T_0}{T} = 1 + \frac{(k-1)}{2}M^2 \tag{14.37}$$

For an isentropic process,

$$\left(\frac{T_0}{T}\right)^{k/(k-1)} = \frac{P_0}{P} \qquad \left(\frac{T_0}{T}\right)^{1/(k-1)} = \frac{\rho_0}{\rho}$$

Therefore,

$$\frac{P_0}{P} = \left[1 + \frac{(k-1)}{2}M^2\right]^{k/(k-1)} \tag{14.38}$$

$$\frac{\rho_0}{\rho} = \left[1 + \frac{(k-1)}{2}M^2\right]^{1/(k-1)} \tag{14.39}$$

Values of P/P_0, ρ/ρ_0, and T/T_0 are given as a function of M in Table A.18 for the value $k = 1.40$. In the software version of Table A.18, both M and k are input variables.

The conditions at the throat of the nozzle can be found by noting that $M = 1$ at the throat. The properties at the throat are denoted by an asterisk (*). Therefore,

$$\frac{T^*}{T_0} = \frac{2}{k+1} \tag{14.40}$$

$$\frac{P^*}{P_0} = \left(\frac{2}{k+1}\right)^{k/(k-1)} \tag{14.41}$$

$$\frac{\rho^*}{\rho_0} = \left(\frac{2}{k+1}\right)^{1/(k-1)} \tag{14.42}$$

These properties at the throat of a nozzle when $M = 1$ are frequently referred to as critical pressure, critical temperature, and critical density and the ratios given by Eqs. 14.40,

TABLE 14.1

Critical Pressure, Density, and Temperature Ratios
for Isentropic Flow of an Ideal Gas

	k = 1.1	k = 1.2	k = 1.3	k = 1.4	k = 1.67
P^*/P_0	0.5847	0.5644	0.5457	0.5283	0.4867
ρ^*/ρ_0	0.6139	0.6209	0.6276	0.6340	0.6497
T^*/T_0	0.9524	0.9091	0.8696	0.8333	0.7491

14.41, and 14.42 are referred to as the critical-temperature ratio, critical-pressure ratio, and critical-density ratio. Table 14.1 gives these ratios for various values of k.

14.7 MASS RATE OF FLOW OF AN IDEAL GAS THROUGH AN ISENTROPIC NOZZLE

We now turn our attention to a consideration of the mass rate of flow per unit area, $\dot{m}/\mathscr{A}$, in a nozzle. From the continuity equation we proceed as follows:

$$\frac{\dot{m}}{\mathscr{A}} = \rho \mathbf{V} = \frac{P\mathbf{V}}{RT}\sqrt{\frac{kT_0}{kT_0}}$$

$$= \frac{P\mathbf{V}}{\sqrt{kRT}}\sqrt{\frac{k}{R}}\sqrt{\frac{T_0}{T}}\sqrt{\frac{1}{T_0}}$$

$$= \frac{PM}{\sqrt{T_0}}\sqrt{\frac{k}{R}}\sqrt{1+\frac{k-1}{2}M^2} \tag{14.43}$$

By substituting Eq. 14.38 into Eq. 14.42 the flow per unit area can be expressed in terms of stagnation pressure, stagnation temperature, Mach number, and gas properties.

$$\frac{\dot{m}}{\mathscr{A}} = \frac{P_0}{\sqrt{T_0}}\sqrt{\frac{k}{R}} \times \frac{M}{\left(1+\dfrac{k-1}{2}M^2\right)^{(k+1)/2(k-1)}} \tag{14.44}$$

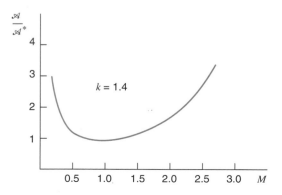

FIGURE 14.10 Area ratio as a function of Mach number for a reversible, adiabatic nozzle.

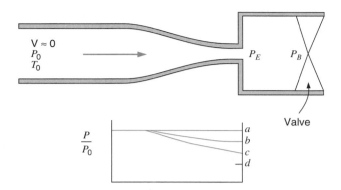

FIGURE 14.11 Pressure ratio as a function of back pressure for a convergent nozzle.

At the throat, $M = 1$, and therefore the flow per unit area at the throat, $\dot{m}/\mathscr{A}^*$, can be found by setting $M = 1$ in Eq. 14.44.

$$\frac{\dot{m}}{\mathscr{A}^*} = \frac{P_0}{\sqrt{T_0}}\sqrt{\frac{k}{R}} \times \frac{1}{\left(\dfrac{k+1}{2}\right)^{(k+1)/2(k-1)}} \tag{14.45}$$

The area ratio $\mathscr{A}/\mathscr{A}^*$ can be obtained by dividing Eq. 14.45 by Eq. 14.44.

$$\frac{\mathscr{A}}{\mathscr{A}^*} = \frac{1}{M}\left[\left(\frac{2}{k+1}\right)\left(1 + \frac{k-1}{2}M^2\right)\right]^{(k+1)/2(k-1)} \tag{14.46}$$

The area ratio $\mathscr{A}/\mathscr{A}^*$ is the ratio of the area at the point where the Mach number is M to the throat area, and values of $\mathscr{A}/\mathscr{A}^*$ as a function of Mach number are given in Table A.18 in the Appendix. Figure 14.10 shows a plot of $\mathscr{A}/\mathscr{A}^*$ vs. M, which is in accordance with our previous conclusion that a subsonic nozzle is converging and a supersonic nozzle is diverging.

The final point to be made regarding the isentropic flow of an ideal gas through a nozzle involves the effect of varying the back pressure (the pressure outside the nozzle exit) on the mass rate of flow.

Consider first a convergent nozzle as shown in Fig. 14.11, which also shows the pressure ratio P/P_0 along the length of the nozzle. The conditions upstream are the stagnation conditions, which are assumed to be constant. The pressure at the exit plane of the nozzle is designated P_E, and the back pressure P_B. Let us consider how the mass rate of

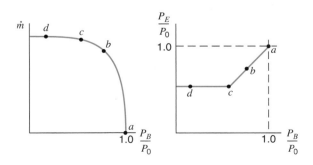

FIGURE 14.12 Mass rate of flow and exit pressure as a function of back pressure for a convergent nozzle.

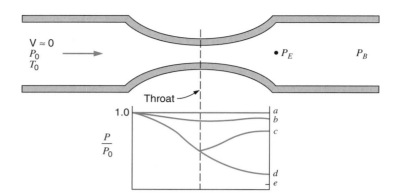

FIGURE 14.13
Nozzle pressure ratio as a function of back pressure for a reversible, convergent–divergent nozzle.

flow $\dot{m}$ and the exit plane pressure P_E/P_0 vary as the back pressure P_B is decreased. These quantities are plotted in Fig. 14.12.

When $P_B/P_0 = 1$ there is of course no flow, and $P_E/P_0 = 1$ as designated by point a. Next let the back pressure P_B be lowered to that designated by point b, so that P_B/P_0 is greater than the critical-pressure ratio. The mass rate of flow has a certain value and $P_E = P_B$. The exit Mach number is less than 1. Next let the back pressure be lowered to the critical pressure, designated by point c. The Mach number at the exit is now unity, and P_E is equal to P_B. When P_B is decreased below the critical pressure, designated by point d, there is no further increase in the mass rate of flow, and P_E remains constant at a value equal to the critical pressure, and the exit Mach number is unity. The drop in pressure from P_E to P_B takes place outside the nozzle exit. Under these conditions the nozzle is said to be choked, which means that for given stagnation conditions the nozzle is passing the maximum possible mass flow.

Consider next a convergent–divergent nozzle in a similar arrangement, Fig. 14.13. Point a designates the condition when $P_B = P_0$ and there is no flow. When P_B is decreased to the pressure indicated by point b, so that P_B/P_0 is less than 1 but considerably greater than the critical-pressure ratio, the velocity increases in the convergent section, but $M < 1$ at the throat. Therefore, the diverging section acts as a subsonic diffuser in which the pressure increases and velocity decreases. Point c designates the back pressure at which $M = 1$ at the throat, but the diverging section acts as a subsonic diffuser (with $M = 1$ at the inlet) in which the pressure increases and velocity decreases. Point d designates one other back pressure that permits isentropic flow, and in this case the diverging section acts as a supersonic nozzle, with a decrease in pressure and an increase in velocity. Between the back pressures designated by points c and d, an isentropic solution is not possible, and shock waves will be present. This matter is discussed in the section that follows. When the back pressure is decreased below that designated by point d, the exit-plane pressure P_E remains constant, the drop in pressure from P_E to P_B takes place outside the nozzle. This is designated by point e.

EXAMPLE 14.6 A convergent nozzle has an exit area of 500 mm². Air enters the nozzle with a stagnation pressure of 1000 kPa and a stagnation temperature of 360 K. Determine the mass rate of flow for back pressures of 800 kPa, 528 kPa, and 300 kPa, assuming isentropic flow.

Analysis and Solution

For air $k = 1.4$ and Table A.18 may be used. The critical-pressure ratio, $P*/P_0$, is 0.528. Therefore, for a back pressure of 528 kPa, $M = 1$ at the nozzle exit and the nozzle is choked. Decreasing the back pressure below 528 kPa will not increase the flow.

For a back pressure of 528 kPa,

$$\frac{T^*}{T} = 0.8333 \qquad T^* = 300 \text{ K}$$

At the exit

$$\mathbf{V} = c = \sqrt{kRT}$$

$$= \sqrt{1.4 \times 0.287 \times 300 \times 1000} = 347.2 \text{ m/s}$$

$$\rho^* = \frac{P^*}{RT^*} = \frac{528}{0.287 \times 300} = 6.1324 \text{ kg/m}^3$$

$$\dot{m} = \rho \mathscr{A} \mathbf{V}$$

Applying this relation to the throat section

$$\dot{m} = 6.1324 \times 500 \times 10^{-6} \times 347.2 = 1.0646 \text{ kg/s}$$

For a back pressure of 800 kPa, $P_E/P_0 = 0.8$ (subscript E designates the properties in the exit plane). From Table A.18

$$M_E = 0.573 \qquad T_E / T_0 = 0.9381$$

$$T_E = 337.7 \text{ K}$$

$$c_E = \sqrt{kRT_E} = \sqrt{1.4 \times 0.287 \times 337.7 \times 1000} = 368.4 \text{ m/s}$$

$$\mathbf{V}_E = M_E c_E = 211.1 \text{ m/s}$$

$$\rho_E = \frac{P_E}{RT_E} = \frac{800}{0.287 \times 337.7} = 8.2542 \text{ kg/m}^3$$

$$\dot{m} = \rho \mathscr{A} \mathbf{V}$$

Applying this relation to the exit section,

$$\dot{m} = 8.2542 \times 500 \times 10^{-6} \times 211.1 = 0.8712 \text{ kg/s}$$

For a back pressure less than the critical pressure, which in this case is 528 kPa, the nozzle is choked and the mass rate of flow is the same as that for the critical pressure. Therefore, for an exhaust pressure of 300 kPa, the mass rate of flow is 1.0646 kg/s.

EXAMPLE 14.7   A converging–diverging nozzle has an exit area to throat area ratio of 2. Air enters this nozzle with a stagnation pressure of 1000 kPa and a stagnation temperature of 360 K. The throat area is 500 mm^2. Determine the mass rate of flow, exit pressure, exit temperature, exit Mach number, and exit velocity for the following conditions:

(a) Sonic velocity at the throat, diverging section acting as a nozzle. (Corresponds to point d in Fig. 14.13.)

(b) Sonic velocity at the throat, diverging section acting as a diffuser. (Corresponds to point c in Fig. 14.13.)

Analysis and Solution

(a) In Table A.18 of the Appendix we find that there are two Mach numbers listed for $\mathscr{A}/\mathscr{A}^* = 2$. One of these is greater than unity and one is less than unity. When the diverging section acts as a supersonic nozzle we use the value for $M > 1$. The following are from Table A.18.

$$\frac{\mathscr{A}_E}{\mathscr{A}_0} = 2.0 \qquad M_E = 2.197 \qquad \frac{P_E}{P_0} = 0.0939 \qquad \frac{T_E}{T_0} = 0.5089$$

Therefore,

$$P_E = 0.0939(1000) = 93.9 \text{ kPa}$$

$$T_E = 0.5089(360) = 183.2 \text{ K}$$

$$c_E = \sqrt{kRT_E} = \sqrt{1.4 \times 0.287 \times 183.2 \times 1000} = 271.3 \text{ m/s}$$

$$\mathbf{V}_E = M_E c_E = 2.197(271.3) = 596.1 \text{ m/s}$$

The mass rate of flow can be determined by considering either the throat section or the exit section. However, in general it is preferable to determine the mass rate of flow from conditions at the throat. Since in this case $M = 1$ at the throat, the calculation is identical to the calculation for the flow in the convergent nozzle of Example 14.6 when it is choked.

(b) The following are from Table A.18.

$$\frac{\mathscr{A}_E}{\mathscr{A}^*} = 2.0 \qquad M = 0.308 \qquad \frac{P_E}{P_0} = 0.936 \qquad \frac{T_E}{T_0} = 0.9812$$

$$P_E = 0.936(1000) = 936 \text{ kPa}$$

$$T_E = 0.9812(360) = 353.3 \text{ K}$$

$$c_E = \sqrt{kRT_E} = \sqrt{1.4 \times 0.287 \times 353.3 \times 1000} = 376.8 \text{ m/s}$$

$$\mathbf{V}_E = M_E c_E = 0.308(376.8) = 116 \text{ m/s}$$

Since $M = 1$ at the throat, the mass rate of flow is the same as in (a), which is also equal to the flow in the convergent nozzle of Example 14.6 when it is choked.

In the example above a solution assuming isentropic flow is not possible if the back pressure is between 936 kPa and 93.9 kPa. If the back pressure is in this range there will be either a normal shock in the nozzle or oblique shock waves outside the nozzle. The matter of normal shock waves is considered in the following section.

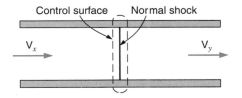

FIGURE 14.14 One-dimensional normal shock.

14.8 NORMAL SHOCK IN AN IDEAL GAS FLOWING THROUGH A NOZZLE

A shock wave involves an extremely rapid and abrupt change of state. In a normal shock this change of state takes place across a plane normal to the direction of the flow. Figure 14.14 shows a control surface that includes such a normal shock. We can now determine the relations that govern the flow. Assuming steady-state, steady-flow we can write the following relations, where subscripts x and y denote the conditions upstream and downstream of the shock, respectively. Note that no heat and work across the control surface.

First law:

$$h_x + \frac{\mathbf{V}_x^2}{2} = h_y + \frac{\mathbf{V}_y^2}{2} = h_{0x} = h_{0y} \tag{14.47}$$

Continuity equation:

$$\frac{\dot{m}}{\mathscr{A}} = \rho_x \mathbf{V}_x = \rho_y \mathbf{V}_y \tag{14.48}$$

Momentum equation:

$$\mathscr{A}(P_x - P_y) = \dot{m}(\mathbf{V}_y - \mathbf{V}_x) \tag{14.49}$$

Second law: Since the process is adiabatic

$$s_y - s_x \geq 0 \tag{14.50}$$

The energy and continuity equations can be combined to give an equation that when plotted on the h–s diagram is called the Fanno line. Similarly, the momentum and continuity equations can be combined to give an equation the plot of which on the h–s diagram is known as the Rayleigh line. Both of these lines are shown on the h–s diagram of Fig. 14.15. It can be shown that the point of maximum entropy on each line, points a and b, corresponds to $M = 1$. The lower part of each line corresponds to supersonic velocities, and the upper part to subsonic velocities.

The two points where all three equations are satisfied are points x and y, x being in the supersonic region and y in the subsonic region. Since the second law requires that $s_y - s_x \geq 0$ in an adiabatic process, we conclude that the normal shock can proceed only from x to y. This means that the velocity changes from supersonic ($M > 1$) before the shock to subsonic ($M < 1$) after the shock.

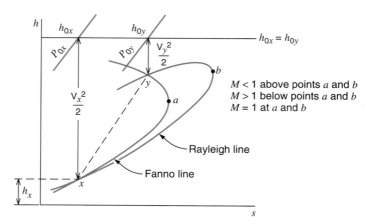

FIGURE 14.15 End states for a one-dimensional normal shock on an enthalpy–entropy diagram.

The equations governing normal shock waves will now be developed. If we assume constant specific heats we conclude from Eq. 14.47, the energy equation, that

$$T_{0x} = T_{0y} \qquad (14.51)$$

That is, there is no change in stagnation temperature across a normal shock. Introducing Eq. 14.37

$$\frac{T_{0x}}{T_x} = 1 + \frac{k-1}{2} M_x^2 \qquad \frac{T_{0y}}{T_y} = 1 + \frac{k-1}{2} M_y^2$$

and substituting into Eq. 14.51 we have

$$\frac{T_y}{T_x} = \frac{1 + \dfrac{k-1}{2} M_x^2}{1 + \dfrac{k-1}{2} M_y^2} \qquad (14.52)$$

The equation of state, the definition of Mach number, and the relation $c = \sqrt{kRT}$ can be introduced into the continuity equation as follows:

$$\rho_x \mathbf{V}_x = \rho_y \mathbf{V}_y$$

But

$$\rho_x = \frac{P_x}{RT_x} \qquad \rho_y = \frac{P_y}{RT_y}$$

$$\frac{T_y}{T_x} = \frac{P_y \mathbf{V}_y}{P_x \mathbf{V}_x} = \frac{P_y M_y c_y}{P_x M_x c_x} = \frac{P_y M_y \sqrt{T_y}}{P_x M_x \sqrt{T_x}}$$

$$= \left(\frac{P_y}{P_x}\right)^2 \left(\frac{M_y}{M_x}\right)^2 \qquad (14.53)$$

Combining Eqs. 14.52 and 14.53, which involves combining the energy equation and the continuity equation, gives the equation of the Fanno line.

$$\frac{P_y}{P_x} = \frac{M_x \sqrt{1 + \dfrac{k-1}{2} M_x^2}}{M_y \sqrt{1 + \dfrac{k-1}{2} M_y^2}} \qquad (14.54)$$

The momentum and continuity equations can be combined as follows to give the equation of the Rayleigh line.

$$P_x - P_y = \frac{\dot{m}}{\mathscr{A}}(\mathbf{V}_y - \mathbf{V}_x) = \rho_y \mathbf{V}_y^2 - \rho_x \mathbf{V}_x^2$$

$$P_x + \rho_x \mathbf{V}_x^2 = P_y + \rho_y \mathbf{V}_y^2$$

$$P_x + \rho_x M_x^2 c_x^2 = P_y + \rho_y M_y^2 c_y^2$$

$$P_x + \frac{P_x M_x^2}{RT_x}(kRT_x) = P_y + \frac{P_y M_y^2}{RT_y}(kRT_y)$$

$$P_x(1 + kM_x^2) = P_y(1 + kM_y^2)$$

$$\frac{P_y}{P_x} = \frac{1 + kM_x^2}{1 + kM_y^2} \tag{14.55}$$

Equations 14.54 and 14.55 can be combined to give the following equation relating M_x and M_y.

$$M_y^2 = \frac{M_x^2 + \dfrac{2}{k-1}}{\dfrac{2k}{k-1} M_x^2 - 1} \tag{14.56}$$

Table A.19 gives the normal shock functions, which include M_y as a function of M_x. This table applies to an ideal gas with a value $k = 1.40$. In the software version of Table A.19, both M and k are input variables. Note that M_x is always supersonic and M_y is always subsonic, which agrees with the previous statement that in a normal shock the velocity changes from supersonic to subsonic. These tables also give the pressure, density, temperature, and stagnation pressure ratios across a normal shock as a function of M_x. These are found from Eqs. 14.52 and 14.53 and the equation of state. Note that there is always a drop in stagnation pressure across a normal shock and an increase in the static pressure.

EXAMPLE 14.8　Consider the convergent–divergent nozzle of Example 14.7 in which the diverging section acts as a supersonic nozzle (Fig. 14.16). Assume that a normal shock stands in the exit plane of the nozzle. Determine the static pressure and temperature and the stagnation pressure just downstream of the normal shock.

　　　　Sketch:　Fig. 14.16.

Analysis and Solution

From Table A.19

$$M_x = 2.197 \qquad M_y = 0.547 \qquad \frac{P_y}{P_x} = 5.46 \qquad \frac{T_y}{T_x} = 1.854 \qquad \frac{P_{0y}}{P_{0x}} = 0.630$$

$$P_y = 5.46 \times P_x = 5.46(93.9) = 512.7 \text{ kPa}$$

$$T_y = 1.854 \times T_x = 1.854(183.2) = 339.7 \text{ K}$$

$$P_{0y} = 0.630 \times P_{0x} = 0.630(1000) = 630 \text{ kPa}$$

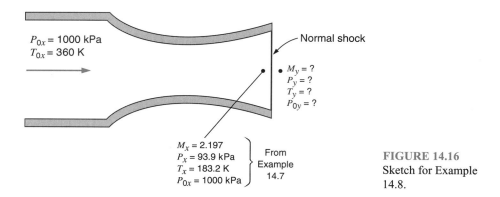

P_{0x} = 1000 kPa
T_{0x} = 360 K

Normal shock

M_y = ?
P_y = ?
T_y = ?
P_{0y} = ?

M_x = 2.197
P_x = 93.9 kPa
T_x = 183.2 K
P_{0x} = 1000 kPa

From
Example
14.7

FIGURE 14.16

Sketch for Example
14.8.

In the light of this example we can conclude the discussion concerning the flow
through a convergent–divergent nozzle. Figure 14.13 is repeated here as Fig. 14.17 for
convenience, except that points f, g, and h have been added. Consider point d. We have
already noted that with this back pressure the exit plane pressure P_E is just equal to the
back pressure P_B, and isentropic flow is maintained in the nozzle. Let the back pressure
be raised to that designated by point f. The exit-plane pressure P_E is not influenced by
this increase in back pressure, and the increase in pressure from P_E to P_B takes place outside the nozzle. Let the back pressure be raised to that designated by point g, which is
just sufficient to cause a normal shock to stand in the exit plane of the nozzle. The exit-
plane pressure P_E (downstream of the shock) is equal to the back pressure P_B, and $M < 1$
leaving the nozzle. This is the case in Example 14.8. Now let the back pressure be raised
to that corresponding to point h. As the back pressure is raised from g to h the normal
shock moves into the nozzle as indicated. Since $M < 1$ downstream of the normal shock,
the diverging part of the nozzle that is downstream of the shock acts as a subsonic diffuser. As the back pressure is increased from h to c the shock moves further upstream
and disappears at the nozzle throat where the back pressure corresponds to c. This is reasonable since there are no supersonic velocities involved when the back pressure corresponds to c, and hence no shock waves are possible.

FIGURE 14.17
Nozzle pressure ratio
as a function of back
pressure for a conver-
gent–divergent nozzle.

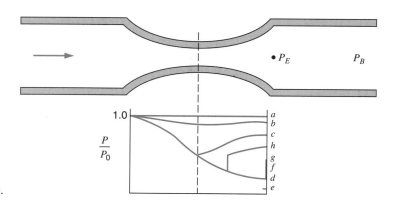

EXAMPLE 14.9   Consider the convergent–divergent nozzle of Examples 14.7 and 14.8. Assume that there is a normal shock wave standing at the point where $M = 1.5$. Determine the exit-plane pressure, temperature, and Mach number. Assume isentropic flow except for the normal shock (Fig. 14.18).

Sketch: Fig. 14.18.

Analysis and Solution

The properties at point x can be determined from Table A.18, because the flow is isentropic to point x.

$$M_x = 1.5 \qquad \frac{P_x}{P_{0x}} = 0.2724 \qquad \frac{T_x}{T_{0x}} = 0.6897 \qquad \frac{\mathscr{A}_x}{\mathscr{A}_x^*} = 1.1762$$

Therefore,

$$P_x = 0.2724(1000) = 272.4 \text{ kPa}$$

$$T_x = 0.6897(360) = 248.3 \text{ K}$$

The properties at point y can be determined from the normal shock functions, Table A.19.

$$M_y = 0.7011 \qquad \frac{P_y}{P_x} = 2.4583 \qquad \frac{T_y}{T_x} = 1.320 \qquad \frac{P_{0y}}{P_{0x}} = 0.9298$$

$$P_y = 2.4583 P_x = 2.4583(272.4) = 669.6 \text{ kPa}$$

$$T_y = 1.320 T_x = 1.320(248.3) = 327.8 \text{ K}$$

$$P_{0y} = 0.9298 P_{0x} = 0.9298(1000) = 929.8 \text{ kPa}$$

Since there is no change in stagnation temperature across a normal shock,

$$T_{0x} = T_{0y} = 360 \text{ K}$$

From y to E the diverging section acts as a subsonic diffuser. In solving this problem it is convenient to think of the flow at y as having come from an isentropic nozzle having a throat area $\mathscr{A}_y^*$. Such a hypothetical nozzle is shown by the dotted line. From the table of isentropic flow functions, Table A.18, we find the following for $M_y = 0.7011$.

$$M_y = 0.7011 \qquad \frac{\mathscr{A}_y}{\mathscr{A}_y^*} = 1.0938 \qquad \frac{P_y}{P_{0y}} = 0.7202 \qquad \frac{T_y}{T_{0y}} = 0.9105$$

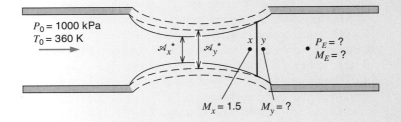

FIGURE 14.18
Sketch for Example 14.9.

From the statement of the problem

$$\frac{\mathcal{A}_E}{\mathcal{A}_x^*} = 2.0$$

Also, since the flow from y to E is isentropic,

$$\frac{\mathcal{A}_E}{\mathcal{A}_E^*} = \frac{\mathcal{A}_E}{\mathcal{A}_y^*} = \frac{\mathcal{A}_E}{\mathcal{A}_x^*} \times \frac{\mathcal{A}_x^*}{\mathcal{A}_x} \times \frac{\mathcal{A}_x}{\mathcal{A}_y} \times \frac{\mathcal{A}_y}{\mathcal{A}_y^*}$$

$$= \frac{\mathcal{A}_E}{\mathcal{A}_y^*} = 2.0 \times \frac{1}{1.1762} \times 1 \times 1.0938 = 1.860$$

From the table of isentropic flow functions for $\mathcal{A}/\mathcal{A}^* = 1.860$ and $M < 1$

$$M_E = 0.339 \qquad \frac{P_E}{P_{0E}} = 0.9222 \qquad \frac{T_E}{T_{0E}} = 0.9771$$

$$\frac{P_E}{P_{0E}} = \frac{P_E}{P_{0y}} = 0.9222$$

$$P_E = 0.9222(P_{0y}) = 0.9222(929.8) = 857.5 \text{ kPa}$$

$$T_E = 0.9771(T_{0E}) = 0.9771(360) = 351.7 \text{ K}$$

In conclusion it should be pointed out that in considering the normal shock we have ignored the effect of viscosity and thermal conductivity, which are certain to be present. The actual shock wave will occur over some finite thickness. However, the development as given here gives a very good qualitative picture of normal shocks, and also provides a basis for fairly accurate quantitative results.

14.9 FLOW OF A VAPOR THROUGH A NOZZLE

In this section we consider a vapor to be a substance that is in the gaseous phase but with limited superheat. Therefore, the vapor will probably deviate significantly from the ideal gas relations, and the possibility of condensation must be considered. The most familiar example is the flow of steam through the nozzles of a steam turbine.

The principles that have been developed for the isentropic flow of an ideal gas apply also to the isentropic flow of a vapor. However, because the vapor deviates from the ideal gas relationships, the appropriate tables of thermodynamic properties must be used. Further, the possibility of condensation must be borne in mind, as indicated in Fig. 14.19. If steam expands isentropically from state 1 to state 2, and if equilibrium is maintained throughout the nozzle, condensation would begin at point a, and at pressures below this a mixture of liquid droplets and vapor would be present. In an actual nozzle the formation of the liquid tends to be delayed due to an effect known as supersaturation. This will be considered in a later paragraph.

Let us consider first the isentropic flow of steam in a nozzle without condensation. The value of the specific-heat ratio k for steam varies, but $k = 1.3$ is a good approxima-

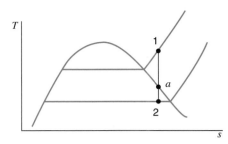

FIGURE 14.19 Reversible, adiabatic expansion of a vapor.

tion over a considerable range. Therefore, the critical pressure ratio, P^*/P_0, can be found by the relation

$$\frac{P^*}{P_0} = \left(\frac{2}{k+1}\right)^{k/k-1} = 0.545 \tag{14.57}$$

This is also the value given in Table 14.1. Knowing, therefore, the critical-pressure ratio, the throat area for a given flow can be calculated. The exit area of the nozzle can be calculated in a similar manner. The example below illustrates this procedure.

EXAMPLE 14.10   Steam at stagnation pressure of 800 kPa and a stagnation temperature of 350°C expands in a nozzle to 200 kPa. Determine the throat area and exit area required for a flow of 3 kg/s, assuming reversible adiabatic flow.

Analysis and Solution

First consider the properties and flow at the throat.

$$\frac{P^*}{P_0} = 0.545 \qquad P^* = 436 \text{ kPa}$$

$$s^* = s_0 = 7.4089 \text{ kJ / kg K} \quad h_0 = 3161.7 \text{ kJ / kg}$$

$$T^* = 268.7°\text{C} \qquad h^* = 3001.4 \text{ kJ / kg}$$

$$h_0 = h^* + \frac{\mathbf{V}^{*2}}{2} \qquad \mathbf{V}^* = \sqrt{2(h_0 - h^*)}$$

$$\mathbf{V}^* = \sqrt{2 \times 1000(3161.7 - 3001.4)} = 566.2 \text{ m / s}$$

$$v^* = 0.5724 \text{ m}^3 / \text{kg (from the steam tables)}$$

$$\dot{m}v^* = \mathcal{A}^* \mathbf{V}^*$$

$$\mathcal{A}^* = \frac{3 \times 0.5724}{566.2} = 3.033 \times 10^{-3} \text{ m}^2$$

At the nozzle exit

$$P_E = 200 \text{ kPa} \qquad s_E = s_0 = 7.4089 \text{ kJ / kg K}$$

Therefore,

$$T_E = 178.5°C$$

$$h_E = 2826.7 \text{ kJ / kg} \qquad v_e = 1.0284 \text{ m}^3 / \text{kg}$$

$$\mathbf{V}_E = \sqrt{2(h_0 - h_E)} = \sqrt{2 \times 1000(3161.7 - 2826.7)} = 818.5 \text{ m / s}$$

$$\dot{m}v_E = \mathscr{A}_E \mathbf{V}_E$$

$$\mathscr{A}_E = \frac{3 \times 1.0284}{818.5} = 3.769 \times 10^{-3} \text{ m}^2$$

For steam initially saturated the critical-pressure ratio is usually taken as

$$\frac{P*}{P_0} = 0.577 \tag{14.58}$$

If the flow with initially saturated steam is assumed to be isentropic there will be a certain amount of liquid entrained with the vapor. Calculation of throat and exit areas under most conditions is similar to the example above.

When the steam flowing through a nozzle is initially superheated, there will be a certain point in the nozzle where the steam becomes saturated. However, as noted in Section 13.5 in connection with a discussion of metastable equilibrium, if the point at which condensation would occur under equilibrium conditions (point a in Fig. 13.9) occurs in the divergent section of the nozzle, a condition of metastable equilibrium exists. That is, the formation of droplets is delayed, and the vapor temperature is less than the saturation temperature for the given pressure. This is frequently referred to as supersaturation.

This phenomenon of supersaturation is observed only in the diverging portions of the nozzle. In a nozzle in which supersaturation occurs the rate of mass flow might be slightly greater than that obtained in an isentropic flow without supersaturation.

Supersaturation may also be present in a supersonic wind tunnel. An abrupt condensation of the moisture in the air would disturb the flow pattern, and for this reason most of the moisture is removed from the air to be used in the tunnel.

A similar phenomenon may occur in the flow of a saturated liquid through a nozzle or valve. When a saturated liquid flows through a nozzle or valve some of it will vaporize as the pressure is reduced. In practice the formation of the vapor is delayed, because a metastable state exists.

14.10 NOZZLE AND DIFFUSER COEFFICIENTS

Up to this point we have considered only isentropic flow and normal shocks. As was pointed out in Chapter 7, isentropic flow through a nozzle provides a standard to which the performance of an actual nozzle can be compared. For nozzles, the three important parameters by which actual flow can be compared to the ideal flow are nozzle efficiency, velocity coefficient, and discharge coefficient. These are defined as follows:

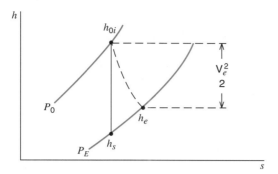

FIGURE 14.20 Enthalpy–entropy diagram showing the effects of irreversibility in a nozzle.

The nozzle efficiency η_N is defined as

$$\eta_N = \frac{\text{Actual kinetic energy at nozzle exit}}{\text{Kinetic energy at nozzle exit with isentropic flow to same exit pressure}} \quad (14.59)$$

The efficiency can be defined in terms of properties. On the h–s diagram of Fig. 14.20 state $0i$ represents the stagnation state of the fluid entering the nozzle; state e represents the actual state at the nozzle exit; and state s represents the state that would have been achieved at the nozzle exit if the flow had been reversible and adiabatic to the same exit pressure. Therefore, in terms of these states the nozzle efficiency is

$$\eta_N = \frac{h_{0i} - h_e}{h_{0i} - h_s}$$

Nozzle efficiencies vary in general from 90 to 99%. Large nozzles usually have higher efficiencies than small nozzles, and nozzles with straight axes have higher efficiencies than nozzles with curved axes. The irreversibilities, which cause the departure from isentropic flow, are primarily due to frictional effects, and are confined largely to the boundary layer. The rate of change of cross-sectional area along the nozzle axis (that is, the nozzle contour) is an important parameter in the design of an efficient nozzle, particularly in the divergent section. Detailed consideration of this matter is beyond the scope of this text, and the reader is referred to standard references on the subject.

The velocity coefficient C_V is defined as

$$C_V = \frac{\text{Actual velocity at nozzle exit}}{\text{Velocity at nozzle exit with isentropic flow and same exit pressure}} \quad (14.60)$$

It follows that the velocity coefficient is equal to the square root of the nozzle efficiency

$$C_V = \sqrt{\eta_N} \quad (14.61)$$

The coefficient of discharge C_D is defined by the relation

$$C_D = \frac{\text{Actual mass rate of flow}}{\text{Mass rate of flow with isentropic flow}}$$

In determining the mass rate of flow with isentropic conditions, the actual back pressure is used if the nozzle is not choked. If the nozzle is choked, the isentropic mass rate of flow is based on isentropic flow and sonic velocity at the minimum section (that is, sonic velocity at the exit of a convergent nozzle and at the throat of a convergent–divergent nozzle).

The performance of a diffuser is usually given in terms of diffuser efficiency, which is best defined wih the aid of an *h–s* diagram. On the *h–s* diagram of Fig. 14.21 states 1 and 01 are the actual and stagnation states of the fluid entering the diffuser. States 2 and 02 are the actual and stagnation states of the fluid leaving the diffuser. State 3 is not attained in the diffuser, but it is the state that has the same entropy as the initial state and the pressure of the isentropic stagnation state leaving the diffuser. The efficiency of the diffuser η_D is defined as

$$\eta_D = \frac{\Delta h_s}{\mathbf{V}_1^2 / 2} = \frac{h_3 - h_1}{h_{01} - h_1} = \frac{h_3 - h_1}{h_{02} - h_1} \tag{14.62}$$

If we assume an ideal gas with constant specific heat this reduces to

$$\eta_D = \frac{T_3 - T_1}{T_{02} - T_1} = \frac{\dfrac{(T_3 - T_1)}{T_1} T_1}{\dfrac{\mathbf{V}_1^2}{2 C_{po}}}$$

$$C_{po} = \frac{kR}{k-1} \qquad T_1 = \frac{c_1^2}{kR} \qquad \mathbf{V}_1^2 = M_1^2 c_1^2 \qquad \frac{T_3}{T_1} = \left(\frac{P_{02}}{P_1}\right)^{(k-1)/k}$$

Therefore,

$$\eta_D = \frac{\left(\dfrac{P_{02}}{P_1}\right)^{(k-1)/k} - 1}{\dfrac{k-1}{2} M_1^2}$$

$$\left(\frac{P_{02}}{P_1}\right)^{(k-1)/k} = \left(\frac{P_{01}}{P_1}\right)^{(k-1)/k} \times \left(\frac{P_{02}}{P_{01}}\right)^{(k-1)/k}$$

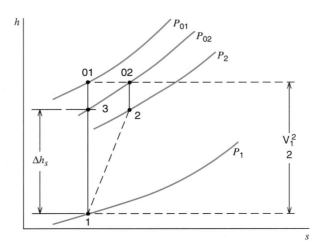

FIGURE 14.21 Enthalpy–entropy diagram showing the definition of diffuser efficiency.

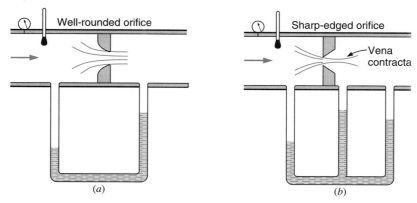

FIGURE 14.22 Nozzles and orifices as flow-measuring devices.

$$\left(\frac{P_{02}}{P_1}\right)^{(k-1)/k} = \left(1+\frac{k-1}{2}M_1^2\right)\left(\frac{P_{02}}{P_{01}}\right)^{(k-1)/k}$$

$$\eta_D = \frac{\left(1+\dfrac{k-1}{2}M_1^2\right)\left(\dfrac{P_{02}}{P_{01}}\right)^{(k-1)/k}-1}{\dfrac{k-1}{2}M_1^2} \qquad (14.63)$$

14.11 NOZZLES AND ORIFICES AS FLOW-MEASURING DEVICES

The mass rate of flow of a fluid flowing in a pipe is frequently determined by measuring the pressure drop across a nozzle or orifice in the line, as shown in Fig. 14.22. The ideal process for such a nozzle or orifice is assumed to be isentropic flow through a nozzle that has the measured pressure drop from inlet to exit and a minimum cross-sectional area equal to the minimum area of the nozzle or orifice. The actual flow is related to the ideal flow by the coefficient of discharge, which is defined by Eq. 14.62.

The pressure difference measured across an orifice depends upon the location of the pressure taps as indicated in Fig. 14.22. Since the ideal flow is based on the measured pressure difference, it follows that the coefficient of discharge depends on the locations of the pressure taps. Also, the coefficient of discharge for a sharp-edged orifice is considerably less than that for a well-rounded nozzle, primarily due to a contraction of the stream, known as the vena contracta, as it flows through a sharpedged orifice.

There are two approaches to determining the discharge coefficient of a nozzle or orifice. One is to follow a standard design procedure, such as the ones established by the American Society of Mechanical Engineers,[1] and use the coefficient of discharge given for a particular design. A more accurate method is to calibrate a given nozzle or orifice,

---

[1]*Fluid Meters, Their Theory and Application,* ASME, 1959. *Flow Measurement,* ASME, 1959.

and determine the discharge coefficient for a given installation by accurately measuring the actual mass rate of flow. The procedure to be followed will depend on the accuracy desired and other factors involved (such as time, expense, availability of calibration facilities) in a given situation.

For incompressible fluids flowing through an orifice the ideal flow for a given pressure drop can be found by the procedure outlined in Section 14.4. Actually, it is advantageous to combine Eqs. 14.24 and 14.20 to give the following relation, which is valid for reversible flow.

$$v(P_2 - P_1) + \frac{\mathbf{V}_2^2 - \mathbf{V}_1^2}{2} = v(P_2 - P_1) + \frac{\mathbf{V}_2^2 - (\mathcal{A}_2/\mathcal{A}_1)^2 \mathbf{V}_2^2}{2} = 0 \tag{14.64}$$

or

$$v(P_2 - P_1) + \frac{\mathbf{V}_2^2}{2}\left[1 - \left(\frac{\mathcal{A}_2}{\mathcal{A}_1}\right)^2\right] = 0$$

$$\mathbf{V}_2 = \sqrt{\frac{2v(P_1 - P_2)}{[1 - (\mathcal{A}_2/\mathcal{A}_1)^2]}} \tag{14.65}$$

For an ideal gas it is frequently advantageous to use the following simplified procedure when the pressure drop across an orifice or nozzle is small. Consider the nozzle shown in Fig. 14.23. From the first law we conclude that

$$h_i + \frac{\mathbf{V}_i^2}{2} = h_e + \frac{\mathbf{V}_e^2}{2}$$

Assuming constant specific heat, this reduces to

$$\frac{\mathbf{V}_e^2 - \mathbf{V}_i^2}{2} = h_i - h_e = C_{po}(T_i - T_e)$$

Let ΔP and ΔT be the decrease in pressure and temperature across the nozzle. Since we are considering reversible adiabatic flow we note that

$$\frac{T_e}{T_i} = \left(\frac{P_e}{P_i}\right)^{(k-1)/k}$$

or

$$\frac{T_i - \Delta T}{T_i} = \left(\frac{P_i - \Delta P}{P_i}\right)^{(k-1)/k}$$

$$1 - \frac{\Delta T}{T_i} = \left(1 - \frac{\Delta P}{P_i}\right)^{(k-1)/k}$$

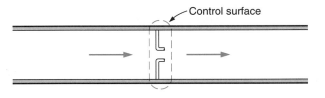

FIGURE 14.23 Analysis of a nozzle as a flow-measuring device.

Using the binomial expansion on the right side of the equation we have

$$1 - \frac{\Delta T}{T_i} = 1 - \frac{k-1}{k}\frac{\Delta P}{P_i} - \frac{k-1}{2k^2}\frac{\Delta P^2}{P_i^2}\cdots$$

If $\Delta P/P_i$ is small this reduces to

$$\frac{\Delta T}{T_i} = \frac{k-1}{k}\frac{\Delta P}{P_i}$$

Substituting this into the first-law equation we have

$$\frac{\mathbf{V}_e^2 - \mathbf{V}_i^2}{2} = C_{po}\frac{k-1}{k}\Delta P\frac{T_i}{P_i}$$

But for an ideal gas

$$C_{po} = \frac{kR}{k-1} \qquad \text{and} \qquad v_i = R\frac{T_i}{P_i}$$

Therefore,

$$\frac{\mathbf{V}_e^2 - \mathbf{V}_i^2}{2} = v_i\Delta P$$

which is the same as Eq. 14.64, which was developed for incompressible flow. Therefore, when the pressure drop across a nozzle or orifice is small, the flow can be calculated with high accuracy by assuming incompressible flow.

The Pitot tube, Fig. 14.24, is an important instrument for measuring the velocity of a fluid. In calculating the flow with a Pitot tube it is assumed that the fluid is decelerated isentropically in front of the Pitot tube, and therefore the stagnation pressure of the free stream can be measured.

Applying the first law to this process we have

$$h + \frac{\mathbf{V}^2}{2} = h_0$$

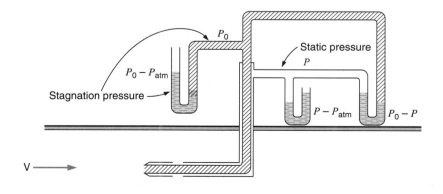

FIGURE 14.24 Schematic arrangement of a Pitot tube.

If we assume incompressible flow for this isentropic process, the first law reduces to (because $T \, ds = dh - v \, dp$)

$$\frac{\mathbf{V}^2}{2} = h_0 - h = v(P_0 - P)$$

or

$$\mathbf{V} = \sqrt{2v(P_0 - P)} \tag{14.66}$$

If we consider the compressible flow of an ideal gas with constant specific heat, the velocity can be found from the relation

$$\frac{\mathbf{V}^2}{2} = h_0 - h = C_{po}(T_0 - T) = C_{po}T\left(\frac{T_0}{T} - 1\right)$$

$$= C_{po}T\left[\left(\frac{P_0}{P}\right)^{(k-1)/k} - 1\right] \tag{14.67}$$

It is of interest to know the error introduced by assuming incompressible flow when using the Pitot tube to measure the velocity of an ideal gas. To do so we introduce Eq. 14.38 and rearrange it as follows:

$$\frac{P_0}{P} = \left(1 + \frac{k-1}{2}M^2\right)^{k/(k-1)} = \left[1 + \left(\frac{k-1}{2}\right)\left(\frac{\mathbf{V}^2}{c^2}\right)\right]^{k/(k-1)} \tag{14.68}$$

But

$$\frac{\mathbf{V}^2}{2} + C_{po}T = C_{po}T_0$$

$$\frac{\mathbf{V}^2}{2} + \frac{kRc^2}{(k-1)kR} = \frac{kRc_0^2}{(k-1)kR}$$

$$1 + \frac{2c^2}{(k-1)\mathbf{V}^2} = \frac{2c_0^2}{(k-1)\mathbf{V}^2} \quad \text{where} \quad c_0 = \sqrt{kRT_0}$$

$$\frac{c^2}{\mathbf{V}^2} = \frac{k-1}{2}\left[\left(\frac{2}{k-1}\right)\left(\frac{c_0^2}{\mathbf{V}^2}\right) - 1\right] = \frac{c_0^2}{\mathbf{V}^2} - \frac{k-1}{2}$$

or

$$\frac{c^2}{\mathbf{V}^2} = \frac{c_0^2}{\mathbf{V}^2} - \frac{k-1}{2} \tag{14.69}$$

Substituting this into Eq. 14.68 and rearranging

$$\frac{P}{P_0} = \left[1 - \frac{k-1}{2}\left(\frac{\mathbf{V}}{c_0}\right)^2\right]^{k/(k-1)} \tag{14.70}$$

Expanding this equation by the binomial theorem, and including terms through $(\mathbf{V}/c_0)^4$, we have

$$\frac{P}{P_0} = 1 - \frac{k}{2}\left(\frac{\mathbf{V}}{c_0}\right)^2 + \frac{k}{8}\left(\frac{\mathbf{V}}{c_0}\right)^4$$

TABLE 14.2

V/c_0	Approximate Room-Temperature Velocity, m/s	Error in Pressure for a Given Velocity, %	Error in Velocity for a Given Pressure, %
0.0	0	0	0
0.1	35	0.25	-0.13
0.2	70	1.0	-0.5
0.3	105	2.25	-1.2
0.4	140	4.0	-2.1
0.5	175	6.25	-3.3

On rearranging this we have

$$\frac{P_0 - P}{\rho_0 \mathbf{V}^2 / 2} = 1 - \frac{1}{4}\left(\frac{\mathbf{V}}{c_0}\right)^2 \tag{14.71}$$

For incompressible flow the corresponding equation is

$$\frac{P_0 - P}{\rho_0 \mathbf{V}^2 / 2} = 1$$

Therefore, the second term on the right side of Eq. 14.71 represents the error involved if incompressible flow is assumed. The error in pressure for a given velocity and the error in velocity for a given pressure that would result from assuming incompressible flow are given in Table 14.2.

14.12 FLOW THROUGH BLADE PASSAGES

We will now apply the principles we have been considering in this chapter to the flow of fluids through blade passages. We will limit our consideration to uniform, parallel, one-dimensional flow.

There are a number of different devices that involve flow through blade passages. Some of the familiar ones are steam and gas turbines, axial-flow compressors, and certain types of blowers, pumps, and torque converters. Some of these devices, such as the steam turbine, are made up of a number of stages. One stage consists of a nozzle (or blade passages that act as a nozzle) and the moving row of blades that follows. Sometimes the blades are also called buckets, but in this text the term blade will be used.

In the analysis of such problems it is necessary to discuss both absolute and relative velocities. By absolute velocity we mean the velocity a stationary observer would detect. The relative velocity is the velocity an observer traveling with the blade would detect, which is designated with a subscript R. The velocity of the blade is designated $\mathbf{V}_B$.

It is advantageous to show these velocities on a vector diagram. Figure 14.25 shows typical vector diagrams for both a turbine and a compressor. In each case the first sketch shows the velocity vectors in relation to the blades, and the second sketch shows how these vector diagrams are usually drawn. In these diagrams $\mathbf{V}_1$ represents the velocity of the fluid entering the blade passage, and α designates the angle at which it enters. $\mathbf{V}_{1R}$ represents the relative velocity of the fluid entering the blade passage and β the angle at which it enters. Similarly, $\mathbf{V}_2$ and $\mathbf{V}_{2R}$ represent the velocity and relative velocity of the fluid leaving at the angles δ and γ, respectively.

In analyzing the flow through a blade passage it is convenient to consider the control surface shown in Fig. 14.26, which is shown from both a stationary and a moving observer's point of view.

Let us begin the analysis by writing the first law for both cases, assuming steady, adiabatic flow through the blade passage. The stationary observer is aware that work is being done on the blade, and the steady-flow equation for the first law is

$$h_1 + \frac{\mathbf{V}_{1R}^2}{2} = h_2 + \frac{\mathbf{V}_{2R}^2}{2} \tag{14.72}$$

The moving observer is not aware of any work being done on the blade, and in this case the first law is

$$h_1 + \frac{\mathbf{V}_{1R}^2}{2} = h_2 + \frac{\mathbf{V}_{2R}^2}{2} \tag{14.73}$$

By combining Eqs. 14.72 and 14.73, we can derive the relation

$$w = \frac{(\mathbf{V}_1^2 - \mathbf{V}_{1R}^2) - (\mathbf{V}_2^2 - \mathbf{V}_{2R}^2)}{2} \tag{14.74}$$

Applying the second law of thermodynamics to this process we conclude, since the process is adiabatic, that

$$s_2 \geq s_1$$

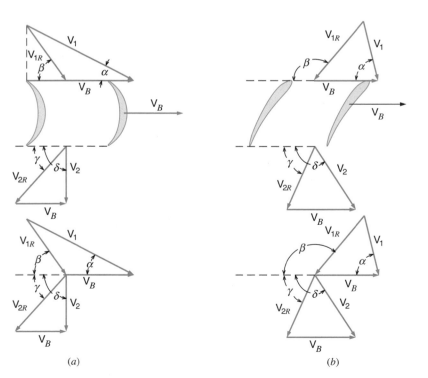

FIGURE 14.25
Velocity vector diagrams for (a) a turbine and (b) a compressor.

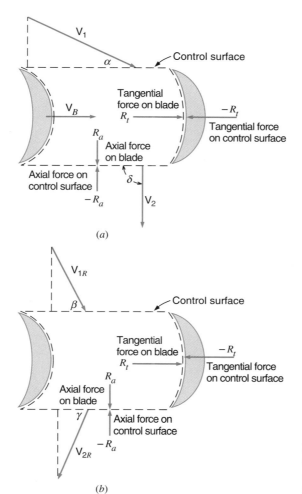

FIGURE 14.26 Analysis of forces on a turbine blade. (*a*) Stationary observer. (*b*) Observer moving with the blade.

We will next consider the momentum equation, and evaluate both the tangential force R_t and the axial force R_a on the blade. For our sign convention we assume that tangential velocities and forces in the direction of the movement of the blade are positive and axial velocities and forces in the direction of the axial component of the velocity are positive.

Let us first apply the momentum equation in the tangential direction to the control volume as viewed by the stationary observer. In doing so we make the reasonable assumption that the pressure forces in the tangential direction balance. Then, using Eq. 14.19

$$-R_t = \dot{m}(-\mathbf{V}_2 \cos \delta - \mathbf{V}_1 \cos \alpha)$$

or

$$R_t = \dot{m}(\mathbf{V}_1 \cos \alpha + \mathbf{V}_2 \cos \delta) \tag{14.75}$$

Similarly, for the control volume as viewed by the moving observer we have

$$-R_t = \dot{m}(-\mathbf{V}_{2R} \cos \gamma - \mathbf{V}_{1R} \cos \beta)$$

or

$$R_t = \dot{m}(\mathbf{V}_{1R} \cos \beta + \mathbf{V}_{2R} \cos \gamma) \tag{14.76}$$

Equation 14.76 could have been developed from Eq. 14.75 and from the geometry of the vector diagram by noting that

$$\mathbf{V}_1 \cos \alpha - \mathbf{V}_B = \mathbf{V}_{1R} \cos \beta$$

and

$$\mathbf{V}_2 \cos \delta + \mathbf{V}_B = \mathbf{V}_{2R} \cos \gamma$$

Therefore,

$$\mathbf{V}_1 \cos \alpha + \mathbf{V}_2 \cos \delta = \mathbf{V}_{1R} \cos \beta + \mathbf{V}_{2R} \cos \gamma \tag{14.77}$$

The work done by the control volume on the blade in time dt as the blade moves a distance dx is

$$\delta W = R_t \, dx$$

The rate at which work is done, or the power, is

$$\frac{\delta W}{dt} = R_t \frac{dx}{dt} = R_t \mathbf{V}_B$$

Substituting for R_t from Eqs. 14.75 and 14.76

$$\frac{\delta W}{dt} = \dot{W}_{\text{c.v.}} = \dot{m} \mathbf{V}_B (\mathbf{V}_1 \cos \alpha + \mathbf{V}_2 \cos \delta)$$

$$\frac{\delta W}{dt} = \dot{W}_{\text{c.v.}} = \dot{m} \mathbf{V}_B (\mathbf{V}_{1R} \cos \beta + \mathbf{V}_{2R} \cos \gamma) \tag{14.78}$$

The work w per kilogram of fluid flowing is therefore

$$\dot{W}_{\text{c.v.}} / \dot{m} = w = \mathbf{V}_B (\mathbf{V}_1 \cos \alpha + \mathbf{V}_2 \cos \delta) \tag{14.79}$$

or

$$\dot{W}_{\text{c.v.}} / \dot{m} = w = \mathbf{V}_B (\mathbf{V}_{1R} \cos \beta + \mathbf{V}_{2R} \cos \gamma) \tag{14.80}$$

The axial thrust on the blade can be found by applying the momentum equation in the axial direction. We must recognize that the pressure at the inlet section may be different from the pressure at the exit section. Consider first the control volume as viewed by the stationary observer. Again we use Eq. 14.19.

$$R_a = \dot{m} (\mathbf{V}_2 \sin \delta - \mathbf{V}_1 \sin \alpha) - (P_1 \mathscr{A}_1 - P_2 \mathscr{A}_2) \tag{14.81}$$

As viewed by the moving observer we have

$$R_a = \dot{m} (\mathbf{V}_{2R} \sin \gamma - \mathbf{V}_{1R} \sin \beta) - (P_1 \mathscr{A}_1 - P_2 \mathscr{A}_2) \tag{14.82}$$

Since there is no motion in the axial direction there is no work involved. However, adequate thrust bearings must be provided to handle the axial force. It is important to note in Eqs. 14.81 and 14.82 that if the pressure changes across the moving blade, the axial forces will be significantly influenced. This will be discussed more fully in the next section when the difference between an impulse and a reaction stage is considered.

EXAMPLE 14.11   Steam enters the blade passage of one stage of a steam turbine with a velocity of 550 m/s and at an angle (α) of 20°. The steam leaves the blade (as seen by the moving observer) at an angle (γ) of 50°. There is no pressure change across the blade and there are no irreversibilities (which means that there is no change in the magnitude of the relative velocity during the flow through the blade passage). Determine the work per kilogram of steam and the axial thrust. The blade speed is 250 m/s.

Analysis and Solution

We first draw the vector diagram as shown in Fig. 14.27.

$$\mathbf{V}_{1R}\sin\beta = \mathbf{V}_1\sin\alpha = 550\sin20° = 550(0.342) = 188.1 \text{ m/s}$$
$$\mathbf{V}_{1R}\cos\beta + \mathbf{V}_B = \mathbf{V}_1\cos\alpha = 550\cos20° = 550(0.9397) = 516.8 \text{ m/s}$$
$$\mathbf{V}_{1R}\cos\beta = 516.8 - 250 = 266.8 \text{ m/s}$$
$$\tan\beta = \frac{\mathbf{V}_{1R}\sin\beta}{\mathbf{V}_{1R}\cos\beta} = \frac{188.1}{266.8} = 0.7050$$
$$\beta = 35.185°$$
$$\mathbf{V}_{1R}\sin\beta = 188.1$$
$$\mathbf{V}_{1R} = \frac{188.1}{\sin35.185°} = \frac{188.1}{0.5762} = 326.4 \text{ m/s}$$
$$\mathbf{V}_{2R} = \mathbf{V}_{1R} = 326.4 \text{ m/s}$$
$$\mathbf{V}_2\sin\delta = \mathbf{V}_{2R}\sin\gamma = 326.4\sin50° = 326.4(0.7660) = 250.1 \text{ m/s}$$
$$\mathbf{V}_2\cos\delta + \mathbf{V}_B = \mathbf{V}_{2R}\cos\gamma = 326.4\cos50° = 326.4(0.6428) = 209.8 \text{ m/s}$$
$$\mathbf{V}_2\cos\delta = 209.8 - 250 = -40.2 \text{ m/s}$$

This implies that we draw our vector diagram slightly wrong and it should have been as shown in Fig. 14.28

$$\cot\delta = \frac{\mathbf{V}_2\cos\delta}{\mathbf{V}_2\sin\delta} = \frac{-40.2}{250.1} = -0.1607$$
$$\delta = 99.13°$$
$$\mathbf{V}_2 = \sqrt{(\mathbf{V}_2\sin\delta)^2 + (\mathbf{V}_2\cos\delta)^2}$$
$$= \sqrt{(250.1)^2 + (40.2)^2} = 253.3 \text{ m/s}$$

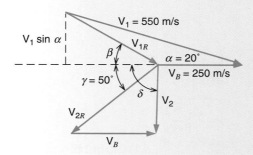

FIGURE 14.27 Sketch for Example 14.11.

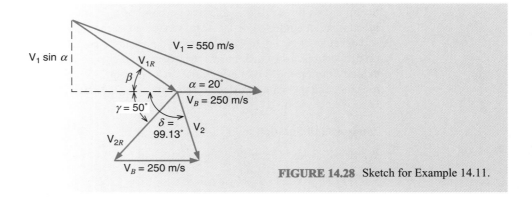

FIGURE 14.28 Sketch for Example 14.11.

14.13 IMPULSE AND REACTION STAGES FOR TURBINES

Two terms used in connection with turbines are impulse stage and reaction stage. A turbine stage is defined as the fixed nozzle or blade and the moving blades that follow it. Figure 14.29*a* shows an impulse stage. In an impulse stage the entire pressure drop takes place in a stationary nozzle, and the pressure remains constant as the fluid flows through the blade passage. There is a decrease in the kinetic energy of the fluid as it flows through the blade passage, and the enthalpy will increase due to irreversibilities associated with the fluid flow.

In the pure reaction stage the entire pressure drop occurs as the fluid flows through the moving blades. Thus, the moving blade acts as a nozzle, and the blade passage must

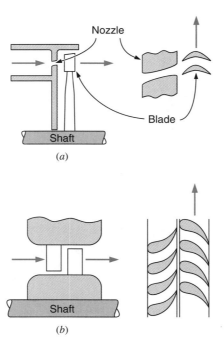

FIGURE 14.29 Schematic arrangement for (*a*) an impulse stage and (*b*) a reaction stage.

have the proper contour for a nozzle (converging if the exit pressure is greater than the critical pressure and converging-diverging if the exit pressure is less than the critical pressure). In the pure reaction stage the only purpose of the stationary blade is to direct the fluid into the moving blade at the proper angle and velocity.

The pure reaction stage is used relatively infrequently. Rather, most turbines that are classified as reaction turbines have a pressure and enthalpy drop in both the fixed and moving blades. The degree of reaction is defined as the fraction of the enthalpy drop that occurs in the moving blades. Thus, in the 50% reaction stage, which is very commonly used, half of the enthalpy drop across the stage occurs in the fixed blade and the other half in the moving blade.

A few comparisons between the impulse and reaction blade may be noted here. Since there is a pressure drop across both the fixed and moving blades in the reaction stage, there is a tendency for leakage to occur across the tip of the blade. Therefore, very close clearances must be maintained at the blade tips. Further, the pressure drop across the moving blade of the reaction stage gives rise to axial forces that must be balanced in order to prevent axial motion in reaction turbines. In an impulse stage it is possible to utilize only part of the periphery for admission of steam. This is usually termed partial admission. In fact, in turbines that utilize an impulse stage for the first stage the power output can be controlled by opening and closing nozzles. The main advantage of the reaction stage is that lower fluid velocities are utilized and higher efficiencies can be attained at lower velocities. This will be discussed more fully in the following sections.

14.14 SOME FURTHER CONSIDERATIONS OF IMPULSE STAGES

Figure 14.30 shows a typical velocity diagram for an impulse stage. The blade velocity coefficient k_B is defined as the ratio of the relative velocity leaving the blade to the relative velocity entering the blade. That is,

$$k_B = \frac{\mathbf{V}_{2R}}{\mathbf{V}_{1R}} \tag{14.83}$$

In the ideal impulse stage $\mathbf{V}_{2R} = \mathbf{V}_{1R}$ and $k_B = 1$, but in the actual impulse stage $\mathbf{V}_{2R}$ will be less than $\mathbf{V}_{1R}$ due to irreversibilities in the fluid as it flows through the blade passage.

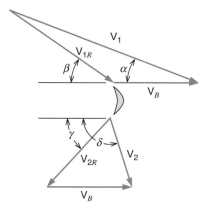

FIGURE 14.30 A velocity vector diagram for an impulse blade.

The blade efficiency η_B is defined as the ratio of the actual work per kilogram of fluid flowing to the kinetic energy of the fluid entering the blade passage.

$$\eta_B = \frac{w}{\mathbf{V}_1^2 / 2} \tag{14.84}$$

Thus, a blade efficiency of 100% means that the work is exactly equal to the kinetic energy of the fluid entering the blade, and the kinetic energy of the fluid leaving the blade is zero.

Quite obviously, the fluid must have some axial velocity if it is to flow out the blade passage. However, it is of interest to consider a turbine having a zero axial velocity and to determine the blade-speed ratio that will give an efficiency of 100%. The blade-speed ratio is the ratio of the blade speed $\mathbf{V}_B$ to the velocity of the fluid leaving the nozzle $\mathbf{V}_1$ and the ratio is designated $\mathbf{V}_B/\mathbf{V}_1$.

Figure 14.31a shows a reversible impulse stage that has a very small entrance and blade exit angle, and a blade-speed ratio of 0.5. It is evident that as the angles α and γ (the inlet and blade exit angles) approach zero the exit velocity $\mathbf{V}_2$ will also approach zero. Such a stage would have an efficiency of 100 percent. Figures 14.31b and c show reversible impulse stages with essentially zero angles, but with blade-speed ratios considerably less and considerably greater than 0.5. In both of these the exit velocity is large, and therefore the blade efficiency is considerably less than 100%. We conclude, therefore, that for a reversible, zero-angle impulse turbine the optimum blade-speed ratio is 0.5. The curve marked impulse stage on Fig. 14.34 shows the variation of blade efficiency with blade-speed ratio for a reversible stage.

An expression for the blade efficiency as a function of the angles α and γ (the inlet angle and blade exit angles), the blade-speed ratio, and the blade velocity coefficient can be derived.

From Eq. 14.79

$$w = \mathbf{V}_B (\mathbf{V}_1 \cos \alpha + \mathbf{V}_2 \cos \delta)$$

$$\mathbf{V}_{2R} = k_B \mathbf{V}_{1R}$$

From the vector diagram we note that

$$\mathbf{V}_2 \cos \delta = \mathbf{V}_{2R} \cos \gamma - \mathbf{V}_B = k_B \mathbf{V}_{1R} \cos \gamma - \mathbf{V}_B$$

$$\mathbf{V}_{1R} = \frac{\mathbf{V}_1 \cos \alpha - \mathbf{V}_B}{\cos \beta}$$

Therefore,

$$w = \mathbf{V}_B \left(\mathbf{V}_1 \cos \alpha + k_B \cos \gamma \frac{\mathbf{V}_1 \cos \alpha - \mathbf{V}_B}{\cos \beta} - \mathbf{V}_B \right)$$

$$= \mathbf{V}_B \left[(\mathbf{V}_1 \cos \alpha - \mathbf{V}_B) \left(1 + \frac{k_B \cos \gamma}{\cos \beta} \right) \right]$$

$$\eta_B = \frac{w}{\mathbf{V}_1^2 / 2} = \frac{2\mathbf{V}_B}{\mathbf{V}_1^2} (\mathbf{V}_1 \cos \alpha - \mathbf{V}_B) \left(1 + k_B \frac{\cos \gamma}{\cos \beta} \right)$$

$$= \frac{2\mathbf{V}_B^2}{\mathbf{V}_1^2} \left(\frac{\mathbf{V}_1}{\mathbf{V}_B} \cos \alpha - 1 \right) \left(1 + k_B \frac{\cos \gamma}{\cos \beta} \right)$$

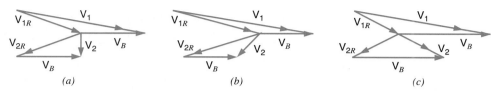

(a)   *(b)*   *(c)*

FIGURE 14.31 A velocity vector diagram illustrating optimum blade-speed ratio.

But

$$\cos\beta = \frac{\mathbf{V}_1\cos\alpha - \mathbf{V}_B}{\sqrt{(\mathbf{V}_1\cos\alpha - \mathbf{V}_B)^2 + (\mathbf{V}_1\sin\alpha)^2}}$$

$$= \frac{\cos\alpha - \mathbf{V}_B/\mathbf{V}_1}{\sqrt{(\cos\alpha - \mathbf{V}_B/\mathbf{V}_1)^2 + \sin^2\alpha}}$$

$$\eta_B = 2\frac{\mathbf{V}_B}{\mathbf{V}_1}\left(\cos\alpha - \frac{\mathbf{V}_B}{\mathbf{V}_1}\right)\left[1 + \frac{k_B\cos\gamma\sqrt{(\cos\alpha - \mathbf{V}_B/\mathbf{V}_1)^2 + \sin^2\alpha}}{\cos\alpha - \mathbf{V}_B/\mathbf{V}_1}\right] \quad (14.85)$$

This relation gives the blade efficiency as a function of the blade-speed ratio $\mathbf{V}_B/\mathbf{V}_1$, the angles α and γ, and the blade velocity coefficient k_B. From this relation it can be shown that for the usual values of angles and blade velocity coefficient, the optimum efficiency is obtained with blade velocity coefficient of approximately 0.5, which was the value obtained for the reversible, zero-angle impulse stage.

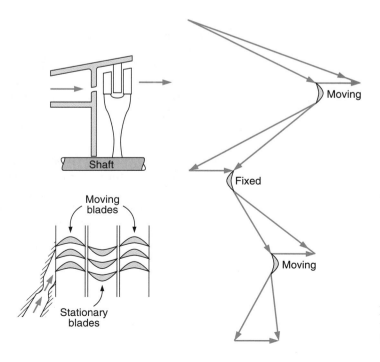

FIGURE 14.32 A velocity-compounded impulse stage.

This leads to consideration of the velocity-compounded turbine. Suppose one was attempting to build a turbine having one impulse stage, and that it was to receive steam at 700 kPa, 250°C, and exhaust at a pressure of 15 kPa. The velocity of the steam leaving the nozzle would be about 1100 m/s. To have maximum efficiency a blade-speed ratio of 0.5 must be used, and this would require a blade speed of 550 m/s. Blade speeds of this magnitude result in very high stresses due to centrifugal force, and further, the irreversibilities associated with fluid flow increase as the fluid velocity increases.

For these reasons velocity compounding is used, and a schematic diagram and a vector diagram for a velocity-compounded stage are shown in Fig. 14.32. The stationary blade serves to change the direction of the fluid so that it will enter the second row of moving blades properly. It can be shown that for a reversible, zero-angle turbine having two moving rows the blade-speed ratio for optimum efficiency is ¼. Sometimes velocity-compounded turbines are built with as many as three moving rows.

Velocity-compounded turbines have a lower efficiency than comparable turbines that have a number of impulse or reaction stages. However, they are relatively inexpensive and are therefore frequently used in small sizes where initial cost is more important than efficiency. A velocity-compounded stage is also frequently used as the first stage in large steam turbines in order to have a large pressure and temperature drop before the steam enters the moving blades.

14.15 SOME FURTHER CONSIDERATIONS OF REACTION STAGES

First a remark should be made regarding the blade efficiency of a reaction stage. Both the fixed blades and moving blades act as a nozzle. Therefore, the efficiency of the moving blades can be defined in the same way as the efficiency of a nozzle if the relative velocities are used.

The second matter to consider is the blade-speed ratio for maximum efficiency of a reaction stage. Let us consider a reversible reaction stage in which the isentropic enthalpy drops across the fixed and moving blades are equal (that is, a 50% reaction stage). For a zero-angle turbine it is evident from Fig. 14.33 that for zero kinetic energy leaving the blade, a blade-speed ratio of 1.0 is required. Thus, maximum efficiency in a reaction stage is obtained for a blade-speed ratio of 1.0.

This leads us to a comparison of the impulse stage with the reaction stage regarding the blade-speed ratio for maximum efficiency.

If we compare reversible impulse and reaction stages for a given velocity from the stationary nozzle or blade, it is evident that for maximum efficiency the blade velocity for the reaction stage would be twice the blade velocity of the impulse stage. However, this is not a good comparison because the reaction stage would have a greater enthalpy drop (due to the enthalpy drop in the moving row) than the impulse stage.

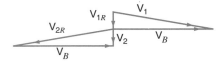

FIGURE 14.33 A velocity vector diagram for a reaction stage.

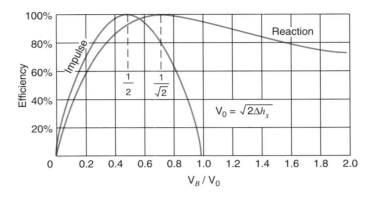

FIGURE 14.34
Efficiency of an ideal impulse and reaction stage as a function of $\mathbf{V}_B/\mathbf{V}_0$.

A better comparison can be made by considering equal enthalpy drops across the stage. Let this enthalpy drop be denoted Δh_s, and let the velocity $\mathbf{V}_0$ be defined by the relation

$$\mathbf{V}_0 = \sqrt{2\Delta h_s}$$

That is, $\mathbf{V}_0$ would be the velocity leaving the nozzle of the reversible impulse stage that has this enthalpy drop. That is, for the impulse stage

$$\mathbf{V}_1 = \mathbf{V}_0$$

For the reversible 50% reaction stage the velocity leaving the fixed blades would be

$$\mathbf{V}_1 = \sqrt{\tfrac{1}{2} \times 2\Delta h_s} = \frac{\mathbf{V}_0}{\sqrt{2}}$$

For maximum efficiency in the impulse stage, $\mathbf{V}_B/\mathbf{V}_1 = 0.5$. Therefore, for the impulse stage

$$\frac{\mathbf{V}_B}{\mathbf{V}_1} = \frac{\mathbf{V}_B}{\mathbf{V}_0} = 0.5$$

For maximum efficiency in the reaction stage, $\mathbf{V}_B/\mathbf{V}_1 = 1.0$. Therefore, for the reaction stage

$$\frac{\mathbf{V}_B}{\mathbf{V}_1} = \frac{\mathbf{V}_B \sqrt{2}}{\mathbf{V}_0} \qquad \frac{\mathbf{V}_B}{\mathbf{V}_1} = \frac{1}{\sqrt{2}}$$

Thus, for a given enthalpy drop per stage, the reaction stage requires a higher blade speed for maximum efficiency than the impulse stage. Conversely, for a given blade speed the reaction turbine requires more stages than an impulse stage, and in this case the fluid velocity is lower in the reaction stage than in the impulse stage.

Figure 14.34 shows the blade efficiency as a function of $\mathbf{V}_B/\mathbf{V}_0$ for an impulse stage and reaction stage.

PROBLEMS

14.1 Steam leaves a nozzle with a pressure of 500 kPa, a temperature of 350°C, and a velocity of 250 m/s. What is the isentropic stagnation pressure and temperature?

14.2 The products of combustion of a jet engine leave the engine with a velocity relative to the plane of 400 m/s, a temperature of 480°C, and a pressure of 75 kPa.

Assuming that $k = 1.32$, $C_p = 1.15$ kJ/kg K for the products, determine the stagnation pressure and temperature of the products relative to the airplane.

14.3 Air leaves a compressor in a pipe with a stagnation temperature and pressure of 150°C, 300 kPa, and a velocity of 125 m/s. The pipe has a cross-sectional area of 0.02 m². Determine the static temperature and pressure and the mass flow rate.

14.4 (Adv.) Atmospheric air is at 20°C, 100 kPa with zero velocity. An adiabatic reversible compressor takes atmospheric air in through a pipe with cross-sectional area of 0.1 m² at a rate of 1 kg/s. It is compressed up to a measured stagnation pressure of 500 kPa and leaves through a pipe with cross-sectional area of 0.01 m². What is the required compressor work and the air velocity, static pressure and temperature in the exit pipeline?

14.5 A water cannon sprays 1 kg/s liquid water at a velocity of 100 m/s horizontally out from a nozzle. It is driven by a pump that receives the water from a tank at 15°C, 100 kPa. The elevation is the same for all parts of this system.

   a. Find the nozzle exit area.

   b. Find the required pressure out of the pump (neglect the kinetic energy of the water flowing in the pump and hose to the nozzle).

   c. Find the horizontal force needed to hold the cannon in place.

14.6 An irrigation pump takes water from a lake and discharges it through a nozzle as shown in Fig. P14.6. At the pump exit the pressure is 700 kPa, and the temperature is 20°C. The nozzle is located 10 m above the pump and the atmospheric pressure is 100 kPa. Assuming reversible flow through the system determine the velocity of the water leaving the nozzle.

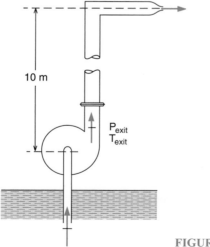

FIGURE P14.6

14.7 A water turbine using nozzles is located at the bottom of Hoover Dam 175 m below the surface of Lake Mead. The water enters the nozzles at a stagnation pressure corresponding to the column of water above it minus 20% due to friction. The temperature is 15°C and the water leaves at standard atmospheric pressure. If the flow through the nozzle is reversible and adiabatic, determine the velocity and kinetic energy per kilogram of water leaving the nozzle.

14.8 A convergent-divergent nozzle has a throat diameter of 0.05 m and an exit diameter of 0.1 m. The inlet stagnation state is 500 kPa, 500 K. Find the back pressure that will lead to the maximum possible flow rate and the mass flow rate if

   a. Air is flowing through it.

   b. Hydrogen is flowing through it.

   c. Carbon dioxide is flowing through it.

14.9 Air is expanded in a nozzle from 2 MPa, 600 K, to 200 kPa. The mass rate of flow through the nozzle is 5 kg/s. Assume the flow to be reversible and adiabatic.

   a. Determine specific volume, velocity, Mach number, and cross-sectional area for each 200 kPa decrease of pressure, and plot these as a function of pressure.

   b. Determine the throat and exit areas for the nozzle.

14.10 Consider the nozzle of Problem 14.9 and determine what back pressure will cause a normal shock to stand in the exit plane of the nozzle. What is the mass flow rate under these conditions?

14.11 At what Mach number will the normal shock occur in the nozzle of Problem 14.9 if the back pressure is 1.4 MPa?

14.12 Consider the nozzle of Problem 14.9. What back pressure will be required to cause subsonic flow throughout the entire nozzle with $M = 1$ at the throat?

14.13 Determine the mass flow rate through the nozzle of Problem 14.9 for a back pressure of 1.9 MPa.

14.14 At what Mach number will the normal shock occur in the nozzle of Problem 14.8 flowing with air if the back pressure is halfway between the pressures at c and d in Fig. 14.17?

14.15 A convergent-divergent nozzle has a throat area of 100 mm² and an exit area of 175 mm². The inlet flow is helium at a total pressure of 1 MPa, stagnation temperature of 375 K. What is the back pressure that will give sonic condition at the throat, but subsonic everywhere else?

14.16 A nozzle is designed assuming reversible adiabatic flow with an exit Mach number of 2.6 while flowing air with a stagnation pressure and temperature of 2 MPa and 150°C, respectively. The mass flow rate is 5 kg/s, and k may be assumed to be 1.40 and constant.

   a. Determine the exit pressure, temperature, and area, and the throat area.

   b. Suppose that the back pressure at the nozzle exit is raised to 1.4 MPa, and that the flow remains isentropic except for a normal shock wave. Determine the exit Mach number and temperature, and the mass flow rate through the nozzle.

14.17 A jet plane travels through the air with a speed of 1000 km/h at an altitude of 6 km, where the pressure is 40 kPa and the temperature is −12°C. Consider the inlet diffuser of the engine where air leaves with a velocity of 100 m/s. Determine the pressure and temperature leaving the diffuser, and the ratio of inlet to exit area of the diffuser, assuming the flow to be reversible and adiabatic.

14.18 A 1-m³ insulated tank contains air at 1 MPa, 560 K. The tank is now discharged through a small convergent nozzle to the atmosphere at 100 kPa. The nozzle has an exit area of 2×10^{-5} m².

a. Find the initial mass flow rate out of the tank.

b. Find the mass flow rate when half the mass has been discharged.

c. Find the mass of air in the tank and the mass flow rate out of the tank when the nozzle flow changes to become subsonic.

14.19 A 1-m^3 uninsulated tank contains air at 1 MPa, 560 K. The tank is now discharged through a small convergent nozzle to the atmosphere at 100 kPa while heat transfer from some source keeps the air temperature in the tank at 560 K. The nozzle has an exit area of 2×10^{-5} m^2.

a. Find the initial mass flow rate out of the tank.

b. Find the mass flow rate when half the mass has been discharged.

c. Find the mass of air in the tank and the mass flow rate out of the tank when the nozzle flow changes to become subsonic.

14.20 The products of combustion enter a nozzle of a jet engine at a total pressure of 125 kPa, and a total temperature of 650°C. The atmospheric pressure is 45 kPa. The nozzle is convergent, and the mass flow rate is 25 kg/s. Assume the flow is reversible and adiabatic. Determine the exit area of the nozzle.

14.21 Air is expanded in a nozzle from 700 kPa, 200°C, to 150 kPa in a nozzle having an efficiency of 90%. The mass flow rate is 4 kg/s. Determine the exit area of the nozzle, the exit velocity, and the increase of entropy per kilogram of air. Compare these results with those of a reversible adiabatic nozzle.

14.22 Repeat Problem 14.17 assuming a diffuser efficiency of 80%.

14.23 Consider the diffuser of a supersonic aircraft flying at $M = 1.4$ at such an altitude that the temperature is −20°C, and the atmospheric pressure is 50 kPa. Consider two possible ways in which the diffuser might operate, and for each case calculate the throat area required for a flow of 50 kg/s.

a. The diffuser operates as a reversible adiabatic diffuser with subsonic exit velocity.

b. A normal shock stands at the entrance to the diffuser. Except for the normal shock the flow is reversible and adiabatic, and the exit velocity is subsonic. This is shown in Fig. P14.23. Assume a convergent-divergent diffuser with $M = 1$ at the throat.

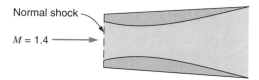

FIGURE P14.23

14.24 Air enters a diffuser with a velocity of 200 m/s, a static pressure of 70 kPa, and a temperature of −6°C. The velocity leaving the diffuser is 60 m/s and the static pressure at the diffuser exit is 80 kPa. Determine the static temperature at the diffuser exit and the diffuser efficiency. Compare the stagnation pressures at the inlet and the exit.

14.25 Steam at a pressure of 1 MPa and temperature of 400°C expands in a nozzle to a pressure of 200 kPa. The nozzle efficiency is 90% and the mass flow rate is 10 kg/s. Determine the nozzle exit area and the exit velocity.

14.26 Steam at 800 kPa, 350°C flows through a convergent-divergent nozzle that has a throat area of 350 mm². The pressure at the exit plane is 150 kPa and the exit velocity is 800 m/s. The flow from the nozzle entrance to the throat is reversible and adiabatic. Determine the exit area of the nozzle, the overall nozzle efficiency, and the entropy generation in the process.

14.27 A sharp-edged orifice is used to measure the flow of air in a pipe. The pipe diameter is 100 mm and the diameter of the orifice is 25 mm. Upstream of the orifice, the absolute pressure is 150 kPa and the temperature is 35°C. The pressure drop across the orifice is 15 kPa, and the coefficient of discharge is 0.62. Determine the mass flow rate in the pipeline.

14.28 A critical nozzle is used for the accurate measurement of the flow rate of air. Exhaust from a car engine is diluted with air so its temperature is 50°C at a total pressure of 100 kPa. It flows through the nozzle with throat area of 700 mm² by suction from a blower. Find the needed suction pressure that will lead to critical flow in the nozzle, the mass flow rate and the blower work, assuming the blower exit is at atmospheric pressure 100 kPa.

14.29 A convergent nozzle is used to measure the flow of air to an engine. The atmosphere is at 100 kPa, 25°C. The nozzle used has a minimum area of 2000 mm² and the coefficient of discharge is 0.95. A pressure difference across the nozzle is measured to 2.5 kPa. Find the mass flow rate assuming incompressible flow. Also find the mass flow rate assuming compressible adiabatic flow.

14.30 Steam at 600 kPa, 300°C is fed to a set of convergent nozzles in a steam turbine. The total nozzle exit area is 0.005 m² and they have a discharge coefficient of 0.94. The mass flow rate should be estimated from the measurement of the pressure drop across the nozzles, which is measured to be 200 kPa. Determine the mass flow rate.

14.31 The coefficient of discharge of a sharp-edged orifice is determined at one set of conditions by use of an accurately calibrated gasometer. The orifice has a diameter of 20 mm and the pipe diameter is 50 mm. The absolute upstream pressure is 200 kPa and the pressure drop across the orifice is 82 mm of mercury. The temperature of the air entering the orifice is 25°C and the mass flow rate measured with the gasometer is 2.4 kg/min. What is the coefficient of discharge of the orifice at these conditions?

14.32 Consider the flow of air through an impulse blade passage. The air enters the blade passage at an angle of 18° with a velocity of 460 m/s, a pressure of 110 kPa, and a temperature of 90°C. The blade velocity is 250 m/s and the air leaves the blade passage at an angle of 45° relative to the blade. The mass flow rate is 10 kg/s, and it is assumed that the flow is reversible and adiabatic.

a. Draw the velocity diagram to scale.

b. What is the power output of the turbine?

c. If the blade wheel to which the blades are attached has a diameter of 0.3 m, what is the length of each blade?

14.33 Steam enters an impulse turbine in which all processes are assumed to be reversible and adiabatic. The inlet pressure is 700 kPa, the inlet temperature is 450°C and the exhaust pressure is 110 kPa. The steam leaves the nozzle and enters the turbine at an angle of 20°. The blade speed ratio is 0.5 and the blade exit angle is 50°. Determine the blade efficiency of this turbine.

14.34 Steam enters an impulse turbine having the same inlet conditions as Problem 14.33. The nozzle efficiency is 92% and the blade velocity coefficient is 0.96 with a blade speed ratio of 0.5. Determine the work per kilogram of steam and draw the velocity diagram to scale.

14.35 Consider a two-stage reversible, velocity-compounded impulse turbine that has the same inlet conditions as in Problem 14.33. Assume a blade velocity coefficient of 0.25, and an inlet angle of 20°. Both the moving and the fixed blades have equal inlet and exit angles (i.e., the angle at which the steam enters relative to the blade is equal to the relative angle at which it leaves). Determine the blade efficiency of the turbine and draw velocity diagram to scale.

14.36 Consider a single-stage reaction turbine having equal enthalpy drops across the fixed and moving blades. Consider the same inlet conditions and exit pressure as in Problem 14.33. All processes are reversible and adiabatic. The blade inlet angle is 20° and the blade-speed ratio is 0.9.

a. What is the pressure at the exit of the fixed blades?

b. Draw the velocity diagram to scale.

c. What is the net work per kilogram of steam flowing through the turbine?

14.37 A single-stage air turbine having 50% reaction operates with inlet pressure and temperature of 350 kPa, 1000 K, and exhaust pressure of 100 kPa. The fixed-blade exit angle is 18°, and exhaust from the turbine is in the axial direction. The blade-speed ratio is 0.9, and all processes are reversible and adiabatic. Determine

a. All velocities and angles, and draw the velocity vector diagram.

b. The work per kilogram of air flowing through the turbine.

English Unit Problems

14.38E Air leaves the compressor of a jet engine at a temperature of 300 F, a pressure of 45 lbf/in.^{2}, and a velocity of 400 ft/s. Determine the isentropic stagnation temperature and pressure.

14.39E Steam leaves a nozzle with a velocity of 800 ft/s. The stagnation pressure is 100 lbf/in.^{2}, and the stagnation temperature is 500 F. What is the static pressure and temperature?

14.40E A water turbine using nozzles is located at the bottom of Hoover Dam 575 ft below the surface of Lake Mead. The water enters the nozzles at a stagnation pressure corresponding to the column of water above it minus 20% due to friction. The temperature is 60 F and the water leaves at standard atmospheric pressure. If the flow through the nozzle is reversible and adiabatic, determine the velocity and kinetic energy per kilogram of water leaving the nozzle.

14.41E A 40-ft^3 insulated tank contains air at 150 lbf/in.^{2}, 1000 R. The tank is now discharged through a small convergent nozzle to the atmosphere at 14.7 lbf/in.^{2}. The nozzle has an exit area of 0.03 in.^{2}.

a. Find the initial mass flow rate out of the tank.

      b. Find the mass flow rate when half the mass has been discharged.

      c. Find the mass of air in the tank and the mass flow rate out of the tank when the nozzle flow changes to become subsonic.

14.42E A nozzle is designed assuming reversible adiabatic flow with an exit Mach number of 2.6 while flowing air with a stagnation pressure and temperature of 300 lbf/in.^{2} and 300 F, respectively. The mass flow rate is 10 lbm/s, and k may be assumed to be 1.40 and constant.

      a. Determine the exit pressure, temperature, and area, and the throat area.

      b. Suppose that the back pressure at the nozzle exit is raised to 200 lbf/in.^{2}, and that the flow remains isentropic except for a normal shock wave. Determine the exit Mach number and temperature, and the mass flow rate through the nozzle.

14.43E A jet plane travels through the air with a speed of 600 mi/h at an altitude of 20 000 ft, where the pressure is 5.75 lbf/in.^{2} and the temperature is 25 F. Consider the diffuser of the engine where air leaves at with a velocity of 300 ft/s. Determine the pressure and temperature leaving the diffuser, and the ratio of inlet to exit area of the diffuser, assuming the flow to be reversible and adiabatic.

14.44E The products of combustion enter a nozzle of a jet engine at a total pressure of 18 lbf/in.^{2}, and a total temperature of 1200 F. The atmospheric pressure is 6.75 lbf/in.^{2}. The nozzle is convergent, and the mass flow rate is 50 lbm/s. Assume the flow is reversible and adiabatic. Determine the exit area of the nozzle.

14.45E Repeat Problem 14.43 assuming a diffuser efficiency of 80%.

14.46E A 50-ft^3 uninsulated tank contains air at 150 lbf/in.^{2}, 1000 R. The tank is now discharged through a small convergent nozzle to the atmosphere at 14.7 lbf/in.^{2} while heat transfer from some source keeps the air temperature in the tank at 1000 R. The nozzle has an exit area of 2×10^{-4} ft^2.

      a. Find the initial mass flow rate out of the tank.

      b. Find the mass flow rate when half the mass has been discharged.

      c. Find the mass of air in the tank and the mass flow rate out of the tank when the nozzle flow changes to become subsonic.

14.47E Air enters a diffuser with a velocity of 600 ft/s, a static pressure of 10 lbf/in.^{2}, and a temperature of 20 F. The velocity leaving the diffuser is 200 ft/s and the static pressure at the diffuser exit is 11.7 lbf/in.^{2}. Determine the static temperature at the diffuser exit and the diffuser efficiency. Compare the stagnation pressures at the inlet and the exit.

14.48E Steam at 100 lbf/in.^{2}, 600 F flows through a convergent-divergent nozzle that has a throat area of 0.5 in.^{2}. The pressure at the exit plane is 20 lbf/in.^{2} and the exit velocity is 2600 ft/s. The flow from the nozzle entrance to the throat is reversible and adiabatic. Determine the exit area of the nozzle, the overall nozzle efficiency, and the entropy generation in the process.

14.49E Steam enters an impulse turbine in which all processes are assumed to be reversible and adiabatic. The inlet pressure is 100 lbf/in.^{2}, the inlet temperature is

800 F and the exhaust pressure is 16 lbf/in.^{2}. The steam leaves the nozzle and enters the turbine at an angle of 20°. The blade speed ratio is 0.5 and the blade exit angle is 50°. Determine the blade efficiency of this turbine.

14.50E Steam enters an impulse turbine having the same inlet conditions in Problem 14.49. The nozzle efficiency is 92% and the blade velocity coefficient is 0.96 with a blade speed ratio of 0.5. Determine the work per pound-mass of steam and draw the velocity diagram to scale.

COMPUTER, DESIGN, AND OPEN-ENDED PROBLEMS

14.51 Make a program that calculates the stagnation pressure and temperature from a static pressure, temperature and velocity. Assume the fluid is air with constant specific heats. If the inverse relation is sought one of the three properties in the flow must be given, include that also.

14.52 Make a program to solve Problem 14.9 assuming constant specific heats for air.

14.53 Use the menu-driven software to solve Problem 14.26. Find from the menu-driven steam tables the ratio of specific heats at the inlet and the speed of sound from its definition in Eq. 14.28.

14.54 (Adv.) Make a program that will track the process in time as described in Problems 14.18 and 14.19. Investigate the time it takes to bring the tank pressure to 125 kPa as a function of the size of the nozzle exit area. Plot several of the key variables as function of time.

14.55 Make a computer program to do the necessary calculations for Problem 14.32. Let the angles be input variables and assume constant specific heats for the air.

14.56 Write a program to solve Problem 14.33 using the supplied software for the properties of steam.

14.57 Write a program to solve Problem 14.35 using the supplied software for the properties of steam.

14.58 A pump can deliver liquid water at an exit pressure of 400 kPa using 0.5 kW of power. Assume the inlet is water at 100 kPa, 15°C and that the pipe size is the same for the inlet and exit. Design a nozzle to be mounted on the exit line so the water exit velocity is at least 20 m/s. Show the exit velocity and mass flow rate as a function of the nozzle exit area with the same power to the pump.

14.59 In all the problems in the text the efficiency of a pump or compressor has been given as a constant. In reality it is a function of the mass flow rate and the fluid state through the device. Examine in the literature what the characteristics of a real air compressor (blower) are.

14.60 The throttle plate in a carburetor gives a severe restriction to the air flow where at idle it is critical flow. For normal atmospheric conditions estimate then the inlet temperature and pressure to the cylinder of the engine.

14.61 For an experiment in the laboratory an air flow rate should be measured. The range should be from 0.05 to 0.10 kg/s and the flow should be delivered to the experiment at 110 kPa. Size 1 or (2 in parallel) convergent nozzles that sits in a plate. The air is drawn through the nozzle(s) by suction of a blower that delivers the air at 110 kPa. What should be measured and what accuracy can be expected?

14.62 An afterburner in a jet engine adds fuel that is burned after the turbine but before the exit nozzle that accelerates the gases. Examine the effect on a nozzle exit velocity of having a higher inlet temperature, but same pressure as without the afterburner. Are these nozzles operating with subsonic or supersonic flow?

14.63 Examine the nozzle flow for the steam turbine in Problem 14.33. It was stated that the flow was adiabatic and reversible. If the nozzle receives the steam at the given inlet and has the given outlet pressure what are the limits for the outlet velocity? Discuss the factors that have an influence and check for possible shocks standing in the nozzle.

14.64 Investigate the literature for analysis that must be done to do the design of a steam or gas turbine stage. In most of the analysis presented we neglected pressure drops due to friction and shocks in the blade passage. Make a list of the most important things to consider for the design of the turbine stage.

14.65 For the maximum efficiency the normal velocity was considered very small when the analysis was done. Investigate a real situation and estimate the area's and velocities for the flow in the normal (axial) direction in the case of a steam turbine (Problem 14.33) and an air turbine (Problem 14.37).

TABLE A.1SI *Thermodynamic Properties of Water (SI Units)*
TABLE A.1.1SI *Saturated Water: Temperature Table (SI Units)*

Temp. °C T	Press. kPa, MPa P	Specific Volume, m³/kg		Internal Energy, kJ/kg			Enthalpy, kJ/kg			Entropy, kJ/kg K		
		Sat. Liquid v_f	Sat. Vapor v_g	Sat. Liquid u_f	Evap. u_{fg}	Sat. Vapor u_g	Sat. Liquid h_f	Evap. h_{fg}	Sat. Vapor h_g	Sat. Liquid s_f	Evap. s_{fg}	Sat. Vapor s_g
0.01	0.6113	0.001000	206.132	0.00	2375.3	2375.3	0.00	2501.3	2501.3	0.0000	9.1562	9.1562
5	0.8721	0.001000	147.118	20.97	2361.3	2382.2	20.98	2489.6	2510.5	0.0761	8.9496	9.0257
10	1.2276	0.001000	106.377	41.99	2347.2	2389.2	41.99	2477.7	2519.7	0.1510	8.7498	8.9007
15	1.7051	0.001001	77.925	62.98	2333.1	2396.0	62.98	2465.9	2528.9	0.2245	8.5569	8.7813
20	2.3385	0.001002	57.790	83.94	2319.0	2402.9	83.94	2454.1	2538.1	0.2966	8.3706	8.6671
25	3.1691	0.001003	43.359	104.86	2304.9	2409.8	104.87	2442.3	2547.2	0.3673	8.1905	8.5579
30	4.2461	0.001004	32.893	125.77	2290.8	2416.6	125.77	2430.5	2556.2	0.4369	8.0164	8.4533
35	5.6280	0.001006	25.216	146.65	2276.7	2423.4	146.66	2418.6	2565.3	0.5052	7.8478	8.3530
40	7.3837	0.001008	19.523	167.53	2262.6	2430.1	167.54	2406.7	2574.3	0.5724	7.6845	8.2569
45	9.5934	0.001010	15.258	188.41	2248.4	2436.8	188.42	2394.8	2583.2	0.6386	7.5261	8.1647
50	12.350	0.001012	12.032	209.30	2234.2	2443.5	209.31	2382.7	2592.1	0.7037	7.3725	8.0762
55	15.758	0.001015	9.568	230.19	2219.9	2450.1	230.20	2370.7	2600.9	0.7679	7.2234	7.9912
60	19.941	0.001017	7.671	251.09	2205.5	2456.6	251.11	2358.5	2609.6	0.8311	7.0784	7.9095
65	25.033	0.001020	6.197	272.00	2191.1	2463.1	272.03	2346.2	2618.2	0.8934	6.9375	7.8309
70	31.188	0.001023	5.042	292.93	2176.6	2469.5	292.96	2333.8	2626.8	0.9548	6.8004	7.7552
75	38.578	0.001026	4.131	313.87	2162.0	2475.9	313.91	2321.4	2635.3	1.0154	6.6670	7.6824
80	47.390	0.001029	3.407	334.84	2147.4	2482.2	334.88	2308.8	2643.7	1.0752	6.5369	7.6121
85	57.834	0.001032	2.828	355.82	2132.6	2488.4	355.88	2296.0	2651.9	1.1342	6.4102	7.5444
90	70.139	0.001036	2.361	376.82	2117.7	2494.5	376.90	2283.2	2660.1	1.1924	6.2866	7.4790
95	84.554	0.001040	1.982	397.86	2102.7	2500.6	397.94	2270.2	2668.1	1.2500	6.1659	7.4158
100	0.10135	0.001044	1.6729	418.91	2087.6	2506.5	419.02	2257.0	2676.0	1.3068	6.0480	7.3548

TABLE A.1SI (Continued) Thermodynamic Properties of Water (SI Units)
TABLE A.1.1SI Saturated Water: Temperature Table (SI Units)

Temp. °C T	Press. MPa P	Specific Volume, m³/kg		Internal Energy, kJ/kg			Enthalpy, kJ/kg			Entropy, kJ/kg K		
		Sat. Liquid v_f	Sat. Vapor v_g	Sat. Liquid u_f	Evap. u_{fg}	Sat. Vapor u_g	Sat. Liquid h_f	Evap. h_{fg}	Sat. Vapor h_g	Sat. Liquid s_f	Evap. s_{fg}	Sat. Vapor s_g
105	0.12082	0.001047	1.4194	440.00	2072.3	2512.3	440.13	2243.7	2683.8	1.3629	5.9328	7.2958
110	0.14328	0.001052	1.2102	461.12	2057.0	2518.1	461.27	2230.2	2691.5	1.4184	5.8202	7.2386
115	0.16906	0.001056	1.0366	482.28	2041.4	2523.7	482.46	2216.5	2699.0	1.4733	5.7100	7.1832
120	0.19853	0.001060	0.8919	503.48	2025.8	2529.2	503.69	2202.6	2706.3	1.5275	5.6020	7.1295
125	0.2321	0.001065	0.77059	524.72	2009.9	2534.6	524.96	2188.5	2713.5	1.5812	5.4962	7.0774
130	0.2701	0.001070	0.66850	546.00	1993.9	2539.9	546.29	2174.2	2720.5	1.6343	5.3925	7.0269
135	0.3130	0.001075	0.58217	567.34	1977.7	2545.0	567.67	2159.6	2727.3	1.6869	5.2907	6.9777
140	0.3613	0.001080	0.50885	588.72	1961.3	2550.0	589.11	2144.8	2733.9	1.7390	5.1908	6.9298
145	0.4154	0.001085	0.44632	610.16	1944.7	2554.9	610.61	2129.6	2740.3	1.7906	5.0926	6.8832
150	0.4759	0.001090	0.39278	631.66	1927.9	2559.5	632.18	2114.3	2746.4	1.8417	4.9960	6.8378
155	0.5431	0.001096	0.34676	653.23	1910.8	2564.0	653.82	2098.6	2752.4	1.8924	4.9010	6.7934
160	0.6178	0.001102	0.30706	674.85	1893.5	2568.4	675.53	2082.6	2758.1	1.9426	4.8075	6.7501
165	0.7005	0.001108	0.27269	696.55	1876.0	2572.5	697.32	2066.2	2763.5	1.9924	4.7153	6.7078
170	0.7917	0.001114	0.24283	718.31	1858.1	2576.5	719.20	2049.5	2768.7	2.0418	4.6244	6.6663
175	0.8920	0.001121	0.21680	740.16	1840.0	2580.2	741.16	2032.4	2773.6	2.0909	4.5347	6.6256
180	1.0022	0.001127	0.19405	762.08	1821.6	2583.7	763.21	2015.0	2778.2	2.1395	4.4461	6.5857
185	1.1227	0.001134	0.17409	784.08	1802.9	2587.0	785.36	1997.1	2782.4	2.1878	4.3586	6.5464
190	1.2544	0.001141	0.15654	806.17	1783.8	2590.0	807.61	1978.8	2786.4	2.2358	4.2720	6.5078
195	1.3978	0.001149	0.14105	828.36	1764.4	2592.8	829.96	1960.0	2790.0	2.2835	4.1863	6.4697
200	1.5538	0.001156	0.12736	850.64	1744.7	2595.3	852.43	1940.7	2793.2	2.3308	4.1014	6.4322
205	1.7230	0.001164	0.11521	873.02	1724.5	2597.5	875.03	1921.0	2796.0	2.3779	4.0172	6.3951
210	1.9063	0.001173	0.10441	895.51	1703.9	2599.4	897.75	1900.7	2798.5	2.4247	3.9337	6.3584
215	2.1042	0.001181	0.09479	918.12	1682.9	2601.1	920.61	1879.9	2800.5	2.4713	3.8507	6.3221
220	2.3178	0.001190	0.08619	940.85	1661.5	2602.3	943.61	1858.5	2802.1	2.5177	3.7683	6.2860

Temp. °C T	Press. MPa P	Specific Volume, m³/kg		Internal Energy, kJ/kg			Enthalpy, kJ/kg			Entropy, kJ/kg K		
		Sat. Liquid v_f	Sat. Vapor v_g	Sat. Liquid u_f	Evap. u_{fg}	Sat. Vapor u_g	Sat. Liquid h_f	Evap. h_{fg}	Sat. Vapor h_g	Sat. Liquid s_f	Evap. s_{fg}	Sat. Vapor s_g
225	2.5477	0.001199	0.07849	963.72	1639.6	2603.3	966.77	1836.5	2803.3	2.5639	3.6863	6.2502
230	2.7949	0.001209	0.07158	986.72	1617.2	2603.9	990.10	1813.8	2803.9	2.6099	3.6047	6.2146
235	3.0601	0.001219	0.06536	1009.88	1594.2	2604.1	1013.61	1790.5	2804.1	2.6557	3.5233	6.1791
240	3.3442	0.001229	0.05976	1033.19	1570.8	2603.9	1037.31	1766.5	2803.8	2.7015	3.4422	6.1436
245	3.6482	0.001240	0.05470	1056.69	1546.7	2603.4	1061.21	1741.7	2802.9	2.7471	3.3612	6.1083
250	3.9730	0.001251	0.05013	1080.37	1522.0	2602.4	1085.34	1716.2	2801.5	2.7927	3.2802	6.0729
255	4.3195	0.001263	0.04598	1104.26	1496.7	2600.9	1109.72	1689.8	2799.5	2.8382	3.1992	6.0374
260	4.6886	0.001276	0.04220	1128.37	1470.6	2599.0	1134.35	1662.5	2796.9	2.8837	3.1181	6.0018
265	5.0813	0.001289	0.03877	1152.72	1443.9	2596.6	1159.27	1634.3	2793.6	2.9293	3.0368	5.9661
270	5.4987	0.001302	0.03564	1177.33	1416.3	2593.7	1184.49	1605.2	2789.7	2.9750	2.9551	5.9301
275	5.9418	0.001317	0.03279	1202.23	1387.9	2590.2	1210.05	1574.9	2785.0	3.0208	2.8730	5.8937
280	6.4117	0.001332	0.03017	1227.43	1358.7	2586.1	1235.97	1543.6	2779.5	3.0667	2.7903	5.8570
285	6.9094	0.001348	0.02777	1252.98	1328.4	2581.4	1262.29	1511.0	2773.3	3.1129	2.7069	5.8198
290	7.4360	0.001366	0.02557	1278.89	1297.1	2576.0	1289.04	1477.1	2766.1	3.1593	2.6227	5.7821
295	7.9928	0.001384	0.02354	1305.21	1264.7	2569.9	1316.27	1441.8	2758.0	3.2061	2.5375	5.7436
300	8.5810	0.001404	0.02167	1331.97	1231.0	2563.0	1344.01	1404.9	2748.9	3.2533	2.4511	5.7044
305	9.2018	0.001425	0.01995	1359.22	1195.9	2555.2	1372.33	1366.4	2738.7	3.3009	2.3633	5.6642
310	9.8566	0.001447	0.01835	1387.03	1159.4	2546.4	1401.29	1326.0	2727.3	3.3492	2.2737	5.6229
315	10.547	0.001472	0.01687	1415.44	1121.1	2536.6	1430.97	1283.5	2714.4	3.3981	2.1821	5.5803
320	11.274	0.001499	0.01549	1444.55	1080.9	2525.5	1461.45	1238.6	2700.1	3.4479	2.0882	5.5361
330	12.845	0.001561	0.012996	1505.24	993.7	2498.9	1525.29	1140.6	2665.8	3.5506	1.8909	5.4416
340	14.586	0.001638	0.010797	1570.26	894.3	2464.5	1594.15	1027.9	2622.0	3.6593	1.6763	5.3356
350	16.514	0.001740	0.008813	1641.81	776.6	2418.4	1670.54	893.4	2563.9	3.7776	1.4336	5.2111
360	18.651	0.001892	0.006945	1725.19	626.3	2351.5	1760.48	720.5	2481.0	3.9146	1.1379	5.0525
370	21.028	0.002213	0.004926	1843.84	384.7	2228.5	1890.37	441.8	2332.1	4.1104	0.6868	4.7972
374.14	22.089	0.003155	0.003155	2029.58	0	2029.6	2099.26	0	2099.3	4.4297	0	4.4297

TABLE A.1.2SI *Saturated Water: Pressure Table (SI Units)*

Press. kPa P	Temp. °C T	Specific Volume, m³/kg Sat. Liquid v_f	Sat. Vapor v_g	Internal Energy, kJ/kg Sat. Liquid u_f	Evap. u_{fg}	Sat. Vapor u_g	Enthalpy, kJ/kg Sat. Liquid h_f	Evap. h_{fg}	Sat. Vapor h_g	Entropy, kJ/kg K Sat. Liquid s_f	Evap. s_{fg}	Sat. Vapor s_g
0.6113	0.01	0.001000	206.132	0	2375.3	2375.3	0.00	2501.3	2501.3	0	9.1562	9.1562
1.0	6.98	0.001000	129.208	29.29	2355.7	2385.0	29.29	2484.9	2514.2	0.1059	8.8697	8.9756
1.5	13.03	0.001001	87.980	54.70	2338.6	2393.3	54.70	2470.6	2525.3	0.1956	8.6322	8.8278
2.0	17.50	0.001001	67.004	73.47	2326.0	2399.5	73.47	2460.0	2533.5	0.2607	8.4629	8.7236
2.5	21.08	0.001002	54.254	88.47	2315.9	2404.4	88.47	2451.6	2540.0	0.3120	8.3311	8.6431
3.0	24.08	0.001003	45.665	101.03	2307.5	2408.5	101.03	2444.5	2545.5	0.3545	8.2231	8.5775
4.0	28.96	0.001004	34.800	121.44	2293.7	2415.2	121.44	2432.9	2554.4	0.4226	8.0520	8.4746
5.0	32.88	0.001005	28.193	137.79	2282.7	2420.5	137.79	2423.7	2561.4	0.4763	7.9187	8.3950
7.5	40.29	0.001008	19.238	168.76	2261.7	2430.5	168.77	2406.0	2574.8	0.5763	7.6751	8.2514
10.0	45.81	0.001010	14.674	191.79	2246.1	2437.9	191.81	2392.8	2584.6	0.6492	7.5010	8.1501
15.0	53.97	0.001014	10.022	225.90	2222.8	2448.7	225.91	2373.1	2599.1	0.7548	7.2536	8.0084
20.0	60.06	0.001017	7.649	251.35	2205.4	2456.7	251.38	2358.3	2609.7	0.8319	7.0766	7.9085
25.0	64.97	0.001020	6.204	271.88	2191.2	2463.1	271.90	2346.3	2618.2	0.8930	6.9383	7.8313
30.0	69.10	0.001022	5.229	289.18	2179.2	2468.4	289.21	2336.1	2625.3	0.9439	6.8247	7.7686
40.0	75.87	0.001026	3.993	317.51	2159.5	2477.0	317.55	2319.2	2636.7	1.0258	6.6441	7.6700
50.0	81.33	0.001030	3.240	340.42	2143.4	2483.8	340.47	2305.4	2645.9	1.0910	6.5029	7.5939
75.0	91.77	0.001037	2.217	384.29	2112.4	2496.7	384.36	2278.6	2663.0	1.2129	6.2434	7.4563
MPa												
0.100	99.62	0.001043	1.6940	417.33	2088.7	2506.1	417.44	2258.0	2675.5	1.3025	6.0568	7.3593
0.125	105.99	0.001048	1.3749	444.16	2069.3	2513.5	444.30	2241.1	2685.3	1.3739	5.9104	7.2843
0.150	111.37	0.001053	1.1593	466.92	2052.7	2519.6	467.08	2226.5	2693.5	1.4335	5.7897	7.2232
0.175	116.06	0.001057	1.0036	486.78	2038.1	2524.9	486.97	2213.6	2700.5	1.4848	5.6868	7.1717
0.200	120.23	0.001061	0.8857	504.47	2025.0	2529.5	504.68	2202.0	2706.6	1.5300	5.5970	7.1271
0.225	124.00	0.001064	0.7933	520.45	2013.1	2533.6	520.69	2191.3	2712.0	1.5705	5.5173	7.0878
0.250	127.43	0.001067	0.7187	535.08	2002.1	2537.2	535.34	2181.5	2716.9	1.6072	5.4455	7.0526

Press. MPa P	Temp. °C T	Specific Volume, m³/kg		Internal Energy, kJ/kg			Enthalpy, kJ/kg			Entropy, kJ/kg K		
		Sat. Liquid v_f	Sat. Vapor v_g	Sat. Liquid u_f	Evap. u_{fg}	Sat. Vapor u_g	Sat. Liquid h_f	Evap. h_{fg}	Sat. Vapor h_g	Sat. Liquid s_f	Evap. s_{fg}	Sat. Vapor s_g
0.275	130.60	0.001070	0.6573	548.57	1992.0	2540.5	548.87	2172.4	2721.3	1.6407	5.3801	7.0208
0.300	133.55	0.001073	0.6058	561.13	1982.4	2543.6	561.45	2163.9	2725.3	1.6717	5.3201	6.9918
0.325	136.30	0.001076	0.5620	572.88	1973.5	2546.3	573.23	2155.8	2729.0	1.7005	5.2646	6.9651
0.350	138.88	0.001079	0.5243	583.93	1965.0	2548.9	584.31	2148.1	2732.4	1.7274	5.2130	6.9404
0.375	141.32	0.001081	0.4914	594.38	1956.9	2551.3	594.79	2140.8	2735.6	1.7527	5.1647	6.9174
0.40	143.63	0.001084	0.4625	604.29	1949.3	2553.6	604.73	2133.8	2738.5	1.7766	5.1193	6.8958
0.45	147.93	0.001088	0.4140	622.75	1934.9	2557.6	623.24	2120.7	2743.9	1.8206	5.0359	6.8565
0.50	151.86	0.001093	0.3749	639.66	1921.6	2561.2	640.21	2108.5	2748.7	1.8606	4.9606	6.8212
0.55	155.48	0.001097	0.3427	655.30	1909.2	2564.5	655.91	2097.0	2752.9	1.8972	4.8920	6.7892
0.60	158.85	0.001101	0.3157	669.88	1897.5	2567.4	670.54	2086.3	2756.8	1.9311	4.8289	6.7600
0.65	162.01	0.001104	0.2927	683.55	1886.5	2570.1	684.26	2076.0	2760.3	1.9627	4.7704	6.7330
0.70	164.97	0.001108	0.2729	696.43	1876.1	2572.5	697.20	2066.3	2763.5	1.9922	4.7158	6.7080
0.75	167.77	0.001111	0.2556	708.62	1866.1	2574.7	709.45	2057.0	2766.4	2.0199	4.6647	6.6846
0.80	170.43	0.001115	0.2404	720.20	1856.6	2576.8	721.10	2048.0	2769.1	2.0461	4.6166	6.6627
0.85	172.96	0.001118	0.2270	731.25	1847.4	2578.7	732.20	2039.4	2771.6	2.0709	4.5711	6.6421
0.90	175.38	0.001121	0.2150	741.81	1838.7	2580.5	742.82	2031.1	2773.9	2.0946	4.5280	6.6225
0.95	177.69	0.001124	0.2042	751.94	1830.2	2582.1	753.00	2023.1	2776.1	2.1171	4.4869	6.6040
1.00	179.91	0.001127	0.19444	761.67	1822.0	2583.6	762.79	2015.3	2778.1	2.1386	4.4478	6.5864
1.10	184.09	0.001133	0.17753	780.08	1806.3	2586.4	781.32	2000.4	2781.7	2.1791	4.3744	6.5535
1.20	187.99	0.001139	0.16333	797.27	1791.6	2588.8	798.64	1986.2	2784.8	2.2165	4.3067	6.5233
1.30	191.64	0.001144	0.15125	813.42	1777.5	2590.9	814.91	1972.7	2787.6	2.2514	4.2438	6.4953
1.40	195.07	0.001149	0.14084	828.68	1764.1	2592.8	830.29	1959.7	2790.0	2.2842	4.1850	6.4692
1.50	198.32	0.001154	0.13177	843.14	1751.3	2594.5	844.87	1947.3	2792.1	2.3150	4.1298	6.4448
1.75	205.76	0.001166	0.11349	876.44	1721.4	2597.8	878.48	1918.0	2796.4	2.3851	4.0044	6.3895
2.00	212.42	0.001177	0.09963	906.42	1693.8	2600.3	908.77	1890.7	2799.5	2.4473	3.8935	6.3408
2.25	218.45	0.001187	0.08875	933.81	1668.2	2602.0	936.48	1865.2	2801.7	2.5034	3.7938	6.2971

APPENDIX A

TABLE A.1.2SI (Continued) Saturated Water: Pressure Table (SI Units)

Press. MPa P	Temp. °C T	Sat. Liquid v_f	Sat. Vapor v_g	Sat. Liquid u_f	Evap. u_{fg}	Sat. Vapor u_g	Sat. Liquid h_f	Evap. h_{fg}	Sat. Vapor h_g	Sat. Liquid s_f	Evap. s_{fg}	Sat. Vapor s_g
2.50	223.99	0.001197	0.07998	959.09	1644.0	2603.1	962.09	1841.0	2803.1	2.5546	3.7028	6.2574
2.75	229.12	0.001207	0.07275	982.65	1621.2	2603.8	985.97	1817.9	2803.9	2.6018	3.6190	6.2208
3.00	233.90	0.001216	0.06668	1004.76	1599.3	2604.1	1008.41	1795.7	2804.1	2.6456	3.5412	6.1869
3.25	238.38	0.001226	0.06152	1025.62	1578.4	2604.0	1029.60	1774.4	2804.0	2.6866	3.4685	6.1551
3.50	242.60	0.001235	0.05707	1045.41	1558.3	2603.7	1049.73	1753.7	2803.4	2.7252	3.4000	6.1252
4.0	250.40	0.001252	0.049778	1082.28	1520.0	2602.3	1087.29	1714.1	2801.4	2.7963	3.2737	6.0700
5.0	263.99	0.001286	0.039441	1147.78	1449.3	2597.1	1154.21	1640.1	2794.3	2.9201	3.0532	5.9733
6.0	275.64	0.001319	0.032440	1205.41	1384.3	2589.7	1213.32	1571.0	2784.3	3.0266	2.8625	5.8891
7.0	285.88	0.001351	0.027370	1257.51	1323.0	2580.5	1266.97	1505.1	2772.1	3.1210	2.6922	5.8132
8.0	295.06	0.001384	0.023518	1305.54	1264.3	2569.8	1316.61	1441.3	2757.9	3.2067	2.5365	5.7431
9.0	303.40	0.001418	0.020484	1350.47	1207.3	2557.8	1363.23	1378.9	2742.1	3.2857	2.3915	5.6771
10.0	311.06	0.001452	0.018026	1393.00	1151.4	2544.4	1407.53	1317.1	2724.7	3.3595	2.2545	5.6140
11.0	318.15	0.001489	0.015987	1433.68	1096.1	2529.7	1450.05	1255.5	2705.6	3.4294	2.1233	5.5527
12.0	324.75	0.001527	0.014263	1472.92	1040.8	2513.7	1491.24	1193.6	2684.8	3.4961	1.9962	5.4923
13.0	330.93	0.001567	0.012780	1511.09	985.0	2496.1	1531.46	1130.8	2662.2	3.5604	1.8718	5.4323
14.0	336.75	0.001611	0.011485	1548.53	928.2	2476.8	1571.08	1066.5	2637.5	3.6231	1.7485	5.3716
15.0	342.24	0.001658	0.010338	1585.58	869.8	2455.4	1610.45	1000.0	2610.5	3.6847	1.6250	5.3097
16.0	347.43	0.001711	0.009306	1622.63	809.1	2431.7	1650.00	930.6	2580.6	3.7460	1.4995	5.2454
17.0	352.37	0.001770	0.008365	1660.16	744.8	2405.0	1690.25	856.9	2547.2	3.8078	1.3698	5.1776
18.0	357.06	0.001840	0.007490	1698.86	675.4	2374.3	1731.97	777.1	2509.1	3.8713	1.2330	5.1044
19.0	361.54	0.001924	0.006657	1739.87	598.2	2338.1	1776.43	688.1	2464.5	3.9387	1.0841	5.0227
20.0	365.81	0.002035	0.005834	1785.47	507.6	2293.1	1826.18	583.6	2409.7	4.0137	0.9132	4.9269
21.0	369.89	0.002206	0.004953	1841.97	388.7	2230.7	1888.30	446.4	2334.7	4.1073	0.6942	4.8015
22.0	373.80	0.002808	0.003526	1973.16	108.2	2081.4	2034.92	124.0	2159.0	4.3307	0.1917	4.5224
22.09	374.14	0.003155	0.003155	2029.58	0	2029.6	2099.26	0	2099.3	4.4297	0	4.4297

TABLE A.1.3SI *Superheated Vapor Water (SI Units)*

T	P = 10 kPa (45.81)				P = 50 kPa (81.33)				P = 100 kPa (99.62)			
	v	u	h	s	v	u	h	s	v	u	h	s
Sat.	14.674	2437.9	2584.6	8.1501	3.240	2483.8	2645.9	7.5939	1.6940	2506.1	2675.5	7.3593
50	14.869	2443.9	2592.6	8.1749	—	—	—	—	—	—	—	—
100	17.196	2515.5	2687.5	8.4479	3.418	2511.6	2682.5	7.6947	1.6958	2506.6	2676.2	7.3614
150	19.513	2587.9	2783.0	8.6881	3.889	2585.6	2780.1	7.9400	1.9364	2582.7	2776.4	7.6133
200	21.825	2661.3	2879.5	8.9037	4.356	2659.8	2877.6	8.1579	2.1723	2658.0	2875.3	7.8342
250	24.136	2736.0	2977.3	9.1002	4.821	2735.0	2976.0	8.3555	2.4060	2733.7	2974.3	8.0332
300	26.445	2812.1	3076.5	9.2812	5.284	2811.3	3075.5	8.5372	2.6388	2810.4	3074.3	8.2157
400	31.063	2968.9	3279.5	9.6076	6.209	2968.4	3278.9	8.8641	3.1026	2967.8	3278.1	8.5434
500	35.679	3132.3	3489.0	9.8977	7.134	3131.9	3488.6	9.1545	3.5655	3131.5	3488.1	8.8341
600	40.295	3302.5	3705.4	10.1608	8.058	3302.2	3705.1	9.4177	4.0278	3301.9	3704.7	9.0975
700	44.911	3479.6	3928.7	10.4028	8.981	3479.5	3928.5	9.6599	4.4899	3479.2	3928.2	9.3398
800	49.526	3663.8	4159.1	10.6281	9.904	3663.7	4158.9	9.8852	4.9517	3663.5	4158.7	9.5652
900	54.141	3855.0	4396.4	10.8395	10.828	3854.9	4396.3	10.0967	5.4135	3854.8	4396.1	9.7767
1000	58.757	4053.0	4640.6	11.0392	11.751	4052.9	4640.5	10.2964	5.8753	4052.8	4640.3	9.9764
1100	63.372	4257.5	4891.2	11.2287	12.674	4257.4	4891.1	10.4858	6.3370	4257.3	4890.9	10.1658
1200	67.987	4467.9	5147.8	11.4090	13.597	4467.8	5147.7	10.6662	6.7986	4467.7	5147.6	10.3462
1300	72.603	4683.7	5409.7	11.5810	14.521	4683.6	5409.6	10.8382	7.2603	4683.5	5409.5	10.5182

T	P = 200 kPa (120.23)				P = 300 kPa (133.55)				P = 400 kPa (143.63)			
	v	u	h	s	v	u	h	s	v	u	h	s
Sat.	0.88573	2529.5	2706.6	7.1271	0.60582	2543.6	2725.3	6.9918	0.46246	2553.6	2738.5	6.8958
150	0.95964	2576.9	2768.8	7.2795	0.63388	2570.8	2761.0	7.0778	0.47084	2564.5	2752.8	6.9299
200	1.08034	2654.4	2870.5	7.5066	0.71629	2650.7	2865.5	7.3115	0.53422	2646.8	2860.5	7.1706
250	1.19880	2731.2	2971.0	7.7085	0.79636	2728.7	2967.6	7.5165	0.59512	2726.1	2964.2	7.3788
300	1.31616	2808.6	3071.8	7.8926	0.87529	2806.7	3069.3	7.7022	0.65484	2804.8	3066.7	7.5661
400	1.54930	2966.7	3276.5	8.2217	1.03151	2965.5	3275.0	8.0329	0.77262	2964.4	3273.4	7.8984
500	1.78139	3130.7	3487.0	8.5132	1.18669	3130.0	3486.0	8.3250	0.88934	3129.2	3484.9	8.1912
600	2.01297	3301.4	3704.0	8.7769	1.34136	3300.8	3703.2	8.5892	1.00555	3300.2	3702.4	8.4557
700	2.24426	3478.8	3927.7	9.0194	1.49573	3478.4	3927.1	8.8319	1.12147	3477.9	3926.5	8.6987
800	2.47539	3663.2	4158.3	9.2450	1.64994	3662.9	4157.8	9.0575	1.23722	3662.5	4157.4	8.9244

TABLE A.1.3SI (Continued) Superheated Vapor Water (SI Units)

T	P = 200 kPa (120.23)				P = 300 kPa (133.55)				P = 400 kPa (143.63)			
	v	u	h	s	v	u	h	s	v	u	h	s
900	2.70643	3854.5	4395.8	9.4565	1.80406	3854.2	4395.4	9.2691	1.35288	3853.9	4395.1	9.1361
1000	2.93740	4052.5	4640.0	9.6563	1.95812	4052.3	4639.7	9.4689	1.46847	4052.0	4639.4	9.3360
1100	3.16834	4257.0	4890.7	9.8458	2.11214	4256.8	4890.4	9.6585	1.58404	4256.5	4890.1	9.5255
1200	3.39927	4467.5	5147.3	10.0262	2.26614	4467.2	5147.1	9.8389	1.69958	4467.0	5146.8	9.7059
1300	3.63018	4683.2	5409.3	10.1982	2.42013	4683.0	5409.0	10.0109	1.81511	4682.8	5408.8	9.8780

T	P = 500 kPa (151.86)				P = 600 kPa (158.85)				P = 800 kPa (170.43)			
	v	u	h	s	v	u	h	s	v	u	h	s
Sat.	0.37489	2561.2	2748.7	6.8212	0.31567	2567.4	2756.8	6.7600	0.24043	2576.8	2769.1	6.6627
200	0.42492	2642.9	2855.4	7.0592	0.35202	2638.9	2850.1	6.9665	0.26080	2630.6	2839.2	6.8158
250	0.47436	2723.5	2960.7	7.2708	0.39383	2720.9	2957.2	7.1816	0.29314	2715.5	2950.0	7.0384
300	0.52256	2802.9	3064.2	7.4598	0.43437	2801.0	3061.6	7.3723	0.32411	2797.1	3056.4	7.2327
350	0.57012	2882.6	3167.6	7.6328	0.47424	2881.1	3165.7	7.5463	0.35439	2878.2	3161.7	7.4088
400	0.61728	2963.2	3271.8	7.7937	0.51372	2962.0	3270.2	7.7078	0.38426	2959.7	3267.1	7.5715
500	0.71093	3128.4	3483.8	8.0872	0.59199	3127.6	3482.7	8.0020	0.44331	3125.9	3480.6	7.8672
600	0.80406	3299.6	3701.7	8.3521	0.66974	3299.1	3700.9	8.2673	0.50184	3297.9	3699.4	8.1332
700	0.89691	3477.5	3926.0	8.5952	0.74720	3477.1	3925.4	8.5107	0.56007	3476.2	3924.3	8.3770
800	0.98959	3662.2	4157.0	8.8211	0.82450	3661.8	4156.5	8.7367	0.61813	3661.1	4155.7	8.6033
900	1.08217	3853.6	4394.7	9.0329	0.90169	3853.3	4394.4	8.9485	0.67610	3852.8	4393.6	8.8153
1000	1.17469	4051.8	4639.1	9.2328	0.97883	4051.5	4638.8	9.1484	0.73401	4051.0	4638.2	9.0153
1100	1.26718	4256.3	4889.9	9.4224	1.05594	4256.1	4889.6	9.3381	0.79188	4255.6	4889.1	9.2049
1200	1.35964	4466.8	5146.6	9.6028	1.13302	4466.5	5146.3	9.5185	0.84974	4466.1	5145.8	9.3854
1300	1.45210	4682.5	5408.6	9.7749	1.21009	4682.3	5408.3	9.6906	0.90758	4681.8	5407.9	9.5575

T	P = 1.00 MPa (179.91)				P = 1.20 MPa (187.99)				P = 1.40 MPa (195.07)			
	v	u	h	s	v	u	h	s	v	u	h	s
Sat.	0.19444	2583.6	2778.1	6.5864	0.16333	2588.8	2784.8	6.5233	0.14084	2592.8	2790.0	6.4692
200	0.20596	2621.9	2827.9	6.6939	0.16930	2612.7	2815.9	6.5898	0.14302	2603.1	2803.3	6.4975
250	0.23268	2709.9	2942.6	6.9246	0.19235	2704.2	2935.0	6.8293	0.16350	2698.3	2927.2	6.7467
300	0.25794	2793.2	3051.2	7.1228	0.21382	2789.2	3045.8	7.0316	0.18228	2785.2	3040.4	6.9533
350	0.28247	2875.2	3157.7	7.3010	0.23452	2872.2	3153.6	7.2120	0.20026	2869.1	3149.5	7.1359
400	0.30659	2957.3	3263.9	7.4650	0.25480	2954.9	3260.7	7.3773	0.21780	2952.5	3257.4	7.3025
500	0.35411	3124.3	3478.4	7.7621	0.29463	3122.7	3476.3	7.6758	0.25215	3121.1	3474.1	7.6026

(Superheated water — continuation of blocks from preceding page, rows $T = 600$–1300)

T	v	u	h	s	v	u	h	s	v	u	h	s
600	0.40109	3296.8	3697.9	8.0289	0.33393	3295.6	3696.3	7.9434	0.28596	3294.4	3694.8	7.8710
700	0.44779	3475.4	3923.1	8.2731	0.37294	3474.5	3922.0	8.1881	0.31947	3473.6	3920.9	8.1160
800	0.49432	3660.5	4154.8	8.4996	0.41177	3659.8	4153.9	8.4149	0.35281	3659.1	4153.0	8.3431
900	0.54075	3852.2	4392.9	8.7118	0.45051	3851.6	4392.2	8.6272	0.38606	3851.0	4391.5	8.5555
1000	0.58712	4050.5	4637.6	8.9119	0.48919	4050.0	4637.0	8.8274	0.41924	4049.5	4636.4	8.7558
1100	0.63345	4255.1	4888.5	9.1016	0.52783	4254.6	4888.0	9.0171	0.45239	4254.1	4887.5	8.9456
1200	0.67977	4465.6	5145.4	9.2821	0.56646	4465.1	5144.9	9.1977	0.48552	4464.6	5144.4	9.1262
1300	0.72608	4681.3	5407.4	9.4542	0.60507	4680.9	5406.9	9.3698	0.51864	4680.4	5406.5	9.2983

T	P = 1.60 MPa (201.40)				P = 1.80 MPa (207.15)				P = 2.00 MPa (212.42)			
	v	u	h	s	v	u	h	s	v	u	h	s
Sat.	0.12380	2595.9	2794.0	6.4217	0.11042	2598.4	2797.1	6.3793	0.09963	2600.3	2799.5	6.3408
225	0.13287	2644.6	2857.2	6.5518	0.11673	2636.6	2846.7	6.4807	0.10377	2628.3	2835.8	6.4146
250	0.14184	2692.3	2919.2	6.6732	0.12497	2686.0	2911.0	6.6066	0.11144	2679.6	2902.5	6.5452
300	0.15862	2781.0	3034.8	6.8844	0.14021	2776.8	3029.2	6.8226	0.12547	2772.6	3023.5	6.7663
350	0.17456	2866.0	3145.4	7.0693	0.15457	2862.9	3141.2	7.0099	0.13857	2859.8	3137.0	6.9562
400	0.19005	2950.1	3254.2	7.2373	0.16847	2947.7	3250.9	7.1793	0.15120	2945.2	3247.6	7.1270
500	0.22029	3119.5	3471.9	7.5389	0.19550	3117.8	3469.7	7.4824	0.17568	3116.2	3467.6	7.4316
600	0.24998	3293.3	3693.2	7.8080	0.22199	3292.1	3691.7	7.7523	0.19960	3290.9	3690.1	7.7023
700	0.27937	3472.7	3919.7	8.0535	0.24818	3471.9	3918.6	7.9983	0.22323	3471.0	3917.5	7.9487
800	0.30859	3658.4	4152.1	8.2808	0.27420	3657.7	4151.3	8.2258	0.24668	3657.0	4150.4	8.1766
900	0.33772	3850.5	4390.8	8.4934	0.30012	3849.9	4390.1	8.4386	0.27004	3849.3	4389.4	8.3895
1000	0.36678	4049.0	4635.8	8.6938	0.32598	4048.4	4635.2	8.6390	0.29333	4047.9	4634.6	8.5900
1100	0.39581	4253.7	4887.0	8.8837	0.35180	4253.2	4886.4	8.8290	0.31659	4252.7	4885.9	8.7800
1200	0.42482	4464.2	5143.9	9.0642	0.37761	4463.7	5143.4	9.0096	0.33984	4463.2	5142.9	8.9606
1300	0.45382	4679.9	5406.0	9.2364	0.40340	4679.4	5405.6	9.1817	0.36306	4679.0	5405.1	9.1328

T	P = 2.50 MPa (223.99)				P = 3.00 MPa (233.90)				P = 3.50 MPa (242.60)			
	v	u	h	s	v	u	h	s	v	u	h	s
Sat.	0.07998	2603.1	2803.1	6.2574	0.06668	2604.1	2804.1	6.1869	0.05707	2603.7	2803.4	6.1252
225	0.08027	2605.6	2806.3	6.2638	—	—	—	—	—	—	—	—
250	0.08700	2662.5	2880.1	6.4084	0.07058	2644.0	2855.8	6.2871	0.05873	2623.7	2829.2	6.1748
300	0.09890	2761.6	3008.8	6.6437	0.08114	2750.0	2993.5	6.5389	0.06842	2738.0	2977.5	6.4460
350	0.10976	2851.8	3126.2	6.8402	0.09053	2843.7	3115.3	6.7427	0.07678	2835.3	3104.0	6.6578
400	0.12010	2939.0	3239.3	7.0147	0.09936	2932.7	3230.8	6.9211	0.08453	2926.4	3222.2	6.8404
450	0.13014	3025.4	3350.8	7.1745	0.10787	3020.4	3344.0	7.0833	0.09196	3015.3	3337.2	7.0051
500	0.13998	3112.1	3462.0	7.3233	0.11619	3107.9	3456.5	7.2337	0.09918	3103.7	3450.9	7.1571
600	0.15930	3288.0	3686.2	7.5960	0.13243	3285.0	3682.3	7.5084	0.11324	3282.1	3678.4	7.4338
700	0.17832	3468.8	3914.6	7.8435	0.14838	3466.6	3911.7	7.7571	0.12699	3464.4	3908.8	7.6837

TABLE A.1.3SI (Continued) Superheated Vapor Water (SI Units)

T	P = 2.50 MPa (223.99)				P = 3.00 MPa (233.90)				P = 3.50 MPa (242.60)			
	v	u	h	s	v	u	h	s	v	u	h	s
800	0.19716	3655.3	4148.2	8.0720	0.16414	3653.6	4146.0	7.9862	0.14056	3651.8	4143.8	7.9135
900	0.21590	3847.9	4387.6	8.2853	0.17980	3846.5	4385.9	8.1999	0.15402	3845.0	4384.1	8.1275
1000	0.23458	4046.7	4633.1	8.4860	0.19541	4045.4	4631.6	8.4009	0.16743	4044.1	4630.1	8.3288
1100	0.25322	4251.5	4884.6	8.6761	0.21098	4250.3	4883.3	8.5911	0.18080	4249.1	4881.9	8.5191
1200	0.27185	4462.1	5141.7	8.8569	0.22652	4460.9	5140.5	8.7719	0.19415	4459.8	5139.3	8.7000
1300	0.29046	4677.8	5404.0	9.0291	0.24206	4676.6	5402.8	8.9442	0.20749	4675.5	5401.7	8.8723

T	P = 4.00 MPa (250.40)				P = 4.50 MPa (257.48)				P = 5.00 MPa (263.99)			
	v	u	h	s	v	u	h	s	v	u	h	s
Sat.	0.04978	2602.3	2801.4	6.0700	0.04406	2600.0	2798.3	6.0198	0.03944	2597.1	2794.3	5.9733
275	0.05457	2667.9	2886.2	6.2284	0.04730	2650.3	2863.1	6.1401	0.04141	2631.2	2838.3	6.0543
300	0.05884	2725.3	2960.7	6.3614	0.05135	2712.0	2943.1	6.2827	0.04532	2697.9	2924.5	6.2083
350	0.06645	2826.6	3092.4	6.5820	0.05840	2817.8	3080.6	6.5130	0.05194	2808.7	3068.4	6.4492
400	0.07341	2919.9	3213.5	6.7689	0.06475	2913.3	3204.7	6.7046	0.05781	2906.6	3195.6	6.6458
450	0.08003	3010.1	3330.2	6.9362	0.07074	3004.9	3323.2	6.8745	0.06330	2999.6	3316.1	6.8185
500	0.08643	3099.5	3445.2	7.0900	0.07651	3095.2	3439.5	7.0300	0.06857	3090.9	3433.8	6.9758
600	0.09885	3279.1	3674.4	7.3688	0.08765	3276.0	3670.5	7.3109	0.07869	3273.0	3666.5	7.2588
700	0.11095	3462.1	3905.9	7.6198	0.09847	3459.9	3903.0	7.5631	0.08849	3457.7	3900.1	7.5122
800	0.12287	3650.1	4141.6	7.8502	0.10911	3648.4	4139.4	7.7942	0.09811	3646.6	4137.2	7.7440
900	0.13469	3843.6	4382.3	8.0647	0.11965	3842.1	4380.6	8.0091	0.10762	3840.7	4378.8	7.9593
1000	0.14645	4042.9	4628.7	8.2661	0.13013	4041.6	4627.2	8.2108	0.11707	4040.3	4625.7	8.1612
1100	0.15817	4248.0	4880.6	8.4566	0.14056	4246.8	4879.3	8.4014	0.12648	4245.6	4878.0	8.3519
1200	0.16987	4458.6	5138.1	8.6376	0.15098	4457.4	5136.9	8.5824	0.13587	4456.3	5135.7	8.5330
1300	0.18156	4674.3	5400.5	8.8099	0.16139	4673.1	5399.4	8.7548	0.14526	4672.0	5398.2	8.7055

T	P = 6.00 MPa (275.64)				P = 7.00 MPa (285.88)				P = 8.00 MPa (295.06)			
	v	u	h	s	v	u	h	s	v	u	h	s
Sat.	0.03244	2589.7	2784.3	5.8891	0.02737	2580.5	2772.1	5.8132	0.02352	2569.8	2757.9	5.7431
300	0.03616	2667.2	2884.2	6.0673	0.02947	2632.1	2838.4	5.9304	0.02426	2590.9	2785.0	5.7905
350	0.04223	2789.6	3043.0	6.3334	0.03524	2769.3	3016.0	6.2282	0.02995	2747.7	2987.3	6.1300
400	0.04739	2892.8	3177.2	6.5407	0.03993	2878.6	3158.1	6.4477	0.03432	2863.8	3138.3	6.3633
450	0.05214	2988.9	3301.8	6.7192	0.04416	2977.9	3287.0	6.6326	0.03817	2966.7	3272.0	6.5550
500	0.05665	3082.2	3422.1	6.8802	0.04814	3073.3	3410.3	6.7974	0.04175	3064.3	3398.3	6.7239

T	v	u	h	s	v	u	h	s	v	u	h	s
550	0.06101	3174.6	3540.6	7.0287	0.05195	3167.2	3530.9	6.9486	0.04516	3159.8	3521.0	6.8778
600	0.06525	3266.9	3658.4	7.1676	0.05565	3260.7	3650.3	7.0894	0.04845	3254.4	3642.0	7.0205
700	0.07352	3453.2	3894.3	7.4234	0.06283	3448.6	3888.4	7.3476	0.05481	3444.0	3882.5	7.2812
800	0.08160	3643.1	4132.7	7.6566	0.06981	3639.6	4128.3	7.5822	0.06097	3636.1	4123.8	7.5173
900	0.08958	3837.8	4375.3	7.8727	0.07669	3835.0	4371.8	7.7991	0.06702	3832.1	4368.3	7.7350
1000	0.09749	4037.8	4622.7	8.0751	0.08350	4035.3	4619.8	8.0020	0.07301	4032.8	4616.9	7.9384
1100	0.10536	4243.3	4875.4	8.2661	0.09027	4240.9	4872.8	8.1933	0.07896	4238.6	4870.3	8.1299
1200	0.11321	4454.0	5133.3	8.4473	0.09703	4451.7	5130.9	8.3747	0.08489	4449.4	5128.5	8.3115
1300	0.12106	4669.6	5396.0	8.6199	0.10377	4667.3	5393.7	8.5472	0.09080	4665.0	5391.5	8.4842
	P = 9.00 MPa (303.40)				*P = 10.00 MPa (311.06)*				*P = 12.50 MPa (327.89)*			
Sat.	0.02048	2557.8	2742.1	5.6771	0.01803	2544.4	2724.7	5.6140	0.01350	2505.1	2673.8	5.4623
325	0.02327	2646.5	2855.9	5.8711	0.01986	2610.4	2809.0	5.7568	—	—	—	—
350	0.02580	2724.4	2956.5	6.0361	0.02242	2699.2	2923.4	5.9442	0.01613	2624.6	2826.2	5.7117
400	0.02993	2848.4	3117.8	6.2853	0.02641	2832.4	3096.5	6.2119	0.02000	2789.3	3039.3	6.0416
450	0.03350	2955.1	3256.6	6.4843	0.02975	2943.3	3240.8	6.4189	0.02299	2912.4	3199.8	6.2718
500	0.03677	3055.1	3386.1	6.6575	0.03279	3045.8	3373.6	6.5965	0.02560	3021.7	3341.7	6.4617
550	0.03987	3152.2	3511.0	6.8141	0.03564	3144.5	3500.9	6.7561	0.02801	3124.9	3475.1	6.6289
600	0.04285	3248.1	3633.7	6.9588	0.03837	3241.7	3625.3	6.9028	0.03029	3225.4	3604.0	6.7810
650	0.04574	3343.7	3755.3	7.0943	0.04101	3338.2	3748.3	7.0397	0.03248	3324.4	3730.4	6.9218
700	0.04857	3439.4	3876.5	7.2221	0.04358	3434.7	3870.5	7.1687	0.03460	3422.9	3855.4	7.0536
800	0.05409	3632.5	4119.4	7.4597	0.04859	3629.0	4114.9	7.4077	0.03869	3620.0	4103.7	7.2965
900	0.05950	3829.2	4364.7	7.6782	0.05349	3826.3	4361.2	7.6272	0.04267	3819.1	4352.5	7.5181
1000	0.06485	4030.3	4613.9	7.8821	0.05832	4027.8	4611.0	7.8315	0.04658	4021.6	4603.8	7.7237
1100	0.07016	4236.3	4867.7	8.0739	0.06312	4234.0	4865.1	8.0236	0.05045	4228.2	4858.8	7.9165
1200	0.07544	4447.2	5126.2	8.2556	0.06789	4444.9	5123.8	8.2054	0.05430	4439.3	5118.0	8.0987
1300	0.08072	4662.7	5389.2	8.4283	0.07265	4660.4	5387.0	8.3783	0.05813	4654.8	5381.4	8.2717
	P = 15 MPa (342.24)				*P = 17.5 MPa (354.75)*				*P = 20 MPa (365.81)*			
Sat.	.010338	2455.4	2610.5	5.3097	.0079204	2390.2	2528.8	5.1418	.0058342	2293.1	2409.7	4.9269
350	.011470	2520.4	2692.4	5.4420	.0017139	1632.0	1662.0	3.7612	.0016640	1612.3	1645.6	3.7280
400	.015649	2740.7	2975.4	5.8810	.0124477	2685.0	2902.8	5.7212	.0099423	2619.2	2818.1	5.5539
450	.018446	2879.5	3156.2	6.1403	.0151740	2844.2	3109.7	6.0182	.0126953	2806.2	3060.1	5.9016
500	.020800	2996.5	3308.5	6.3442	.0173585	2970.3	3274.0	6.2382	.0147683	2942.8	3238.2	6.1400
550	.022927	3104.7	3448.6	6.5198	.0192877	3083.8	3421.4	6.4229	.0165553	3062.3	3393.5	6.3347
600	.024911	3208.6	3582.3	6.6775	.0210640	3191.5	3560.1	6.5866	.0181781	3174.0	3537.6	6.5048
650	.026797	3310.4	3712.3	6.8223	.0227372	3296.0	3693.9	6.7356	.0196929	3281.5	3675.3	6.6582

TABLE A.1.3SI (Continued) Superheated Vapor Water (SI Units)

T		P = 15 MPa (342.24)				P = 17.5 MPa (354.75)				P = 20 MPa (365.81)		
	v	u	h	s	v	u	h	s	v	u	h	s
700	.028612	3410.9	3840.1	6.9572	.0243365	3398.8	3824.7	6.8736	.0211311	3386.5	3809.1	6.7993
800	.032096	3611.0	4092.4	7.2040	.0273849	3601.9	4081.1	7.1245	.0238532	3592.7	4069.8	7.0544
900	.035457	3811.9	4343.8	7.4279	.0303071	3804.7	4335.1	7.3507	.0264463	3797.4	4326.4	7.2830
1000	.038748	4015.4	4596.6	7.6347	.0331580	4009.3	4589.5	7.5588	.0289666	4003.1	4582.5	7.4925
1100	.042001	4222.6	4852.6	7.8282	.0359695	4216.9	4846.4	7.7530	.0314471	4211.3	4840.2	7.6874
1200	.045233	4433.8	5112.3	8.0108	.0387605	4428.3	5106.6	7.9359	.0339071	4422.8	5101.0	7.8706
1300	.048455	4649.1	5375.9	8.1839	.0415417	4643.5	5370.5	8.1093	.0363574	4638.0	5365.1	8.0441

T		P = 25 MPa				P = 30 MPa				P = 35 MPa		
	v	u	h	s	v	u	h	s	v	u	h	s
375	.001973	1798.6	1847.9	4.0319	.001789	1737.8	1791.4	3.9303	.001700	1702.9	1762.4	3.8721
400	.006004	2430.1	2580.2	5.1418	.002790	2067.3	2151.0	4.4728	.002100	1914.0	1987.5	4.2124
425	.007882	2609.2	2806.3	5.4722	.005304	2455.1	2614.2	5.1503	.003428	2253.4	2373.4	4.7747
450	.009162	2720.7	2949.7	5.6743	.006735	2619.3	2821.4	5.4423	.004962	2498.7	2672.4	5.1962
500	.011124	2884.3	3162.4	5.9592	.008679	2820.7	3081.0	5.7904	.006927	2751.9	2994.3	5.6281
550	.012724	3017.5	3335.6	6.1764	.010168	2970.3	3275.4	6.0342	.008345	2920.9	3213.0	5.9025
600	.014138	3137.9	3491.4	6.3602	.011446	3100.5	3443.9	6.2330	.009527	3062.0	3395.5	6.1178
650	.015433	3251.6	3637.5	6.5229	.012596	3221.0	3598.9	6.4057	.010575	3189.8	3559.9	6.3010
700	.016647	3361.4	3777.6	6.6707	.013661	3335.8	3745.7	6.5606	.011533	3309.9	3713.5	6.4631
800	.018913	3574.3	4047.1	6.9345	.015623	3555.6	4024.3	6.8332	.013278	3536.8	4001.5	6.7450
900	.021045	3783.0	4309.1	7.1679	.017448	3768.5	4291.9	7.0717	.014883	3754.0	4274.9	6.9886
1000	.023102	3990.9	4568.5	7.3801	.019196	3978.8	4554.7	7.2867	.016410	3966.7	4541.1	7.2063
1100	.025119	4200.2	4828.2	7.5765	.020903	4189.2	4816.3	7.4845	.017895	4178.3	4804.6	7.4056
1200	.027115	4412.0	5089.9	7.7604	.022589	4401.3	5079.0	7.6691	.019360	4390.7	5068.4	7.5910
1300	.029101	4626.9	5354.4	7.9342	.024266	4616.0	5344.0	7.8432	.020815	4605.1	5333.6	7.7652

T		P = 40 MPa				P = 50 MPa				P = 60 MPa		
	v	u	h	s	v	u	h	s	v	u	h	s
375	.0016406	1677.1	1742.7	3.8289	.0015593	1638.6	1716.5	3.7638	.0015027	1609.4	1699.5	3.7140
400	.0019077	1854.5	1930.8	4.1134	.0017309	1788.0	1874.6	4.0030	.0016335	1745.3	1843.4	3.9317
425	.0025319	2096.8	2198.1	4.5028	.0020071	1959.6	2060.0	4.2733	.0018165	1892.7	2001.7	4.1625
450	.0036931	2365.1	2512.8	4.9459	.0024862	2159.6	2283.9	4.5883	.0020850	2053.9	2179.0	4.4119
500	.0056225	2678.4	2903.3	5.4699	.0038924	2525.5	2720.1	5.1725	.0029557	2390.5	2567.9	4.9320

T	P = 40 MPa				P = 50 MPa				P = 60 MPa			
	v	u	h	s	v	u	h	s	v	u	h	s
600	.0080943	3022.6	3346.4	6.0113	.0061123	2942.0	3247.6	5.8177	.0048345	2861.1	3151.2	5.6451
650	.0090636	3158.0	3520.6	6.2054	.0069657	3093.6	3441.8	6.0342	.0055953	3028.8	3364.6	5.8829
700	.0099415	3283.6	3681.3	6.3750	.0077274	3230.5	3616.9	6.2189	.0062719	3177.3	3553.6	6.0824
800	.0115228	3517.9	3978.8	6.6662	.0090761	3479.8	3933.6	6.5290	.0074588	3441.6	3889.1	6.4110
900	.0129626	3739.4	4257.9	6.9150	.0102831	3710.3	4224.4	6.7882	.0085083	3681.0	4191.5	6.6805
1000	.0143238	3954.6	4527.6	7.1356	.0114113	3930.5	4501.1	7.0146	.0094800	3906.4	4475.2	6.9126
1100	.0156426	4167.4	4793.1	7.3364	.0124966	4145.7	4770.6	7.2183	.0104091	4124.1	4748.6	7.1194
1200	.0169403	4380.1	5057.7	7.5224	.0135606	4359.1	5037.2	7.4058	.0113167	4338.2	5017.2	7.3082
1300	.0182292	4594.3	5323.5	7.6969	.0146159	4572.8	5303.6	7.5807	.0122155	4551.4	5284.3	7.4837

TABLE A.1.4SI *Compressed Liquid Water (SI Units)*

T	P = 5.00 MPa (263.99)				P = 10.00 MPa (311.06)				P = 15.00 MPa (342.24)			
	v	u	h	s	v	u	h	s	v	u	h	s
Sat.	.0012859	1147.78	1154.21	2.9201	.0014524	1393.00	1407.53	3.3595	.0016581	1585.58	1610.45	3.6847
0	.0009977	0.03	5.02	0.0001	.0009952	0.10	10.05	0.0003	.0009928	0.15	15.04	0.0004
20	.0009995	83.64	88.64	0.2955	.0009972	83.35	93.32	0.2945	.0009950	83.05	97.97	0.2934
40	.0010056	166.93	171.95	0.5705	.0010034	166.33	176.36	0.5685	.0010013	165.73	180.75	0.5665
60	.0010149	250.21	255.28	0.8284	.0010127	249.34	259.47	0.8258	.0010105	248.49	263.65	0.8231
80	.0010268	333.69	338.83	1.0719	.0010245	332.56	342.81	1.0687	.0010222	331.46	346.79	1.0655
100	.0010410	417.50	422.71	1.3030	.0010385	416.09	426.48	1.2992	.0010361	414.72	430.26	1.2954
120	.0010576	501.79	507.07	1.5232	.0010549	500.07	510.61	1.5188	.0010522	498.39	514.17	1.5144
140	.0010768	586.74	592.13	1.7342	.0010737	584.67	595.40	1.7291	.0010707	582.64	598.70	1.7241
160	.0010988	672.61	678.10	1.9374	.0010953	670.11	681.07	1.9316	.0010918	667.69	684.07	1.9259
180	.0011240	759.62	765.24	2.1341	.0011199	756.63	767.83	2.1274	.0011159	753.74	770.48	2.1209
200	.0011530	848.08	853.85	2.3254	.0011480	844.49	855.97	2.3178	.0011433	841.04	858.18	2.3103
220	.0011866	938.43	944.36	2.5128	.0011805	934.07	945.88	2.5038	.0011748	929.89	947.52	2.4952
240	.0012264	1031.34	1037.47	2.6978	.0012187	1025.94	1038.13	2.6872	.0012114	1020.82	1038.99	2.6770

TABLE A.1.4SI (Continued) Compressed Liquid Water (SI Units)

T	P = 5.00 MPa (263.99)				P = 10.00 MPa (311.06)				P = 15.00 MPa (342.24)			
	v	u	h	s	v	u	h	s	v	u	h	s
260	.0012748	1127.92	1134.30	2.8829	.0012645	1121.03	1133.68	2.8698	.0012550	1114.59	1133.41	2.8575
280					.0013216	1220.90	1234.11	3.0547	.0013084	1212.47	1232.09	3.0392
300					.0013972	1328.34	1342.31	3.2468	.0013770	1316.58	1337.23	3.2259
320									.0014724	1431.05	1453.13	3.4246
340									.0016311	1567.42	1591.88	3.6545

T	P = 20 MPa (365.81)				P = 30 MPa				P = 50 MPa			
	v	u	h	s	v	u	h	s	v	u	h	s
Sat.	.0020353	1785.47	1826.18	4.0137	—				—			
0	.0009904	0.20	20.00	0.0004	.0009856	0.25	29.82	0.0001	.0009766	0.20	49.03	-0.0014
20	.0009928	82.75	102.61	0.2922	.0009886	82.16	111.82	0.2898	.0009804	80.98	130.00	0.2847
40	.0009992	165.15	185.14	0.5646	.0009951	164.01	193.87	0.5606	.0009872	161.84	211.20	0.5526
60	.0010084	247.66	267.82	0.8205	.0010042	246.03	276.16	0.8153	.0009962	242.96	292.77	0.8051
80	.0010199	330.38	350.78	1.0623	.0010156	328.28	358.75	1.0561	.0010073	324.32	374.68	1.0439
100	.0010337	413.37	434.04	1.2917	.0010290	410.76	441.63	1.2844	.0010201	405.86	456.87	1.2703
120	.0010496	496.75	517.74	1.5101	.0010445	493.58	524.91	1.5017	.0010348	487.63	539.37	1.4857
140	.0010678	580.67	602.03	1.7192	.0010621	576.86	608.73	1.7097	.0010515	569.76	622.33	1.6915
160	.0010885	665.34	687.11	1.9203	.0010821	660.81	693.27	1.9095	.0010703	652.39	705.91	1.8890
180	.0011120	750.94	773.18	2.1146	.0011047	745.57	778.71	2.1024	.0010912	735.68	790.24	2.0793
200	.0011387	837.70	860.47	2.3031	.0011302	831.34	865.24	2.2892	.0011146	819.73	875.46	2.2634
220	.0011693	925.89	949.27	2.4869	.0011590	918.32	953.09	2.4710	.0011408	904.67	961.71	2.4419
240	.0012046	1015.94	1040.04	2.6673	.0011920	1006.84	1042.60	2.6489	.0011702	990.69	1049.20	2.6158
260	.0012462	1108.53	1133.45	2.8459	.0012303	1097.38	1134.29	2.8242	.0012034	1078.06	1138.23	2.7860
280	.0012965	1204.69	1230.62	3.0248	.0012755	1190.69	1228.96	2.9985	.0012415	1167.19	1229.26	2.9536
300	.0013596	1306.10	1333.29	3.2071	.0013304	1287.89	1327.80	3.1740	.0012860	1258.66	1322.95	3.1200
320	.0014437	1415.66	1444.53	3.3978	.0013997	1390.64	1432.63	3.3538	.0013388	1353.23	1420.17	3.2867
340	.0015683	1539.64	1571.01	3.6074	.0014919	1501.71	1546.47	3.5425	.0014032	1451.91	1522.07	3.4556
360	.0018226	1702.78	1739.23	3.8770	.0016265	1626.57	1675.36	3.7492	.0014838	1555.97	1630.16	3.6290
380	—			—	.0018691	1781.35	1837.43	4.0010	.0015883	1667.13	1746.54	3.8100

TABLE A.1.5SI *Saturated Solid–Saturated Vapor Water (SI Units)*

Temp. °C T	Press. kPa P	Specific Volume, m³/kg		Internal Energy, kJ/kg			Enthalpy, kJ/kg			Entropy, kJ/kg K		
		Sat. Solid $v_i \times 10^{-3}$	Sat. Vapor v_g	Sat. Solid u_i	Evap. u_{ig}	Sat. Vapor u_g	Sat. Solid h_i	Evap. h_{ig}	Sat. Vapor h_g	Sat. Solid s_i	Evap. s_{ig}	Sat. Vapor s_g
0.01	0.6113	1.0908	206.153	−333.40	2708.7	2375.3	−333.40	2834.7	2501.3	−1.2210	10.3772	9.1562
0	0.6108	1.0908	206.315	−333.42	2708.7	2375.3	−333.42	2834.8	2501.3	−1.2211	10.3776	9.1565
−2	0.5177	1.0905	241.663	−337.61	2710.2	2372.5	−337.61	2835.3	2497.6	−1.2369	10.4562	9.2193
−4	0.4376	1.0901	283.799	−341.78	2711.5	2369.8	−341.78	2835.7	2494.0	−1.2526	10.5358	9.2832
−6	0.3689	1.0898	334.139	−345.91	2712.9	2367.0	−345.91	2836.2	2490.3	−1.2683	10.6165	9.3482
−8	0.3102	1.0894	394.414	−350.02	2714.2	2364.2	−350.02	2836.6	2486.6	−1.2839	10.6982	9.4143
−10	0.2601	1.0891	466.757	−354.09	2715.5	2361.4	−354.09	2837.0	2482.9	−1.2995	10.7809	9.4815
−12	0.2176	1.0888	553.803	−358.14	2716.8	2358.7	−358.14	2837.3	2479.2	−1.3150	10.8648	9.5498
−14	0.1815	1.0884	658.824	−362.16	2718.0	2355.9	−362.16	2837.6	2475.5	−1.3306	10.9498	9.6192
−16	0.1510	1.0881	785.907	−366.14	2719.2	2353.1	−366.14	2837.9	2471.8	−1.3461	11.0359	9.6898
−18	0.1252	1.0878	940.183	−370.10	2720.4	2350.3	−370.10	2838.2	2468.1	−1.3617	11.1233	9.7616
−20	0.10355	1.0874	1128.113	−374.03	2721.6	2347.5	−374.03	2838.4	2464.3	−1.3772	11.2120	9.8348
−22	0.08535	1.0871	1357.864	−377.93	2722.7	2344.7	−377.93	2838.6	2460.6	−1.3928	11.3020	9.9093
−24	0.07012	1.0868	1639.753	−381.80	2723.7	2342.0	−381.80	2838.7	2456.9	−1.4083	11.3935	9.9852
−26	0.05741	1.0864	1986.776	−385.64	2724.8	2339.2	−385.64	2838.9	2453.2	−1.4239	11.4864	10.0625
−28	0.04684	1.0861	2415.201	−389.45	2725.8	2336.4	−389.45	2839.0	2449.5	−1.4394	11.5808	10.1413
−30	0.03810	1.0858	2945.228	−393.23	2726.8	2333.6	−393.23	2839.0	2445.8	−1.4550	11.6765	10.2215
−32	0.03090	1.0854	3601.823	−396.98	2727.8	2330.8	−396.98	2839.1	2442.1	−1.4705	11.7733	10.3028
−34	0.02499	1.0851	4416.253	−400.71	2728.7	2328.0	−400.71	2839.1	2438.4	−1.4860	11.8713	10.3853
−36	0.02016	1.0848	5430.116	−404.40	2729.6	2325.2	−404.40	2839.1	2434.7	−1.5014	11.9704	10.4690
−38	0.01618	1.0844	6707.022	−408.06	2730.5	2322.4	−408.06	2839.0	2431.0	−1.5168	12.0714	10.5546
−40	0.01286	1.0841	8366.396	−411.70	2731.3	2319.6	−411.70	2838.9	2427.2	−1.5321	12.1768	10.6447

TABLE A.2SI *Thermodynamic Properties of Ammonia*
TABLE A.2.1SI *Saturated Ammonia (SI Units)*

Temp. °C	Abs. Press. kPa P	Specific Volume, m³/kg			Enthalpy, kJ/kg			Entropy, kJ/kg K		
		Sat. Liquid v_f	Evap. v_{fg}	Sat. Vapor v_g	Sat. Liquid h_f	Evap. h_{fg}	Sat. Vapor h_g	Sat. Liquid s_f	Evap. s_{fg}	Sat. Vapor s_g
−50	40.86	0.001424	2.62524	2.62667	−43.76	1416.34	1372.57	−0.1916	6.3470	6.1553
−48	45.94	0.001429	2.35297	2.35440	−35.04	1410.95	1375.90	−0.1528	6.2666	6.1139
−46	51.52	0.001434	2.11359	2.11503	−26.31	1405.50	1379.19	−0.1142	6.1875	6.0733
−44	57.66	0.001439	1.90262	1.90406	−17.56	1400.00	1382.44	−0.0759	6.1095	6.0336
−42	64.38	0.001444	1.71625	1.71769	−8.79	1394.44	1385.65	−0.0378	6.0326	5.9948
−40	71.72	0.001450	1.55124	1.55269	0	1388.82	1388.82	0	5.9568	5.9568
−38	79.74	0.001455	1.40482	1.40627	8.81	1383.13	1391.94	0.0376	5.8820	5.9196
−36	88.48	0.001460	1.27461	1.27607	17.64	1377.39	1395.03	0.0749	5.8082	5.8831
−34	97.98	0.001465	1.15857	1.16004	26.49	1371.58	1398.07	0.1120	5.7353	5.8473
−32	108.29	0.001471	1.05496	1.05643	35.36	1365.70	1401.06	0.1489	5.6634	5.8123
−30	119.46	0.001476	0.96226	0.96374	44.26	1359.76	1404.01	0.1856	5.5924	5.7780
−28	131.54	0.001482	0.87916	0.88064	53.17	1353.74	1406.92	0.2220	5.5223	5.7443
−26	144.59	0.001487	0.80453	0.80602	62.11	1347.66	1409.77	0.2582	5.4530	5.7113
−24	158.65	0.001493	0.73738	0.73887	71.07	1341.51	1412.58	0.2942	5.3846	5.6788
−22	173.80	0.001498	0.67685	0.67835	80.05	1335.29	1415.34	0.3301	5.3170	5.6470
−20	190.08	0.001504	0.62220	0.62371	89.05	1329.00	1418.05	0.3657	5.2501	5.6158
−18	207.56	0.001510	0.57277	0.57428	98.08	1322.64	1420.71	0.4011	5.1840	5.5851
−16	226.29	0.001516	0.52800	0.52951	107.12	1316.20	1423.32	0.4363	5.1187	5.5550
−14	246.35	0.001522	0.48737	0.48889	116.19	1309.68	1425.88	0.4713	5.0541	5.5254
−12	267.79	0.001528	0.45045	0.45197	125.29	1303.09	1428.38	0.5061	4.9901	5.4963
−10	290.67	0.001534	0.41684	0.41837	134.41	1296.42	1430.83	0.5408	4.9269	5.4676
−8	315.08	0.001540	0.38621	0.38775	143.55	1289.67	1433.22	0.5753	4.8642	5.4395
−6	341.07	0.001546	0.35824	0.35979	152.72	1282.84	1435.56	0.6095	4.8023	5.4118
−4	368.72	0.001553	0.33268	0.33423	161.91	1275.93	1437.84	0.6437	4.7409	5.3846
−2	398.10	0.001559	0.30928	0.31084	171.12	1268.94	1440.06	0.6776	4.6801	5.3577

Temp. °C	Abs. Press. kPa P	Specific Volume, m³/kg			Enthalpy, kJ/kg			Entropy, kJ/kg K		
		Sat. Liquid v_f	Evap. v_{fg}	Sat. Vapor v_g	Sat. Liquid h_f	Evap. h_{fg}	Sat. Vapor h_g	Sat. Liquid s_f	Evap. s_{fg}	Sat. Vapor s_g
0	429.29	0.001566	0.28783	0.28940	180.36	1261.86	1442.22	0.7114	4.6199	5.3313
2	462.34	0.001573	0.26815	0.26972	189.63	1254.69	1444.32	0.7450	4.5603	5.3053
4	497.35	0.001579	0.25005	0.25163	198.93	1247.43	1446.35	0.7785	4.5012	5.2796
6	534.39	0.001586	0.23341	0.23499	208.25	1240.08	1448.32	0.8118	4.4426	5.2543
8	573.54	0.001593	0.21807	0.21966	217.60	1232.63	1450.23	0.8449	4.3845	5.2294
10	614.87	0.001600	0.20392	0.20553	226.97	1225.10	1452.07	0.8779	4.3269	5.2048
12	658.48	0.001608	0.19086	0.19247	236.38	1217.46	1453.84	0.9108	4.2698	5.1805
14	704.43	0.001615	0.17878	0.18040	245.81	1209.72	1455.53	0.9435	4.2131	5.1565
16	752.81	0.001623	0.16761	0.16923	255.28	1201.88	1457.16	0.9760	4.1568	5.1328
18	803.71	0.001630	0.15725	0.15888	264.77	1193.94	1458.71	1.0085	4.1009	5.1094
20	857.22	0.001638	0.14764	0.14928	274.30	1185.89	1460.18	1.0408	4.0455	5.0863
22	913.41	0.001646	0.13872	0.14037	283.85	1177.73	1461.58	1.0730	3.9904	5.0634
24	972.38	0.001654	0.13043	0.13208	293.44	1169.45	1462.89	1.1050	3.9357	5.0407
26	1034.21	0.001663	0.12272	0.12438	303.07	1161.06	1464.13	1.1370	3.8813	5.0182
28	1099.00	0.001671	0.11553	0.11720	312.72	1152.55	1465.27	1.1688	3.8272	4.9960
30	1166.83	0.001680	0.10883	0.11051	322.42	1143.92	1466.33	1.2005	3.7735	4.9740
32	1237.80	0.001688	0.10258	0.10427	332.14	1135.16	1467.30	1.2321	3.7200	4.9521
34	1312.00	0.001697	0.09675	0.09845	341.91	1126.27	1468.17	1.2635	3.6669	4.9304
36	1389.52	0.001707	0.09129	0.09300	351.71	1117.25	1468.95	1.2949	3.6140	4.9089
38	1470.46	0.001716	0.08619	0.08790	361.55	1108.09	1469.64	1.3262	3.5613	4.8875
40	1554.92	0.001725	0.08141	0.08313	371.43	1098.79	1470.22	1.3574	3.5088	4.8662
42	1642.98	0.001735	0.07693	0.07866	381.35	1089.34	1470.69	1.3885	3.4566	4.8451
44	1734.75	0.001745	0.07272	0.07447	391.31	1079.75	1471.06	1.4195	3.4045	4.8240
46	1830.33	0.001755	0.06878	0.07053	401.32	1070.00	1471.32	1.4504	3.3526	4.8030
48	1929.82	0.001766	0.06507	0.06684	411.38	1060.09	1471.46	1.4813	3.3009	4.7822
50	2033.32	0.001777	0.06159	0.06336	421.48	1050.01	1471.49	1.5121	3.2493	4.7613

TABLE A.2.2SI Superheated Ammonia (SI Units)

Abs. Press. kPa (Sat. T)		−20	−10	0	10	20	30	40	50	60	70	80	100
							Temperature, °C						
50	v	2.4463	2.5471	2.6474	2.7472	2.8466	2.9458	3.0447	3.1435	3.2421	3.3406	3.4390	—
(−46.53)	h	1434.6	1455.7	1476.9	1498.1	1519.3	1540.6	1562.0	1583.5	1605.1	1626.9	1648.8	—
	s	6.3187	6.4006	6.4795	6.5556	6.6293	6.7008	6.7703	6.8379	6.9038	6.9682	7.0312	—
75	v	1.6222	1.6905	1.7582	1.8255	1.8924	1.9591	2.0255	2.0917	2.1577	2.2237	2.2895	—
(−39.16)	h	1431.7	1453.3	1474.8	1496.2	1517.7	1539.2	1560.7	1582.4	1604.1	1626.0	1648.0	—
	s	6.1120	6.1954	6.2756	6.3527	6.4272	6.4993	6.5693	6.6373	6.7036	6.7683	6.8315	—
100	v	1.2101	1.2621	1.3136	1.3647	1.4153	1.4657	1.5158	1.5658	1.6156	1.6652	1.7148	1.8137
(−33.59)	h	1428.8	1450.8	1472.6	1494.4	1516.1	1537.7	1559.5	1581.2	1603.1	1625.1	1647.1	1691.7
	s	5.9626	6.0477	6.1291	6.2073	6.2826	6.3553	6.4258	6.4943	6.5609	6.6258	6.6892	6.8120
125	v	0.9627	1.0051	1.0468	1.0881	1.1290	1.1696	1.2100	1.2502	1.2903	1.3302	1.3700	1.4494
(−29.06)	h	1425.9	1448.3	1470.5	1492.5	1514.4	1536.3	1558.2	1580.1	1602.1	1624.1	1646.3	1691.0
	s	5.8446	5.9314	6.0141	6.0933	6.1694	6.2428	6.3138	6.3827	6.4496	6.5149	6.5785	6.7017
150	v	0.7977	0.8336	0.8689	0.9037	0.9381	0.9723	1.0062	1.0398	1.0734	1.1068	1.1401	1.2065
(−25.21)	h	1422.9	1445.7	1468.3	1490.6	1512.8	1534.8	1556.9	1578.9	1601.0	1623.2	1645.4	1690.2
	s	5.7465	5.8349	5.9189	5.9992	6.0761	6.1502	6.2217	6.2910	6.3583	6.4238	6.4877	6.6112
200	v	—	0.6193	0.6465	0.6732	0.6995	0.7255	0.7513	0.7769	0.8023	0.8275	0.8527	0.9028
(−18.85)	h	—	1440.6	1463.8	1486.8	1509.4	1531.9	1554.3	1576.6	1598.9	1621.3	1643.7	1688.8
	s	—	5.6791	5.7659	5.8484	5.9270	6.0025	6.0751	6.1453	6.2133	6.2794	6.3437	6.4679
250	v	—	0.4905	0.5129	0.5348	0.5563	0.5774	0.5983	0.6190	0.6396	0.6600	0.6803	0.7206
(−13.65)	h	—	1435.3	1459.3	1482.9	1506.0	1529.0	1551.7	1574.3	1596.8	1619.4	1641.9	1687.3
	s	—	5.5544	5.6441	5.7288	5.8093	5.8861	5.9599	6.0309	6.0997	6.1663	6.2312	6.3561
300	v	—	—	0.4238	0.4425	0.4608	0.4787	0.4964	0.5138	0.5311	0.5483	0.5653	0.5992
(−9.22)	h	—	—	1454.7	1478.9	1502.6	1525.9	1549.0	1571.9	1594.7	1617.5	1640.2	1685.8
	s	—	—	5.5420	5.6290	5.7113	5.7896	5.8645	5.9365	6.0060	6.0732	6.1385	6.2642

Temperature, °C

kPa (Sat. T)		−20	−10	0	10	20	30	40	50	60	70	80	100
350 (−5.34)	v	—	—	0.3601	0.3765	0.3925	0.4081	0.4235	0.4386	0.4536	0.4685	0.4832	0.5124
	h	—	—	1449.9	1474.9	1499.1	1522.9	1546.3	1569.5	1592.6	1615.5	1638.4	1684.3
	s	—	—	5.4532	5.5427	5.6270	5.7068	5.7828	5.8557	5.9259	5.9938	6.0596	6.1860
400 (−1.87)	v	—	—	0.3123	0.3270	0.3413	0.3552	0.3688	0.3823	0.3955	0.4086	0.4216	0.4473
	h	—	—	1445.1	1470.7	1495.6	1519.8	1543.6	1567.1	1590.4	1613.6	1636.7	1682.8
	s	—	—	5.3741	5.4663	5.5525	5.6338	5.7111	5.7850	5.8560	5.9244	5.9907	6.1179
450 (1.27)	v	—	—	—	0.2885	0.3014	0.3140	0.3263	0.3384	0.3503	0.3620	0.3737	0.3967
	h	—	—	—	1466.5	1492.0	1516.7	1540.9	1564.7	1588.2	1611.6	1634.9	1681.3
	s	—	—	—	5.3972	5.4855	5.5685	5.6470	5.7219	5.7936	5.8627	5.9295	6.0575

kPa (Sat. T)		20	30	40	50	60	70	80	100	120	140	160	180
500 (4.15)	v	0.2695	0.2810	0.2923	0.3033	0.3141	0.3248	0.3353	0.3562	0.3768	0.3972	—	—
	h	1488.3	1513.5	1538.1	1562.3	1586.1	1609.6	1633.1	1679.8	1726.6	1773.8	—	—
	s	5.4244	5.5090	5.5889	5.6647	5.7373	5.8070	5.8744	6.0031	6.1253	6.2422	—	—
600 (9.29)	v	0.2215	0.2315	0.2412	0.2506	0.2598	0.2689	0.2778	0.2955	0.3128	0.3300	—	—
	h	1480.8	1507.1	1532.5	1557.3	1581.6	1605.7	1629.5	1676.8	1724.0	1771.5	—	—
	s	5.3156	5.4037	5.4862	5.5641	5.6383	5.7094	5.7778	5.9081	6.0314	6.1491	—	—
700 (13.81)	v	0.1872	0.1961	0.2046	0.2129	0.2210	0.2289	0.2367	0.2521	0.2671	0.2819	—	—
	h	1473.0	1500.4	1526.7	1552.2	1577.1	1601.6	1625.8	1673.7	1721.4	1769.2	—	—
	s	5.2196	5.3115	5.3968	5.4770	5.5529	5.6254	5.6949	5.8268	5.9512	6.0698	—	—
800 (17.86)	v	0.1614	0.1695	0.1772	0.1846	0.1919	0.1990	0.2059	0.2195	0.2328	0.2459	0.2589	—
	h	1464.9	1493.5	1520.8	1547.0	1572.5	1597.5	1622.1	1670.6	1718.7	1766.9	1815.3	—
	s	5.1328	5.2287	5.3171	5.3996	5.4774	5.5513	5.6219	5.7555	5.8811	6.0006	6.1150	—
900 (21.53)	v	—	0.1487	0.1558	0.1626	0.1692	0.1756	0.1819	0.1942	0.2061	0.2179	0.2295	—
	h	—	1486.5	1514.7	1541.7	1567.9	1593.3	1618.4	1667.5	1716.1	1764.5	1813.2	—
	s	—	5.1530	5.2447	5.3296	5.4093	5.4847	5.5565	5.6919	5.8187	5.9389	6.0541	—
1000 (24.91)	v	—	0.1321	0.1387	0.1450	0.1511	0.1570	0.1627	0.1739	0.1848	0.1955	0.2060	0.2164
	h	—	1479.1	1508.5	1536.3	1563.1	1589.1	1614.6	1664.3	1713.4	1762.2	1811.2	1860.5
	s	—	5.0826	5.1778	5.2654	5.3471	5.4240	5.4971	5.6342	5.7622	5.8834	5.9992	6.1105

TABLE A.2.2SI (Continued) Superheated Ammonia (SI Units)

Abs. Press. kPa (Sat. T)		Temperature, °C											
		40	50	60	70	80	100	120	140	160	180	200	220
1200 (30.95)	v	0.1129	0.1185	0.1238	0.1289	0.1339	0.1435	0.1527	0.1618	0.1707	0.1795	—	—
	h	1495.4	1525.1	1553.3	1580.5	1606.8	1658.0	1708.0	1757.5	1807.1	1856.9	—	—
	s	5.0564	5.1497	5.2357	5.3159	5.3916	5.5325	5.6631	5.7860	5.9031	6.0156	—	—
1400 (36.26)	v	0.0943	0.0994	0.1042	0.1088	0.1132	0.1217	0.1299	0.1378	0.1455	0.1532	—	—
	h	1481.6	1513.4	1543.1	1571.5	1598.8	1651.4	1702.5	1752.8	1802.9	1853.2	—	—
	s	4.9463	5.0462	5.1370	5.2209	5.2994	5.4443	5.5775	5.7023	5.8208	5.9343	—	—
1600 (41.03)	v	—	0.0851	0.0895	0.0937	0.0977	0.1054	0.1127	0.1197	0.1266	0.1334	—	—
	h	—	1501.0	1532.5	1562.3	1590.7	1644.8	1696.9	1748.0	1798.7	1849.5	—	—
	s	—	4.9510	5.0472	5.1351	5.2167	5.3659	5.5018	5.6286	5.7485	5.8631	—	—
1800 (45.37)	v	—	0.0738	0.0780	0.0819	0.0857	0.0927	0.0993	0.1057	0.1119	0.1180	—	—
	h	—	1487.9	1521.4	1552.7	1582.2	1638.0	1691.2	1743.1	1794.5	1845.7	—	—
	s	—	4.8614	4.9637	5.0561	5.1410	5.2948	5.4337	5.5624	5.6838	5.7995	—	—
2000 (49.36)	v	—	0.0647	0.0687	0.0725	0.0760	0.0825	0.0886	0.0945	0.1002	0.1057	—	—
	h	—	1473.9	1509.8	1542.7	1573.5	1631.1	1685.5	1738.2	1790.2	1842.0	—	—
	s	—	4.7754	4.8848	4.9821	5.0707	5.2294	5.3714	5.5022	5.6251	5.7420	—	—

TABLE A.3.1SI *Saturated R-12 (SI Units)*

Temp. °C	Abs. Press. MPa P	Specific Volume, m³/kg			Enthalpy, kJ/kg			Entropy, kJ/kg K		
		Sat. Liquid v_f	Evap. v_{fg}	Sat. Vapor v_g	Sat. Liquid h_f	Evap. h_{fg}	Sat. Vapor h_g	Sat. Liquid s_f	Evap. s_{fg}	Sat. Vapor s_g
−90	0.00284	0.000608	4.414937	4.415545	−43.284	189.748	146.464	−0.20863	1.03593	0.82730
−85	0.00424	0.000612	3.036704	3.037316	−39.005	187.737	148.731	−0.18558	0.99771	0.81213
−80	0.00617	0.000617	2.137728	2.138345	−34.721	185.740	151.018	−0.16312	0.96155	0.79843
−75	0.00879	0.000622	1.537030	1.537651	−30.430	183.751	153.321	−0.14119	0.92725	0.78606
−70	0.01227	0.000627	1.126654	1.127280	−26.128	181.764	155.636	−0.11977	0.89465	0.77489
−65	0.01680	0.000632	0.840534	0.841166	−21.814	179.774	157.960	−0.09880	0.86361	0.76480
−60	0.02262	0.000637	0.637274	0.637911	−17.485	177.775	160.289	−0.07827	0.83397	0.75570
−55	0.02998	0.000642	0.490358	0.491000	−13.141	175.762	162.621	−0.05815	0.80563	0.74748
−50	0.03915	0.000648	0.382457	0.383105	−8.779	173.730	164.951	−0.03841	0.77848	0.74007
−45	0.05044	0.000654	0.302029	0.302682	−4.400	171.676	167.276	−0.01903	0.75241	0.73338
−40	0.06417	0.000659	0.241251	0.241910	0	169.595	169.595	0	0.72735	0.72735
−35	0.08071	0.000666	0.194732	0.195398	4.420	167.482	171.903	0.01871	0.70322	0.72193
−30	0.10041	0.000672	0.158703	0.159375	8.862	165.335	174.197	0.03711	0.67993	0.71704
−25	0.12368	0.000679	0.130487	0.131166	13.327	163.149	176.476	0.05522	0.65742	0.71264
−20	0.15093	0.000685	0.108162	0.108847	17.816	160.920	178.736	0.07306	0.63563	0.70869
−15	0.18260	0.000693	0.090326	0.091018	22.331	158.643	180.974	0.09063	0.61450	0.70513
−10	0.21912	0.000700	0.075946	0.076646	26.874	156.314	183.188	0.10796	0.59397	0.70194
−5	0.26096	0.000708	0.064255	0.064963	31.446	153.928	185.375	0.12506	0.57400	0.69907
0	0.30861	0.000716	0.054673	0.055389	36.052	151.479	187.531	0.14196	0.55453	0.69649
5	0.36255	0.000724	0.046761	0.047485	40.694	148.961	189.654	0.15865	0.53551	0.69416
10	0.42330	0.000733	0.040180	0.040914	45.375	146.365	191.740	0.17517	0.51689	0.69206
15	0.49137	0.000743	0.034671	0.035413	50.100	143.684	193.784	0.19154	0.49862	0.69015
20	0.56729	0.000752	0.030028	0.030780	54.874	140.909	195.783	0.20777	0.48064	0.68841
25	0.65162	0.000763	0.026091	0.026854	59.702	138.028	197.730	0.22388	0.46292	0.68680
30	0.74490	0.000774	0.022734	0.023508	64.592	135.028	199.620	0.23991	0.44539	0.68530
35	0.84772	0.000786	0.019855	0.020641	69.551	131.896	201.446	0.25587	0.42800	0.68387
40	0.96065	0.000798	0.017373	0.018171	74.587	128.613	203.200	0.27179	0.41068	0.68248
45	1.08432	0.000811	0.015220	0.016032	79.712	125.160	204.872	0.28771	0.39338	0.68109
50	1.21932	0.000826	0.013344	0.014170	84.936	121.514	206.450	0.30366	0.37601	0.67967
55	1.36630	0.000841	0.011701	0.012542	90.274	117.645	207.920	0.31967	0.35849	0.67817
60	1.52592	0.000858	0.010253	0.011111	95.743	113.521	209.264	0.33580	0.34073	0.67653
65	1.69884	0.000877	0.008971	0.009847	101.362	109.099	210.460	0.35209	0.32262	0.67471

TABLE A.3SI (Continued) Thermodynamic Properties of Refrigerant-12 (Dichlorodifluoromethane)

TABLE A.3.1SI Saturated R-12 (SI Units)

Temp. °C	Abs. Press. MPa P	Specific Volume, m³/kg			Enthalpy, kJ/kg			Entropy, kJ/kg K		
		Sat. Liquid v_f	Evap. v_{fg}	Sat. Vapor v_g	Sat. Liquid h_f	Evap. h_{fg}	Sat. Vapor h_g	Sat. Liquid s_f	Evap. s_{fg}	Sat. Vapor s_g
70	1.88578	0.000897	0.007828	0.008725	107.155	104.326	211.481	0.36861	0.30401	0.67262
75	2.08745	0.000920	0.006802	0.007723	113.153	99.136	212.288	0.38543	0.28474	0.67017
80	2.30460	0.000946	0.005875	0.006821	119.394	93.437	212.832	0.40265	0.26457	0.66722
85	2.53802	0.000976	0.005029	0.006005	125.932	87.107	213.039	0.42040	0.24320	0.66361
90	2.78850	0.001012	0.004246	0.005258	132.841	79.961	212.802	0.43887	0.22018	0.65905
95	3.05689	0.001056	0.003508	0.004563	140.235	71.707	211.942	0.45833	0.19477	0.65310
100	3.34406	0.001113	0.002790	0.003903	148.314	61.810	210.124	0.47928	0.16564	0.64492
105	3.65093	0.001197	0.002045	0.003242	157.521	49.047	206.568	0.50285	0.12970	0.63254
110	3.97846	0.001364	0.001098	0.002462	169.550	28.444	197.995	0.53334	0.07423	0.60758
112	4.11548	0.001792	0	0.001792	183.418	0	183.418	0.56888	0	0.56888

TABLE A.3.2SI Superheated Refrigerant-12 (SI Units)

Temp. °C	0.05 MPa			0.10 MPa			0.15 MPa		
	v m³/kg	h kJ/kg	s kJ/kg K	v m³/kg	h kJ/kg	s kJ/kg K	v m³/kg	h kJ/kg	s kJ/kg K
−20	0.341859	181.170	0.79172	0.167702	179.987	0.74064	—	—	—
−10	0.356228	186.889	0.81388	0.175223	185.839	0.76331	0.114826	184.753	0.73240
0	0.370509	192.705	0.83557	0.182648	191.765	0.78541	0.119980	190.800	0.75495
10	0.384717	198.614	0.85681	0.189995	197.770	0.80700	0.125051	196.906	0.77690
20	0.398864	204.617	0.87764	0.197277	203.855	0.82812	0.130053	203.077	0.79832
30	0.412960	210.710	0.89808	0.204507	210.018	0.84879	0.135000	209.314	0.81924
40	0.427014	216.891	0.91814	0.211692	216.262	0.86905	0.139900	215.621	0.83971
50	0.441031	223.160	0.93784	0.218839	222.583	0.88892	0.144761	221.998	0.85975
60	0.455018	229.512	0.95720	0.225956	228.982	0.90842	0.149589	228.446	0.87940
70	0.468980	235.946	0.97623	0.233045	235.457	0.92757	0.154391	234.963	0.89867
80	0.482919	242.460	0.99494	0.240112	242.007	0.94638	0.159168	241.549	0.91759
90	0.496839	249.050	1.01334	0.247160	248.629	0.96487	0.163926	248.204	0.93617

Temp. °C	v m³/kg	h kJ/kg	s kJ/kg K	v m³/kg	h kJ/kg	s kJ/kg K	v m³/kg	h kJ/kg	s kJ/kg K
	0.20 MPa			**0.25 MPa**			**0.30 MPa**		
0	0.088609	189.805	0.73249	0.069752	188.779	0.71437	0.057150	187.718	0.69894
10	0.092550	196.020	0.75484	0.073024	195.109	0.73713	0.059984	194.173	0.72215
20	0.096419	202.281	0.77657	0.076219	201.468	0.75920	0.062735	200.636	0.74458
30	0.100229	208.597	0.79775	0.079351	207.866	0.78066	0.065419	207.119	0.76633
40	0.103990	214.971	0.81843	0.082432	214.309	0.80157	0.068049	213.635	0.78747
50	0.107710	221.405	0.83866	0.085470	220.803	0.82198	0.070636	220.191	0.80808
60	0.111397	227.902	0.85846	0.088474	227.351	0.84193	0.073186	226.793	0.82820
70	0.115056	234.462	0.87786	0.091449	233.956	0.86147	0.075706	233.444	0.84787
80	0.118691	241.087	0.89689	0.094399	240.620	0.88061	0.078200	240.147	0.86712
90	0.122305	247.775	0.91556	0.097328	247.341	0.89937	0.080673	246.904	0.88599
100	0.125902	254.525	0.93390	0.100239	254.122	0.91779	0.083127	253.716	0.90449
110	0.129484	261.338	0.95191	0.103135	260.962	0.93588	0.085566	260.582	0.92265

Temp. °C	v m³/kg	h kJ/kg	s kJ/kg K	v m³/kg	h kJ/kg	s kJ/kg K	v m³/kg	h kJ/kg	s kJ/kg K
	0.40 MPa			**0.50 MPa**			**0.60 MPa**		
20	0.045837	198.906	0.72043	0.035646	197.077	0.70043	—	—	—
30	0.047971	205.577	0.74281	0.037464	203.963	0.72352	0.030422	202.263	0.70679
40	0.050046	212.250	0.76447	0.039215	210.810	0.74574	0.031966	209.307	0.72965
50	0.052072	218.939	0.78549	0.040912	217.643	0.76722	0.033450	216.300	0.75163
60	0.054059	225.653	0.80595	0.042566	224.479	0.78806	0.034887	223.268	0.77287
70	0.056014	232.401	0.82591	0.044185	231.330	0.80832	0.036286	230.231	0.79346
80	0.057941	239.188	0.84540	0.045775	238.206	0.82807	0.037653	237.201	0.81348
90	0.059846	246.017	0.86447	0.047341	245.112	0.84735	0.038996	244.188	0.83299
100	0.061731	252.892	0.88315	0.048886	252.054	0.86621	0.040316	251.200	0.85204
110	0.063601	259.815	0.90145	0.050415	259.035	0.88467	0.041619	258.242	0.87066
120	0.065456	266.786	0.91941	0.051929	266.057	0.90276	0.042907	265.318	0.88889
130	0.067299	273.806	0.93704	0.053430	273.123	0.92050	0.044181	272.431	0.90675

Temp. °C	v m³/kg	h kJ/kg	s kJ/kg K	v m³/kg	h kJ/kg	s kJ/kg K	v m³/kg	h kJ/kg	s kJ/kg K
	0.70 MPa			**0.80 MPa**			**0.90 MPa**		
40	0.026761	207.732	0.71529	0.022830	206.074	0.70210	0.019744	204.320	0.68972
50	0.028100	214.903	0.73783	0.024068	213.446	0.72527	0.020912	211.921	0.71361
60	0.029387	222.017	0.75951	0.025247	220.720	0.74744	0.022012	219.373	0.73633
70	0.030632	229.099	0.78045	0.026380	227.934	0.76878	0.023062	226.730	0.75808
80	0.031843	236.171	0.80076	0.027477	235.114	0.78940	0.024073	234.028	0.77905
90	0.033028	243.244	0.82051	0.028545	242.279	0.80941	0.025051	241.290	0.79932

TABLE A.3.2SI (Continued) Superheated Refrigerant-12 (SI Units)

Temp. °C	v m³/kg	h kJ/kg	s kJ/kg K	v m³/kg	h kJ/kg	s kJ/kg K	v m³/kg	h kJ/kg	s kJ/kg K
	0.70 MPa			0.80 MPa			0.90 MPa		
100	0.034189	250.330	0.83976	0.029588	249.443	0.82887	0.026005	248.537	0.81901
110	0.035332	257.436	0.85855	0.030612	256.616	0.84784	0.026937	255.781	0.83817
120	0.036459	264.568	0.87693	0.031619	263.806	0.86636	0.027852	263.032	0.85685
130	0.037572	271.730	0.89492	0.032612	271.019	0.88448	0.028751	270.298	0.87510
140	0.038673	278.925	0.91254	0.033592	278.259	0.90221	0.029639	277.585	0.89295
150	0.039765	286.155	0.92984	0.034563	285.529	0.91960	0.030515	284.896	0.91043

Temp. °C	v m³/kg	h kJ/kg	s kJ/kg K	v m³/kg	h kJ/kg	s kJ/kg K	v m³/kg	h kJ/kg	s kJ/kg K
	1.00 MPa			1.20 MPa			1.40 MPa		
50	0.018366	210.317	0.70259	0.014483	206.813	0.68165	—	—	—
60	0.019410	217.970	0.72591	0.015463	214.964	0.70649	0.012579	211.613	0.68806
70	0.020397	225.485	0.74814	0.016368	222.851	0.72982	0.013448	219.984	0.71281
80	0.021341	232.910	0.76946	0.017221	230.568	0.75198	0.014247	228.059	0.73601
90	0.022251	240.278	0.79004	0.018032	238.171	0.77321	0.014997	235.940	0.75802
100	0.023133	247.612	0.80996	0.018812	245.699	0.79366	0.015710	243.692	0.77907
110	0.023993	254.931	0.82931	0.019567	253.180	0.81344	0.016393	251.355	0.79934
120	0.024835	262.246	0.84816	0.020301	260.632	0.83265	0.017053	258.961	0.81893
130	0.025661	269.567	0.86655	0.021018	268.072	0.85133	0.017695	266.530	0.83795
140	0.026474	276.902	0.88452	0.021721	275.509	0.86955	0.018321	274.078	0.85644
150	0.027275	284.255	0.90211	0.022412	282.952	0.88735	0.018934	281.618	0.87447
160	0.028068	291.632	0.91933	0.023093	290.408	0.90477	0.019535	289.158	0.89208

Temp. °C	v m³/kg	h kJ/kg	s kJ/kg K	v m³/kg	h kJ/kg	s kJ/kg K	v m³/kg	h kJ/kg	s kJ/kg K
	1.60 MPa			1.80 MPa			2.00 MPa		
70	0.011208	216.810	0.69641	0.009406	213.208	0.67992	—	—	—
80	0.011984	225.344	0.72092	0.010187	222.363	0.70622	0.008704	219.024	0.69143
90	0.012698	233.563	0.74387	0.010884	231.007	0.73036	0.009406	228.226	0.71713
100	0.013366	241.575	0.76564	0.011525	239.332	0.75297	0.010035	236.936	0.74079
110	0.014000	249.448	0.78646	0.012126	247.446	0.77443	0.010615	245.336	0.76300
120	0.014608	257.225	0.80649	0.012697	255.417	0.79497	0.011159	253.528	0.78411
130	0.015195	264.937	0.82586	0.013244	263.288	0.81474	0.011676	261.577	0.80433
140	0.015765	272.606	0.84465	0.013772	271.090	0.83385	0.012172	269.526	0.82380
150	0.016320	280.250	0.86293	0.014284	278.847	0.85240	0.012651	277.405	0.84265

Temp. °C	1.60 MPa v m³/kg	h kJ/kg	s kJ/kg K	1.80 MPa v m³/kg	h kJ/kg	s kJ/kg K	2.00 MPa v m³/kg	h kJ/kg	s kJ/kg K
160	0.016864	287.880	0.88076	0.014784	286.574	0.87045	0.013116	285.237	0.86094
170	0.017398	295.506	0.89816	0.015272	294.284	0.88805	0.013570	293.037	0.87874
180	0.017923	303.136	0.91519	0.015752	301.988	0.90524	0.014013	300.819	0.89611

Temp. °C	2.50 MPa v m³/kg	h kJ/kg	s kJ/kg K	3.00 MPa v m³/kg	h kJ/kg	s kJ/kg K	3.50 MPa v m³/kg	h kJ/kg	s kJ/kg K
90	0.006595	219.736	0.68284	—	—	—	—	—	—
100	0.007264	230.029	0.71081	0.005231	220.723	0.67755	—	—	—
110	0.007837	239.453	0.73573	0.005886	232.256	0.70806	0.004324	222.360	0.67559
120	0.008351	248.379	0.75873	0.006419	242.398	0.73420	0.004959	235.086	0.70840
130	0.008827	256.986	0.78035	0.006887	251.825	0.75788	0.005456	245.865	0.73548
140	0.009273	265.377	0.80091	0.007313	260.818	0.77991	0.005884	255.728	0.75965
150	0.009697	273.616	0.82062	0.007709	269.521	0.80072	0.006270	265.053	0.78195
160	0.010104	281.748	0.83961	0.008083	278.024	0.82059	0.006626	274.027	0.80291
170	0.010497	289.802	0.85799	0.008439	286.384	0.83967	0.006961	282.759	0.82284
180	0.010879	297.802	0.87584	0.008782	294.640	0.85809	0.007279	291.319	0.84194
190	0.011250	305.764	0.89322	0.009114	302.820	0.87594	0.007584	299.752	0.86035
200	0.011614	313.701	0.91018	0.009436	310.946	0.89330	0.007878	308.092	0.87816

Temp. °C	4.00 MPa v m³/kg	h kJ/kg	s kJ/kg K	5.00 MPa v m³/kg	h kJ/kg	s kJ/kg K
120	0.003736	225.180	0.67769	0.001369	176.303	0.54710
130	0.004325	238.691	0.71164	0.002501	216.458	0.64811
140	0.004781	249.930	0.73918	0.003139	235.004	0.69359
150	0.005172	260.124	0.76357	0.003585	248.416	0.72568
160	0.005522	269.710	0.78596	0.003950	259.910	0.75253
170	0.005845	278.903	0.80694	0.004268	270.400	0.77648
180	0.006147	287.825	0.82685	0.004555	280.276	0.79851
190	0.006434	296.552	0.84590	0.004821	289.740	0.81917
200	0.006708	305.136	0.86424	0.005071	298.916	0.83877
210	0.006972	313.614	0.88197	0.005308	307.882	0.85753
220	0.007228	322.013	0.89917	0.005535	316.690	0.87557
230	0.007477	330.352	0.91592	0.005753	325.380	0.89301

TABLE A.4SI *Thermodynamic Properties of Refrigerant-22 (Chlorodifluoromethane)*
TABLE A.4.1SI *Saturated Refrigerant-22 (SI Units)*

Temp. °C	Abs. Press. MPa P	Specific Volume, m³/kg			Enthalpy, kJ/kg			Entropy, kJ/kg K		
		Sat. Liquid v_f	Evap. v_{fg}	Sat. Vapor v_g	Sat. Liquid h_f	Evap. h_{fg}	Sat. Vapor h_g	Sat. Liquid s_f	Evap. s_{fg}	Sat. Vapor s_g
−70	0.0205	0.000670	0.940268	0.940938	−30.607	249.425	218.818	−0.1401	1.2277	1.0876
−65	0.0280	0.000676	0.704796	0.705472	−25.658	246.925	221.267	−0.1161	1.1862	1.0701
−60	0.0375	0.000682	0.536470	0.537152	−20.652	244.354	223.702	−0.0924	1.1463	1.0540
−55	0.0495	0.000689	0.414138	0.414827	−15.585	241.703	226.117	−0.0689	1.1079	1.0390
−50	0.0644	0.000695	0.323862	0.324557	−10.456	238.965	228.509	−0.0457	1.0708	1.0251
−45	0.0827	0.000702	0.256288	0.256990	−5.262	236.132	230.870	−0.0227	1.0349	1.0122
−40	0.1049	0.000709	0.205036	0.205745	0	233.198	233.197	0	1.0002	1.0002
−35	0.1317	0.000717	0.165683	0.166400	5.328	230.156	235.484	0.0225	0.9664	0.9889
−30	0.1635	0.000725	0.135120	0.135844	10.725	227.001	237.726	0.0449	0.9335	0.9784
−25	0.2010	0.000733	0.111126	0.111859	16.191	223.727	239.918	0.0670	0.9015	0.9685
−20	0.2448	0.000741	0.092102	0.092843	21.728	220.327	242.055	0.0890	0.8703	0.9593
−15	0.2957	0.000750	0.076876	0.077625	27.334	216.798	244.132	0.1107	0.8398	0.9505
−10	0.3543	0.000759	0.064581	0.065340	33.012	213.132	246.144	0.1324	0.8099	0.9422
−5	0.4213	0.000768	0.054571	0.055339	38.762	209.323	248.085	0.1538	0.7806	0.9344
0	0.4976	0.000778	0.046357	0.047135	44.586	205.364	249.949	0.1751	0.7518	0.9269
5	0.5838	0.000789	0.039567	0.040356	50.485	201.246	251.731	0.1963	0.7235	0.9197
10	0.6807	0.000800	0.033914	0.034714	56.463	196.960	253.423	0.2173	0.6956	0.9129
15	0.7891	0.000812	0.029176	0.029987	62.523	192.495	255.018	0.2382	0.6680	0.9062
20	0.9099	0.000824	0.025179	0.026003	68.670	187.836	256.506	0.2590	0.6407	0.8997
25	1.0439	0.000838	0.021787	0.022624	74.910	182.968	257.877	0.2797	0.6137	0.8934
30	1.1919	0.000852	0.018890	0.019742	81.250	177.869	259.119	0.3004	0.5867	0.8871
35	1.3548	0.000867	0.016401	0.017269	87.700	172.516	260.216	0.3210	0.5598	0.8809
40	1.5335	0.000884	0.014251	0.015135	94.272	166.877	261.149	0.3417	0.5329	0.8746
45	1.7290	0.000902	0.012382	0.013284	100.982	160.914	261.896	0.3624	0.5058	0.8682
50	1.9423	0.000922	0.010747	0.011669	107.851	154.576	262.428	0.3832	0.4783	0.8615
55	2.1744	0.000944	0.009308	0.010252	114.905	147.800	262.705	0.4042	0.4504	0.8546
60	2.4266	0.000969	0.008032	0.009001	122.180	140.497	262.678	0.4255	0.4217	0.8472
65	2.6999	0.000997	0.006890	0.007887	129.729	132.547	262.276	0.4472	0.3920	0.8391

Temp. °C	Abs. Press. MPa P	Sat. Liquid v_f	Evap. v_{fg}	Sat. Vapor v_g	Sat. Liquid h_f	Evap. h_{fg}	Sat. Vapor h_g	Sat. Liquid s_f	Evap. s_{fg}	Sat. Vapor s_g
70	2.9959	0.001030	0.005859	0.006889	137.625	123.772	261.397	0.4695	0.3607	0.8302
75	3.3161	0.001069	0.004914	0.005983	145.986	113.902	259.888	0.4927	0.3272	0.8198
80	3.6623	0.001118	0.004031	0.005149	155.011	102.475	257.486	0.5173	0.2902	0.8075
85	4.0368	0.001183	0.003175	0.004358	165.092	88.598	253.690	0.5445	0.2474	0.7918
90	4.4425	0.001282	0.002282	0.003564	177.204	70.037	247.241	0.5767	0.1929	0.7695
95	4.8835	0.001521	0.001030	0.002551	196.359	34.925	231.284	0.6273	0.0949	0.7222
96.006	4.9773	0.001906	0	0.001906	212.546	0	212.546	0.6708	0	0.6708

TABLE A.4.2SI Superheated Refrigerant-22 (SI Units)

Temp. °C	v m³/kg	h kJ/kg	s kJ/kg K	v m³/kg	h kJ/kg	s kJ/kg K	v m³/kg	h kJ/kg	s kJ/kg K
	0.05 MPa			0.10 MPa			0.15 MPa		
−40	0.440633	234.724	1.07616	0.216331	233.337	1.00523	—	—	—
−30	0.460641	240.602	1.10084	0.226754	239.359	1.03052	0.148723	238.078	0.98773
−20	0.480543	246.586	1.12495	0.237064	245.466	1.05513	0.155851	244.319	1.01288
−10	0.500357	252.676	1.14855	0.247279	251.665	1.07914	0.162879	250.631	1.03733
0	0.520095	258.874	1.17166	0.257415	257.956	1.10261	0.169823	257.022	1.06116
10	0.539771	265.180	1.19433	0.267485	264.345	1.12558	0.176699	263.496	1.08444
20	0.559393	271.594	1.21659	0.277500	270.831	1.14809	0.183516	270.057	1.10721
30	0.578970	278.115	1.23846	0.287467	277.416	1.17017	0.190284	276.709	1.12952
40	0.598507	284.743	1.25998	0.297394	284.101	1.19187	0.197011	283.452	1.15140
50	0.618011	291.478	1.28114	0.307287	290.887	1.21320	0.203702	290.289	1.17289
60	0.637485	298.319	1.30199	0.317149	297.772	1.23418	0.210362	297.220	1.19402
70	0.656935	305.265	1.32253	0.326986	304.757	1.25484	0.216997	304.246	1.21479
80	0.676362	312.314	1.34278	0.336801	311.842	1.27519	0.223608	311.368	1.23525
90	0.695771	319.465	1.36275	0.346596	319.026	1.29524	0.230200	318.584	1.25540
	0.20 MPa			0.25 MPa			0.30 MPa		
−20	0.115203	243.140	0.98184	—	—	—	—	—	—
−10	0.120647	249.574	1.00676	0.095280	248.492	0.98231	0.078344	247.382	0.96170
0	0.126003	256.069	1.03098	0.099689	255.097	1.00695	0.082128	254.104	0.98677
10	0.131286	262.633	1.05458	0.104022	261.755	1.03089	0.085832	260.861	1.01106

TABLE A.4.2SI *Superheated Refrigerant-22 (SI Units)*

Temp. °C	v m³/kg	h kJ/kg	s kJ/kg K	v m³/kg	h kJ/kg	s kJ/kg K	v m³/kg	h kJ/kg	s kJ/kg K
	0.20 MPa			0.25 MPa			0.30 MPa		
20	0.136509	269.273	1.07763	0.108292	268.476	1.05421	0.089469	267.667	1.03468
30	0.141681	275.992	1.10016	0.112508	275.267	1.07699	0.093051	274.531	1.05771
40	0.146809	282.796	1.12224	0.116681	282.132	1.09927	0.096588	281.460	1.08019
50	0.151902	289.686	1.14390	0.120815	289.076	1.12109	0.100085	288.460	1.10220
60	0.156963	296.664	1.16516	0.124918	296.102	1.14250	0.103550	295.535	1.12376
70	0.161997	303.731	1.18607	0.128993	303.212	1.16353	0.106986	302.689	1.14491
80	0.167008	310.890	1.20663	0.133044	310.409	1.18420	0.110399	309.924	1.16569
90	0.171999	318.139	1.22687	0.137075	317.692	1.20454	0.113790	317.241	1.18612
100	0.176972	325.480	1.24681	0.141089	325.063	1.22456	0.117164	324.643	1.20623
110	0.181931	332.912	1.26646	0.145086	332.522	1.24428	0.120522	332.129	1.22603
	0.40 MPa			0.50 MPa			0.60 MPa		
0	0.060131	252.051	0.95359	—	—	—	—	—	—
10	0.063060	259.023	0.97866	0.049355	257.108	0.95223	0.040180	255.109	0.92945
20	0.065915	266.010	1.00291	0.051751	264.295	0.97717	0.042280	262.517	0.95517
30	0.068710	273.029	1.02646	0.054081	271.483	1.00128	0.044307	269.888	0.97989
40	0.071455	280.092	1.04938	0.056358	278.690	1.02467	0.046276	277.250	1.00378
50	0.074160	287.209	1.07175	0.058590	285.930	1.04743	0.048198	284.622	1.02695
60	0.076830	294.386	1.09362	0.060786	293.215	1.06963	0.050081	292.020	1.04950
70	0.079470	301.630	1.11504	0.062951	300.552	1.09133	0.051931	299.456	1.07149
80	0.082085	308.944	1.13605	0.065090	307.949	1.11257	0.053754	306.938	1.09298
90	0.084679	316.332	1.15668	0.067206	315.410	1.13340	0.055553	314.475	1.11403
100	0.087254	323.796	1.17695	0.069303	322.939	1.15386	0.057332	322.071	1.13466
110	0.089813	331.339	1.19690	0.071384	330.539	1.17395	0.059094	329.731	1.15492
120	0.092358	338.961	1.21654	0.073450	338.213	1.19373	0.060842	337.458	1.17482
130	0.094890	346.664	1.23588	0.075503	345.963	1.21319	0.062576	345.255	1.19441
	0.70 MPa			0.80 MPa			0.90 MPa		
20	0.035487	260.667	0.93565	0.030366	258.737	0.91787	0.026355	256.713	0.90132
30	0.037305	268.240	0.96105	0.032034	266.533	0.94402	0.027915	264.760	0.92831
40	0.039059	275.769	0.98549	0.033632	274.243	0.96905	0.029397	272.670	0.95398

Temp. °C	0.70 MPa v m³/kg	h kJ/kg	s kJ/kg K	0.80 MPa v m³/kg	h kJ/kg	s kJ/kg K	0.90 MPa v m³/kg	h kJ/kg	s kJ/kg K
50	0.040763	283.282	1.00910	0.035175	281.907	0.99314	0.030819	280.497	0.97859
60	0.042424	290.800	1.03201	0.036674	289.553	1.01644	0.032193	288.278	1.00230
70	0.044052	298.339	1.05431	0.038136	297.202	1.03906	0.033528	296.042	1.02526
80	0.045650	305.912	1.07606	0.039568	304.868	1.06108	0.034832	303.807	1.04757
90	0.047224	313.527	1.09732	0.040974	312.565	1.08257	0.036108	311.590	1.06930
100	0.048778	321.192	1.11815	0.042359	320.303	1.10359	0.037363	319.401	1.09052
110	0.050313	328.914	1.13856	0.043725	328.087	1.12417	0.038598	327.251	1.11128
120	0.051834	336.696	1.15861	0.045076	335.925	1.14437	0.039817	335.147	1.13162
130	0.053341	344.541	1.17832	0.046413	343.821	1.16420	0.041022	343.094	1.15158
140	0.054836	352.454	1.19770	0.047738	351.778	1.18369	0.042215	351.097	1.17119
150	0.056321	360.435	1.21679	0.049052	359.799	1.20288	0.043398	359.159	1.19047

Temp. °C	1.00 MPa v m³/kg	h kJ/kg	s kJ/kg K	1.20 MPa v m³/kg	h kJ/kg	s kJ/kg K	1.40 MPa v m³/kg	h kJ/kg	s kJ/kg K
30	0.024600	262.912	0.91358	—	—	—	—	—	—
40	0.025995	271.042	0.93996	0.020851	267.602	0.91411	0.017120	263.861	0.89010
50	0.027323	279.046	0.96512	0.022051	276.011	0.94055	0.018247	272.766	0.91809
60	0.028601	286.973	0.98928	0.023191	284.263	0.96570	0.019299	281.401	0.94441
70	0.029836	294.859	1.01260	0.024282	292.415	0.98981	0.020295	289.858	0.96942
80	0.031038	302.727	1.03520	0.025336	300.508	1.01305	0.021248	298.202	0.99339
90	0.032213	310.599	1.05718	0.026359	308.570	1.03556	0.022167	306.473	1.01649
100	0.033364	318.488	1.07861	0.027357	316.623	1.05744	0.023058	314.703	1.03884
110	0.034495	326.405	1.09955	0.028334	324.682	1.07875	0.023926	322.916	1.06056
120	0.035609	334.360	1.12004	0.029292	332.762	1.09957	0.024775	331.128	1.08172
130	0.036709	342.360	1.14014	0.030236	340.871	1.11994	0.025608	339.354	1.10238
140	0.037797	350.410	1.15986	0.031166	349.019	1.13990	0.026426	347.603	1.12259
150	0.038873	358.514	1.17924	0.032084	357.210	1.15949	0.027233	355.885	1.14240
160	0.039940	366.677	1.19831	0.032993	365.450	1.17873	0.028029	364.206	1.16183

Temp. °C	1.60 MPa v m³/kg	h kJ/kg	s kJ/kg K	1.80 MPa v m³/kg	h kJ/kg	s kJ/kg K	2.00 MPa v m³/kg	h kJ/kg	s kJ/kg K
50	0.015351	269.262	0.89689	0.013052	265.423	0.87625	—	—	—
60	0.016351	278.358	0.92461	0.014028	275.097	0.90573	0.012135	271.563	0.88729
70	0.017284	287.171	0.95068	0.014921	284.331	0.93304	0.013008	281.310	0.91612
80	0.018167	295.797	0.97546	0.015755	293.282	0.95876	0.013811	290.640	0.94292
90	0.019011	304.301	0.99920	0.016546	302.046	0.98323	0.014563	299.697	0.96821
100	0.019825	312.725	1.02209	0.017303	310.683	1.00669	0.015277	308.571	0.99232

TABLE A.4.2SI (Continued) Superheated Refrigerant-22 (SI Units)

Temp. °C	1.60 MPa			1.80 MPa			2.00 MPa		
	v m³/kg	h kJ/kg	s kJ/kg K	v m³/kg	h kJ/kg	s kJ/kg K	v m³/kg	h kJ/kg	s kJ/kg K
110	0.020614	321.103	1.04424	0.018032	319.239	1.02932	0.015960	317.322	1.01546
120	0.021382	329.457	1.06576	0.018738	327.745	1.05123	0.016619	325.991	1.03780
130	0.022133	337.805	1.08673	0.019427	336.224	1.07253	0.017258	334.610	1.05944
140	0.022869	346.162	1.10721	0.020099	344.695	1.09329	0.017881	343.201	1.08049
150	0.023592	354.540	1.12724	0.020759	353.172	1.11356	0.018490	351.783	1.10102
160	0.024305	362.945	1.14688	0.021407	361.666	1.13340	0.019087	360.369	1.12107
170	0.025008	371.386	1.16614	0.022045	370.186	1.15284	0.019673	368.970	1.14070
180	0.025703	379.869	1.18507	0.022675	378.738	1.17193	0.020251	377.595	1.15995

Temp. °C	2.50 MPa			3.00 MPa			3.50 MPa		
	v m³/kg	h kJ/kg	s kJ/kg K	v m³/kg	h kJ/kg	s kJ/kg K	v m³/kg	h kJ/kg	s kJ/kg K
70	0.009459	272.677	0.87476	—	—	—	—	—	—
80	0.010243	283.332	0.90537	0.007747	274.530	0.86780	0.005765	262.739	0.82489
90	0.010948	293.338	0.93332	0.008465	286.042	0.89995	0.006597	277.268	0.86548
100	0.011598	302.935	0.95939	0.009098	296.663	0.92881	0.007257	289.504	0.89872
110	0.012208	312.261	0.98405	0.009674	306.744	0.95547	0.007829	300.640	0.92818
120	0.012788	321.400	1.00760	0.010211	316.470	0.98053	0.008346	311.129	0.95520
130	0.013343	330.412	1.03023	0.010717	325.955	1.00435	0.008825	321.196	0.98049
140	0.013880	339.336	1.05210	0.011200	335.270	1.02718	0.009276	330.976	1.00445
150	0.014400	348.205	1.07331	0.011665	344.467	1.04918	0.009704	340.554	1.02736
160	0.014907	357.040	1.09395	0.012114	353.584	1.07047	0.010114	349.989	1.04940
170	0.015402	365.860	1.11408	0.012550	362.647	1.09116	0.010510	359.324	1.07071
180	0.015887	374.679	1.13376	0.012976	371.679	1.11131	0.010894	368.590	1.09138
190	0.016364	383.508	1.15303	0.013392	380.695	1.13099	0.011268	377.810	1.11151
200	0.016834	392.354	1.17192	0.013801	389.708	1.15024	0.011634	387.004	1.13115

Temp. °C	4.00 MPa			5.00 MPa			6.00 MPa		
	v m³/kg	h kJ/kg	s kJ/kg K	v m³/kg	h kJ/kg	s kJ/kg K	v m³/kg	h kJ/kg	s kJ/kg K
90	0.005037	265.629	0.82544	—	—	—	—	—	—
100	0.005804	280.997	0.86721	0.003334	253.042	0.78005	—	—	—
110	0.006405	293.748	0.90094	0.004255	275.919	0.84064	0.002432	243.278	0.74674
120	0.006924	305.273	0.93064	0.004851	291.362	0.88045	0.003333	272.385	0.82185
130	0.007391	316.080	0.95778	0.005335	304.469	0.91337	0.003899	290.253	0.86675

Temp. °C	m³/kg	kJ/kg	kJ/kg K	m³/kg	kJ/kg	kJ/kg K	m³/kg	kJ/kg	kJ/kg K
	4.00 MPa			5.00 MPa			6.00 MPa		
140	0.007822	326.422	0.98312	0.005757	316.379	0.94256	0.004345	304.757	0.90230
150	0.008226	336.446	1.00710	0.006139	327.563	0.96931	0.004728	317.633	0.93310
160	0.008610	346.246	1.02999	0.006493	338.266	0.99431	0.005071	329.553	0.96094
170	0.008978	355.885	1.05199	0.006826	348.633	1.01797	0.005386	340.849	0.98673
180	0.009332	365.409	1.07324	0.007142	358.760	1.04057	0.005680	351.715	1.01098
190	0.009675	374.853	1.09386	0.007444	368.713	1.06230	0.005958	362.271	1.03402
200	0.010009	384.240	1.11391	0.007735	378.537	1.08328	0.006222	372.602	1.05609
210	0.010335	393.593	1.13347	0.008018	388.268	1.10363	0.006477	382.764	1.07734
220	0.010654	402.925	1.15259	0.008292	397.932	1.12343	0.006722	392.801	1.09790

TABLE A.5SI *Thermodynamic Properties of Refrigerant-134a (1,1,2-tetrafluoroethane)*
TABLE A.5.1SI *Saturated R-134a (SI Units)*

Temp. °C	Abs. Press. MPa P	Specific Volume, m³/kg			Enthalpy, kJ/kg			Entropy, kJ/kg K		
		Sat. Liquid v_f	Evap. v_{fg}	Sat. Vapor v_g	Sat. Liquid h_f	Evap. h_{fg}	Sat. Vapor h_g	Sat. Liquid s_f	Evap. s_{fg}	Sat. Vapor s_g
−33	0.0737	0.000718	0.25574	0.25646	157.417	220.491	377.908	0.8346	0.9181	1.7528
−30	0.0851	0.000722	0.22330	0.22402	161.118	218.683	379.802	0.8499	0.8994	1.7493
−26.25	0.1013	0.000728	0.18947	0.19020	165.802	216.360	382.162	0.8690	0.8763	1.7453
−25	0.1073	0.000730	0.17956	0.18029	167.381	215.569	382.950	0.8754	0.8687	1.7441
−20	0.1337	0.000738	0.14575	0.14649	173.744	212.340	386.083	0.9007	0.8388	1.7395
−15	0.1650	0.000746	0.11932	0.12007	180.193	209.004	389.197	0.9258	0.8096	1.7354
−10	0.2017	0.000755	0.098454	0.099209	186.721	205.564	392.285	0.9507	0.7812	1.7319
−5	0.2445	0.000764	0.081812	0.082576	193.324	202.016	395.340	0.9755	0.7534	1.7288
0	0.2940	0.000773	0.068420	0.069193	200.000	198.356	398.356	1.0000	0.7262	1.7262
5	0.3509	0.000783	0.057551	0.058334	206.751	194.572	401.323	1.0243	0.6995	1.7239
10	0.4158	0.000794	0.048658	0.049451	213.580	190.652	404.233	1.0485	0.6733	1.7218
15	0.4895	0.000805	0.041326	0.042131	220.492	186.582	407.075	1.0725	0.6475	1.7200
20	0.5728	0.000817	0.035238	0.036055	227.493	182.345	409.838	1.0963	0.6220	1.7183
25	0.6663	0.000829	0.030148	0.030977	234.590	177.920	412.509	1.1201	0.5967	1.7168
30	0.7710	0.000843	0.025865	0.026707	241.790	173.285	415.075	1.1437	0.5716	1.7153
35	0.8876	0.000857	0.022237	0.023094	249.103	168.415	417.518	1.1673	0.5465	1.7139
40	1.0171	0.000873	0.019147	0.020020	256.539	163.282	419.821	1.1909	0.5214	1.7123

TABLE A.5SI (Continued) Thermodynamic Properties of Refrigerant-134a (1,1,1,2-tetrafluoroethane)
TABLE A.5.1SI Saturated R-134a (SI Units)

Temp. °C	Abs. Press. MPa P	Specific Volume, m³/kg			Enthalpy, kJ/kg			Entropy, kJ/kg K		
		Sat. Liquid v_f	Evap. v_{fg}	Sat. Vapor v_g	Sat. Liquid h_f	Evap. h_{fg}	Sat. Vapor h_g	Sat. Liquid s_f	Evap. s_{fg}	Sat. Vapor s_g
45	1.1602	0.000890	0.016499	0.017389	264.110	157.852	421.962	1.2145	0.4962	1.7106
50	1.3180	0.000908	0.014217	0.015124	271.830	152.085	423.915	1.2381	0.4706	1.7088
55	1.4915	0.000928	0.012237	0.013166	279.718	145.933	425.650	1.2619	0.4447	1.7066
60	1.6818	0.000951	0.010511	0.011462	287.794	139.336	427.130	1.2857	0.4182	1.7040
65	1.8898	0.000976	0.008995	0.009970	296.088	132.216	428.305	1.3099	0.3910	1.7009
70	2.1169	0.001005	0.007653	0.008657	304.642	124.468	429.110	1.3343	0.3627	1.6970
75	2.3644	0.001038	0.006453	0.007491	313.513	115.939	429.451	1.3592	0.3330	1.6923
80	2.6337	0.001078	0.005368	0.006446	322.794	106.395	429.189	1.3849	0.3013	1.6862
85	2.9265	0.001128	0.004367	0.005495	332.644	95.440	428.084	1.4117	0.2665	1.6782
90	3.2448	0.001195	0.003412	0.004606	343.380	82.295	425.676	1.4404	0.2266	1.6670
95	3.5914	0.001297	0.002432	0.003729	355.834	64.984	420.818	1.4733	0.1765	1.6498
101.15	4.0640	0.001969	0	0.001969	390.977	0	390.977	1.5658	0	1.5658

TABLE A.5.2SI Superheated R-134a

Temp. °C	0.10 MPa			0.15 MPa			0.20 MPa		
	v m³/kg	h kJ/kg	s kJ/kg K	v m³/kg	h kJ/kg	s kJ/kg K	v m³/kg	h kJ/kg	s kJ/kg K
−25	0.19400	383.212	1.75058	—	—	—	—	—	—
−20	0.19860	387.215	1.76655	—	—	—	—	—	—
−10	0.20765	395.270	1.79775	0.13603	393.839	1.76058	0.10013	392.338	1.73276
0	0.21652	403.413	1.82813	0.14222	402.187	1.79171	0.10501	400.911	1.76474
10	0.22527	411.668	1.85780	0.14828	410.602	1.82197	0.10974	409.500	1.79562
20	0.23393	420.048	1.88689	0.15424	419.111	1.85150	0.11436	418.145	1.82563
30	0.24250	428.564	1.91545	0.16011	427.730	1.88041	0.11889	426.875	1.85491
40	0.25102	437.223	1.94355	0.16592	436.473	1.90879	0.12335	435.708	1.88357
50	0.25948	446.029	1.97123	0.17168	445.350	1.93669	0.12776	444.658	1.91171

Temp. °C	v m³/kg	h kJ/kg	s kJ/kg K	v m³/kg	h kJ/kg	s kJ/kg K	v m³/kg	h kJ/kg	s kJ/kg K
	0.10 MPa			0.15 MPa			0.20 MPa		
60	0.26791	454.986	1.99853	0.17740	454.366	1.96416	0.13213	453.735	1.93937
70	0.27631	464.096	2.02547	0.18308	463.525	1.99125	0.13646	462.946	1.96661
80	0.28468	473.359	2.05208	0.18874	472.831	2.01798	0.14076	472.296	1.99346
90	0.29303	482.777	2.07837	0.19437	482.285	2.04438	0.14504	481.788	2.01997
100	0.30136	492.349	2.10437	0.19999	491.888	2.07046	0.14930	491.424	2.04614

Temp. °C	v m³/kg	h kJ/kg	s kJ/kg K	v m³/kg	h kJ/kg	s kJ/kg K	v m³/kg	h kJ/kg	s kJ/kg K
	0.25 MPa			0.30 MPa			0.40 MPa		
0	0.082 637	399.579	1.74284	—	—	—	—	—	—
10	0.086 584	408.357	1.77440	0.071 110	407.171	1.75637	0.051 681	404.651	1.72611
20	0.090 408	417.151	1.80492	0.074 415	416.124	1.78744	0.054 362	413.965	1.75844
30	0.094 139	425.997	1.83460	0.077 620	425.096	1.81754	0.056 926	423.216	1.78947
40	0.097 798	434.925	1.86357	0.080 748	434.124	1.84684	0.059 402	432.465	1.81949
50	0.101 401	443.953	1.89195	0.083 816	443.234	1.87547	0.061 812	441.751	1.84868
60	0.104 958	453.094	1.91980	0.086 838	452.442	1.90354	0.064 169	451.104	1.87718
70	0.108 480	462.359	1.94720	0.089 821	461.763	1.93110	0.066 484	460.545	1.90510
80	0.111 972	471.754	1.97419	0.092 774	471.206	1.95823	0.068 767	470.088	1.93252
90	0.115 440	481.285	2.00080	0.095 702	480.777	1.98495	0.071 022	479.745	1.95948
100	0.118 888	490.955	2.02707	0.098 609	490.482	2.01131	0.073 254	489.523	1.98604
110	0.122 318	500.766	2.05302	0.101 498	500.324	2.03734	0.075 468	499.428	2.01223
120	0.125 734	510.720	2.07866	0.104 371	510.304	2.06305	0.077 665	509.464	2.03809

Temp. °C	v m³/kg	h kJ/kg	s kJ/kg K	v m³/kg	h kJ/kg	s kJ/kg K	v m³/kg	h kJ/kg	s kJ/kg K
	0.50 MPa			0.60 MPa			0.70 MPa		
20	0.042 256	411.645	1.73420	—	—	—	—	—	—
30	0.044 457	421.221	1.76632	0.036 094	419.093	1.74610	0.030 069	416.809	1.72770
40	0.046 557	430.720	1.79715	0.037 958	428.881	1.77786	0.031 781	426.933	1.76056
50	0.048 581	440.205	1.82696	0.039 735	438.589	1.80838	0.033 392	436.895	1.79187
60	0.050 547	449.718	1.85596	0.041 447	448.279	1.83791	0.034 929	446.782	1.82201
70	0.052 467	459.290	1.88426	0.043 108	457.994	1.86664	0.036 410	456.655	1.85121
80	0.054 351	468.942	1.91199	0.044 730	467.764	1.89471	0.037 848	466.554	1.87964
90	0.056 205	478.690	1.93921	0.046 319	477.611	1.92220	0.039 251	476.507	1.90743
100	0.058 035	488.546	1.96598	0.047 883	487.550	1.94920	0.040 627	486.535	1.93467
110	0.059 845	498.518	1.99235	0.049 426	497.594	1.97576	0.041 980	496.654	1.96143
120	0.061 639	508.613	2.01836	0.050 951	507.750	2.00193	0.043 314	506.875	1.98777
130	0.063 418	518.835	2.04403	0.052 461	518.026	2.02774	0.044 633	517.207	2.01372
140	0.065 184	529.187	2.06940	0.053 958	528.425	2.05322	0.045 938	527.656	2.03932

TABLE A.5.2SI (Continued) Superheated R-134a

Temp. °C	v m³/kg	h kJ/kg	s kJ/kg K	v m³/kg	h kJ/kg	s kJ/kg K	v m³/kg	h kJ/kg	s kJ/kg K
	0.80 MPa			0.90 MPa			1.00 MPa		
40	0.027 113	424.860	1.74457	0.023 446	422.642	1.72943	0.020 473	420.249	1.71479
50	0.028 611	435.114	1.77680	0.024 868	433.235	1.76273	0.021 849	431.243	1.74936
60	0.030 024	445.223	1.80761	0.026 192	443.595	1.79431	0.023 110	441.890	1.78181
70	0.031 375	455.270	1.83732	0.027 447	453.835	1.82459	0.024 293	452.345	1.81273
80	0.032 678	465.308	1.86616	0.028 649	464.025	1.85387	0.025 417	462.703	1.84248
90	0.033 944	475.375	1.89427	0.029 810	474.216	1.88232	0.026 497	473.027	1.87131
100	0.035 180	485.499	1.92177	0.030 940	484.441	1.91010	0.027 543	483.361	1.89938
110	0.036 392	495.698	1.94874	0.032 043	494.726	1.93730	0.028 561	493.736	1.92682
120	0.037 584	505.988	1.97525	0.033 126	505.088	1.96399	0.029 556	504.175	1.95371
130	0.038 760	516.379	2.00135	0.034 190	515.542	1.99025	0.030 533	514.694	1.98013
140	0.039 921	526.880	2.02708	0.035 241	526.096	2.01611	0.031 495	525.305	2.00613
150	0.041 071	537.496	2.05247	0.036 278	536.760	2.04161	0.032 444	536.017	2.03175
	1.20 MPa			1.40 MPa			1.60 MPa		
50	0.017 243	426.845	1.72373	—			—		
60	0.018 439	438.210	1.75837	0.015 032	434.079	1.73597	0.012 392	429.322	1.71349
70	0.019 530	449.179	1.79081	0.016 083	445.720	1.77040	0.013 449	441.888	1.75066
80	0.020 548	459.925	1.82168	0.017 040	456.944	1.80265	0.014 378	453.722	1.78466
90	0.021 512	470.551	1.85135	0.017 931	467.931	1.83333	0.015 225	465.145	1.81656
100	0.022 436	481.128	1.88009	0.018 775	478.790	1.86282	0.016 015	476.333	1.84695
110	0.023 329	491.702	1.90805	0.019 583	489.589	1.89139	0.016 763	487.390	1.87619
120	0.024 197	502.307	1.93537	0.020 362	500.379	1.91918	0.017 479	498.387	1.90452
130	0.025 044	512.965	1.96214	0.021 118	511.192	1.94634	0.018 169	509.371	1.93211
140	0.025 874	523.697	1.98844	0.021 856	522.054	1.97296	0.018 840	520.376	1.95908
150	0.026 691	534.514	2.01431	0.022 579	532.984	1.99910	0.019 493	531.427	1.98551
160	0.027 495	545.426	2.03980	0.023 289	543.994	2.02481	0.020 133	542.542	2.01147
170	0.028 289	556.443	2.06494	0.023 988	555.097	2.05015	0.020 761	553.735	2.03702

Temp. °C	v m³/kg	h kJ/kg	s kJ/kg K	v m³/kg	h kJ/kg	s kJ/kg K	v m³/kg	h kJ/kg	s kJ/kg K
	1.80 MPa			2.0 MPa			2.50 MPa		
70	0.011 341	437.562	1.73085	0.009 581	432.531	1.71011	—	—	—
80	0.012 273	450.202	1.76717	0.010 550	446.304	1.74968	0.007 221	433.797	1.70180
90	0.013 099	462.164	1.80057	0.011 374	458.951	1.78500	0.008 157	449.499	1.74567
100	0.013 854	473.741	1.83202	0.012 111	470.996	1.81772	0.008 907	463.279	1.78311
110	0.014 560	485.095	1.86205	0.012 789	482.693	1.84866	0.009 558	476.129	1.81709
120	0.015 230	496.325	1.89098	0.013 424	494.187	1.87827	0.010 148	488.457	1.84886
130	0.015 871	507.498	1.91905	0.014 028	505.569	1.90686	0.010 694	500.474	1.87904
140	0.016 490	518.659	1.94639	0.014 608	516.900	1.93463	0.011 208	512.307	1.90804
150	0.017 091	529.841	1.97314	0.015 168	528.224	1.96171	0.011 698	524.037	1.93609
160	0.017 677	541.068	1.99936	0.015 712	539.571	1.98821	0.012 169	535.722	1.96338
170	0.018 251	552.357	2.02513	0.016 242	550.963	2.01421	0.012 624	547.399	1.99004
180	0.018 814	563.724	2.05049	0.016 762	562.418	2.03977	0.013 066	559.098	2.01614
190	0.019 369	575.177	2.07549	0.017 272	573.950	2.06494	0.013 498	570.841	2.04177

Temp. °C	v m³/kg	h kJ/kg	s kJ/kg K	v m³/kg	h kJ/kg	s kJ/kg K	v m³/kg	h kJ/kg	s kJ/kg K
	3.0 MPa			3.50 MPa			4.0 MPa		
90	0.005 755	436.193	1.69950	—	—	—	—	—	—
100	0.006 653	453.731	1.74717	0.004 839	440.433	1.70386	—	—	—
110	0.007 339	468.500	1.78623	0.005 667	459.211	1.75355	0.004 277	446.844	1.71480
120	0.007 924	482.043	1.82113	0.006 289	474.697	1.79346	0.005 005	465.987	1.76415
130	0.008 446	494.915	1.85347	0.006 813	488.771	1.82881	0.005 559	481.865	1.80404
140	0.008 926	507.388	1.88403	0.007 279	502.079	1.86142	0.006 027	496.295	1.83940
150	0.009 375	519.618	1.91328	0.007 706	514.928	1.89216	0.006 444	509.925	1.87200
160	0.009 801	531.704	1.94151	0.008 103	527.496	1.92151	0.006 825	523.072	1.90271
170	0.010 208	543.713	1.96892	0.008 480	539.890	1.94980	0.007 181	535.917	1.93203
180	0.010 601	555.690	1.99565	0.008 839	552.185	1.97724	0.007 517	548.573	1.96028
190	0.010 982	567.670	2.02180	0.009 185	564.430	2.00397	0.007 837	561.117	1.98766
200	0.011 353	579.678	2.04745	0.009 519	576.665	2.03010	0.008 145	573.601	2.01432

TABLE A.6SI *Thermodynamic Properties of Nitrogen*
TABLE A.6.1SI *Saturated Nitrogen (SI Units)*

Temp. K	Abs. Press. MPa P	Specific Volume, m³/kg			Enthalpy, kJ/kg			Entropy, kJ/kg K		
		Sat. Liquid v_f	Evap. v_{fg}	Sat. Vapor v_g	Sat. Liquid h_f	Evap. h_{fg}	Sat. Vapor h_g	Sat. Liquid s_f	Evap. s_{fg}	Sat. Vapor s_g
63.148	0.01252	0.001150	1.480099	1.481249	−150.911	215.392	64.482	2.4234	3.4108	5.8342
65	0.01741	0.001160	1.092665	1.093825	−147.172	213.384	66.212	2.4816	3.2829	5.7646
70	0.03858	0.001191	0.525015	0.526206	−137.088	207.788	70.700	2.6307	2.9683	5.5991
75	0.07610	0.001223	0.280499	0.281722	−126.949	201.816	74.867	2.7700	2.6909	5.4609
77.348	0.101325	0.001240	0.215145	0.216385	−122.150	198.839	76.689	2.8326	2.5707	5.4033
80	0.13699	0.001259	0.162485	0.163744	−116.689	195.319	78.630	2.9014	2.4415	5.3429
85	0.22903	0.001299	0.100204	0.101503	−106.252	188.149	81.898	3.0266	2.2136	5.2401
90	0.36066	0.001343	0.064803	0.066146	−95.577	180.137	84.560	3.1466	2.0016	5.1482
95	0.54082	0.001393	0.043398	0.044792	−84.593	171.075	86.482	3.2627	1.8009	5.0636
100	0.77881	0.001452	0.029764	0.031216	−73.199	160.691	87.493	3.3761	1.6070	4.9831
105	1.08423	0.001522	0.020673	0.022195	−61.238	148.597	87.359	3.4883	1.4153	4.9036
110	1.46717	0.001610	0.014342	0.015952	−48.446	134.165	85.719	3.6017	1.2197	4.8215
115	1.93875	0.001729	0.009717	0.011445	−34.308	116.212	81.904	3.7204	1.0106	4.7310
120	2.51248	0.001915	0.006083	0.007998	−17.605	91.930	74.324	3.8536	0.7661	4.6197
125	3.20886	0.002353	0.002530	0.004883	6.677	48.762	55.438	4.0395	0.3901	4.4296
126.193	3.39780	0.003194	0	0.003194	29.791	0	29.791	4.2193	0	4.2193

TABLE A.6.2SI *Superheated Nitrogen (SI Units)*

Abs. Press. MPa		100	125	150	175	200	225	250	275	300
					Temperature, K					
0.1	v	0.291030	0.367236	0.442612	0.517577	0.592311	0.666904	0.741404	0.815839	0.890229
	h	101.938	128.423	154.695	180.860	206.967	233.039	259.090	285.128	311.160
	s	5.6944	5.9308	6.1225	6.2838	6.4232	6.5461	6.6559	6.7551	6.8457
0.2	v	0.142521	0.181711	0.220007	0.257878	0.295515	0.333008	0.370408	0.407743	0.445033
	h	100.238	127.294	153.876	180.236	206.476	232.644	258.766	284.861	310.938
	s	5.4775	5.7191	5.9130	6.0755	6.2157	6.3390	6.4491	6.5486	6.6393
0.5	v	0.053062	0.070328	0.086429	0.102059	0.117442	0.132677	0.147817	0.162892	0.177921
	h	94.460	123.776	151.376	178.349	204.998	231.458	257.799	284.063	310.276
	s	5.1660	5.4282	5.6296	5.7959	5.9383	6.0629	6.1740	6.2741	6.3653
1.0	v	—	0.033064	0.041876	0.050120	0.058093	0.065911	0.073631	0.081285	0.088893
	h	—	117.397	147.062	175.156	202.522	229.482	256.194	282.743	309.182
	s	—	5.1872	5.4039	5.5772	5.7234	5.8504	5.9630	6.0642	6.1562
2.0	v	—	0.014030	0.019541	0.024153	0.028436	0.032548	0.036558	0.040500	0.044395
	h	—	101.541	137.779	168.584	197.528	225.543	253.014	280.140	307.034
	s	—	4.8887	5.1541	5.3443	5.4989	5.6310	5.7467	5.8502	5.9438
4.0	v	—	—	0.008231	0.011185	0.013650	0.015912	0.018063	0.020145	0.022179
	h	—	—	115.595	154.672	187.417	217.740	246.800	275.098	302.898
	s	—	—	4.8379	5.0797	5.2548	5.3978	5.5203	5.6282	5.7250

TABLE A.6.2SI(Continued) *Superheated Nitrogen (SI Units)*

Abs. Press. MPa		Temperature, K										
		150	175	200	225	250	275	300	350	400
6.0	v	0.004421	0.006909	0.008771	0.010412	0.011937	0.013393	0.014803	0.017532	0.020187
	h	87.298	139.945	177.293	210.125	240.822	270.294	298.988	354.951	409.83
	s	4.5685	4.8956	5.0955	5.2503	5.3797	5.4921	5.5920	5.7646	5.9112
8.0	v	0.002914	0.004861	0.006387	0.007701	0.008905	0.010042	0.011135	0.013236	0.015264
	h	61.924	125.326	167.469	202.833	235.145	265.761	295.318	352.511	408.237
	s	4.3522	4.7460	4.9717	5.1385	5.2748	5.3916	5.4945	5.6709	5.8197
10.0	v	0.002388	0.003752	0.005014	0.006112	0.007113	0.008053	0.008952	0.01067	0.01232
	h	48.659	112.363	158.353	196.022	229.841	261.532	291.902	350.260	406.790
	s	4.2290	4.6233	4.8697	5.0474	5.1901	5.3109	5.4167	5.5967	5.7477
15.0	v	0.001955	0.002598	0.003365	0.004109	0.004804	0.005461	0.006088	0.007280	0.008416
	h	36.805	91.928	140.599	181.908	218.586	252.470	284.565	345.466	403.791
	s	4.0790	4.4191	4.6796	4.8745	5.0292	5.1585	5.2702	5.4581	5.6139
20.0	v	0.001782	0.002187	0.002687	0.003213	0.003729	0.004226	0.004704	0.005617	0.006487
	h	33.644	83.317	130.168	172.324	210.434	245.699	279.007	341.856	401.649
	s	3.9960	4.3024	4.5529	4.7517	4.9124	5.0469	5.1629	5.3568	5.5166

TABLE A.7SI *Thermodynamic Properties of Methane*
TABLE A.7.1SI *Saturated Methane (SI Units)*

Temp. K	Abs. Press. MPa P	Specific Volume, m³/kg			Enthalpy, kJ/kg			Entropy, kJ/kg K		
		Sat. Liquid v_f	Evap. v_{fg}	Sat. Vapor v_g	Sat. Liquid h_f	Evap. h_{fg}	Sat. Vapor h_g	Sat. Liquid s_f	Evap. s_{fg}	Sat. Vapor s_g
90.685	0.01169	0.00221	3.97955	3.98176	-358.1	543.1	185.1	4.226	5.989	10.216
95	0.01983	0.00224	2.44824	2.45048	-343.7	537.2	193.4	4.381	5.654	10.035
100	0.03441	0.00228	1.47657	1.47885	-326.8	529.8	202.9	4.554	5.298	9.851
105	0.05643	0.00231	0.93791	0.94022	-309.7	521.8	212.2	4.721	4.970	9.691
110	0.08820	0.00235	0.62219	0.62454	-292.3	513.3	221.0	4.882	4.666	9.548
115	0.13232	0.00239	0.42808	0.43048	-274.7	504.1	229.4	5.037	4.384	9.421
120	0.19158	0.00244	0.30371	0.30615	-257.0	494.2	237.2	5.187	4.118	9.305
125	0.26896	0.00249	0.22110	0.22359	-239.0	483.4	244.5	5.332	3.868	9.200
130	0.36760	0.00254	0.16448	0.16702	-220.7	471.7	251.0	5.473	3.629	9.102
135	0.49072	0.00259	0.12457	0.12717	-202.1	458.9	256.8	5.611	3.399	9.011
140	0.64165	0.00265	0.09574	0.09839	-183.2	444.8	261.7	5.746	3.177	8.924
145	0.82379	0.00272	0.07444	0.07716	-163.7	429.4	265.7	5.879	2.961	8.841
150	1.04065	0.00279	0.05838	0.06117	-143.7	412.3	268.5	6.011	2.748	8.759
155	1.29580	0.00288	0.04604	0.04892	-123.1	393.3	270.2	6.141	2.537	8.679
160	1.59296	0.00297	0.03638	0.03935	-101.6	372.0	270.3	6.272	2.325	8.597
165	1.93607	0.00309	0.02868	0.03176	-79.1	347.8	268.7	6.405	2.108	8.512
170	2.32936	0.00322	0.02241	0.02563	-55.2	320.0	264.8	6.540	1.882	8.422
175	2.77762	0.00339	0.01718	0.02058	-29.3	287.2	257.9	6.681	1.641	8.322
180	3.28655	0.00362	0.01266	0.01628	-0.5	246.8	246.2	6.833	1.371	8.204
185	3.86361	0.00398	0.00845	0.01243	33.8	192.1	225.9	7.009	1.038	8.048
190	4.52082	0.00499	0.00298	0.00796	92.2	79.8	172.0	7.305	0.420	7.725
190.551	4.59920	0.00615	0	0.00615	129.7	0	129.7	7.500	0	7.500

TABLE A.7.2SI *Superheated Methane (SI Units)*

Abs. Press. MPa		Temperature, K									
		150	175	200	225	250	275	300	350	400	450
0.05	v	1.5433	1.8054	2.0665	2.3270	2.5872	2.8472	3.1069	3.6262	4.1451	—
	h	308.5	360.8	413.2	465.8	518.9	572.9	628.1	742.9	865.4	—
	s	10.5170	10.8399	11.1196	11.3674	11.5914	11.7972	11.9891	12.3429	12.6697	—
0.10	v	0.7659	0.8984	1.0299	1.1609	1.2915	1.4219	1.5521	1.8123	2.0721	—
	h	306.8	359.6	412.2	465.0	518.3	572.4	627.6	742.6	865.1	—
	s	10.1504	10.4759	10.7570	11.0058	11.2303	11.4365	11.6286	11.9829	12.3099	—
0.50	v	0.1433	0.1726	0.2006	0.2280	0.2550	0.2817	0.3083	0.3611	0.4137	—
	h	292.3	349.1	404.1	458.5	512.9	567.8	623.7	739.6	862.8	—
	s	9.2515	9.6021	9.8959	10.1520	10.3812	10.5906	10.7850	11.1422	11.4710	—
1.00	v	0.0643	0.0815	0.0968	0.1113	0.1254	0.1392	0.1528	0.1798	0.2064	—
	h	270.6	334.9	393.5	450.1	506.0	562.0	618.8	735.9	860.0	—
	s	8.7902	9.1871	9.5006	9.7672	10.0028	10.2164	10.4138	10.7748	11.1059	—
1.50	v	—	0.0508	0.0621	0.0724	0.0822	0.0917	0.1010	0.1193	0.1373	—
	h	—	318.8	382.3	441.4	499.0	556.2	613.8	732.3	857.2	—
	s	—	8.9121	9.2514	9.5303	9.7730	9.9911	10.1916	10.5565	10.8899	—
2.00	v	—	0.0350	0.0446	0.0529	0.0606	0.0680	0.0751	0.0891	0.1027	—
	h	—	300.0	370.2	432.4	491.8	550.3	608.9	728.6	854.3	—
	s	—	8.6839	9.0596	9.3532	9.6036	9.8266	10.0303	10.3992	10.7349	—

TABLE A.7.2SI (Continued) *Superheated Methane (SI Units)*

Abs. Press. MPa		Temperature, K									
		100	125	200	225	250	275	300	350	400	450
3.00	v	—	—	0.0269	0.0333	0.0390	0.0442	0.0492	0.0589	0.0682	0.0774
	h	—	—	342.7	413.3	477.1	538.3	598.8	721.2	848.8	983.5
	s	—	—	8.7492	9.0823	9.3512	9.5848	9.7954	10.1726	10.5130	10.8303
4.00	v	—	—	0.0176	0.0235	0.0281	0.0324	0.0363	0.0438	0.0510	0.0580
	h	—	—	308.2	392.4	461.6	526.1	588.7	713.9	843.2	979.2
	s	—	—	8.4675	8.8653	9.1574	9.4031	9.6212	10.0071	10.3523	10.6725
5.00	v	—	—	0.0114	0.0175	0.0216	0.0252	0.0286	0.0348	0.0406	0.0463
	h	—	—	258.3	369.3	445.6	513.6	578.6	706.7	837.8	975.0
	s	—	—	8.1459	8.6728	8.9945	9.2540	9.4802	9.8751	10.2251	10.5483
6.00	v	—	—	0.0061	0.0135	0.0173	0.0205	0.0234	0.0288	0.0338	0.0386
	h	—	—	160.3	343.7	428.8	500.9	568.4	699.5	832.4	970.9
	s	—	—	7.6125	8.4907	8.8502	9.1253	9.3601	9.7643	10.1192	10.4453
8.00	v	—	—	0.0041	0.0085	0.0120	0.0147	0.0171	0.0213	0.0252	0.0289
	h	—	—	88.5	285.0	393.9	475.4	548.1	685.4	822.0	962.9
	s	—	—	7.2069	8.1344	8.5954	8.9064	9.1598	9.5831	9.9477	10.2796
10.00	v	—	—	0.0038	0.0059	0.0089	0.0113	0.0133	0.0169	0.0201	0.0231
	h	—	—	72.2	229.3	358.6	450.1	528.4	671.8	811.9	955.3
	s	—	—	7.0862	7.8245	8.3716	8.7210	8.9936	9.4362	9.8104	10.1480

TABLE A.8SI *Critical Constants (SI Units)*

Substance	Formula	Molec. Weight	Temp. K	Press. MPa	Vol. m³/kmol	Acentric Factor
Ammonia	NH_3	17.031	405.5	11.35	0.0725	0.250
Argon	Ar	39.948	150.8	4.87	0.0749	0.001
Bromine	Br_2	159.808	588	10.30	0.1272	0.108
Carbon dioxide	CO_2	44.01	304.1	7.38	0.0939	0.239
Carbon monoxide	CO	28.01	132.9	3.50	0.0932	0.066
Chlorine	Cl_2	70.906	416.9	7.98	0.1238	0.090
Deuterium (normal)	D_2	4.032	38.4	1.66	—	−0.160
Fluorine	F_2	37.997	144.3	5.22	0.0663	0.054
Helium	He	4.003	5.19	0.227	0.0574	−0.365
Helium³	He	3.017	3.31	0.114	0.0729	−0.473
Hydrogen (normal)	H_2	2.016	33.2	1.30	0.0651	−0.218
Krypton	Kr	83.80	209.4	5.50	0.0912	0.005
Neon	Ne	20.183	44.4	2.76	0.0416	−0.029
Nitric oxide	NO	30.006	180	6.48	0.0577	0.588
Nitrogen	N_2	28.013	126.2	3.39	0.0898	0.039
Nitrogen dioxide	NO_2	46.006	431	10.1	0.1678	0.834
Nitrous oxide	N_2O	44.013	309.6	7.24	0.0974	0.165
Oxygen	O_2	31.999	154.6	5.04	0.0734	0.025
Sulfur dioxide	SO_2	64.063	430.8	7.88	0.1222	0.256
Water	H_2O	18.015	647.3	22.12	0.0571	0.344
Xenon	Xe	131.30	289.7	5.84	0.1184	0.008
Acetylene	C_2H_2	26.038	308.3	6.14	0.1127	0.190
Benzene	C_6H_6	78.114	562.2	4.89	0.2590	0.212
n-Butane	C_3H_{10}	58.124	425.2	3.80	0.2550	0.199
Carbon tetrachloride	CCL_4	153.823	556.4	4.56	0.2759	0.193
Chlorodifluoroethane[a] (142b)	CH_3CCLF_2	100.495	410.3	4.25	0.2310	0.250
Chlorodifluoromethane (22)	$CHCLF_2$	86.469	369.3	4.97	0.1656	0.221
Chloroform	$CHCL_3$	119.378	536.4	5.37	0.2389	0.218
Dichlorodifluoromethane (12)	CCL_2F_2	120.914	385.0	4.14	0.2167	0.204
Dichlorofluoroethane[a] (141)	CH_3CCL_2F	116.95	481.5	4.54	0.2520	0.215
Dichlorofluoromethane (21)	$CHCL_2F$	102.923	451.6	5.18	0.1964	0.210
Dichlorotrifluoroethane[a] (123)	$CHCL_2CF_3$	152.93	456.9	3.67	0.2781	0.282
Difluoroethane[a] (152a)	CHF_2CH_3	66.05	386.4	4.52	0.1795	0.275
Ethane	C_2H_6	30.070	305.4	4.88	0.1483	0.099
Ethyl alcohol	C_2H_5OH	46.069	513.9	6.14	0.1671	0.644
Ethylene	C_2H_4	28.054	282.4	5.04	0.1304	0.089
n-Heptane	C_7H_{16}	100.205	540.3	2.74	0.4320	0.349
n-Hexane	C_6H_{14}	86.178	507.5	3.01	0.3700	0.299
Methane	CH_4	16.043	190.4	4.60	0.0992	0.011
Methyl alcohol	CH_3OH	32.042	512.6	8.09	0.1180	0.556

TABLE A.8SI (Continued) *Critical Constants (SI Units)*

Substance	Formula	Molec. Weight	Temp. K	Press. MPa	Vol. m³/kmol	Acentric Factor
Methyl chloride	CH_3CL	50.488	416.3	6.70	0.1389	0.153
n-Octane	C_8H_{18}	114.232	568.8	2.49	0.4920	0.398
n-Pentane	C_5H_{16}	72.151	469.7	3.37	0.3040	0.251
Propane	C_3H_8	44.094	369.8	4.25	0.2030	0.153
Propene	C_3H_6	42.081	364.9	4.60	0.1810	0.144
Propyne	C_3H_4	40.065	402.4	5.63	0.1640	0.215
Tetrafluoroethane[a] (134a)	CF_3CH_2F	102.03	374.2	4.06	0.1980	0.327

Source: R. C. Reid, J. M. Prausnitz, and B. E. Poling, *The Properties of Gases and Liquids*, fourth edition, McGraw-Hill Book Company, New York, 1987.

[a]Data from M. O. McLinden, NIST Thermophysics Division, 1989.

TABLE A.9SI *Properties of Various Solids and Liquids at 25°C (SI Units)*

Solid	C_p kJ/kg K	ρ, kg/m³	Liquid	C_p kJ/kg K	ρ, kg/m³
Aluminum	0.9	2700	Ammonia	4.8	602
Concrete	0.65	2300	Benzene	1.72	879
Copper	0.386	8900	Butane	2.469	556
Glass	0.8	2300	Ethanol	2.456	783
Granite	1.017	2700	Glycerine	2.40	1200
Graphite	0.711	2500	Iso-octane	2.1	692
Iron	0.450	7840	Mercury	0.139	13560
Lead	0.128	11310	Methanol	2.55	787
Rubber (soft)	1.84	1100	Oil (light)	1.8	910
Sand (dry)	0.8	1450–1750	Propane	2.54	510
Silver	0.235	10470	R-12	0.971	1310
Steel (AISI302)	0.48	8050	R-134a	1.43	1206
Tin	0.217	5730	Water	4.184	997
Wood (most)	1.76	350–700			

TABLE A.10SI *Properties of Various Ideal Gases at 300 K (SI Units)*

Gas	Chemical Formula	Molecular Mass	R kJ/kg K	C_{po} kJ/kg K	C_{vo} kJ/kg K	k
Acetylene	C_2H_2	26.038	0.3193	1.6986	1.3793	1.231
Air		28.97	0.2870	1.0035	0.7165	1.400
Ammonia	NH_3	17.031	0.48819	2.1300	1.6418	1.297
Argon	Ar	39.948	0.20813	0.5203	0.3122	1.667
Butane	C_4H_{10}	58.124	0.14304	1.7164	1.5734	1.091
Carbon dioxide	CO2	44.01	0.18892	0.8418	0.6529	1.289
Carbon monoxide	CO	28.01	0.29683	1.0413	0.7445	1.400

TABLE A.10SI (Continued) *Properties of Various Ideal Gases at 300 K (SI Units)*

Gas	Chemical Formula	Molecular Mass	R kJ/kg K	C_{po} kJ/kg K	C_{vo} kJ/kg K	k
Ethane	C_2H_6	30.07	0.27650	1.7662	1.4897	1.186
Ethanol	C_2H_5OH	46.069	0.18048	1.427	1.246	1.145
Ethylene	C_2H_4	28.054	0.29637	1.5482	1.2518	1.237
Helium	He	4.003	2.07703	5.1926	3.1156	1.667
Hydrogen	H_2	2.016	4.12418	14.2091	10.0849	1.409
Methane	CH_4	16.04	0.51835	2.2537	1.7354	1.299
Methanol	CH_3OH	32.042	0.25948	1.4050	1.1455	1.227
Neon	Ne	20.183	0.41195	1.0299	0.6179	1.667
Nitrogen	N_2	28.013	0.29680	1.0416	0.7448	1.400
Nitrous oxide	N_2O	44.013	0.18891	0.8793	0.6904	1.274
n-octane	C_8H_{18}	114.23	0.07279	1.7113	1.6385	1.044
Oxygen	O_2	31.999	0.25983	0.9216	0.6618	1.393
Propane	C_3H_8	44.097	0.18855	1.6794	1.4909	1.126
Steam	H_2O	18.015	0.46152	1.8723	1.4108	1.327
Sulfur dioxide	SO_2	64.059	0.12979	0.6236	0.4938	1.263
Sulfur trioxide	SO_3	80.058	0.10386	0.6346	0.5307	1.196

TABLE A.11SI *Constant-Pressure Specific Heats of Various Ideal Gases (SI Units)*

C_{p0}=kJ/kmol K $\theta=T$(Kelvin)/100

Gas		Range K	Max Error %
N_2	$\overline{C}_{p0} = 39.060 - 512.79\,\theta^{-1.5} + 1072.7\,\theta^{-2} - 820.40\,\theta^{-3}$	300–3500	0.43
O_2	$\overline{C}_{p0} = 37.432 + 0.020\,102\,\theta^{1.5} - 178.57\,\theta^{-1.5} + 236.88\,\theta^{-2}$	300–3500	0.30
H_2	$\overline{C}_{p0} = 56.505 - 702.74\,\theta^{-0.75} + 1165.0\,\theta^{-1} - 560.70\,\theta^{-1.5}$	300–3500	0.60
CO	$\overline{C}_{p0} = 69.145 - 0.704\,63\,\theta^{0.75} - 200.77\,\theta^{-0.5} + 176.76\,\theta^{-0.75}$	300–3500	0.42
OH	$\overline{C}_{p0} = 81.546 - 59.350\,\theta^{0.25} + 17.329\,\theta^{0.75} - 4.2660\,\theta$	300–3500	0.43
NO	$\overline{C}_{p0} = 59.283 - 1.7096\,\theta^{0.5} - 70.613\,\theta^{-0.5} + 74.889\,\theta^{-1.5}$	300–3500	0.34
H_2O	$\overline{C}_{p0} = 143.05 - 183.54\,\theta^{0.25} + 82.751\,\theta^{0.5} - 3.6989\,\theta$	300–3500	0.43
CO_2	$\overline{C}_{p0} = -3.7357 + 30.529\,\theta^{0.5} - 4.1034\,\theta + 0.024\,198\,\theta^2$	300–3500	0.19
NO_2	$\overline{C}_{p0} = 46.045 + 216.10\,\theta^{-0.5} - 363.66\,\theta^{-0.75} + 232.550\,\theta^{-2}$	300–3500	0.26
CH_4	$\overline{C}_{p0} = -672.87 + 439.74\,\theta^{0.25} - 24.875\,\theta^{0.75} + 323.88\,\theta^{-0.5}$	300–2000	0.15
C_2H_4	$\overline{C}_{p0} = -95.395 + 123.15\,\theta^{0.5} - 35.641\,\theta^{0.75} + 182.77\,\theta^{-3}$	300–2000	0.07
C_2H_6	$\overline{C}_{p0} = 6.895 + 17.26\,\theta - 0.6402\,\theta^2 + 0.007\,28\,\theta^3$	300–1500	0.83
C_3H_8	$\overline{C}_{p0} = -4.042 + 30.46\,\theta - 1.571\,\theta^2 + 0.031\,71\,\theta^3$	300–1500	0.40
C_4H_{10}	$\overline{C}_{p0} = 3.954 + 37.12\,\theta - 1.833\,\theta^2 + 0.034\,98\,\theta^3$	300–1500	0.54

Source: From T.C. Scott and R.E. Sonntag. University of Michigan, unpublished 1971, except C_2H_6, C_3H_8, and C_4H_{10} from K.A. Kobe, Petroleum Refiner, 28, No. 2, 113 (1949).

TABLE A.12SI *Ideal-Gas Properties of Air, SI Units Standard Entropy at 0.1 MPa (1 bar) Pressure*

T K	u kJ/kg	h kJ/kg	s^o kJ/kg K	P_r	v_r
200	142.768	200.174	6.46260	0.27027	493.466
220	157.071	220.218	6.55812	0.37700	389.150
240	171.379	240.267	6.64535	0.51088	313.274
260	185.695	260.323	6.72562	0.67573	256.584
280	200.022	280.390	6.79998	0.87556	213.257
290	207.191	290.430	6.83521	0.98990	195.361
298.15	213.036	298.615	6.86305	1.09071	182.288
300	214.364	300.473	6.86926	1.11458	179.491
320	228.726	320.576	6.93413	1.39722	152.728
340	243.113	340.704	6.99515	1.72814	131.200
360	257.532	360.863	7.05276	2.11226	113.654
380	271.988	381.060	7.10735	2.55479	99.1882
400	286.487	401.299	7.15926	3.06119	87.1367
420	301.035	421.589	7.20875	3.63727	77.0025
440	315.640	441.934	7.25607	4.28916	68.4088
460	330.306	462.340	7.30142	5.02333	61.0658
480	345.039	482.814	7.34499	5.84663	54.7479
500	359.844	503.360	7.38692	6.76629	49.2777
520	374.726	523.982	7.42736	7.78997	44.5143
540	389.689	544.686	7.46642	8.92569	40.3444
560	404.736	565.474	7.50422	10.18197	36.6765
580	419.871	586.350	7.54084	11.56771	33.4358
600	435.097	607.316	7.57638	13.09232	30.5609
620	450.415	628.375	7.61090	14.76564	28.0008
640	465.828	649.528	7.64448	16.59801	25.7132
660	481.335	670.776	7.67717	18.60025	23.6623
680	496.939	692.120	7.70903	20.78367	21.8182
700	512.639	713.561	7.74010	23.16010	20.1553
720	528.435	735.098	7.77044	25.74188	18.6519
740	544.328	756.731	7.80008	28.54188	17.2894
760	560.316	778.460	7.82905	31.57347	16.0518
780	576.400	800.284	7.85740	34.85061	14.9250
800	592.577	822.202	7.88514	38.38777	13.8972
850	633.422	877.397	7.95207	48.46828	11.6948
900	674.824	933.152	8.01581	60.51977	9.91692
950	716.756	989.436	8.07667	74.81519	8.46770
1000	759.189	1046.221	8.13493	91.65077	7.27604
1050	802.095	1103.478	8.19081	111.3467	6.28845
1100	845.445	1161.180	8.24449	134.2478	5.46408
1150	881.211	1219.298	8.29616	160.7245	4.77141

T K	u kJ/kg	h kJ/kg	$s°$ kJ/kg K	P_r	v_r
1100	845.445	1161.180	8.24449	134.2478	5.46408
1120	862.903	1184.379	8.26539	144.3878	5.17272
1140	880.426	1207.642	8.28598	155.1245	4.90068
1160	898.012	1230.969	8.30626	166.4834	4.64642
1180	915.660	1254.357	8.32625	178.4908	4.40857
1200	933.367	1277.805	8.34596	191.1736	4.18586
1250	977.888	1336.677	8.39402	226.0192	3.68804
1300	1022.751	1395.892	8.44046	265.7145	3.26257
1350	1067.936	1455.429	8.48539	310.7426	2.89711
1400	1113.426	1515.270	8.52891	361.6192	2.58171
1450	1159.202	1575.398	8.57111	418.8942	2.30831
1500	1205.253	1635.800	8.61208	483.1554	2.07031
1550	1251.547	1696.446	8.65185	554.9577	1.86253
1600	1298.079	1757.329	8.69051	634.9670	1.68035
1650	1344.834	1818.436	8.72811	723.8560	1.52007
1700	1391.801	1879.755	8.76472	822.3320	1.37858
1750	1438.970	1941.275	8.80039	931.1376	1.25330
1800	1486.331	2002.987	8.83516	1051.051	1.14204
1850	1533.873	2064.882	8.86908	1182.888	1.04294
1900	1581.591	2126.951	8.90219	1327.498	0.95445
1950	1629.474	2189.186	8.93452	1485.772	0.87521
2000	1677.518	2251.581	8.96611	1658.635	0.80410
2050	1725.714	2314.128	8.99699	1847.077	0.74012
2100	1774.057	2376.823	9.02721	2052.109	0.68242
2150	1822.541	2439.659	9.05678	2274.789	0.63027
2200	1871.161	2502.630	9.08573	2516.217	0.58305
2250	1919.912	2565.733	9.11409	2777.537	0.54020
2300	1968.790	2628.962	9.14189	3059.939	0.50124
2350	2017.789	2692.313	9.16913	3364.658	0.46576
2400	2066.907	2755.782	9.19586	3692.974	0.43338
2450	2116.138	2819.366	9.22208	4046.215	0.40378
2500	2165.480	2883.059	9.24781	4425.759	0.37669
2550	2214.929	2946.859	9.27308	4833.031	0.35185
2600	2264.481	3010.763	9.29790	5269.505	0.32903
2650	2314.133	3074.767	9.32228	5736.707	0.30805
2700	2363.883	3138.868	9.34625	6236.215	0.28872
2750	2413.727	3203.064	9.36980	6769.657	0.27089
2800	2463.663	3267.351	9.39297	7338.715	0.25443
2850	2513.687	3331.726	9.41576	7945.124	0.23921
2900	2563.797	3396.188	9.43818	8590.676	0.22511
2950	2613.990	3460.733	9.46025	9277.216	0.21205
3000	2664.265	3525.359	9.48198	10006.645	0.19992

TABLE A.13SI *Ideal-Gas Properties of Various Substances (SI Units), Entropies at 0.1-MPa (1-bar) Pressure*

	Nitrogen, Diatomic (N_2) $\overline{h}^{\circ}_{f,\,298} = 0$ kJ/kmol $M = 28.013$		Nitrogen, Monatomic (N) $\overline{h}^{\circ}_{f,\,298} = 472\,680$ kJ/kmol $M = 14.007$	
T K	$(\overline{h}-\overline{h}^{\circ}_{298})$ kJ/kmol	$\overline{s}^{\circ}$ kJ/kmol K	$(\overline{h}-\overline{h}^{\circ}_{298})$ kJ/kmol	$\overline{s}^{\circ}$ kJ/kmol K
0	−8670	0	−6197	0
100	−5768	159.812	−4119	130.593
200	−2857	179.985	−2040	145.001
298	0	191.609	0	153.300
300	54	191.789	38	153.429
400	2971	200.181	2117	159.409
500	5911	206.740	4196	164.047
600	8894	212.177	6274	167.837
700	11937	216.865	8353	171.041
800	15046	221.016	10431	173.816
900	18223	224.757	12510	176.265
1000	21463	228.171	14589	178.455
1100	24760	231.314	16667	180.436
1200	28109	234.227	18746	182.244
1300	31503	236.943	20825	183.908
1400	34936	239.487	22903	185.448
1500	38405	241.881	24982	186.883
1600	41904	244.139	27060	188.224
1700	45430	246.276	29139	189.484
1800	48979	248.304	31218	190.672
1900	52549	250.234	33296	191.796
2000	56137	252.075	35375	192.863
2200	63362	255.518	39534	194.845
2400	70640	258.684	43695	196.655
2600	77963	261.615	47860	198.322
2800	85323	264.342	52033	199.868
3000	92715	266.892	56218	201.311
3200	100134	269.286	60420	202.667
3400	107577	271.542	64646	203.948
3600	115042	273.675	68902	205.164
3800	122526	275.698	73194	206.325
4000	130027	277.622	77532	207.437
4400	145078	281.209	86367	209.542
4800	160188	284.495	95457	211.519
5200	175352	287.530	104843	213.397
5600	190572	290.349	114550	215.195
6000	205848	292.984	124590	216.926

TABLE A.13SI (Continued) *Ideal-Gas Properties of Various Substances (SI Units), Entropies at 0.1-MPa (1-bar) Pressure*

	Oxygen, Diatomic (O_2) $\bar{h}^o_{f,298} = 0$ kJ/kmol $M = 31.999$		Oxygen, Monatomic (O) $\bar{h}^o_{f,298} = 249\ 170$ kJ/kmol $M = 16.00$	
T K	$(\bar{h}-\bar{h}^o_{298})$ kJ/kmol	$\bar{s}^o$ kJ/kmol K	$(\bar{h}-\bar{h}^o_{298})$ kJ/kmol	$\bar{s}^o$ kJ/kmol K
0	−8683	0	−6725	0
100	−5777	173.308	−4518	135.947
200	−2868	193.483	−2186	152.153
298	0	205.148	0	161.059
300	54	205.329	41	161.194
400	3027	213.873	2207	167.431
500	6086	220.693	4343	172.198
600	9245	226.450	6462	176.060
700	12499	231.465	8570	179.310
800	15836	235.920	10671	182.116
900	19241	239.931	12767	184.585
1000	22703	243.579	14860	186.790
1100	26212	246.923	16950	188.783
1200	29761	250.011	19039	190.600
1300	33345	252.878	21126	192.270
1400	36958	255.556	23212	193.816
1500	40600	258.068	25296	195.254
1600	44267	260.434	27381	196.599
1700	47959	262.673	29464	197.862
1800	51674	264.797	31547	199.053
1900	55414	266.819	33630	200.179
2000	59176	268.748	35713	201.247
2200	66770	272.366	39878	203.232
2400	74453	275.708	44045	205.045
2600	82225	278.818	48216	206.714
2800	90080	281.729	52391	208.262
3000	98013	284.466	56574	209.705
3200	106022	287.050	60767	211.058
3400	114101	289.499	64971	212.332
3600	122245	291.826	69190	213.538
3800	130447	294.043	73424	214.682
4000	138705	296.161	77675	215.773
4400	155374	300.133	86234	217.812
4800	172240	303.801	94873	219.691
5200	189312	307.217	103592	221.435
5600	206618	310.423	112391	223.066
6000	224210	313.457	121264	224.597

TABLE A.13SI (Continued) *Ideal-Gas Properties of Various Substances (SI Units), Entropies at 0.1-MPa (1-bar) Pressure*

	Carbon Dioxide (CO_2) $\bar{h}^\circ_{f,298}$ = -393 522 kJ/kmol M = 44.01		Carbon Monoxide (CO) $\bar{h}^\circ_{f,298}$ = -110 527 kJ/kmol M = 28.01	
T K	$(\bar{h}-\bar{h}^\circ_{298})$ kJ/kmol	$\bar{s}^\circ$ kJ/kmol K	$(\bar{h}-\bar{h}^\circ_{298})$ kJ/kmol	$\bar{s}^\circ$ kJ/kmol K
0	−9364	0	−8671	0
100	−6457	179.010	−5772	165.852
200	−3413	199.976	−2860	186.024
298	0	213.794	0	197.651
300	69	214.024	54	197.831
400	4003	225.314	2977	206.240
500	8305	234.902	5932	212.833
600	12906	243.284	8942	218.321
700	17754	250.752	12021	223.067
800	22806	257.496	15174	227.277
900	28030	263.646	18397	231.074
1000	33397	269.299	21686	234.538
1100	38885	274.528	25031	237.726
1200	44473	279.390	28427	240.679
1300	50148	283.931	31867	243.431
1400	55895	288.190	35343	246.006
1500	61705	292.199	38852	248.426
1600	67569	295.984	42388	250.707
1700	73480	299.567	45948	252.866
1800	79432	302.969	49529	254.913
1900	85420	306.207	53128	256.860
2000	91439	309.294	56743	258.716
2200	103562	315.070	64012	262.182
2400	115779	320.384	71326	265.361
2600	128074	325.307	78679	268.302
2800	140435	329.887	86070	271.044
3000	152853	334.170	93504	273.607
3200	165321	338.194	100962	276.012
3400	177836	341.988	108440	278.279
3600	190394	345.576	115938	280.422
3800	202990	348.981	123454	282.454
4000	215624	352.221	130989	284.387
4400	240992	358.266	146108	287.989
4800	266488	363.812	161285	291.290
5200	292112	368.939	176510	294.337
5600	317870	373.711	191782	297.167
6000	343782	378.180	207105	299.809

TABLE A.13SI (Continued) *Ideal-Gas Properties of Various Substances (SI Units), Entropies at 0.1-MPa (1-bar) Pressure*

T K	Water (H_2O) $\bar{h}^o_{f, 298}$ = -241 826 kJ/kmol M = 18.015 $(\bar{h}-\bar{h}^o_{298})$ kJ/kmol	$\bar{s}^o$ kJ/kmol K	Hydroxyl (OH) $\bar{h}^o_{f, 298}$ = 38 987 kJ/kmol M = 17.007 $(\bar{h}-\bar{h}^o_{298})$ kJ/kmol	$\bar{s}^o$ kJ/kmol K
0	−9904	0	−9172	0
100	−6617	152.386	−6140	149.591
200	−3282	175.488	−2975	171.592
298	0	188.835	0	183.709
300	62	189.043	55	183.894
400	3450	198.787	3034	192.466
500	6922	206.532	5991	199.066
600	10499	213.051	8943	204.448
700	14190	218.739	11902	209.008
800	18002	223.826	14881	212.984
900	21937	228.460	17889	216.526
1000	26000	232.739	20935	219.735
1100	30190	236.732	24024	222.680
1200	34506	240.485	27159	225.408
1300	38941	244.035	30340	227.955
1400	43491	247.406	33567	230.347
1500	48149	250.620	36838	232.604
1600	52907	253.690	40151	234.741
1700	57757	256.631	43502	236.772
1800	62693	259.452	46890	238.707
1900	67706	262.162	50311	240.556
2000	72788	264.769	53763	242.328
2200	83153	269.706	60751	245.659
2400	93741	274.312	67840	248.743
2600	104520	278.625	75018	251.614
2800	115463	282.680	82268	254.301
3000	126548	286.504	89585	256.825
3200	137756	290.120	96960	259.205
3400	149073	293.550	104388	261.456
3600	160484	296.812	111864	263.592
3800	171981	299.919	119382	265.625
4000	183552	302.887	126940	267.563
4400	206892	308.448	142165	271.191
4800	230456	313.573	157522	274.531
5200	254216	318.328	173002	277.629
5600	278161	322.764	188598	280.518
6000	302295	326.926	204309	283.227

TABLE A.13SI (Continued) *Ideal-Gas Properties of Various Substances (SI Units), Entropies at 0.1-MPa (1-bar) Pressure*

T K	Hydrogen (H_2) $\bar{h}^o_{f,298} = 0$ kJ/kmol $M = 2.016$ $(\bar{h}-\bar{h}^o_{298})$ kJ/kmol	$\bar{s}^o$ kJ/kmol K	Hydrogen, Monatomic (H) $\bar{h}^o_{f,298} = 217\,999$ kJ/kmol $M = 1.008$ $(\bar{h}-\bar{h}^o_{298})$ kJ/kmol	$\bar{s}^o$ kJ/kmol K
0	−8467	0	−6197	0
100	−5467	100.727	−4119	92.009
200	−2774	119.410	−2040	106.417
298	0	130.678	0	114.716
300	53	130.856	38	114.845
400	2961	139.219	2117	120.825
500	5883	145.738	4196	125.463
600	8799	151.078	6274	129.253
700	11730	155.609	8353	132.457
800	14681	159.554	10431	135.233
900	17657	163.060	12510	137.681
1000	20663	166.225	14589	139.871
1100	23704	169.121	16667	141.852
1200	26785	171.798	18746	143.661
1300	29907	174.294	20825	145.324
1400	33073	176.637	22903	146.865
1500	36281	178.849	24982	148.299
1600	39533	180.946	27060	149.640
1700	42826	182.941	29139	150.900
1800	46160	184.846	31218	152.089
1900	49532	186.670	33296	153.212
2000	52942	188.419	35375	154.279
2200	59865	191.719	39532	156.260
2400	66915	194.789	43689	158.069
2600	74082	197.659	47847	159.732
2800	81355	200.355	52004	161.273
3000	88725	202.898	56161	162.707
3200	96187	205.306	60318	164.048
3400	103736	207.593	64475	165.308
3600	111367	209.773	68633	166.497
3800	119077	211.856	72790	167.620
4000	126864	213.851	76947	168.687
4400	142658	217.612	85261	170.668
4800	158730	221.109	93576	172.476
5200	175057	224.379	101890	174.140
5600	191607	227.447	110205	175.681
6000	208332	230.322	118519	177.114

TABLE A.13SI (Continued) *Ideal-Gas Properties of Various Substances (SI Units), Entropies at 0.1-MPa (1-bar) Pressure*

T K	Nitric Oxide (NO) $\bar{h}^\circ_{f,298} = 90\ 291$ kJ/kmol $M = 30.006$		Nitrogen Dioxide (NO$_2$) $\bar{h}^\circ_{f,298} = 33\ 100$ kJ/kmol $M = 46.005$	
	$(\bar{h}-\bar{h}^\circ_{298})$ kJ/kmol	$\bar{s}^\circ$ kJ/kmol K	$(\bar{h}-\bar{h}^\circ_{298})$ kJ/kmol	$\bar{s}^\circ$ kJ/kmol K
0	−9192	0	−10186	0
100	−6073	177.031	−6861	202.563
200	−2951	198.747	−3495	225.852
298	0	210.759	0	240.034
300	55	210.943	68	240.263
400	3040	219.529	3927	251.342
500	6059	226.263	8099	260.638
600	9144	231.886	12555	268.755
700	12308	236.762	17250	275.988
800	15548	241.088	22138	282.513
900	18858	244.985	27180	288.450
1000	22229	248.536	32344	293.889
1100	25653	251.799	37606	298.904
1200	29120	254.816	42946	303.551
1300	32626	257.621	48351	307.876
1400	36164	260.243	53808	311.920
1500	39729	262.703	59309	315.715
1600	43319	265.019	64846	319.289
1700	46929	267.208	70414	322.664
1800	50557	269.282	76008	325.861
1900	54201	271.252	81624	328.898
2000	57859	273.128	87259	331.788
2200	65212	276.632	98578	337.182
2400	72606	279.849	109948	342.128
2600	80034	282.822	121358	346.695
2800	87491	285.585	132800	350.934
3000	94973	288.165	144267	354.890
3200	102477	290.587	155756	358.597
3400	110000	292.867	167262	362.085
3600	117541	295.022	178783	365.378
3800	125099	297.065	190316	368.495
4000	132671	299.007	201860	371.456
4400	147857	302.626	224973	376.963
4800	163094	305.940	248114	381.997
5200	178377	308.998	271276	386.632
5600	193703	311.838	294455	390.926
6000	209070	314.488	317648	394.926

TABLE A.14 The Reduced Second and Third Virial Coefficients and Force Constants for the Lennard–Jones (6-12) Potential

TABLE A.14.1
The Reduced Second and Third Coefficients and their Derivatives

T^*	B^*	$B_1^* = T^* \dfrac{dB^*}{dT^*}$	C^*	$C_1^* = T^* \dfrac{dC^*}{dT^*}$
0.3	−27.88061	76.60701		
0.4	−13.79885	30.26698		
0.5	−8.72022	16.92367		
0.6	−6.19798	11.24883		
0.7	−4.71004	8.25711	−3.44223	29.02471
0.8	−3.73423	6.45414	−0.87753	11.80911
0.9	−3.04712	5.26492	0.06579	5.05023
1.0	−2.53809	4.42826	0.42600	2.12100
1.1	−2.14638	3.81063	0.55670	0.76761
1.2	−1.83595	3.33749	0.59235	0.12051
1.3	−1.58411	2.96421	0.58821	−0.18965
1.4	−1.37585	2.66262	0.56823	−0.33189
1.5	−1.20089	2.41414	0.54307	−0.38813
1.6	−1.05191	2.20602	0.51748	−0.39994
1.7	−0.92362	2.02926	0.49348	−0.38906
1.8	−0.81203	1.87733	0.47183	−0.36719
1.9	−0.71415	1.74537	0.45267	−0.34065
2.0	−0.62763	1.62972	0.43590	−0.31290
2.2	−0.48171	1.43663	0.40861	−0.26013
2.4	−0.36358	1.28190	0.38797	−0.21492
2.6	−0.26613	1.15517	0.37228	−0.17792
2.8	−0.18451	1.04948	0.36022	−0.14821
3.0	−0.11523	0.96000	0.35084	−0.12454
3.2	−0.05579	0.88328	0.34342	−0.10574
3.4	−0.00428	0.81676	0.33748	−0.09081
3.6	0.04072	0.75854	0.33264	−0.07895
3.8	0.08033	0.70716	0.32863	−0.06955
4.0	0.11542	0.66148	0.32526	−0.06209
4.2	0.14668	0.62060	0.32238	−0.05619
4.4	0.17469	0.58381	0.31988	−0.05154
4.6	0.19990	0.55051	0.31767	−0.04789
4.8	0.22268	0.52024	0.31569	−0.04506
5.0	0.24334	0.49260	0.31390	−0.04288
6.0	0.32290	0.38397	0.30661	−0.03831
7.0	0.37609	0.30826	0.30069	−0.03899
8.0	0.41343	0.25248	0.29533	−0.04152
9.0	0.44060	0.20970	0.29027	−0.04456
10.0	0.46088	0.17587	0.28541	−0.04758
20.0	0.52538	0.02866	0.24609	−0.06402
30.0	0.52693	-0.01749	0.21930	−0.06728

TABLE A.14.2
Force Constants from Experimental Virial Coefficient Data

Substance	ε/k, K	b_0, m³/kmol
Ne	35.8	0.0262
Ar	119.0	0.0502
Kr	173.0	0.0583
Xe	225.3	0.0854
N_2	95.05	0.0635
O_2	117.5	0.0578
CO	100.2	0.0675
NO	131.0	0.0402
CO_2	186.0	0.118
N_2O	193.0	0.118
CH_4	148.1	0.0698
CF_4	152.0	0.131

TABLE A.15 *Generalized Three Parameter (T_r, P_r, ω) Compressibility Factor Tables.*
TABLE A.15.1 *Generalized Saturation Pressure, Compressibility Factor, Enthalpy Departure, Entropy Departure, Fugacity Coefficient Table of a Simple Fluid*

T_r	$\ln(P_r)$	Z_f	Z_g	$\left(\dfrac{h^*-h}{RT_c}\right)_f$	$\left(\dfrac{h^*-h}{RT_c}\right)_g$	$\left(\dfrac{s_p^*-s_p}{R}\right)_f$	$\left(\dfrac{s_p^*-s_p}{R}\right)_g$	$\ln\left(\dfrac{f}{P}\right)$
0.30	−13.14053	0.00000	0.99998	6.04616	0.00002	20.09953	0.00005	−0.00002
0.32	−11.89025	0.00000	0.99993	5.99061	0.00007	18.06562	0.00014	−0.00007
0.34	−10.79655	0.00001	0.99983	5.93515	0.00018	16.51869	0.00035	−0.00017
0.36	−9.83281	0.00002	0.99963	5.87895	0.00040	16.01837	0.00076	−0.00037
0.38	−8.97801	0.00003	0.99927	5.82177	0.00085	15.31191	0.00150	−0.00073
0.40	−8.21540	0.00006	0.99865	5.76367	0.00163	14.41129	0.00272	−0.00134
0.42	−7.53140	0.00012	0.99770	5.70481	0.00291	13.58378	0.00463	−0.00230
0.44	−6.91492	0.00022	0.99628	5.64539	0.00489	12.82092	0.00741	−0.00371
0.46	−6.35683	0.00038	0.99430	5.58558	0.00781	12.13784	0.01129	−0.00568
0.48	−5.84950	0.00061	0.99165	5.52553	0.01191	11.50344	0.01648	−0.00832
0.50	−5.38653	0.00095	0.98820	5.46534	0.01745	10.91801	0.02317	−0.01173
0.52	−4.96253	0.00141	0.98389	5.40509	0.02472	10.37636	0.03155	−0.01600
0.54	−4.57289	0.00204	0.97862	5.34481	0.03399	9.87361	0.04176	−0.02118
0.56	−4.21367	0.00286	0.97233	5.28447	0.04552	9.40554	0.05395	−0.02734
0.58	−3.88146	0.00390	0.96498	5.22403	0.05958	8.96810	0.06823	−0.03449
0.60	−3.57331	0.00521	0.95652	5.16343	0.07641	8.55883	0.08470	−0.04265
0.62	−3.28666	0.00682	0.94695	5.10255	0.09626	8.17415	0.10344	−0.05182
0.64	−3.01926	0.00877	0.93623	5.04126	0.11938	7.81156	0.12455	−0.06198
0.66	−2.76913	0.01110	0.92436	4.97940	0.14601	7.46873	0.14811	−0.07312
0.68	−2.53452	0.01384	0.91133	4.91678	0.17640	7.14358	0.17423	−0.08519
0.70	−2.31388	0.01705	0.89711	4.85318	0.21084	6.83415	0.20302	−0.09817
0.72	−2.10584	0.02077	0.88170	4.78832	0.24961	6.53867	0.23465	−0.11203
0.74	−1.90915	0.02504	0.86506	4.72190	0.29309	6.25549	0.26934	−0.12674
0.76	−1.72272	0.02994	0.84712	4.65355	0.34170	5.98304	0.30734	−0.14227
0.78	−1.54554	0.03552	0.82782	4.58283	0.39595	5.71981	0.34902	−0.15861
0.80	−1.37672	0.04186	0.80704	4.50922	0.45650	5.46432	0.39486	−0.17576
0.82	−1.21545	0.04906	0.78463	4.43203	0.52418	5.21508	0.44552	−0.19373
0.84	−1.06097	0.05725	0.76036	4.35043	0.60010	4.97049	0.50187	−0.21254
0.86	−0.91263	0.06658	0.73394	4.26329	0.68574	4.72880	0.56514	−0.23224
0.88	−0.76980	0.07727	0.70491	4.16906	0.78319	4.48793	0.63709	−0.25290
0.90	−0.63192	0.08961	0.67262	4.06548	0.89550	4.24517	0.72040	−0.27461
0.92	−0.49847	0.10407	0.63605	3.94904	1.02744	3.99668	0.81928	−0.29750
0.94	−0.36898	0.12143	0.59347	3.81358	1.18719	3.73613	0.94120	−0.32176
0.96	−0.24301	0.14328	0.54146	3.64643	1.39116	3.45088	1.10146	−0.34766
0.98	−0.12014	0.17412	0.47112	3.41136	1.68360	3.10521	1.34235	−0.37560
1.00	0.00000	0.29010	0.29010	2.58438	2.58438	2.17799	2.17799	−0.40639

TABLE A.15.2 Compressibility Factor of a Simple Fluid, Z

T_r\\P_r	0.10	0.20	0.40	0.60	0.80	1.00	1.20	1.40	1.70	2.00	2.50	3.00	5.00	7.00	10.00
0.30	0.0290	0.0579	0.1158	0.1737	0.2315	0.2892	0.3470	0.4047	0.4911	0.5775	0.7213	0.8648	1.4366	2.0048	2.8507
0.40	0.0239	0.0477	0.0953	0.1429	0.1904	0.2379	0.2853	0.3327	0.4036	0.4744	0.5921	0.7095	1.1758	1.6373	2.3211
0.50	0.0207	0.0413	0.0825	0.1236	0.1647	0.2056	0.2465	0.2873	0.3483	0.4092	0.5103	0.6110	1.0094	1.4017	1.9801
0.60	0.0186	0.0371	0.0741	0.1109	0.1476	0.1842	0.2207	0.2571	0.3115	0.3657	0.4554	0.5446	0.8959	1.2398	1.7440
0.70	0.0172	0.0344	0.0687	0.1027	0.1366	0.1703	0.2038	0.2372	0.2869	0.3364	0.4181	0.4991	0.8161	1.1241	1.5729
0.75	0.9165	0.0336	0.0670	0.1001	0.1330	0.1656	0.1981	0.2303	0.2784	0.3260	0.4046	0.4823	0.7854	1.0787	1.5047
0.80	0.9319	0.8539	0.0661	0.0985	0.1307	0.1626	0.1942	0.2255	0.2721	0.3182	0.3942	0.4690	0.7598	1.0400	1.4456
0.85	0.9436	0.8810	0.0661	0.0983	0.1301	0.1614	0.1924	0.2230	0.2684	0.3132	0.3868	0.4591	0.7388	1.0071	1.3943
0.90	0.9528	0.9015	0.7800	0.1006	0.1321	0.1630	0.1935	0.2235	0.2678	0.3114	0.3828	0.4527	0.7220	0.9793	1.3496
0.95	0.9600	0.9174	0.8206	0.6967	0.1410	0.1705	0.1998	0.2288	0.2717	0.3138	0.3827	0.4501	0.7092	0.9561	1.3108
1.00	0.9659	0.9300	0.8509	0.7574	0.6353	0.2901	0.2237	0.2459	0.2839	0.3229	0.3880	0.4522	0.7004	0.9372	1.2772
1.05	0.9707	0.9401	0.8743	0.8002	0.7130	0.6026	0.4437	0.3246	0.3182	0.3452	0.4014	0.4604	0.6956	0.9222	1.2481
1.10	0.9747	0.9485	0.8930	0.8323	0.7649	0.6880	0.5984	0.5003	0.4086	0.3953	0.4277	0.4770	0.6950	0.9110	1.2232
1.15	0.9780	0.9554	0.9081	0.8576	0.8032	0.7443	0.6803	0.6129	0.5227	0.4760	0.4718	0.5042	0.6987	0.9033	1.2021
1.20	0.9808	0.9611	0.9205	0.8779	0.8330	0.7858	0.7363	0.6856	0.6135	0.5605	0.5295	0.5425	0.7069	0.8990	1.1844
1.30	0.9852	0.9702	0.9396	0.9083	0.8764	0.8438	0.8111	0.7784	0.7316	0.6908	0.6467	0.6344	0.7358	0.8998	1.1580
1.40	0.9884	0.9768	0.9534	0.9298	0.9062	0.8827	0.8595	0.8367	0.8043	0.7753	0.7387	0.7202	0.7761	0.9112	1.1419
1.50	0.9909	0.9818	0.9636	0.9456	0.9278	0.9103	0.8933	0.8768	0.8536	0.8328	0.8052	0.7887	0.8200	0.9297	1.1339
1.60	0.9928	0.9856	0.9714	0.9575	0.9439	0.9308	0.9180	0.9059	0.8889	0.8738	0.8537	0.8410	0.8617	0.9518	1.1320
1.80	0.9955	0.9910	0.9823	0.9739	0.9659	0.9583	0.9511	0.9444	0.9353	0.9275	0.9176	0.9118	0.9297	0.9961	1.1391
2.00	0.9972	0.9944	0.9892	0.9842	0.9796	0.9754	0.9715	0.9680	0.9635	0.9599	0.9561	0.9550	0.9772	1.0328	1.1516
2.50	0.9994	0.9989	0.9981	0.9975	0.9971	0.9969	0.9970	0.9973	0.9982	0.9996	1.0031	1.0080	1.0395	1.0866	1.1763
3.00	1.0004	1.0008	1.0018	1.0030	1.0043	1.0057	1.0074	1.0091	1.0121	1.0153	1.0215	1.0284	1.0635	1.1075	1.1848
3.50	1.0008	1.0017	1.0035	1.0055	1.0075	1.0097	1.0120	1.0143	1.0181	1.0221	1.0292	1.0368	1.0723	1.1138	1.1834
4.00	1.0010	1.0021	1.0043	1.0066	1.0090	1.0115	1.0140	1.0166	1.0207	1.0249	1.0323	1.0401	1.0747	1.1136	1.1773
5.00	1.0012	1.0024	1.0048	1.0073	1.0098	1.0124	1.0150	1.0176	1.0217	1.0259	1.0331	1.0405	1.0722	1.1064	1.1611

TABLE A.15.3 Enthalpy Departure of a Simple Fluid, $(h^* - h)/RT_c$

T_r\P_r	0.10	0.20	0.40	0.60	0.80	1.00	1.20	1.40	1.70	2.00	2.50	3.00	5.00	7.00	10.00
0.30	6.040	6.034	6.022	6.011	5.999	5.987	5.975	5.963	5.945	5.927	5.898	5.868	5.748	5.628	5.446
0.40	5.757	5.751	5.738	5.726	5.713	5.700	5.687	5.675	5.655	5.636	5.604	5.572	5.442	5.311	5.113
0.50	5.459	5.453	5.440	5.427	5.414	5.401	5.388	5.375	5.355	5.336	5.303	5.270	5.135	4.999	4.791
0.60	5.159	5.153	5.141	5.129	5.116	5.104	5.091	5.079	5.060	5.041	5.008	4.976	4.842	4.704	4.492
0.70	4.853	4.848	4.839	4.828	4.818	4.808	4.797	4.786	4.769	4.752	4.723	4.693	4.566	4.432	4.221
0.75	0.183	4.687	4.679	4.672	4.664	4.655	4.646	4.637	4.622	4.607	4.581	4.554	4.434	4.303	4.095
0.80	0.160	0.345	4.507	4.504	4.499	4.494	4.488	4.481	4.470	4.459	4.437	4.413	4.303	4.178	3.974
0.85	0.141	0.300	4.308	4.313	4.316	4.316	4.316	4.314	4.309	4.302	4.287	4.269	4.173	4.056	3.857
0.90	0.126	0.264	0.596	4.074	4.094	4.108	4.118	4.125	4.130	4.132	4.129	4.119	4.043	3.935	3.744
0.95	0.113	0.235	0.516	0.885	3.763	3.825	3.865	3.893	3.922	3.939	3.955	3.958	3.910	3.815	3.634
1.00	0.103	0.212	0.455	0.750	1.151	2.584	3.441	3.560	3.653	3.706	3.757	3.782	3.774	3.695	3.526
1.05	0.094	0.192	0.407	0.654	0.955	1.359	2.034	2.831	3.243	3.398	3.521	3.583	3.632	3.575	3.420
1.10	0.086	0.175	0.367	0.581	0.827	1.120	1.487	1.955	2.609	2.965	3.231	3.353	3.484	3.453	3.315
1.15	0.079	0.160	0.334	0.523	0.732	0.968	1.239	1.550	2.059	2.479	2.888	3.091	3.329	3.329	3.211
1.20	0.073	0.148	0.305	0.474	0.657	0.857	1.076	1.315	1.704	2.079	2.537	2.807	3.166	3.202	3.107
1.30	0.063	0.127	0.259	0.399	0.545	0.698	0.860	1.029	1.293	1.560	1.964	2.274	2.825	2.942	2.899
1.40	0.055	0.110	0.224	0.341	0.463	0.588	0.716	0.848	1.050	1.253	1.576	1.857	2.486	2.679	2.692
1.50	0.048	0.097	0.196	0.297	0.400	0.505	0.611	0.719	0.883	1.046	1.309	1.549	2.175	2.421	2.486
1.60	0.043	0.086	0.173	0.261	0.350	0.440	0.531	0.622	0.759	0.894	1.114	1.318	1.904	2.177	2.285
1.80	0.034	0.068	0.137	0.206	0.275	0.344	0.413	0.481	0.583	0.683	0.844	0.996	1.476	1.751	1.908
2.00	0.028	0.056	0.111	0.167	0.222	0.276	0.330	0.384	0.463	0.541	0.665	0.782	1.167	1.411	1.577
2.50	0.018	0.035	0.070	0.104	0.137	0.170	0.203	0.234	0.281	0.326	0.398	0.465	0.687	0.838	0.954
3.00	0.011	0.023	0.045	0.067	0.088	0.109	0.129	0.149	0.177	0.205	0.248	0.288	0.415	0.495	0.545
3.50	0.007	0.015	0.029	0.043	0.056	0.069	0.081	0.093	0.111	0.127	0.152	0.174	0.239	0.270	0.264
4.00	0.005	0.009	0.017	0.026	0.033	0.041	0.048	0.054	0.064	0.072	0.085	0.095	0.116	0.110	0.061
5.00	0.001	0.001	0.002	0.003	0.004	0.004	0.004	0.004	0.003	0.001	-0.003	-0.009	-0.045	-0.101	-0.213

TABLE A.15.4 *Entropy Departure of a Simple Fluid,* $(s_P^* - s_P)/R$

T_r\\P_r	0.10	0.20	0.40	0.60	0.80	1.00	1.20	1.40	1.70	2.00	2.50	3.00	5.00	7.00	10.00
0.30	9.319	8.635	7.961	7.574	7.304	7.099	6.935	6.799	6.633	6.497	6.319	6.182	5.847	5.683	5.578
0.40	8.506	7.821	7.144	6.755	6.483	6.275	6.109	5.970	5.799	5.660	5.475	5.330	4.967	4.772	4.619
0.50	7.842	7.156	6.479	6.089	5.816	5.608	5.441	5.302	5.130	4.989	4.802	4.656	4.282	4.074	3.899
0.60	7.294	6.610	5.933	5.544	5.273	5.066	4.900	4.762	4.591	4.451	4.266	4.120	3.747	3.537	3.353
0.70	6.823	6.140	5.467	5.082	4.814	4.610	4.446	4.310	4.143	4.007	3.826	3.684	3.322	3.117	2.935
0.75	0.164	5.917	5.248	4.866	4.600	4.399	4.238	4.104	3.940	3.807	3.630	3.491	3.138	2.939	2.761
0.80	0.134	0.294	5.026	4.649	4.388	4.191	4.034	3.904	3.744	3.615	3.444	3.310	2.970	2.777	2.605
0.85	0.111	0.239	4.785	4.418	4.166	3.976	3.825	3.701	3.548	3.425	3.262	3.135	2.812	2.629	2.463
0.90	0.094	0.199	0.463	4.145	3.912	3.738	3.599	3.484	3.344	3.231	3.081	2.963	2.663	2.491	2.334
0.95	0.080	0.168	0.377	0.671	3.556	3.433	3.326	3.235	3.119	3.023	2.893	2.790	2.520	2.361	2.215
1.00	0.069	0.144	0.315	0.532	0.847	2.178	2.893	2.893	2.843	2.784	2.690	2.609	2.380	2.239	2.105
1.05	0.060	0.124	0.267	0.439	0.656	0.965	1.523	2.185	2.444	2.483	2.461	2.415	2.242	2.121	2.001
1.10	0.053	0.108	0.230	0.371	0.537	0.742	1.012	1.368	1.855	2.081	2.191	2.202	2.104	2.007	1.903
1.15	0.047	0.096	0.201	0.319	0.452	0.607	0.790	1.007	1.366	1.649	1.885	1.968	1.966	1.897	1.810
1.20	0.042	0.085	0.177	0.277	0.389	0.512	0.651	0.807	1.063	1.308	1.587	1.727	1.827	1.789	1.722
1.30	0.033	0.068	0.140	0.217	0.298	0.385	0.478	0.576	0.732	0.891	1.127	1.299	1.554	1.581	1.556
1.40	0.027	0.056	0.114	0.174	0.237	0.303	0.372	0.442	0.552	0.663	0.839	0.990	1.303	1.386	1.402
1.50	0.023	0.046	0.094	0.143	0.194	0.246	0.299	0.353	0.436	0.520	0.654	0.777	1.088	1.208	1.260
1.60	0.019	0.039	0.079	0.120	0.162	0.204	0.247	0.290	0.356	0.421	0.528	0.628	0.913	1.050	1.130
1.80	0.014	0.029	0.058	0.088	0.117	0.147	0.177	0.207	0.252	0.296	0.369	0.438	0.661	0.799	0.908
2.00	0.011	0.022	0.044	0.067	0.089	0.111	0.134	0.156	0.189	0.221	0.274	0.325	0.497	0.620	0.733
2.50	0.006	0.013	0.026	0.038	0.051	0.064	0.076	0.088	0.106	0.124	0.153	0.181	0.281	0.361	0.453
3.00	0.004	0.008	0.017	0.025	0.033	0.041	0.049	0.057	0.068	0.080	0.098	0.116	0.181	0.236	0.303
3.50	0.003	0.006	0.012	0.017	0.023	0.029	0.034	0.040	0.048	0.056	0.068	0.081	0.126	0.166	0.216
4.00	0.002	0.004	0.009	0.013	0.017	0.021	0.025	0.029	0.035	0.041	0.050	0.059	0.093	0.123	0.162
5.00	0.001	0.003	0.005	0.008	0.010	0.013	0.015	0.018	0.021	0.025	0.031	0.036	0.057	0.075	0.100

TABLE A.15.5 Fugacity Coefficient of a Simple Fluid, $\ln(f/P)$

T_r\\P_r	0.10	0.20	0.40	0.60	0.80	1.00	1.20	1.40	1.70	2.00	2.50	3.00	5.00	7.00	10.00
0.30	-10.815	-11.479	-12.114	-12.462	-12.692	-12.857	-12.981	-13.078	-13.185	-13.261	-13.340	-13.378	-13.313	-13.076	-12.576
0.40	-5.887	-6.556	-7.202	-7.559	-7.800	-7.975	-8.110	-8.216	-8.339	-8.431	-8.535	-8.599	-8.638	-8.506	-8.164
0.50	-3.077	-3.749	-4.401	-4.766	-5.012	-5.194	-5.335	-5.448	-5.581	-5.682	-5.803	-5.883	-5.989	-5.923	-5.682
0.60	-1.304	-1.979	-2.635	-3.003	-3.254	-3.440	-3.586	-3.703	-3.842	-3.950	-4.082	-4.173	-4.323	-4.304	-4.133
0.70	-0.110	-0.786	-1.445	-1.816	-2.069	-2.258	-2.407	-2.527	-2.670	-2.782	-2.922	-3.021	-3.202	-3.215	-3.095
0.75	-0.080	-0.332	-0.991	-1.363	-1.618	-1.808	-1.957	-2.078	-2.223	-2.336	-2.478	-2.580	-2.773	-2.799	-2.699
0.80	-0.066	0.137	-0.608	-0.981	-1.236	-1.426	-1.576	-1.698	-1.844	-1.959	-2.103	-2.206	-2.409	-2.445	-2.362
0.85	-0.055	-0.113	-0.284	-0.656	-0.911	-1.102	-1.252	-1.374	-1.521	-1.636	-1.782	-1.887	-2.097	-2.142	-2.074
0.90	-0.046	-0.095	-0.198	-0.382	-0.636	-0.826	-0.976	-1.098	-1.245	-1.360	-1.506	-1.613	-1.829	-1.881	-1.826
0.95	-0.039	-0.080	-0.166	-0.261	-0.405	-0.594	-0.742	-0.864	-1.009	-1.124	-1.270	-1.376	-1.596	-1.655	-1.610
1.00	-0.034	-0.068	-0.140	-0.218	-0.303	-0.406	-0.548	-0.667	-0.809	-0.923	-1.067	-1.173	-1.394	-1.457	-1.422
1.05	-0.029	-0.059	-0.120	-0.185	-0.254	-0.329	-0.414	-0.511	-0.644	-0.753	-0.893	-0.997	-1.217	-1.284	-1.256
1.10	-0.025	-0.051	-0.103	-0.158	-0.215	-0.276	-0.340	-0.410	-0.517	-0.615	-0.747	-0.847	-1.063	-1.131	-1.110
1.15	-0.022	-0.044	-0.089	-0.136	-0.184	-0.235	-0.287	-0.341	-0.425	-0.507	-0.626	-0.719	-0.929	-0.997	-0.981
1.20	-0.019	-0.038	-0.078	-0.118	-0.159	-0.202	-0.245	-0.289	-0.357	-0.425	-0.527	-0.612	-0.811	-0.879	-0.867
1.30	-0.015	-0.030	-0.060	-0.090	-0.121	-0.152	-0.183	-0.215	-0.262	-0.309	-0.384	-0.450	-0.619	-0.682	-0.674
1.40	-0.012	-0.023	-0.046	-0.070	-0.093	-0.117	-0.140	-0.163	-0.198	-0.232	-0.287	-0.336	-0.472	-0.527	-0.521
1.50	-0.009	-0.018	-0.036	-0.055	-0.073	-0.091	-0.109	-0.126	-0.152	-0.178	-0.218	-0.255	-0.361	-0.406	-0.397
1.60	-0.007	-0.014	-0.029	-0.043	-0.057	-0.071	-0.085	-0.098	-0.118	-0.137	-0.168	-0.196	-0.277	-0.310	-0.298
1.80	-0.005	-0.009	-0.018	-0.027	-0.035	-0.044	-0.052	-0.060	-0.072	-0.083	-0.100	-0.116	-0.159	-0.174	-0.152
2.00	-0.003	-0.006	-0.011	-0.016	-0.022	-0.027	-0.032	-0.036	-0.043	-0.049	-0.058	-0.067	-0.086	-0.086	-0.055
2.50	-0.001	-0.001	-0.002	-0.003	-0.004	-0.005	-0.005	-0.006	-0.006	-0.006	-0.006	-0.005	0.006	0.026	0.072
3.00	0.000	0.001	0.002	0.003	0.004	0.005	0.006	0.007	0.009	0.011	0.016	0.020	0.042	0.070	0.121
3.50	0.001	0.002	0.003	0.005	0.007	0.009	0.011	0.013	0.016	0.019	0.025	0.031	0.058	0.089	0.141
4.00	0.001	0.002	0.004	0.006	0.009	0.011	0.013	0.016	0.019	0.023	0.029	0.036	0.064	0.095	0.146
5.00	0.001	0.002	0.005	0.007	0.010	0.012	0.015	0.017	0.021	0.025	0.031	0.038	0.066	0.096	0.143

TABLE A.15.6 *The Lee-Kesler Equation of State, The Lee-Kesler Generalized Equation of State, Eq. 10.63, is*

$$Z = \frac{P_r v_r'}{T_r} = 1 + \frac{B}{v_r'} + \frac{C}{v_r'^2} + \frac{D}{v_r'^5} + \frac{c_4}{T_r^3 v_r'^2} \left(\beta + \frac{\gamma}{v_r'^2} \right) \exp \left(-\frac{\gamma}{v_r'^2} \right)$$

$$B = b_1 - \frac{b_2}{T_r} - \frac{b_3}{T_r^2} - \frac{b_4}{T_r^3}$$

$$C = c_1 - \frac{c_2}{T_r} + \frac{c_3}{T_r^3}$$

$$D = d_1 + \frac{d_2}{T_r}$$

in which

$$T_r = \frac{T}{T_c}, \qquad P_r = \frac{P}{P_c}, \qquad v_r' = \frac{v}{RT_c / P_c}$$

The two sets of constants are as follows:

Constant	Simple Fluids	Reference Fluid
b_1	0.118 119 3	0.202 657 9
b_2	0.265 728	0.331 511
b_3	0.154 790	0.027 655
b_4	0.030 323	0.203 488
c_1	0.023 674 4	0.031 338 5
c_2	0.018 698 4	0.050 361 8
c_3	0.0	0.016 901
c_4	0.042 724	0.041 577
$d_1 \times 10^4$	0.155 488	0.487 36
$d_2 \times 10^4$	0.623 689	0.074 033 6
β	0.653 92	1.226
γ	0.060 167	0.037 54

TABLE A.16SI *Enthalpy of Formation, Gibbs Function of Formation, and Absolute Entropy of Various Substances at 25°C, 100 kPa Pressure.*

Substance	Formula	M	State	$\overline{h}_f^0$ kJ/kmol	$\overline{g}_f^0$ kJ/kmol	$\overline{s}_f^0$ kJ/kmol K
Water	H_2O	18.015	gas	−241 826	−228 582	188.834
Water	H_2O	18.015	liq	−285 830	−237 141	69.950
Hydrogen peroxide	H_2O_2	34.015	gas	−136 106	−105 445	232.991
Ozone	O_3	47.998	gas	+142 674	+163 184	238.932
Carbon (graphite)	C	12.011	solid	0	0	5.740
Carbon monoxide	CO	28.011	gas	−110 527	−137 163	197.653
Carbon dioxide	CO_2	44.010	gas	−393 522	−394 389	213.795
Methane	CH_4	16.043	gas	−74 873	−50 768	186.251
Acetylene	C_2H_2	26.038	gas	+226 731	+209 200	200.958
Ethene	C_2H_4	28.054	gas	+52 467	+68 421	219.330
Ethane	C_2H_6	30.070	gas	−84 740	−32 885	229.597
Propene	C_3H_6	42.081	gas	+20 430	+62 825	267.066
Propane	C_3H_8	44.094	gas	−103 900	−23 393	269.917
Butane	C_4H_{10}	58.124	gas	−126 200	−15 970	306.647
Pentane	C_5H_{12}	72.151	gas	−146 500	−8 208	348.945
Benzene	C_6H_6	78.114	gas	+82 980	+129 765	269.562
Hexane	C_6H_{14}	86.178	gas	−167 300	+28	387.979
Heptane	C_7H_{16}	100.205	gas	−187 900	+8 227	427.805
n-Octane	C_8H_{18}	114.232	gas	−208 600	+16 660	466.514
n-Octane	C_8H_{18}	114.232	liq	−250 105	+6 741	360.575
Methanol	CH_3OH	32.042	gas	−201 300	−162 551	239.709
Ethanol	C_2H_5OH	46.069	gas	−235 000	−168 319	282.444
Ammonia	NH_3	17.031	gas	−45 720	−16 128	192.572
T-T-Diesel	$C_{14.4}H_{24.9}$	198.06	liq	−174 000	+178 919	525.90
Sulfur	S	32.06	solid	0	0	32.056
Sulfur dioxide	SO_2	64.059	gas	−296 842	−300 125	248.212
Sulfur trioxide	SO_3	80.058	gas	−395 765	−371 016	256.769
Nitrogen oxide	N_2O	44.013	gas	+82 050	+104 179	219.957
Nitromethane	CH_3NO_2	61.04	liq	−113 100	−14 439	171.80

TABLE A.17SI *Logarithms to the Base e of the Equilibrium Constant K*

For the reaction $v_A A + v_b B \rightleftarrows v_C C + v_D D$, the equilibrium constant K is defined as

$$K = \frac{y_C^{v_C} y_D^{v_D}}{y_A^{v_A} y_B^{v_B}} \left(\frac{P}{P^\circ}\right)^{v_C + v_D - v_A - v_B}, P^\circ = 0.1 \text{ MPa}$$

Temp K	$H_2 \rightleftarrows 2H$	$O_2 \rightleftarrows 2O$	$N_2 \rightleftarrows 2N$	$2H_2O \rightleftarrows 2H2 + O_2$	$2H_2O \rightleftarrows H_2 + 2OH$	$2CO_2 \rightleftarrows 2CO + O_2$	$N_2 + O_2 \rightleftarrows 2NO$	$N_2 + 2O_2 \rightleftarrows 2NO_2$
298	−164.003	−186.963	−367.528	−184.420	−212.075	−207.529	−69.868	−41.355
500	−92.830	−105.623	−213.405	−105.385	−120.331	−115.234	−40.449	−30.725
1000	−39.810	−45.146	−99.146	−46.321	−51.951	−47.052	−18.709	−23.039
1200	−30.878	−35.003	−80.025	−36.363	−40.467	−35.736	−15.082	−21.752
1400	−24.467	−27.741	−66.345	−29.222	−32.244	−27.679	−12.491	−20.826
1600	−19.638	−22.282	−56.069	−23.849	−26.067	−21.656	−10.547	−20.126
1800	−15.868	−18.028	−48.066	−19.658	−21.258	−16.987	−9.035	−19.577
2000	−12.841	−14.619	−41.655	−16.299	−17.406	−13.266	−7.825	−19.136
2200	−10.356	−11.826	−36.404	−13.546	−14.253	−10.232	−6.836	−18.773
2400	−8.280	−9.495	−32.023	−11.249	−11.625	−7.715	−6.012	−18.470
2600	−6.519	−7.520	−28.313	−9.303	−9.402	−5.594	−5.316	−18.214
2800	−5.005	−5.826	−25.129	−7.633	−7.496	−3.781	−4.720	−17.994
3000	−3.690	−4.356	−22.367	−6.184	−5.845	−2.217	−4.205	−17.805
3200	−2.538	−3.069	−19.947	−4.916	−4.401	−0.853	−3.755	−17.640
3400	−1.519	−1.932	−17.810	−3.795	−3.128	0.346	−3.359	−17.496
3600	−0.611	−0.922	−15.909	−2.799	−1.996	1.408	−3.008	−17.369
3800	0.201	−0.017	−14.205	−1.906	−0.984	2.355	−2.694	−17.257
4000	0.934	0.798	−12.671	−1.101	−0.074	3.204	−2.413	−17.157
4500	2.483	2.520	−9.423	0.602	1.847	4.985	−1.824	−16.953
5000	3.724	3.898	−6.816	1.972	3.383	6.397	−1.358	−16.797
5500	4.739	5.027	−4.672	3.098	4.639	7.542	−0.980	−16.678
6000	5.587	5.969	−2.876	4.040	5.684	8.488	−0.671	−16.588

Source: Consistent with thermodynamic data in *JANAF Thermochemical Tables*, third edition, Thermal Group, Dow Chemical U.S.A., Midland, MI 1985.

TABLE A.18 *One-Dimensional Isentropic Compressible-Flow Functions for an Ideal Gas with Constant Specific Heat and Molecular Weight and $k = 1.4$*

M	M*	A/A*	P/P_o	ρ/ρ_o	T/T_o
0.0	0.00000	∞	1.00000	1.00000	1.00000
0.1	0.10944	5.82183	0.99303	0.99502	0.99800
0.2	0.21822	2.96352	0.97250	0.98028	0.99206
0.3	0.32572	2.03506	0.93947	0.95638	0.98232
0.4	0.43133	1.59014	0.89561	0.92427	0.96899
0.5	0.53452	1.33984	0.84302	0.88517	0.95238
0.6	0.63481	1.18820	0.78400	0.84045	0.93284
0.7	0.73179	1.09437	0.72093	0.79158	0.91075
0.8	0.82514	1.03823	0.65602	0.73999	0.88652
0.9	0.91460	1.00886	0.59126	0.68704	0.86059
1.0	1.0000	1.00000	0.52828	0.63394	0.83333
1.1	1.0812	1.00793	0.46835	0.58170	0.80515
1.2	1.1583	1.03044	0.41238	0.53114	0.77640
1.3	1.2311	1.06630	0.36091	0.48290	0.74738
1.4	1.2999	1.11493	0.31424	0.43742	0.71839
1.5	1.3646	1.17617	0.27240	0.39498	0.68966
1.6	1.4254	1.25023	0.23527	0.35573	0.66138
1.7	1.4825	1.33761	0.20259	0.31969	0.63371
1.8	1.5360	1.43898	0.17404	0.28682	0.60680
1.9	1.5861	1.55526	0.14924	0.25699	0.58072
2.0	1.6330	1.68750	0.12780	0.23005	0.55556
2.1	1.6769	1.83694	0.10935	0.20580	0.53135
2.2	1.7179	2.00497	0.93522E-01	0.18405	0.50813
2.3	1.7563	2.19313	0.79973E-01	0.16458	0.48591
2.4	1.7922	2.40310	0.68399E-01	0.14720	0.46468
2.5	1.8257	2.63672	0.58528E-01	0.13169	0.44444
2.6	1.8571	2.89598	0.50115E-01	0.11787	0.42517
2.7	1.8865	3.18301	0.42950E-01	0.10557	0.40683
2.8	1.9140	3.50012	0.36848E-01	0.94626E-01	0.38941
2.9	1.9398	3.84977	0.31651E-01	0.84889E-01	0.37286
3.0	1.9640	4.23457	0.27224E-01	0.76226E-01	0.35714
3.5	2.0642	6.78962	0.13111E-01	0.45233E-01	0.28986
4.0	2.1381	10.7188	0.65861E-02	0.27662E-01	0.23810
4.5	2.1936	16.5622	0.34553E-02	0.17449E-01	0.19802
5.0	2.2361	25.0000	0.18900E-02	0.11340E-01	0.16667
6.0	2.2953	53.1798	0.63336E-03	0.51936E-02	0.12195
7.0	2.3333	104.143	0.24156E-03	0.26088E-02	0.09259
8.0	2.3591	190.109	0.10243E-03	0.14135E-02	0.07246
9.0	2.3772	327.189	0.47386E-04	0.81504E-03	0.05814
10.0	2.3905	535.938	0.23563E-04	0.49482E-03	0.04762
∞	2.4495	∞	0.0	0.0	0.0

TABLE A.19 *One-Dimensional Normal Shock Functions for an Ideal Gas with Constant Specific Heat and Molecular Weight and k = 1.4*

M_x	M_y	P_y/P_x	ρ_y/ρ_x	T_y/T_x	P_{0y}/P_{0x}	P_{0y}/P_x
1.00	1.00000	1.0000	1.0000	1.0000	1.00000	1.8929
1.10	0.91177	1.2450	1.1691	1.0649	0.99893	2.1328
1.20	0.84217	1.5133	1.3416	1.1280	0.99280	2.4075
1.30	0.78596	1.8050	1.5157	1.1909	0.97937	2.7136
1.40	0.73971	2.1200	1.6897	1.2547	0.95819	3.0492
1.50	0.70109	2.4583	1.8621	1.3202	0.92979	3.4133
1.60	0.66844	2.8200	2.0317	1.3880	0.89520	3.8050
1.70	0.64054	3.2050	2.1977	1.4583	0.85572	4.2238
1.80	0.61650	3.6133	2.3592	1.5316	0.81268	4.6695
1.90	0.59562	4.0450	2.5157	1.6079	0.76736	5.1418
2.00	0.57735	4.5000	2.6667	1.6875	0.72087	5.6404
2.10	0.56128	4.9783	2.8119	1.7705	0.67420	6.1654
2.20	0.54706	5.4800	2.9512	1.8569	0.62814	6.7165
2.30	0.53441	6.0050	3.0845	1.9468	0.58329	7.2937
2.40	0.52312	6.5533	3.2119	2.0403	0.54014	7.8969
2.50	0.51299	7.1250	3.3333	2.1375	0.49901	8.5261
2.60	0.50387	7.7200	3.4490	2.2383	0.46012	9.1813
2.70	0.49563	8.3383	3.5590	2.3429	0.42359	9.8624
2.80	0.48817	8.9800	3.6636	2.4512	0.38946	10.569
2.90	0.48138	9.6450	3.7629	2.5632	0.35773	11.302
3.00	0.47519	10.333	3.8571	2.6790	0.32834	12.061
4.00	0.43496	18.500	4.5714	4.0469	0.13876	21.068
5.00	0.41523	29.000	5.0000	5.8000	0.06172	32.653
10.00	0.38758	116.50	5.7143	20.387	0.00304	129.22
∞	0.37796	∞	6.0000	∞	0.0	∞

TABLE A.20 *Atomic Masses, and Melting and Boiling Points of the Elements Based on the Assigned Relative Atomic Mass of $^{12}C = 12$*

Name	Symbol	Atomic Number	Atomic Mass	Melting Point, °C	Boiling Point, °C
Actinium	Ac	89	227.028[a]	1050	3200 ± 300
Aluminum	Al	13	26.9815	660.37	2467
Americium	Am	95	(243)	994 ± 4	2607
Antimony (stibium)	Sb	51	121.75	630.74	1750
Argon	Ar	18	39.948	−189.2	−185.7
Arsenic	As	33	74.9216	817 (28 atm)	613 (sub)
Astatine	At	85	(210	302	337
Barium	Ba	56	137.33	725	1640
Berkelium	Bk	97	(247)		
Beryllium	Be	4	9.012 18	1278 ± 5	2970 (5 mm)
Bismuth	Bi	83	208.980	271.3	1560 ± 5
Boron	B	5	10.81	2079	2550 (sub)
Bromine	Br	35	79.904	−7.2	58.78
Cadmium	Cd	48	112.41	320.9	765
Caesium (cesium)	Cs	55	132.905	28.40 ± 0.01	669.3
Calcium	Ca	20	40.08	839 ± 2	1484
Californium	Cf	98	(251)		
Carbon	C	6	12.011	3652 (sub)	t
Cerium	Ce	58	140.12	798 ± 2	3257
Cesium (caesium)	Cs	55	132.9054	28.40 ± 0.01	669.3
Chlorine	Cl	17	35.453	−100.98	−34.6
Chromium	Cr	24	51.996	1857 ± 20	2672
Cobalt	Co	27	51.9332	1495	2870
Copper (cuprum)	Cu	29	63.546	1083.4 ± 0.2	2567
Curium	Cm	96	(247)	1340 ± 40	
Dysprosium	Dy	66	162.50	1409	2335
Einsteinium	Es	99	(252)		
Erbium	Er	68	167.26	1522	2510
Europium	Eu	63	151.96	822 ± 5	1597
Fermium	Fm	100	(257)		
Fluorine	F	9	18.9984	−219.62	−188.14
Francium	Fr	87	(223)	(27)	(677)
Gadolinium	Gd	64	157.25	1311 ± 1	3233
Gallium	Ga	31	69.72	29.78	2403
Germanium	Ge	32	72.59	937.4	2830
Gold (aurum)	Au	79	196.967	1064.434	3080
Hafnium	Hf	72	178.49	2227 ± 20	4602

[a]The values apply to elements as they exist in materials of terrestrial origin and to certain artificial elements.

Source: *Handbook of Chemistry and Physics*, sixty-seventh edition, 1986–1987, CRC Press, Boca Raton, FL, 1986.

TABLE A.20 (Continued) *Atomic Masses, and Melting and Boiling Points of the Elements Based on the Assigned Relative Atomic Mass of $^{12}C = 12$*

Name	Symbol	Atomic Number	Atomic Mass	Melting Point, °C	Boiling Point, °C
Helium	He	2	4.002 60	$-272.2^{26\ atm}$	-268.934
Holmium	Ho	67	164.930	1470	2720
Hydrogen	H	1	1.007 94	-259.14	-252.87
Indium	In	49	114.82	156.61	2080
Iodine	I	53	126.905	113.5	184.35
Iridium	Ir	77	192.22	2410	4130
Iron (ferrum)	Fe	26	55.847	1535	2750
Krypton	Kr	36	83.80	-156.6	-152.30 ± 0.10
Lanthanum	La	57	138.906	920 ± 5	3454
Lawrencium	Lr	103	(260)		
Lead (plumbum)	Pb	82	207.2	327.502	1740
Lithium	Li	3	6.941	180.54	1342
Lutetium	Lu	71	174.967	1656 ± 5	3315
Magnesium	Mg	12	24.305	648.8 ± 0.5	1090
Manganese	Mn	25	54.9380	1244 ± 3	1962
Mendelevium	Md	101	(258)		
Mercury (hydrargyrum)	Hg	80	200.59	-38.87	356.58
Molybdenum	Mo	42	95.94	2617	4612
Neodymium	Nd	60	144.24	1010	3127
Neon	Ne	10	20.1179	-248.67	-246.048
Neptunium	Np	93	237.048	640 ± 1	3902
Nickel	Ni	28	58.69	1453	2732
Niobium (columbium)	Nb	41	92.9064	2468 ± 10	4742
Nitrogen	N	7	14.0067	-209.86	-195.8
Nobelium	No	102	(259)		
Osmium	Os	76	190.2	3045 ± 30	5027 ± 100
Oxygen	O	8	15.9994	-218.4	-182.962
Palladium	Pd	46	106.42	1554	3140
Phosphorus	P	15	30.9738	44.1 (white)	280 (white)
Platinum	Pt	78	195.08	1772	3827 ± 100
Plutonium	Pu	94	(244)	641	3232
Polonium	Po	84	(209)	254	962
Potassium (kalium)	K	19	39.0983	63.25	759.9
Praseodymium	Pr	59	140.908	931 ± 4	3212
Promethium	Pm	61	(145)	1080 (approx)	2460 (?)
Protactinium	Pa	91	231.0359	1600	

TABLE A.20 (Continued) *Atomic Masses, and Melting and Boiling Points of the Elements Based on the Assigned Relative Atomic Mass of $^{12}C = 12$*

Name	Symbol	Atomic Number	Atomic Mass	Melting Point, °C	Boiling Point, °C
Radium	Ra	88	226.025	700	1140
Radon	Rn	86	(222)	−71	−61.8
Rhenium	Re	75	186.207	3180	5627 (est.)
Rhodium	Rh	45	102.906	1965 ± 3	3727 ± 100
Rubidium	Rb	37	85.4678	38.89	686
Ruthenium	Ru	44	101.07	2310	3900
Samarium	Sm	62	150.36	1072 ± 5	1778
Scandium	Sc	21	44.9559	1539	2832
Selenium	Se	34	78.96	217	684.9 ± 1.0
Silicon	Si	14	28.0855	1410	2355
Silver (argentum)	Ag	47	107.868	961.93	2212
Sodium (natrium)	Na	11	22.9898	97.81 ± 0.03	882.9
Strontium	Sr	38	87.62	769	1384
Sulfur	S	16	32.06	112.8	444.674
Tantalum	Ta	73	180.9479	2996	5425 ± 100
Technetium	Tc	43	(98)	2172	4877
Tellurium	Te	52	127.60	449.5 ± 0.3	989.8 ± 3.8
Terbium	Tb	65	158.925	1360 ± 4	3041
Thallium	Tl	81	204.383	303.5	1457 ± 10
Thorium	Th	90	232.038	1750	4790 (approx)
Thulium	Tm	69	168.934	1545 ± 15	1727
Tin (stannum)	Sn	50	118.71	231.9681	2270
Titanium	Ti	22	47.88	1660 ± 10	3287
Tungsten (wolfram)	W	74	183.85	3410 ± 20	5660
Unnihexium	(Unh)	106	(263)		
Unnilpentium	(Unp)	105	(262)		
Unnilquadium	(Unq)	104	(261)		
Unnilseptium	(Uns)	107	(262)		
Uranium	U	92	238.029	1132 ± 0.8	3818
Vanadium	V	23	50.9415	1890 ± 10	3380
Wolfram (see tungsten)					
Xenon	Xe	54	131.29	−111.9	−107.1 ± 3
Ytterbium	Yb	70	173.04	824 ± 5	1193
Yttrium	Y	39	88.9059	1523 ± 8	3337
Zinc	Zn	30	65.39	419.58	907
Zirconium	Zr	40	91.224	1852 ± 2	4377

TABLE A.1E *Thermodynamic Properties of Water (English Units)*
TABLE A.1.1E *Saturated Water: Temperature Table (English Units)*

Temp. F T	Press. lbf/in.² P	Specific Volume, ft³/lbm		Internal Energy, Btu/lbm			Enthalpy, Btu/lbm			Entropy, Btu/lbm R		
		Sat. Liquid v_f	Sat. Vapor v_g	Sat. Liquid u_f	Evap. u_{fg}	Sat. Vapor u_g	Sat. Liquid h_f	Evap. h_{fg}	Sat. Vapor h_g	Sat. Liquid s_f	Evap. s_{fg}	Sat. Vapor s_g
32	0.08867	0.016022	3301.9	0.00	1021.2	1021.2	0.00	1075.4	1075.4	0.00000	2.1869	2.1869
35	0.09993	0.016021	2947.5	2.99	1019.2	1022.2	2.99	1073.7	1076.7	0.00607	2.1703	2.1764
40	0.12167	0.016020	2445.1	8.01	1015.8	1023.8	8.01	1070.9	1078.9	0.01617	2.1430	2.1591
45	0.14750	0.016021	2037.0	13.03	1012.5	1025.5	13.03	1068.1	1081.1	0.02617	2.1161	2.1423
50	0.17805	0.016024	1704.0	18.05	1009.1	1027.2	18.05	1065.2	1083.3	0.03606	2.0898	2.1259
60	0.2563	0.016034	1206.7	28.08	1002.4	1030.4	28.08	1059.6	1087.7	0.05554	2.0388	2.0943
70	0.3632	0.016051	867.60	38.09	995.6	1033.7	38.09	1054.0	1092.0	0.07462	1.9896	2.0642
80	0.5074	0.016072	632.69	48.08	988.9	1037.0	48.08	1048.3	1096.4	0.09331	1.9423	2.0356
90	0.6989	0.016099	467.60	58.06	982.2	1040.2	58.06	1042.7	1100.7	0.11163	1.8966	2.0083
100	0.9504	0.016130	349.98	68.04	975.4	1043.5	68.04	1037.0	1105.0	0.12962	1.8526	1.9822
110	1.2765	0.016166	265.07	78.01	968.7	1046.7	78.01	1031.3	1109.3	0.14728	1.8101	1.9574
120	1.6947	0.016205	203.03	87.99	961.9	1049.9	87.99	1025.5	1113.5	0.16464	1.7690	1.9336
130	2.2254	0.016247	157.16	97.96	955.1	1053.0	97.97	1019.8	1117.7	0.18171	1.7292	1.9109
140	2.892	0.016293	122.87	107.95	948.2	1056.2	107.96	1014.0	1121.9	0.19850	1.6907	1.8892
150	3.722	0.016343	96.977	117.94	941.3	1059.3	117.95	1008.1	1126.1	0.21502	1.6533	1.8683
160	4.745	0.016395	77.224	127.94	934.4	1062.3	127.95	1002.2	1130.1	0.23129	1.6171	1.8484
170	5.997	0.016450	62.015	137.94	927.4	1065.4	137.96	996.2	1134.2	0.24731	1.5819	1.8292
180	7.515	0.016509	50.199	147.96	920.4	1068.3	147.98	990.2	1138.2	0.26309	1.5478	1.8109
190	9.344	0.016570	40.942	157.99	913.3	1071.3	158.02	984.1	1142.1	0.27865	1.5146	1.7932
200	11.530	0.016634	33.631	168.03	906.2	1074.2	168.07	977.9	1145.9	0.29399	1.4822	1.7762
210	14.126	0.016702	27.813	178.09	898.9	1077.0	178.13	971.6	1149.7	0.30912	1.4507	1.7599
212	14.699	0.016715	26.797	180.10	897.5	1077.6	180.15	970.3	1150.5	0.31212	1.4445	1.7567
220	17.189	0.016772	23.149	188.16	891.7	1079.8	188.21	965.3	1153.5	0.32404	1.4201	1.7441
230	20.780	0.016845	19.385	198.25	884.3	1082.6	198.31	958.8	1157.1	0.33878	1.3901	1.7289
240	24.968	0.016922	16.326	208.36	876.9	1085.3	208.43	952.3	1160.7	0.35333	1.3609	1.7142
250	29.823	0.017001	13.825	218.48	869.4	1087.9	218.58	945.6	1164.2	0.36771	1.3324	1.7001
260	35.422	0.017084	11.767	228.64	861.8	1090.5	228.75	938.8	1167.6	0.38191	1.3044	1.6864
270	41.848	0.017170	10.065	238.81	854.1	1092.9	238.95	932.0	1170.9	0.39596	1.2771	1.6731
280	49.189	0.017259	8.650	249.02	846.4	1095.4	249.17	924.9	1174.1	0.40985	1.2504	1.6602
290	57.535	0.017352	7.466	259.25	838.5	1097.7	259.43	917.8	1177.2	0.42359	1.2241	1.6477
300	66.985	0.017448	6.471	269.51	830.5	1100.0	269.73	910.4	1180.2	0.43719	1.1984	1.6356
310	77.641	0.017548	5.631	279.80	822.3	1102.1	280.06	903.0	1183.0	0.45065	1.1731	1.6238
320	89.609	0.017652	4.919	290.13	814.1	1104.2	290.43	895.3	1185.8	0.46399	1.1483	1.6122
330	103.00	0.017760	4.312	300.50	805.7	1106.2	300.84	887.5	1188.4	0.47720	1.1238	1.6010
340	117.94	0.017871	3.792	310.90	797.1	1108.0	311.29	879.5	1190.8	0.49030	1.0997	1.5900

TABLE A.1.1E (Continued) *Saturated Water: Temperature Table (English Units)*

Temp. F T	Press. lbf/in.² P	Specific Volume, ft³/lbm		Internal Energy, Btu/lbm			Enthalpy, Btu/lbm			Entropy, Btu/lbm R		
		Sat. Liquid v_f	Sat. Vapor v_g	Sat. Liquid u_f	Evap. u_{fg}	Sat. Vapor u_g	Sat. Liquid h_f	Evap. h_{fg}	Sat. Vapor h_g	Sat. Liquid s_f	Evap. s_{fg}	Sat. Vapor s_g
350	134.54	0.017987	3.346	321.35	788.4	1109.8	321.80	871.3	1193.1	0.50328	1.0760	1.5793
360	152.93	0.018108	2.961	331.83	779.6	1111.4	332.35	862.9	1195.2	0.51616	1.0526	1.5688
370	173.24	0.018233	2.628	342.37	770.6	1112.9	342.95	854.2	1197.2	0.52893	1.0295	1.5584
380	195.61	0.018363	2.339	352.95	761.4	1114.3	353.61	845.4	1199.0	0.54161	1.0067	1.5483
390	220.17	0.018498	2.087	363.58	752.0	1115.5	364.33	836.2	1200.6	0.55420	0.9841	1.5383
400	247.08	0.018638	1.8660	374.26	742.4	1116.6	375.11	826.8	1201.9	0.56671	0.9617	1.5284
410	276.48	0.018784	1.6725	385.00	732.6	1117.6	385.96	817.2	1203.1	0.57914	0.9395	1.5187
420	308.52	0.018936	1.5023	395.80	722.5	1118.3	396.89	807.2	1204.1	0.59150	0.9175	1.5090
430	343.37	0.019094	1.3520	406.67	712.2	1118.9	407.89	796.9	1204.8	0.60380	0.8957	1.4995
440	381.18	0.019260	1.2191	417.61	701.7	1119.3	418.97	786.3	1205.3	0.61604	0.8740	1.4900
450	422.13	0.019433	1.1011	428.63	690.9	1119.5	430.15	775.4	1205.5	0.62823	0.8523	1.4805
460	466.38	0.019613	0.9961	439.73	679.8	1119.5	441.42	764.1	1205.5	0.64038	0.8308	1.4711
470	514.11	0.019803	0.9024	450.92	668.4	1119.4	452.80	752.4	1205.2	0.65250	0.8093	1.4618
480	565.50	0.020002	0.8186	462.21	656.7	1118.9	464.30	740.3	1204.6	0.66460	0.7878	1.4524
490	620.74	0.020211	0.74353	473.60	644.7	1118.3	475.92	727.8	1203.7	0.67669	0.7663	1.4430
500	680.02	0.020432	0.67605	485.11	632.3	1117.4	487.68	714.8	1202.4	0.68877	0.7447	1.4335
520	811.48	0.020911	0.56042	508.53	606.2	1114.8	511.67	687.2	1198.9	0.71299	0.7015	1.4144
540	961.51	0.021452	0.46579	532.56	578.4	1111.0	536.38	657.5	1193.8	0.73736	0.6576	1.3950
560	1131.9	0.022068	0.38764	557.35	548.4	1105.8	561.97	625.0	1187.0	0.76203	0.6129	1.3749
580	1324.4	0.022782	0.32250	583.05	515.9	1098.9	588.63	589.3	1178.0	0.78714	0.5668	1.3539
600	1541.1	0.023625	0.26765	609.91	480.1	1090.0	616.64	549.7	1166.3	0.81294	0.5187	1.3317
620	1784.5	0.024646	0.22090	638.26	440.2	1078.5	646.40	505.0	1151.4	0.83975	0.4677	1.3075
640	2057.2	0.025934	0.18044	668.68	394.5	1063.2	678.55	453.3	1131.9	0.86808	0.4122	1.2803
660	2362.6	0.027665	0.14457	702.24	340.0	1042.3	714.34	391.1	1105.5	0.89896	0.3493	1.2483
680	2705.1	0.030313	0.11125	741.70	269.3	1011.0	756.87	309.8	1066.6	0.93497	0.2718	1.2068
700	3090.5	0.036648	0.07435	801.66	145.9	947.6	822.61	167.5	990.1	0.99006	0.1444	1.1345
705.44	3203.8	0.050531	0.05053	872.56	0	872.6	902.52	0	902.5	1.05801	0	1.0580

TABLE A.1.2E *Saturated Steam: Pressure Table (English Units)*

Press. lbf/in.² P	Temp. F T	Specific Volume, ft³/lbm		Internal Energy, Btu/lbm			Enthalpy, Btu/lbm			Entropy, Btu/lbm R		
		Sat. Liquid v_f	Sat. Vapor v_g	Sat. Liquid u_f	Evap. u_{fg}	Sat. Vapor u_g	Sat. Liquid h_f	Evap. h_{fg}	Sat. Vapor h_g	Sat. Liquid s_f	Evap. s_{fg}	Sat. Vapor s_g
1	101.70	0.016136	333.58	69.73	974.3	1044.0	69.73	1036.0	1105.7	0.13264	1.8453	1.9779
2	126.03	0.016230	173.75	94.01	957.8	1051.8	94.01	1022.1	1116.1	0.17497	1.7448	1.9198
3	141.43	0.016300	118.72	109.37	947.2	1056.6	109.38	1013.1	1122.5	0.20087	1.6852	1.8861
4	152.93	0.016358	90.64	120.86	939.3	1060.2	120.88	1006.4	1127.3	0.21981	1.6426	1.8624
5	162.20	0.016407	73.53	130.14	932.9	1063.0	130.15	1000.9	1131.0	0.23484	1.6093	1.8441
6	170.02	0.016450	61.985	137.97	927.4	1065.4	137.98	996.2	1134.2	0.24734	1.5819	1.8292
8	182.83	0.016526	47.347	150.80	918.4	1069.2	150.83	988.4	1139.3	0.26752	1.5383	1.8058
10	193.19	0.016590	38.424	161.19	911.0	1072.2	161.22	982.1	1143.3	0.28356	1.5041	1.7877
14.696	211.99	0.016715	26.803	180.09	897.5	1077.6	180.13	970.4	1150.5	0.31210	1.4446	1.7567
15	213.02	0.016723	26.295	181.13	896.8	1077.9	181.18	969.7	1150.9	0.31365	1.4414	1.7550
20	227.96	0.016830	20.091	196.18	885.8	1082.0	196.25	960.1	1156.4	0.33578	1.3962	1.7320
25	240.07	0.016922	16.306	208.43	876.9	1085.3	208.51	952.2	1160.7	0.35344	1.3607	1.7141
30	250.34	0.017004	13.748	218.83	869.2	1088.0	218.92	945.4	1164.3	0.36819	1.3314	1.6996
35	259.29	0.017078	11.900	227.92	862.4	1090.3	228.03	939.3	1167.4	0.38091	1.3064	1.6873
40	267.26	0.017146	10.501	236.02	856.3	1092.3	236.15	933.9	1170.0	0.39212	1.2845	1.6767
45	274.45	0.017209	9.403	243.35	850.7	1094.0	243.50	928.8	1172.3	0.40216	1.2651	1.6673
50	281.03	0.017268	8.518	250.07	845.5	1095.6	250.23	924.2	1174.4	0.41127	1.2476	1.6589
55	287.09	0.017325	7.789	256.27	840.8	1097.0	256.45	919.9	1176.3	0.41961	1.2317	1.6513
60	292.73	0.017378	7.177	262.04	836.3	1098.3	262.24	915.8	1178.0	0.42731	1.2170	1.6444
65	298.00	0.017429	6.657	267.45	832.1	1099.5	267.66	911.9	1179.6	0.43448	1.2035	1.6380
70	302.95	0.017477	6.208	272.55	828.1	1100.6	272.77	908.3	1181.0	0.44118	1.1909	1.6320
75	307.63	0.017524	5.818	277.36	824.3	1101.6	277.60	904.8	1182.4	0.44747	1.1791	1.6265
80	312.06	0.017569	5.474	281.93	820.6	1102.6	282.19	901.4	1183.6	0.45342	1.1679	1.6214
85	316.28	0.017613	5.170	286.29	817.2	1103.4	286.57	898.2	1184.8	0.45905	1.1574	1.6165
90	320.31	0.017655	4.898	290.45	813.8	1104.3	290.75	895.1	1185.8	0.46440	1.1475	1.6119
95	324.16	0.017696	4.654	294.44	810.6	1105.0	294.75	892.1	1186.9	0.46950	1.1380	1.6075
100	327.85	0.017736	4.434	298.27	807.5	1105.8	298.60	889.2	1187.8	0.47437	1.1290	1.6034
110	334.82	0.017813	4.051	305.51	801.6	1107.1	305.87	883.7	1189.6	0.48352	1.1122	1.5957
120	341.30	0.017886	3.730	312.26	796.0	1108.3	312.66	878.5	1191.1	0.49199	1.0966	1.5886
130	347.37	0.017957	3.457	318.59	790.8	1109.3	319.03	873.5	1192.5	0.49987	1.0822	1.5821
140	353.08	0.018024	3.221	324.57	785.7	1110.3	325.04	868.7	1193.8	0.50725	1.0688	1.5760
150	358.47	0.018089	3.016	330.23	781.0	1111.2	330.73	864.2	1194.9	0.51420	1.0562	1.5704

TABLE A.1.2E (Continued) *Saturated Steam: Pressure Table (English Units)*

Press. lbf/in.² P	Temp. F T	Specific Volume, ft³/lbm		Internal Energy, Btu/lbm			Enthalpy, Btu/lbm			Entropy, Btu/lbm		
		Sat. Liquid v_f	Sat. Vapor v_g	Sat. Liquid u_f	Evap. u_{fg}	Sat. Vapor u_g	Sat. Liquid h_f	Evap. h_{fg}	Sat. Vapor h_g	Sat. Liquid s_f	Evap. s_{fg}	Sat. Vapor s_g
160	363.59	0.018152	2.836	335.61	776.4	1112.0	336.15	859.8	1196.0	0.52076	1.0443	1.5650
170	368.47	0.018213	2.676	340.75	772.0	1112.7	341.32	855.6	1196.9	0.52698	1.0330	1.5600
180	373.12	0.018273	2.533	345.67	767.7	1113.4	346.27	851.5	1197.8	0.53290	1.0224	1.5553
190	377.58	0.018331	2.405	350.38	763.6	1114.0	351.03	847.5	1198.6	0.53855	1.0122	1.5507
200	381.86	0.018387	2.289	354.92	759.6	1114.6	355.60	843.7	1199.3	0.54396	1.0025	1.5464
250	401.03	0.018653	1.8448	375.37	741.4	1116.7	376.23	825.9	1202.1	0.56800	0.9594	1.5274
300	417.42	0.018896	1.5441	393.01	725.1	1118.1	394.06	809.8	1203.9	0.58832	0.9232	1.5115
350	431.81	0.019124	1.3267	408.65	710.3	1119.0	409.89	795.0	1204.9	0.60602	0.8917	1.4978
400	444.69	0.019340	1.1619	422.77	696.7	1119.4	424.21	781.2	1205.5	0.62177	0.8638	1.4856
450	456.39	0.019547	1.0326	435.71	683.9	1119.6	437.34	768.2	1205.6	0.63600	0.8385	1.4745
500	467.12	0.019748	0.92828	447.69	671.7	1119.4	449.51	755.8	1205.3	0.64901	0.8154	1.4645
550	477.06	0.019942	0.84230	458.88	660.2	1119.1	460.91	743.9	1204.8	0.66104	0.7941	1.4551
600	486.33	0.020133	0.77016	469.40	649.1	1118.5	471.64	732.4	1204.1	0.67225	0.7742	1.4464
700	503.22	0.020505	0.65579	488.85	628.2	1117.0	491.50	710.5	1202.0	0.69266	0.7378	1.4305
800	518.36	0.020870	0.56906	506.58	608.4	1115.0	509.67	689.6	1199.3	0.71099	0.7050	1.4160
900	532.11	0.021231	0.50093	523.00	589.6	1112.6	526.54	669.5	1196.0	0.72772	0.6750	1.4027
1000	544.74	0.021590	0.44591	538.37	571.5	1109.9	542.36	650.0	1192.4	0.74318	0.6471	1.3903
1200	567.36	0.022318	0.36232	566.69	536.8	1103.5	571.64	612.3	1183.9	0.77120	0.5961	1.3673
1400	587.25	0.023070	0.30156	592.63	503.3	1096.0	598.61	575.5	1174.1	0.79640	0.5497	1.3461
1600	605.05	0.023863	0.25515	616.91	470.5	1087.4	623.98	538.9	1162.9	0.81960	0.5062	1.3258
1800	621.20	0.024715	0.21831	640.02	437.6	1077.7	648.26	502.1	1150.4	0.84140	0.4645	1.3059
2000	635.99	0.025648	0.18813	662.38	404.3	1066.6	671.87	464.4	1136.3	0.86224	0.4238	1.2861
2500	668.30	0.028602	0.13059	717.62	313.4	1031.0	730.85	360.6	1091.4	0.91302	0.3196	1.2326
3000	695.52	0.034300	0.08402	783.37	185.4	968.8	802.41	213.0	1015.4	0.97312	0.1844	1.1575
3203.6	705.44	0.050531	0.05053	872.56	0.0	872.6	902.52	0.0	902.5	1.05801	0.0000	1.0580

TABLE A.1.3E Superheated Vapor Water (English Units)

T	P = 1.00 (101.70)				P = 5.00 (162.20)				P = 10.0 (193.19)			
	v	u	h	s	v	u	h	s	v	u	h	s
Sat.	333.58	1044.0	1105.7	1.9779	73.531	1063.0	1131.0	1.8441	38.424	1072.2	1143.3	1.7877
200	392.51	1077.5	1150.1	2.0507	78.147	1076.2	1148.6	1.8715	38.848	1074.7	1146.6	1.7927
240	416.42	1091.2	1168.3	2.0775	83.001	1090.3	1167.1	1.8987	41.320	1089.0	1165.5	1.8205
280	440.32	1105.0	1186.5	2.1028	87.831	1104.3	1185.5	1.9244	43.768	1103.3	1184.3	1.8467
320	464.19	1118.9	1204.8	2.1269	92.645	1118.3	1204.0	1.9487	46.200	1117.6	1203.0	1.8713
360	488.05	1132.9	1223.2	2.1499	97.447	1132.4	1222.6	1.9719	48.620	1131.8	1221.8	1.8948
400	511.91	1147.0	1241.7	2.1720	102.24	1146.6	1241.2	1.9941	51.032	1146.1	1240.5	1.9171
440	535.76	1161.2	1260.4	2.1932	107.03	1160.9	1259.9	2.0154	53.438	1160.5	1259.3	1.9385
500	571.53	1182.8	1288.5	2.2235	114.21	1182.5	1288.2	2.0458	57.039	1182.2	1287.7	1.9690
600	631.13	1219.3	1336.1	2.2706	126.15	1219.1	1335.8	2.0930	63.027	1218.8	1335.5	2.0164
700	690.72	1256.7	1384.5	2.3142	138.09	1256.5	1384.3	2.1367	69.006	1256.3	1384.0	2.0601
800	750.30	1294.9	1433.7	2.3549	150.01	1294.7	1433.5	2.1774	74.978	1294.6	1433.3	2.1009
1000	869.45	1373.9	1534.8	2.4294	173.86	1373.8	1534.7	2.2520	86.912	1373.7	1534.6	2.1755
1200	988.60	1456.7	1639.6	2.4967	197.70	1456.6	1639.5	2.3192	98.837	1456.5	1639.4	2.2428
1400	1107.74	1543.1	1748.1	2.5584	221.54	1543.1	1748.1	2.3809	110.759	1543.0	1748.0	2.3045

T	P = 14.696 (211.99)				P = 20 (227.96)				P = 40 (267.26)			
	v	u	h	s	v	u	h	s	v	u	h	s
Sat.	26.803	1077.6	1150.5	1.7567	20.091	1082.0	1156.4	1.7320	10.501	1092.3	1170.0	1.6767
240	27.999	1087.9	1164.0	1.7764	20.475	1086.5	1162.3	1.7405	—	—	—	—
280	29.687	1102.4	1183.1	1.8030	21.734	1101.4	1181.8	1.7676	10.711	1097.3	1176.6	1.6857
320	31.358	1116.8	1202.1	1.8280	22.976	1116.0	1201.0	1.7929	11.360	1112.8	1196.9	1.7124
360	33.018	1131.2	1221.0	1.8516	24.206	1130.6	1220.1	1.8168	11.996	1128.0	1216.8	1.7373
400	34.668	1145.6	1239.9	1.8741	25.427	1145.1	1239.2	1.8395	12.623	1143.0	1236.4	1.7606
440	36.313	1160.0	1258.8	1.8956	26.642	1159.6	1258.2	1.8611	13.242	1157.8	1255.8	1.7827
500	38.772	1181.8	1287.3	1.9262	28.456	1181.5	1286.8	1.8919	14.164	1180.1	1284.9	1.8140
600	42.857	1218.6	1335.2	1.9737	31.466	1218.3	1334.8	1.9395	15.685	1217.3	1333.4	1.8621
700	46.932	1256.1	1383.8	2.0175	34.466	1255.9	1383.5	1.9834	17.196	1255.1	1382.4	1.9063
800	51.001	1294.4	1433.1	2.0584	37.460	1294.3	1432.9	2.0243	18.701	1293.7	1432.1	1.9474
1000	59.128	1373.6	1534.4	2.1330	43.437	1373.5	1534.3	2.0989	21.700	1373.1	1533.7	2.0222

TABLE A.1.3E (Continued) *Superheated Vapor Water (English Units)*

T	P = 14.696 (211.99)				P = 20 (227.96)				P = 40 (267.26)			
	v	u	h	s	v	u	h	s	v	u	h	s
1200	67.247	1456.5	1639.3	2.2003	49.406	1456.4	1639.2	2.1663	24.690	1456.1	1638.9	2.0897
1400	75.361	1543.0	1747.9	2.2620	55.371	1542.9	1747.8	2.2280	27.677	1542.7	1747.6	2.1515
1600	83.473	1633.2	1860.2	2.3194	61.333	1633.1	1860.1	2.2854	30.660	1633.0	1859.9	2.2089

T	P = 60 (292.73)				P = 80 (312.06)				P = 100 (327.86)			
	v	u	h	s	v	u	h	s	v	u	h	s
Sat.	7.177	1098.3	1178.0	1.6444	5.474	1102.6	1183.6	1.6214	4.434	1105.8	1187.8	1.6034
320	7.485	1109.5	1192.6	1.6633	5.544	1105.9	1188.0	1.6270	—	—	—	—
360	7.924	1125.3	1213.3	1.6893	5.886	1122.5	1209.7	1.6541	4.662	1119.7	1205.9	1.6259
400	8.353	1140.8	1233.5	1.7134	6.217	1138.5	1230.6	1.6790	4.934	1136.2	1227.5	1.6517
440	8.775	1156.0	1253.4	1.7360	6.541	1154.1	1251.0	1.7022	5.199	1152.2	1248.5	1.6755
500	9.399	1178.6	1283.0	1.7678	7.017	1177.2	1281.1	1.7346	5.587	1175.7	1279.1	1.7085
600	10.425	1216.3	1332.1	1.8165	7.794	1215.3	1330.7	1.7838	6.216	1214.2	1329.3	1.7582
700	11.440	1254.4	1381.4	1.8609	8.561	1253.6	1380.3	1.8285	6.834	1252.8	1379.2	1.8033
800	12.448	1293.0	1431.2	1.9022	9.321	1292.4	1430.4	1.8700	7.445	1291.8	1429.6	1.8449
1000	14.454	1372.7	1533.2	1.9773	10.831	1372.3	1532.6	1.9453	8.657	1371.9	1532.1	1.9204
1200	16.452	1455.8	1638.5	2.0448	12.333	1455.5	1638.1	2.0129	9.861	1455.2	1637.7	1.9882
1400	18.445	1542.5	1747.3	2.1067	13.830	1542.3	1747.0	2.0749	11.060	1542.0	1746.7	2.0502
1600	20.436	1632.8	1859.7	2.1641	15.324	1632.6	1859.5	2.1323	12.257	1632.4	1859.3	2.1076
1800	22.426	1726.7	1975.7	2.2178	16.818	1726.5	1975.5	2.1861	13.452	1726.4	1975.3	2.1614
2000	24.415	1824.0	2095.1	2.2685	18.310	1823.9	2094.9	2.2367	14.647	1823.7	2094.8	2.2120

T	P = 120 (341.30)				P = 140 (353.08)				P = 160 (363.59)			
	v	u	h	s	v	u	h	s	v	u	h	s
Sat.	3.730	1108.3	1191.1	1.5886	3.221	1110.3	1193.8	1.5760	2.836	1112.0	1196.0	1.5650
360	3.844	1116.6	1202.0	1.6021	3.259	1113.5	1197.9	1.5812	0.018	331.8	332.4	0.5161
400	4.079	1133.8	1224.4	1.6287	3.466	1131.4	1221.2	1.6088	3.007	1128.8	1217.8	1.5910
450	4.360	1154.3	1251.2	1.6590	3.713	1152.4	1248.6	1.6399	3.228	1150.5	1246.1	1.6230
500	4.633	1174.2	1277.1	1.6868	3.952	1172.7	1275.1	1.6682	3.440	1171.1	1273.0	1.6518
550	4.900	1193.8	1302.6	1.7127	4.184	1192.5	1300.9	1.6944	3.646	1191.2	1299.2	1.6784
600	5.164	1213.2	1327.8	1.7371	4.412	1212.1	1326.4	1.7191	3.848	1211.0	1325.0	1.7033
700	5.682	1252.0	1378.2	1.7825	4.860	1251.2	1377.1	1.7648	4.243	1250.4	1376.0	1.7494
800	6.195	1291.2	1428.7	1.8243	5.301	1290.5	1427.9	1.8068	4.631	1289.9	1427.0	1.7916

T	P = 120 (341.30)				P = 140 (353.08)				P = 160 (363.59)			
	v	u	h	s	v	u	h	s	v	u	h	s
1000	7.208	1371.5	1531.5	1.9000	6.173	1371.0	1531.0	1.8827	5.397	1370.6	1530.4	1.8677
1200	8.213	1454.9	1637.3	1.9679	7.036	1454.6	1636.9	1.9507	6.154	1454.3	1636.5	1.9358
1400	9.214	1541.8	1746.4	2.0300	7.895	1541.6	1746.1	2.0128	6.906	1541.4	1745.8	1.9980
1600	10.212	1632.3	1859.0	2.0875	8.752	1632.1	1858.8	2.0704	7.656	1631.9	1858.6	2.0556
1800	11.209	1726.2	1975.1	2.1412	9.607	1726.1	1975.0	2.1242	8.405	1725.9	1974.8	2.1094
2000	12.205	1823.6	2094.6	2.1919	10.461	1823.5	2094.5	2.1748	9.153	1823.3	2094.3	2.1601

T	P = 180 (373.12)				P = 200 (381.86)				P = 225 (391.86)			
	v	u	h	s	v	u	h	s	v	u	h	s
Sat.	2.533	1113.4	1197.8	1.5553	2.289	1114.6	1199.3	1.5464	2.043	1115.8	1200.8	1.5364
400	2.648	1126.2	1214.4	1.5749	2.361	1123.5	1210.8	1.5600	2.073	1119.9	1206.2	1.5427
450	2.850	1148.5	1243.4	1.6078	2.548	1146.4	1240.7	1.5938	2.245	1143.8	1237.3	1.5779
500	3.042	1169.6	1270.9	1.6372	2.724	1168.0	1268.8	1.6238	2.405	1165.9	1266.1	1.6087
550	3.228	1190.0	1297.5	1.6641	2.893	1188.6	1295.7	1.6512	2.558	1187.0	1293.5	1.6366
600	3.409	1210.0	1323.5	1.6893	3.058	1208.9	1322.0	1.6767	2.707	1207.5	1320.2	1.6624
700	3.763	1249.6	1374.9	1.7357	3.379	1248.8	1373.8	1.7234	2.995	1247.7	1372.4	1.7095
800	4.110	1289.3	1426.2	1.7781	3.693	1288.6	1425.3	1.7659	3.276	1287.8	1424.2	1.7523
900	4.453	1329.4	1477.7	1.8175	4.003	1328.9	1477.0	1.8055	3.553	1328.3	1476.2	1.7920
1000	4.793	1370.2	1529.8	1.8544	4.310	1369.8	1529.3	1.8425	3.827	1369.2	1528.6	1.8292
1200	5.467	1454.0	1636.1	1.9227	4.918	1453.7	1635.7	1.9109	4.369	1453.4	1635.2	1.8977
1400	6.137	1541.2	1745.6	1.9849	5.521	1540.9	1745.3	1.9732	4.906	1540.7	1744.9	1.9600
1600	6.804	1631.7	1858.4	2.0425	6.123	1631.6	1858.1	2.0308	5.441	1631.3	1857.9	2.0177
1800	7.470	1725.8	1974.6	2.0963	6.722	1725.6	1974.4	2.0847	5.975	1725.4	1974.2	2.0716
2000	8.135	1823.2	2094.1	2.1470	7.321	1823.0	2094.0	2.1354	6.507	1822.8	2093.8	2.1223

T	P = 250 (401.03)				P = 275 (409.52)				P = 300 (417.42)			
	v	u	h	s	v	u	h	s	v	u	h	s
Sat.	1.8448	1116.7	1202.1	1.5274	1.6813	1117.5	1203.1	1.5191	1.5441	1118.1	1203.9	1.5115
450	2.0018	1141.1	1233.7	1.5632	1.8026	1138.3	1230.0	1.5494	1.6361	1135.4	1226.2	1.5365
500	2.1498	1163.8	1263.3	1.5948	1.9407	1161.7	1260.4	1.5820	1.7662	1159.5	1257.5	1.5701
550	2.2903	1185.3	1291.3	1.6233	2.0708	1183.6	1289.0	1.6110	1.8878	1181.8	1286.6	1.5997
600	2.4258	1206.1	1318.3	1.6494	2.1958	1204.7	1316.4	1.6375	2.0041	1203.2	1314.5	1.6266
650	2.5581	1226.5	1344.8	1.6739	2.3174	1225.3	1343.2	1.6623	2.1168	1224.1	1341.6	1.6516
700	2.6879	1246.7	1371.1	1.6970	2.4365	1245.7	1369.7	1.6856	2.2269	1244.6	1368.3	1.6751
800	2.9426	1287.0	1423.2	1.7401	2.6696	1286.2	1422.1	1.7289	2.4421	1285.4	1421.0	1.7187
900	3.1929	1327.6	1475.3	1.7799	2.8983	1327.0	1474.5	1.7689	2.6528	1326.3	1473.6	1.7589

TABLE A.1.3E (Continued) Superheated Vapor Water (English Units)

T	P = 250 (401.03)				P = 275 (409.52)				P = 300 (417.42)			
	v	u	h	s	v	u	h	s	v	u	h	s
1000	3.4402	1368.7	1527.9	1.8172	3.1239	1368.7	1527.2	1.8063	2.8604	1367.7	1526.4	1.7964
1200	3.9291	1453.0	1634.8	1.8858	3.5696	1452.6	1634.3	1.8751	3.2700	1452.2	1633.8	1.8653
1400	4.4136	1540.4	1744.6	1.9483	4.0108	1540.1	1744.2	1.9376	3.6751	1539.8	1743.8	1.9279
1600	4.8957	1631.1	1857.6	2.0060	4.4495	1630.9	1857.3	1.9954	4.0777	1630.7	1857.0	1.9857
1800	5.3763	1725.2	1974.0	2.0599	4.8868	1725.0	1973.7	2.0493	4.4790	1724.8	1973.5	2.0396
2000	5.8562	1822.7	2093.6	2.1106	5.3234	1822.5	2093.4	2.1000	4.8794	1822.3	2093.2	2.0904

T	P = 350 (431.81)				P = 400 (444.69)				P = 450 (456.39)			
	v	u	h	s	v	u	h	s	v	u	h	s
Sat.	1.3267	1119.0	1204.9	1.4978	1.1619	1119.4	1205.5	1.4856	1.0326	1119.6	1205.6	1.4745
450	1.3733	1129.2	1218.2	1.5125	1.1745	1122.6	1209.6	1.4901	—	—	—	—
500	1.4913	1154.9	1251.5	1.5481	1.2843	1150.1	1245.2	1.5282	1.1226	1145.0	1238.5	1.5097
550	1.5998	1178.3	1281.9	1.5790	1.3834	1174.6	1277.0	1.5605	1.2146	1170.7	1271.9	1.5436
600	1.7025	1200.3	1310.6	1.6068	1.4760	1197.3	1306.6	1.5892	1.2996	1194.3	1302.5	1.5732
650	1.8013	1221.6	1338.3	1.6323	1.5646	1219.1	1334.9	1.6153	1.3803	1216.6	1331.5	1.6000
700	1.8975	1242.5	1365.4	1.6562	1.6503	1240.4	1362.5	1.6396	1.4580	1238.2	1359.6	1.6248
800	2.0846	1283.8	1418.8	1.7004	1.8163	1282.1	1416.6	1.6844	1.6077	1280.5	1414.4	1.6700
900	2.2670	1325.0	1471.8	1.7409	1.9776	1323.7	1470.1	1.7252	1.7524	1322.4	1468.3	1.7113
1000	2.4463	1366.6	1525.0	1.7787	2.1357	1365.5	1523.6	1.7632	1.8941	1364.4	1522.2	1.7495
1200	2.7993	1451.5	1632.8	1.8478	2.4462	1450.7	1631.8	1.8327	2.1715	1450.0	1630.8	1.8192
1400	3.1476	1539.3	1743.1	1.9106	2.7520	1538.7	1742.4	1.8956	2.4443	1538.1	1741.7	1.8823
1600	3.4935	1630.2	1856.5	1.9685	3.0553	1629.8	1855.9	1.9535	2.7146	1629.3	1855.4	1.9403
1800	3.8380	1724.5	1973.0	2.0225	3.3573	1724.1	1972.6	2.0076	2.9834	1723.7	1972.1	1.9944
2000	4.1817	1822.0	2092.8	2.0732	3.6585	1821.6	2092.4	2.0584	3.2515	1821.3	2092.0	2.0452

T	P = 500 (467.12)				P = 600 (486.33)				P = 700 (503.22)			
	v	u	h	s	v	u	h	s	v	u	h	s
Sat.	0.9283	1119.4	1205.3	1.4645	0.7702	1118.5	1204.1	1.4464	0.6558	1117.0	1202.0	1.4305
500	0.9924	1139.7	1231.5	1.4922	0.7947	1128.0	1216.2	1.4592	0.7275	1149.0	1243.2	1.4723
550	1.0792	1166.7	1266.6	1.5279	0.8749	1158.2	1255.4	1.4990	0.7929	1177.5	1280.2	1.5081
600	1.1583	1191.1	1298.3	1.5585	0.9456	1184.5	1289.5	1.5320	0.8520	1203.0	1313.4	1.5387
650	1.2327	1214.0	1328.0	1.5860	1.0109	1208.6	1320.9	1.5609	0.9073	1226.9	1344.4	1.5660

T	$P = 500$ (467.12)				$P = 600$ (486.33)				$P = 700$ (503.22)			
	v	u	h	s	v	u	h	s	v	u	h	s
700	1.3040	1236.0	1356.7	1.6112	1.0728	1231.5	1350.6	1.5871	1.0109	1272.0	1402.9	1.6145
800	1.4407	1278.8	1412.1	1.6571	1.1900	1275.4	1407.6	1.6343	1.1089	1315.6	1459.3	1.6575
900	1.5723	1321.0	1466.5	1.6986	1.3021	1318.4	1462.9	1.6766	1.2036	1358.9	1514.9	1.6970
1000	1.7008	1363.3	1520.7	1.7371	1.4108	1361.2	1517.8	1.7155	1.2960	1402.4	1570.2	1.7337
1100	1.8271	1406.0	1575.1	1.7731	1.5173	1404.2	1572.7	1.7519	1.3868	1446.1	1625.8	1.7682
1200	1.9518	1449.2	1629.8	1.8071	1.6222	1447.7	1627.8	1.7861				
1400	2.1981	1537.6	1741.0	1.8704	1.8289	1536.4	1739.5	1.8497	1.5652	1535.3	1738.1	1.8321
1600	2.4419	1628.9	1854.8	1.9285	2.0330	1628.0	1853.7	1.9080	1.7409	1627.1	1852.6	1.8905
1800	2.6843	1723.3	1971.7	1.9826	2.2357	1722.6	1970.8	1.9622	1.9152	1721.8	1969.9	1.9449
2000	2.9259	1820.9	2091.6	2.0335	2.4375	1820.2	2090.8	2.0131	2.0887	1819.5	2090.1	1.9958

T	$P = 800$ (518.36)				$P = 1000$ (544.74)				$P = 1250$ (572.56)			
	v	u	h	s	v	u	h	s	v	u	h	s
Sat.	0.5691	1115.0	1199.3	1.4160	0.4459	1109.9	1192.4	1.3903	0.3454	1101.7	1181.6	1.3619
550	0.6154	1138.8	1229.9	1.4469	0.4534	1114.8	1198.7	1.3965	0.3786	1129.0	1216.6	1.3954
600	0.6776	1170.1	1270.4	1.4861	0.5140	1153.7	1248.8	1.4450	0.4267	1167.2	1266.0	1.4409
650	0.7324	1197.2	1305.6	1.5186	0.5637	1184.7	1289.1	1.4822	0.4670	1198.4	1306.4	1.4766
700	0.7829	1222.1	1338.0	1.5471	0.6080	1212.0	1324.5	1.5135	0.5030	1226.1	1342.4	1.5070
750	0.8306	1245.7	1368.6	1.5729	0.6490	1237.2	1357.3	1.5412	0.5364	1251.8	1375.8	1.5341
800	0.8764	1268.5	1398.2	1.5969	0.6878	1261.2	1388.5	1.5664				
900	0.9640	1312.9	1455.6	1.6408	0.7610	1307.3	1448.1	1.6120	0.5984	1300.0	1438.4	1.5819
1000	1.0482	1356.7	1511.9	1.6807	0.8305	1352.2	1505.9	1.6530	0.6563	1346.4	1498.2	1.6244
1100	1.1300	1400.5	1567.8	1.7178	0.8976	1396.8	1562.9	1.6908	0.7116	1392.0	1556.6	1.6631
1200	1.2102	1444.6	1623.8	1.7525	0.9630	1441.5	1619.7	1.7260	0.7652	1437.5	1614.5	1.6990
1400	1.3674	1534.2	1736.6	1.8167	1.0905	1531.9	1733.7	1.7909	0.8690	1529.0	1730.0	1.7647
1600	1.5218	1626.2	1851.5	1.8754	1.2152	1624.4	1849.3	1.8499	0.9699	1622.1	1846.5	1.8242
1800	1.6749	1721.0	1969.0	1.9298	1.3384	1719.5	1967.2	1.9046	1.0693	1717.6	1964.9	1.8791
2000	1.8271	1818.8	2089.3	1.9808	1.4608	1817.4	2087.7	1.9557	1.1678	1815.7	2085.8	1.9304

T	$P = 1500$ (596.38)				$P = 1750$ (617.30)				$P = 2000$ (635.99)			
	v	u	h	s	v	u	h	s	v	u	h	s
Sat.	0.2769	1091.8	1168.7	1.3358	0.2268	1080.2	1153.7	1.3109	0.1881	1066.6	1136.3	1.2861
600	0.2816	1096.6	1174.8	1.3416	—	—	—	—	—	—	—	—
650	0.3329	1147.0	1239.3	1.4012	0.2627	1122.5	1207.6	1.3603	0.2057	1091.1	1167.2	1.3141
700	0.3716	1183.4	1286.6	1.4429	0.3022	1166.7	1264.6	1.4106	0.2487	1147.7	1239.8	1.3782
750	0.4049	1214.1	1326.5	1.4766	0.3341	1201.3	1309.5	1.4485	0.2803	1187.3	1291.1	1.4216
800	0.4350	1241.8	1362.5	1.5058	0.3622	1231.3	1348.6	1.4801	0.3071	1220.1	1333.8	1.4562

TABLE A.1.3E (Continued) Superheated Vapor Water (English Units)

T	P = 1500 (596.38)				P = 1750 (617.30)				P = 2000 (635.99)			
	v	u	h	s	v	u	h	s	v	u	h	s
850	0.4631	1267.7	1396.2	1.5321	0.3878	1258.8	1384.4	1.5080	0.3312	1249.5	1372.0	1.4860
900	0.4897	1292.5	1428.5	1.5562	0.4119	1284.8	1418.2	1.5334	0.3534	1276.8	1407.6	1.5126
1000	0.5400	1340.4	1490.3	1.6001	0.4569	1334.3	1482.3	1.5789	0.3945	1328.1	1474.1	1.5598
1100	0.5876	1387.2	1550.3	1.6398	0.4990	1382.2	1543.8	1.6197	0.4325	1377.2	1537.2	1.6017
1200	0.6334	1433.4	1609.3	1.6765	0.5392	1429.3	1603.9	1.6570	0.4685	1425.2	1598.6	1.6398
1400	0.7213	1526.1	1726.3	1.7431	0.6158	1523.1	1722.6	1.7245	0.5368	1520.2	1718.8	1.7082
1600	0.8064	1619.9	1843.7	1.8031	0.6896	1617.6	1840.9	1.7850	0.6020	1615.4	1838.2	1.7692
1800	0.8899	1715.7	1962.7	1.8582	0.7617	1713.8	1960.5	1.8404	0.6656	1712.0	1958.3	1.8248
2000	0.9725	1814.0	2083.9	1.9096	0.8330	1812.3	2082.0	1.8919	0.7284	1810.6	2080.1	1.8765

T	P = 2500 (668.30)				P = 3000 (695.52)				P = 3500			
	v	u	h	s	v	u	h	s	v	u	h	s
Sat.	0.13059	1031.0	1091.4	1.2326	0.08402	968.8	1015.4	1.1575				
650	—	—	—	—	—	—	—	—	0.02491	663.5	679.6	0.8629
700	0.16840	1098.7	1176.6	1.3073	0.09772	1003.9	1058.1	1.1944	0.03058	759.5	779.3	0.9506
750	0.20302	1155.2	1249.1	1.3686	0.14831	1114.7	1197.1	1.3122	0.10460	1058.4	1126.1	1.2440
800	0.22907	1195.7	1301.7	1.4112	0.17572	1167.6	1265.2	1.3675	0.13626	1134.7	1223.0	1.3226
850	0.25126	1229.5	1345.8	1.4456	0.19731	1207.6	1317.2	1.4080	0.15819	1183.4	1285.9	1.3716
900	0.27118	1259.9	1385.4	1.4752	0.21598	1241.8	1361.7	1.4413	0.17625	1222.4	1336.5	1.4095
950	0.28960	1288.2	1422.2	1.5018	0.23284	1272.7	1402.0	1.4705	0.19214	1256.4	1380.8	1.4416
1000	0.30694	1315.2	1457.2	1.5262	0.24847	1301.7	1439.6	1.4967	0.20663	1287.6	1421.4	1.4699
1100	0.33933	1366.8	1523.8	1.5704	0.27720	1356.2	1510.0	1.5434	0.23282	1345.2	1496.0	1.5193
1200	0.36958	1416.7	1587.7	1.6101	0.30365	1408.0	1576.6	1.5847	0.25657	1399.1	1565.3	1.5624
1300	0.39836	1465.7	1650.0	1.6465	0.32855	1458.5	1640.8	1.6223	0.27871	1451.1	1631.6	1.6012

T	P = 2500 (668.30)				P = 3000 (695.52)				P = 3500			
	v	u	h	s	v	u	h	s	v	u	h	s
1400	0.42609	1514.2	1711.3	1.6804	0.35236	1508.1	1703.7	1.6571	0.29972	1501.9	1696.1	1.6368
1600	0.47947	1610.8	1832.6	1.7424	0.39782	1606.3	1827.1	1.7201	0.33953	1601.7	1821.6	1.7010
1800	0.53116	1708.2	1954.0	1.7986	0.44156	1704.5	1949.6	1.7769	0.37759	1700.8	1945.4	1.7583
2000	0.58199	1807.2	2076.4	1.8506	0.48441	1803.9	2072.8	1.8291	0.41474	1800.6	2069.2	1.8108

T	P = 4000				P = 5000				P = 6000			
	v	u	h	s	v	u	h	s	v	u	h	s
650	0.02447	657.7	675.8	0.8574	0.02377	648.0	670.0	0.8482	0.02322	640.0	665.8	0.8404
700	0.02867	742.1	763.3	0.9345	0.02676	721.8	746.5	0.9156	0.02563	708.1	736.5	0.9028
750	0.06332	960.7	1007.6	1.1395	0.03364	821.4	852.6	1.0049	0.02978	788.6	821.7	0.9746
800	0.10523	1095.0	1172.9	1.2740	0.05933	987.2	1042.1	1.1583	0.03942	896.9	940.6	1.0708
850	0.12833	1156.5	1251.5	1.3352	0.08556	1092.7	1171.9	1.2596	0.05818	1018.8	1083.4	1.1820
900	0.14623	1201.5	1309.7	1.3789	0.10385	1155.1	1251.1	1.3190	0.07588	1102.9	1187.2	1.2599
950	0.16152	1239.2	1358.8	1.4143	0.11853	1202.2	1311.8	1.3629	0.09009	1162.0	1262.0	1.3140
1000	0.17520	1272.9	1402.6	1.4449	0.13120	1242.0	1363.4	1.3988	0.10207	1209.1	1322.4	1.3561
1100	0.19954	1333.9	1481.6	1.4973	0.15302	1310.6	1452.2	1.4577	0.12219	1286.4	1422.1	1.4222
1200	0.22129	1390.1	1553.9	1.5423	0.17199	1371.6	1530.8	1.5066	0.13928	1352.7	1507.3	1.4752
1300	0.24137	1443.7	1622.4	1.5823	0.18919	1428.6	1603.7	1.5493	0.15453	1413.3	1584.9	1.5206
1400	0.26029	1495.7	1688.4	1.6188	0.20517	1483.2	1673.0	1.5876	0.16854	1470.5	1657.6	1.5608
1600	0.29586	1597.1	1816.1	1.6841	0.23480	1587.9	1805.2	1.6551	0.19421	1578.7	1794.3	1.6307
1800	0.32964	1697.1	1941.1	1.7420	0.26261	1689.7	1932.7	1.7142	0.21801	1682.4	1924.5	1.6910
2000	0.36251	1797.3	2065.6	1.7948	0.28947	1790.8	2058.6	1.7676	0.24087	1784.3	2051.7	1.7450

TABLE A.1.4E Compressed Liquid Water (English Units)

T	P = 500 (467.12)				P = 1000 (544.74)				P = 1500 (596.39)			
	v	u	h	s	v	u	h	s	v	u	h	s
Sat.	0.019748	447.69	449.51	0.64901	0.021590	538.37	542.36	0.74318	0.023461	604.95	611.46	0.80821
32	0.015994	0.00	1.48	0.00000	0.015967	0.02	2.98	0.00004	0.015939	0.04	4.47	0.00006
50	0.015998	18.02	19.50	0.03599	0.015971	17.98	20.94	0.03592	0.015946	17.95	22.37	0.03583
100	0.016106	67.87	69.36	0.12931	0.016081	67.70	70.67	0.12900	0.016057	67.53	71.99	0.12869
150	0.016318	117.66	119.17	0.21455	0.016293	117.37	120.39	0.21409	0.016268	117.10	121.61	0.21362
200	0.016607	167.64	169.18	0.29339	0.016580	167.25	170.32	0.29279	0.016554	166.86	171.46	0.29220
250	0.016971	217.99	219.56	0.36700	0.016941	217.46	220.60	0.36626	0.016910	216.95	221.65	0.36553
300	0.017416	268.91	270.52	0.43640	0.017379	268.24	271.45	0.43550	0.017343	267.57	272.39	0.43462
350	0.017954	320.70	322.36	0.50247	0.017909	319.83	323.14	0.50139	0.017865	318.97	323.93	0.50032
400	0.018608	373.68	375.40	0.56603	0.018549	372.55	375.98	0.56471	0.018493	371.45	376.58	0.56341
450	0.019420	428.39	430.19	0.62797	0.019340	426.89	430.47	0.62630	0.019264	425.43	430.78	0.62468
500	—	—	—	—	0.020357	483.77	487.54	0.68736	0.020245	481.76	487.38	0.68524
550	—	—	—	—	—	—	—	—	0.021579	542.08	548.07	0.74686

T	P = 2000 (635.99)				P = 3000 (695.52)				P = 5000			
	v	u	h	s	v	u	h	s	v	u	h	s
Sat.	0.025648	662.38	671.87	0.86224	0.034300	783.37	802.41	0.97312	—	—	—	—
32	0.015912	0.06	5.95	0.00008	0.015859	0.09	8.89	0.00009	0.015755	0.11	14.69	-0.00002
50	0.015920	17.91	23.80	0.03574	0.015870	17.83	26.64	0.03555	0.015773	17.66	32.26	0.03507
100	0.016034	67.36	73.30	0.12838	0.015987	67.03	75.91	0.12776	0.015897	66.40	81.11	0.12650
200	0.016527	166.48	172.60	0.29161	0.016476	165.74	174.88	0.29044	0.016376	164.31	179.46	0.28817
300	0.017308	266.92	273.33	0.43374	0.017240	265.65	275.22	0.43203	0.017110	263.24	279.07	0.42874
400	0.018439	370.38	377.20	0.56214	0.018334	368.31	378.49	0.55969	0.018141	364.46	381.24	0.55504
450	0.019191	424.03	431.13	0.62312	0.019052	421.35	431.93	0.62010	0.018802	416.44	433.83	0.61449
500	0.020139	479.84	487.29	0.68320	0.019944	476.24	487.31	0.67935	0.019603	469.80	487.93	0.67238
550	0.021407	539.24	547.16	0.74400	0.021103	534.07	545.79	0.73874	0.020602	525.24	544.30	0.72963
600	0.023302	605.37	613.99	0.80856	0.022736	596.99	609.62	0.80041	0.021914	583.96	604.23	0.78755

TABLE A.1.5E *Saturated Solid–Saturated Vapor Water (English Units)*

Temp. F T	Press. lbf/in.^{2} P	Specific Volume ft^3/lbm		Internal Energy Btu/lbm			Enthalpy Btu/lbm			Entropy Btu/lbm R		
		Sat. Solid v_i	Sat. Vapor $v_g \times 10^{-3}$	Sat. Solid u_i	Evap. u_{ig}	Sat. Vapor u_g	Sat. Solid h_i	Evap. h_{ig}	Sat. Vapor h_g	Sat. Solid s_i	Evap. s_{ig}	Sat. Vapor s_g
32.02	0.08866	0.017473	3.302	−143.34	1164.5	1021.2	−143.34	1218.7	1075.4	−0.2916	2.4786	2.1869
32	0.08859	0.01747	3.305	−143.35	1164.5	1021.2	−143.35	1218.7	1075.4	−0.2917	2.4787	2.1870
30	0.08083	0.01747	3.607	−144.35	1164.9	1020.5	−144.35	1218.8	1074.5	−0.2938	2.4891	2.1953
25	0.06406	0.01746	4.505	−146.84	1165.7	1018.9	−146.84	1219.1	1072.3	−0.2990	2.5154	2.2164
20	0.05051	0.01745	5.655	−149.31	1166.5	1017.2	−149.31	1219.4	1070.1	−0.3042	2.5422	2.2380
15	0.03963	0.01745	7.133	−151.75	1167.3	1015.6	−151.75	1219.6	1067.9	−0.3093	2.5695	2.2601
10	0.03093	0.01744	9.043	−154.16	1168.1	1013.9	−154.16	1219.8	1065.7	−0.3145	2.5973	2.2827
5	0.02402	0.01743	11.522	−156.56	1168.8	1012.2	−156.56	1220.0	1063.5	−0.3197	2.6256	2.3059
0	0.01855	0.01742	14.761	−158.93	1169.5	1010.6	−158.93	1220.2	1061.2	−0.3248	2.6544	2.3296
−5	0.01424	0.01742	19.019	−161.27	1170.2	1008.9	−161.27	1220.3	1059.0	−0.3300	2.6839	2.3539
−10	0.01086	0.01741	24.657	−163.59	1170.8	1007.3	−163.59	1220.4	1056.8	−0.3351	2.7140	2.3788
−15	0.00823	0.01740	32.169	−165.89	1171.5	1005.6	−165.89	1220.5	1054.6	−0.3403	2.7447	2.4044
−20	0.00620	0.01740	42.238	−168.16	1172.1	1003.9	−168.16	1220.5	1052.4	−0.3455	2.7761	2.4307
−25	0.00464	0.01739	55.782	−170.40	1172.7	1002.3	−170.40	1220.6	1050.2	−0.3506	2.8081	2.4575
−30	0.00346	0.01738	74.046	−172.63	1173.2	1000.6	−172.63	1220.6	1048.0	−0.3557	2.8406	2.4849
−35	0.00256	0.01737	98.890	−174.82	1173.8	998.9	−174.82	1220.6	1045.7	−0.3608	2.8737	2.5129
−40	0.00187	0.01737	134.017	−177.00	1174.3	997.3	−177.00	1220.5	1043.5	−0.3659	2.9084	2.5425

TABLE A.2E *Thermodynamic Properties of Ammonia*
TABLE A.2.1E *Saturated Ammonia (English Units)*

Temp. F	Abs. Press. lbf/in.² P	Specific Volume, ft³/lbm Sat. Liquid v_f	Evap. v_{fg}	Sat. Vapor v_g	Enthalpy, Btu/lbm Sat. Liquid h_f	Evap. h_{fg}	Sat. Vapor h_g	Entropy, Btu/lbm R Sat. Liquid s_f	Evap. s_{fg}	Sat. Vapor s_g
−60	5.547	0.02277	44.7335	44.756	−20.89	610.19	589.30	−0.0510	1.5267	1.4758
−55	6.535	0.02288	38.3792	38.402	−15.69	606.99	591.29	−0.0380	1.4999	1.4619
−50	7.664	0.02299	33.0685	33.091	−10.48	603.73	593.26	−0.0252	1.4737	1.4485
−45	8.948	0.02311	28.6093	28.632	−5.25	600.43	595.19	−0.0126	1.4480	1.4354
−40	10.403	0.02322	24.8485	24.872	0	597.08	597.08	0	1.4228	1.4228
−35	12.046	0.02334	21.6629	21.686	5.26	593.68	598.95	0.0124	1.3980	1.4105
−30	13.894	0.02345	18.9535	18.977	10.54	590.23	600.77	0.0248	1.3737	1.3985
−25	15.967	0.02357	16.6399	16.663	15.84	586.72	602.56	0.0370	1.3498	1.3869
−20	18.282	0.02369	14.6567	14.680	21.15	583.16	604.31	0.0492	1.3264	1.3756
−15	20.861	0.02382	12.9505	12.974	26.49	579.54	606.03	0.0612	1.3033	1.3645
−10	23.725	0.02394	11.4773	11.501	31.84	575.86	607.70	0.0731	1.2807	1.3538
−5	26.896	0.02407	10.2011	10.225	37.21	572.12	609.33	0.0850	1.2584	1.3434
0	30.397	0.02420	9.0918	9.116	42.60	568.33	610.92	0.0967	1.2364	1.3332
5	34.251	0.02433	8.1245	8.149	48.00	564.47	612.47	0.1084	1.2148	1.3232
10	38.483	0.02446	7.2784	7.303	53.43	560.54	613.97	0.1200	1.1936	1.3135
15	43.120	0.02460	6.5363	6.561	58.88	556.56	615.43	0.1315	1.1726	1.3040
20	48.186	0.02474	5.8833	5.908	64.34	552.51	616.85	0.1429	1.1519	1.2948
25	53.709	0.02488	5.3073	5.332	69.83	548.39	618.21	0.1542	1.1315	1.2857
30	59.717	0.02502	4.7978	4.823	75.33	544.20	619.53	0.1654	1.1114	1.2769
35	66.239	0.02517	4.3460	4.371	80.86	539.93	620.80	0.1766	1.0916	1.2682
40	73.303	0.02532	3.9443	3.970	86.41	535.60	622.01	0.1877	1.0720	1.2597
45	80.940	0.02548	3.5863	3.612	91.98	531.19	623.17	0.1987	1.0526	1.2513
50	89.180	0.02564	3.2666	3.292	97.58	526.70	624.28	0.2097	1.0335	1.2431
55	98.055	0.02580	2.9803	3.006	103.20	522.12	625.32	0.2206	1.0145	1.2351
60	107.596	0.02597	2.7234	2.749	108.84	517.47	626.31	0.2314	0.9958	1.2272
65	117.836	0.02614	2.4924	2.519	114.51	512.73	627.24	0.2422	0.9773	1.2194
70	128.808	0.02631	2.2843	2.311	120.21	507.90	628.10	0.2529	0.9589	1.2118
75	140.546	0.02649	2.0964	2.123	125.93	502.97	628.90	0.2635	0.9407	1.2042

Temp. F	Abs. Press. lbf/in.² P	Sat. Liquid v_f	Evap. v_{fg}	Sat. Vapor v_g	Sat. Liquid h_f	Evap. h_{fg}	Sat. Vapor h_g	Sat. Liquid s_f	Evap. s_{fg}	Sat. Vapor s_g
80	153.084	0.02668	1.9264	1.953	131.68	497.95	629.63	0.2741	0.9227	1.1968
85	166.457	0.02687	1.7724	1.799	137.45	492.83	630.29	0.2846	0.9048	1.1895
90	180.700	0.02706	1.6325	1.660	143.26	487.61	630.87	0.2951	0.8871	1.1822
95	195.850	0.02726	1.5053	1.533	149.10	482.28	631.37	0.3055	0.8695	1.1750
100	211.943	0.02747	1.3894	1.417	154.97	476.83	631.80	0.3159	0.8520	1.1679
105	229.016	0.02768	1.2836	1.311	160.87	471.27	632.14	0.3263	0.8346	1.1609
110	247.107	0.02790	1.1869	1.215	166.80	465.59	632.40	0.3366	0.8173	1.1539
115	266.254	0.02813	1.0983	1.126	172.78	459.78	632.56	0.3468	0.8001	1.1469
120	286.496	0.02836	1.0171	1.045	178.79	453.84	632.63	0.3571	0.7829	1.1400
125	307.872	0.02860	0.9425	0.971	184.84	447.76	632.60	0.3673	0.7658	1.1331

TABLE A.2.2E *Superheated Ammonia (English Units)*

Abs. Press. lbf/in.² (Sat. T)		Temperature, F											
		0	20	40	60	80	100	120	140	160	180	200	220
10 (−41.32)	v	28.567	29.881	31.185	32.481	33.770	35.055	36.336	37.613	38.889	40.162	41.434	—
	h	617.9	628.1	638.3	648.6	658.8	669.1	679.4	689.8	700.2	710.7	721.3	—
	s	1.475	1.496	1.517	1.537	1.557	1.575	1.593	1.611	1.628	1.645	1.661	—
15 (−27.26)	v	18.909	19.804	20.688	21.564	22.434	23.298	24.159	25.017	25.872	26.726	27.577	—
	h	616.2	626.7	637.2	647.5	657.9	668.3	678.7	689.1	699.6	710.2	720.8	—
	s	1.425	1.447	1.468	1.489	1.508	1.527	1.545	1.563	1.580	1.597	1.613	—
20 (−16.61)	v	14.077	14.763	15.439	16.105	16.765	17.420	18.071	18.719	19.364	20.007	20.649	21.290
	h	614.5	625.3	635.9	646.5	657.0	667.5	678.0	688.5	699.1	709.7	720.4	731.1
	s	1.388	1.411	1.433	1.454	1.473	1.492	1.511	1.529	1.546	1.563	1.579	1.595
25 (−7.93)	v	11.177	11.738	12.288	12.829	13.363	13.893	14.418	14.940	15.459	15.976	16.492	17.006
	h	612.8	623.9	634.7	645.5	656.1	666.7	677.3	687.9	698.5	709.2	719.9	730.7
	s	1.359	1.383	1.405	1.426	1.446	1.465	1.484	1.502	1.519	1.536	1.553	1.569

TABLE A.2.2E *Superheated Ammonia (English Units)*

Abs. Press. lbf/in.² (Sat. T)		Temperature, F											
		0	20	40	60	80	100	120	140	160	180	200	220
30 (−0.54)	v	9.242	9.721	10.187	10.645	11.095	11.541	11.982	12.420	12.855	13.289	13.721	14.151
	h	611.1	622.4	633.5	644.4	655.2	665.9	676.6	687.3	698.0	708.7	719.5	730.3
	s	1.335	1.359	1.382	1.403	1.424	1.443	1.462	1.480	1.497	1.514	1.531	1.547
35 (5.92)	v	—	8.279	8.686	9.084	9.475	9.861	10.242	10.620	10.996	11.369	11.741	12.111
	h	—	620.9	632.2	643.3	654.3	665.1	675.9	686.7	697.4	708.2	719.0	729.9
	s	—	1.339	1.362	1.384	1.404	1.424	1.443	1.461	1.479	1.496	1.513	1.529
40 (11.69)	v	—	7.196	7.560	7.913	8.260	8.600	8.937	9.270	9.601	9.929	10.256	10.582
	h	—	619.4	631.0	642.3	653.4	664.3	675.2	686.0	696.9	707.7	718.5	729.4
	s	—	1.321	1.344	1.366	1.387	1.407	1.426	1.445	1.463	1.480	1.497	1.513
45 (16.91)	v	—	6.354	6.683	7.002	7.314	7.620	7.922	8.220	8.516	8.809	9.101	9.392
	h	—	617.8	629.7	641.2	652.4	663.5	674.5	685.4	696.3	707.2	718.1	729.0
	s	—	1.304	1.329	1.351	1.372	1.393	1.412	1.430	1.448	1.465	1.482	1.499
50 (21.69)	v	—	—	5.981	6.273	6.557	6.836	7.110	7.380	7.648	7.914	8.177	8.440
	h	—	—	628.4	640.1	651.5	662.7	673.8	684.8	695.7	706.7	717.6	728.6
	s	—	—	1.314	1.337	1.359	1.379	1.399	1.417	1.435	1.453	1.469	1.486
60 (30.23)	v	—	—	4.928	5.179	5.422	5.659	5.891	6.120	6.346	6.569	6.791	7.012
	h	—	—	625.7	637.8	649.6	661.1	672.3	683.5	694.6	705.6	716.7	727.7
	s	—	—	1.289	1.313	1.335	1.356	1.375	1.394	1.413	1.430	1.447	1.464

		60	80	100	120	140	160	180	200	240	280	320	360
70 (37.71)	v	4.396	4.610	4.817	5.020	5.219	5.415	5.609	5.801	6.182	6.558	—	—
	h	635.5	647.6	659.4	670.9	682.2	693.4	704.6	715.7	738.0	760.5	—	—
	s	1.291	1.314	1.335	1.356	1.375	1.393	1.411	1.428	1.461	1.492	—	—

Temperature, F

Abs. Press. lbf/in² (Sat. T)		60	80	100	120	140	160	180	200	240	280	320	360
80	v	3.808	4.001	4.186	4.367	4.544	4.717	4.889	5.059	5.394	5.725	—	—
	h	633.2	645.6	657.7	669.4	680.9	692.3	703.6	714.8	737.2	759.8	—	—
(44.40)	s	1.272	1.296	1.318	1.338	1.358	1.376	1.394	1.412	1.445	1.476	—	—
90	v	3.350	3.526	3.695	3.858	4.018	4.174	4.329	4.481	4.782	5.078	—	—
	h	630.7	643.6	655.9	667.9	679.6	691.1	702.5	713.8	736.4	759.1	—	—
(50.48)	s	1.255	1.279	1.301	1.322	1.342	1.361	1.379	1.397	1.430	1.462	—	—
100	v	2.983	3.146	3.301	3.451	3.597	3.740	3.880	4.019	4.291	4.560	—	—
	h	628.3	641.5	654.2	666.4	678.2	689.9	701.4	712.9	735.6	758.4	—	—
(56.05)	s	1.239	1.264	1.287	1.308	1.328	1.347	1.366	1.383	1.417	1.449	—	—
140	v	—	2.164	2.287	2.403	2.514	2.622	2.727	2.830	3.031	3.228	3.421	—
	h	—	632.7	646.8	660.0	672.7	685.0	697.1	709.0	732.4	755.6	779.0	—
(74.77)	s	—	1.212	1.237	1.261	1.282	1.302	1.321	1.340	1.374	1.407	1.437	—
180	v	—	—	1.719	1.818	1.910	1.999	2.085	2.168	2.330	2.487	2.641	—
	h	—	—	638.7	653.3	666.9	680.0	692.6	704.9	729.0	752.8	776.5	—
(89.76)	s	—	—	1.197	1.222	1.245	1.267	1.287	1.306	1.341	1.374	1.406	—
220	v	—	—	—	1.443	1.524	1.602	1.676	1.747	1.884	2.016	2.145	2.208
	h	—	—	—	646.0	660.8	674.7	687.9	700.8	725.6	749.9	774.1	786.1
(102.40)	s	—	—	—	1.189	1.214	1.237	1.258	1.278	1.314	1.348	1.380	1.395
240	v	—	—	—	1.301	1.379	1.452	1.522	1.588	1.716	1.839	1.958	2.017
	h	—	—	—	642.2	657.6	671.9	685.6	698.7	723.9	748.5	772.8	785.0
(108.07)	s	—	—	—	1.174	1.200	1.223	1.245	1.265	1.302	1.337	1.369	1.384
260	v	—	—	—	1.180	1.256	1.325	1.391	1.454	1.575	1.689	1.801	1.855
	h	—	—	—	638.2	654.3	669.1	683.1	696.5	722.2	747.0	771.6	783.8
(113.40)	s	—	—	—	1.159	1.186	1.211	1.233	1.253	1.291	1.326	1.358	1.374
280	v	—	—	—	1.076	1.149	1.216	1.279	1.339	1.453	1.561	1.665	1.717
	h	—	—	—	634.0	650.9	666.2	680.6	694.3	720.4	745.6	770.3	782.6
(118.43)	s	—	—	—	1.145	1.173	1.198	1.221	1.242	1.281	1.316	1.348	1.364

TABLE A.3E *Thermodynamic Properties of Refrigerant-12 (Dichlorodifluoromethane)*
TABLE A.3.1E *Saturated R-12 (English Units)*

Temp. F	Abs. Press. lbf/in.² P	Specific Volume, ft³/lbm			Enthalpy, Btu/lbm			Entropy, Btu/lbm R		
		Sat. Liquid v_f	Evap. v_{fg}	Sat. Vapor v_g	Sat. Liquid h_f	Evap. h_{fg}	Sat. Vapor h_g	Sat. Liquid s_f	Evap. s_{fg}	Sat. Vapor s_g
−130	0.41224	0.009736	70.7205	70.730	−18.609	81.577	62.968	−0.04983	0.24743	0.19760
−120	0.64190	0.009816	46.7309	46.741	−16.565	80.617	64.052	−0.04372	0.23732	0.19359
−110	0.97034	0.009899	31.7667	31.777	−14.518	79.663	65.146	−0.03779	0.22781	0.19002
−100	1.4280	0.009985	22.1537	22.164	−12.466	78.714	66.248	−0.03200	0.21883	0.18683
−90	2.0509	0.010073	15.8109	15.821	−10.409	77.765	67.355	−0.02637	0.21034	0.18398
−80	2.8807	0.010164	11.5228	11.533	−8.345	76.812	68.467	−0.02086	0.20230	0.18143
−70	3.9651	0.010259	8.5584	8.5687	−6.273	75.853	69.580	−0.01548	0.19465	0.17916
−60	5.3575	0.010357	6.4670	6.4774	−4.192	74.885	70.694	−0.01021	0.18735	0.17714
−50	7.1168	0.010459	4.9637	4.9742	−2.101	73.906	71.805	−0.00506	0.18039	0.17533
−40	9.3076	0.010564	3.8645	3.8750	0	72.913	72.913	0	0.17373	0.17373
−30	11.999	0.010674	3.0478	3.0585	2.112	71.903	74.015	0.00496	0.16733	0.17229
−20	15.267	0.010788	2.4322	2.4429	4.236	70.874	75.110	0.00983	0.16119	0.17102
−10	19.189	0.010906	1.9618	1.9727	6.372	69.824	76.196	0.01462	0.15527	0.16989
0	23.849	0.011030	1.5978	1.6089	8.521	68.751	77.271	0.01932	0.14956	0.16888
10	29.335	0.011160	1.3129	1.3241	10.684	67.651	78.335	0.02395	0.14403	0.16798
20	35.736	0.011296	1.0875	1.0988	12.863	66.522	79.385	0.02852	0.13867	0.16719
30	43.148	0.011438	0.90736	0.91880	15.058	65.361	80.419	0.03301	0.13347	0.16648
40	51.667	0.011588	0.76198	0.77357	17.273	64.163	81.436	0.03745	0.12840	0.16586
50	61.394	0.011746	0.64363	0.65538	19.508	62.926	82.433	0.04184	0.12346	0.16530
60	72.433	0.011913	0.54647	0.55839	21.766	61.643	83.409	0.04618	0.11861	0.16479
70	84.888	0.012089	0.46609	0.47818	24.051	60.309	84.359	0.05048	0.11385	0.16434
80	98.870	0.012277	0.39908	0.41135	26.365	58.917	85.282	0.05475	0.10917	0.16392
90	114.49	0.012478	0.34281	0.35529	28.713	57.461	86.174	0.05900	0.10453	0.16353
100	131.86	0.012693	0.29525	0.30794	31.100	55.929	87.029	0.06323	0.09993	0.16315
110	151.11	0.012924	0.25477	0.26769	33.531	54.313	87.844	0.06745	0.09534	0.16279
120	172.35	0.013174	0.22009	0.23326	36.013	52.598	88.610	0.07168	0.09073	0.16241
130	195.71	0.013447	0.19019	0.20364	38.553	50.768	89.322	0.07593	0.08609	0.16202
140	221.32	0.013746	0.16424	0.17799	41.162	48.805	89.967	0.08020	0.08138	0.16159
150	249.31	0.014078	0.14156	0.15564	43.850	46.684	90.534	0.08453	0.07657	0.16110

Temp. F	Abs. Press. lbf/in.² P	Sat. Liquid v_f	Evap. v_{fg}	Sat. Vapor v_g	Sat. Liquid h_f	Evap. h_{fg}	Sat. Vapor h_g	Sat. Liquid s_f	Evap. s_{fg}	Sat. Vapor s_g
160	279.82	0.014449	0.12159	0.13604	46.633	44.373	91.006	0.08893	0.07160	0.16053
170	313.00	0.014871	0.10385	0.11873	49.529	41.831	91.359	0.09342	0.06643	0.15985
180	349.00	0.015360	0.087943	0.10330	52.562	38.999	91.561	0.09804	0.06096	0.15900
190	387.98	0.015942	0.073476	0.089418	55.769	35.792	91.561	0.10284	0.05509	0.15793
200	430.09	0.016659	0.060068	0.076728	59.203	32.075	91.278	0.10789	0.04862	0.15651
210	475.52	0.017601	0.047242	0.064843	62.959	27.599	90.558	0.11332	0.04121	0.15453
220	524.43	0.018986	0.034154	0.053140	67.246	21.790	89.037	0.11943	0.03206	0.15149
230	577.03	0.021854	0.017581	0.039435	72.894	12.229	85.122	0.12739	0.01773	0.14512
233.6 (critical)	596.90	0.028700	0	0.028700	78.855	0	78.855	0.13587	0	0.13587

TABLE A.3.2.E *Superheated Refrigerant-12 (English Units)*

Temp. F	5.00 lbf/in.² v ft³/lbm	h Btu/lbm	s Btu/lbm R	10.00 lbf/in.² v ft³/lbm	h Btu/lbm	s Btu/lbm R	15.00 lbf/in.² v ft³/lbm	h Btu/lbm	s Btu/lbm R
0	8.0612	78.582	0.19663	3.9810	78.246	0.18471	2.6201	77.902	0.17751
20	8.4265	81.309	0.20244	4.1691	81.014	0.19061	2.7494	80.712	0.18349
40	8.7903	84.090	0.20812	4.3556	83.828	0.19635	2.8770	83.561	0.18931
60	9.1528	86.922	0.21367	4.5408	86.689	0.20197	3.0031	86.451	0.19498
80	9.5142	89.806	0.21912	4.7248	89.596	0.20746	3.1281	89.383	0.20051
100	9.8747	92.738	0.22445	4.9079	92.548	0.21283	3.2521	92.357	0.20593
120	10.2344	95.717	0.22968	5.0903	95.546	0.21809	3.3754	95.373	0.21122
140	10.5936	98.743	0.23481	5.2721	98.586	0.22325	3.4981	98.429	0.21640
160	10.9523	101.812	0.23985	5.4533	101.669	0.22830	3.6202	101.525	0.22148
180	11.3105	104.925	0.24479	5.6341	104.793	0.23326	3.7419	104.661	0.22646
200	11.6684	108.079	0.24964	5.8145	107.957	0.23813	3.8632	107.835	0.23135
220	12.0260	111.272	0.25441	5.9946	111.159	0.24291	3.9841	111.046	0.23614

Temp. F	20.00 lbf/in.² v ft³/lbm	h Btu/lbm	s Btu/lbm R	25.00 lbf/in.² v ft³/lbm	h Btu/lbm	s Btu/lbm R	30.00 lbf/in.² v ft³/lbm	h Btu/lbm	s Btu/lbm R
20	2.0391	80.403	0.17829	1.6125	80.088	0.17414	1.3278	79.765	0.17065
40	2.1373	83.289	0.18419	1.6932	83.012	0.18012	1.3969	82.730	0.17671
60	2.2340	86.210	0.18992	1.7723	85.965	0.18591	1.4644	85.716	0.18257

TABLE A.3.2.E (Continued) *Superheated Refrigerant-12 (English Units)*

Temp. F	20.00 lbf/in.² v ft³/lbm	h Btu/lbm	s Btu/lbm R	25.00 lbf/in.² v ft³/lbm	h Btu/lbm	s Btu/lbm R	30.00 lbf/in.² v ft³/lbm	h Btu/lbm	s Btu/lbm R
60	2.2340	86.210	0.18992	1.7723	85.965	0.18591	1.4644	85.716	0.18257
80	2.3295	89.168	0.19550	1.8502	88.950	0.19155	1.5306	88.729	0.18826
100	2.4241	92.164	0.20095	1.9271	91.968	0.19704	1.5957	91.770	0.19379
120	2.5179	95.198	0.20628	2.0032	95.021	0.20240	1.6600	94.843	0.19918
140	2.6110	98.270	0.21149	2.0786	98.110	0.20763	1.7237	97.948	0.20445
160	2.7036	101.380	0.21659	2.1535	101.234	0.21276	1.7868	101.086	0.20960
180	2.7957	104.528	0.22159	2.2279	104.393	0.21778	1.8494	104.258	0.21463
200	2.8874	107.712	0.22649	2.3019	107.588	0.22269	1.9116	107.464	0.21957
220	2.9789	110.932	0.23130	2.3757	110.817	0.22752	1.9735	110.702	0.22440
240	3.0700	114.186	0.23602	2.4491	114.080	0.23225	2.0351	113.973	0.22915

Temp. F	35.00 lbf/in.² v ft³/lbm	h Btu/lbm	s Btu/lbm R	40.00 lbf/in.² v ft³/lbm	h Btu/lbm	s Btu/lbm R	50.00 lbf/in.² v ft³/lbm	h Btu/lbm	s Btu/lbm R
40	1.1850	82.442	0.17375	1.0258	82.148	0.17112	0.80248	81.540	0.16655
60	1.2442	85.463	0.17968	1.0789	85.206	0.17712	0.84713	84.676	0.17271
80	1.3021	88.504	0.18542	1.1306	88.277	0.18292	0.89025	87.811	0.17862
100	1.3589	91.570	0.19100	1.1812	91.367	0.18854	0.93216	90.953	0.18434
120	1.4148	94.663	0.19643	1.2309	94.480	0.19401	0.97313	94.110	0.18988
140	1.4701	97.785	0.20172	1.2798	97.620	0.19933	1.0133	97.286	0.19527
160	1.5248	100.938	0.20689	1.3282	100.788	0.20453	1.0529	100.485	0.20051
180	1.5789	104.122	0.21195	1.3761	103.985	0.20961	1.0920	103.708	0.20563
200	1.6327	107.338	0.21690	1.4236	107.212	0.21457	1.1307	106.958	0.21064
220	1.6862	110.586	0.22175	1.4707	110.469	0.21944	1.1690	110.235	0.21553
240	1.7394	113.865	0.22651	1.5176	113.757	0.22420	1.2070	113.539	0.22032
260	1.7923	117.175	0.23117	1.5642	117.074	0.22888	1.2448	116.871	0.22502

Temp. F	60.00 lbf/in.² v ft³/lbm	h Btu/lbm	s Btu/lbm R	70.00 lbf/in.² v ft³/lbm	h Btu/lbm	s Btu/lbm R	80.00 lbf/in.² v ft³/lbm	h Btu/lbm	s Btu/lbm R
60	0.69210	84.126	0.16892	0.58088	83.552	0.16556	—	—	—
80	0.72964	87.330	0.17497	0.61458	86.832	0.17175	0.52795	86.316	0.16885
100	0.76588	90.528	0.18079	0.64685	90.091	0.17768	0.55734	89.640	0.17489
120	0.80110	93.731	0.18641	0.67803	93.343	0.18339	0.58556	92.945	0.18070

Temp. F	60.00 lbf/in² v ft³/lbm	h Btu/lbm	s Btu/lbm R	70.00 lbf/in² v ft³/lbm	h Btu/lbm	s Btu/lbm R	80.00 lbf/in² v ft³/lbm	h Btu/lbm	s Btu/lbm R
140	0.83551	96.945	0.19186	0.70836	96.597	0.18891	0.61286	96.242	0.18629
160	0.86928	100.176	0.19716	0.73800	99.862	0.19427	0.63943	99.542	0.19170
180	0.90252	103.427	0.20233	0.76708	103.141	0.19948	0.66543	102.851	0.19696
200	0.93531	106.700	0.20736	0.79571	106.439	0.20455	0.69095	106.174	0.20207
220	0.96775	109.997	0.21229	0.82397	109.756	0.20951	0.71609	109.513	0.20706
240	0.99989	113.319	0.21710	0.85191	113.096	0.21435	0.74090	112.872	0.21193
260	1.03176	116.666	0.22182	0.87959	116.459	0.21909	0.76544	116.251	0.21669
280	1.06342	120.039	0.22644	0.90705	119.846	0.22373	0.78975	119.652	0.22135

Temp. F	90.00 lbf/in² v ft³/lbm	h Btu/lbm	s Btu/lbm R	100.00 lbf/in² v ft³/lbm	h Btu/lbm	s Btu/lbm R	125.00 lbf/in² v ft³/lbm	h Btu/lbm	s Btu/lbm R
100	0.48749	89.175	0.17234	0.43138	88.694	0.16996	0.32943	87.407	0.16455
120	0.51346	92.536	0.17824	0.45562	92.116	0.17597	0.35086	91.008	0.17087
140	0.53845	95.879	0.18391	0.47881	95.507	0.18172	0.37098	94.537	0.17686
160	0.56268	99.216	0.18938	0.50118	98.884	0.18726	0.39015	98.023	0.18258
180	0.58629	102.557	0.19469	0.52291	102.257	0.19262	0.40857	101.484	0.18807
200	0.60941	105.905	0.19984	0.54413	105.633	0.19782	0.42642	104.934	0.19338
220	0.63213	109.267	0.20486	0.56492	109.018	0.20287	0.44380	108.380	0.19853
240	0.65451	112.644	0.20976	0.58538	112.415	0.20780	0.46081	111.829	0.20353
260	0.67662	116.040	0.21455	0.60554	115.828	0.21261	0.47750	115.287	0.20840
280	0.69849	119.456	0.21923	0.62546	119.258	0.21731	0.49394	118.756	0.21316
300	0.72016	122.892	0.22381	0.64518	122.707	0.22191	0.51016	122.238	0.21780
320	0.74166	126.349	0.22830	0.66472	126.176	0.22641	0.52619	125.737	0.22235

Temp. F	150.00 lbf/in² v ft³/lbm	h Btu/lbm	s Btu/lbm R	175.00 lbf/in² v ft³/lbm	h Btu/lbm	s Btu/lbm R	200.00 lbf/in² v ft³/lbm	h Btu/lbm	s Btu/lbm R
120	0.28007	89.800	0.16629	—	—	—	—	—	—
140	0.29845	93.498	0.17256	0.24595	92.373	0.16859	0.20579	91.137	0.16480
160	0.31566	97.112	0.17849	0.26198	96.142	0.17478	0.22121	95.100	0.17130
180	0.33200	100.675	0.18415	0.27697	99.823	0.18062	0.23535	98.921	0.17737
200	0.34769	104.206	0.18958	0.29120	103.447	0.18620	0.24860	102.652	0.18311
220	0.36285	107.720	0.19483	0.30485	107.036	0.19156	0.26117	106.325	0.18860
240	0.37761	111.226	0.19992	0.31804	110.605	0.19674	0.27323	109.962	0.19387
260	0.39203	114.732	0.20485	0.33087	114.162	0.20175	0.28489	113.576	0.19896
280	0.40617	118.242	0.20967	0.34339	117.717	0.20662	0.29623	117.178	0.20390
300	0.42008	121.761	0.21436	0.35567	121.273	0.21137	0.30730	120.775	0.20870
320	0.43379	125.290	0.21894	0.36773	124.835	0.21599	0.31815	124.373	0.21337
340	0.44733	128.833	0.22343	0.37963	128.407	0.22052	0.32881	127.974	0.21793

TABLE A.3.2.E (Continued) Superheated Refrigerant-12 (English Units)

Temp. F	250.00 lbf/in.² v ft³/lbm	h Btu/lbm	s Btu/lbm R	300.00 lbf/in.² v ft³/lbm	h Btu/lbm	s Btu/lbm R	400.00 lbf/in.² v ft³/lbm	h Btu/lbm	s Btu/lbm R
160	0.16249	92.717	0.16462	—	—	—	—	—	—
180	0.17605	96.925	0.17130	0.13482	94.556	0.16537	—	—	—
200	0.18824	100.930	0.17747	0.14697	98.975	0.17217	0.09101	93.718	0.16092
220	0.19952	104.809	0.18326	0.15774	103.136	0.17838	0.10316	99.046	0.16888
240	0.21014	108.607	0.18877	0.16761	107.140	0.18419	0.11300	103.735	0.17568
260	0.22027	112.351	0.19404	0.17685	111.043	0.18969	0.12163	108.105	0.18183
280	0.23001	116.060	0.19913	0.18562	114.879	0.19495	0.12949	112.286	0.18756
300	0.23944	119.747	0.20405	0.19402	118.670	0.20000	0.13680	116.343	0.19298
320	0.24862	123.420	0.20882	0.20214	122.430	0.20489	0.14372	120.318	0.19814
340	0.25759	127.088	0.21346	0.21002	126.171	0.20963	0.15032	124.235	0.20310
360	0.26639	130.754	0.21799	0.21770	129.900	0.21423	0.15668	128.112	0.20789
380	0.27504	134.423	0.22241	0.22522	133.624	0.21872	0.16285	131.961	0.21253

Temp. F	500.00 lbf/in.² v ft³/lbm	h Btu/lbm	s Btu/lbm R	600.00 lbf/in.² v ft³/lbm	h Btu/lbm	s Btu/lbm R	700.00 lbf/in.² v ft³/lbm	h Btu/lbm	s Btu/lbm R
220	0.06421	92.397	0.15683	—	—	—	—	—	—
240	0.07762	99.218	0.16672	0.04749	91.024	0.15335	0.01997	71.902	0.12528
260	0.08705	104.526	0.17421	0.06192	99.741	0.16566	0.04022	91.889	0.15344
280	0.09492	109.277	0.18072	0.07086	105.637	0.17374	0.05245	100.963	0.16589
300	0.10190	113.729	0.18666	0.07806	110.729	0.18053	0.06047	107.202	0.17421
320	0.10829	117.997	0.19221	0.08433	115.420	0.18663	0.06693	112.526	0.18113
340	0.11426	122.143	0.19746	0.09002	119.871	0.19227	0.07255	117.393	0.18730
360	0.11992	126.205	0.20247	0.09529	124.167	0.19757	0.07763	121.986	0.19297
380	0.12533	130.207	0.20730	0.10025	128.355	0.20262	0.08232	126.401	0.19829
400	0.13054	134.166	0.21196	0.10498	132.466	0.20746	0.08674	130.692	0.20334
420	0.13559	138.096	0.21648	0.10952	136.523	0.21213	0.09093	134.895	0.20818
440	0.14051	142.004	0.22087	0.11391	140.539	0.21664	0.09495	139.033	0.21283

TABLE A.4.1E *Saturated Refrigerant-22 (English Units)*

Temp. F	Abs. Press. lbf/in.² P	Specific Volume, ft³/lbm			Enthalpy, Btu/lbm			Entropy, Btu/lbm R		
		Sat. Liquid v_f	Evap. v_{fg}	Sat. Vapor v_g	Sat. Liquid h_f	Evap. h_{fg}	Sat. Vapor h_g	Sat. Liquid s_f	Evap. s_{fg}	Sat. Vapor s_g
−100	2.398	0.010664	18.4219	18.4326	−14.564	107.935	93.371	−0.03734	0.30008	0.26274
−90	3.423	0.010771	13.2243	13.2351	−12.216	106.759	94.544	−0.03091	0.28878	0.25787
−80	4.782	0.010881	9.6840	9.69487	−9.838	105.548	95.710	−0.02457	0.27799	0.25342
−70	6.552	0.010995	7.2208	7.23180	−7.429	104.297	96.868	−0.01832	0.26764	0.24932
−60	8.818	0.011113	5.4733	5.48442	−4.988	103.001	98.014	−0.01214	0.25770	0.24556
−50	11.674	0.011235	4.2111	4.22238	−2.512	101.656	99.144	−0.00604	0.24813	0.24209
−40	15.222	0.011363	3.2844	3.29572	0	100.257	100.257	0	0.23888	0.23888
−30	19.573	0.011495	2.5934	2.60489	2.547	98.801	101.348	0.00598	0.22994	0.23591
−20	24.845	0.011634	2.0709	2.08257	5.131	97.285	102.415	0.01189	0.22126	0.23315
−10	31.162	0.011778	1.6707	1.68248	7.751	95.704	103.455	0.01776	0.21282	0.23058
0	38.657	0.011930	1.3603	1.37227	10.409	94.056	104.465	0.02356	0.20461	0.22817
10	47.464	0.012088	1.1170	1.12904	13.104	92.338	105.442	0.02932	0.19659	0.22592
20	57.727	0.012255	0.92405	0.936309	15.837	90.545	106.383	0.03503	0.18876	0.22379
30	69.591	0.012431	0.76965	0.782082	18.609	88.674	107.284	0.04070	0.18108	0.22178
40	83.206	0.012618	0.64491	0.657527	21.421	86.720	108.142	0.04632	0.17355	0.21986
50	98.727	0.012815	0.54324	0.556059	24.275	84.678	108.953	0.05190	0.16614	0.21803
60	116.312	0.013025	0.45969	0.472719	27.172	82.540	109.712	0.05745	0.15882	0.21627
70	136.123	0.013251	0.39048	0.403734	30.116	80.298	110.414	0.06296	0.15159	0.21456
80	158.326	0.013492	0.33271	0.346206	33.109	77.943	111.052	0.06846	0.14442	0.21288
90	183.094	0.013754	0.28413	0.297887	36.158	75.461	111.619	0.07394	0.13728	0.21122
100	210.604	0.014038	0.24298	0.257021	39.267	72.838	112.105	0.07942	0.13014	0.20956
110	241.042	0.014350	0.20787	0.222220	42.446	70.052	112.498	0.08491	0.12296	0.20787
120	274.604	0.014694	0.17768	0.192379	45.705	67.077	112.782	0.09042	0.11571	0.20613
130	311.496	0.015080	0.15153	0.166606	49.059	63.877	112.936	0.09598	0.10832	0.20431
140	351.944	0.015518	0.12866	0.144176	52.528	60.403	112.931	0.10163	0.10072	0.20235
150	396.194	0.016025	0.10846	0.124484	56.143	56.585	112.728	0.10739	0.09281	0.20020
160	444.525	0.016627	0.090384	0.107011	59.948	52.316	112.263	0.11334	0.08442	0.19776
170	497.259	0.017367	0.073912	0.091279	64.019	47.419	111.438	0.11959	0.07531	0.19490
180	554.783	0.018332	0.058458	0.076790	68.498	41.570	110.068	0.12635	0.06498	0.19133
190	617.590	0.019733	0.043104	0.062837	73.711	34.023	107.734	0.13409	0.05237	0.18646
200	686.356	0.022436	0.025002	0.047438	80.862	21.990	102.853	0.14460	0.03333	0.17794
204.81	721.906	0.030525	0	0.030525	91.378	0	91.378	0.16022	0	0.16022

TABLE A.4.2E *Superheated Refrigerant-22 (English Units)*

Temp. F	5.00 lbf/in.² v ft³/lbm	h Btu/lbm	s Btu/lbm R	10.00 lbf/in.² v ft³/lbm	h Btu/lbm	s Btu/lbm R	15.00 lbf/in.² v ft³/lbm	h Btu/lbm	s Btu/lbm R
−20	10.8034	103.885	0.27238	5.3460	103.526	0.25588	3.5261	103.159	0.24596
0	11.3114	106.735	0.27872	5.6060	106.414	0.26230	3.7037	106.088	0.25248
20	11.8177	109.643	0.28491	5.8643	109.356	0.26856	3.8794	109.065	0.25882
40	12.3227	112.611	0.29097	6.1212	112.353	0.27468	4.0537	112.091	0.26500
60	12.8265	115.638	0.29691	6.3769	115.404	0.28067	4.2268	115.168	0.27103
80	13.3293	118.724	0.30274	6.6316	118.512	0.28654	4.3989	118.298	0.27694
100	13.8313	121.867	0.30845	6.8855	121.674	0.29229	4.5701	121.480	0.28273
120	14.3327	125.069	0.31407	7.1387	124.892	0.29794	4.7406	124.715	0.28841
140	14.8335	128.327	0.31960	7.3913	128.165	0.30349	4.9105	128.003	0.29399
160	15.3337	131.642	0.32504	7.6434	131.493	0.30895	5.0799	131.344	0.29947
180	15.8336	135.012	0.33039	7.8951	134.875	0.31432	5.2489	134.737	0.30486
200	16.3331	138.437	0.33566	8.1464	138.310	0.31961	5.4174	138.183	0.31016
220	16.8322	141.916	0.34086	8.3974	141.799	0.32482	5.5857	141.680	0.31538
240	17.3312	145.449	0.34598	8.6481	145.339	0.32995	5.7537	145.229	0.32053

Temp. F	20.00 lbf/in.² v ft³/lbm	h Btu/lbm	s Btu/lbm R	25.00 lbf/in.² v ft³/lbm	h Btu/lbm	s Btu/lbm R	30.00 lbf/in.² v ft³/lbm	h Btu/lbm	s Btu/lbm R
0	2.7521	105.756	0.24535	2.1808	105.419	0.23970	1.7997	105.076	0.23497
20	2.8867	108.769	0.25177	2.2908	108.469	0.24619	1.8933	108.165	0.24154
40	3.0198	111.826	0.25801	2.3992	111.558	0.25250	1.9853	111.286	0.24792
60	3.1516	114.930	0.26410	2.5063	114.689	0.25864	2.0760	114.445	0.25412
80	3.2823	118.082	0.27005	2.6123	117.865	0.26464	2.1655	117.645	0.26016
100	3.4122	121.284	0.27588	2.7175	121.087	0.27050	2.2542	120.888	0.26606
120	3.5414	124.536	0.28159	2.8219	124.357	0.27624	2.3421	124.176	0.27183
140	3.6700	127.840	0.28719	2.9257	127.675	0.28187	2.4294	127.510	0.27748
160	3.7981	131.194	0.29269	3.0289	131.043	0.28739	2.5162	130.891	0.28303
180	3.9257	134.599	0.29810	3.1318	134.460	0.29282	2.6025	134.320	0.28848
200	4.0529	138.055	0.30342	3.2342	137.926	0.29815	2.6884	137.797	0.29383
220	4.1799	141.562	0.30865	3.3363	141.443	0.30340	2.7739	141.323	0.29909
240	4.3065	145.119	0.31381	3.4382	145.008	0.30857	2.8592	144.897	0.30427
260	4.4329	148.725	0.31889	3.5397	148.622	0.31367	2.9443	148.518	0.30938

Temp. F	40.00 lbf/in.² v ft³/lbm	40.00 lbf/in.² h Btu/lbm	40.00 lbf/in.² s Btu/lbm R	50.00 lbf/in.² v ft³/lbm	50.00 lbf/in.² h Btu/lbm	50.00 lbf/in.² s Btu/lbm R	60.00 lbf/in.² v ft³/lbm	60.00 lbf/in.² h Btu/lbm	60.00 lbf/in.² s Btu/lbm R
20	1.3959	107.541	0.23399	1.0968	106.897	0.22788	—	—	—
40	1.4676	110.732	0.24051	1.1564	110.163	0.23455	0.94863	109.577	0.22950
60	1.5378	113.950	0.24682	1.2145	113.443	0.24099	0.99871	112.923	0.23607
80	1.6068	117.199	0.25296	1.2714	116.745	0.24722	1.0475	116.281	0.24241
100	1.6749	120.485	0.25893	1.3272	120.075	0.25328	1.0952	119.658	0.24855
120	1.7423	123.810	0.26477	1.3822	123.438	0.25918	1.1421	123.061	0.25453
140	1.8090	127.176	0.27048	1.4366	126.838	0.26495	1.1882	126.495	0.26035
160	1.8751	130.585	0.27607	1.4903	130.276	0.27059	1.2338	129.963	0.26604
180	1.9407	134.039	0.28156	1.5436	133.755	0.27611	1.2788	133.468	0.27161
200	2.0060	137.538	0.28694	1.5965	137.276	0.28153	1.3235	137.012	0.27706
220	2.0709	141.082	0.29223	1.6491	140.840	0.28686	1.3678	140.596	0.28241
240	2.1356	144.673	0.29744	1.7013	144.448	0.29209	1.4118	144.221	0.28767
260	2.2000	148.310	0.30257	1.7533	148.100	0.29723	1.4556	147.889	0.29284
280	2.2641	151.992	0.30761	1.8051	151.797	0.30230	1.4991	151.600	0.29792

Temp. F	70.00 lbf/in.² v ft³/lbm	70.00 lbf/in.² h Btu/lbm	70.00 lbf/in.² s Btu/lbm R	80.00 lbf/in.² v ft³/lbm	80.00 lbf/in.² h Btu/lbm	80.00 lbf/in.² s Btu/lbm R	100.00 lbf/in.² v ft³/lbm	100.00 lbf/in.² h Btu/lbm	100.00 lbf/in.² s Btu/lbm R
40	0.79981	108.972	0.22507	0.68782	108.347	0.22107	—	—	—
60	0.84429	112.391	0.23178	0.72820	111.843	0.22793	0.56498	110.700	0.22117
80	0.88736	115.807	0.23823	0.76708	115.323	0.23450	0.59818	114.319	0.22801
100	0.92932	119.234	0.24446	0.80477	118.801	0.24083	0.63003	117.911	0.23454
120	0.97038	122.679	0.25051	0.84152	122.290	0.24696	0.66084	121.492	0.24083
140	1.0107	126.148	0.25639	0.87751	125.796	0.25290	0.69081	125.077	0.24691
160	1.0504	129.647	0.26213	0.91286	129.326	0.25869	0.72011	128.674	0.25281
180	1.0896	133.178	0.26774	0.94770	132.885	0.26435	0.74885	132.290	0.25855
200	1.1284	136.745	0.27323	0.98209	136.476	0.26987	0.77712	135.931	0.26415
220	1.1669	140.350	0.27862	1.0161	140.101	0.27529	0.80501	139.599	0.26963
240	1.2050	143.993	0.28390	1.0498	143.763	0.28060	0.83257	143.299	0.27500
260	1.2428	147.677	0.28909	1.0833	147.463	0.28581	0.85985	147.032	0.28026
280	1.2805	151.401	0.29419	1.1165	151.202	0.29094	0.88689	150.800	0.28542
300	1.3179	155.167	0.29922	1.1495	154.981	0.29598	0.91373	154.605	0.29050

Temp. F	125.00 lbf/in.² v ft³/lbm	125.00 lbf/in.² h Btu/lbm	125.00 lbf/in.² s Btu/lbm R	150.00 lbf/in.² v ft³/lbm	150.00 lbf/in.² h Btu/lbm	150.00 lbf/in.² s Btu/lbm R	175.00 lbf/in.² v ft³/lbm	175.00 lbf/in.² h Btu/lbm	175.00 lbf/in.² s Btu/lbm R
80	0.46219	112.989	0.22104	0.37055	111.556	0.21484	—	—	—
100	0.48962	116.743	0.22787	0.39534	115.504	0.22202	0.32731	114.180	0.21668
120	0.51584	120.456	0.23439	0.41870	119.368	0.22880	0.34884	118.221	0.22378

TABLE A.4.2E (Continued) *Superheated Refrigerant-22 (English Units)*

Temp. F	125.00 lbf/in.²			150.00 lbf/in.²			175.00 lbf/in.²		
	v ft³/lbm	h Btu/lbm	s Btu/lbm R	v ft³/lbm	h Btu/lbm	s Btu/lbm R	v ft³/lbm	h Btu/lbm	s Btu/lbm R
140	0.54112	124.148	0.24065	0.44097	123.181	0.23527	0.36910	122.172	0.23048
160	0.56565	127.835	0.24670	0.46241	126.968	0.24148	0.38843	126.069	0.23687
180	0.58957	131.528	0.25256	0.48319	130.744	0.24748	0.40702	129.935	0.24301
200	0.61299	135.234	0.25827	0.50342	134.520	0.25329	0.42502	133.787	0.24894
220	0.63600	138.959	0.26383	0.52322	138.305	0.25895	0.44256	137.637	0.25469
240	0.65867	142.708	0.26927	0.54265	142.107	0.26446	0.45970	141.493	0.26028
260	0.68104	146.485	0.27459	0.56177	145.929	0.26984	0.47651	145.363	0.26573
280	0.70316	150.292	0.27980	0.58062	149.775	0.27512	0.49305	149.251	0.27106
300	0.72507	154.130	0.28493	0.59925	153.649	0.28028	0.50936	153.162	0.27628
320	0.74679	158.003	0.28996	0.61770	157.553	0.28536	0.52546	157.098	0.28139
340	0.76835	161.911	0.29491	0.63597	161.489	0.29034	0.54140	161.062	0.28641

Temp. F	200.00 lbf/in.²			250.00 lbf/in.²			300.00 lbf/in.²		
	v ft³/lbm	h Btu/lbm	s Btu/lbm R	v ft³/lbm	h Btu/lbm	s Btu/lbm R	v ft³/lbm	h Btu/lbm	s Btu/lbm R
100	0.27553	112.750	0.21165	—	—	—	—	—	—
120	0.29595	117.004	0.21911	0.22038	114.297	0.21036	—	—	—
140	0.31487	121.114	0.22608	0.23795	118.818	0.21803	0.18520	116.199	0.21042
160	0.33270	125.134	0.23268	0.25401	123.141	0.22512	0.20062	120.938	0.21820
180	0.34972	129.100	0.23898	0.26902	127.340	0.23179	0.21460	125.436	0.22534
200	0.36609	133.034	0.24503	0.28325	131.461	0.23814	0.22759	129.784	0.23204
220	0.38196	136.953	0.25089	0.29688	135.533	0.24422	0.23984	134.035	0.23839
240	0.39741	140.867	0.25656	0.31002	139.576	0.25008	0.25154	138.225	0.24446
260	0.41251	144.787	0.26209	0.32278	143.604	0.25576	0.26280	142.375	0.25031
280	0.42733	148.719	0.26747	0.33522	147.629	0.26127	0.27370	146.503	0.25597
300	0.44190	152.668	0.27274	0.34740	151.659	0.26665	0.28431	150.621	0.26146
320	0.45627	156.637	0.27790	0.35935	155.699	0.27190	0.29467	154.738	0.26681
340	0.47046	160.631	0.28296	0.37111	159.756	0.27704	0.30483	158.861	0.27203
360	0.48449	164.652	0.28792	0.38270	163.831	0.28207	0.31482	162.995	0.27714
380	0.49839	168.700	0.29280	0.39416	167.929	0.28701	0.32465	167.145	0.28214
400	0.51218	172.779	0.29760	0.40549	172.052	0.29186	0.33436	171.314	0.28705

Temp. F	400.00 lbf/in.²			500.00 lbf/in.²			600.00 lbf/in.²		
	v ft³/lbm	h Btu/lbm	s Btu/lbm R	v ft³/lbm	h Btu/lbm	s Btu/lbm R	v ft³/lbm	h Btu/lbm	s Btu/lbm R
160	0.13053	115.516	0.20459	—	—	—	—	—	—
180	0.14460	121.014	0.21332	0.098744	115.064	0.20054	—	—	—
200	0.15674	126.023	0.22104	0.11220	121.432	0.21034	0.079113	115.116	0.19810
220	0.16770	130.755	0.22810	0.12319	126.954	0.21859	0.091943	122.320	0.20887
240	0.17785	135.315	0.23471	0.13285	132.051	0.22598	0.10197	128.295	0.21753
260	0.18740	139.761	0.24098	0.14164	136.894	0.23281	0.11061	133.705	0.22516
280	0.19650	144.131	0.24697	0.14984	141.572	0.23922	0.11839	138.790	0.23213
300	0.20524	148.451	0.25273	0.15757	146.141	0.24532	0.12558	143.668	0.23863
320	0.21369	152.741	0.25831	0.16495	150.634	0.25115	0.13233	148.407	0.24479
340	0.22191	157.012	0.26372	0.17206	155.077	0.25678	0.13875	153.050	0.25067
360	0.22992	161.275	0.26898	0.17893	159.487	0.26223	0.14490	157.628	0.25633
380	0.23776	165.538	0.27412	0.18561	163.877	0.26752	0.15083	162.160	0.26179
400	0.24546	169.807	0.27914	0.19212	168.257	0.27267	0.15658	166.662	0.26709
420	0.25303	174.087	0.28407	0.19851	172.634	0.27771	0.16219	171.146	0.27225
440	0.26049	178.381	0.28889	0.20477	177.014	0.28263	0.16766	175.621	0.27727

Temp. F	700.00 lbf/in.²			800.00 lbf/in.²			900.00 lbf/in.²		
	v ft³/lbm	h Btu/lbm	s Btu/lbm R	v ft³/lbm	h Btu/lbm	s Btu/lbm R	v ft³/lbm	h Btu/lbm	s Btu/lbm R
220	0.067104	116.048	0.19748	0.042152	104.391	0.17883	—	—	—
240	0.078820	123.792	0.20872	0.059974	118.024	0.19865	0.043352	109.908	0.18568
260	0.087856	130.088	0.21759	0.070151	125.882	0.20973	0.055694	120.852	0.20113
280	0.095590	135.733	0.22533	0.078163	132.339	0.21858	0.064298	128.530	0.21165
300	0.10253	141.007	0.23237	0.085058	138.131	0.22631	0.071320	135.010	0.22030
320	0.10891	146.046	0.23892	0.091252	143.538	0.23333	0.077446	140.872	0.22792
340	0.11489	150.925	0.24510	0.096959	148.696	0.23987	0.082984	146.359	0.23487
360	0.12057	155.695	0.25099	0.10231	153.686	0.24603	0.088104	151.600	0.24134
380	0.12599	160.387	0.25664	0.10737	158.557	0.25190	0.092910	156.672	0.24745
400	0.13122	165.025	0.26210	0.11222	163.344	0.25753	0.097471	161.623	0.25328
420	0.13628	169.625	0.26739	0.11688	168.071	0.26297	0.10184	166.487	0.25888
440	0.14120	174.201	0.27253	0.12140	172.756	0.26824	0.10604	171.289	0.26427
460	0.14601	178.761	0.27755	0.12579	177.412	0.27335	0.11011	176.046	0.26950
480	0.15071	183.315	0.28245	0.13007	182.049	0.27834	0.11406	180.771	0.27458
500	0.15532	187.867	0.28724	0.13425	186.675	0.28321	0.11792	185.475	0.27954

TABLE A.5E *Thermodynamic Properties of Refrigerant-134a (1,1,1,2-tetrafluoroethane)*
TABLE A.5.1E *Saturated R-134a (English Units)*

Temp. F	Abs. Press. lbf/in.² P	Specific Volume, ft³/lbm			Enthalpy, Btu/lbm			Entropy, Btu/lbm R		
		Sat. Liquid v_f	Evap. v_{fg}	Sat. Vapor v_g	Sat. Liquid h_f	Evap. h_{fg}	Sat. Vapor h_g	Sat. Liquid s_f	Evap. s_{fg}	Sat. Vapor s_g
−25	11.4028	0.011526	3.8551	3.8666	68.382	94.451	162.833	0.2010	0.2173	0.4183
−20	13.0097	0.011592	3.4049	3.4164	69.863	93.724	163.587	0.2044	0.2132	0.4175
−10	16.7602	0.011729	2.6804	2.6921	72.868	92.220	165.089	0.2111	0.2051	0.4162
0	21.3150	0.011873	2.1339	2.1458	75.924	90.658	166.582	0.2178	0.1972	0.4150
10	26.7864	0.012024	1.7162	1.7282	79.024	89.039	168.064	0.2244	0.1896	0.4140
30	40.9608	0.012350	1.1398	1.1522	85.344	85.632	170.976	0.2375	0.1749	0.4124
35	45.2708	0.012437	1.0339	1.0464	86.948	84.742	171.690	0.2408	0.1713	0.4121
40	49.9206	0.012526	0.93952	0.9520	88.563	83.834	172.397	0.2440	0.1678	0.4118
45	54.9278	0.012619	0.85509	0.8677	90.188	82.910	173.097	0.2472	0.1643	0.4115
50	60.3101	0.012714	0.77942	0.7921	91.823	81.966	173.789	0.2504	0.1608	0.4112
55	66.0855	0.012812	0.71145	0.72426	93.469	81.002	174.471	0.2536	0.1574	0.4110
60	72.2724	0.012913	0.65024	0.66316	95.127	80.017	175.144	0.2568	0.1540	0.4108
65	78.8898	0.013017	0.59502	0.60803	96.797	79.010	175.807	0.2600	0.1506	0.4105
70	85.9569	0.013125	0.54508	0.55820	98.479	77.979	176.457	0.2631	0.1472	0.4103
75	93.4933	0.013237	0.49983	0.51307	100.174	76.922	177.096	0.2663	0.1439	0.4101
80	101.5189	0.013353	0.45875	0.47210	101.882	75.838	177.720	0.2694	0.1405	0.4099
85	110.0543	0.013474	0.42138	0.43485	103.605	74.725	178.330	0.2725	0.1372	0.4097
90	119.1203	0.013599	0.38731	0.40091	105.342	73.582	178.924	0.2757	0.1339	0.4095
95	128.7381	0.013730	0.35621	0.36994	107.095	72.406	179.501	0.2788	0.1305	0.4093
100	138.9297	0.013866	0.32775	0.34162	108.864	71.194	180.059	0.2819	0.1272	0.4091
110	161.1238	0.014156	0.27772	0.29188	112.455	68.657	181.113	0.2882	0.1205	0.4087
120	185.8877	0.014475	0.23540	0.24988	116.123	65.949	182.072	0.2945	0.1138	0.4082
130	213.4166	0.014830	0.19933	0.21416	119.876	63.043	182.919	0.3008	0.1069	0.4077
140	243.9180	0.015227	0.16837	0.18360	123.729	59.904	183.633	0.3071	0.0999	0.4070
150	277.6140	0.015679	0.14156	0.15724	127.698	56.487	184.185	0.3135	0.0927	0.4062

Temp. F	Abs. Press. lbf/in.² P	Specific Volume, ft³/lbm Sat. Liquid v_f	Evap. v_{fg}	Sat. Vapor v_g	Enthalpy, Btu/lbm Sat. Liquid h_f	Evap. h_{fg}	Sat. Vapor h_g	Entropy, Btu/lbm R Sat. Liquid s_f	Evap. s_{fg}	Sat. Vapor s_g
160	314.7458	0.016203	0.11813	0.13434	131.807	52.728	184.535	0.3200	0.0851	0.4051
170	355.5800	0.016827	0.097399	0.11423	136.094	48.532	184.626	0.3267	0.0771	0.4037
180	400.4193	0.017599	0.078726	0.096325	140.623	43.737	184.360	0.3336	0.0684	0.4020
190	449.6228	0.018615	0.061411	0.080026	145.514	38.043	183.557	0.3409	0.0586	0.3995
200	503.6514	0.020125	0.044361	0.064487	151.069	30.724	181.792	0.3491	0.0466	0.3957
210	563.2197	0.023379	0.023989	0.047368	158.654	18.735	177.389	0.3601	0.0280	0.3881
214.07	589.4334	0.031532	0	0.031532	168.090	0	168.090	0.3740	0	0.3740

TABLE A.5.2E *Superheated R = 134a (English Units)*

Temp. F	10 lbf/in.² v ft³/lbm	h Btu/lbm	s Btu/lbm R	15 lbf/in.² v ft³/lbm	h Btu/lbm	s Btu/lbm R	20 lbf/in.² v ft³/lbm	h Btu/lbm	s Btu/lbm R
−20	4.4879	163.924	0.42323	—	—	—	—	—	—
0	4.7168	167.657	0.43154	3.1033	167.193	0.42288	2.2955	166.712	0.41647
20	4.9417	171.448	0.43961	3.2586	171.059	0.43111	2.4164	170.658	0.42488
40	5.1637	175.306	0.44749	3.4109	174.974	0.43911	2.5340	174.634	0.43300
60	5.3836	179.238	0.45521	3.5610	178.951	0.44691	2.6492	178.658	0.44089
80	5.6019	183.249	0.46278	3.7093	182.997	0.45455	2.7627	182.741	0.44860
100	5.8189	187.340	0.47022	3.8563	187.117	0.46204	2.8748	186.890	0.45615
120	6.0350	191.515	0.47755	4.0024	191.314	0.46941	2.9859	191.111	0.46356
140	6.2503	195.772	0.48477	4.1476	195.591	0.47666	3.0962	195.407	0.47085
160	6.4650	200.114	0.49189	4.2922	199.948	0.48381	3.2058	199.780	0.47802
180	6.6791	204.540	0.49892	4.4364	204.387	0.49086	3.3149	204.233	0.48509
200	6.8929	209.049	0.50586	4.5801	208.907	0.49782	3.4236	208.765	0.49207
220	7.1064	213.641	0.51272	4.7234	213.509	0.50469	3.5319	213.376	0.49895
240	7.3195	218.315	0.51950	4.8665	218.192	0.51148	3.6399	218.068	0.50576

Temp. F	30 lbf/in.² v ft³/lbm	h Btu/lbm	s Btu/lbm R	40 lbf/in.² v ft³/lbm	h Btu/lbm	s Btu/lbm R	50 lbf/in.² v ft³/lbm	h Btu/lbm	s Btu/lbm R
20	1.5725	169.818	0.41565	—	—	—	—	—	—
40	1.6559	173.928	0.42404	1.2157	173.182	0.41731	—	—	—
60	1.7367	178.053	0.43214	1.2796	177.420	0.42563	1.0045	176.756	0.42032
80	1.8155	182.214	0.44000	1.3413	181.668	0.43365	1.0563	181.099	0.42852

TABLE A.5.2E (Continued) *Superheated R = 134a (English Units)*

Temp. F	v ft³/lbm	h Btu/lbm	s Btu/lbm R	v ft³/lbm	h Btu/lbm	s Btu/lbm R	v ft³/lbm	h Btu/lbm	s Btu/lbm R
	30 lbf/in.²			40 lbf/in.²			50 lbf/in.²		
100	1.8929	186.426	0.44766	1.4015	185.947	0.44143	1.1062	185.451	0.43644
120	1.9691	190.697	0.45516	1.4604	190.271	0.44902	1.1549	189.834	0.44413
140	2.0445	195.034	0.46251	1.5184	194.651	0.45645	1.2026	194.260	0.45164
160	2.1192	199.441	0.46974	1.5757	199.094	0.46374	1.2495	198.741	0.45899
180	2.1933	203.921	0.47686	1.6324	203.604	0.47090	1.2957	203.282	0.46620
200	2.2670	208.477	0.48387	1.6886	208.185	0.47795	1.3415	207.889	0.47329
220	2.3403	213.109	0.49079	1.7444	212.838	0.48490	1.3869	212.565	0.48027
240	2.4133	217.818	0.49761	1.7999	217.566	0.49176	1.4319	217.311	0.48716
260	2.4860	222.605	0.50436	1.8552	222.369	0.49853	1.4766	222.131	0.49395

Temp. F	v ft³/lbm	h Btu/lbm	s Btu/lbm R	v ft³/lbm	h Btu/lbm	s Btu/lbm R	v ft³/lbm	h Btu/lbm	s Btu/lbm R
	60 lbf/in.²			70 lbf/in.²			80 lbf/in.²		
60	0.82042	176.057	0.41574	0.68820	175.318	0.41165	—	—	—
80	0.86571	180.506	0.42415	0.72913	179.886	0.42027	0.62624	179.237	0.41675
100	0.90906	184.939	0.43221	0.76789	184.407	0.42850	0.66170	183.855	0.42516
120	0.95098	189.384	0.44001	0.80509	188.920	0.43642	0.69545	188.441	0.43321
140	0.99182	193.860	0.44760	0.84112	193.449	0.44410	0.72793	193.028	0.44099
160	1.0318	198.380	0.45502	0.87626	198.012	0.45159	0.75947	197.636	0.44855
180	1.0712	202.954	0.46229	0.91071	202.621	0.45891	0.79027	202.282	0.45593
200	1.1100	207.589	0.46942	0.94461	207.284	0.46609	0.82050	206.975	0.46315
220	1.1484	212.288	0.47644	0.97807	212.007	0.47314	0.85026	211.724	0.47024
240	1.1865	217.054	0.48335	1.0112	216.795	0.48008	0.87964	216.532	0.47721
260	1.2243	221.891	0.49016	1.0440	221.649	0.48692	0.90871	221.405	0.48408
280	1.2618	226.799	0.49689	1.0765	226.573	0.49367	0.93752	226.345	0.49085
300	1.2991	231.780	0.50353	1.1088	231.567	0.50033	0.96611	231.353	0.49753

Temp. F	v ft³/lbm	h Btu/lbm	s Btu/lbm R	v ft³/lbm	h Btu/lbm	s Btu/lbm R	v ft³/lbm	h Btu/lbm	s Btu/lbm R
	90 lbf/in.²			100 lbf/in.²			125 lbf/in.²		
80	0.54574	178.554	0.41349	0.48088	177.833	0.41039	—	—	—
100	0.57881	183.281	0.42209	0.51219	182.682	0.41922	0.39104	181.059	0.41262
120	0.60997	187.947	0.43028	0.54138	187.437	0.42756	0.41714	186.077	0.42143
140	0.63976	192.595	0.43816	0.56909	192.151	0.43556	0.44133	190.982	0.42975

Temp. F	90 lbf/in.² v ft³/lbm	90 lbf/in.² h Btu/lbm	90 lbf/in.² s Btu/lbm R	100 lbf/in.² v ft³/lbm	100 lbf/in.² h Btu/lbm	100 lbf/in.² s Btu/lbm R	125 lbf/in.² v ft³/lbm	125 lbf/in.² h Btu/lbm	125 lbf/in.² s Btu/lbm
160	0.66854	197.252	0.44580	0.59569	196.859	0.44328	0.46421	195.836	0.43771
180	0.69653	201.937	0.45324	0.62147	201.585	0.45079	0.48610	200.675	0.44540
200	0.72392	206.662	0.46051	0.64661	206.343	0.45811	0.50726	205.525	0.45286
220	0.75082	211.436	0.46764	0.67123	211.145	0.46528	0.52785	210.402	0.46015
240	0.77732	216.267	0.47465	0.69545	216.000	0.47232	0.54798	215.318	0.46727
260	0.80351	221.159	0.48154	0.71933	220.911	0.47924	0.56774	220.282	0.47427
280	0.82942	226.115	0.48834	0.74293	225.884	0.48606	0.58720	225.299	0.48115
300	0.85510	231.138	0.49504	0.76629	230.921	0.49278	0.60641	230.375	0.48792
320	0.88059	236.229	0.50165	0.78946	236.025	0.49941	0.62540	235.513	0.49459

Temp. F	150 lbf/in.² v ft³/lbm	150 lbf/in.² h Btu/lbm	150 lbf/in.² s Btu/lbm R	175 lbf/in.² v ft³/lbm	175 lbf/in.² h Btu/lbm	175 lbf/in.² s Btu/lbm R	200 lbf/in.² v ft³/lbm	200 lbf/in.² h Btu/lbm	200 lbf/in.² s Btu/lbm
120	0.33316	184.573	0.41586	0.27189	182.883	0.41054	—	—	—
140	0.35543	189.719	0.42459	0.29330	188.341	0.41980	0.24585	186.818	0.41519
160	0.37606	194.748	0.43284	0.31260	193.583	0.42840	0.26449	192.328	0.42423
180	0.39552	199.719	0.44073	0.33049	198.711	0.43654	0.28139	197.642	0.43267
200	0.41413	204.673	0.44836	0.34739	203.783	0.44435	0.29712	202.852	0.44069
220	0.43210	209.633	0.45577	0.36355	208.837	0.45190	0.31200	208.011	0.44839
240	0.44955	214.617	0.46299	0.37915	213.896	0.45924	0.32624	213.154	0.45585
260	0.46661	219.638	0.47007	0.39430	218.979	0.46640	0.34000	218.305	0.46311
280	0.48333	224.704	0.47701	0.40909	224.097	0.47341	0.35337	223.479	0.47020
300	0.49978	229.821	0.48384	0.42359	229.259	0.48030	0.36643	228.688	0.47715
320	0.51601	234.995	0.49056	0.43786	234.471	0.48707	0.37923	233.941	0.48397
340	0.53204	240.228	0.49719	0.45192	239.738	0.49374	0.39181	239.242	0.49069
360	0.54791	245.524	0.50373	0.46580	245.062	0.50032	0.40421	244.598	0.49730

Temp. F	250 lbf/in.² v ft³/lbm	250 lbf/in.² h Btu/lbm	250 lbf/in.² s Btu/lbm R	300 lbf/in.² v ft³/lbm	300 lbf/in.² h Btu/lbm	300 lbf/in.² s Btu/lbm R	350 lbf/in.² v ft³/lbm	350 lbf/in.² h Btu/lbm	350 lbf/in.² s Btu/lbm
160	0.19551	189.462	0.41620	0.14666	185.845	0.40783	—	—	—
180	0.21168	195.284	0.42545	0.16374	192.528	0.41845	0.12745	189.131	0.41104
200	0.22612	200.843	0.43401	0.17793	198.592	0.42778	0.14249	196.007	0.42163
220	0.23942	206.259	0.44210	0.19050	204.347	0.43638	0.15497	202.236	0.43093
240	0.25191	211.598	0.44984	0.20202	209.934	0.44448	0.16602	208.139	0.43949
260	0.26381	216.905	0.45732	0.21279	215.428	0.45222	0.17613	213.863	0.44756
280	0.27524	222.206	0.46458	0.22302	220.877	0.45969	0.18557	219.487	0.45527

TABLE A.5.2.E (Continued) *Superheated R = 134a (English Units)*

Temp. F	v ft³/lbm	h Btu/lbm	s Btu/lbm R	v ft³/lbm	h Btu/lbm	s Btu/lbm R	v ft³/lbm	h Btu/lbm	s Btu/lbm R
	250 lbf/in.²			300 lbf/in.²			350 lbf/in.²		
300	0.28632	227.519	0.47167	0.23282	226.311	0.46694	0.19452	225.059	0.46270
320	0.29710	232.860	0.47861	0.24230	231.751	0.47401	0.20310	230.611	0.46992
340	0.30764	238.237	0.48542	0.25150	237.212	0.48092	0.21138	236.165	0.47695
360	0.31799	243.658	0.49212	0.26049	242.705	0.48771	0.21941	241.736	0.48383
380	0.32816	249.128	0.49871	0.26930	248.237	0.49438	0.22725	247.335	0.49058
400	0.33819	254.651	0.50521	0.27794	253.814	0.50094	0.23492	252.970	0.49721
	400 lbf/in.²			500 lbf/in.²			600 lbf/in.²		
180	0.096581	184.411	0.40204	—	—	—	—	—	—
200	0.11456	192.920	0.41515	0.066633	182.544	0.39688	—	—	—
220	0.12766	199.861	0.42552	0.086749	193.804	0.41372	0.050668	182.226	0.39480
240	0.13864	206.187	0.43470	0.099037	201.622	0.42506	0.070300	195.560	0.41418
260	0.14840	212.196	0.44317	0.10888	208.473	0.43472	0.081478	204.022	0.42611
280	0.15735	218.028	0.45116	0.11743	214.861	0.44348	0.090262	211.274	0.43605
300	0.16572	223.759	0.45880	0.12515	220.994	0.45166	0.097804	217.964	0.44497
320	0.17366	229.438	0.46618	0.13230	226.977	0.45943	0.10457	224.342	0.45326
340	0.18126	235.094	0.47335	0.13903	232.874	0.46690	0.11080	230.537	0.46111
360	0.18860	240.750	0.48033	0.14543	238.725	0.47413	0.11664	236.621	0.46862
380	0.19572	246.422	0.48717	0.15158	244.559	0.48116	0.12218	242.643	0.47588
400	0.20266	252.119	0.49387	0.15753	250.393	0.48803	0.12748	248.633	0.48293

TABLE A.8E *Critical Constants (English Units)*

Substance	Formula	M	Temp. R	Pressure lbf/in.²	Volume ft³/lb-mole	Acentric Factor
Ammonia	NH_3	17.031	729.9	1646	1.1613	0.250
Argon	Ar	39.948	271.4	706	1.1998	0.001
Bromine	Br_2	159.808	1058.4	1494	2.0375	0.108
Carbon dioxide	CO_2	44.010	547.4	1070	1.5041	0.239
Carbon monoxide	CO	28.010	239.2	508	1.4929	0.066
Chlorine	Cl_2	70.906	750.4	1157	1.9831	0.090
Deuterium (normal)	D_2	4.032	69.1	241	0.0000	-0.160
Fluorine	F_2	37.997	259.7	757	1.0620	0.054
Helium	He	4.003	9.34	32.9	0.9195	-0.365
Helium³	He	3.017	5.96	16.5	1.1677	-0.473
Hydrogen (normal)	H_2	2.016	59.76	188.6	1.0428	-0.218
Krypton	Kr	83.800	376.9	798	1.4609	0.005
Neon	Ne	20.183	79.92	400	0.6664	-0.029
Nitric oxide	NO	30.006	324.0	940	0.9243	0.588
Nitrogen	N_2	28.013	227.2	492	1.4385	0.039
Nitrogen dioxide	NO_2	46.006	775.8	1465	2.6879	0.834
Nitrous oxide	N_2O	44.013	557.3	1050	1.5602	0.165
Oxygen	O_2	31.999	278.3	731	1.1758	0.025
Sulfur dioxide	SO_2	64.063	775.4	1143	1.9575	0.256
Water	H_2O	18.015	1165.1	3208	0.9147	0.344
Xenon	Xe	131.300	521.5	847	1.8966	0.008
Acetylene	C_2H_2	26.038	554.9	891	1.8053	0.190
Benzene	C_6H_6	78.114	1012.0	709	4.1488	0.212
n-Butane	C_4H_{10}	58.124	765.4	551	4.0847	0.199
Carbon tetrachloride	CCl_4	153.823	1001.5	661	4.4195	0.193
Chlorodifluoroethane[a]	CH_3CClF_2	100.495	738.5	616	3.7003	0.250
Chlorodifluoromethane	$CHClF_2$	86.469	664.7	721	2.6527	0.221
Chloroform	$CHCl_3$	119.378	965.5	779	3.8268	0.218
Dichlorodifluoromethane	CCl_2F_2	120.914	693.0	600	3.4712	0.204
Dichlorofluoroethane[a]	CH_3CCl_2F	116.950	866.7	658	4.0367	0.215
Dichlorofluoromethane	$CHCl_2F$	102.923	812.9	751	3.1460	0.210
Dichlorotrifluoroethane[a]	$CHCl_2CF_3$	152.930	822.4	532	4.4547	0.282
Difluoroethane[a]	CHF_2CH_3	66.050	695.5	656	2.8753	0.275
Ethane	C_2H_6	30.070	549.7	708	2.3755	0.099
Ethyl alcohol	C_2H_5OH	46.069	925.0	891	2.6767	0.644
Ethylene	C_2H_4	28.054	508.3	731	2.0888	0.089
n-Heptane	C_7H_{16}	100.205	972.5	397	6.9200	0.349
n-Hexane	C_6H_{14}	86.178	913.5	437	5.9268	0.299
Methane	CH_4	16.043	342.7	667	1.5890	0.011

TABLE A.8E (Continued) *Critical Constants (English Units)*

Substance	Formula	M	Temp. R	Pressure lbf/in.^{2}	Volume ft^3/lb-mole	Acentric Factor
Methyl alcohol	CH_3OH	32.042	922.7	1173	1.8902	0.556
Methyl chloride	CH_3Cl	50.488	749.3	972	2.2250	0.153
n-Octane	C_8H_{18}	114.232	1023.8	361	7.8811	0.398
n-Pentane	C_5H_{16}	72.151	845.5	489	4.8696	0.251
Propane	C_3H_8	44.094	665.6	616	4.2517	0.153
Propene	C_3H_6	42.081	656.8	667	2.8993	0.144
Propyne	C_3H_4	40.065	724.3	817	2.6270	0.215
Tetrafluoroethane	CF_3CH_2F	102.030	673.6	589	3.1717	0.327

Source: R. C. Reid, J. M. Prausnitz, and B. E. Poling, *The Properties of Gases and Liquids*, fourth edition, McGraw-Hill Book Company, New York, 1987.

[a]Data from M. O. McLinden, NIST Thermophysics Division, 1989.

TABLE A.9E *Properties of Various Solids and Liquids at 80 F (English Units)*

Solid	C_p Btu/lbm R	ρ, lbm/ft^3	Liquid	C_p Btu/lbm R	ρ, lbm/ft^3
Aluminum	0.215	170	Ammonia	1.146	38
Concrete	0.155	144	Benzene	0.41	55
Copper	0.092	555	Butane	0.590	35
Glass	0.8	144	Ethanol	0.587	49
Granite	0.243	170	Glycerine	0.573	75
Graphite	0.170	155	Iso-octane	0.50	43
Iron	0.107	490	Mercury	0.033	847
Lead	0.030	705	Methanol	0.609	49
Rubber (soft)	0.439	70	Oil (light)	0.430	57
Sand (dry)	0.190	90–110	Propane	0.61	32
Silver	0.056	655	R-12	0.232	82
Steel (AISI302)	0.115	503	R-134a	0.342	75
Tin	0.052	360	Water	1.000	62
Wood (most)	0.420	22–45			

TABLE A.10E *Properties of Various Ideal Gases at 80 F (English Units)*

Gas	Chemical Formula	Molecular Mass	R, ft-lbf/ lbm R	C_{po}, Btu/ lbm R	C_{vo}, Btu/ lbm R	k
Acetylene	C_2H_2	26.038	59.34	0.406	0.329	1.231
Air		28.97	53.34	0.240	0.171	1.400
Ammonia	NH_3	17.031	90.72	0.509	0.392	1.297
Argon	Ar	39.948	38.68	0.1253	0.0756	1.667
Butane	C_4H_{10}	58.124	26.58	0.415	0.381	1.091
Carbon dioxide	CO_2	44.01	35.10	0.203	0.158	1.289

TABLE A.10E (*Continued*) *Properties of Various Ideal Gases at 80 F (English Units)*

Gas	Chemical Formula	Molecular Mass	R, ft-lbf/ lbm R	C_{po}, Btu/ lbm R	C_{vo}, Btu/ lbm R	k
Carbon monoxide	CO	28.01	55.16	0.249	0.178	1.400
Ethane	C_2H_6	30.07	51.38	0.427	0.361	1.186
Ethanol	C_2H_5OH	46.069	33.54	0.341	0.298	1.145
Ethylene	C_2H_4	28.054	55.07	0.411	0.340	1.237
Helium	He	4.003	386.0	1.25	0.753	1.667
Hydrogen	H_2	2.016	766.4	3.43	2.44	1.409
Methane	CH_4	16.04	96.35	0.532	0.403	1.299
Methanol	CH_3OH	32.042	48.22	0.336	0.274	1.227
Neon	Ne	20.183	76.55	0.246	0.1477	1.667
Nitrogen	N_2	28.013	55.15	0.248	0.177	1.400
Nitrous oxide	N_2O	44.013	35.10	0.210	0.165	1.274
n-octane	C_8H_{18}	114.23	13.53	0.409	0.392	1.044
Oxygen	O_2	31.999	48.28	0.219	0.157	1.393
Propane	C_3H_8	44.097	35.04	0.407	0.362	1.126
Steam	H_2O	18.015	85.76	0.445	0.335	1.327
Sulfur dioxide	SO_2	64.059	24.12	0.149	0.118	1.263
Sulfur trioxide	SO_3	80.058	19.30	0.152	0.127	1.196

TABLE A.11E *Constant-Pressure Specific Heats of Various Ideal Gases (English Units)*

$$C_{p0} = \frac{Btu}{lb\ mole\ R} \qquad \theta = \frac{T(Rankine)}{180}$$

Gas		Range R	Max Error %
N_2	$\bar{C}_{po} = 9.3355 - 122.56\ \theta^{-1.5} + 256.38\ \theta^{-2} - 196.08\ \theta^{-3}$	540–6300	0.43
O_2	$\bar{C}_{po} = 8.9465 + 4.8044 \times 10^{-3}\ \theta^{1.5} - 42.679\ \theta^{-1.5} + 56.615\ \theta^{-2}$	540–6300	0.30
H_2	$\bar{C}_{po} = 13.505 - 167.96\ \theta^{-0.75} + 278.44\ \theta^{-1} - 134.01\ \theta^{-1.5}$	540–6300	0.60
CO	$\bar{C}_{po} = 16.526 - 0.16841\ \theta^{0.75} - 47.985\ \theta^{-0.5} + 42.246\ \theta^{-0.75}$	540–6300	0.42
OH	$\bar{C}_{po} = 19.490 - 14.185\ \theta^{0.25} + 4.1418\ \theta^{0.75} - 1.0196\ \theta$	540–6300	0.43
NO	$\bar{C}_{po} = 14.169 - 0.40861\ \theta^{0.5} - 16877\ \theta^{-0.5} + 17.889\ \theta^{-1.5}$	540–6300	0.34
H_2O	$\bar{C}_{po} = 34.190 - 43.868\ \theta^{0.25} + 19.778\ \theta^{0.5} - 0.88407\ \theta$	540–6300	0.43
CO_2	$\bar{C}_{po} = -0.89286 + 7.2967\ \theta^{0.5} - 0.98074\ \theta + 5.7835 \times 10^{-3}\ \theta^2$	540–6300	0.19
NO_2	$\bar{C}_{po} = 11.005 + 51.650\ \theta^{-0.5} - 86.916\ \theta^{-0.75} + 55.580\ \theta^{-2}$	540–6300	0.26
CH_4	$\bar{C}_{po} = -160.82 + 105.10\ \theta^{0.25} - 5.9452\ \theta^{0.75} + 77.408\ \theta^{-0.5}$	540–3600	0.15
C_2H_4	$\bar{C}_{po} = -22.800 + 29.433\ \theta^{0.5} - 8.5185\ \theta^{0.75} + 43.683\ \theta^{-3}$	540–3600	0.07
C_2H_6	$\bar{C}_{po} = 1.648 + 4.124\ \theta - 0.153\ \theta^2 + 1.74 \times 10^{-3}\ \theta^3$	540–2700	0.83
C_3H_8	$\bar{C}_{po} = -0.966 + 7.2790 - 0.3755\ \theta^2 + 7.58 \times 10^{-3}\ \theta^3$	540–2700	0.40
C_4H_{10}	$\bar{C}_{po} = 0.945 + 8.873\ \theta - 0.438\ \theta^2 + 8.36 \times 10^{-3}\ \theta^3$	540–2700	0.54

Source: From T.C. Scott and R.E. Sonntag. University of Michigan, unpublished 1971, except C_2H_6, C_3H_8, and C_4H_{10} from K.A. Kobe, Petroleum Refiner, 28, No. 2, 113 (1949).

TABLE A.12E *Ideal-Gas Properties of Air, English Units Standard Entropy at 1 atm = 101.325 kPa = 14.696 lbf/in.²*

T R	u Btu/lbm	h Btu/lbm	s^0 Btu/lbm R	P_r	v_r
400	68.212	95.634	1.56788	0.39046	379.523
440	75.047	105.212	1.59071	0.54470	299.264
480	81.887	114.794	1.61155	0.73825	240.877
520	88.733	124.383	1.63074	0.97670	197.244
536.67	91.589	128.381	1.63831	1.09071	182.288
540	92.160	129.180	1.63979	1.11458	179.491
560	95.589	133.980	1.64852	1.26592	163.885
600	102.457	143.590	1.66510	1.61217	137.880
640	109.340	153.216	1.68063	2.02204	117.260
680	116.242	162.860	1.69524	2.50257	100.666
720	123.167	172.528	1.70906	3.06119	87.1367
760	130.118	182.221	1.72216	3.70585	75.9775
800	137.099	191.944	1.73463	4.44496	66.6778
840	144.114	201.701	1.74653	5.28751	58.8555
880	151.165	211.494	1.75791	6.24303	52.2211
920	158.255	221.327	1.76884	7.32166	46.5519
960	165.388	231.202	1.77935	8.53415	41.6744
1000	172.564	241.121	1.78947	9.89193	37.4523
1040	179.787	251.086	1.79924	11.40706	33.7768
1080	187.058	261.099	1.80868	13.09232	30.5609
1120	194.378	271.161	1.81783	14.96119	27.7339
1160	201.748	281.273	1.82670	17.02788	25.2381
1200	209.168	291.436	1.83532	19.30735	23.0259
1240	216.640	301.650	1.84369	21.81531	21.0581
1280	224.163	311.915	1.85184	24.56826	19.3017
1320	231.737	322.231	1.85977	27.58348	17.7290
1360	239.362	332.598	1.86751	30.87907	16.3168
1400	247.037	343.016	1.87506	34.47392	15.0451
1440	254.762	353.483	1.88243	38.38777	13.8972
1480	262.537	364.000	1.88964	42.64121	12.8585
1520	270.359	374.565	1.89668	47.25567	11.9165
1560	278.230	385.177	1.90357	52.25344	11.0603
1600	286.146	395.837	1.91032	57.65771	10.2807
1650	296.106	409.224	1.91856	65.02144	9.40126
1700	306.136	422.681	1.92659	73.10700	8.61487
1750	316.232	436.205	1.93444	81.96560	7.90980
1800	326.393	449.794	1.94209	91.65077	7.27604
1850	336.616	463.445	1.94957	102.2183	6.70505

TABLE A.12E (Continued) *Ideal-Gas Properties of Air, English Units Standard Entropy at*
1 atm = 101.325 kPa = 14.696 lbf/in.²

T R	u Btu/lbm	h Btu/lbm	s^0 Btu/lbm R	P_r	v_r
1900	346.901	477.158	1.95689	113.7264	6.18944
1950	357.243	490.928	1.96404	126.2356	5.72284
2000	367.642	504.755	1.97104	139.8090	5.29973
2050	378.096	518.636	1.97790	154.5119	4.91531
2100	388.602	532.570	1.98461	170.4125	4.56538
2150	399.158	546.554	1.99119	187.5812	4.24627
2200	409.764	560.588	1.99765	206.0915	3.95477
2300	431.114	588.793	2.01018	247.4432	3.44359
2400	452.640	617.175	2.02226	295.1096	3.01292
2500	474.330	645.721	2.03391	349.7802	2.64791
2600	496.175	674.421	2.04517	412.1964	2.33683
2700	518.165	703.267	2.05606	483.1554	2.07031
2800	540.286	732.244	2.06659	563.4304	1.84110
2900	562.532	761.345	2.07681	653.9284	1.64296
3000	584.895	790.564	2.08671	755.5802	1.47096
3100	607.369	819.894	2.09633	869.3694	1.32104
3200	629.948	849.328	2.10567	996.3336	1.18988
3300	652.625	878.861	2.11476	1137.566	1.07472
3400	675.396	908.488	2.12361	1294.214	0.97327
3500	698.257	938.204	2.13222	1467.483	0.88360
3600	721.203	968.005	2.14062	1658.635	0.80410
3700	744.230	997.888	2.14880	1869.020	0.73341
3800	767.334	1027.848	2.15679	2100.030	0.67037
3900	790.513	1057.882	2.16459	2353.126	0.61401
4000	813.763	1087.988	2.17221	2629.834	0.56350
4100	837.081	1118.162	2.17967	2931.747	0.51810
4200	860.466	1148.402	2.18695	3260.527	0.47722
4300	883.913	1178.705	2.19408	3617.908	0.44032
4400	907.422	1209.069	2.20106	4005.693	0.40694
4500	930.989	1239.492	2.20790	4425.759	0.37669
4600	954.613	1269.972	2.21460	4880.058	0.34921
4700	978.292	1300.506	2.22117	5370.617	0.32421
4800	1002.023	1331.093	2.22761	5899.541	0.30143
4900	1025.806	1361.732	2.23392	6469.012	0.28062
5000	1049.638	1392.419	2.24012	7081.293	0.26159
5100	1073.518	1423.155	2.24621	7738.728	0.24415
5200	1097.444	1453.936	2.25219	8443.744	0.22815
5300	1121.414	1484.762	2.25806	9198.851	0.21345
5400	1145.428	1515.632	2.26383	10006.645	0.19992

TABLE A.13E *Ideal-Gas Properties of Various Substances (English Units), Entropies at 1 atm Pressure*

T R	Nitrogen, Diatomic (N_2) $\bar{h}^o_{f, 537} = 0$ Btu/lb mol $M = 28.013$		Nitrogen, Monatomic (N) $\bar{h}^o_{f, 537} = 203\ 216$ Btu/lb mol $M = 14.007$	
	$\bar{h}^o - \bar{h}^o_{f,537}$ Btu/lb mol	$\bar{s}^o$ Btu/lbmol/R	$\bar{h}^o - \bar{h}^o_{f,537}$ Btu/lb mol	$\bar{s}^o$ Btu/lbmol/R
0	−3727	0	−2664	0
200	−2341	38.877	−1671	31.689
400	−950	43.695	−679	35.130
537	0	45.739	0	36.589
600	441	46.515	314	37.143
800	1837	48.524	1307	38.571
1000	3251	50.100	2300	39.679
1200	4693	51.414	3293	40.584
1400	6169	52.552	4286	41.349
1600	7681	53.561	5279	42.012
1800	9227	54.472	6272	42.597
2000	10804	55.302	7265	43.120
2200	12407	56.066	8258	43.593
2400	14034	56.774	9251	44.025
2600	15681	57.433	10244	44.423
2800	17345	58.049	11237	44.791
3000	19025	58.629	12230	45.133
3200	20717	59.175	13223	45.454
3400	22421	59.691	14216	45.755
3600	24135	60.181	15209	46.038
3800	25857	60.647	16202	46.307
4000	27587	61.090	17195	46.562
4200	29324	61.514	18189	46.804
4400	31068	61.920	19183	47.035
4600	32817	62.308	20178	47.256
4800	34571	62.682	21174	47.468
5000	36330	63.041	22171	47.672
5500	40745	63.882	24670	48.148
6000	45182	64.654	27186	48.586
6500	49638	65.368	29724	48.992
7000	54109	66.030	32294	49.373
7500	58595	66.649	34903	49.733
8000	63093	67.230	37559	50.076
8500	67603	67.777	40270	50.405
9000	72125	68.294	43040	50.721
9500	76658	68.784	45875	51.028
10000	81203	69.250	48777	51.325

TABLE A.13E (Continued) *Ideal-Gas Properties of Various Substances (English Units), Entropies at 1 atm Pressure*

	Oxygen, Diatomic (O_2) $\overline{h}^\circ_{f, 537} = 0$ Btu/lb mol $M = 31.999$		Oxygen, Monatomic (O) $\overline{h}^\circ_{f, 537} = 107\ 124$ Btu/lb mol $M = 16.00$	
T R	$\overline{h}^\circ - \overline{h}^\circ_{f,537}$ Btu/lb mol	$\overline{s}^\circ$ Btu/lbmol/R	$\overline{h}^\circ - \overline{h}^\circ_{f,537}$ Btu/lb mol	$\overline{s}^\circ$ Btu/lbmol/R
0	−3733	0	−2891	0
200	−2345	42.100	−1829	33.041
400	−955	46.920	−724	36.884
537	0	48.973	0	38.442
600	446	49.758	330	39.023
800	1881	51.819	1358	40.503
1000	3366	53.475	2374	41.636
1200	4903	54.876	3383	42.556
1400	6487	56.096	4387	43.330
1600	8108	57.179	5389	43.999
1800	9761	58.152	6389	44.588
2000	11438	59.035	7387	45.114
2200	13136	59.844	8385	45.589
2400	14852	60.591	9381	46.023
2600	16584	61.284	10378	46.422
2800	18329	61.930	11373	46.791
3000	20088	62.537	12369	47.134
3200	21860	63.109	13364	47.455
3400	23644	63.650	14359	47.757
3600	25441	64.163	15354	48.041
3800	27250	64.652	16349	48.310
4000	29071	65.119	17344	48.565
4200	30904	65.566	18339	48.808
4400	32748	65.995	19334	49.039
4600	34605	66.408	20330	49.261
4800	36472	66.805	21327	49.473
5000	38350	67.189	22325	49.677
5500	43091	68.092	24823	50.153
6000	47894	68.928	27329	50.589
6500	52751	69.705	29847	50.992
7000	57657	70.433	32378	51.367
7500	62608	71.116	34924	51.718
8000	67600	71.760	37485	52.049
8500	72633	72.370	40063	52.362
9000	77708	72.950	42658	52.658
9500	82828	73.504	45270	52.941
10000	87997	74.034	47897	53.210

TABLE A.13E (Continued) *Ideal-Gas Properties of Various Substances (English Units), Entropies at 1 atm Pressure*

	Carbon Dioxide (CO_2) $\bar{h}^o_{f,537} = -169\,184$ Btu/lb mol $M = 44.01$		Carbon Monoxide (CO) $\bar{h}^o_{f,537} = -47\,518$ Btu/lb mol $M = 28.01$	
T R	$\bar{h}^o - \bar{h}^o_{f,537}$ Btu/lb mol	$\bar{s}^o$ Btu/lbmol/R	$\bar{h}^o - \bar{h}^o_{f,537}$ Btu/lb mol	$\bar{s}^o$ Btu/lbmol/R
0	−4026	0	−3728	0
200	−2636	43.466	−2343	40.319
400	−1153	48.565	−951	45.137
537	0	51.038	0	47.182
600	573	52.047	441	47.959
800	2525	54.848	1842	49.974
1000	4655	57.222	3266	51.562
1200	6927	59.291	4723	52.891
1400	9315	61.131	6220	54.044
1600	11798	62.788	7754	55.068
1800	14358	64.295	9323	55.992
2000	16982	65.677	10923	56.835
2200	19659	66.952	12549	57.609
2400	22380	68.136	14197	58.326
2600	25138	69.239	15864	58.993
2800	27926	70.273	17547	59.616
3000	30741	71.244	19243	60.201
3200	33579	72.160	20951	60.752
3400	36437	73.026	22669	61.273
3600	39312	73.847	24395	61.767
3800	42202	74.629	26128	62.236
4000	45105	75.373	27869	62.683
4200	48021	76.084	29614	63.108
4400	50948	76.765	31366	63.515
4600	53885	77.418	33122	63.905
4800	56830	78.045	34883	64.280
5000	59784	78.648	36650	64.641
5500	55739	68.649	39393	61.477
6000	74660	81.360	45548	66.263
6500	82155	82.560	50023	66.979
7000	89682	83.675	54514	67.645
7500	97239	84.718	59020	68.267
8000	104823	85.697	63539	68.850
8500	112434	86.620	68069	69.399
9000	120071	87.493	72610	69.918
9500	127734	88.321	77161	70.410
10000	135426	89.110	81721	70.878

TABLE A.13E (Continued) *Ideal-Gas Properties of Various Substances (English Units), Entropies at 1 atm Pressure*

T R	Water (H_2O) $\bar{h}^o_{f,537} = -103\,966$ Btu/lb mol $M = 18.015$		Hydroxyl (OH) $\bar{h}^o_{f,537} = 16\,761$ Btu/lb mol $M = 17.007$	
	$\bar{h}^o - \bar{h}^o_{f,537}$ Btu/lb mol	$\bar{s}^o$ Btu/lbmol/R	$\bar{h}^o - \bar{h}^o_{f,537}$ Btu/lb mol	$\bar{s}^o$ Btu/lbmol/R
0	−4258	0	−3943	0
200	−2686	37.209	−2484	36.521
400	−1092	42.728	−986	41.729
537	0	45.076	0	43.852
600	509	45.973	452	44.649
800	2142	48.320	1870	46.689
1000	3824	50.197	3280	48.263
1200	5566	51.784	4692	49.549
1400	7371	53.174	6112	50.643
1600	9241	54.422	7547	51.601
1800	11178	55.563	9001	52.457
2000	13183	56.619	10477	53.235
2200	15254	57.605	11978	53.950
2400	17388	58.533	13504	54.614
2600	19582	59.411	15054	55.235
2800	21832	60.245	16627	55.817
3000	24132	61.038	18220	56.367
3200	26479	61.796	19834	56.887
3400	28867	62.520	21466	57.382
3600	31293	63.213	23114	57.853
3800	33756	63.878	24777	58.303
4000	36251	64.518	26455	58.733
4200	38774	65.134	28145	59.145
4400	41325	65.727	29849	59.542
4600	43899	66.299	31563	59.922
4800	46496	66.852	33287	60.289
5000	49114	67.386	35021	60.643
5500	55739	68.649	39393	61.477
6000	62463	69.819	43812	62.246
6500	69270	70.908	48272	62.959
7000	76146	71.927	52767	63.626
7500	83081	72.884	57294	64.250
8000	90069	73.786	61851	64.838
8500	97101	74.639	66434	65.394
9000	104176	75.448	71043	65.921
9500	111289	76.217	75677	66.422
10000	118440	76.950	80335	66.900

TABLE A.13E (Continued) *Ideal-Gas Properties of Various Substances (English Units), Entropies at 1 atm Pressure*

T R	Hydrogen (H_2) $\bar{h}^o_{f,\,537} = 0$ Btu/lb mol $M = 2.016$		Hydrogen, Monatomic (H) $\bar{h}^o_{f,\,537} = 93\ 723$ Btu/lb mol $M = 1.008$	
	$\bar{h}^o - \bar{h}^o_{f,537}$ Btu/lb mol	$\bar{s}^o$ Btu/lbmol/R	$\bar{h}^o - \bar{h}^o_{f,537}$ Btu/lb mol	$\bar{s}^o$ Btu/lbmol/R
0	−3640	0	−2664	0
200	−2224	24.703	−1672	22.473
400	−927	29.193	−679	25.914
537	0	31.186	0	27.373
600	438	31.957	314	27.927
800	1831	33.960	1307	29.355
1000	3225	35.519	2300	30.463
1200	4622	36.797	3293	31.368
1400	6029	37.883	4286	32.134
1600	7448	38.831	5279	32.797
1800	8884	39.676	6272	33.381
2000	10337	40.441	7265	33.905
2200	11812	41.143	8258	34.378
2400	13309	41.794	9251	34.810
2600	14829	42.401	10244	35.207
2800	16372	42.973	11237	35.575
3000	17938	43.512	12230	35.917
3200	19525	44.024	13223	36.238
3400	21133	44.512	14215	36.539
3600	22761	44.977	15208	36.823
3800	24407	45.422	16201	37.091
4000	26071	45.849	17194	37.346
4200	27752	46.260	18187	37.588
4400	29449	46.655	19180	37.819
4600	31161	47.035	20173	38.040
4800	32887	47.403	21166	38.251
5000	34627	47.758	22159	38.454
5500	39032	48.598	24641	38.927
6000	43513	49.378	27124	39.359
6500	48062	50.105	29606	39.756
7000	52678	50.789	32088	40.124
7500	57356	51.434	34571	40.467
8000	62094	52.045	37053	40.787
8500	66889	52.627	39535	41.088
9000	71738	53.182	42018	41.372
9500	76638	53.712	44500	41.640
10000	81581	54.220	46982	41.895

TABLE A.13E (Continued) *Ideal-Gas Properties of Various Substances (English Units), Entropies at 1 atm Pressure*

	Nitric Oxide (NO) $\overline{h}^{\circ}_{f,\,537} = 38\ 818$ Btu/lb mol $M = 30.006$		Nitrogen Dioxide (NO$_2$) $\overline{h}^{\circ}_{f,\,537} = 14\ 230$ Btu/lb mol $M = 46.005$	
T R	$\overline{h}^{o}-\overline{h}^{o}_{f,537}$ **Btu/lb mol**	$\overline{s}^{\,o}$ **Btu/lbmol/R**	$\overline{h}^{o}-\overline{h}^{o}_{f,537}$ **Btu/lb mol**	$\overline{s}^{\,o}$ **Btu/lbmol/R**
0	−3952	0	−4379	0
200	−2224	24.703	−1672	22.473
400	−927	29.193	−679	25.914
537	0	31.186	0	27.373
600	438	31.957	314	27.927
800	1831	33.960	1307	29.355
1000	3225	35.519	2300	30.463
1200	4622	36.797	3293	31.368
1400	6029	37.883	4286	32.134
1600	7448	38.831	5279	32.797
1800	8884	39.676	6272	33.381
2000	10337	40.441	7265	33.905
2200	11812	41.143	8258	34.378
2400	13309	41.794	9251	34.810
2600	14829	42.401	10244	35.207
2800	16372	42.973	11237	35.575
3000	17938	43.512	12230	35.917
3200	19525	44.024	13223	36.238
3400	21133	44.512	14215	36.539
3600	22761	44.977	15208	36.823
3800	24407	45.422	16201	37.091
4000	26071	45.849	17194	37.346
4200	27752	46.260	18187	37.588
4400	29449	46.655	19180	37.819
4600	31161	47.035	20173	38.040
4800	32887	47.403	21166	38.251
5000	34627	47.758	22159	38.454
5500	41726	68.965	63395	84.990
6000	43513	49.378	27124	39.359
6500	48062	50.105	29606	39.756
7000	52678	50.789	32088	40.124
7500	57356	51.434	34571	40.467
8000	62094	52.045	37053	40.787
8500	66889	52.627	39535	41.088
9000	71738	53.182	42018	41.372
9500	76638	53.712	44500	41.640
10000	81581	54.220	46982	41.895

TABLE A.16E *Enthalpy of Formation, Gibbs Function of Formation, and Absolute Entropy of Various Substances at 77 F, 1 atm Pressure*

Substance	Formula	M	State	$\bar{h}_f^o$ Btu/lbmol	$\bar{g}_f^o$ Btu/lbmol	$\bar{s}_f^o$ Btu/lbmol R
Water	H_2O	18.015	gas	−103 966	−98 279	45.076
Water	H_2O	18.015	liq	−122 885	−101 973	16.707
Hydrogen peroxide	H_2O_2	34.015	gas	−58 515	−45 347	55.623
Ozone	O_3	47.998	gas	+61 339	+70 150	57.042
Carbon (graphite)	C	12.011	solid	0	0	1.371
Carbon monoxide	CO	28.011	gas	−47 518	−58 962	47.182
Carbon dioxide	CO_2	44.010	gas	−169 184	−169 556	51.038
Methane	CH_4	16.043	gas	−32 190	−21 841	44.459
Acetylene	C_2H_2	26.038	gas	+97 477	+91 412	47.972
Ethene	C_2H_4	28.054	gas	+22.557	+29 402	52.360
Ethane	C_2H_6	30.070	gas	−36 432	−14 167	54.812
Propene	C_3H_6	42.081	gas	+8 783	+26 981	63.761
Propane	C_3H_8	44.094	gas	−44 669	−10 099	64.442
Butane	C_4H_{10}	58.124	gas	−54 256	−6 922	73.215
Pentane	C_5H_{12}	72.151	gas	−62 984	−3 600	83.318
Benzene	C_6H_6	78.114	gas	+35 675	+55 760	64.358
Hexane	C_6H_{14}	86.178	gas	−71 926	−73	92.641
Heptane	C_7H_{16}	100.205	gas	−80 782	+3 439	102.153
n-Octane	C_8H_{18}	114.232	gas	−89 682	+7 049	111.399
n-Octane	C_8H_{18}	114.232	liq	−107 526	+2 770	86.122
Methanol	CH_3OH	32.042	gas	−86 543	−69 905	57.227
Ethanol	C_2H_5OH	46.069	gas	−101 032	−72 399	67.434
Ammonia	NH_3	17.031	gas	−19 656	−6 948	45.969
T-T-Diesel	$C_{14.4}H_{24.9}$	198.06	liq	−74 807	+76 748	125.609
Sulfur	S	32.06	solid	0	0	7.656
Sulfur dioxide	SO_2	64.059	gas	−127 619	−129 030	59.258
Sulfur trioxide	SO_3	80.058	gas	−170 148	−159 515	61.302
Nitrogen oxide	N_2O	44.013	gas	+35 275	+44 782	52.510
Nitromethane	CH_3NO_2	61.04	liq	−48 624	−6 249	41.034

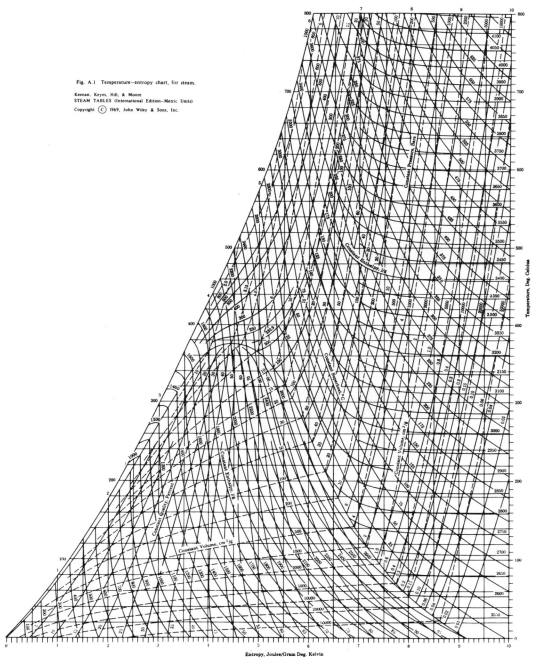

Fig. A.1 Temperature–entropy chart, for steam.

Keenan, Keyes, Hill, & Moore
STEAM TABLES (International Edition–Metric Units)
Copyright © 1969, John Wiley & Sons, Inc.

FIGURE A.1

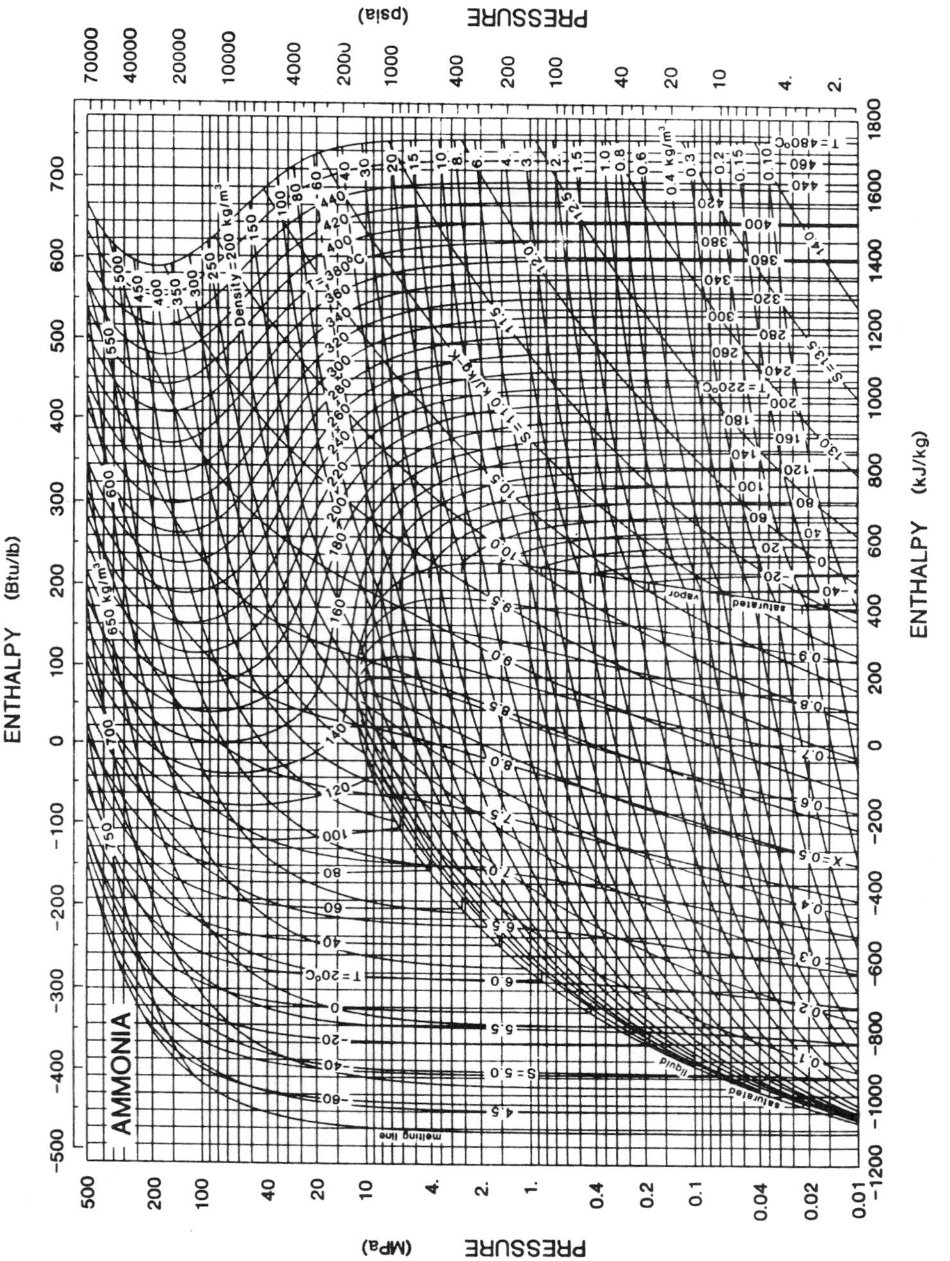

FIGURE A.2

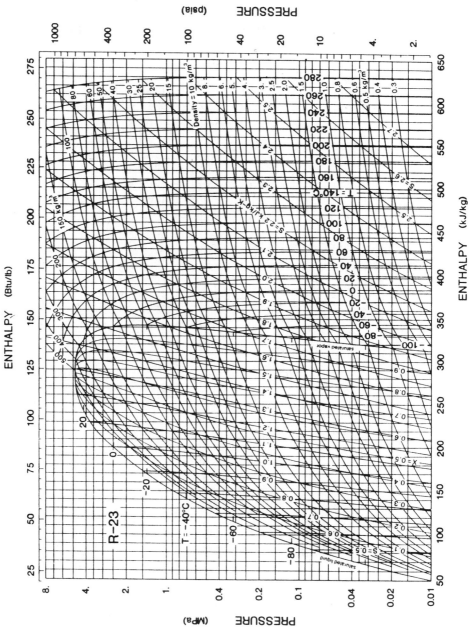

FIGURE A.3

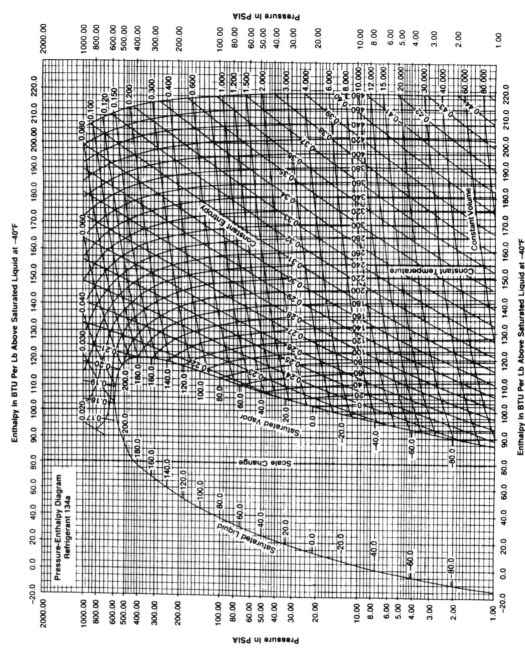

Pressure-Enthalpy Diagram
Refrigerant 134a

FIGURE A.4

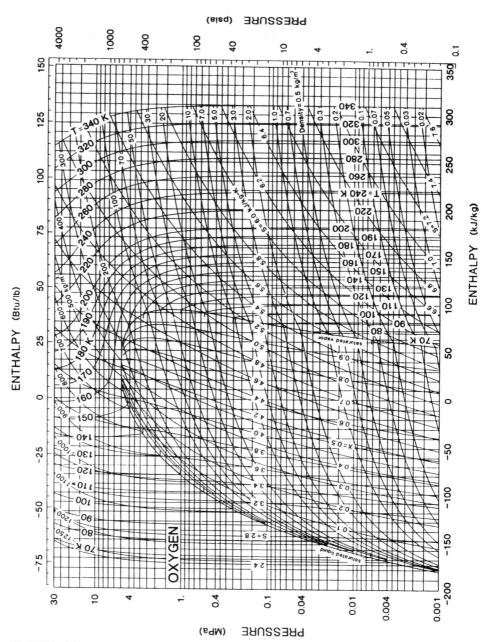

FIGURE A.5

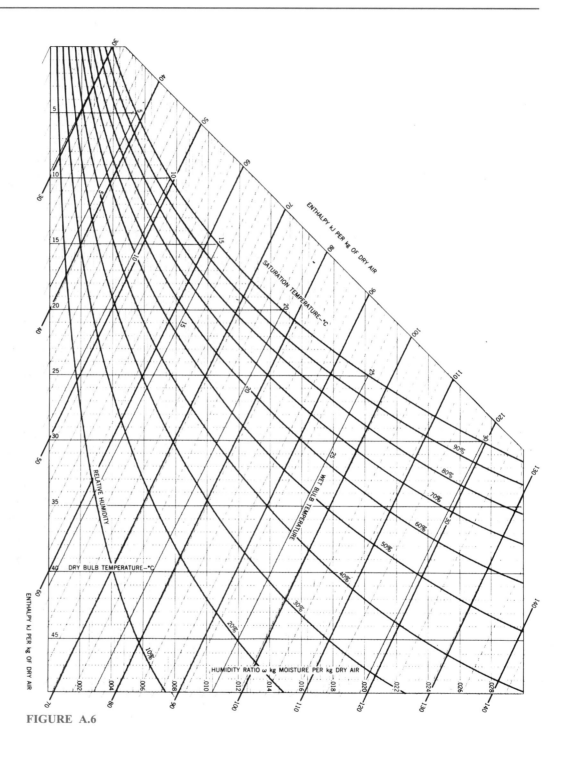

FIGURE A.6

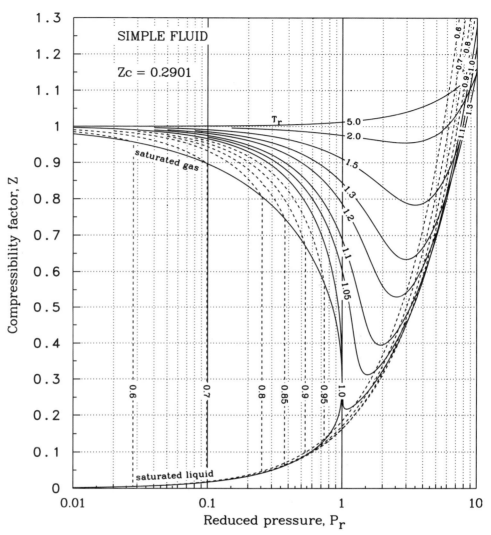

FIGURE A.7

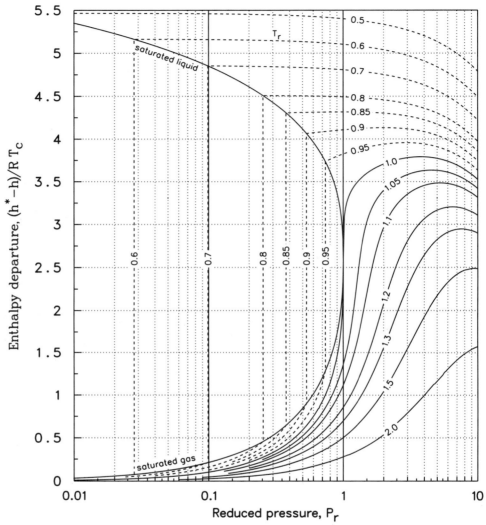

FIGURE A.8

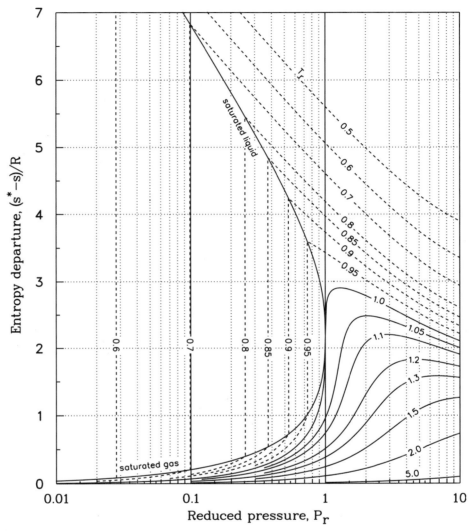

FIGURE A.9

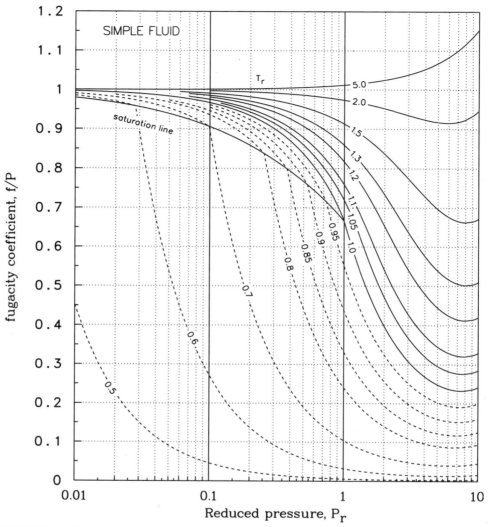

FIGURE A.10

Analysis and engineering judgment of thermodynamic systems often implies a parametric study of the behavior of the system under different conditions. In certain situations it is necessary to follow a transient process from the start to the end. Such tasks are very time consuming if done manually by hand and table look-up. To alleviate this problem, all the properties listed in Appendices A and C are included in the software supplied with this book. Additional refrigerants, mixtures of ideal gases, and the psychrometric chart are added as well as the calculations of chemical equilibrium reactions. The software treats the properties in SI and English units on a mass or mole basis and thus gives a more complete set of tables than printed in Appendix A. It is not a simplified version, but the same software that generated the tables as printed, so the only difference is roundoff in the printed tables.

The software comes in two versions. One is a menu-driven package that works like the tables, and the other is a collection of FORTRAN subroutines that can be called by the users program. Examples and other support programs are supplied as FORTRAN source code to encourage and help the user in applications and the computer homework assignments.

B.1 SUBSTANCES

Substances for which a full set of properties are covered as listed in Appendix A are

Water	Ammonia	Refrigerant-12
Refrigerant-22	Refrigerant-134a	Nitrogen
Methane		

In addition to these substances several refrigerants are available with the total set of choices as

Refrigerant-11	Refrigerant-12	Refrigerant-13
Refrigerant-14	Refrigerant-21	Refrigerant-22
Refrigerant-23	Refrigerant-113	Refrigerant-114
Refrigerant-134a	Refrigerant-500	Refrigerant-502
Refrigerant-C318		

For these substances, the regions covered are the saturated liquid and saturated vapor states and any two-phase liquid–vapor mixture. The superheated vapor region is covered up to a maximum temperature and pressure dependent upon the substance. Compressed liquid is covered for water, ammonia, refrigerant-134a, nitrogen, and methane. Only the water table includes the saturated solid phase and, therefore, also the two-phase saturated solid and saturated vapor mixture. The water tables do not cover the compressed solid nor the solid–liquid interphase (the fusion line).

Common for all the programs are the selection of units and job task to be performed. The following combinations of properties can be used to define a state of the substance

JOB	Input	JOB	Input
1	T and P	5	P and v
2	T and v	6	P and h
3	T and s	7	P and s
4	T and x	8	P and x

If a state is defined by any other combination of properties use the combination that contains one of the properties and iterate on the other. As an example, let a state be given by a pair of properties as (v, s), then use JOB as 2 and guess on T until the state has the desired s.

The units used for the programs can be selected as one of four combinations.

IUNIT	T	P	v or $1/\bar{v}$	u	h	s
1	°C	MPa	m³/kg	kJ/kg	kJ/kg	kJ/kg K
2	K	MPa	kmol/m³	kJ/kmol	kJ/kmol	kJ/kmol K
3	F	psia	ft³/lbm	Btu/lbm	Btu/lbm	Btu/lbm R
4	R	psia	lbmol/ft³	Btu/lbmol	Btu/lbmol	Btu/lbmol R

which gives the most common combinations. In options 2 and 4, the inverse of the specific volume on a mole basis is used similar to density.

B.2 IDEAL GASES

Tables for the ideal gases are done so they follow the tables in Appendix A. Air is an ideal mixture of the ideal gases nitrogen, oxygen and argon as

$$\text{Air} = 0.7803\ N_2 + 0.2099\ O_2 + 0.0098\ Ar$$

with the factors as volume fractions that equals mole fractions. Since the ideal gas properties listed are a function of temperature only, the state is found from one of the table entries as one of the variables

$$\text{Air Table Entry:}\quad T\ u\ h\ P_r\ v_r\ s°$$

Here $s°$ is the standard entropy at 100 kPa for SI units and at 1 atm for the English units. The other ideal gases are as listed in Table A.13.

ID	Gas	ID	Gas	ID	Gas
1	N_2	5	CO_2	9	H_2
2	N	6	CO	10	H
3	O_2	7	H_2O	11	NO
4	O	8	OH	12	NO_2

where the state is found from one of the table entries as $(T, \bar{h}, \bar{s}°)$.

The psychrometric chart is a special application of a binary mixture of water vapor and air. This program uses the air properties as explained above together with the properties of water vapor and saturated liquid water from the steam tables. A set of possible entries are listed below where three pieces of information are necessary (two for the thermodynamic state and one for the mixture composition). The default total pressure is 100 kPa for SI units and 14.696 psia for the English units with an option for changing it.

JOB	Input	JOB	Input
1	T and ω	5	$\tilde{h}$ and ω
2	T and Φ	6	$\tilde{h}$ and Φ
3	T and T_{wet}	7	ω and Φ
4	T and $\tilde{h}$	8	T and T_{dew}

The wet bulb temperature, T_{wet}, is calculated as the adiabatic saturation temperature and the mixture enthalpy is

$$\tilde{h} = h_{\text{air}} + \omega h_v$$

with the same offset as used in the psychrometric chart. This offset sets the enthalpy for air to zero at −20°C and is a constant offset as $h_a = h_a − 253.454$ kJ/kg. As a state may be specified in the input to contain more water than the saturated state, the properties for the saturated state and the excess water ($\omega − \omega_{sat}$) are returned.

B.3 GENERALIZED CHART

The generalized chart program calculates the simple fluid and the reference fluid corrections as listed in Tables A.15 and in Appendix C. They are combined to a single correction with the acentric factor. Since the two-phase region of the simple fluid and the reference fluid are at different pressures for the same temperature, the relation between T_{sat} and P_{sat} is a function of the acentric factor. This feature is built into the program, which makes it different from the tables.

B.4 ADDITIONAL TABLES

Additional tables are included so nearly all the tables in the book are available in the menu-driven software. These tables are listed as follows.

- Table A.8: Critical constants.
- Table A.9: Properties of various solids and liquids.
- Table A.10: Properties of various ideal gases.
- Table A.11: Constant pressure specific heat. The computer version includes the calculation of the heat capacity at any temperature and changes in enthalpy and standard entropy between two temperatures.
- Table A.14: The second and third virial coefficients including their derivatives.
- Table A.16: The enthalpy of formation, Gibbs function of formation and the absolute entropy for a number of substances.
- Table A.17: The equilibrium constants for eight reactions.
- Table A.18: The one-dimensional compressible flow relations.
- Table A.19: The one-dimensional normal shock relations.

B.5 CHEMICAL EQUILIBRIUM

Chemical equilibrium calculations are included in a separate program package. This allows up to four different feed lines into a reactor and determines the final composition and properties for the products of combustion. For each feed line the user selects the composition from a list of species together with the total pressure and the line temperature. The program lists the mole fractions, the mass fractions, the enthalpies, and entropies for each feed line. This part of the program is also useful for problems with mixtures of ideal gases.

The combustion process assumes all the fuel is burned and thus cannot handle rich mixtures (excess fuel) where the chemistry is very complex. Products of combustion is listed for no dissociation and seven dissociation reactions are then included. The user may toggle these equilibrium reactions in or out individually to investigate the effect on the final mixture. This feature makes the program capable of doing many of the multiple simultaneous reactions with ease. As the exit line temperature is set by the user with the enthalpies and entropies available computations of the flame temperature (adiabatic or not) can be done by trial and error. By the use of this program many combustion problems can be analyzed more realistically than feasible by hand calculations.

B.6 THE LIBRARY SUBROUTINES

The subroutines that can be called by a user's FORTRAN program are the same as those used for the menu-driven program. These subroutines are listed below together with the syntax for using them where the parameters JOB and IUNIT have the meaning as explained previously.

Water, The Steam Tables

SUBROUTINE STEAM(JOB,IUNIT,P,V,T,U,H,S,X,IPHASE)

IPHASE	MEANING
0	Inappropriate data input
1	Dense fluid ($T > T_c$)
2	Compressed liquid state
3	Saturated liquid state
4	Saturated liquid–vapor mixture
5	Saturated vapor
6	Superheated vapor
7	Compressed solid (not available)
8	Saturated solid
9	Saturated solid–vapor mixture

Refrigerants. The substances cover a number of refrigerants for which the saturated liquid–vapor region, the superheated vapor and the dense gas region is available. The refrigerant R-134a is listed with the next group of substances since that includes the compressed liquid region.

SUBROUTINE FREON(IDR,JOB,IUNIT,P,V,T,U,H,S,X,IPHASE)

IDR	1	2	3	4	5	6
SUBSTANCE	R-11	R-12	R-13	R-14	R-21	R-22

IDR	7	8	9	10	11	12
SUBSTANCE	R-23	R-113	R-114	R-500	R-502	R-C318

IPHASE	MEANING
0	Inappropriate data input
1	Dense fluid ($T > T_c$)
2	Compressed liquid state (not available)
3	Saturated liquid state
4	Saturated liquid–vapor mixture
5	Saturated vapor
6	Superheated vapor

ARNM The tables for ammonia, R-134a, nitrogen, and methane. The refrigerant R-134a is listed with this group as these tables all include the compressed liquid region.

SUBROUTINE ARNM(ID,JOB,IUNIT,P,V,T,U,H,S,X,IPHASE)

ID	2	5	6	7
SUBSTANCE	Ammonia	R-134a	Nitrogen	Methane

IPHASE	MEANING
0	Inappropriate data input
1	Dense fluid ($T > T_c$)
2	Compressed liquid state
3	Saturated liquid state
4	Saturated liquid–vapor mixture
5	Saturated vapor
6	Superheated vapor

FNAIR This air routine corresponds to the air tables in SI Units.

SUBROUTINE FNAIR(T,U,H,S,PR,VR)

I/O	Var.	Variable and units
Input	T	T Temperature in K
Output	U	u Specific internal energy in kJ/kg
Output	H	h Specific enthalpy in kJ/kg
Output	S	s° Standard entropy in kJ/kg K
Output	PR	P_r Relative pressure, dimensionless
Output	VR	v_r Relative specific volume, dimensionless

IDEAL The ideal gas tables in SI and english units.

SUBROUTINE SUBST(ID,T,H,S,ITUNIT)

I/O	Variable and units
Input	ID, gas identifier integer 1-12, see table in Section B.2
Input	ITUNIT Temperature unit indicator (1-4)
Input	T Temperature in °C, K, F, R according to ITUNIT
Output	h Specific enthalpy in kJ/kg or Btu/lbm
Output	s° Standard entropy in kJ/kg K or Btu/lbm R

ITUNIT	1	2	3	4
T in	°C	K	F	R
h, s in	SI	SI	Engl.	Engl.

LK The generalized charts according to Lee-Kesler. These are split into the single phase region and the two-phase saturated liquid-vapor region. The single phase version is

SUBROUTINE LEEK(TR,PR,W,IPHASE,Z,DH,DS,FP)

I/O	Variable and units
Input	TR, Reduced temperature
Input	PR, Reduced pressure
Input	W, Acentric factor
Output	IPHASE, phase indicator
Output	Z, Compressibility factor
Output	DH, Enthalpy departure, dimensionless
Output	DS, Entropy departure, dimensionless
Output	FP, Fugacity

The two-phase program includes the effect of the acentric factor on the saturation relation between pressure and temperature. For this reason it is done separately as the determination of a two-phase state otherwise must be done with a lengthy iteration procedure.

SUBROUTINE LEEKX(IOPT,TR,PR,W,X,IPHASE,Z,DH,DS,FP)

I/O	Variable and units
Input	IOPT, =1 Temperature given, =2 Pressure given
Input or Output	TR, Reduced temperature
Input or Output	PR, Reduced pressure
Input	W, Acentric factor
Output	Z, Compressibility factor
Output	DH, Enthalpy departure, dimensionless
Output	DS, Entropy departure, dimensionless
Output	FP, Fugacity
Input	X, Quality

The computations of the constants for the Lee–Kesler equation of state as shown with Table C.6 is included for convenience.

Additional comments about the software and its use is provided in a README.DOC file on the diskette. There are also a number of test and example programs in FORTRAN source code to aid in the use of the software package. Finally, we have included a number of utility routines that can be used in the computer assignments so these are more manageable to do.

The generalized compressibility behavior of real substances was discussed in Section 10.9, in which the real behavior was correlated with respect to two parameters, critical pressure and critical temperature. It was mentioned there that a third parameter is often used to improve the accuracy of correlation, one such parameter being the acentric factor ω, which adds a correction factor to the simple-fluid correlation represented by Fig. A.7 or Table A.15. It is observed that simple fluids have a reduced saturation pressure of about 0.1 at a reduced temperature of 0.7. Therefore, the acentric factor for any molecule is defined in terms of its reduced saturation pressure at the reduced temperature of 0.7 as

$$\omega = \frac{\ln P_{r \text{ at } T_r = 0.7}^{\text{sat}}}{2.302585} - 1.0 \tag{C.1}$$

We now introduce notation to identify all of the simple-fluid values listed in Table A.15 (Figs. A.7–A.10) with the superscript (0). The compressibility factor Z for any molecule can now be represented in terms of the simple-fluid value $Z^{(0)}$ plus a first-order correction term in the form

$$Z = Z^{(0)} + \omega Z^{(1)} \tag{C.2}$$

The simple-fluid value is calculated from the Lee–Kesler equation of state, Eq. 10.63, using the simple-fluid constants given in Table A.15. The second set of constants in that table is for a relatively complicated reference fluid, octane, which yields, again from Eq. 10.63, a compressibility factor $Z^{(r)}$. The correction factor $Z^{(1)}$ of Eq. C.2 is now assumed to be a linear weighted average value, such that

$$Z = Z^{(0)} + \frac{\omega}{\omega^{(r)}}[Z^{(r)} - Z^{(0)}] = Z^{(0)} + \omega Z^{(1)} \tag{C.3}$$

$$\omega^{(r)} = 0.3978, \qquad Z^{(1)} = (Z^{(r)} - Z^{(0)})/\omega^{(r)}$$

Values for the correction factor $Z^{(1)}$ as calculated using the generalized empirical equation of state, Eq. 10.63, are listed in Table C.1. No saturation table is included there, since the two-phase saturation requirement depends on the acentric factor, and a separate table would be required for each value of ω. This effect is included when using the software that accompanies this text, such that saturation corrections can be accommodated in that manner.

Generalized enthalpy departures corresponding to the Lee–Kesler equation of state were calculated for simple fluids following the procedure developed in Section 10.11. Those values, given in Table A.15 and Fig. A.8, should now carry a superscript (0), using the present notation. Now, following the same procedure as for Z and using the reference-fluid constants, we obtain values for the reference-fluid enthalpy departure from Eqs. 10.71 and 10.72. The overall enthapy departure at a given state is then the simple-fluid value plus a correction term,

$$\frac{h^* - h}{RT_c} = \left(\frac{h^* - h}{RT_c}\right)^{(0)} + \omega\left(\frac{h^* - h}{RT_c}\right)^{(1)} \tag{C.4}$$

The correction term (1) is taken as the linear weighted average of the (0) and (r) enthalpy values, as was done for the compressibility factor in Eq. C.3. These correction terms are listed in Table C.2, to give more accurate representations for enthalpy calculations.

The same procedure is followed in developing a correction term for entropy departures, using the reference-fluid constants in Eq. 10.63 to evaluate Eq. 10.75. The overall entropy departure at a given state is then the simple-fluid value from Table A.15, or Fig. A.9, plus a correction term,

$$\frac{s_P^* - s_P}{R} = \left(\frac{s_P^* - s_P}{R}\right)^{(0)} + \omega \left(\frac{s_P^* - s_P}{R}\right)^{(1)} \tag{C.5}$$

The correction term (1) is once again taken as the linear weighted average of the (0) and (r) entropy departure values, as was done for the compressibility factor and the enthalpy departure values. These correction terms are listed in Table C.3 to give more accurate representations for entropy departure values than do the simple-fluid correlations alone.

A correction term is calculated for the fugacity coefficient in the same manner as for the other properties, with the result being

$$\ln\frac{f}{P} = \ln\left(\frac{f}{P}\right)^{(0)} + \omega \ln\left(\frac{f}{P}\right)^{(1)} \tag{C.6}$$

The simple fluid term (0) is listed in Table A.15 and shown graphically in Fig. A.10. The correction term (1) is listed in Table C.4.

In using the generalized tables/charts to represent real gas mixture behavior, only one pseudocritical model was discussed in Section 11.8—Kay's rule model as given by Eq. 11.35. A model that is considerably more accurate, and one that also includes the acentric factor correction is the following Lee–Kesler model. For the pseudocritical specific volume and temperature,

$$\bar{v}_{c\,\text{mix}} = \sum_j \sum_k y_j y_k \bar{v}_{c_{jk}} \tag{C.7}$$

$$(\bar{v}_c T_c)_{\text{mix}} = \sum_j \sum_k y_i y_k \bar{v}_{c_{jk}} T_{c_{jk}} \tag{C.8}$$

In these expressions the interaction parameters are

$$\bar{v}_{c_{jk}} = \frac{1}{8}(\bar{v}_{c_j}^{1/3} + \bar{v}_{c_k}^{1/3})^3, \qquad T_{c_{jk}} = (T_{c_j} T_{c_k})^{1/2} \tag{C.9}$$

Once these values have been determined, the pseudocritical pressure is then found from the expressions

$$\omega_{\text{mix}} = \sum_j y_j \omega_j \tag{C.10}$$

$$Z_{c\,\text{mix}} = 0.2905 - 0.085\omega_{\text{mix}} \tag{C.11}$$

and

$$P_{c\,\text{mix}} = Z_{c\,\text{mix}} \bar{R} T_{c\,\text{mix}} / \bar{v}_{c\,\text{mix}} \tag{C.12}$$

These pseudocritical properties, used with the simple-fluid properties from Table A.15 and corrections from Tables C.1–C.4, have been found to give good correlations with properties of mixtures for a wide variety of gas mixtures and compositions.

TABLE C.1 *Generalized Three-Parameter (T_r, P_r, ω) Compressibility Factor Tables Compressibility Factor Correction, $Z^{(1)}$*

T_r\P_r	0.10	0.20	0.40	0.60	0.80	1.00	1.20	1.40	1.70	2.00	2.50	3.00	5.00	7.00	10.00
0.30	-.0081	-.0162	-.0323	-.0484	-.0645	-.0806	-.0967	-.1127	-.1368	-.1608	-.2008	-.2407	-.3996	-.5573	-.7916
0.40	-.0095	-.0190	-.0380	-.0569	-.0758	-.0946	-.1134	-.1321	-.1601	-.1879	-.2340	-.2799	-.4603	-.6365	-.8936
0.50	-.0090	-.0181	-.0360	-.0539	-.0716	-.0893	-.1069	-.1243	-.1504	-.1762	-.2189	-.2611	-.4253	-.5831	-.8099
0.60	-.0082	-.0164	-.0326	-.0487	-.0646	-.0803	-.0960	-.1115	-.1345	-.1572	-.1945	-.2312	-.3718	-.5047	-.6929
0.70	-.0075	-.0148	-.0294	-.0438	-.0579	-.0718	-.0855	-.0990	-.1189	-.1385	-.1703	-.2013	-.3184	-.4270	-.5785
0.75	-.0744	-.0143	-.0282	-.0417	-.0550	-.0681	-.0808	-.0934	-.1118	-.1298	-.1590	-.1872	-.2929	-.3901	-.5250
0.80	-.0487	-.1160	-.0272	-.0401	-.0526	-.0648	-.0767	-.0883	-.1052	-.1217	-.1481	-.1736	-.2682	-.3545	-.4740
0.85	-.0319	-.0715	-.0268	-.0391	-.0509	-.0622	-.0731	-.0837	-.0990	-.1138	-.1375	-.1602	-.2439	-.3201	-.4254
0.90	-.0205	-.0442	-.1118	-.0396	-.0503	-.0604	-.0701	-.0795	-.0929	-.1059	-.1265	-.1463	-.2195	-.2862	-.3788
0.95	-.0126	-.0262	-.0589	-.1110	-.0540	-.0607	-.0678	-.0751	-.0860	-.0967	-.1141	-.1310	-.1943	-.2526	-.3339
1.00	-.0069	-.0140	-.0285	-.0435	-.0588	-.0879	-.0609	-.0652	-.0735	-.0824	-.0972	-.1118	-.1672	-.2185	-.2902
1.05	-.0029	-.0054	-.0092	-.0097	-.0032	0.0220	0.1059	0.0951	-.0117	-.0432	-.0671	-.0838	-.1370	-.1835	-.2476
1.10	0.0001	0.0007	0.0038	0.0106	0.0236	0.0476	0.0897	0.1468	0.1418	0.0698	-.0033	-.0373	-.1021	-.1469	-.2056
1.15	0.0023	0.0052	0.0127	0.0237	0.0396	0.0625	0.0943	0.1345	0.1815	0.1667	0.0906	0.0332	-.0611	-.1084	-.1642
1.20	0.0040	0.0084	0.0190	0.0326	0.0499	0.0719	0.0991	0.1310	0.1780	0.1990	0.1651	0.1095	-.0141	-.0678	-.1231
1.30	0.0061	0.0125	0.0267	0.0429	0.0612	0.0819	0.1048	0.1294	0.1669	0.1991	0.2223	0.2079	0.0875	0.0176	-.0423
1.40	0.0072	0.0147	0.0306	0.0477	0.0661	0.0857	0.1063	0.1276	0.1596	0.1894	0.2259	0.2397	0.1737	0.1008	0.0350
1.50	0.0078	0.0158	0.0323	0.0497	0.0677	0.0864	0.1055	0.1248	0.1535	0.1806	0.2186	0.2433	0.2309	0.1717	0.1058
1.60	0.0080	0.0162	0.0330	0.0501	0.0677	0.0855	0.1035	0.1214	0.1478	0.1729	0.2098	0.2381	0.2631	0.2255	0.1673
1.80	0.0081	0.0162	0.0325	0.0488	0.0652	0.0816	0.0978	0.1137	0.1370	0.1593	0.1932	0.2224	0.2846	0.2871	0.2576
2.00	0.0078	0.0155	0.0310	0.0464	0.0617	0.0767	0.0916	0.1061	0.1273	0.1476	0.1789	0.2069	0.2820	0.3097	0.3096
2.50	0.0068	0.0135	0.0268	0.0399	0.0528	0.0654	0.0778	0.0899	0.1075	0.1245	0.1511	0.1757	0.2542	0.3052	0.3475
3.00	0.0059	0.0117	0.0232	0.0345	0.0456	0.0565	0.0672	0.0776	0.0929	0.1076	0.1310	0.1529	0.2268	0.2817	0.3385
3.50	0.0052	0.0103	0.0204	0.0303	0.0401	0.0497	0.0591	0.0683	0.0818	0.0949	0.1158	0.1356	0.2042	0.2584	0.3194
4.00	0.0046	0.0091	0.0182	0.0270	0.0357	0.0443	0.0527	0.0610	0.0731	0.0849	0.1038	0.1219	0.1857	0.2378	0.2994
5.00	0.0038	0.0075	0.0149	0.0222	0.0294	0.0365	0.0434	0.0503	0.0604	0.0703	0.0863	0.1016	0.1573	0.2047	0.2637

TABLE C.2 Generalized Three-Parameter (T_r, P_r, ω) Compressibility Factor Tables Enthalpy Departure Correction, $((h^*-h)/RT_c)^{(1)}$

T_r\P_r	0.10	0.20	0.40	0.60	0.80	1.00	1.20	1.40	1.70	2.00	2.50	3.00	5.00	7.00	10.00
0.30	11.098	11.095	11.088	11.081	11.074	11.067	11.061	11.054	11.044	11.034	11.017	11.001	10.936	10.873	10.782
0.40	10.120	10.120	10.120	10.120	10.120	10.120	10.120	10.120	10.120	10.121	10.121	10.122	10.127	10.135	10.150
0.50	8.869	8.871	8.875	8.879	8.883	8.887	8.891	8.895	8.902	8.908	8.919	8.931	8.979	9.030	9.111
0.60	7.570	7.573	7.579	7.585	7.592	7.598	7.605	7.611	7.622	7.632	7.650	7.668	7.744	7.825	7.950
0.70	6.356	6.360	6.366	6.373	6.381	6.388	6.396	6.404	6.417	6.430	6.452	6.475	6.573	6.677	6.837
0.75	0.306	5.796	5.803	5.809	5.816	5.824	5.832	5.841	5.854	5.868	5.892	5.918	6.027	6.141	6.317
0.80	0.234	0.542	5.266	5.271	5.277	5.285	5.292	5.301	5.315	5.330	5.356	5.384	5.506	5.632	5.824
0.85	0.182	0.401	4.753	4.754	4.758	4.763	4.771	4.779	4.794	4.810	4.840	4.871	5.008	5.149	5.358
0.90	0.144	0.308	0.751	4.254	4.248	4.249	4.255	4.263	4.279	4.298	4.333	4.371	4.530	4.688	4.916
0.95	0.115	0.241	0.542	0.994	3.737	3.713	3.713	3.723	3.746	3.773	3.822	3.873	4.068	4.248	4.498
1.00	0.093	0.191	0.410	0.675	1.034	2.471	2.952	3.033	3.119	3.186	3.279	3.358	3.615	3.825	4.100
1.05	0.075	0.153	0.318	0.498	0.691	0.877	0.878	1.113	2.027	2.381	2.645	2.800	3.167	3.418	3.722
1.10	0.061	0.123	0.251	0.381	0.507	0.617	0.673	0.631	0.780	1.261	1.853	2.167	2.720	3.023	3.362
1.15	0.050	0.099	0.199	0.296	0.385	0.459	0.503	0.502	0.468	0.604	1.083	1.497	2.275	2.641	3.019
1.20	0.040	0.080	0.158	0.232	0.297	0.349	0.381	0.388	0.358	0.361	0.591	0.934	1.840	2.273	2.692
1.30	0.026	0.052	0.100	0.142	0.177	0.203	0.218	0.221	0.205	0.178	0.182	0.300	1.066	1.592	2.086
1.40	0.016	0.032	0.060	0.083	0.100	0.111	0.115	0.112	0.096	0.070	0.034	0.044	0.504	1.012	1.547
1.50	0.009	0.018	0.032	0.042	0.048	0.049	0.046	0.038	0.018	-0.008	-0.052	-0.078	0.142	0.556	1.080
1.60	0.004	0.007	0.012	0.013	0.011	0.005	-0.004	-0.016	-0.038	-0.065	-0.113	-0.151	-0.082	0.217	0.689
1.80	-0.003	-0.006	-0.015	-0.025	-0.037	-0.051	-0.067	-0.084	-0.113	-0.143	-0.194	-0.241	-0.317	-0.203	0.112
2.00	-0.007	-0.015	-0.030	-0.047	-0.065	-0.085	-0.105	-0.125	-0.157	-0.190	-0.244	-0.295	-0.428	-0.424	-0.255
2.50	-0.012	-0.025	-0.049	-0.075	-0.100	-0.125	-0.150	-0.176	-0.213	-0.250	-0.310	-0.367	-0.552	-0.661	-0.704
3.00	-0.014	-0.029	-0.058	-0.086	-0.114	-0.142	-0.170	-0.198	-0.238	-0.278	-0.342	-0.403	-0.611	-0.763	-0.899
3.50	-0.016	-0.031	-0.062	-0.092	-0.122	-0.152	-0.181	-0.210	-0.253	-0.294	-0.361	-0.425	-0.650	-0.827	-1.015
4.00	-0.016	-0.032	-0.064	-0.096	-0.127	-0.158	-0.188	-0.218	-0.262	-0.306	-0.375	-0.442	-0.680	-0.874	-1.097
5.00	-0.017	-0.034	-0.068	-0.101	-0.133	-0.166	-0.198	-0.229	-0.276	-0.321	-0.395	-0.466	-0.726	-0.947	-1.219

TABLE C.3 Generalized Three-Parameter (T_r, P_r, ω) Compressibility Factor Tables Entropy Departure Correction, $((s_P^* - s_P)/R)$[1]

T_r\\P_r	0.10	0.20	0.40	0.60	0.80	1.00	1.20	1.40	1.70	2.00	2.50	3.00	5.00	7.00	10.00
0.30	16.773	16.754	16.714	16.675	16.637	16.598	16.559	16.520	16.463	16.405	16.310	16.214	15.838	15.469	14.927
0.40	13.980	13.970	13.951	13.932	13.914	13.895	13.876	13.858	13.830	13.803	13.757	13.713	13.540	13.376	13.144
0.50	11.196	11.191	11.181	11.171	11.161	11.151	11.142	11.132	11.118	11.105	11.083	11.063	10.986	10.920	10.836
0.60	8.827	8.823	8.817	8.811	8.806	8.800	8.795	8.790	8.784	8.777	8.768	8.759	8.735	8.723	8.720
0.70	6.954	6.951	6.946	6.941	6.937	6.933	6.930	6.928	6.924	6.922	6.919	6.919	6.928	6.952	7.002
0.75	0.340	6.173	6.167	6.162	6.158	6.155	6.152	6.150	6.148	6.146	6.147	6.149	6.174	6.213	6.285
0.80	0.246	0.578	5.474	5.467	5.462	5.458	5.455	5.453	5.452	5.452	5.455	5.461	5.501	5.555	5.648
0.85	0.183	0.408	4.853	4.841	4.832	4.826	4.822	4.820	4.820	4.822	4.828	4.839	4.898	4.969	5.083
0.90	0.140	0.301	0.744	4.269	4.250	4.238	4.232	4.230	4.231	4.236	4.250	4.267	4.351	4.442	4.578
0.95	0.109	0.228	0.517	0.961	3.697	3.658	3.647	3.646	3.655	3.669	3.697	3.728	3.851	3.966	4.125
1.00	0.086	0.177	0.382	0.632	0.977	2.399	2.868	2.940	3.012	3.067	3.140	3.200	3.387	3.532	3.717
1.05	0.069	0.140	0.292	0.460	0.642	0.820	0.831	1.073	1.951	2.283	2.522	2.655	2.949	3.134	3.348
1.10	0.055	0.112	0.229	0.350	0.470	0.577	0.640	0.620	0.786	1.241	1.786	2.067	2.534	2.767	3.013
1.15	0.045	0.091	0.183	0.275	0.361	0.437	0.489	0.506	0.507	0.654	1.100	1.471	2.138	2.428	2.708
1.20	0.037	0.075	0.149	0.220	0.286	0.343	0.385	0.408	0.414	0.447	0.680	0.991	1.767	2.115	2.430
1.30	0.026	0.052	0.102	0.148	0.190	0.226	0.254	0.275	0.291	0.300	0.351	0.481	1.147	1.569	1.944
1.40	0.019	0.037	0.072	0.104	0.133	0.158	0.178	0.194	0.210	0.220	0.240	0.290	0.730	1.138	1.544
1.50	0.014	0.027	0.053	0.076	0.097	0.115	0.130	0.142	0.156	0.166	0.181	0.206	0.479	0.823	1.222
1.60	0.011	0.021	0.040	0.057	0.073	0.086	0.098	0.108	0.119	0.129	0.142	0.159	0.334	0.604	0.969
1.80	0.006	0.013	0.024	0.035	0.044	0.053	0.060	0.067	0.075	0.083	0.094	0.105	0.195	0.355	0.628
2.00	0.004	0.008	0.016	0.023	0.029	0.035	0.040	0.045	0.052	0.058	0.067	0.077	0.136	0.238	0.434
2.50	0.002	0.004	0.007	0.010	0.014	0.017	0.020	0.022	0.026	0.031	0.037	0.044	0.080	0.130	0.230
3.00	0.001	0.002	0.004	0.006	0.008	0.010	0.012	0.014	0.017	0.020	0.026	0.031	0.058	0.093	0.158
3.50	0.001	0.001	0.003	0.004	0.006	0.007	0.009	0.010	0.013	0.015	0.019	0.024	0.046	0.073	0.122
4.00	0.001	0.001	0.002	0.003	0.005	0.006	0.007	0.008	0.010	0.012	0.016	0.020	0.038	0.060	0.100
5.00	0.000	0.001	0.001	0.002	0.003	0.004	0.005	0.006	0.007	0.009	0.011	0.014	0.028	0.044	0.073

TABLE C.4 Generalized Three-Parameter (T_r, P_r, ω) Compressibility Factor Tables Fugacity Coefficient Correction, $\ln (f/P)^{(1)}$

$T_r \backslash P_r$	0.10	0.20	0.40	0.60	0.80	1.00	1.20	1.40	1.70	2.00	2.50	3.00	5.00	7.00	10.00
0.30	−20.221	−20.229	−20.245	−20.261	−20.278	−20.294	−20.310	−20.326	−20.350	−20.374	−20.414	−20.455	−20.615	−20.774	−21.012
0.40	−11.319	−11.329	−11.348	−11.367	−11.386	−11.404	−11.423	−11.442	−11.471	−11.499	−11.546	−11.592	−11.778	−11.961	−12.231
0.50	−6.542	−6.551	−6.569	−6.587	−6.605	−6.623	−6.641	−6.658	−6.685	−6.711	−6.755	−6.799	−6.971	−7.139	−7.386
0.60	−3.790	−3.798	−3.815	−3.831	−3.847	−3.863	−3.879	−3.895	−3.919	−3.943	−3.982	−4.020	−4.172	−4.318	−4.530
0.70	−2.127	−2.134	−2.149	−2.164	−2.178	−2.193	−2.207	−2.221	−2.242	−2.263	−2.298	−2.331	−2.462	−2.587	−2.765
0.75	−0.068	−1.555	−1.569	−1.583	−1.597	−1.611	−1.624	−1.638	−1.658	−1.677	−1.709	−1.741	−1.862	−1.976	−2.138
0.80	−0.046	−0.099	−1.108	−1.121	−1.135	−1.148	−1.161	−1.173	−1.192	−1.210	−1.240	−1.270	−1.381	−1.485	−1.632
0.85	−0.030	−0.064	−0.739	−0.753	−0.765	−0.778	−0.790	−0.802	−0.820	−0.837	−0.865	−0.892	−0.994	−1.088	−1.221
0.90	−0.020	−0.041	−0.090	−0.458	−0.471	−0.483	−0.495	−0.507	−0.523	−0.539	−0.565	−0.590	−0.682	−0.767	−0.885
0.95	−0.012	−0.025	−0.053	−0.085	−0.237	−0.250	−0.261	−0.272	−0.288	−0.303	−0.326	−0.349	−0.431	−0.505	−0.609
1.00	−0.007	−0.014	−0.028	−0.042	−0.057	−0.072	−0.083	−0.093	−0.106	−0.119	−0.139	−0.158	−0.228	−0.293	−0.383
1.05	−0.003	−0.006	−0.011	−0.015	−0.017	−0.015	−0.005	0.014	0.020	0.015	0.003	−0.011	−0.067	−0.120	−0.197
1.10	0.000	0.000	0.002	0.004	0.009	0.016	0.028	0.047	0.077	0.094	0.101	0.097	0.061	0.019	−0.043
1.15	0.002	0.005	0.010	0.017	0.026	0.037	0.052	0.069	0.100	0.129	0.158	0.170	0.160	0.131	0.083
1.20	0.004	0.008	0.017	0.027	0.039	0.052	0.067	0.085	0.115	0.146	0.188	0.213	0.234	0.220	0.186
1.30	0.006	0.012	0.025	0.039	0.054	0.069	0.086	0.104	0.133	0.163	0.211	0.250	0.327	0.344	0.340
1.40	0.007	0.014	0.029	0.045	0.061	0.078	0.095	0.113	0.141	0.170	0.216	0.259	0.370	0.416	0.440
1.50	0.008	0.016	0.032	0.048	0.065	0.082	0.099	0.117	0.144	0.171	0.216	0.258	0.385	0.453	0.502
1.60	0.008	0.016	0.032	0.049	0.066	0.083	0.100	0.117	0.143	0.169	0.212	0.253	0.386	0.469	0.539
1.80	0.008	0.016	0.032	0.049	0.065	0.081	0.097	0.114	0.138	0.162	0.201	0.239	0.371	0.468	0.566
2.00	0.008	0.016	0.031	0.047	0.062	0.077	0.093	0.108	0.130	0.153	0.189	0.224	0.350	0.450	0.562
2.50	0.007	0.014	0.027	0.040	0.054	0.067	0.080	0.093	0.112	0.131	0.161	0.191	0.300	0.395	0.512
3.00	0.006	0.012	0.023	0.035	0.046	0.058	0.069	0.080	0.097	0.113	0.139	0.165	0.262	0.347	0.458
3.50	0.005	0.010	0.021	0.031	0.041	0.051	0.061	0.070	0.085	0.099	0.123	0.146	0.232	0.309	0.412
4.00	0.005	0.009	0.018	0.027	0.036	0.045	0.054	0.063	0.076	0.089	0.110	0.130	0.208	0.279	0.375
5.00	0.004	0.008	0.015	0.022	0.030	0.037	0.044	0.052	0.062	0.073	0.090	0.107	0.173	0.233	0.317

SOME SELECTED REFERENCES

H. B. Callen, *Thermodynamics and an Introduction to Thermostatics,* Second Edition, John Wiley & Sons, New York, 1985.

G. N. Hatsopoulos and J. H . Keenan, *Principles of General Thermodynamics,* John Wiley & Sons, New York, 1965, 1981.

O. A. Hougen, K. M. Watson, and R. A. Ragatz, *Chemical Process Principles, Part Two: Thermodynamics,* Second Edition, John Wiley & Sons, New York, 1959.

J. R. Howell and R. O. Buckius, *Fundamentals of Engineering Thermodynamics,* Second Edition, McGraw-Hill Book Co., New York, 1992.

J. H. Keenan, *Thermodynamics,* M.I.T. Press, Cambridge, MA, 1970.

J. Kestin, *A Course in Thermodynamics,* Hemisphere Publishing Corp., Washington, D.C., 1966, 1979.

M. J. Moran and H. N. Shapiro, *Fundamentals of Engineering Thermodynamics,* Second Edition, John Wiley & Sons, New York, 1992.

E. F. Obert, *Concepts of Thermodynamics,* McGraw-Hill Book Co., New York, 1960.

H. Reiss, *Methods of Thermodynamics,* Blaisdell Publishing Co., Waltham, MA, 1965.

W. C. Reynolds and H. C. Perkins, *Engineering Thermodynamics,* Second Edition, McGraw-Hill Book Co., New York, 1977.

A. H. Shapiro, *The Dynamics and Thermodynamics of Compressible Fluid Flow,* The Ronald Press Co., New York, 1953.

R. E. Sonntag and G. J. Van Wylen, *Introduction to Thermodynamics, Classical and Statistical,* Third Edition, John Wiley & Sons, New York, 1991.

W. F. Stoecker, *Design of Thermal Systems,* Third Edition, McGraw-Hill Book Co., New York, 1989.

K. Wark, *Thermodynamics,* Fourth Edition, McGraw-Hill Book Co., New York, 1983.

L. C. Woods, *The Thermodynamics of Fluid Systems,* Oxford University Press, London, 1975.

M. W. Zemansky, M. M. Abbott, and H. C. Van Ness, *Basic Engineering Thermodynamics,* McGraw-Hill Book Co., New York, 1966, 1975.

ANSWERS TO SELECTED PROBLEMS

2.3	27.03 m/s^2		3.66	29.0 lbf/in.^{2}
2.6	5 kg; 0.83 kg		3.69	89.423 lbm/s; 1.63%; 46.55%
2.9	1020 N		3.72	417.4 F
2.12	1346 kPa		3.75	9.1 F; 2.6794 ft^3/lbm
2.15	2452 Pa; 500 mm		4.3	0.173 kJ
2.21	184.3 kPa		4.9	−80.4 kJ
2.24	8.33 kg		4.12	0.0963 kJ
2.27	6.2 MPa		4.15	−54.1 kJ
2.30	814.3°C		4.18	117.5 kJ
2.33	385.4 lbf		4.21	−13.4 kJ
2.36	199.1 lbf/in.^{2}		4.24	120.2°C; 0.8857 m^3; 157.5 kJ
2.39	0.274 in.		4.27	40 kJ
2.42	62.3 lbf/in.^{2}		4.30	247 kPa; 33.6 L; 5.8 kJ
3.3	2152 kPa		4.33	a. 81.3°C; 1500 kPa
3.6	0.603 kg			b. 198.3°C; 1500 kPa; 0.13177 m^3/kg; 348.9 kJ
3.9	500 kPa; 906 kPa		4.36	12.46 kJ
3.12	204.4 kPa		4.39	3.93 J
3.15	a. 123.7 kg		4.48	−117.8 Btu
	b. 4.2%		4.51	150 ft lbf
3.27	12.4%; 1.1%		4.54	−10.49 Btu
3.30	a. 5.5607 MPa		4.57	0.43 lbf/in.^{2}; 0.0334; 0
	b. 5.5637 MPa		4.60	18.17 Btu/lbm
	c. 8.904 MPa		5.3	1405 kJ
	d. 5.4816 MPa		5.9	−263.2 kJ
3.33	23.8 MPa		5.12	722 kJ
3.36	0.9991		5.15	a. 99.6°C; 1.0 m^3
3.39	290 kPa			b. 175 kJ; 2434.6 kJ
3.42	1.7159 m^3/h		5.18	a. 0.5903 kg; 0.9695 kg
3.45	1.73 kg; 0.057 81 m^3/kg; 188.0°C; 0.012 04 m^3/kg			b. 352.3°C; −152.1 kJ
3.48	−13.3°C; 0.172 73 m^3/kg		5.21	a. 4.2°C
3.51	a. 0.6829; 0.015 66 m^3 vapor out			b. −115.2 kJ
	b. 0.6829; 0.000 726 m^3 liquid out		5.24	145.9 kJ
3.57	29.4 lbf/in.^{2}		5.27	161.6°C
			5.30	1.677 m^3; 400 kPa; 254.1 kJ; 8840 kJ

5.33	13.76 MJ
5.36	−8390 kJ
5.39	40.6 kJ
5.42	a. 361 kPa
	b. 2080 kJ; 60 kJ
5.45	51.9 L
5.48	211.3 L; 92.2 L; −87.9 kJ
5.51	6.17 kg; 30 kJ; 913.5 kJ
5.54	32.29 MJ
5.57	a. 520.3 kJ/kg
	b. 921.6 kJ/kg
	c. 841.8 kJ/kg
5.60	2.323 kg; 2.489 kg; 905.5 K; 625 kPa
5.63	5048.1 kJ/kg
5.66	540.4 K; 2702.9 kJ
5.69	−1.117 kJ
5.72	a. 1×10^{-5} m^3; 1.16×10^{-5} kg
	b. 2.303 J; 0.9 J
	c. 1.403 J; 13.68 m/s
5.75	a. 524 kPa
	b. −16.6 kJ
	c. 538.6 kJ
5.78	1875 kPa; 66.6 kJ; 1647 kJ
5.84	10.9 m/s; 12.8 m/s
5.87	12.0 kg/s
5.90	0.947; 8.55 mm^2
5.93	−0.98 kW
5.96	0.001 25 kg/s
5.99	0.1766
5.102	316.85 kJ/kg; 306.83 kJ/kg
5.105	18.084 MW
5.108	a. 126.3°C
	b. −2620 kW
5.111	131.1°C
5.114	123 075 kg/h
5.117	a. 410 K; 0.1275 kg
5.120	−379 639 kJ
5.123	764 kPa
5.126	27.24 kg
5.129	6744 kJ
5.132	1.952 kg; −10°C
5.135	3.082 kg; 225 kJ; −819.2 kJ

5.138	290°C
5.144	−191 Btu
5.147	a. 0.3804 lbm; 0.874 lbm
	b. −181 Btu; −105.7 Btu
5.150	69.6 Btu
5.153	4452 Btu
5.156	13.85 Btu
5.159	a. 39.5 lbf/in.^{2}
	b. 6.4 Btu
	c. 9.0 Btu
5.162	641 R
5.165	2206 Btu/lbm
5.168	a. 37 lbf/in.^{2}; −115.7 Btu
	b. 34.4%
5.171	a. 10 in.^{3}; 4.26×10^{-4} lbm
	b. 0.0362 Btu; 0.0142 Btu
	c. 0.022 Btu; 165.9 ft/s
5.174	351.8 Btu
5.177	3.0 ft/s
5.180	24963 ft^3/s
5.183	−0.77 Btu/s
5.186	7.57 lbm/h
5.189	−136.8 Btu/lbm; −127.1 Btu/lbm
5.192	254.6 F
5.195	−201 339 Btu
5.198	589.3 R
5.201	3.92 lbm; 224.9 Btu; −348.1 Btu
6.6	0.595
6.9	0.307
6.12	2.558 kW
6.15	0.051
6.18	0.731
6.24	300 J; 3.33×10^{-8}
6.27	a. 0.91 kW
	b. 45.2°C
6.30	29.8 kW
6.33	49.6 kW
6.36	1493.3 kJ
6.39	0.587
6.48	42.25 Btu/s
7.3	99.6°C; 0.287; 1716.2 kJ/kg; 6.073 kJ/kg K
7.6	3213.7 kJ; 617.1 kJ

7.9   232.6 kJ; 0

7.12 −3.197 kJ; −3.83 kJ

7.15 0.385 m^3

7.18 501 kPa

7.21 487.9°C; 2.534 kJ/K

7.24 0.2422 kJ/K

7.27 874.6 kJ; 6954.8 kJ; 5.27 kJ/K

7.33 2.57 kJ/K

7.36 2.3653 kJ/K

7.39 b. 97.82 kJ, 97.82 kJ; −74.48 kJ, 0; −130.43 kJ,
      −130.43 kJ; 74.48 kJ, 0

7.42 479 mm; 290.5 J

7.45 b. 143.2 K; −623.9 kJ/kg

7.48 0; 0.2947 kJ/K

7.51 0.365 kJ/K

7.54 1.81 kJ; −0.963 kJ

7.60 0.000 47 kJ/K

7.63 0.728 kg/s

7.66 357.6 kPa; 178 mm^2

7.72 a. 989.3 kJ/kg
      b. 510 kPa

7.75 a. 706.0 K; 557.9 kJ/kg
      b. 662.1 K; 539.8 kJ/kg
      c. 706.1 K; 554.1 kJ/kg

7.78 −0.473 kW

7.81 613 m/s

7.84 a. 10.2°C; 10.2°C
      b. 98.5°C; 86.8°C

7.87 2.675 kg; 450 kJ; −1276.3 kJ; 105.8 kPa

7.90 10.868 × 10^5 kg; 680 K; −2.345 × 10^8 kJ; 1248 kPa;
      −2.287 × 10^8 kJ

7.93 a. 2.692 kg
      b. −35.833 kJ/K; +50.449 kJ/K

7.96 410.4 kPa; 757.9 K

7.99 349 kPa; 18.06 kW

7.102 128.6 kPa; 313.1 K

7.105 2687.5 kJ/kg; 0.4992 kJ/kg K; −2495.7 kJ/kg

7.108 0.0499 kW/K

7.111 18 682 kW; 3.612 kW/K

7.114 b. 0.328 kW/K; 4.739 kW/K

7.117 0.92; 0.028 kJ/kg K

7.120 375.5 kJ/kg; 265.55 kJ/kg; 0.877; 0.1657 kJ/kg K

7.123 212 F; 0.262; 775.4 Btu/lbm; 1.4806 Btu/lbm R

7.126 2802.7 Btu; 543.3 Btu

7.129 −4.24 Btu; −5.05 Btu

7.132 85 lbf/in.^{2}

7.135 716.4 Btu; 5842.4 Btu; 2.535 Btu/R

7.138 1.293 Btu/R

7.141 23.94 in.; 0.46 Btu

7.144 5.16 lbm

7.147 b. 421.6 R; −11.8 Btu
      c. 19.4 Btu

7.150 −1.8 Btu/s

7.153 21.6 lbf/in.^{2}; 579.3 R

7.156 1023.9 lbf/in.^{2}; 78 F

7.159 2.1319 × 10^6 lbm; 1221 R; 2.226 × 10^8 Btu; 182.8
      lbf/in.^{2}, −2.158 × 10^8 Btu

7.162 63.13 lbf/in.^{2}; 1349.2 R

7.165 0.837; 0.0397 Btu/lbm R

7.168 3.53 hp

7.171 71.76 lbm/s; 64.93 lbm/s; 1.292 Btu/R s; 2.201
      Btu/R s; 0.153 Btu/R s

8.3   0.504 kW

8.6   1269 kW

8.9   1483.9 kJ/kg; 1636.8 kJ/kg

8.12 635 kJ

8.15 103.3 kJ

8.18 −10 kJ; −10 kJ; −6.14 kJ

8.21 64.6 kJ; 1286 kJ

8.24 −36.1 kJ/kg

8.27 −1576 kJ; 1460.3 kJ

8.30 1.4 MPa; 1.208 m^3; 1085.8 kJ; 1147.6 kJ

8.33 44.5 kJ/kg; 0.95

8.36 a. 299.5 K; 21 533 kJ
      b. 300.3 K; 22 170 kJ

8.39 0.315; 0.672

8.42 0.604; 0.599

8.45 394 389 kJ

8.48 37.87 kJ; −20.93 kJ

8.51 −9.58 kJ; −0.14 kJ

8.54 1.01 kg/s; 0.77

8.57 550 K; 31.2 kJ; 0.816

8.60 961 Btu/s

8.63 87.57 Btu

8.66 2.6 Btu/lbm

8.69 2342.3 Btu

8.72 21.3 Btu/lbm; 0.946

8.75 0.335; 0.70

8.78 332 Btu

8.81 32.5 Btu; −18 Btu

8.84 0.176

9.3 758°C; 4.82 kg/s

9.6 15.23 kW

9.9 1860 kJ/kg; 0; 0.5657

9.12 0.348; 0.362; 0.413; 0.33

9.18 0.2364; 5.55 kJ/kg; 4.51 kJ/kg

9.21 a. 0.438
      b. 0.473
      c. 0.488

9.24 a. 6.53 kg/s
      b. 3.202 kW; 53.22 kW
      c. 304.6 kg/s
      d. 0.977 m

9.27 a. 0.289
      b. 34.8 kJ/kg

9.30 a. 7.82 kg/s
      b. 47.8 kW
      c. 0.333

9.33 a. 8438 kW
      b. 49 196 kW

9.36 a. 3046 kW; 7009 kW; 0.509
      b. 221 kPa

9.39 a. 166.33 MW; 0.399
      b. 0.530

9.42 0.5; 861 K; 250.4 kPa; 201.7 kJ/kg

9.45 478.1 kJ/kg; 1235.5 kJ/kg; 0.613

9.48 163.3 kJ/kg; 133.2 kJ/kg; −133.2 kJ/kg

9.51 545.3 kPa/ 1010.8 m/s

9.54 1028 m/s

9.57 a. 10.07 MPa; 4525 K
      b. 0.541
      c. 1957 kPa

9.60 a. 10.07 MPa; 4525 K
      b. 0.491
      c. 1778 kPa

9.63 a. 2763 kPa

b. 0.783; 0.374

9.66 900 K; 429.9 kJ/kg/ 15.6; 900 K; 460 kJ/kg; 18.1

9.69 106.9 kJ/kg; 134 kJ/kg; 3.95

9.75 15.55 kW

9.78 0.147

9.81 0.258

9.84 a. 61.27 kg/s
      b. 8.746 kg/s
      c. 0.524

9.87 a. 0.79 kg/s
      b. 83.6 kW

9.90 a. 1.433
      b. 1.032

9.93 0.341; 1.8 Btu/lbm; 1254.8 Btu/lbm; 429 Btu/lbm; 827 Btu/lbm

9.96 760 Btu/lbm; 0.563

9.99 0.45; 0.7515

9.102 0.366; 435.9 Btu/lbm

9.105 a. 0.284
       b. 8.2 Btu/lbm

9.108 a. 165 600 hp; 0.396
       b. 0.530

9.111 242.7 Btu/lbm; 449.5 Btu/lbm; 0.427

9.114 3817 R; 791 lbf/in.^{2}

9.117 a. 1450 lbf/in.^{2}; 8128 R
       b. 0.492
       c. 256 lbf/in.^{2}

9.120 0.766

9.123 3.33; 3.30

9.129 a. 120.32 lbm/s
       b. 16.835 lbm/s
       c. 0.525

10.3 215.7 kJ/kg; 3.018 kJ/kg K; 134.8 kJ/kg; 1.226 kJ/kg K

10.6 −150.6 kW

10.15 0.000 198 (°C)$^{-1}$; 0.000 480 (MPa)$^{-1}$; 0.002 977 (°C)$^{-1}$; 0.002 731 (MPa)$^{-1}$

10.18 343 m/s

10.21 451.6 K; 116.6 kJ

10.24 244.6 K

10.33 279.8 kJ/kmol

10.36 −172.6 kJ

10.39   3353 kJ

10.42   85.3 kJ/kg; 362 K

10.45   −981.3 kJ

10.48   −186.7 kJ/kg; −474.2 kJ/kg

10.51   934.2 kJ/kg; 368 K/ 418.7 kJ/kg

10.54   211.5 m/s

10.57   275 K; 0.727

10.60   279 K; 810 kPa

10.63   306; −541 kJ/kg; −123.6 kJ/kg

10.66   304.6 kJ/kg

10.69   0.318; 13.8 kg

10.72   350.8 kJ/kg

10.75   5490 lbf/in.^{2}

10.78   0.000 103 (F)$^{-1}$; 0.000 003 (lbf/in.^{2})$^{-1}$; 0.001 144 (F)$^{-1}$; 0.000 011 (lbf/in.^{2})$^{-1}$

10.81   442.7 R

10.84   0.3433; 29.7 lbm

10.87   −149.8 Btu

10.90   −1.83 Btu/lbm; 0.285

10.93   −79.3 Btu/lbm; −202.7 Btu/lbm

10.96   −704 Btu

10.99   301.2; −233.9 Btu/lbm; −52.6 Btu/lbm

10.102   0.3139; 30.0 lbm

11.3   a. 2.1% H_2, 46.85% CO, 27.61% CO_2, 23.43% N_2

         b. 9.81 × 10^{-4} kg

         c. 0.0719 kJ

11.6   1321 kW

11.9   141 kPa

11.12   279 kPa; 418.6 K; −231 kJ

11.18   0.533 kg

11.21   17%; 16 kJ/kg air; 100%; −151.3 kJ/kg air

11.24   3978 kJ; 9.7%; 0.268 75 kg

11.30   0.0024 kg/kg; −11.94 kJ/kg air; 83%; 0.0106 kg/kg

11.33   a. 0.13

         b. 0.4359 kg

         c. 25.6 kJ

11.36   0.503; 25.8 kJ/kg air

11.39   700 kPa

11.42   b. −12.1 kJ

         c. 0.0138 kJ/K

11.45   a. 0.1318 kg

         b. −389.7 kJ

         c. 0.46 kJ/K

11.48   34.5 kg/s; 0.33 kg/s

11.51   0.7606; 5°C

11.54   0.0545 kg/s; −6.64 kW

11.57   a. −7.21 kJ; −7.21 kJ

         b. −10.86 kJ; −7.79 kJ

         c. −9.93 kJ; −7.81 kJ

11.60   a. 0.0187 m^3

         b. 6.45 kJ/K

11.63   a. 0.2880 m^3/kmol; 0.1740 m^3/kmol

11.66   4.66 kg; 2.314 MPa

11.69   1.569 MPa; 35.8 kJ

11.75   3.02 kg

11.78   1184 Btu/s

11.81   581 R; −49.2 Btu

11.84   0.864 Btu/R s

11.87   78 F; −1.49 Btu

11.90   0.007 16; 15.6 Btu/lbm air; 83%; 0.0106

11.93   −4.18 Btu; 0.0227 lbm; 20.76 lbf/in.^{2}

11.96   47 F; 0.0068; 100%; 1.987 Btu/s; 112 F; 12%; 1 Btu/s

11.99   0.123 lbm; 0.04 lbm/min; 94 F; 9%

11.102   2.53 lbm; 345 lbf/in.^{2}

11.108   23 411 Btu

12.3   1.25; 125%

12.6   2.95 kg/kg

12.9   0.718 kmol air/kmol gas

12.12   a. 43.2°C

         b. 0.0639 kg/kg

12.15   −179 796 kJ/kmol; 1.35

12.18   −778 190 kJ/kmol

12.21   − 1 234 583 kJ

12.24   38.66 kW; −83.27 kW

12.27   6.78; 9.57

12.30   2529 K; 21%

12.33   a. 2909 K

         b. 7400 K

12.36   2460 K; −393 522 kJ/kmol fuel

12.39   1656.6 kPa

12.42   232 009 kJ/kmol gas

12.45   −178 136 kJ/kmol CO; 5741K

12.48	52.9°C; 1.303 kg/kg fuel
12.51	110 630 kJ
12.54	238.3 kPa; -1.613×10^6 kJ/kmol fuel; 4070 kJ/K kmol fuel
12.60	a. 2011 kPa; 666.4 K
	b. 2907 K; 8772 kPa
	c. 152 860 kJ
12.63	923.4 K; 1.318×10^6 kJ/kmol fuel
12.66	−4.081 kW; 0.139
12.69	0.328; 0.414
12.72	a. 1.232 lbm/lbm
	b. 1.492 lbm/lbm
12.75	−42 855 Btu/lb mol
12.78	5045 R
12.81	34.9 Btu/s; −67.45 Btu/s
12.84	a. 6.91
	b. 9.689
12.87	a. 5236 R
	b. 13 300 R
12.90	218.45 lbf/in.^{2}
12.93	− 268 580 Btu; 1001.8 Btu/R
12.96	a. 5.07
	b. 307.99 Btu/R
12.99	2049 R; 8520 Btu/s
13.3	89.6% N_2 gas; 23.3% N_2 liq
13.6	0; −2350 kW
13.9	197 611 kJ
13.15	1444 K
13.18	0.0877
13.21	45.6% CO_2, 21.4% CO, 33.0% O_2
13.24	176 811 kJ
13.27	a. 0.0237
	b. 30.9% CH_4, 30.9% H_2O, 28.7% H_2, 9.5% CO
13.30	0.006 55; −835 974 kJ
13.33	164 245 kJ
13.36	0.27
13.39	0.0024
13.42	40.6% H_2O, 1.9% H_2, 46.3% O_2, 11.2% OH
13.45	1.07
13.48	48.8% H_2O, 5.7% H_2, 7.6% O_2, 8.5% OH, 15.5% CO_2, 13.9% CO
13.54	1.939

13.57	84 616 Btu
13.60	2600 R
13.63	73.7%
13.66	1.3682
13.69	92 070 Btu
13.72	48.7% H_2O, 5.7% H_2, 7.6% O_2, 8.6% OH, 15.5% CO_2, 13.9% CO
14.3	415.4 K; 281 kPa; 5.897 kg/s
14.6	31.72 m/s
14.9	b. 1516 mm^2; 2435 mm^2
14.12	1796 kPa
14.15	906 kPa
14.18	a. 0.034 16 kg/s
	b. 0.014 87 kg/s
	c. 1.895 kg; 0.0082 kg/s
14.21	4457 mm^2; 552 m/s; 0.0539 kJ/kg K
14.24	285.3 K; 0.608; 88.25 kPa; 81.77 kPa
14.27	0.063 65 kg/s
14.30	3.7808 kg/s
14.33	0.798
14.36	a. 306 kPa
	c. 423.8 kJ/kg
14.39	88 lbf/in.^{2}; 466 F
14.42	a. 15.04 lbf/in.^{2}; 323 R; 4.99 in.^{2}; 1.722 in.^{2}
	b. 2.0; 746 R; 10 lbm/s
14.45	7.807 lbf/in.^{2}; 542 R; 0.414
14.48	0.769 in.^{2}; 0.876; 0.0255 Btu/lbm R

INDEX